Electric Vehicles: Theory and Design

Electric Vehicles: Theory and Design

BY YIQING YUAN

INTERNATIONAL®

Warrendale, Pennsylvania, USA

400 Commonwealth Drive
Warrendale, PA 15096-0001 USA
E-mail: CustomerService@sae.org
Phone: 877-606-7323 (inside USA and Canada)
724-776-4970 (outside USA)
Fax: 724-776-0790

Library of Congress Catalog Number 2024938640
http://dx.doi.org/10.4271/9781468607741

ISBN-Print 978-1-4686-0773-4
ISBN-PDF 978-1-4686-0774-1
ISBN-epub 978-1-4686-0775-8

To purchase bulk quantities, please contact: SAE Customer Service

E-mail: CustomerService@sae.org
Phone: 877-606-7323 (inside USA and Canada)
724-776-4970 (outside USA)
Fax: 724-776-0790

Visit the SAE International Bookstore at books.sae.org

Publisher
Sherry Dickinson Nigam

Product Manager
Amanda Zeidan

Production and Manufacturing Associate
Michelle Silberman

Contents

Preface xvii
Acknowledgments xviii
List of Acronyms xix

CHAPTER 1
Dynamic Performance 1

1.1. Indicators of Dynamic Performance 1
1.2. Tractive Effort and Resistance Forces 2
1.2.1. Tractive Effort 2
1.2.2. Resistance Forces 6
1.3. Vehicle Dynamic Equation 17
1.3.1. Equilibrium Diagrams of Driving Force and Running Resistance 18
1.3.2. Acceleration Capacity 19
1.3.3. Gradeability 20
1.4. Tire-Ground Adhesion 20
1.4.1. Adhesion Coefficient and Adhesion Condition 20
1.4.2. Adhesion Coefficient 21
1.4.3. Adhesion-Related Forces 21
1.4.4. Minimal Adhesion Coefficient Requirements 24
1.5. Equilibrium of Power 26
1.6. Dynamic Performance Tests 28
1.6.1. Road Tests 28
1.6.2. Rig Tests 29
References 29

CHAPTER 2

Energy Economy 31

2.1. Indicators of Energy Economy 31

2.2. Estimation of Energy Economy 36

2.3. Factors Impacting Energy Economy 37

2.3.1. Reduction in Vehicle Weight and Size 42

2.3.2. Electric Motor System 43

2.3.3. Drivetrain 43

2.3.4. Body Aerodynamics 43

2.3.5. Tire 44

2.3.6. Thermal Management 44

2.4. Energy Economy Testing 45

References 46

CHAPTER 3

Selection of Dynamic Parameters 47

3.1. Selection of Electric Motor Power 47

3.2. Selection of the Smallest Speed Ratio 49

3.3. Selection of the Largest Speed Ratio 50

3.4. Selection of the Number of Gears and Intermediate Speed Ratios 51

3.5. Determination of Transmission and Final Drive Ratios 54

3.5.1. Determination of the Final Drive Ratio 54

3.5.2. Optimization of Speed Ratios 55

References 58

CHAPTER 4

Braking Performance and Energy Regeneration 61

4.1. Evaluation Indicators for Braking Performance 61

4.2. Wheel Force Analysis during Braking 62

4.2.1. Ground Braking Force 63

4.2.2. Braking Force at the Tire Circumference 63

4.2.3. Relationship between Adhesion and Braking Forces 64

4.2.4. Adhesion Coefficient on Hard-Surface Roads 64

4.3. Braking Efficiency and Constancy 68
4.3.1. Stopping Distance and Deceleration 68
4.3.2. Analysis of Stopping Distance 69
4.3.3. Constancy of Braking Efficiency 70
4.4. Directional Stability in Braking 71
4.4.1. Braking Deviation 71
4.4.2. Sideways Skid and Loss of Steering in Braking 72
4.5. Front and Rear Proportioning of a Braking Force 74
4.5.1. Normal Reaction Forces from the Ground to Front and Rear Wheels 75
4.5.2. Ideal Curve of Front and Rear Braking Force Proportioning 76
4.5.3. Fixed Proportioning of Front and Rear Brake Forces 77
4.5.4. Analysis of the Braking Process of EVs with Fixed Brake Proportioning 78
4.5.5. Adhesion Utilization and Braking Efficiency 81
4.5.6. Requirements for Brake Proportioning 83
4.5.7. ABS 84
4.6. Braking Performance Tests 86
4.6.1. Road Tests 86
4.6.2. Rig Tests 88
4.7. Regenerative Braking 88
4.7.1. Mechanical and Electrical Compound Braking 89
4.7.2. Brake Proportioning for Maximization of Regenerative Energy 91
References 95

CHAPTER 5
Handling and Stability 97

5.1. Cornering Characteristics of Tires 97
5.1.1. Coordinate System 97
5.1.2. Lateral Forces in Cornering 99
5.1.3. Factors Affecting Cornering Behavior 100
5.1.4. Aligning Torque 101
5.1.5. Behavior of a Tire with the Camber Angle 102
5.2. Vehicle Response to the Input of the Front Wheel Angle 104
5.3. Steady-State Response to the Input of the Front Wheel Angle 106
5.3.1. Yaw Velocity Gain 106
5.3.2. Ratio of Turning Radii 108
5.3.3. Static Margin 109
5.3.4. Difference in Sideslip Angles 110

5.4. Instantaneous Response to the Step Input of the Front Wheel Angle 111
5.4.1. Equation of Motion and Its Solution 111
5.4.2. Stability Conditions of Instantaneous Response 115
5.5. Yaw Rate Frequency Response Characteristics 116
5.6. Suspension Effects on Handling Stability 117
5.6.1. Suspension-Related Cornering Behavior 118
5.6.2. Roll Steer 126
5.6.3. Compliance Steer Related to Lateral Force 127
5.6.4. Compliance Steer Related to Aligning Torque 128
5.6.5. Compliance Camber Change 129
5.7. Steering System Effect on Handling 129
5.7.1. Functionality of the Steering System 129
5.7.2. Fundamental Handling Characteristics 132
5.7.3. Steering Effort Characteristics 132
5.7.4. On-Center Handling Characteristics 133
5.7.5. Movement Interference between the Steering System and Suspension during Body Roll 135
5.7.6. Compliance Steer Related to Steering System Stiffness 135
5.8. Drivetrain System Effect on Handling 135
5.8.1. Mechanisms of the Tractive Force Effect on Cornering Characteristics 135
5.8.2. Cornering Stability Control by Tractive and Braking Forces 137
5.8.3. Electronic Control Systems for Handling Stability 139
5.9. Rollover 143
5.9.1. Quasistatic Rollover Analysis of a Rigid Vehicle 143
5.9.2. Quasistatic Rollover Analysis of a Vehicle with Suspension 145
5.9.3. Transient Rollover Analysis 146
5.10. Vehicle Handling Tests 147
5.10.1. Low-Speed Steering Effort Tests 147
5.10.2. Steady-State Steering Characteristics Tests 148
5.10.3. Transient Yaw Response Tests 148
5.10.4. Self-Aligning Tests 149
5.10.5. Pulse Steering Tests 149
5.10.6. On-Center Handling Tests 149
References 149

CHAPTER 6

Ride Comfort 151

6.1. Excitation Sources 152
6.1.1. Electric Motors 152
6.1.2. Drivetrain 155
6.1.3. Wheels and Tires 157
6.1.4. Road Roughness 161

6.2. Vehicle Response to Excitations 165
6.2.1. Dynamic Model of a Vehicle 165
6.2.2. Fundamental Behavior of Sprung Mass 168
6.2.3. Frequency Response to Road Excitation 169
6.2.4. Statistical Characteristics of System Response 171

6.3. Effects of System Parameters on Vibration Response 179

6.4. Human Perception of Vibrations 179

6.5. Evaluation of Ride Quality 180
6.5.1. Basic Evaluation Method 180
6.5.2. Alternative Evaluation Method 181
6.5.3. Sound Quality Evaluation 181

References 183

CHAPTER 7

Trafficability 185

7.1. Indices and Geometric Parameters for Trafficability Evaluation 185
7.1.1. Indices for Trafficability Evaluation Related to Ground Bearing Capacity 185
7.1.2. Structural and Geometric Parameters Related to Trafficability 186

7.2. Physical Properties of Soft Ground 188
7.2.1. Relationship between Shear Stress and Shear Deformation in Soft Ground 188
7.2.2. Relationship between Subsidence and Vertical Loads 190
7.2.3. Effect of the Density of Semifluid Mud and Snow on Trafficability 191

7.3. Drawbar Pull of EVs 191
7.3.1. Traction Force Acting on the Vehicle from Soft Ground 192
7.3.2. Resistance on the Vehicle Traveling on Soft Ground 193
7.3.3. Calculation of the Drawbar Pull 197

7.4. Tractive Trafficability 197
7.5. Conditions for Obstacle Clearance 200
7.5.1. Obstacle Clearance of the Front End 200
7.5.2. Obstacle Clearance of the Underbody 201
7.6. Overcoming Terrain Steps and Trenches 202
7.6.1. FWD Vehicle Overcoming Steps 202
7.6.2. RWD Vehicle Overcoming Steps 204
7.6.3. AWD Vehicle Overcoming Steps 206
7.6.4. Evaluation of Trench Obstacle Capability 208
7.7. Testing of Trafficability 208
7.7.1. Brief Descriptions of Trafficability Tests 208
7.7.2. Measurements of Soil Parameters 209
References 210

CHAPTER 8
Vehicle-Level Design 211

8.1. Categorization and Classification of EVs 211
8.1.1. Classification of Electric Cars 211
8.1.2. Classification of Electric Buses 212
8.1.3. Classification of Trucks 212
8.2. Body Style 212
8.2.1. Body Styles of Cars 212
8.2.2. Body Styles of Buses 217
8.2.3. Body Styles of Trucks 218
8.3. Electric Drive System Layout 220
8.3.1. Number of Axles and Axle Load 220
8.3.2. Basic Principles of Electric Drive System Layout 221
8.4. Selection of Overall Design Parameters 223
8.4.1. Dimensional Parameters 223
8.4.2. Mass-Related Parameters 226
8.4.3. Performance-Related Parameters 228
8.5. Selection of Battery System 231
8.5.1. Battery Type Selection 231
8.5.2. Selection of Battery System Parameters 232
8.5.3. Packing of the Battery System 234

8.6. Selection of Electric Drive Systems 236
8.6.1. Electric Motor Selection 236
8.6.2. Selection of Electric Drive System Parameters 240
8.6.3. Electric Drive System Mounts 244
8.7. Selection of Suspensions 245
8.7.1. Front Suspension 245
8.7.2. Rear Suspension 246
8.8. Selection of Tires 247
8.9. Overall Vehicle Layout Design 247
8.9.1. Battery System Layout 248
8.9.2. Powertrain System Layout 249
8.9.3. Steering System Layout 252
8.9.4. Layout of Suspensions 252
8.9.5. Brake System Layout 253
8.9.6. Layout of Operator Pedals 253
8.9.7. Layout of Luggage Space and the Spare Tire 254
8.9.8. Vehicle Interior Layout 254
8.9.9. Sizing of Vehicle Body 256
8.9.10. Mounting Positions of Seat Belts and Airbags 257
8.10. Motion Correctness Check 258
References 258

CHAPTER 9
Battery System Design 259
9.1. Types of Battery Cells 260
9.2. Battery Pack Architecture 261
9.3. Electrical Topology of Battery Packs 262
9.3.1. Battery Cells Connected in Series 262
9.3.2. Battery Cells Connected in Parallel 263
9.3.3. Battery Cells Connected in Series-Parallel 264
9.3.4. Guidelines for the Design of Pack Topology 265
9.4. Design of Battery Packs 267
9.4.1. Structural Design of Battery Packs 267
9.4.2. Design of a BTMS 273
9.4.3. Design of a BMS 276
References 285

CHAPTER 10

Electric Drive System Design 287

10.1. Electric Drive System Configurations 288

10.2. Design of Electric Motors 289

10.2.1. Topology of Electric Motors 289

10.2.2. Basic Design Parameters and Equations 290

10.2.3. Sizing of Electric Motors 295

10.2.4. Design of a Magnetic Circuit 302

10.2.5. Electromagnetic Design 311

10.2.6. Losses, Temperature Rise, and Cooling 329

10.2.7. Design Guidelines for PMSMs 342

10.3. Design of Electric Motor Controllers 354

10.3.1. Design of Main Control Boards 355

10.3.2. Design of Driver Boards 357

10.3.3. Design of High-Voltage Main Circuits 359

10.3.4. Design Guidelines for SiC Motor Controllers 360

10.4. Design of Drivetrains 374

10.4.1. General Layout Design 379

10.4.2. Determination of Basic Characteristic Parameters 380

10.4.3. Design of Layshaft Gearsets 381

10.4.4. Design of Planetary Gearsets 391

10.4.5. Design of Clutches 410

10.4.6. Design of Synchronizers 416

10.4.7. Design of Shift Actuators 423

10.4.8. Design of Hydraulic and Lubrication Systems 432

10.4.9. Design of Housing 442

References 448

CHAPTER 11

Suspension System Design 453

11.1. Types of Vehicle Suspension System 454

11.1.1. Types of Nonindependent Suspension 456

11.1.2. Types of Independent Suspension 456

11.2. Fundamental Parameters of the Suspension System 457

11.2.1. Static Deflection of Suspension 458

11.2.2. Natural Frequency of Suspension 459

11.2.3. Dynamic Deflection of Suspension 460

11.2.4. Elastic Characteristics of Suspensions 460
11.2.5. Roll Stiffness of Front and Rear Suspensions 461
11.2.6. Stiffness of Primary and Secondary Leaf Springs 461

11.3. Design of Elastic Elements 462
11.3.1. Coil Springs 463
11.3.2. Torsion Bar Springs 465
11.3.3. Leaf Springs 468
11.3.4. Air Springs 472

11.4. Design of Shock Absorbers 477
11.4.1. Hydraulic Shock Absorbers 477
11.4.2. Magnetorheological Shock Absorbers 480
11.4.3. Electrorheological Shock Absorbers 481
11.4.4. Electromagnetic Shock Absorbers 482

11.5. Guide Mechanism Design of Independent Suspension 484
11.5.1. Design Considerations on Alignment Parameters 484
11.5.2. Design of Roll Center 486
11.5.3. Design of the Pitch Center 488
11.5.4. Guide Mechanism Design for McPherson Suspension 489
11.5.5. Guide Mechanism Design for Double-Wishbone Suspension 490

11.6. Design of Stabilizer Bars 494

11.7. Design of Other Suspension Components 496
11.7.1. Control Arms 496
11.7.2. Thrust Rods 497
11.7.3. Joints 497

References 497

CHAPTER 12
Steering System Design 499

12.1. Trapezoidal Steering Mechanism 500
12.1.1. Design of Integral-Type Trapezoidal Linkage 500
12.1.2. Design of Split-Type Trapezoidal Linkage 503

12.2. Design of Steering Gears 505
12.2.1. Motion Transmission in a Steering System 505
12.2.2. Rack and Pinion Steering Gears 507

12.3. Design of Power Steering Systems 512
12.3.1. Types of EPS 512
12.3.2. Design of Boost Characteristics 513

12.4. Design of Other Steering System Components and Features 515
12.4.1. Ball Joints 516
12.4.2. Tie Rods 516
12.4.3. Pitman Arms 516
12.4.4. Steering Dampers 517
12.4.5. Other Safety-Related Components 517
12.5. Steer-by-Wire 518
12.6. All-Wheel Steering 519
12.6.1. Basic Principles of All-Wheel Steering 520
12.6.2. Mechanism of All-Wheel Steering 522
References 524

CHAPTER 13
Braking System Design 525

13.1. Determination of Braking Torques 526
13.2. Design of Service Brakes 527
13.2.1. Disc Brakes 527
13.2.2. Drum Brakes 533
13.3. Design of Park Brakes 542
13.4. Brake Actuation Mechanism 543
13.4.1. Types of Brake Actuation Mechanism 543
13.4.2. Power Brakes 543
13.4.3. Layouts of Brake Line Circuit 544
13.4.4. Design of Hydraulic Actuation System Components 546
13.4.5. Braking Force Adjustment Mechanism 548
13.5. Brake-by-Wire 551
13.5.1. Electrohydraulic Brakes 551
13.5.2. Electromechanical Brakes 553
References 554

CHAPTER 14
Chassis and Body Design 555

14.1. Requirements for the Chassis and Body 556
14.2. Design of an EV Chassis 556
14.2.1. Analysis of Load Cases 556

14.2.2. Design of Vehicle Frames 560
14.2.3. Vehicle Frame Testing 569

14.3. Design of an EV Body 570
14.3.1. Body Styling of EVs 570
14.3.2. Body Construction 572
14.3.3. Aerodynamic Design of Vehicle Bodies 572
14.3.4. Structural Design of Vehicle Bodies 577
14.3.5. Vehicle Body Testing 582

References 583

CHAPTER 15
Control System Design 585

15.1. Hardware Architecture of a VCU 588

15.2. Power-up and Shutdown Procedures 590
15.2.1. Low-Voltage Power-up Procedure 590
15.2.2. Low-Voltage Shutdown Procedure 590
15.2.3. High-Voltage Power-up Procedure during Discharge 591
15.2.4. High-Voltage Shutdown Procedure during Discharge 591
15.2.5. High-Voltage Power-up Procedure during Charge 592
15.2.6. High-Voltage Shutdown Procedure during Charge 592

15.3. Battery Management Strategy 593
15.3.1. Power Management 594
15.3.2. BTM Strategy 595

15.4. Control of Electric Motors 595
15.4.1. Control of Three-Phase IMs 595
15.4.2. Control of Three-Phase PMSMs 601

15.5. Driving Intent and Accelerator Pedal Strategy 602

15.6. Launch Control Strategy 604
15.6.1. Torque Compensation Control Strategy 605
15.6.2. Self-Adaptive Control Strategy 608

15.7. Electric Drive System Control Strategy 612
15.7.1. E-Motor Control Strategy 612
15.7.2. Gear Shift Control Strategy 620

15.8. Regenerative Braking Control Strategy 636

15.9. Electronic Differential Control 644
15.9.1. Two-DOF Model Based on the Ackermann Steering Principle 644
15.9.2. Three-DOF Steering Model 646
15.9.3. Motor Speed-Based EDC 649
15.9.4. Motor Torque-Based EDC 649
15.9.5. Disturbance Rejection Control Methodology 652
15.10. Control Strategies for Auxiliary Systems 652
15.11. Trend of VCU Technology and E/E Architecture 653
References 654

Index 657
About the Author 671

Preface

Time flies; it has been over 14 years since I devoted myself to the field of electric vehicles (EVs). The trickle of vehicle electrification merely a decade ago has already turned into a surging tide today, and in retrospect, I still feel very fortunate about my decision of stepping into this epoch-making trend because it will undoubtedly push mobility away from the smoke and dust produced by fossil fuels forever. I believe that this should also be the moral pursuit of all automotive professionals, especially for those with a combustion discipline background like myself, because they should better understand than ordinary people the damage caused by fossil-fuel-based combustion technology to life on this planet.

EVs are different from the traditional internal combustion engine-powered ones in many aspects from fundamental theory to design principles. As a matter of fact, the vehicle technology has been experiencing the greatest transformation since the dawn of automobile age in the 1880s, toward electrification, autonomous driving, connectivity, and shared mobility. In addition to the battery and propulsion systems, many other subsystems of a vehicle are also being electrified and reinvented to a higher level of intelligence.

However, EV engineers often face difficulties in finding necessary reference books in their daily engineering work. Although there are a number of excellent reference books for the design of internal combustion engine-powered vehicles, much of the contents are obsolete and not applicable to EVs. On the other hand, the knowledge on EVs is fragmented since most of the existing reference books on this topic cover the fundamental theory only on the vehicle system level, without detailed elaboration of subsystems, not to mention related basic concepts, principles, equations, considerations, design essentials, and so on.

This book is written with the aim of presenting up-to-date knowledge on EVs and their essential components in a systematic and in-depth manner. In particular, technical requirements, fundamental principles, design considerations and recommendations, and basic equations of engineering-level precisions are given for automotive engineers to quickly grasp the key points of their engineering tasks and building up their expertise in this field. Statistical data of historical and current production EVs are also presented in the form of tables or charts for readers to better understand about common engineering practices and technical trends. Emphasis is generally placed on the novel technologies that are gaining popularity, with only brief discussions on technologies that are being phased out.

The content of this book is not only applicable to battery electric vehicles (BEVs) but also to fuel cell electric vehicles (FCEVs). It has also certain reference value for the development of plug-in hybrid EVs and range-extended EVs. Although the book is intended for the reference of automotive engineers, students, scholars, and other personnel with educational institutions, other technical organizations may also find its usefulness in their teaching, training, or research activities.

Acknowledgments

I wish to acknowledge my parents, Yunting and Jin, for setting me out on this path of curiosity about the world and my engineering career. I am indebted to my family, in particular, my wife Wendy, my son Rocky, and my daughter-in-law Zoey, for their love and support while I devoted time and effort to this piece of work. I am grateful to my friends and colleagues who have provided valuable comments and advice. Special thanks to Dr. Uri Vandsburger, formerly my advisor and professor at the Mechanical Engineering Department of the Virginia Polytechnic Institute and State University, for his encouragement, inspiration, and enlightenment.

List of Acronyms

A/C	Air conditioning
ABS	Anti-lock brake system
AC	Alternating current
AlNiCo	Alnico alloys
AMT	Automated mechanical transmission
ASF	Audi Space Frame
AT	Automatic transmission
AVE	Advanced vehicle engineering
AWD	All-wheel drive
$BaFe_{12}O_{19}$	Barium ferrite
BDC	DC brushed
BEV	Battery electric vehicle
BIW	Body-in-white
BLDC	Brushless DC
BMS	Battery management system
BP	Backpropagation
BTL	Break through load
BTM	Battery thermal management
BTMS	Battery thermal management system
CAE	Computer-aided engineering
CAN	Controller area network
CAS	Chinese Academy of Science
CFD	Computational fluid dynamics
CFRC	Carbon fiber-reinforced carbon
CFRP	Carbon fiber-reinforced polymer
CG	Center of gravity
CLT	Coreless transformer
CMT	Cold metal transfer
CPCM	Composite phase change materials
CPLD	Complex programmable logic device
CSC	Cell supervising circuit
CTB	Cell-to-body
CTC	Cell-to-chassis
CTP	Cell-to-pack
CTV	Continuously variable transmission
DBC	Direct bonded copper
DC	Direct current
DCT	Dual-clutch transmission
DHM	Digital human models

DOF	Degrees of freedom
DPT	Double pulse tester
DSC	Dynamic stability control
DSPV	Deceleration sensing proportioning valve
DTC	Direct torque control
DYCS	Direct yaw moment control system
E/E	Electric and electronic
ECAS	Electronically controlled air suspension
ECE	Economic Commission of Europe
ECU	Electronic control unit
EDC	Electronic differential control
EHB	Electrohydraulic brake
EHL	Elastohydrodynamic lubrication
EIS	Electrochemical impedance spectroscopy
EKF	Extended Kalman filter
EMB	Electromechanical brakes
EMC	Electromagnetic compatibility
EMF	Electromotive force
EMI	Electromagnetic interference
EPS	Electric power steering
ER	Electrorheological
ESC	Electronic stability control
ESD	Electrostatic discharge
ESP	Electronic stability program
EV	Electric vehicle
EVT	Electric variable transmission
FCEV	Fuel cell electric vehicle
FCV	Fuel cell vehicle
FDS	Flow drill screw
FEA	Finite element analysis
FET	Field effect tube
FFT	Fast Fourier transform
FMVSS	Federal Motor Vehicle Safety Standard
FOC	Field-oriented control
FPCB	Flexible printed circuit boards
FUL	Fault under load
FWD	Front wheel drive
GA	Genetic optimization
GVW	Gross vehicle weight
GVWR	Gross vehicle weight rating
HIC	Head injury criterion
HL	Hydrodynamic lubrication
HP	Heat pipe
HPPC	Hybrid pulse power characterization
HSF	Hard switching fault
ICE	Internal combustion engine
IM	Induction motors
IPM	Interior permanent magnet
ISO	International Organization for Standardization

LFP	Lithium ferrophosphate
LIN	Local Interconnect network
LRR	Low rolling resistance
MCD	Multiconstrained dynamic
MCU	Motor control unit
MEC	Multiaccess edge computing
MIG	Metal inert gas
MIMO	Multi-input multioutput
MIPC	Maximum induced power control
MPGe	Miles per gallon of gasoline equivalent
MPR	Multipurpose vehicle
MR	Magnetorheological
MSD	Manual service disconnect
MT	Manual transmission
NdFeB	Neodymium iron boron
NEDC	New European Driving Cycle
NEVC	National New Energy Vehicle Technology Innovation Center
NMC	Nickel Manganese Cobalt
NSD	Nail safety device
NSP	Neutral steer point
NTC	Negative temperature coefficient
NVH	Noise, vibration, and harshness
OCV	Open circuit voltage
OSD	Overcharge safety device
PCB	Printed circuit board
PCM	Phase change material
PI	Proportional integral
PMSM	Permanent magnet synchronous motors
PRND	Park, reverse, neutral, and drive
PSD	Power spectral density
PSO	Particle swarm optimization
PST	Power shift transmission
PTC	Positive temperature coefficient
PWM	Pulse width modulation
RBF	Radial basis function
RMS	Root-mean-square
RS	Reset-set
RSW	Resistance spot welding
RWD	Rear wheel drive
SI	Severity index
SISO	Single-input multioutput
SMC	Sliding mode control
SmCo	Samarium cobalt
SOA	Service-oriented system architecture
SOC	State of charge
SOH	State of health
SOP	State of power
SOS	State of safety
SPM	Surface permanent magnet

SPR	Self-piercing riveting
$SrFe_{12}O_{19}$	Strontium ferrite
SRM	Switched reluctance motors
SUV	Sport utility vehicle
SVM	Support vector machine
SVPWM	Space vector pulsed width modulation
TCS	Traction control system
TCU	Transmission control unit
TEC	Thermoelectric cooler
TIM	Thermal interface material
TNR	Tone-to-noise ratio
Total TNR	Total tone-to-noise ratio
TRA	Torque roll axis
TSEP	Temperature-sensitive electrical parameters
TVS	Transient voltage suppressor
UDC	Urban driving cycle
UKF	Unscented Kalman filter
VA	Volt-amp
VAR	Volt-amp-reactive
VCC	Vapor compression cycle
VCU	Vehicle control unit
VDC	Vehicle dynamic control
VDV	Vibration dose value
VFD	Variable-frequency drive
VFS	Variable force solenoid
VVVF	Variable voltage and variable frequency
WLTC	World Harmonized Light-Duty Vehicles Test Cycle
WOT	Wide-open throttle

1

Dynamic Performance

The dynamic performance of an EV refers to the average driving speed that it can achieve when it is driven in a straight line on a good road by a longitudinal external force originated from its electric powertrain. As a means of transport, the transport efficiency of an EV largely depends on its dynamic performance. Hence, the dynamic performance is the most fundamental and important aspect of all kinds of EVs.

1.1. Indicators of Dynamic Performance

To obtain the highest possible average speed, the dynamic performance of an EV is mainly assessed by three indicators, namely, its acceleration time, maximum speed, and gradeability.

The acceleration time indicates the ability of an EV to accelerate as it influences the average speed. Usually, the acceleration time from standstill to a certain speed and the acceleration time of overtaking are utilized. Although the acceleration time from 0 to 100 km/h is most frequently quoted, the acceleration time from 0 to 50 km/h or 0 to 60 mph is sometimes employed instead. For today's high-end passenger EVs, the acceleration from 0 to 100 km/h can take less than 3 sec. On the other hand, there is no widely accepted definition for overtaking acceleration time. It often refers to the time required from a certain speed to a higher speed with the accelerator pedal fully depressed (or wide-open throttle, WOT). Furthermore, the ability of acceleration for an EV can be better evaluated by its WOT acceleration curve, that is, the speed versus time graph. The maximum speed is the highest speed that an EV can reach on a good, level road surface typically paved with concrete or asphalt continuously for a certain duration. For some high-end passenger EVs, the maximum speed is nowadays over 200 km/h continuously for quite some time, for example, 30 min or longer. Obviously, the actual maximum speed of the same EV can be higher for a shorter period of time.

The gradeability is the maximum slope of a good road surface that an EV can climb at a certain load. In addition, the maximum speed on a good road surface of a certain slope can be taken as supplementary indicators of gradeability. For instance, the maximum speed on 4, 6, or 12% slope is often cited. For an EV with multiple gears, the gradeability is obviously attained at the first gear. The gradeability requirements vary with the type of EVs, for example, 20% for electric cars, 30% for electric commercial trucks, or 60% for off-road EVs.

All the aforementioned indicators should be measured under calm or breezy conditions.

1.2. Tractive Effort and Resistance Forces

To examine the dynamic performance of an EV, its movement in the direction of motion must be determined. The movement depends on the various external forces acting on the vehicle, including tractive effort and resistance forces. By establishing the equation of motion based on the balance of these forces, the acceleration, maximum speed, and gradeability of an EV can be calculated.

In general, the equation of motion for an EV is as follows:

$$\Sigma F_d = \Sigma F_r \tag{1.1}$$

where

ΣF_d is the sum of all tractive effort forces
ΣF_r is the sum of all resistance forces

Depending on the electric drive system layout and the number of drive wheels, the tractive effort may come from the torque of a single electric motor or multiple motors, transmitted with or without drivetrain(s). The resistance consists of rolling, air drag, slope, and acceleration forces.

1.2.1. Tractive Effort

The tractive effort F_t is the force of the ground reacting to the driving wheels produced by circumferential forces F_a acting on the ground from the driving wheels, as shown in Figure 1.1 [1.1]. The circumferential forces F_a are produced by the torque T_t transmitted to the driving wheels from a single electric motor or multiple electric motors either directly or through a driveline. Apparently, the tractive effort can be expressed by

$$F_t = \frac{T_t}{R} \tag{1.2}$$

where R is the radius of driving wheel. The driving wheel torque is given by

$$T_t = T_e i_g i_0 \eta_T \tag{1.3}$$

where
T_e is the torque of the electric motor(s)
i_g is the gear ratio of transmission
i_0 is the final drive ratio
η_T is the mechanical efficiency of the drivetrain

Therefore, the equation of tractive effort, that is, Equation (1.2), becomes

$$F_t = \frac{T_e i_g i_0 \eta_T}{R} \tag{1.4}$$

FIGURE 1.1 Tractive effort on a wheel.

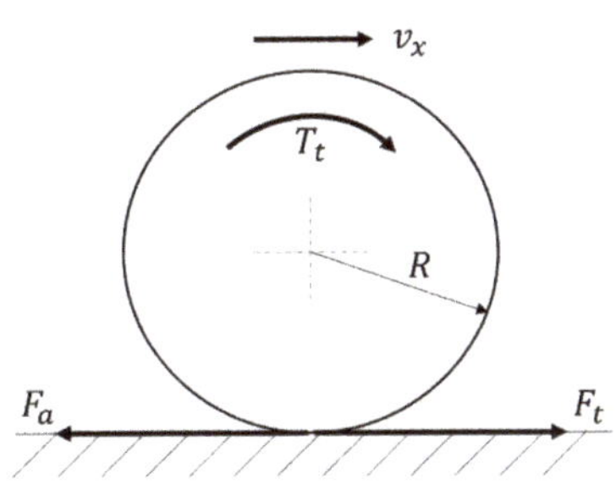

Reprinted with permisssion from Yu, Z.-S. (2009) Automotive Theory. China Machine Press, Beijing.

1.2.1.1. Characteristics of Electric Motors

The characteristics of an electric motor refer to the relationship between its output power P_e and output torque T_e versus speed n_e. Sometimes termed external characteristics, the relationships under peak and rated operating conditions of the electric motor are most frequently used.

The peak characteristics of a typical electric motor are shown in Figure 1.2. The electric motor can operate at any speed ranging from zero to the maximum motor speed n_{emax}. With the speed increasing from zero, the output torque T_e remains constant at T_{emax}, while the output power P_e increases linearly from zero to P_{emax}, until it reaches base speed n_{eb}. Beyond the base speed n_{eb}, the output torque T_e starts to decrease with speed in an inverse proportion, while the output power P_e becomes constant, until it reaches the maximum speed n_{emax}.

FIGURE 1.2 Characteristics of an electric motor.

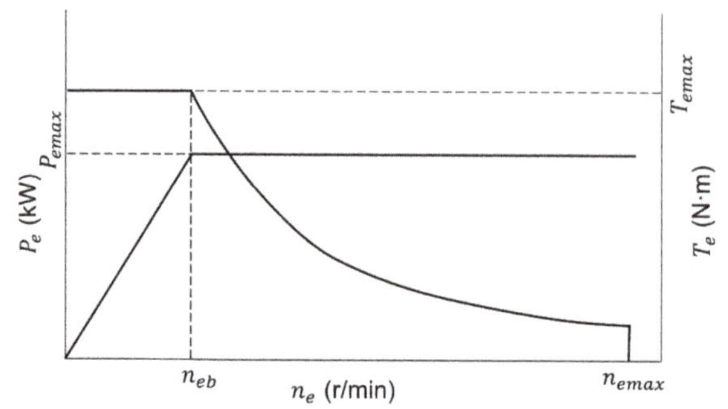

The relationship between the output power P_e and the output torque T_e is as follows:

$$P_e = T_e \omega_e = \frac{\pi T_m n_e}{30{,}000} \tag{1.5}$$

where

P_e is in kW
T_e is in N·m
ω_e is in 1/rad
n_e is in rpm

The external characteristics of an electric motor can be mapped to reflect the variation of output power and torque with speed at different accelerator pedal opening β ranging from approximately 0 to 100%. Obviously, the output torque at various accelerator pedal opening can be described as

$$T_e(\beta) = \beta T_{emax}(n_e) \tag{1.6}$$

where $T_{emax}(n_e)$ is the maximum output torque of the electric motor at the speed n_e. Similarly, for the output power, we have

$$P_e(\beta) = \beta P_{emax}(n_e) \tag{1.7}$$

where $P_{emax}(n_e)$ is the maximum output power of the electric motor at the speed n_e.

1.2.1.2. Mechanical Efficiency of Drivetrain

The mechanical efficiency of drivetrain η_T is defined as the ratio of the power loss to the input power that is transmitted through the drivetrain, that is,

$$\eta_T = \frac{P_i - P_o}{P_i} \tag{1.8}$$

where P_i and P_o are the input power and output power, respectively. As the electric motor operates at a constant speed, the input power of the drivetrain equals the output power of the electric motor, that is, $P_i = P_e$.

The power loss can take place in such components of the drivetrain as transmission, final drive, shaft, universal joint, and so on. In general, the power losses in transmission and final drive are more significant than those in other components. Today, the total power loss in EV drivetrains should be less than 3% in good engineering practice.

The power loss of the drivetrain falls into two categories: mechanical loss and hydraulic loss. Due to the friction between solid surfaces in relative motion, mechanical loss typically occurs in gear pairs, bearings, seals, etc. The mechanical loss depends largely on the load and speed at the friction interfaces. On the contrary, hydraulic loss is related to the drag between a solid surface and a fluid, for instance, gears immersed in lubricants or rotating components exposed to the mixture of air and oil vapor. The hydraulic loss is mainly influenced by fluid properties, temperature, depth of immersion, speed of the rotating component, and so forth.

A systematic perspective has to be adopted when evaluating the mechanical efficiency of the drivetrain. For an EV operating in a given driving cycle, a multiple-speed drivetrain of relatively lower mechanical efficiency could result in a higher energy economy than a single-speed drivetrain, if the former allows the electric motor to operate in its higher efficiency region than the latter. It is, therefore, necessary to take the entire electric drive system into consideration.

1.2.1.3. Wheel Radius

The radius of the wheel not subject to load is termed free radius R_f.

The distance from the center of the wheel to the contact surface between the tire and the road for a stationary EV is called static radius R_s. Because the tire is deformed under radial load, the static radius is always smaller than the free radius.

A rolling radius R_r can be calculated by the number of wheel revolutions N_w and the rolling distance of the wheel S:

$$R_r = \frac{S}{2\pi N_w} \tag{1.9}$$

Obviously, the static radius R_s should be used for dynamic analysis, while the rolling radius R_r is more suitable for kinematic analysis. For simplicity, they are often approximated identical to the wheel radius R, that is,

$$R_s \approx R_r \approx R \tag{1.10}$$

1.2.1.4. Tractive Effort in Different Gears

The maximum tractive effort available for a given EV varies with vehicle speed and gear of transmission, which can be illustrated by so-called tractive effort diagrams. With parameters such as external characteristics of the electric motor, gear ratios, transmission efficiency, wheel radius, etc. available, the tractive effort in each gear can be obtained by Equation (1.4), while the vehicle speed can be calculated by

$$v_x = \frac{3\pi R n_e}{25 i_g i_0} \approx 0.377 \frac{R n_e}{i_g i_0} \tag{1.11}$$

where

v_x is the longitudinal speed in km/h
R is the wheel radius in m
n_e is the electric motor speed in rpm
i_g is the transmission ratio
i_0 is the final drive ratio

1.2.2. Resistance Forces

As an EV traveling at a constant speed on a horizontal road, it must overcome the rolling resistance F_f from the ground and the drag from the air F_d. In addition, for traveling uphill on a ramp, it has to overcome slope resistance F_s, a component of gravity along the ramp. Moreover, an acceleration resistance F_a needs to be overcome for acceleration; this is referred to as acceleration resistance. Thus, the total resistance for an EV to overcome becomes

$$\sum F = F_f + F_d + F_s + F_a \tag{1.12}$$

1.2.2.1. Rolling Resistance

For a driven wheel rolling on a hard road surface, the pressure distribution is asymmetrical in its rolling direction because of the elastic hysteresis in its tire deformation. The pressure in the front is actually higher than that in the back, resulting in an offset distance b between the composite normal force F_z from the ground to the wheel and the center of wheel, as shown in Figure 1.3a [1.2]. The normal force F_z has the same magnitude as the vertical load W but in the opposite direction. The offset distance b increases as the elastic hysteresis of the tire increases. The normal force F_z can be translated to the center of the wheel by adding a rolling resistant moment T_f (see Figure 1.3b). Obviously, we have

$$T_f = F_z b \tag{1.13}$$

FIGURE 1.3 Pressure distribution caused by a driven wheel rolling on a hard road surface. (a) The resultant force and (b) the equivalent.

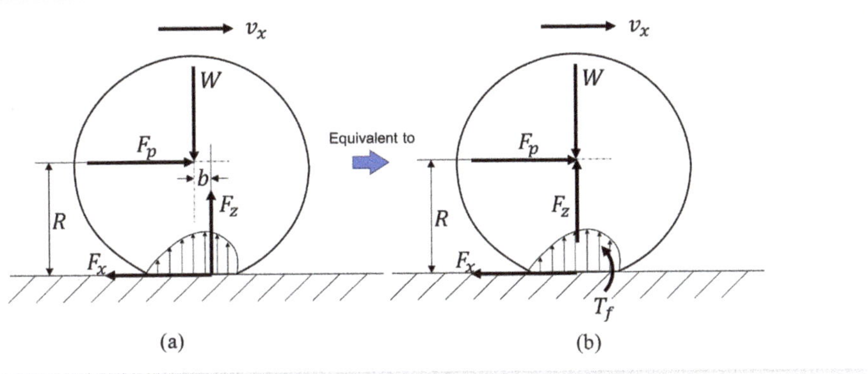

To keep the wheel rolling, a force F_p must be exerted on the wheel center to overcome the rolling resistant moment T_f, that is,

$$F_p = \frac{b}{R} F_z \tag{1.14}$$

Let $f_r = \frac{b}{R}$, we have

$$F_p = f_r F_z = fW \tag{1.15}$$

where f_r is the rolling friction coefficient. For the driven wheel rolling at constant speed, the force F_p should be equal to the rolling resistant force F_f:

$$F_f = \frac{T_f}{R} = f_r W \tag{1.16}$$

For a drive wheel rolling on a hard road surface, the tractive moment T_t must overcome the rolling resistant moment T_f in a similar way to the aforementioned driven wheel scenario, as shown in Figure 1.4. Hence, the net tractive force acting on the wheel from the ground F_x is

$$F_x = \frac{T_t - T_f}{R} = F_t - F_f \tag{1.17}$$

FIGURE 1.4 Pressure distribution of a drive wheel rolling on a hard road surface. (a) The resultant force and (b) the equivalent.

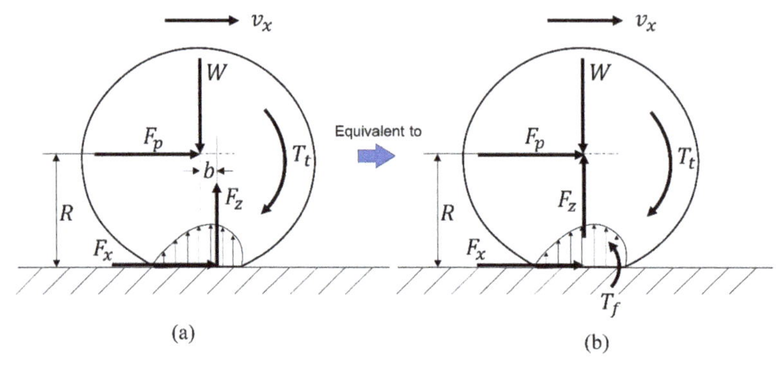

1. Rolling friction coefficient f_r
   The rolling friction coefficient of vehicle tires depends on various factors such as speed, road surface condition, and the structure, material, and tire pressure. The intrinsic complexity makes it only possible to measure the rolling friction coefficient by testing.
2. Effect of speed
   For passenger EV tires, the rolling friction coefficient can be expressed as a function of squared speed:

$$f_r = f_{r0} + f_{r1}v_x^2 \tag{1.18}$$

where

$f_{r0} = 0.015$

$f_{r1} = 7 \times 10^{-6}\ \mathrm{s^2/m^2}$

For commercial EV (truck) tires, the rolling friction coefficient increases linearly with speed:

$$f_r = f_{r0} + f_{r1}v_x \tag{1.19}$$

where

$f_{r0} = 4.1 \times 10^{-3} - 7.6 \times 10^{-3}$

$f_{r1} = 4.1 \times 10^{-5} - 5.6 \times 10^{-5}\ \mathrm{s^2/m^2}$

The speed has a significant effect on rolling resistance. As speed increases, the coefficient slightly increases at low and medium speeds and turns up quickly at high speeds. Once a critical speed is reached, the coefficient increases dramatically because standing waves occur on the tire circumference. Then, the tire bursts in a short time due to overheating.

Factors such as the structure, cord, variety of rubber, and inflation pressure of tires all affect rolling resistance. The radial ply tires, in general, have smaller rolling resistance coefficients than nonradial tires, for example, bias ply tires. The coefficient increases rapidly with a decrease in inflation pressure because greater hysteresis loss results from larger deformation.

3. Effect of road surface
   The rolling resistance coefficients f_{r0} at low to moderate speeds on various road surfaces are listed in Table 1.1 [1.1].

TABLE 1.1 Rolling resistance coefficients of vehicle tires on various road surfaces.

Road surface condition	f_{r0}
Concrete or asphalt, good pavement	0.008–0.018
Concrete or asphalt, average to poor pavement	0.010–0.020
Tarmac, good pavement	0.010–0.013
Gravel, rolled new	0.02
Gravel, loose worn	0.04–0.08
Macadam, good pavement	0.013–0.016
Macadam, average pavement	0.018–0.025
Block, good pavement	0.02
Cobble, small new	0.01–0.015
Cobble, large worn	0.03
Pebble, good pavement	0.025–0.030
Pebble, with potholes	0.035–0.050
Soil, compact and dry	0.025–0.050
Soil, compact after rain	0.050–0.150
Soil, medium hard	0.04–0.08
Muddy soil, rainy season or thawing period	0.100–0.250
Sand, loose	0.2–0.4
Sand, solid	0.04–0.08
Ice	0.014–0.030
Snow, compacted	0.016–0.050
Snow, loose	0.056

EVs are often equipped with low rolling resistance (LRR) tires to maximize their range. LRR tires, in general, incorporate silica instead of carbon black in their tread to reduce low-frequency hysteresis without compromising traction [1.3]. LRR tires can cut the rolling friction coefficient by as much as 43–59% and improve the energy economy of an EV by approximately 5–15% in comparison to regular tires.

4. Effect of load and inflation pressure
   An empirical equation has been developed for estimating the effect of both load and inflation pressure on the rolling friction coefficient:

$$f_r = K\left(5.1\times10^{-3} + \frac{550+0.09F_z}{p} + \frac{1.1+3.88\times10^{-5}F_z}{p}v_x^2\right) \tag{1.20}$$

   where
   - F_z is normal load in N
   - p is inflation pressure in Pa
   - v_x is longitudinal speed in m/s
   - K is 0.8 for radial tires and 1.0 for other types of tires

   Clearly, the rolling friction coefficient f_r rises by an increase in normal load F_z on the tire print. The effect of decreasing inflation pressure p is similar to that of increasing normal load.

5. Effect of tire slip
   A longitudinal slip ratio can be defined to quantify the longitudinal slip developed during driving or braking, that is,

$$s = \frac{\omega_w R - v_x}{v_x} \tag{1.21}$$

   where
   - ω_w is the angular velocity of the tire
   - R is the radius of the tire
   - v_x is the longitudinal velocity of the tire center

Apparently, s is > 0 for driving and s is < 0 for braking.

To accelerate or brake an EV, the longitudinal force developed between the tire and the ground is proportional to the normal force F_z on the tire print:

$$F_x = fF_z \tag{1.22}$$

where f is the longitudinal friction coefficient, also known as longitudinal adhesion coefficient. Figure 1.5 shows the longitudinal friction coefficient f versus slip ratio s [1.4]. The coefficient f is almost linearly proportional to s when $|s| \ll 1$ and then peaks at $|s| \approx 0.1$, where $f = f_{dp}$ for driving and $f = f_{bp}$ for braking. As $|s|$ continues to increase from approximately 0.1, the tire starts to slip and f begins to drop until it reaches a steady value f_{ds} for driving and f_{bs} for braking. The adhesion coefficients of some typical road surfaces are given in Table 1.6 of Section 1.4.2.

FIGURE 1.5 Longitudinal friction coefficient versus slip ratio.

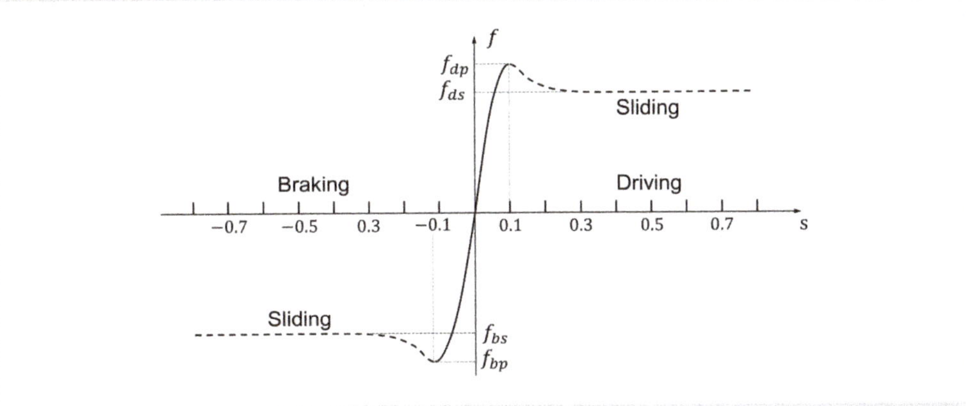

Courtesy of Springer.

For a wheel transferring tractive or braking forces, a certain degree of tire slip relative to the ground can result in additional energy loss in tire rolling. Hence, the rolling friction coefficient f_r for the same tire is typically larger on a drive wheel than that on a driven wheel. Figure 1.6 illustrates the trend of the coefficient f_r varying with the traction coefficient, that is, the ratio of tractive force to normal load [1.3, 1.5]. The coefficient increases quickly with an increase in the tractive force ratio but less significantly for radial tires than for bias tires. Note that a similar trend exists for the rolling friction coefficient as a function of speed. In general, the rolling friction coefficient of radial tires is lower than that of bias tires.

FIGURE 1.6 Variation of the rolling friction coefficient with the traction coefficient.

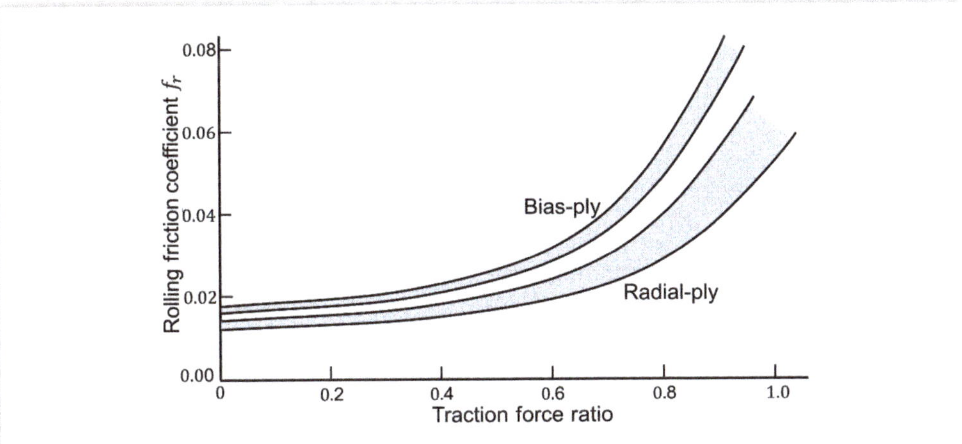

Reprinted with permission from Yu, Z.-S. (2009) Automotive Theory. China Machine Press, Beijing.

Cornering forces, on the other hand, have a similar influence on the rolling friction coefficient f_r. By combining Equations (1.15) and (1.22), the net tractive or braking force on a wheel, F_p, becomes

$$F_p = (f - f_r)F_z \tag{1.23}$$

where $f < 0$ and $f_r > 0$ in the case of braking.

1.2.2.2. Air Resistance

Aerodynamic forces can significantly affect the energy economy, maneuverability, and noise, vibration, and harshness (NVH) quality of an EV in motion. Although the lateral and vertical components of aerodynamic forces contribute to the roll, pitch, and yaw behavior of the EV, only the longitudinal component is taken into account as road load, namely, air resistance.

Air resistance comes from two sources, that is, pressure drag and viscous friction. Pressure drag is the longitudinal component of the resultant force of air pressure acting on the surface of an EV. Viscous friction, on the other hand, is the longitudinal component of the resultant force of tangential viscosity force of the air acting on the surface of an EV. The pressure drag is, in general, produced by these mechanisms: first, the exterior shape of the main body; second, the interference of protrusions such as rearview mirrors, door handles, and drive shafts; third, the internal circulation such as cooling system and ventilation; and fourth, the induction due to the longitudinal component of lift or down forces. Figure 1.7 illustrates the normal pressure distribution in several cross sections of a typical electric sedan with a small grille at the bottom of its front end [1.1]. However, the front grille is generally not deemed a necessity for EVs.

FIGURE 1.7 Normal pressure distribution on the body surface. (a) In vertical longitudinal zero plane and (b) in horizontal planes.

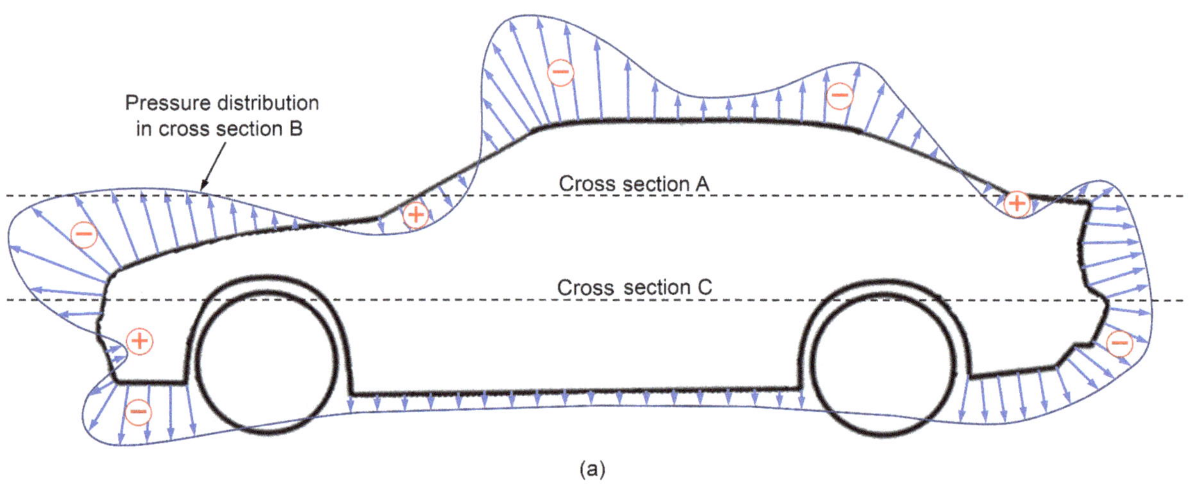

FIGURE 1.7 (Continued) Normal pressure distribution on the body surface. (a) In vertical longitudinal zero plane and (b) in horizontal planes.

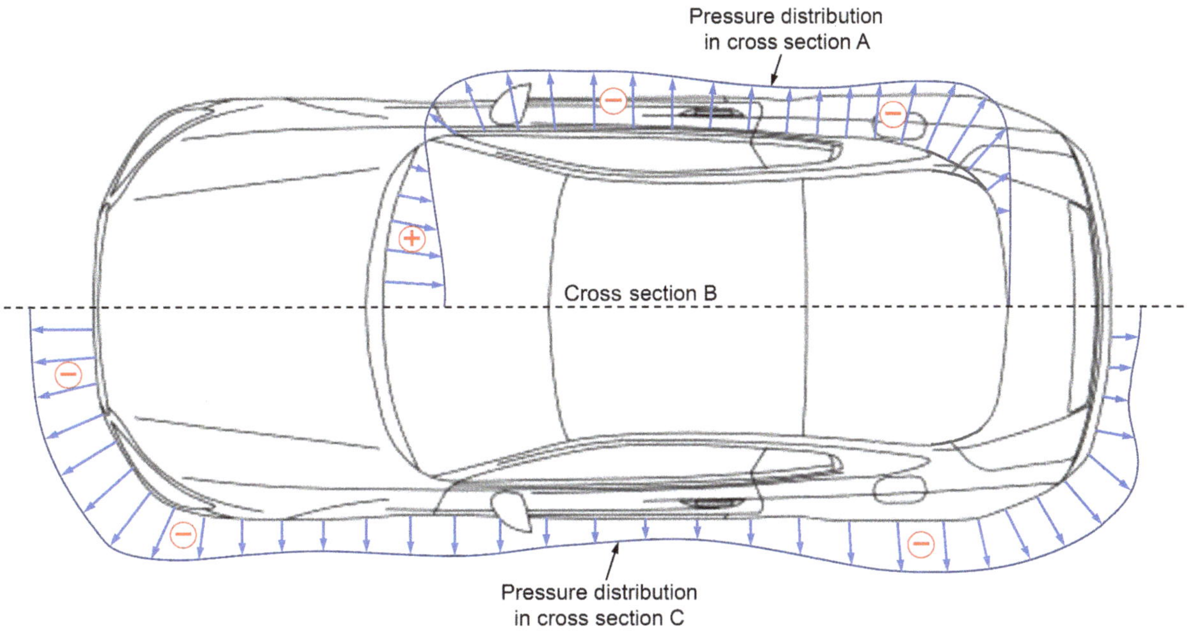

For a regular car, the rough proportion of these resistances is 55–58% of shape resistance, 10–14% of interference resistance, 11–12% of internal circulation resistance, 7–9% of induction resistance, and 9–10% of friction resistance. Clearly, the exterior shape of the vehicle body is the most important contributor to air resistance.

The air resistance usually takes the form of a positive proportion to the dynamic pressure of the air flow relative to the EV, that is,

$$F_d = \frac{1}{2}\rho C_D A v_r^2 \tag{1.24}$$

where

ρ is air density

C_D is the air resistance coefficient

A is the frontal area, that is, the projected area of the vehicle in the direction of motion

v_r is air velocity relative to the vehicle

Reduction in either C_D or A can result in decreased air resistance for energy economy improvement. The frontal area A, however, is often subject to interior space limitations. Thus, the minimization of the air resistance coefficient C_D becomes one of the main engineering targets in EV design. The C_D values of well-known EV models are listed in Table 1.2.

TABLE 1.2 Drag coefficients of popular EV models.

Manufacturer	Model year	Body style	C_D
Ford	2022 Ford Mustang Mach-E	Crossover	0.30
General Motor	1996 EV1	Sedan	0.19
	2017 Chevy Bolt	Coupe	0.308
Lucid	2022 Air	Sedan	0.21
Rivian	2022 R1S	SUV	0.28
Tesla	2021 Model S	Liftback car	0.208
	2017 Model 3	Sedan	0.23
	2018 Model X	SUV	0.24
	2020 Model Y	Crossover	0.23
Mitsubishi	2011 i-MiEV	Hatchback car	0.35
Nissan	2017 Leaf	Sedan	0.28
	2022 Ariya	Crossover	0.297
Subaru	2023 Solterra	Crossover	0.28
Toyota	2022 Prius Prime	Hatchback car	0.25
	2023 bZ4X	SUV	0.28
Hyundai	2016 Ioniq	Liftback car	0.25
	2020 Kona EV	SUV	0.32
	2022 Ioniq 5	SUV	0.288
	2022 Ioniq 6	Sedan	0.21
Kia	2020 Soul EV	SUV	0.35
	2022 Niro EV	Crossover	0.29
	2022 EV6	Crossover	0.28
Audi	2022 e-tron GT	Sedan	0.24
	2020 e-tron Sportback	Crossover	0.26
	2022 e-tron Quattro	SUV	0.28
BMW	2014 i3	Hatchback	0.30
	2015 i8	Coupe	0.26
	2022 i4 eDrive40 Gran Coupe	Coupe	0.24
	2022 ix xDrive50	Crossover	0.25
Mercedes-Benz	2021 EQS	Sedan	0.20
Porsche	2022 Taycan	Sedan	0.22
	2022 Porsche Taycan Turbo 4S	Shooting brake (Sport Turismo)	0.26
NIO	2021 ET7	Sedan	0.208
Polestar (Volvo)	2022 Polestar 2	Liftback car	0.278
Volvo	2022 C40 Recharge	SUV	0.329

The rules generally followed to lower the C_D value in designing EVs are given as follows:

1. Tilt the hood forward and down. Round the angle at the junctions between faces. Position windshield as level as possible with a smooth transition to the roof. Reshape the water channel on the A pillars. Round the front strut. Adjust the front fascia and mount a suitable spoiler (air dam) under the bumper in the front end. Eliminate air intake grille.
2. Minimize and streamline the lamps, rear view mirrors, and their attachment arms, as well as other protrusions.

3. Flush the side window with the body. Adjust the angle of the rear window.
4. Tune the wheel openings. Flush the wheel cover with the tire. Place wheel spoilers in front of tires to reduce turbulence.
5. Lean the entire body of the EV forward by 1–2°.
6. Make the projection of the body in the horizontal plane drum shaped in the middle, semicircular at front, and tapered slightly toward center at the rear.
7. Style the afterbody with a hatchback or fastback with a rear spoiler whenever possible. Raise the trunk cover and shorten its length for a notchback, with a duck tail design.
8. Make the underbody flat, and preferably cover it with a smooth panel. Raise the panel slightly upward from the mid-body or rear wheels toward the back. Use a diffuser to tune air coming off the underside.
9. Design the thermal management system carefully, in particular the positions of the air inlet and outlet if ram air is used for cooling purposes.

Table 1.3 lists the typical fractions that body components or features of an EV contributing to C_D.

TABLE 1.3 Contributions of body components or features to the drag coefficient.

Component or feature		Fraction of contribution to C_D (%)	
Body	Forebody	11.9	65.5
	Afterbody	33.3	
	Underbody	14.3	
	Skin friction	5.95	
Protuberance	External mirrors	2.4	28.6
	Wheels and wheel wells	21.4	
	Window recesses	2.4	
	Drip rails	2.4	
Internal drag	Cooling system	5.95	5.95

It is worth noting that the C_D value of an EV actually varies with ground clearance, pitch angle, and the strength of lateral wind.

Table 1.4 summarizes the air resistance coefficients and frontal areas of various types of EVs.

TABLE 1.4 Air resistance coefficient and frontal area of various types of EVs.

Type of EV	C_D	A (m^2)
Cars	0.19–0.31	1.7–2.0
SUVs	0.24–0.35	2.34–2.65
Vans	0.33–0.35	2.9–3.4
Pickup trucks	0.42–0.46	2.0–3.3
Buses	0.5–0.8	4–7
Trucks	0.6–1.0	3–10.5

1.2.2.3. Grading Resistance

Grading resistance is the component along the ramp of the gravity of the EV traveling up a slope:

$$F_s = mg\sin\alpha \tag{1.25}$$

where

m is the mass of EV
g is the acceleration of gravity
α is the slope angle

The road slope i_s is often expressed as the ratio of the height of slope H to the length of bottom L as a percentile, that is,

$$i_s = \frac{100H}{L}\% = \tan\alpha \tag{1.26}$$

It is apparent from Equation (1.25) that the grading resistance grows with an increase in the slope angle. As the grading resistance becomes large, the vehicle speed must be reduced to gain enough tractive force to overcome the resistance due to the limited power from the electric motor. In fact, the speed limit is set according to the road slope, although there are some differences from nation to nation, as listed in Table 1.5.

TABLE 1.5 Speed limits set according to road slope in some nations.

	Speed limit (km/h)	120	100	80	60
Road slope (%)	The US (Hilly)	4	4	5	6
	The US (Mountainous)	5	6	7	8
	China	3	4	5	6
	Japan	2	3	4	5
	Australia	4-6	4-6	5-7	7-9
	France	—	5	6	7

The combination of grading resistance and rolling resistance is referred to as road resistance because they are both road-related. The total resistance of air resistance and road resistance is usually called road load F_r:

$$F_r = F_f + F_s + F_d \tag{1.27}$$

1.2.2.4. Acceleration Resistance

The inertial force required to accelerate the mass of an EV is referred to as acceleration resistance. The mass consists of translation mass and rotational mass. The moment required to accelerate the rotational inertia is, in general, converted into a translational force by introducing a rotational inertia conversion factor δ for the convenience of calculation, that is,

$$F_a = \delta m \frac{dv}{dt} \tag{1.28}$$

where
- m is vehicle mass
- v is vehicle velocity

δ is related to the inertia of the drivetrain and wheels, as well as the ratios of the drivetrain by the following expression:

$$\delta = 1 + \frac{\sum I_w}{mR^2} + \frac{I_i i_g^2 i_0^2 \eta_T}{mR^2} \tag{1.29}$$

where
- I_w is the inertia of all wheels
- I_i is the drivetrain inertia equivalent to the input shaft
- i_g is the transmission ratio
- i_0 is the final drive ratio
- η_T is transmission efficiency

Clearly, $\delta > 1$, with the typical value of δ for EVs ranging from approximately 1.05 to 1.45.

1.3. Vehicle Dynamic Equation

According to the driving resistances analyzed previously, the one-dimensional vehicle dynamic equation of an EV can be expressed as

$$F_t = F_f + F_s + F_d + F_a \tag{1.30}$$

where F_t is the tractive effort force.

Substitution of aforementioned driving resistances yields

$$\frac{T_t}{R} = mgf\cos\alpha + mg\sin\alpha + \frac{1}{2}\rho C_D A v^2 + \delta m \frac{dv}{dt} \tag{1.31}$$

Considering the tractive effort torque T_t from an electric motor through the drivetrain and $\rho = 1.2258$ kg/m^3 for air at the sea level, Equation (1.31) becomes

$$\frac{T_e i_g i_0 \eta_T}{R} = mgf\cos\alpha + mg\sin\alpha + \frac{C_D A}{21.15}v^2 + \delta m\frac{dv}{dt} \tag{1.32}$$

where

T_e is the output torque of the electric motor
i_g is the transmission ratio
i_0 is the final drive ratio
η_T is the mechanical efficiency of the drivetrain
R is the tire radius

In addition, the following relationship holds between electric motor speed ω_e and vehicle speed v:

$$\omega_e = i_g i_o \omega_w = \frac{i_g i_o v}{R} \tag{1.33}$$

where ω_w is the angular speed of wheel.

1.3.1. Equilibrium Diagrams of Driving Force and Running Resistance

Figure 1.8 is the equilibrium diagram of driving force and running resistance of a two-speed EV. Rolling resistance does not vary with the velocity when a constant friction coefficient is assumed. Air resistance, however, increases with the square of the velocity. The diagram helps evaluate the dynamic performance of an EV. The maximum velocity in the continuous operation of an EV, v_{max}, can be determined directly from the diagram, which is the intersection of the rated driving force curve $F_{t,rated}$ with the running resistance curve $F_f + F_d$. The extra force over resistance is available from the rated driving force for accelerating the EV at any velocity below v_{max}. On the contrary, the extra force from the peak driving force is only available for a short-time acceleration (typically 30 sec to 3 min) for the velocity above v_{max}.

FIGURE 1.8 Equilibrium diagram of driving force and running resistance of a two-speed EV.

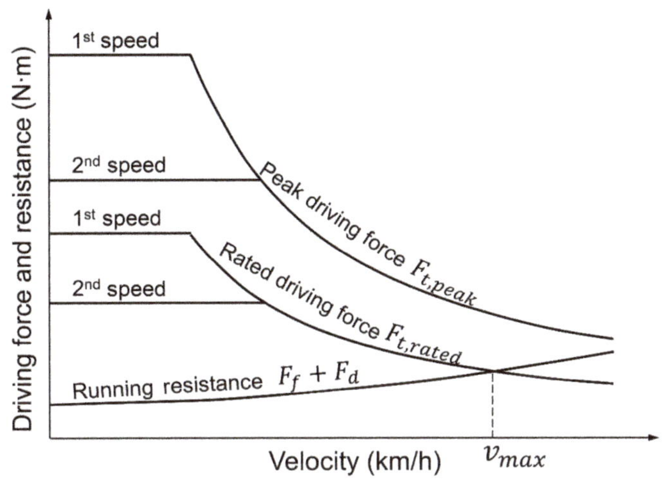

1.3.2. Acceleration Capacity

The acceleration capacity of EV is evaluated on a well-paved flat road. By rearranging Equations (1.30), (1.31), and (1.32), the acceleration can be obtained as follows:

$$a = \frac{dv}{dt} = \frac{1}{\delta m}\left(F_t - F_f - F_d\right) \tag{1.34}$$

Obviously, the maximum acceleration of an EV should be evaluated with the acceleration pedal fully depressed. When a multispeed transmission is utilized, the maximum acceleration at 1st speed can be greater than that at 2nd gear at low velocities, but they become identical as the tractive effort force is eventually limited by the available peak power at higher velocities.

Acceleration time is actually more frequently utilized to quantity acceleration capacity for its ease of measurement. The acceleration time of 0–100 km/h, 0–50 km/h, or 0–60 mph is most often quoted. Per Equations (1.30) through (1.32), the acceleration time from 0 to V km/h is

$$t_{0-V} = \int_0^V \frac{1}{a} dv = \frac{1}{3.6}\int_0^V \frac{\delta m}{\frac{3600\eta_T P_{max}(v)}{v} - \left(mgf + \frac{C_d A v^2}{21.15}\right)} dv \tag{1.35}$$

where

t_{0-V} is in sec

$P_{max}(v)$ is the maximum power of electric motor at the vehicle velocity v

Per Equation (1.33), as a function of electric motor speed, the speed ratio of drivetrain and wheel radius, vehicle velocity is also indirectly related to transmission shift schedules if a multispeed transmission is equipped.

1.3.3. Gradeability

Gradeability refers to the maximum grade an EV can climb by overcoming the slope resistance with the extra force on top of $F_f + F_d$, which can also be computed by assuming $\frac{dv}{dt} = 0$ in the driving equation, that is, Equation (1.32):

$$\alpha = \arcsin\left(\frac{T_t i_g i_0 \eta_T}{mgR} - f\cos\alpha - \frac{C_D A v^2}{21.15mg}\right) \tag{1.36}$$

Sometimes the maximum velocity of an EV for a given slope needs to be examined, thus

$$v = \sqrt{\frac{21.15}{C_D A}\left(\frac{T_t i_g i_0 \eta_T}{R} - mgf\cos\alpha - mg\sin\alpha\right)} \tag{1.37}$$

1.4. Tire-Ground Adhesion

1.4.1. Adhesion Coefficient and Adhesion Condition

The dynamic performance of an EV depends largely on its electric drive system, that is, the electric motor(s) and drivetrain. This is, however, only true if the tire–ground adhesion is sufficient, for example, good tires on dry concrete pavements. As the adhesion is poor on wet pavement, the large driving force may cause the wheels to slip and spin abruptly. The decreased reaction force from the ground due to slip can result in degraded acceleration performance. Thus, the driving performance of an EV is limited not only by the driving force but also by the tire–ground adhesion condition.

Adhesion friction F_{ad} is the upper limit of tangential reaction force from the ground to the tire, which is proportional to the normal reaction force F_z from the road surface to the driving tire:

$$F_{ad} = fF_z \tag{1.38}$$

where f is the adhesion coefficient, which depends on both the tire and the road surface. To prevent the driving wheels from slip, the so-called adhesion condition must be held:

$$\frac{F_x}{F_z} \le f \tag{1.39}$$

where F_x is the tractive effort force on the driving wheel. By defining a ratio C_f as

$$C_f = \frac{F_x}{F_z} \tag{1.40}$$

The adhesion condition expressed by Equation (1.38) becomes

$$C_f \le f \tag{1.41}$$

The ratio C_f of driving wheels is an important indicator of the adhesion behavior of an EV, as it represents the smallest adhesion coefficient needed for driving with the maximum torque available without slip.

1.4.2. Adhesion Coefficient

The adhesion coefficient depends largely on factors such as the type of road surface and its condition, driving speed, and state of wheel motion. The common engineering practice is to take the average value of adhesion coefficient in motion analysis. The adhesion coefficients of some typical road surfaces are listed in Table 1.6.

TABLE 1.6 Adhesion coefficients of some typical road surfaces.

Road surface	Adhesion coefficient *f*
Concrete or asphalt, good pavement, dry	0.7–0.9
Concrete or asphalt, good pavement, wet	0.45–0.8
Macadam, dry	0.6–0.7
Gravel	0.55–0.6
Compact soil, wet	0.2–0.4
Compact soil, dry	0.5–0.6
Compacted snow	0.15–0.2
Ice	0.07–0.1

1.4.3. Adhesion-Related Forces

The ground reaction force on drive wheels depends mainly on the overall structure of the vehicle, body shape, slope of the road, and driving conditions. Figure 1.9 illustrates the related forces acting on a front-wheel drive (FWD) EV going up a grade in acceleration [1.5]. The normal reaction force acting on the front and rear wheels from the ground is as follows, respectively:

$$F_{z1} = mg\left(\frac{L_2}{L}\cos\alpha - \frac{h_g}{L}\sin\alpha\right) - \left(\frac{mh_g}{L} + \frac{\Sigma I_w}{LR} \pm \frac{I_{iy}i_g i_0}{LR}\right)\frac{\mathrm{d}v}{\mathrm{d}t} - \frac{mgRf_r}{L}\cos\alpha - F_{zw1} \tag{1.42}$$

$$F_{z2} = mg\left(\frac{L_1}{L}\cos\alpha + \frac{h_g}{L}\sin\alpha\right) + \left(\frac{mh_g}{L} + \frac{\Sigma I_w}{LR} \pm \frac{I_{iy}i_g i_0}{LR}\right)\frac{dv}{dt} + \frac{mgRf_r}{L}\cos\alpha - F_{zw2} \quad (1.43)$$

where

L is the wheelbase

α is the slope angle

h_g is the centroid height

R is the tire radius

I_w is the wheel inertia

I_{iy} is the rotating inertia about y axis of all components in the drivetrain equivalent to the input shaft

i_g is the transmission ratio

i_0 is the final drive ratio

f_r is the rolling friction coefficient

F_{zw1} and F_{zw2} are the aerodynamic lift force on vehicle body above front and rear tire-prints, respectively

FIGURE 1.9 Forces acting on an FWD EV going up a grade.

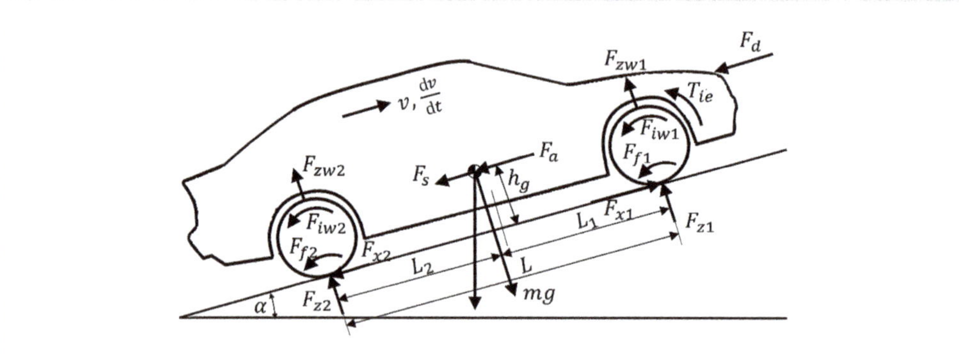

The first term on the right is the normal rection force from the ground due to static axle load. The second term on the right is the normal rection force from the ground dynamically produced by the inertia force and resistance moment during acceleration. In particular, $\pm\frac{I_{iy}i_g i_0}{LR}\frac{dv}{dt}$ is the resistance moment of the lateral mounted electric motor and drivetrain, whereas "+" is for the motor shaft rotating at the same direction as wheels and "−" is for the motor shaft rotating at the opposite direction. This resistance moment is typically small and negligible. The third term on the right is the contribution of rolling resistance moment, that is, $\frac{mgRf_r}{L}\cos\alpha$, which is also small and negligible most of the time. The aerodynamic lift force is produced by the difference in air flow velocity passing the top and bottom of the EV. The lift is often decomposed into front and rear ones acting on the front and rear tire prints:

$$F_{zw1} = \frac{1}{2}C_{Lf}\rho A v_r^2 \quad (1.44)$$

$$F_{zw2} = \frac{1}{2} C_{Lr} \rho A v_r^2 \tag{1.45}$$

where C_{Lf} and C_{Lr} are front and rear aerodynamic lift coefficients, respectively, which can be measured by tests; v_r is the air velocity relative to the EV.

In Table 1.7, the front and rear aerodynamic lift coefficients of some typical body styles are listed [1.1].

TABLE 1.7 Front and rear aerodynamic lift coefficients of typical body styles.

Body style	C_{Lf}	C_{Lr}
Sedan	0.136	0.246
Coupe	0.133	0.279
Fastback	0.135	0.118
Station wagon	0.116	0.063

Reprinted with permisssion from Yu, Z.-S. (2009) Automotive Theory. China Machine Press, Beijing.

By neglecting the resistance moment of rotating inertia and rolling resistance moment, the normal reaction force acting on the front and rear wheels from the ground expressed in Equations (1.42) and (1.43) can be reduced to

$$F_{z1} = mg\left(\frac{L_2}{L}\cos\alpha - \frac{h_g}{L}\sin\alpha\right) - \frac{mh_g}{L}\frac{\mathrm{d}v}{\mathrm{d}t} - F_{zw1} \tag{1.46}$$

$$F_{z2} = mg\left(\frac{L_1}{L}\cos\alpha + \frac{h_g}{L}\sin\alpha\right) + \frac{mh_g}{L}\frac{\mathrm{d}v}{\mathrm{d}t} - F_{zw2} \tag{1.47}$$

For a front-wheel driving EV, the tangential reaction force acting on the front wheels from the ground is

$$F_{x1} = F_{f2} + F_d + F_s + F_a' \tag{1.48}$$

where

F_{f2} is the rolling resistance force acting on rear wheels, $F_{f2} = \frac{T_{f2}}{R}$

F_d is aerodynamic resistance acting on the body

F_s is slope resistance

F_a' is the inertia force due to the linear acceleration of vehicle mass, with $F_a' = m\frac{\mathrm{d}v}{\mathrm{d}t}$

Note that in Equation (1.45), the moment of inertia of wheels is ignored.

Similarly, for a rear-wheel driving EV, the tangential reaction force acting on the rear wheels from the ground is

$$F_{x2} = F_{f1} + F_d + F_s + F_a' \tag{1.49}$$

where F_{f1} is the rolling resistance force acting on front wheels and $F_{f1} = \frac{T_{f1}}{R}$.

For a four-wheel drive EV, the tangential reaction forces acting on the front and rear wheels from the ground are, respectively,

$$F_{x1} = (1-\varphi)\left(mg\sin\alpha + m\frac{dv}{dt}\right) \tag{1.50}$$

$$F_{x2} = \varphi\left(mg\sin\alpha + m\frac{dv}{dt}\right) \tag{1.51}$$

where φ is the rear torque split ratio, with $\varphi = \frac{T_{t2}}{T_{t1} + T_{t2}}$. Here, T_{t1} and T_{t2} are drive torques at front and rear axles, respectively.

1.4.4. Minimal Adhesion Coefficient Requirements

A minimal adhesion coefficient is required to make full use of the tractive effort of the electric drive system for a given EV, which varies with speed under a straight driving condition. For a tractive wheel, the required minimal adhesion coefficient can be defined as

$$f_{min} = \frac{F_x}{F_z} \tag{1.52}$$

Hence, based on Equations (1.46) and (1.48), the required minimal adhesion coefficient for an FWD EV becomes

$$f_{min,f} = \frac{F_{x1}}{F_{z1}} = \frac{F_{f2} + F_d + F_s + F_a'}{mg\left(\frac{L_2}{L}\cos\alpha - \frac{h_g}{L}\sin\alpha\right) - \frac{mh_g}{L}\frac{dv}{dt} - F_{zw1}} \tag{1.53}$$

Likewise, based on Equations (1.47) and (1.49), the required minimal adhesion coefficient for a rear-wheel drive (RWD) EV is given by

$$f_{min,r} = \frac{F_{x2}}{F_{z2}} = \frac{F_{f1} + F_d + F_s + F_a'}{mg\left(\frac{L_1}{L}\cos\alpha + \frac{h_g}{L}\sin\alpha\right) + \frac{mh_g}{L}\frac{dv}{dt} - F_{zw2}} \tag{1.54}$$

As accelerating or grading at low speeds, large minimal adhesion coefficients are required because of a greater tractive effort force. Or, at extremely high speeds on level road, large minimal adhesion coefficients are required as well.

With rolling friction and aerodynamic resistances neglected, the minimal adhesion coefficient required for FWD in Equation (1.53) is reduced to

$$f_{min,f} = \frac{i_{se}}{\frac{L_2}{L} - \frac{h_g}{L} i_{se}} \tag{1.55}$$

$$i_{se} = i_s + \frac{1}{g\cos\alpha}\frac{dv}{dt} \tag{1.56}$$

where i_{se} is the equivalent slope taking acceleration resistance [i.e., the second term in Equation (1.47)] into account and i_{se} is an indicator of the acceleration and gradeability capacity of an EV.

Similarly, the minimal adhesion coefficient required for RWD in Equation (1.54) is reduced to

$$f_{min,r} = \frac{i_{se}}{\frac{L_1}{L} + \frac{h_g}{L} i_{se}} \tag{1.57}$$

For four-wheel drive EVs, the minimal adhesion coefficients required for front and rear wheels, respectively, are

$$f_{min,f} = \frac{(1-\varphi) i_{se}}{\frac{L_2}{L} - \frac{h_g}{L} i_{se}} \tag{1.58}$$

$$f_{min,r} = \frac{\varphi i_{se}}{\frac{L_2}{L} + \frac{h_g}{L} i_{se}} \tag{1.59}$$

Most of the four-wheel drive EVs utilize a front axle and a rear axle, instead of a transfer case. This enables independent adjustment of driving torques between front and rear axles according to operating conditions. Conversion of tractive effort into driving force can be maximized only if front and rear driving forces reach the allowable limits of adhesion friction simultaneously. Thus, the following condition must be satisfied:

$$i_{se} = f \tag{1.60}$$

With much higher equivalent slopes, four-wheel drive EVs usually have better gradeability and acceleration capacity than RWD or FWD EVs.

Assuming both road slope and acceleration are small for an EV operating at high speeds, the minimal adhesion coefficient required for FWD EVs in Equation (1.53) is

$$f_{min,f} = \frac{F_{f2} + F_d}{g\left(\frac{L_2}{L}\cos\alpha - \frac{h_g}{L}\sin\alpha\right) - F_{zw1}} \tag{1.61}$$

Also, the minimal adhesion coefficient required for RWD EVs in Equation (1.54) is

$$f_{min,r} = \frac{F_{f1} + F_d}{mg\left(\frac{L_1}{L}\cos\alpha + \frac{h_g}{L}\sin\alpha\right) - F_{zw2}} \tag{1.62}$$

As the vehicle speed increases, the minimal adhesion coefficient required for an RWD EV increases only moderately for a speed lower than 250 km/h, within the range of a well-paved concrete road; however, it quickly exceeds the range once the speed reaches 300 km/h, due to a quick rise in the rear aerodynamic lift coefficient C_{Lr}.

Reduction in the aerodynamic lift coefficient helps cut the minimal adhesion coefficient requirement and enhance the instability of handling and dynamic performance. Some relevant design considerations include body shape improvements, addition of such auxiliary aerodynamic features as a spoiler, and amendment of the overall layout for adjusting front and rear axle loads. The principle involved in the previous discussion applies to FWD and four-wheel drive scenarios as well.

1.5. Equilibrium of Power

Under the driving condition, the driving force and resistance are in equilibrium, as discussed in Section 1.3. In addition, propulsion power and resistance power are in equilibrium as well. That means, at every moment of driving, the power from electric motor(s) is always equal to the power consumption in overcoming all resistances and in making up associated mechanical losses in motion transmission.

The power balance equation of an EV is as follows:

$$P_e = P_f + P_d + P_s + P_a \tag{1.63}$$

where

P_e is output power of electric motor(s)
P_f is rolling resistance power
P_d is aerodynamic resistance power
P_s is slope resistance power
P_a is acceleration resistance power

According to Equation (1.31), Equation (1.62) can be expressed as

$$P_e = \frac{1}{3600\eta_T}\left(mgfv + mgv\sin\alpha + \frac{C_D A}{21.15}v^3 + \delta mv\frac{dv}{dt}\right) \tag{1.64}$$

where P_e is in kW.

Figure 1.10 illustrates a diagram of power balance for a three-speed EV. The power of the electric motor varies with the velocity of the EV for different gears, while the peak power does not change with gears. The operating range in terms of velocity is, in general, narrow in low gear and broad in high gear. Both rolling resistance power and aerodynamic resistance power increase with velocity. At a high velocity, the aerodynamic resistance power consumption becomes dominant as it is proportional to the third power of velocity. The maximum velocity v_{max} is located at the intersection of the motor power curve and resistance power consumption curve.

FIGURE 1.10 Diagram of power balance for a three-speed EV.

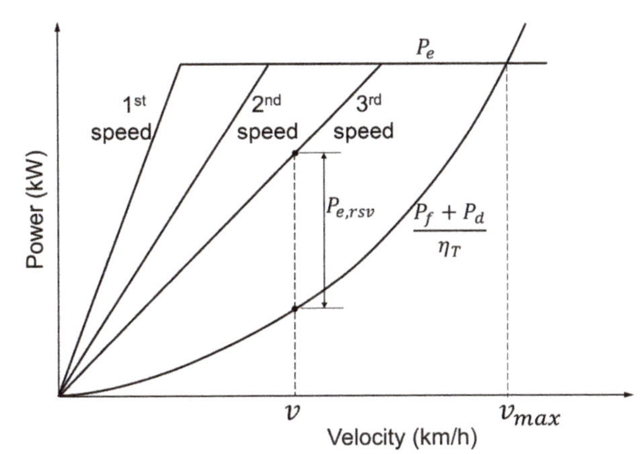

For the EV operating at a given velocity v, the extra power over resistance power consumption is

$$P_{e,rsv} = P_e - \frac{1}{\eta_T}\left(P_f + P_d\right) \tag{1.65}$$

where $P_{e,rsv}$ is referred to as reserved power, as shown in Figure 1.10. By enlarging the opening of the acceleration pedal, the reserved power can be made use of partially or fully. An increase in reserved power helps improve the dynamic performance.

1.6. Dynamic Performance Tests

In general, dynamic performance tests for EVs consist of road tests and rig tests. Of these tests, the most important ones are briefly described as follows.

The acceleration behavior test on level road. The purpose of this test is to determine the maximum acceleration on level road at various states of charge (SOCs) of the battery.

1. Gradeability test. The purpose of this test is to determine the maximum gradient that an EV can climb.
2. Climbing speed test. The purpose of this test is to determine the maximum speed that can be kept on the road of various gradients.
3. Top speed test. The purpose of this test is to determine the maximum speed that can be kept for a certain duration of time.

1.6.1. Road Tests

The road tests are typically performed on a hard-surface, dry, level road. The wind speed should not exceed the upper limit. The test vehicle must be preloaded to a specified weight (e.g., gross vehicle weight, GVW), fully charged, and soaked in a specified range of ambient temperature for certain hours. The tire pressure of the test vehicle must be inflated according to specifications.

In the acceleration test, the test vehicle is accelerated from standstill to nearly maximum speed on a straight road with the acceleration pedal fully depressed for several different battery SOCs. With the speed versus time trace measured in the tests, the acceleration at any given speed can be calculated.

In the gradeability test, the test vehicle is driven a certain distance or duration on a road of certain gradient. The gradeability is then obtained by calculation based on the measurement data of vehicle weight, tractive force, or the actual slope, gear ratios, etc.

In the climbing speed test, the test is run in a similar way to that in the acceleration test. With the speed versus time trace measured in the tests, the acceleration at any given speed can be calculated. Then, the gradient for any given climbing speed can be computed accordingly.

In the top speed test, the test vehicle is driven and kept at the maximum speed for a certain distance. The top speed can be calculated according to the distance and time duration measured in the tests.

1.6.2. Rig Tests

Rig tests are typically conducted on chassis dynamometers, whereas the vehicle speed is controlled by the speeds of drums and rollers.

In the acceleration test, the test vehicle is accelerated from standstill to nearly maximum speed with the acceleration pedal fully depressed for several different battery SOCs. With the speed versus time trace measured in the tests, the acceleration at any given speed can be calculated.

In the climbing speed test, the test vehicle is accelerated to a stable speed and kept for a certain time duration on a chassis dynamometer with preselected load. The power consumption of the dynamometer can be measured for climbing speed calculation.

References

[1.1]. Yu, Z., *Automobile Theory*, 3rd ed. (Beijing: China Machine Press, 2000), 1-36, ISBN:7-111-02076-6.

[1.2]. Bastow, D., Howard, G., and Whitehead, J., *Car Suspension and Handling*, 4th ed. (Warrendale: Society of Automotive Engineers, Inc., 2004), ISBN:0-7680-0872-7.

[1.3]. Wayback Machine, "Tyres-Online: The Benefits of Silica in Tyre Design," archived February 2013, https://web.archive.org/web/20130204065301/http://www.tyrepriceadvisor.co.uk/en/top-tyre-brands/falken-tyres.

[1.4]. Jazar, R., *Vehicle Dynamics: Theory and Application* (New York: Springer, 2008), 121, doi:10.1007/978-0-387-74244-1.

[1.5]. Gillespie, T., *Fundamentals of Vehicle Dynamics* (Warrendale: Society of Automotive Engineers, Inc., 1992), 21-118, ISBN-10:1560911999, ISBN-13:978-1560911999.

2

Energy Economy

The energy economy of an EV refers to its ability to operate with the least energy consumption under the premise of satisfaction of dynamic performance requirements.

Good energy economy is of ultimate importance in the design of EVs because it helps not only reduce the direct costs of usage and the indirect emission in energy production but also extends the maximum range of the vehicle.

2.1. Indicators of Energy Economy

The energy economy of an EV is generally measured by electric energy consumption per hundred kilometer of driving distance. Thus, the energy economy is usually expressed by

$$C_{100} = \frac{100E}{D} \tag{2.1}$$

where

C_{100} is the energy economy in kWh/100 km
E is the electricity consumption in kWh
D is the driving distance in km

The simplest indicator of energy economy is evaluated at a constant speed of 40 or 60 km/h. The constant speed operating condition, however, cannot completely reflect real-world driving situations, in particular without scenarios such as acceleration, deceleration, and stop frequently encountered in urban traffic. Therefore, the energy economy can be better evaluated based on some driving cycles which are designed according to statistics of actual operating data of road vehicles.

The New European Driving Cycle (NEDC) consists of four repeated Economic Commission of Europe (ECE)-15 urban driving cycles (UDCs) and one extra-urban driving cycle (EUDC), as shown in Figure 2.1a.

The World Harmonized Light-Duty Vehicles Test Cycle (WLTC) is composed of four different subparts, namely, low, medium, high, and extra-high, simulating urban, suburban, rural, and highway scenarios, respectively, as shown in Figure 2.1b.

The energy economy of an EV can also be measured by miles per gallon of gasoline equivalent (MPGe), which is calculated by

$$\text{MPGe} = \frac{\psi D}{E} \tag{2.2}$$

where D is the total miles driven; E is the total energy of all fuels consumed, for example, electric energy in kWh or hydrogen in kg; and ψ is the energy of 1 gal of gasoline, with $\psi = 33.7$ kWh/gal for electric energy or $\psi = 1.011$ kg/gal for hydrogen.

The US Environmental Protection Agency (EPA) Federal Test Procedure for MPGe measurement comprises city driving cycle (FTP-75), highway driving cycle (HWFET), aggressive driving cycle (US06), and optional air conditioning test cycle (SC03), as shown in Figure 2.1c, 2.1d, 2.1e, and 2.1f, respectively.

The energy economy tests are typically performed on chassis dynamometers in a laboratory environment.

The energy economy of some well-known BEV and fuel cell vehicle (FCV) models is compared in Table 2.1.

FIGURE 2.1 Various driving cycles for the evaluation of energy economy. (a) NEDC, (b) WLTC, (c) FTP-7, (d) HWFET, (e) US06, and (f) SC03.

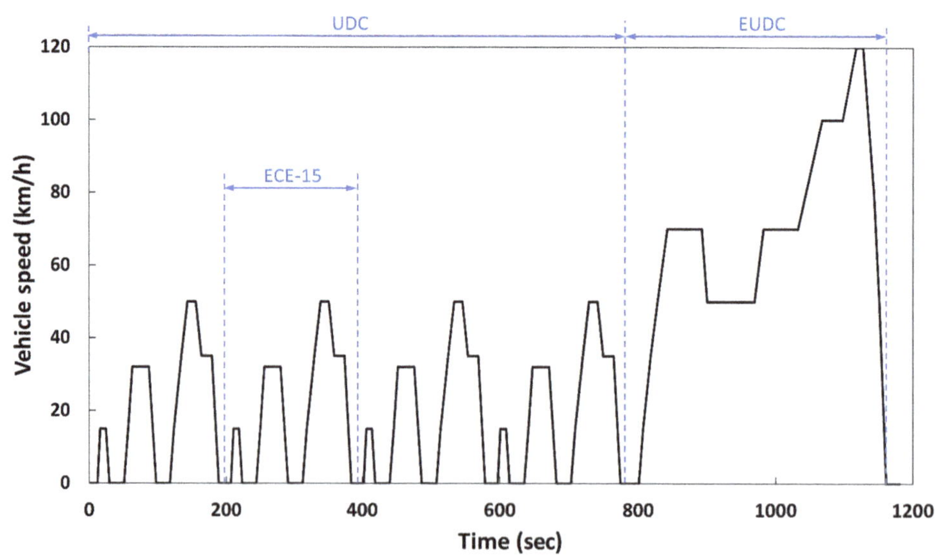

(a)

FIGURE 2.1 **(Continued)** Various driving cycles for the evaluation of energy economy. (a) NEDC, (b) WLTC, (c) FTP-7, (d) HWFET, (e) US06, and (f) SC03.

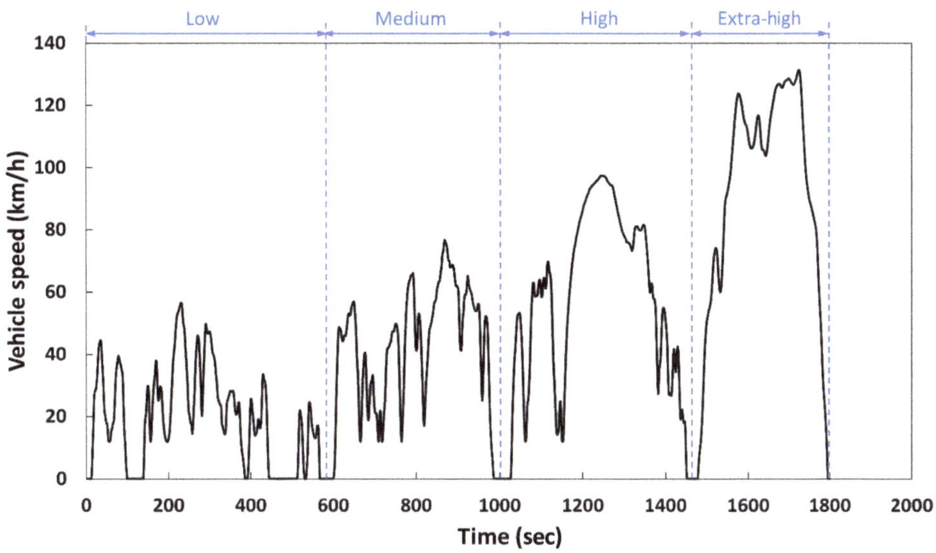

(b)

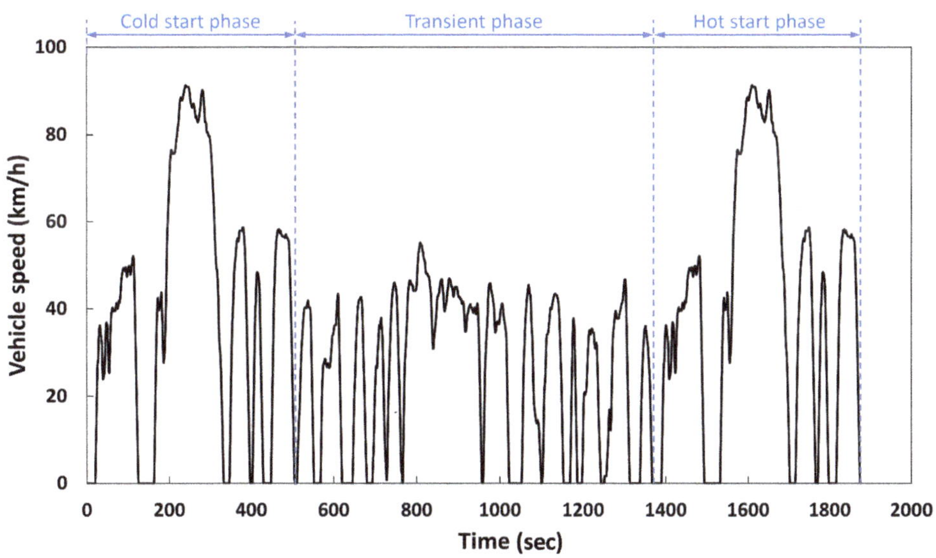

(c)

FIGURE 2.1 (Continued) Various driving cycles for the evaluation of energy economy. (a) NEDC, (b) WLTC, (c) FTP-7, (d) HWFET, (e) US06, and (f) SC03.

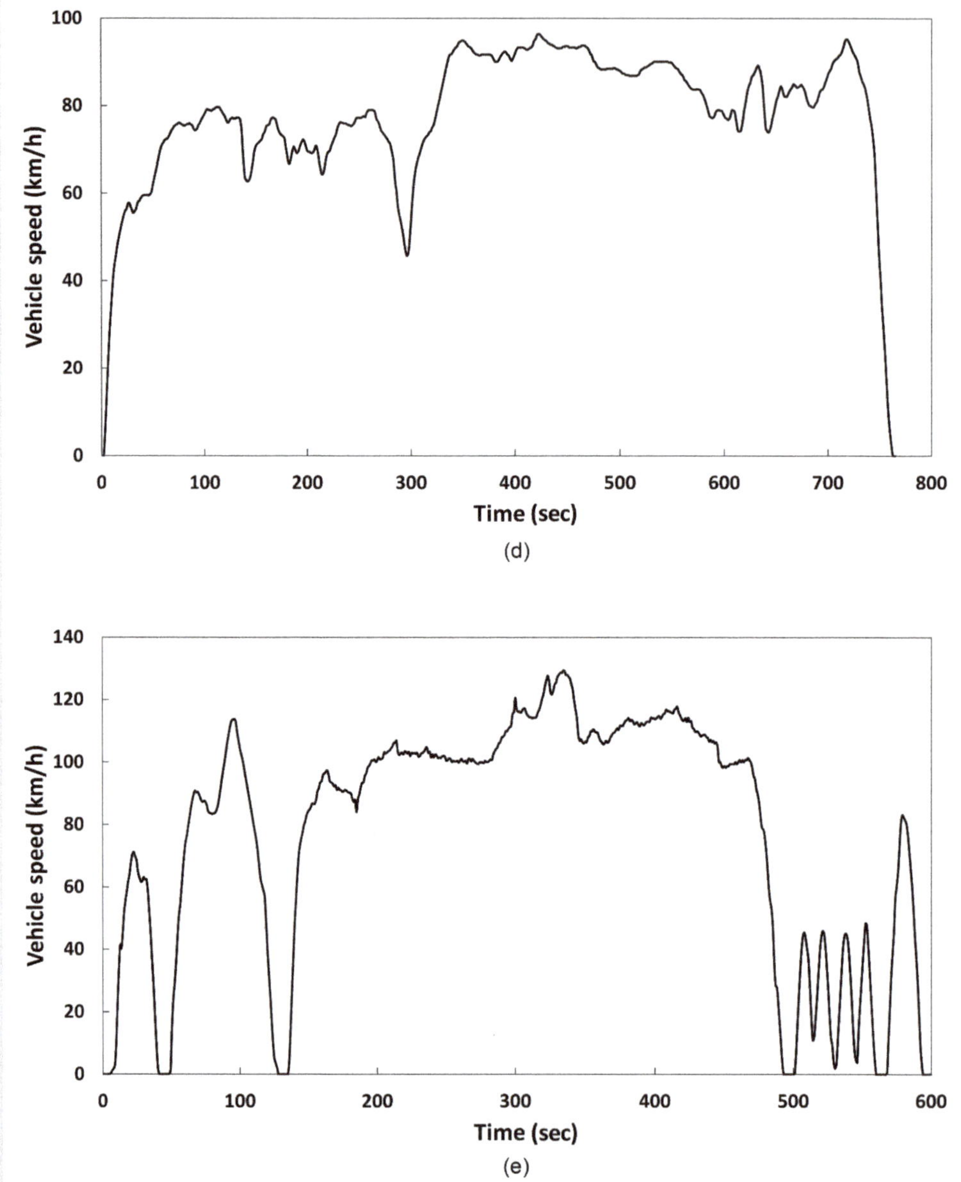

FIGURE 2.1 **(Continued)** Various driving cycles for the evaluation of energy economy. (a) NEDC, (b) WLTC, (c) FTP-7, (d) HWFET, (e) US06, and (f) SC03.

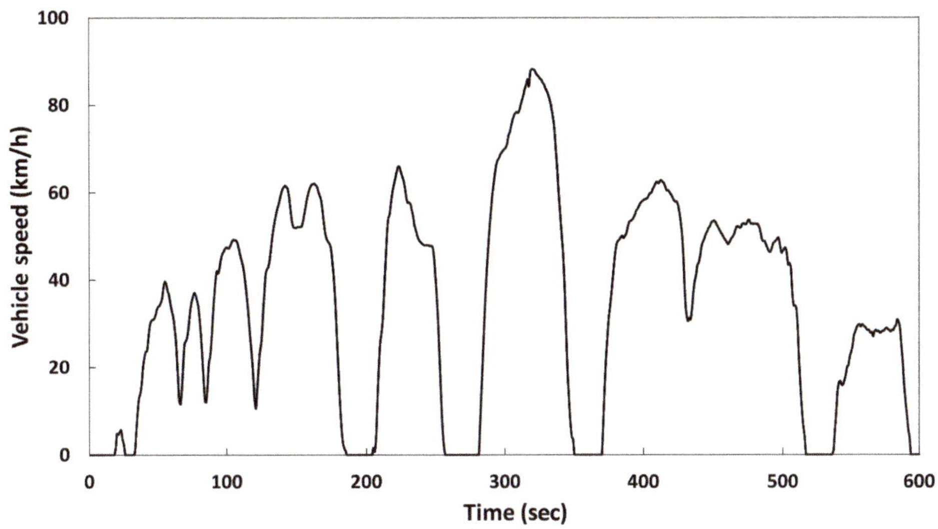

(f)

TABLE 2.1 Comparison of energy economy of popular BEV and FCV models.

Type	Vehicle model	Model year	Total battery capacity (kWh)	Energy economy: Combined mpg-e	Combined kWh/100 km	City mpg-e	City kWh/100 km	Highway mpg-e	Highway kWh/100 km
BEV	Chevrolet Bolt	2022	65	120	17.5	131	16	109	19.2
	Ford Mustang Mach-E	2022	91	93	22.5	99	21.1	86	24.4
	Nissan Leaf	2023	40	111	18.9	123	17.0	99	21.1
	Tesla Model S Plaid	2022	100	116	18.1	119	17.6	112	18.7
	Tesla Model 3	2022	75	131	16	134	15.6	126	16.6
	Tesla Model X	2022	100	102	20.5	107	19.6	97	21.6
	Tesla Model Y	2022	75	125	16.8	131	16	117	17.9
	Lucid Air	2022	112	131	16	130	16.1	132	15.9
	Rivian R1S	2022	135	69	30.4	73	28.7	65	32.2
	Toyota bZ4X	2022	71.4	119	17.6	131	16	107	19.6
	Hyundai Ioniq 5	2022	58	114	18.4	132	15.9	98	21.4
	Audi e-tron GT	2022	93.4	82	25.5	81	25.9	83	25.2
	BMW ix xDrive50	2022	105.2	86	24.4	86	24.4	87	24.1
	Mercedes-Benz EQS	2021	108	97	21.6	97	21.6	97	21.6
FCV	Honda Clarity Fuel Cell	2017	—	67	—	68	—	66	—
	Hyundai Tucson Fuel Cell	2017	—	49	—	48	—	50	—
	Toyota Mirai	2017	—	66	—	66	—	66	—

2.2. Estimation of Energy Economy

It is often needed to estimate the energy economy of an EV in the design and development process based on the efficiency map of electric motors.

Figure 2.2 illustrates an example of a combined efficiency map of an electric motor and its controller. For a given operating point (T_e, ω_e) on the map, the combined efficiency η_e can be obtained by two-dimensional intrapolation. Provided the time history of vehicle speed $v(t)$ as in a driving cycle, the output torque T_e and speed ω_e of an electric motor can be calculated by Equations (1.32) and (1.33). Note that an electric motor may operate in any of the four quadrants. It has two different operating modes depending on the combination of the directions of its angular velocity and torque, as listed in Table 2.2.

FIGURE 2.2 Combined efficiency map of an electric motor and its controller.

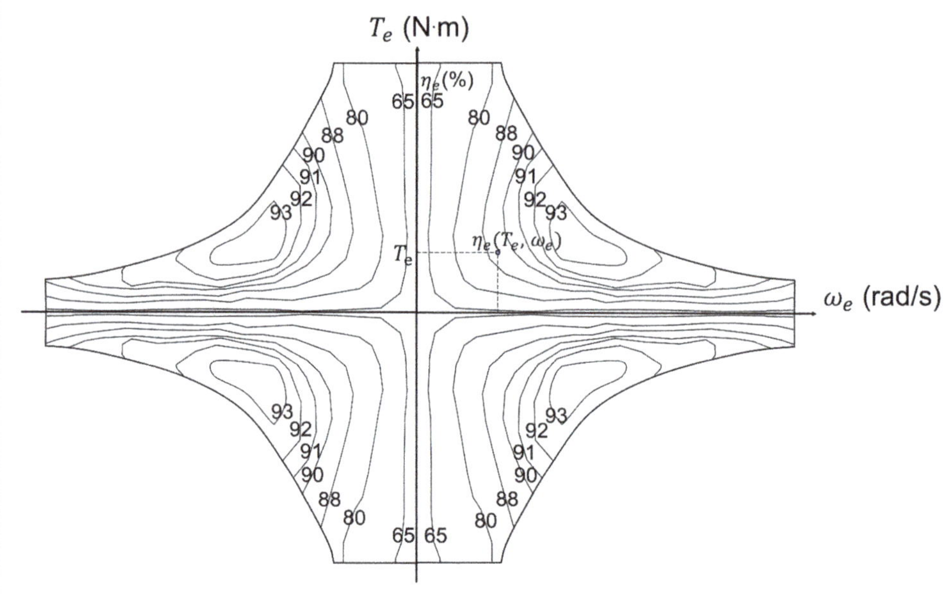

TABLE 2.2 Various operating modes of an electric motor.

Quadrant	Operating mode	Direction of velocity	Direction of torque
1	Driving	$\omega_e > 0$	$T_e > 0$
2	Regenerative braking	$\omega_e < 0$	$T_e > 0$
3	Driving	$\omega_e < 0$	$T_e < 0$
4	Regenerative braking	$\omega_e > 0$	$T_e < 0$

The energy consumption of electric motors can be expressed by

$$E_e = \int_0^t P_e dt = \int_0^t \frac{T_e \omega_e}{\eta_e} dt \tag{2.3}$$

Moreover, the energy consumption of the battery is

$$E_b = \int_0^t P_b dt \tag{2.4}$$

Since the discharging efficiency of a battery may differ from its charging efficiency, the battery power P_b is

$$P_b = \begin{cases} \dfrac{T_e \omega_e}{\eta_e \eta_{b,dc}} & \text{if} \quad sign(T_e) = sign(\omega_e) \\ T_e \omega_e \eta_e \eta_{b,ch} & \text{if} \quad sign(T_e) \neq sign(\omega_e) \end{cases} \tag{2.5}$$

where

$\eta_{b,dc}$ is the discharging efficiency of the battery
$\eta_{b,ch}$ is the charging efficiency of the battery

2.3. Factors Impacting Energy Economy

One of the most important factors that affect the energy economy of an EV is the electric drive system. Technical characteristics such as the motor type, peak efficiency, ratio of high-efficiency area to the entire operatable area of the electric motor, and the efficiency of the drivetrain contribute to the energy economy, to some extent. In addition, the energy economy is also related to the operating conditions, for example, acceleration, deceleration, regenerative braking, and usage of air conditioning (A/C) or heating systems. Furthermore, the energy economy of an EV can be significantly affected by many other factors such as its body style, type of tires, effectiveness of regenerative braking, and thermal management.

The energy consumption balance maps of a typical EV in city and highway operating cycles are shown in Figure 2.3a and 2.3b, respectively [2.1]. Due to considerable losses in the electric motor, motor control unit (MCU), driveline, and accessories, only a portion of the energy from the battery reaches the wheels to propel the vehicle, which is 78% in city driving cycles and 81% in highway driving cycles. Note that the loss and efficiency values are approximate means based on state-of-the-art technology.

FIGURE 2.3 Energy consumption balance maps of a typical EV. (a) City and (b) highway.

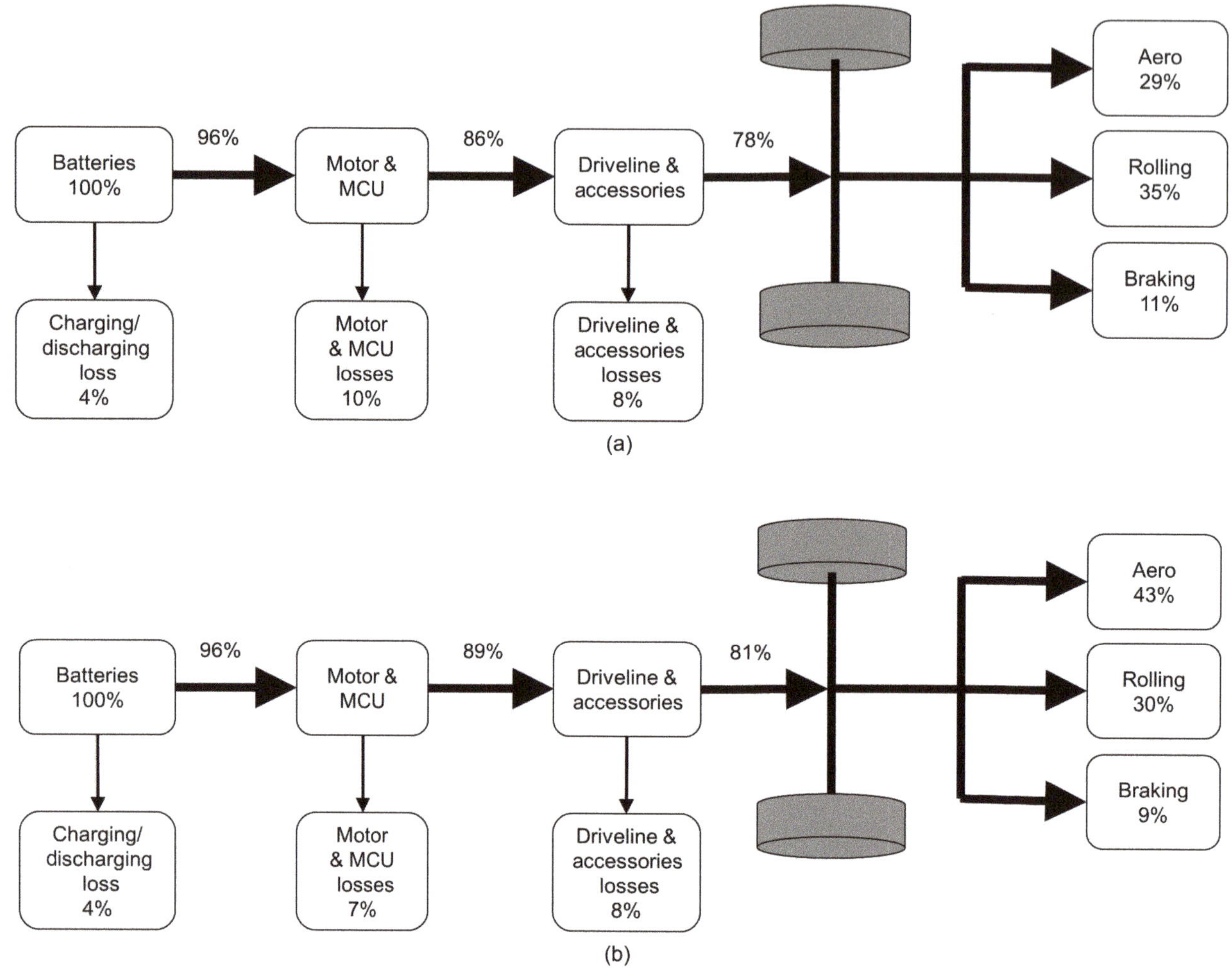

Courtesy of Dan Lauber.

Another metric for the assessment of energy economy is the so-called well-to-wheel efficiency, which takes the efficiency in power generation, transmission, and distribution through the grid into account. However, since the efficiency of power generation varies substantially from one energy source to another, the well-to-wheel efficiency can be quite different for different energy sources. The energy balance maps for a typical EV in city driving cycles using electricity generated by coal, natural gas, wind, sunlight, nuclear, hydro potential, and terrestrial heat are shown in Figure 2.4a through 2.4g, respectively. For example, only 28% of the energy from coal reached wheels for propulsion because of the significant losses in power generation and the transmission process.

FIGURE 2.4 Well-to-wheel energy balance maps of a typical EV using electricity generated from various energy sources. (a) Coal, (b) natural gas, (c) wind, (d) sunlight, (e) uranium/thorium, (f) hydro potential, and (g) terrestrial heat.

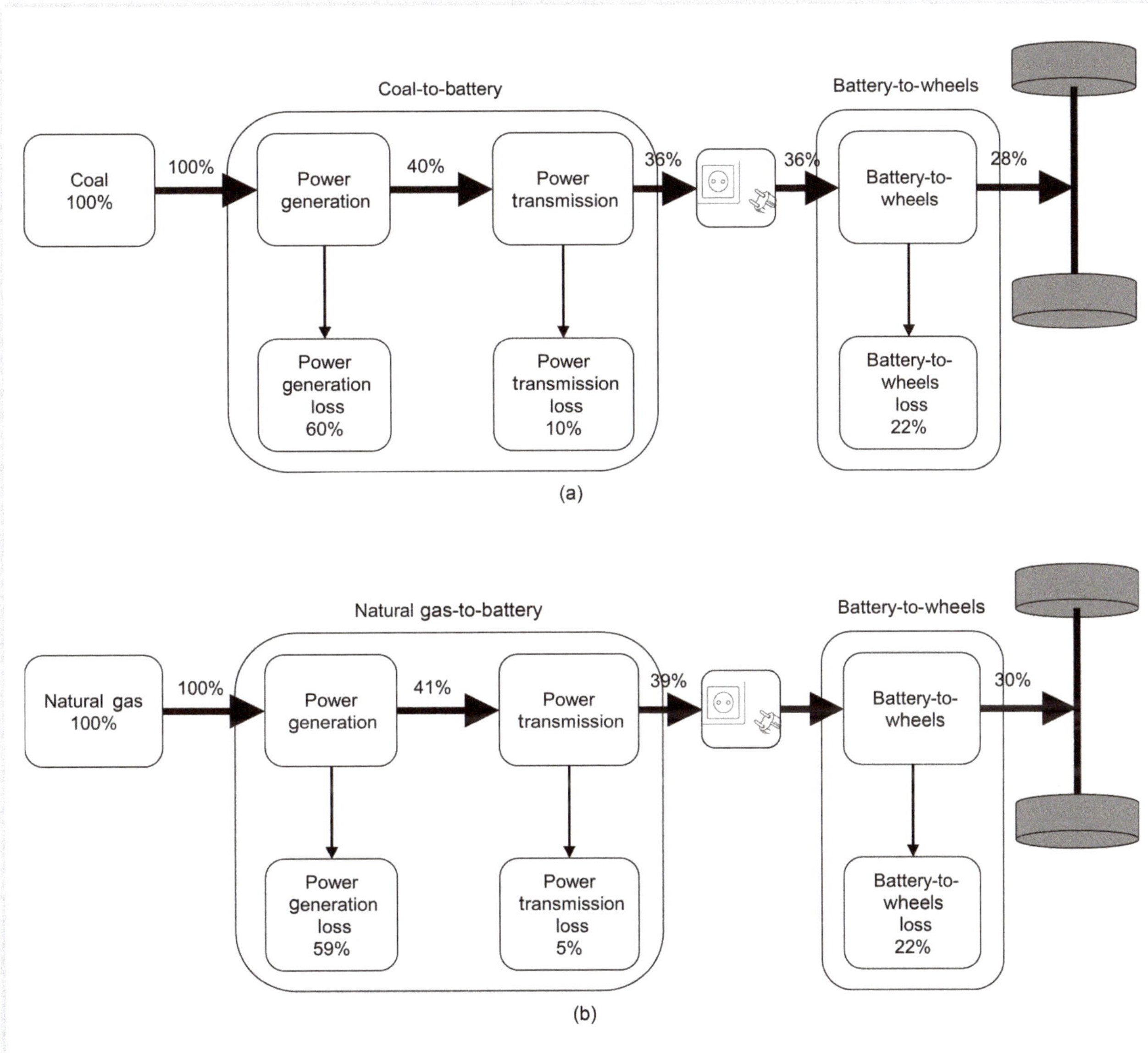

FIGURE 2.4 **(Continued)** Well-to-wheel energy balance maps of a typical EV using electricity generated from various energy sources. (a) Coal, (b) natural gas, (c) wind, (d) sunlight, (e) uranium/thorium, (f) hydro potential, and (g) terrestrial heat.

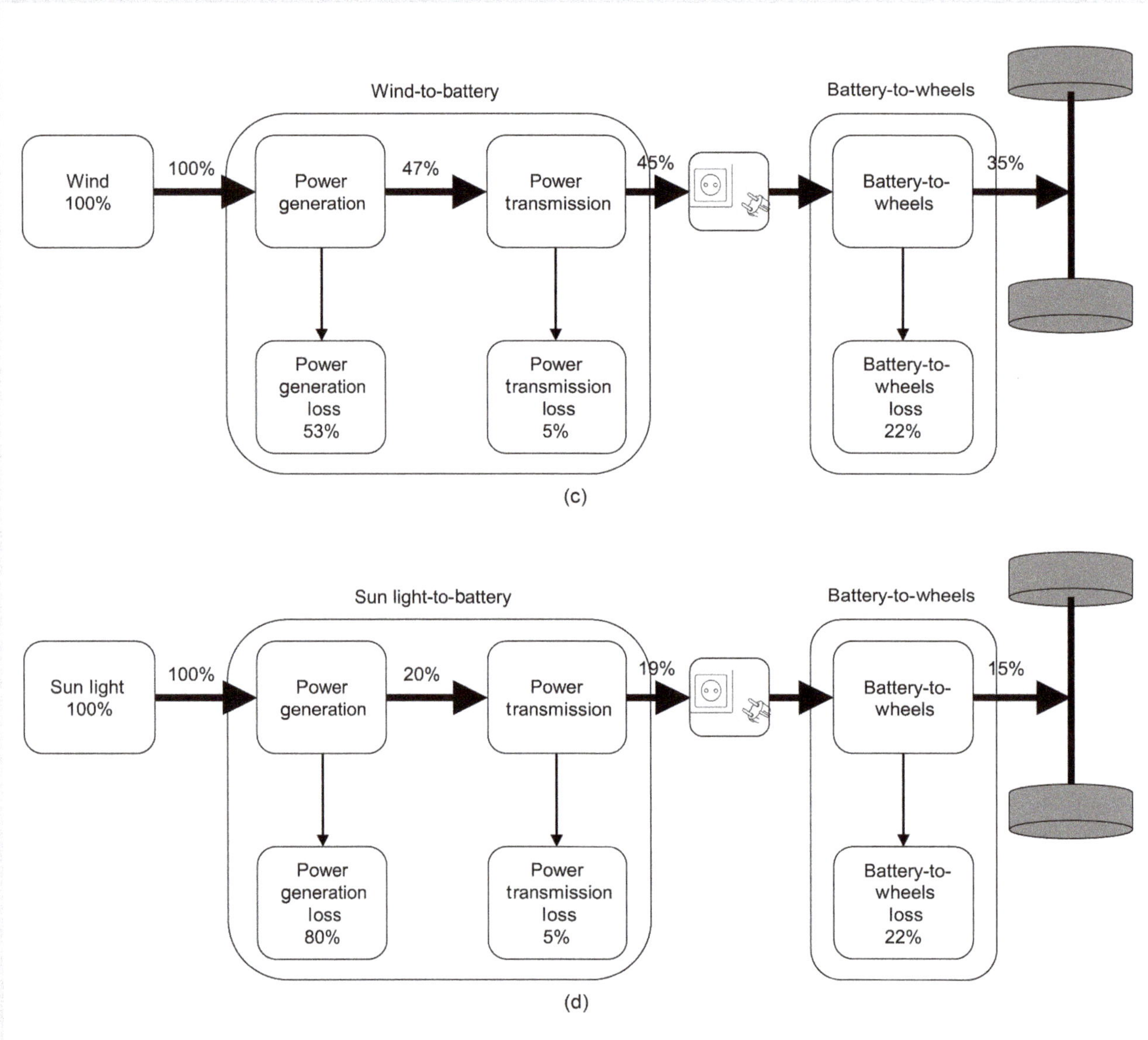

FIGURE 2.4 (Continued) Well-to-wheel energy balance maps of a typical EV using electricity generated from various energy sources. (a) Coal, (b) natural gas, (c) wind, (d) sunlight, (e) uranium/thorium, (f) hydro potential, and (g) terrestrial heat.

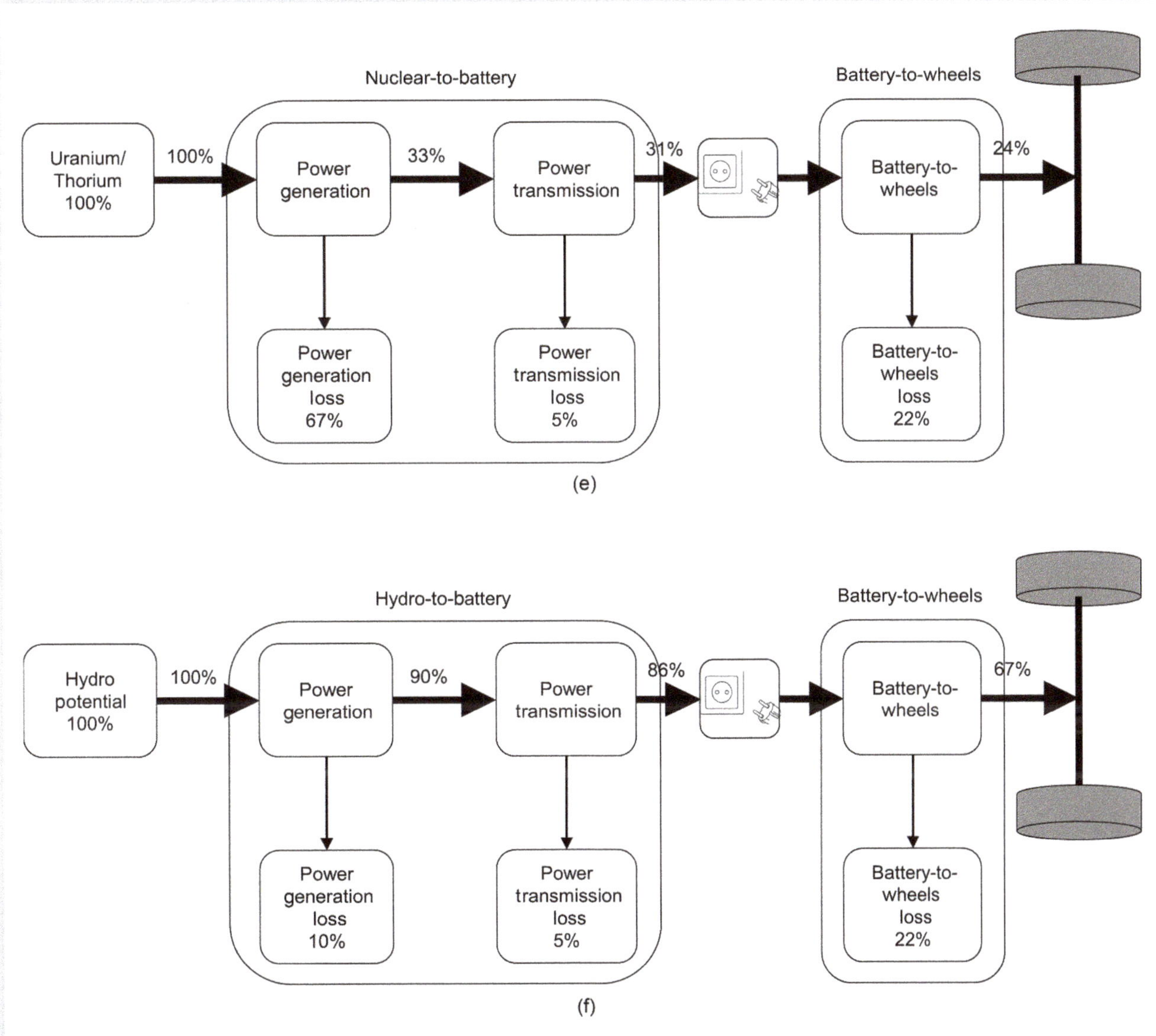

FIGURE 2.4 (Continued) Well-to-wheel energy balance maps of a typical EV using electricity generated from various energy sources. (a) Coal, (b) natural gas, (c) wind, (d) sunlight, (e) uranium/thorium, (f) hydro potential, and (g) terrestrial heat.

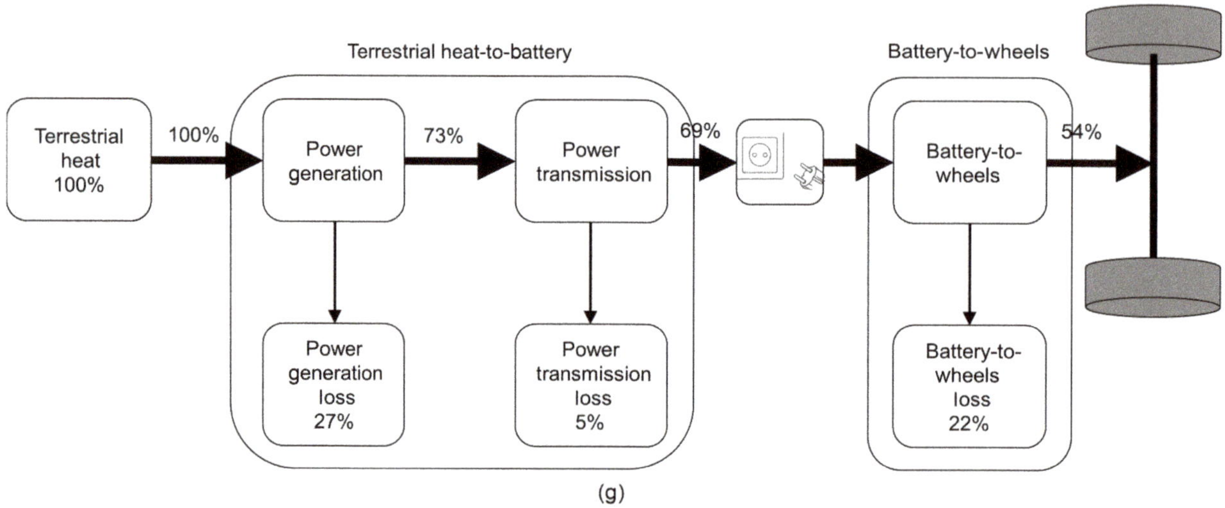

Despite of their deficiencies in energy utilization, the EVs are still the most efficient mobility by far in comparison to traditional vehicles or FCVs. In the year 2020, using 100% renewable electricity, the well-to-wheel efficiency is as high as 77% for a battery EV, in comparison to 33% for a hydrogen FCV, or 16–20% for a conventional vehicle burning liquid fuel [2.2].

The design considerations for energy economy improvements of EVs are discussed now.

2.3.1. Reduction in Vehicle Weight and Size

Lightweight design helps reduce the power consumption needed for overcoming rolling friction, grading, and acceleration resistances. A 10% weight reduction for EVs can improve the electric range by 13.7% [2.3]. Another investigation indicates that for light-duty EVs, 100 kg of weight reduction can result in approximately 0.5–0.7 kWh/100 km of energy savings [2.4], depending on driving cycles. The application of lightweight materials such as aluminum, magnesium, and carbon fiber composites and optimization of the structural design of EVs are the primary engineering solutions to weight reduction. Chassis and bodies made of aluminum are becoming common engineering practices. With issues in mass production being resolved, parts and components made of composite materials, for example, carbon fiber, are gaining popularity as well.

Reduction in vehicle size can improve the energy economy of EVs as well. A decrease in the frontal area can reduce aerodynamic resistance directly. It is not always pragmatic, however, to make vehicle dimensions smaller because of the requirements for interior space and cargo capacity.

2.3.2. Electric Motor System

The electric motor system significantly affects the energy economy of an EV because its efficiency can vary in a wide range, for example, 65–95%. The efficiency of electric motor systems generally involves two aspects: the maximum efficiency of electric motors, MCUs, and the combined efficiency of the system, and the ratio of high-efficiency area to entire operatable area. New technologies such as SiC power electronics, dual-side cooling packaging, and voltage boosting can help further improve the efficiency of MCUs.

Some key efficiency indicators of state-of-the-art electric motor systems in mass production are listed in Table 2.3.

TABLE 2.3 Key efficiency indicators of state-of-the-art electric motor systems in mass production.

Key efficiency indicator	Value (%)
Maximum efficiency of electric motors	97.5
Maximum efficiency of MCUs	99.7
Maximum efficiency of system	95
$A_{\eta \geq 90\%}/A_{op}$*	92

* $A_{\eta \geq 90\%}$ is the operatable area in which system efficiency is larger than 90%, and A_{op} is the entire operatable area of the electric motor system.

2.3.3. Drivetrain

The improvement of mechanical transmission efficiency can help an EV to gain energy economy. A transmission efficiency of 97% is routine nowadays, while some carefully designed drivetrains can reach 98% or higher. Continued efforts are mainly being made to reduce power losses in bearings, clutches, oil pumps, hydraulic circuits, and lubrication systems.

Today, single-speed transmissions are still the norm in the industry, multispeed transmissions, however, can better work with electric motor systems than single-speed transmissions in certain conditions by allowing electric motor systems to shift operating points to those of higher efficiencies. A two-speed transmission is generally reckoned adequate for application in EVs, though three-speed and four-speed transmissions are also utilized in some cases, especially for heavy-duty EVs.

2.3.4. Body Aerodynamics

Reduction in the drag coefficient is an effective way to improve energy. Some investigations suggest that an increase of nearly 10% in the driving range can occur as the drag coefficient drops from 0.3 to 0.2 [2.5]. Refer to Section 14.3.3.3 for design rules to reduce the drag coefficient.

On the other hand, passive and active aerodynamic body accessories can also be beneficial to energy economy.

Examples of passive accessories are wheel spoilers and roof spoilers. Wheel spoilers can reduce the drag coefficient by altering the flow around wheels. Roof spoilers in the rear can reduce both drag and aerodynamic lift.

In addition, examples of active aerodynamic body accessories include active roof spoilers, active air dams, active ride height, and adaptive suspension. Active rear spoilers can pop up to reduce lift at higher speeds. Air dams are spoilers on the front of a vehicle, which are typically positioned under or integrated with the front bumper to improve the drag coefficient of an EV body or to generate downforce. Active air dams can be dropped lower at higher speeds to reduce the drag coefficient. Active ride height lowers the vehicle at speed using actively controlled suspensions, reducing the underbody drag coefficient and lift. Adaptive suspensions can dynamically adjust to compensate for differences in forward and rear loads, for altering the lift on the body.

If we now look at the suite of aerodynamic systems available to improve the highway driving range, grill shutters and air dams are expected to equip most EVs in 2025, while roof spoilers are only expected on some battery electric sport utility vehicles (SUVs) to improve aerodynamics. However, battery electric sports cars are also expected to be equipped with active roof spoilers to improve road handling at higher speeds.

2.3.5. Tire

Tires used for EVs must meet requirements for strength, wearproof, durability, and dynamic and economical performances. It is generally agreed that radial tires have the best overall performance.

It is estimated that all-electric passenger EVs can use approximately 23% of their energy to overcome rolling resistance. To minimize rolling resistance, LRR tires are often used in EVs. As mentioned in Chapter 1, LRR tires can cut the rolling friction coefficient by as much as 43–59% and improve the energy economy of an EV by approximately 5–15%, in comparison to regular tires [2.6].

Nevertheless, it is not acceptable to compromise traction and safety in exchange for only a slight improvement in the energy economy. The shortcomings with LRR tires, for example, long stopping distances at high speeds and lack of grip in the corners, have to be addressed for large-scale applications.

2.3.6. Thermal Management

Effectively managing heat across an EV from a systematic perspective can provide an improved energy economy and driving range.

Thermal management is of particular importance for lithium-ion batteries widely used in today's EVs because the performance and life of these batteries are quite sensitive to temperature. It is suggested that the operating temperature should range from 15 to 35°C and the maximum temperature difference from module to module should be below 5°C to avoid adverse effects on batteries [2.7]. At present, the dominant battery cooling media are air, liquid, and phase change material (PCM). Direct liquid cooling, especially liquid immersion cooling, appears to be promising due to its high cooling efficiency [2.7]. Battery thermal management (BTM) systems can be classified in terms of thermal cycle options [2.8]. The BTM systems with vapor compression cycle (VCC), for example, cabin air cooling system, secondary loop liquid cooling system, and direct refrigerant two-phase cooling system, are applied in most EVs nowadays. This type of system can take advantage of existing A/C systems to cool or heat batteries. It consumes, however, a considerable amount of additional energy in pumps, compressors, condensers, heaters, etc., with an undesired impact on A/C performance [2.8]. On the other hand, the instances of BTM systems without VCC are PCM cooling systems, heat pipe (HP) cooling systems, and thermoelectric element cooling systems. Although not widely applied in EVs yet, this type of system is being studied extensively due to their great potential in performance and energy consumption [2.8].

The driving range of an EV can be reduced dramatically due to significant energy drain from auxiliary electrical loads, particularly the cabin heating and A/C loads. Studies have shown that thermal subsystem loads can reduce the driving range by as much as 45% under ambient temperatures below −10°C [2.9]. The driving range can be considerably improved by recovering the waste heat from battery and power electronics, etc., and utilizing a heat pump. It is indicated that the range improvement of 15–18% at severe cold ambient conditions is feasible [2.9].

2.4. Energy Economy Testing

The energy economy tests for EVs are typically performed on proving ground roads or on test rigs in a laboratory environment. The results are presented in the form of energy consumption, for example, kWh/100 km or MPGe, and driving range.

The road tests in the proving ground are typically performed on a hard surface, dry, level, paved loop or on a straight road of an adequate length. The wind speed should not exceed the upper limit. The test vehicle must be preloaded to a specified weight (e.g., GVW), fully charged, and soaked in a specified range of ambient temperature for certain hours. The tire pressure of the test vehicle must be inflated according to specification. A certain existing milage prior to tests is sometimes required for the test vehicle to pass the running-in stage.

The test usually starts by fully charging the battery for the first time. Then, the test vehicle is driven according to a given driving cycle or at a specified constant speed. The deviations in speed and time must be kept within certain limits from the standard driving cycle data. The most frequently adopted driving cycles are described in Section 2.1. The test is terminated per indicator of an on-board instrument or failure to follow the specified driving cycle. The test vehicle is fully charged once again within a certain time after the test per the specified charging procedure. The driving range is simply the total distance driven in the test. The energy consumption rate can be calculated by the distance measured in the test and the electric energy for charging the battery.

For tests conducted on a chassis dynamometer, an appropriate equivalent inertia has to be selected for the dynamometer according to the weight of the test vehicle, for simulating acceleration resistance. Moreover, the road load of the test vehicle is also needed for setting up the dynamometer. Here, the road load for chassis dynamometer tests refers to the sum of rolling friction and aerodynamic resistances only. Slope resistance is excluded because driving cycles for energy economy testing typically do not involve variations in altitude. The road load F_r can be obtained from coast-down data collected in proving ground tests, which is often presented as a polynomial function of speed:

$$F_r(v) = A + Bv + Cv^2 \tag{2.6}$$

where

F_r is in N

v is the vehicle speed in km/h

A, B, and C are coefficients obtained by regression analysis (e.g., least-squares fitting) of coast-down data

The dynamometer is then set up to simulate the road load according to speed. FWD or RWD EVs are usually tested on a two-wheel chassis dynamometer, while four-wheel drive EVs are tested on a four-wheel chassis dynamometer.

The coast-down tests are commonly performed on a flat road under light wind conditions. The procedure involves three steps: accelerate the test vehicle to a certain speed and then put the transmission in neutral; begin coast down in a straight line; and record the time trace of vehicle velocity.

References

[2.1]. Larber, D., "Electric Vehicle 101—An Introduction," MIT Electric Vehicle Team, accessed August 2023, https://web.mit.edu/evt/EVs%20101%20-%2011-13-09(web).ppt.

[2.2]. European Federation for Transport and Environment, "Cars: Battery Electric Most Efficient by Far," accessed May 2023, https://cleantechnica.com/files/2020/06/most-efficient-automobiles.png.

[2.3]. Kan, Y., Shida, R., Takahashi, J., and Uzawa, K., "Energy Saving Effect of Light-Weight Electric Vehicle Using CFRP on Transportation Sector," Paper Presented at in *JISSE-10 2007*, Tokyo, Japan, November 27–30, 2007.

[2.4]. Helms, H. and Kräck, J., "Energy Savings by Light-Weighting—2016 Update," Commissioned by the International Aluminium Institute Supported by European Aluminium, Heidelberg, December 2016.

[2.5]. Linley, D. and Todeschini, G., "Analysis of Factors Which Estimate the Maximum Distance Achievable by an Electric Vehicle," ResearchGate, April 2, 2019.

[2.6]. North American Council for Freight Efficiency (NACFE), "Confidence Report: Low Rolling Resistance Tires," North American Council for Freight Efficiency Report, August 13, 2015.

[2.7]. Xia, G., Cao, L., and Bi, G., "A Review on Battery Thermal Management in Electric Vehicle Application," *Journal of Power Sources* 367 (2017): 90-105, doi:10.1016/j.jpowsour.2017.09.046.

[2.8]. Kim, J., Oh, J., and Lee, H., "Review on Battery Thermal Management System for Electric Vehicles," *Applied Thermal Engineering* 149 (2019): 192-212, doi:10.1016/j.applthermaleng.2018.12.020.

[2.9]. Chowdhury, S., Leitzel, L., Zima, M., Santacesaria, M. et al., "Total Thermal Management of Battery Electric Vehicles (BEVs)," SAE Technical Paper 2018-37-0026 (2018), doi:https://doi.org/10.4271/2018-37-0026.

3

Selection of Dynamic Parameters

The parameters for electric drive systems refer to the system power and transmission ratio in this chapter. Because of their direct impacts on the dynamic performance and energy economy of an EV, these parameters have to be carefully selected to satisfy the essential requirements of dynamic performance and energy economy, with drivability taken into consideration as well. Drivability stands for the driver's perception of the EV, which largely depends on its response to the driver's expectations in terms of movement, noise, vibration, and so on.

3.1. Selection of Electric Motor Power

The selection of electric motor power involves both peak power and rated power.

Peak power determines the performance of an EV, which can only be sustained for a short period of time, typically from 30 sec to 3 min. Adequate peak power would provide an EV with a satisfactory performance in overtaking and short-slope climbing at high speeds.

Given the overtaking speed v_{ot} for a flat road, the peak power must exceed the road load, that is,

$$P_{e1} = \frac{1}{3600\eta_T}\left(mgfv_{ot} + \frac{C_D A}{21.15}v_{ot}^3\right) \tag{3.1}$$

Provided the speed v_{ss} for climbing a short slope α_{ss}, the peak power must be greater than

$$P_{e2} = \frac{1}{3600\eta_T}\left(mgfv_{ss} + mgv_{ss}\sin\alpha + \frac{C_D A}{21.15} v_{ss}^3 \right) \tag{3.2}$$

The peak power of the electric motor can be selected as the larger one between P_{e1} and P_{e2}, namely,

$$P_{e,peak} = \max\left(P_{e1}, P_{e2}\right) \tag{3.3}$$

The rated power determines the performance of the EV that can be sustained for a long time. Adequate rated power would allow an EV to have satisfactory performance in operating at its nominal top speed.

Assuming the nominal top speed is v_{max}, the rated power must be no less than the road load:

$$P_{e,rate} \geq \frac{1}{3600\eta_T}\left(mgfv_{max} + \frac{C_D A}{21.15} v_{max}^3 \right) \tag{3.4}$$

Statistics of retrospective EV data are also a good reference for selection of electric motor power. Figure 3.1 illustrates the specific power of a number of production EVs. The specific power, in general, increases as GVW increases. The top speed ranges from 140 to 250 km/h for these production EVs, with 140 to 160 km/h for most economy EVs and over 200 km/h for luxury EVs.

FIGURE 3.1 Specific power of a number of production EVs.

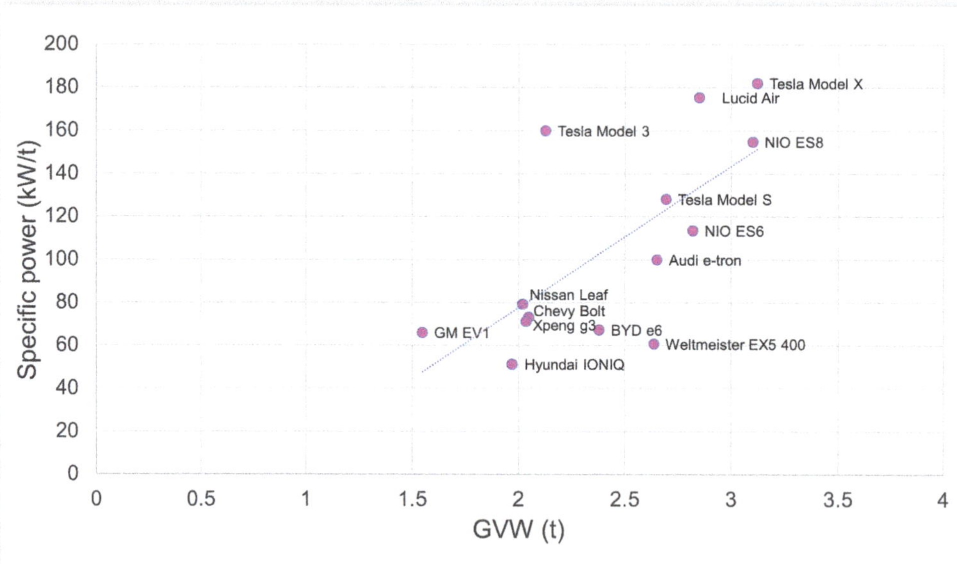

3.2. Selection of the Smallest Speed Ratio

For a transmission of n speeds, the smallest speed ratio is the ratio of highest gear i_n.

$$i_n = i_{gn} i_0 \tag{3.5}$$

where
i_{gn} is the transmission ratio of nth gear
i_0 is the final drive ratio

The selection of the smallest speed ratio i_n requires that the electric motor should not exceed the maximum speed at the top speed of the EV. Thus,

$$i_n \le \frac{3\pi N_{max} R_d}{25 v_{max}} \tag{3.6}$$

where
N_{max} is the maximum speed of the electric motor
R_d is the dynamic wheel radius
v_{max} is the top speed of the EV

Moreover, the selection of the smallest speed ratio i_n should consider the power utilization factor. In the diagram of power balance, assuming all three smallest speed ratios i_{n1}, i_{n2}, and i_{n3} satisfy Equation (3.6) and $i_{n1} \le i_{n2} \le i_{n3}$, then the intersection of the road load curve $(P_f + P_d)/\eta_T$ with any of the three electric motor power lines must fall in the constant power region. Otherwise, the power utilization factor of the electric motor is inadequate, implying the ratio is too large. In this example, the speed ratio i_{n3} gives the greatest extra power over road load and, thus, the best dynamic performance. The speed ratio i_{n1} would result in the smallest extra power and, therefore, the worst performance, while the extra power and performance of the speed ratio i_{n3} go in between those of ratios i_{n1} and i_{n3}.

Finally, the trade-off between the energy economy and dynamic performance should be taken into account. For a few candidates of smallest speed ratios $i_{n1}, i_{n2}, \ldots, i_{nk}$, assuming $i_{n1} < i_{n2} < \cdots < i_{nk}$, then an energy economy versus acceleration curve can be plotted, as shown in Figure 3.2 [3.1]. Apparently, the selection of a ratio around i_{n1} would result in a good energy economy but poor acceleration time, while a ratio near i_{nk} would give just the opposite result. Therefore, the selection of a ratio in the middle of the curve can yield compromised economy and performance.

FIGURE 3.2 Energy economy versus acceleration for an EV.

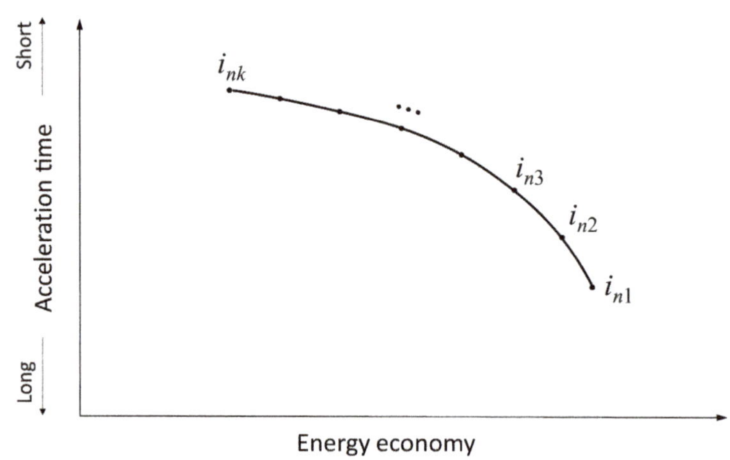

Some planetary gear-based transmissions have a direct drive gear, that is, $i_{gn} = 1$. The mechanical efficiency is typically high in this gear because there is no relative motion between the sun gear and planetary gears or between planetary gears and the ring gear. Furthermore, a transmission gear with a ratio $i_{gn} < 1$ is referred to as an overdrive gear. Whether an EV should apply an overdrive gear depends on the ratio of the maximum speed of the electric motor to the speed of its high-efficiency region.

The energy economy of an EV can be maximized as its operating points at highway cruising speeds (e.g., 80–120 km/h) are located in the high-efficiency region of the electric motor.

3.3. Selection of the Largest Speed Ratio

For a multispeed transmission, the largest speed ratio is the first gear ratio i_1. Gradeability, acceleration, and tire–ground adhesion must be considered in the selection of the largest transmission ratio.

Given the gradeability α_{max}, the first gear ratio i_1 must satisfy the following relationship:

$$i_1 = i_{g1} i_0 \geq \frac{\left(m_a g \sin\alpha_{max} + m_a g f \cos\alpha_{max} + \dfrac{C_d A v_{min}^2}{21.15} \right) R_d}{T_{peak} \eta_T} \tag{3.7}$$

where

i_{g1} is the transmission ratio
i_0 is the final drive ratio
m_a is the gross vehicle mass in kg
f is the rolling friction coefficient
α_{max} is the maximum slope angle in rad
C_d is the drag coefficient
A is the frontal area in m^2
v_{min} is the lowest acceptable climbing speed in km/h, typically equal to 10 km/h
η_T is the mechanical efficiency of transmission
R_d is the dynamic wheel radius in m
T_{peak} is the peak torque of the electric motor in N·m

In addition to the maximum slope angle, there are often minimum speeds required for driving on several different slope angles. For each required minimum speed for particular slope angles, substitute them for v_{max} and α_{max} in Equation (3.5) to obtain the corresponding first gear ratio i_1. Then, take the maximum of all i_1 as the first gear ratio i_1.

Next, the first gear ratio i_1 has to satisfy acceleration requirements. Thus,

$$i_1 = i_{g1} i_0 \geq \frac{\left(\delta m_a \dfrac{dv}{dt} + m_a g f + \dfrac{C_d A v^2}{21.15} \right) R_d}{T_{peak} \eta_T} \tag{3.8}$$

where δ is the conversion factor for rotating mass. Since the shift schedule is typically not available at this point, it can be postulated that the transmission only operates in the first gear in the entire course of acceleration. By solving Equation (3.8), the lower bound of the first gear ratio that satisfies given requirements, for example, acceleration time from 0 to 50 km/h, can be obtained. Once all speed ratios are selected and the corresponding shift schedules established, these ratios must be checked with Equation (1.35) to verify if all the acceleration requirements are met. If not, modify speed ratios and shift schedules with a larger i_1, until all acceleration requirements are satisfied.

Finally, tire–ground adhesion must be checked with Equation (1.53) or Equation (1.54) to make sure it meets the requirements of acceleration and gradeability. If not, it might be necessary to make design changes to the total layout of the EV for improvement in its tire–ground adhesion.

3.4. Selection of the Number of Gears and Intermediate Speed Ratios

The number of gears in the drivetrain of an EV affects its energy economy and dynamic performance. In addition, the appropriate number of gears for an EV is also related to the segment of the EV model, GVW, characteristics of the electric motor system, and so forth.

Today, most EVs are equipped with single-speed transmissions, that is, fixed-ratio reducers. However, some models of two- to four-speed automated mechanical transmissions (AMTs) are in production for passenger and commercial EV applications. Powershift transmissions, such as dual-clutch transmissions (DCTs) and planetary gear transmissions, are in development as well.

Extensive investigations have indicated that multispeed transmissions can improve gradeability, top speed, acceleration, and energy economy for various EV applications, as listed in Table 3.1 [3.1].

TABLE 3.1 Effect of multispeed transmissions on dynamic performance and energy economy of EVs.

No. of speeds	Type	Improvement in 0–50 km/h* 0–60 km/h** acceleration (%)	Improvement in 0–100 km/h acceleration (%)	Improvement in gradeability (%)	Improvement in top speed (%)	Driving cycle	Reduction in energy consumption (%)	Improvement in driving range (%)	Ref.
2	AMT	13.27*	4.05	14.60	22.04	EUDC	5.36	—	[3.2]
2	N/A	4.68*	-11.82	—	20.76	ECE	6.60	7.10	[3.3]
2	N/A	—	32.24	—	—	NEDC	2.71	—	[3.4]
2	I-AMT	8.30*	19.60	—	31.10	NEDC	11.40	—	[3.5]
						UDDS	7.31	—	
						1015	8.61	—	
2	SED	7.70*	3.50	37.40	11.90	NEDC	10.10	—	[3.6]
						FTP75	7.00	—	
						UDDS	6.69	—	
2	DCT	—	—	—	—	EUDC	5-10	—	[3.7]
		—	—	—	—	NEDC	—	3.63	[3.8]
2	DCT	0**	0	0	57.1	HWFET	—	16.4	[3.9]
						ECE	—	2.8	
2	N/A	11.8-12.8*	4.03-10.2	86.1-86.9	13.8-15.2	NEDC	1.94-3.88	—	[3.10]
						FTP	4.49-15.0	—	
2	MST	—	—	—	—	SFUDS	—	7.73-13.09	[3.11]
2	N/A	37.9*	17.9	40	27.6	NEDC	9.8	—	[3.12]
						WLTC	14.5	—	
2	N/A	15.5**	1.39	86.2	94.1	NEDC	—	4.17	[3.13]
2	DCT	—	12	6.25	21	NEDC	1.4	—	[3.14]

For a given EV, the largest speed ratio of a multispeed transmission can be greater than the ratio of a single-speed transmission, which obviously improves the acceleration and gradeability if the same electric motor is used. On the other hand, the smallest ratio of a multispeed transmission can be less than the ratio of a single-speed transmission, which increases the top speed. A multispeed transmission enables the electric motor to operate more frequently in its high-efficiency region than a single-speed transmission.

The ratio steps between adjacent speed ratios affect the shift time and shift quality. In general, a ratio step smaller than 3.4 is recommended to make shift time and shift quality acceptable. A large ratio step can cause difficulties in the shift process, especially in the prolonged time required for speed synchronization between the input and output sides of the shift device.

The effect of multispeed transmissions on dynamic performance and energy economy is usually more prominent in commercial or heavy-duty EVs than light-duty passenger EVs. As a rule of thumb, single- or two-speed transmissions are the most appropriate for light-duty passenger EVs, while two- to four-speed transmissions are proper choices for commercial or heavy-duty EVs.

The intermediate speed ratios in between the smallest and largest speed ratios are selected with different methodologies.

To enhance shift comfort, progressive ratio steps are typically applied for light-passenger EVs, in which gear steps decrease as the gear shifts to high. The intermediate ratio i_k can be estimated by [3.15]

$$i_k = i_n \varphi_1^{n-k} \varphi_2^{0.5(n-k)(n-k-1)} \tag{3.9}$$

where

φ_1 is the basic ratio change factor

φ_2 is the progression factor

Typically, $\varphi_1 = 1.1 - 1.7$ and $\varphi_2 = 1.0 - 1.2$ [3.15]. In general, the maximum speed of the electric motor remains approximately constant, while synchronization speed rises with an increase in gears. The differences between maximum vehicle speeds of adjacent gears do not vary much with gears.

On the other hand, geometric speed steps are commonly applied for commercial or heavy-duty EVs to improve their dynamic performance. The intermediate ratio i_k can be estimated by [3.15]

$$i_k = i_n \varphi^{n-k} \tag{3.10}$$

where $\varphi = \sqrt[n-1]{\sigma_{ratio}}$. The ratio spread σ_{ratio} is defined as $\sigma_{ratio} = i_1/i_n$. Both the maximum speed of the electric motor and synchronization speed remain roughly constant with increased gears. The differences between maximum vehicle speeds of adjacent gears increase as the gear shifts from low to high.

For transmissions of a fewer number of gears, for example, three- to five-speed, the geometric speed steps are not strictly applied. Instead, the speed steps often drop with an increase in gears to balance the usage of different gears, namely,

$$\frac{i_{g1}}{i_{g2}} \geq \frac{i_{g2}}{i_{g3}} \geq \frac{i_{g3}}{i_{g4}} \geq \frac{i_{g4}}{i_{g5}} \tag{3.11}$$

Once all the speed ratios are selected, a gear plan can be put together, which is also referred to as the saw profile diagram. The saw profile diagram of a geometrically stepped four-speed transmission for an EV is illustrated in Figure 3.3 [3.16], with the earliest down-shift points for the electric motor not exceeding the desired operating speed.

FIGURE 3.3 Saw profile diagram of a four-speed EV with geometric ratio steps.

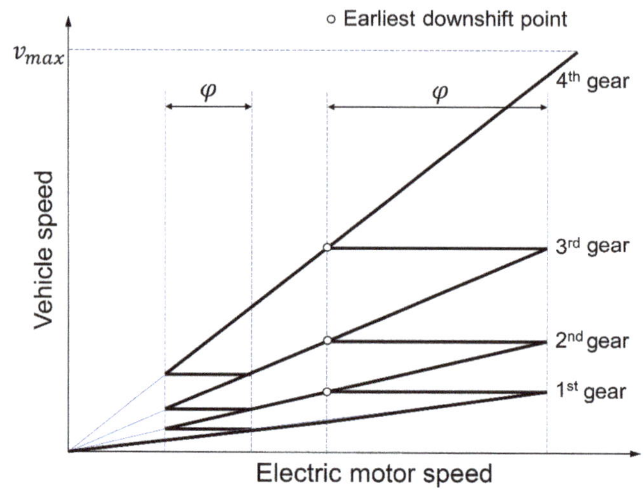

Courtesy of Springer.

3.5. Determination of Transmission and Final Drive Ratios

With the preliminary ratios being selected, as described in the previous sections, the final parameters of the electric drive system can be determined by further analysis and calculation of dynamic performance and energy economy.

3.5.1. Determination of the Final Drive Ratio

The final drive ratio can be determined by computing the dynamic performance and energy economy under the assumption of holding other parameters of the electric drive system unchanged. A curve of energy economy versus acceleration time can be plotted for a group of candidates of the final drive ratio $i_{0,1} < i_{0,2} < i_{0,3} < \cdots < i_{0,n}$. A final drive ratio near $i_{0,1}$ may result in good economy and poor performance, while a final drive ratio near $i_{0,n}$ may produce just the opposite. Consequently, a final drive ratio around the middle of the curve yields a balanced economy and performance.

The final drive made of a single-stage gear train typically has a ratio from 2.0 to 7.0. Otherwise, a two-stage gear train can be utilized for a final drive ratio greater than 7.0. For passenger vehicles, the final drive ratios usually range from 3.0 to 5.5 if it is made of cylindrical gears, or from 2.5 to 5.0, if it is made of bevel gears. Nevertheless, a larger final drive ratio can sometimes be considered for passenger EVs because only a couple of speed ratios are used and the smallest ratio is quite large. An oversized final drive ratio, however, can cause difficulties in packaging it in the vehicle or preserving an appropriate ground clearance.

For a given electric motor, proper transmission and final drive ratios can be selected from several available transmissions and final drive ratios. Alternatively, a suitable electric motor and final drive ratio can be selected from a few available electric motors and final drive ratios.

Furthermore, an appropriate combination can be selected from a number of available electric motors, transmissions, and final drive ratios. An example of three envelopes of energy economy versus acceleration time curves for three electric motors mating with provided two-speed, three-speed, and four-speed transmissions with various final drive ratios is illustrated in Figure 3.4.

FIGURE 3.4 Curves for selecting from e-motors, final drive ratios, and two- to four-speed transmissions.

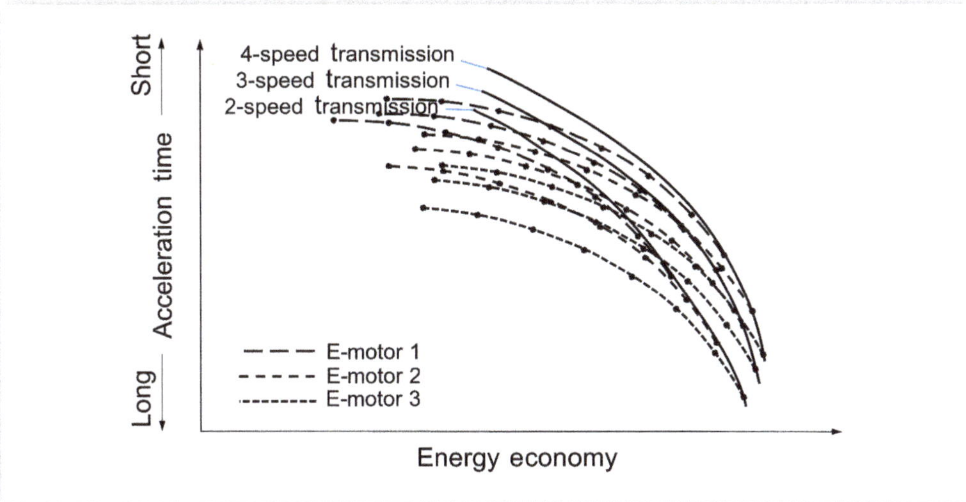

3.5.2. Optimization of Speed Ratios

Based on the preliminary speed ratios obtained by the selection methods described previously, optimization methods can be used to determine speed ratios in a more precise manner.

Since the energy economy of an EV can be characterized by the energy consumption of the battery, the objective function for the optimization can be written as

$$\min E_{bus} = \min \int_0^{t_{cyc}} P_{bus}(T_e, n_e)\,dt \tag{3.12}$$

where

E_{bus} is the total electric energy consumption at the DC bus
P_{bus} is the electric power consumption at the DC bus
t_{cyc} is the duration of the driving cycle
T_e is the electric motor torque
n_e is the electric motor speed

In Equation (3.12),

$$P_{bus} = \frac{P_e(n_e, T_e)}{\eta_e(n_e, T_e)} \tag{3.13}$$

where

P_e is the output power of the electric motor
η_e is the combined efficiency of the electric motor and motor controller, and

$$P_e = \frac{\pi n_e T_e}{30} \tag{3.14}$$

By discretizing Equation (3.12), the minimum total electric energy consumption can be solved to obtain the optimal speed ratios:

$$\min E_{bus} = \min \sum_{j=1}^{N} P_{bus}(T_e, n_e)\Delta t \tag{3.15}$$

where

N is the number of sampling points in the driving cycle
Δt is the time step at each sampling point

According to the discussion in Sections 3.2 and 3.3, the speed ratios fall into certain ranges that are set by requirements such as acceleration, gradeability, and top speed. This constitutes a problem of searching for the optimal solution in a two-dimensional domain. Two optimization algorithms, namely, particle swarm optimization (PSO) and genetic optimization (GA), are available for solving such an optimization problem, as discussed subsequently.

3.5.2.1. PSO Optimization Algorithm

PSO offers a good likelihood to obtain the globally optimal solution and high computational efficiency and feasibility.

Based on PSO, a swarm of population $S = \{x_1, x_2, \ldots, x_j, \ldots, x_m\}$ consisting of m particles can be formed, where m particles represent speed ratios and x_j stands for the position of the jth particle. The position is updated according to the following equations through the dimension j:

$$x_j^{k+1} = v_j^{k+1} + x_j^k \tag{3.16}$$

$$v_j^{k+1} = wv_j^k + c_1 r_1 (p_j - x_j^k) + c_n r_n (p_g - x_j^k) \tag{3.17}$$

where j is the label of particles and $j = 1, 2, \ldots, m$; k is the number of iterations; v_j^k is the velocity of particles; $v_{min} \leq v_j^k \leq v_{max}$ to prevent them from escaping the search domain; p_j is the best point obtained by self-searching, whose fitness value is p_{best}; p_g is the best point obtained by population searching, with g_{best} being the fitness value and g being the corresponding label for the particle; w is the inertia weigh factor, which should be selected by considering both convergence speed and accuracy; c_1 and c_n are learning factors, and $c_1 = c_n = 2$; and r_1 and r_n are arbitrary numbers between 0 and 1 [3.3].

The fitness value of each particle can be calculated by substituting x_j into the objective function of Equation (3.15) for judging the quality of the particle search point. A high-quality search point can be indicated by a small fitness value because the optimization goal is to minimize the input power consumption of the electric motor controller.

The procedures of PSO are briefly described as follows [3.17]:

1. Initialize the position and speed of m particles, that is, x_j^1 and v_j^1, within the search domain.
2. Compute the fitness value and initialize the individual best value p_{best} and global best value g_{best}.
3. Update p_j, p_{best}, p_g, and g_{best} using Equations (3.16) and (3.16), until the termination conditions are met.

3.5.2.2. GA Optimization Algorithm

GA not only allows a highly efficient search for the global optimal solution with good robustness but also supports parallel computing.

The procedure of GA optimization is as follows:

1. Coding
   Set the code string length in the optimization model to the number of speed ratios that need to be optimized. For n speed ratios, the length of the coding string $l_n = n$, and the coding string can be expressed as $s = \{i_1, i_2, \ldots, i_n\}$, where i_1, i_2, …, i_n are the speed ratios of the first gear, second gear, …, and n^{th} gear, respectively.
2. Selection of population size
   A small population size may affect the diversity of the population, resulting in early convergence. In contrast, a large population size would reduce search speed. A population size of 40–400 is, in general, deemed appropriate.
3. Initialization of population
   Although the initial population can be generated randomly, appropriate selection of the initial population is a better approach as it can greatly reduce search time and avoid premature. The rule is to keep the distance between individuals as much as possible, to avoid falling into the local optimum due to insufficient diversity. If the best possible solution or near-optimal solution is known, it should be included as much as possible.
4. Crossover operation
   The crossover operation has to ensure that excellent genes can be retained to the next generation during the genetic process, thereby providing the possibility for new individuals to generate better-quality genes. Therefore, the crossover probability can be selected from a moderately large value, generally between 0.4 and 0.9.

5. Selection operation
According to the principle of natural selection for survival of the fittest, individuals with higher fitness in the population are selected to be inherited by the next generation, so that the optimal solution in the evolutionary process is retained. Usually, the genetic generation gap is 0.9.
6. Mutation operation
The mutation operation can improve the local search ability and effectively avoid premature convergence. According to a certain probability, one or more genes of an individual are changed to generate a new subindividual. Generally speaking, the variation probability is 0.001–0.1, with 0.01–0.1 being the most common.
7. Calculation of the fitness function
The objective function is transformed into a fitness function. Then, the probability of individual inheritance to the next generation is determined.
8. Judgement of termination conditions
There are two commonly used termination conditions: set the maximum evolutionary generation, generally from 100 to 1000 generations, and terminate when the population has matured and no longer has obvious changes according to the degree of group convergence.

References

[3.1]. Yuan, Y., *Electric Vehicle Transmissions: Structure and Control* (Beijing: China Machine Press, 2020), ISBN:978-7-111-66809-1.

[3.2]. Huang, W., Wang, Y., Feng, K., and Zhang, J., "Development of Two-Speed Automatic Transmission for Electric Vehicle," *Automotive Technology* 10 (2001): 17-21.

[3.3]. Qin, D., Zhou, B., Hu, M., Hu, J. et al., "Parameters Design of Powertrain System of Electric Vehicle with Two-Speed Gearbox," *Journal of Chongqing University* 34, no. 1 (2011): 1-6.

[3.4]. Ren, Q., Crolla, D., and Morris, A., "Effect of Transmission Design on Electric Vehicle (EV) Performance," Paper Presented at in *IEEE Vehicle Power and Propulsion Conference*, Dearborn, September 7–10, 2009, doi:10.1109/VPPC.2009.5289707.

[3.5]. Gao, B., Liang, Q., Xiang, Y. et al., "Gear Ratio Optimization and Shift Control of 2-Speed I-AMT in Electric Vehicle," *Mechanical Systems and Signal Processing* 50-51 (2015): 615-631, doi:10.1016/j.ymssp.2014.05.045.

[3.6]. Stubbs, B., "Multi-Speed Transmissions for Electric Vehicles," Panel Presentation at in *Niche Vehicle Symposium*, Gaydon, UK, March 2012.

[3.7]. Green Car Congress, "New Two-Speed Electric Vehicle Transmission for Improved Performance," accessed May 2009, https://www.greencarcongress.com/2009/05/vocis-20090512.html.

[3.8]. Zhou, X., Walker, P., and Zhang, N., "Performance Improvement of a Two Speed EV through Combined Gear Ratio and Shift Schedule Optimization," SAE Technical Paper 2013-01-1477 (2013), doi:https://doi.org/10.4271/2013-01-1477.

[3.9]. Ruan, J., Walker, P., and Zhang, N., "A Comparative Study Energy Consumption and Costs of Battery Electric Vehicle Transmissions," *Applied Energy* 165 (2016): 119-134, doi:10.1016/j.apenergy.2015.12.081.

[3.10]. Sorniotti, A., Boscolo, M., Turner, A., and Cavallino, C., "Optimization of a Multi-Speed Electric Axle as a Function of the Electric Motor Properties," Paper Presented at in *IEEE Vehicle Power and Propulsion Conference*, Lille, France, September 2010, doi:10.1109/VPPC.2010.5729120.

[3.11]. Liauw, Y., Roozegar, M., Zou, T., and Morozov, A., "Range Model of Electric Vehicles with Multi-Speed Transmissions," Paper Presented at in *ASME International Design Engineering Technical Conferences and Computers and Information in Engineering Conference*, Quebec City, Quebec, Canada, August 26–29, 2018, doi:10.1115/DETC2018-85119.

[3.12]. Renţea, C., Oprean, M., Băţăuş, M., and Frăţilă, G., "The Influence of Multi-Speed Transmissions on Electric Vehicles Energy Consumption," *IOP Conference Series: Materials Science and Engineering* 564 (2019): 012107, doi:10.1088/1757-899X/564/1/012107.

[3.13]. Zhou, X., Walker, P., Zhang, N., Zhu, B. et al., "The Influence of Transmission Ratios Selection on Electric Vehicle Motor Performance," Paper Presented at in *ASME International Mechanical Engineering Congress and Exposition*, Houston, November 9–15, 2012, doi:10.1115/IMECE2012-85906.

[3.14]. Li, C., Zhao, L., Li, Y., and Chen, W., "Parameter Optimization of 2-Speed EV Powertrain for Energy Consumption," *Computer Integrated Manufacturing Systems* 24, no. 6 (2018): 1327-1336, doi:10.13196/j.cims.2018.06.001.

[3.15]. Naunheimer, H., Bertsche, B., Ryborz, J., and Novak, W., *Automotive Transmissions Fundamentals, Selection, Design and Application*, 2nd ed. (New York: Springer, 2010), 73-139, doi:10.1007/978-3-642-16214-5?nosfx=y.

[3.16]. Lechner, G. and Naunheimer, H., *Automotive Transmissions: Fundamentals, Selection, Design and Application* (New York: Springer, 1999), 110-111, ISBN:3-540-65903-X.

[3.17]. Yang, Y., Zhang, J., Wu, S., and Wang, P., "Gear Ratio Optimization of 2-Speed AMT for Electric Vehicle by PSO Algorithm," *Drive System Technique* 30, no. 1 (2016): 3-8, doi:10.3969/j.issn.1006-8244.a0818.

4

Braking Performance and Energy Regeneration

Braking performance refers to the capacity of an EV to stop at short distances while maintaining its stability in the driving direction or to keep a certain speed while traveling down a long slope.

Braking performance is of crucial importance to EVs because it is directly related to traffic safety. Traffic accidents often result from long stopping distances, sideslips during emergency braking, overheated brakes, and so on. On the other hand, the kinetic energy of an EV can be recovered partially in the braking process, termed regenerative braking, to boost the energy economy of an EV. It is also desired, therefore, to maximize the amount of energy recovered in regenerative braking without compromising braking performance.

4.1. Evaluation Indicators for Braking Performance

The braking capacity of an EV can be evaluated by the following four criteria [4.1]:

1. Braking effectiveness
   As the most fundamental indicator for braking performance evaluation, braking effectiveness refers to the distance or deceleration of an EV traveling before it comes to a full stop when its brakes are applied at a certain initial speed.

2. Constancy of braking effectiveness
   The kinetic energy of an EV is partially converted into thermal energy during the braking process, causing a dramatic increase in brake temperature. In other circumstances, the brakes can get wet if an EV travels through the flood. Therefore, the constancy of braking effectiveness depends on the resistance of the brake system to thermal or water decay.
3. Directional stability during braking
   Directional stability during braking refers to the capacity of an EV not deviating from its original path due to running off, sideslip, or loss of steering in the process.
4. Effectiveness of regenerative braking
   The effectiveness of regenerative braking can be evaluated by the ratio of the recovered energy to the total kinetic energy of the EV before braking.

Part of the requirements for braking performance in the regulations of different regions is listed in Table 4.1 [4.1].

TABLE 4.1 Requirements for braking performance in the regulations of different regions.

Items	US	European Union	China	Japan
Regulation no.	Federal 135	EU 71/320	GB7258-1998	R13H
Test pavement	Skid no. 81	Good adhesion	Adhesion coefficient ≥ 0.7	High, low, split adhesion
Load	Light	A driver or full load	Any load	Load or no load
Initial speed before braking	96.5 km/h (60 mph)	80 km/h	50 km/h	100 km/h or 80% V_{max}
Braking stability	Deflection ≤ 3.66 m	No deflection, no locking	Deflection ≤ 2.5 m	N/A
Stopping distance or deceleration	≤65.8 m	≤50.7 m, ≥5.8 m/s²	≤20 m	≤70 m, ≥4.82 m/s²
Pedal force	66.7-667 N	<490 N	<500 N	<500 N

Reprinted with permission from Yu, Z.-S. (2009) Automotive Theory. China Machine Press, Beijing.

4.2. Wheel Force Analysis during Braking

An EV can decelerate only when it is subject to external forces opposite to its traveling direction, which are typically from the ground or air. Because air resistance is relatively small, the external force comes primarily from the ground, which is called a ground braking force. Both braking deceleration and stopping distance are proportional to the ground braking force. Hence, the ground braking force has a dominant effect on the braking performance of an EV.

4.2.1. Ground Braking Force

Forces acting on a wheel traveling on a well-paved hard surface during braking are shown in Figure 4.1, where rolling resistance moment, inertia force, and inertia moment are neglected [4.1].

FIGURE 4.1 Forces acting on a braked wheel on a well-paved hard surface.

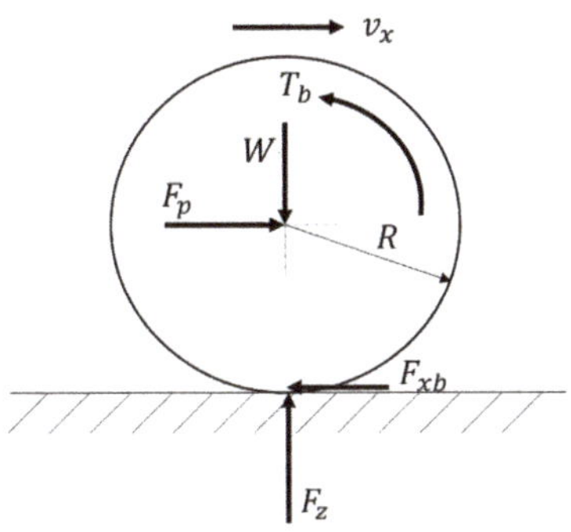

Assuming that there is no slip at the wheel–road interface, the ground braking force F_{xb} is given by

$$F_{xb} = \frac{T_b}{R} \tag{4.1}$$

where

- T_b is the braking moment produced between the friction pads and the brake drum or the brake rotor about the center of the wheel
- R is the wheel radius
- W is the vertical load on the wheel
- F_p is the propulsion force from the shaft to the wheel
- F_z is the normal reaction force from the ground to the wheel

The ground braking force depends on two forces: the friction force between friction pads and brake drum or rotor, and the adhesion force between the tire and the ground.

4.2.2. Braking Force at the Tire Circumference

The braking force at the tire circumference, denoted by F_b, is the force required to overcome the frictional moment of a brake and the braking torque exerted through the driveline by the electric motor. It is equivalent to the force needed to push the tire circumference to turn when taking the vehicle off the ground and pressing the brake pedal. Apparently,

$$F_b = \frac{T_b + T_{eb}}{R} \tag{4.2}$$

where T_{eb} is the braking torque applied on the wheel by the electric motor.

Clearly, the braking force at the tire circumference for an EV mainly depends on the structural parameters of brakes, the dynamic parameters of the electric motor and drivetrain, and the wheel radius. The structural parameters of brakes include the configuration, size, and the friction coefficient between the pad and the drum or rotor. The dynamic parameters of the electric motor and drivetrain involve the torque and power rating of the electric motor, and speed ratios of the drivetrain. For traditional hydraulic or air brake systems, the braking force at the tire circumference is proportional to the brake pedal force, that is, the hydraulic or air pressure in the brake system.

4.2.3. Relationship between Adhesion and Braking Forces

In braking, wheels can be simplified into two operating conditions, namely, rolling or slipping, depending on whether they lock or not.

When both the braking force F_b and brake friction moment are small, the ground braking force F_{xb} between the ground and the tire is adequate to overcome the braking force F_b and make the wheel rolling. Thus, the ground braking force F_{xb} is equal to the brake force F_b and grows proportionally with the increase in the braking force F_b [4.1]. As a limited reaction force against sliding friction, the ground braking force F_{xb} cannot exceed the ground adhesion force F_f in value, that is,

$$F_{xb} \leq F_f = fF_z \tag{4.3}$$

where f is the adhesion coefficient. As the braking force F_b reaches F_{bf}, the ground braking force reaches its maximum $F_{xb,max}$, which is equal to the ground adhesion force F_f, namely,

$$F_{xb,max} = F_f = fF_z \tag{4.4}$$

At this point, the wheels lock and start to slip. With a further increase in the braking force F_b, the ground braking force F_{xb} stops increasing and remains equal to ground adhesion F_f. Meanwhile, the wheels continue to slip due to lockup.

For an EV equipped with an anti-lock brake system (ABS), the braking force F_b undergoes fluctuation between 0 and a certain value F_b'. Apparently, F_b' depends on the brake pedal force or displacement, and $F_b' > F_{bf}$. As a result, the wheels only lock intermittently to prevent the EV from slipping and going out of control.

Depending on the braking force produced by the brake and the electric motor in the first place, the ground braking force is also subject to the ground–tire adhesion condition. Therefore, adequate ground braking force can be attained only if both braking force and ground–tire adhesion force are sufficient.

4.2.4. Adhesion Coefficient on Hard-Surface Roads

4.2.4.1. Longitudinal Behavior

Careful observations of tire marks left on the ground reveal that the wheels actually go through a transient development from rolling to lockup in the braking process [4.1].

When the braking force is small, the tire mark has almost the same shape as the tire tread pattern when the wheels are in sheer rolling. The velocity of the wheel center, v_w, is approximated by

$$v_x \approx \omega_w R_0 \tag{4.5}$$

where

ω_w is the angular velocity of the wheel

R_0 is the wheel radius in rolling subject to the zero ground braking force

With the increase in the braking force, the tire pattern marks can still be recognized, but the pattern becomes gradually blurred in the direction of motion. It is implied that, instead of simple rolling, a certain degree of relative slip occurs between the tread and the ground. In other words, the wheel is in mixed states of rolling and slipping. The velocity of wheel center u_w becomes

$$v_x > \omega_w R_0 \tag{4.6}$$

With a further increase in the severity of braking, that is, braking ratio, the portion of slipping augments; therefore,

$$v_x \gg \omega_w R_0 \tag{4.7}$$

When the braking force is large, a thick black mark is formed, but the tire pattern is no longer recognizable. The wheels lock completely and go into pure slip, namely,

$$\omega_w = 0 \tag{4.8}$$

Based on the previous discussion, the wheels undergo less rolling and more slipping with the increased braking ratio. As defined in Equation (1.21), the longitudinal slip ratio s in a braking process is [4.2]

$$s = \frac{v_x - \omega_w R_0}{v_x} \tag{4.9}$$

Obviously, the wheels are in sheer rolling when $s = 0$, in pure slip when $s = 1$, and in partial rolling and slipping when $0 < s < 1$.

To quantify the ground braking force F_{xb}, the braking force coefficient, also known as braking coefficient, φ_b, can be defined as

$$\varphi_b = \frac{F_{xb}}{F_z} \tag{4.10}$$

where F_z is the vertical load.

Figure 4.2 illustrates a typical curve of the braking force coefficient φ_b versus the slip ratio s [4.3]. With increasing s, φ_b rises fast and almost linearly from 0 to A, and there is actually no relative motion at the tire–ground interface, despite $s > 0$. This is a result of enlarged rolling radius of the tire, as the tread in the front about to come in contact with the ground is stretched slightly under the action of the ground braking force. In fact, the rolling radius of the tire increases with the increasing ground braking force, that is, $v_x = \omega_w R > \omega_w R_0$. Hence, $s > 0$ according to Equation (4.9). From A to P, local relative motion takes place in the part of the contact area between the tire and the ground, slowing down the increase in φ_b, until it peaks at P. The braking force coefficient at P, φ_{bp}, is referred to as the peak adhesion coefficient, which typically occurs around $s = 0.15 - 0.2$. Since the dynamic friction coefficient is smaller than the static friction coefficient, φ_b decreases monotonically with a further increase in s from P to S. The braking force coefficient at S, φ_{bs}, is called the slip adhesion coefficient, where $s = 1$. The difference between φ_{bp} and φ_{bs} is small on dry roads but significant on wet roads. Typically, $\frac{\varphi_{bs}}{\varphi_{bp}} = \frac{1}{3} - 1$ [4.2].

FIGURE 4.2 Braking force coefficient versus slip ratio.

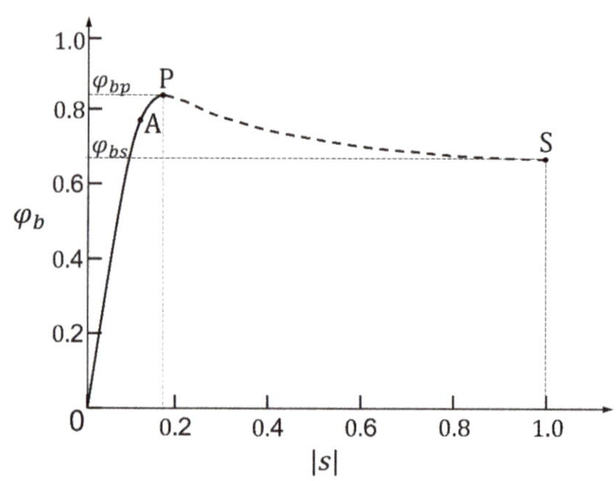

A number of road surface, tire, and speed-related factors contribute to the adhesion coefficient of tires, as described here.

Effect of Road Surface on the Adhesion Coefficient. The adhesion coefficients of road surfaces of various materials and conditions are listed in Table 1.6 [4.1]. The adhesion capacity increases with the increase in the tire–ground contact area. Hence, the use of radial tires, increase in tread width, and reduction in inflation pressure can all help elevate the adhesion coefficient, to some extent.

The adhesion capacity declines substantially on wet road surfaces, particularly in two scenarios: First, when it just starts raining, the small amount of raindrops can get mixed with dust, dirt, or grease on the road surface. The rolling tire cannot squeeze out the mixture film at the tire–road interface due to its high viscosity, causing the adhesion to diminish. Second, when driven through standing water or flood at relatively high speeds, the vehicle can get lifted completely by the water film without touching the road surface, a phenomenon called dynamic hydroplaning. In addition, viscous hydroplaning refers to a vehicle driven on a wet road covered with a layer of dust, oil, or grease. Rubber hydroplaning, on the other hand, refers to the situation where the rubber begins to melt due to the lockup of tires in heavy braking, as the trapped water under the tire turns into superheated steam at high pressure.

The tire comes in direct contact with the road surface as it rolls through standing water at low speeds because the water can be displaced completely from under the tire. As speed increases, a transition condition occurs when only the aggregate in the road comes into contact with the tire. At higher speeds, the tire is lifted off the road surface totally and starts to slide on top of water since the water cannot escape quickly enough from under the tire. The hydroplaning speed v_h can be estimated by

$$v_h = 6.348\sqrt{p_t} \tag{4.11}$$

where

p_t is the inflation pressure of the tire in kPa

v_h is in km/h

Adhesion Coefficients of Different Types of Tires. As hydroplaning occurs, the adhesion coefficient behaves differently for different types of tires. On most road surfaces, the tire with properly designed tread patterns can improve the adhesion coefficient considerably. Apparently, tread wear reduces the tread depth and adhesion capacity of a tire dramatically.

Effect of Speed on the Adhesion Coefficient. For a given slip ratio, the adhesion coefficient generally declines with an increase in speed. The effect of speed on the adhesion coefficient can be enlightened by an example of a tire rolling on a dry bituminous surface with fine stone aggregates. When the vehicle speed is lower than 32 km/h, the adhesion coefficient peaks first and then declines with an increase in the slip ratio; however, when the vehicle speed exceeds 64 km/h, the adhesion coefficient hardly peaks before decreasing only slightly with an increase in the slip ratio; when the vehicle speed reaches 96 km/h, the adhesion coefficient grows moderately with an increase in the slip ratio at a slip ratio of nearly 0.1 and only drops slightly when the wheel is almost locked with the slip ratio very close to 1.

4.2.4.2. Effect of Sideslip on Adhesion

In reality, a lateral deviation or slip often happens in braking when a tire is subject to a lateral force. The characteristics of the braking force coefficient φ_b versus slip ratio $|s|$ under various sideslip angles α are illustrated in Figure 4.3 [4.3]. By defining a lateral force coefficient φ_l as the ratio of lateral force to vertical load, its relationship with the slip ratio $|s|$ for various sideslip angles α is presented in Figure 4.4 [4.3]. For a given sideslip angle α, the lateral force coefficient φ_l becomes larger as the slip ratio $|s|$ decreases, implying stronger resistance of tires to sideslip and loss of steering. Thus, an ABS can elevate both braking force coefficient and lateral force coefficient by keeping the slip ratio at a low level in braking, which improves the braking performance and lateral stability of the EV.

FIGURE 4.3 Braking force coefficient versus slip ratio for various sideslip angles.

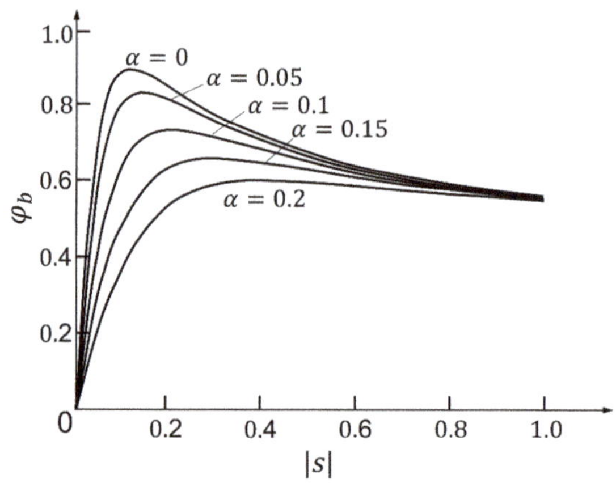

FIGURE 4.4 Lateral force coefficient versus slip ratio for various sideslip angles.

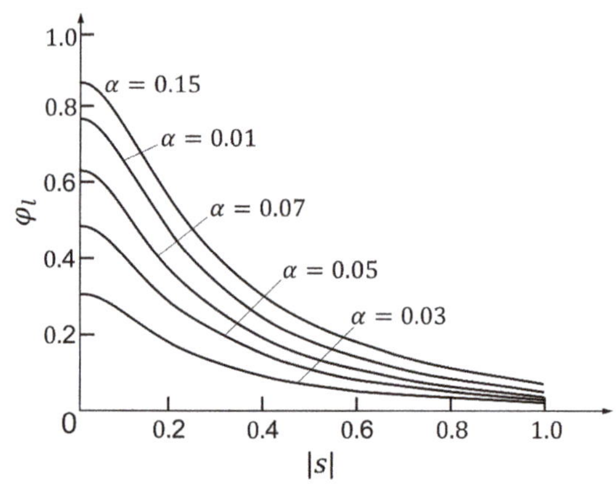

4.3. Braking Efficiency and Constancy

The braking efficiency of an EV refers to its ability to slow down rapidly until it stops, which is commonly indicated by stopping distance and deceleration.

4.3.1. Stopping Distance and Deceleration

Directly related to the driving safety of an EV, stopping distance refers to the distance that the EV drives from the moment when the brake pedal is depressed until it comes to a full stop. The stopping distance of an EV mainly depends on factors such as brake pedal opening, adhesion condition of the road surface, vehicle load, brake temperature, and the role that traction motors play in the braking process.

Braking deceleration is the time derivative of vehicle velocity in braking, which depends on the ground braking force. Hence, the maximum deceleration is given by [4.1]

$$a_{bmax} = \varphi_b g \tag{4.12}$$

For an EV equipped with ABS, the maximum deceleration is subject to the limit imposed by the peak adhesion coefficient:

$$a_{bmax} = \varphi_{bp} g \tag{4.13}$$

For an EV without ABS, the maximum deceleration in the case that both front and rear wheels are locked is limited by the slip adhesion coefficient:

$$a_{bmax} = \varphi_{bs} g \tag{4.14}$$

As the instantaneous deceleration varies with time in a complex manner, a time-averaged deceleration is often employed instead to evaluate the braking performance:

$$\bar{a} = \frac{1}{t_2 - t_1} \int_{t_1}^{t_2} a(t)\mathrm{d}t \tag{4.15}$$

4.3.2. Analysis of Stopping Distance

A typical braking process can be simplified into four stages [4.1]. The process starts at the moment $t = 0$ when the driver observes a situation and, after a time delay of τ_{11}, realizes an emergency stop is necessary. Then, the driver starts to depress the brake pedal after another time delay of τ_{12}. The combined delay $\tau_1 = \tau_{11} + \tau_{12}$ is called driver reaction time, typically ranging from 0.3 to 1.0 sec. After that, the brake pedal opening α_b rises dramatically and soon reaches the maximum. However, brake deceleration takes place only after a delay of τ_{21} due to the clearance and hysteresis in brake systems. It takes additional time τ_{22} for the brake deceleration to reach its maximum. The duration $\tau_2 = \tau_{21} + \tau_{22}$, or brake action time, typically lasts 0.2–0.9 sec. The deceleration remains almost constant in the third stage, with a duration of τ_3. Eventually, the deceleration disappears quickly after the driver releases the brake pedal, and it takes as much time as τ_4 to diminish. In general, τ_4 ranges from 0.2 to 1.0 sec. A prolonged τ_4 can cause a delay in launching the EV after braking.

The stopping distance refers to the distance from where the brake pedal is depressed to where the EV finally stops. Apparently, the total stopping distance

$$s = s_2 + s_3 \tag{4.16}$$

where s_2 and s_3 are the distances that the EV travels in time duration τ_2 and τ_3, respectively. We have

$$s_2 = v_0\tau_{21} + v_0\tau_{22} - \frac{1}{6}a_{bmax}\tau_{22}^2 \tag{4.17}$$

where

v_0 is the initial velocity of the EV

a_{bmax} is the maximum of stopping deceleration

$$s_3 = \frac{v_0^2}{2a_{bmax}} - \frac{v_0\tau_{22}}{2} + \frac{a_{bmax}\tau_{22}^2}{8} \tag{4.18}$$

Substitution of Equations (4.17) and (4.18) into Equation (4.16) yields the total stopping distance:

$$s = v_0\left(\tau_{21} + \frac{\tau_{22}}{2}\right) + \frac{v_0^2}{2a_{bmax}} - \frac{a_{bmax}\tau_{22}^2}{24} \tag{4.19}$$

Because τ_{22} is typically small, the last term in Equation (4.19) can often be neglected. Hence,

$$s \approx v_0\left(\tau_{21} + \frac{\tau_{22}}{2}\right) + \frac{v_0^2}{2a_{bmax}} \tag{4.20}$$

Equation (4.20) implies that the stopping distance mainly depends on the initial velocity, maximum braking deceleration, and time needed for the actuation of brakes.

The brake actuation time varies with the brake system configuration. The actuation time can be as short as 0.1 sec for hydraulic brake systems compared with 0.3–0.9 sec for air brake systems. The brake takes longer time to actuate for a vehicle with a trailer, which ranges from 0.4 to 2.0 sec.

4.3.3. Constancy of Braking Efficiency

Constancy of braking efficiency refers to the brake system's performance in preventing brake fade. Brake temperature often exceeds 300°C and sometimes even reaches 600–700°C during heavy or sustained braking, for example, when driving the vehicle down a long slope. In particular, brake temperature rises quickly when the initial vehicle speed is high. With the increase in brake temperature, the friction torque drops significantly, resulting in a prolonged stopping distance. This phenomenon is called brake fade. The other type of brake fade is caused by water that enters the brake when vehicle wading.

The antifade performance of brakes is generally evaluated by the extent of braking efficiency being retained in a series of repeated brake applications. Most countries have regulations and standards specifying the percentile braking efficiency that must be retained after a given number of repeated brake applications of a certain severity. Higher antifade performance is often required for EVs operating at high speeds or in mountainous regions. Some countries enforce a secondary brake system to ensure braking efficiency in mountainous regions.

The antifade performance of brakes is related to their configuration and the material of friction pairs. The regular brake drums and rotors are made of cast iron, while the friction pads are made of semimetallic, nonasbestos organic, or ceramic materials. The temperature of brake friction pairs is about 200°C in normal braking, with a friction coefficient of 0.3–0.4.

With a further increase in temperature, the friction coefficient of some friction pairs reduces dramatically, resulting in brake fade. The brake fluid temperature may also rise because of poor design or misuse of brakes. Evaporation of the brake fluid can cause brake failure if its temperature exceeds the boiling point of the brake fluid.

The braking efficiency factor K_{ef} is defined as the brake friction force produced by the unit brake cylinder force. The braking efficiency factor of configurations such as bidirectional servo brakes and dual leading shoe brakes is relatively high. However, their braking efficiency factor varies nonlinearly with the friction coefficient, which can deteriorate braking stability. On the other hand, despite their lower braking efficiency factors, the disc brakes have much better braking stability than the brakes of other configurations. Hence, the disc brakes have been widely used in various types of EVs.

Quick recovery from short-term fade of braking efficiency is also an important metric of the constancy of braking efficiency.

4.4. Directional Stability in Braking

The directional stability of an EV refers to its ability to follow expected straight or curved trajectories in braking. This is of crucial importance to EVs because any situation such as loss of steering of front wheels, skid sideways of rear wheels, or deviations during braking can cause the EV to run out of control and result in traffic accidents.

When the front wheel steering is lost, the vehicle either moves by itself in the tangential direction, instead of following the original turn on a curved road, or moves straight forward with no response to the turning steering wheel. Or, during braking, a lateral movement may develop at one or two of the axles, that is, skid. The most dangerous skid occurs at the rear axle when a vehicle is braked at high speeds because it may run into irregular sharp gyratory motion and eventually become out of control. The movement of a vehicle toward the left or the right by itself during braking is called braking deviation. Loss of steering is related to the skid of the rear axle in that they typically do not happen at the same time. Moreover, sideways skid and braking deviation are also interconnected in that severe braking deviation can result in a skid of the rear axle in some cases, while those EVs prone to a sideways skid tend to worsen braking deviation.

4.4.1. Braking Deviation

In general, braking deviation results either from unequal braking forces between left and right wheels, particularly between the front ones, or from the interference between the steering rod and the guide rod system of suspension.

The inequality of braking forces between the left and right wheels can be quantified by

$$\varepsilon = \frac{F_{fl} - F_{fs}}{F_{fl}} \tag{4.21}$$

where F_{fl} and F_{fs} are the larger and smaller braking forces, respectively. Braking deviation can be quantified by the lateral displacement of the vehicle body and the yaw angle. The yaw angle is the angle between the longitudinal line and the driving direction in braking.

Figure 4.5 is the analysis of the braking deviation due to unequal braking forces between the left and right front wheels of an EV [4.1]. For simplicity, it is assumed that the steering wheel is not turned in the process, and the centrifugal force and inertia moment of the centroid of the vehicle in circular motion are negligible with no sideways skid. Provided that the braking force of the right wheel is greater than that of the left wheel, the right ground braking force is, therefore, larger than that of the left ground braking force, that is, $F_{xfr} > F_{xfl}$. The front and rear axles are subject to lateral reaction forces from the ground F_{yf} and F_{yr}, respectively. Thus, the moment about steering kingpin axis resulted by F_{xfr} is greater than that by F_{xfl}. Although the steering wheel is not turned at this moment, the clearance and elasticity of the components of the steering system still allow the front wheels to turn a little angle toward the right, resulting in deviation. Moreover, the lateral reaction force F_{yr} produces an extra moment in the same direction due to the presence of a kingpin caster angle, which adds to the turning angle toward the right as well. Both the lateral displacement and the yaw angle increase with the inequality of braking forces, and the deviation worsens when the rear wheels are locked.

FIGURE 4.5 Braking deviation due to unequal braking forces between the left and right front wheels.

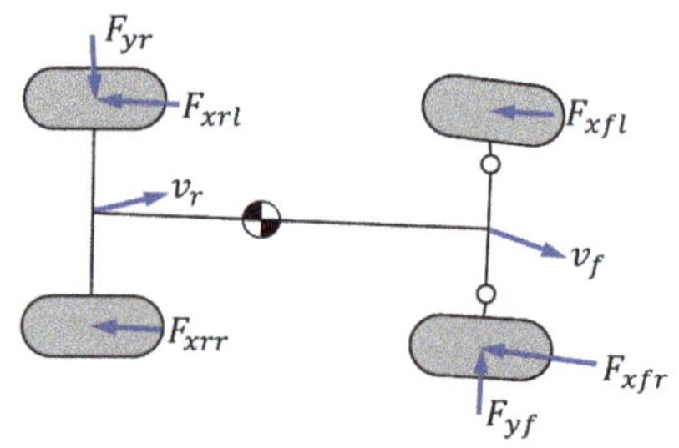

Reprinted with permisssion from Yu, Z.-S. (2009) Automotive Theory. China Machine Press, Beijing.

On the other hand, the interference between the steering rod and the guide rod system of suspension can cause deviation toward either the left or the right direction, depending on which side the steering wheel is mounted on the EV. This is typically caused by the excessive height of the ball joint connecting the upper control arms to the steering knuckle over the centerline of the front axle, and the insufficient stiffness of the front suspension.

4.4.2. Sideways Skid and Loss of Steering in Braking

If locked up earlier than front wheels in braking, rear wheels may enter the sideways skid condition. Sideways skid, especially at rear wheels, can result in dramatic circular motion of the vehicle and even make it turn around. In contrast, the sideways skid of rear wheels can be prevented if they are locked no earlier than front wheels. Nevertheless, steering can be lost due to the lockup of front wheels.

4.4.2.1. Effect of Wheel Lockup Sequence on Directional Stability

The yaw angle is a good indicator for a sideways skid in the braking process. In characterizing the relationship between the yaw angle and the vehicle velocity under different wheel lockup conditions, the following three scenarios are considered here:

1. When braking forces are only applied to rear wheels, the yaw angle increases radically with an increase in velocity. The yaw angle can even exceed 180° at relatively high speeds, meaning that the vehicle can turn around and head in the opposite direction. This is true whether the vehicle travels forward or backward.
2. When braking forces are only applied to front wheels, the yaw angle is quite small and barely increases with vehicle velocity. The motion of the vehicle follows more or less a straight line, with a maximum yaw angle of less than 10°. However, the vehicle may fail to make turns due to the loss of steering when front wheels are locked. Similar behaviors are observed when the vehicle travels backward.
3. When braking forces are applied on both front and rear wheels, the vehicle moves nearly straight ahead if the duration between lockup of front and rear wheels is short, for example, less than 0.5 sec. Nevertheless, if both rear wheels are locked more than a certain duration (e.g., 0.5 sec) earlier than front wheels, then a severe sideways skid could occur at the rear axle. Note that the sideways skid does not occur if only one of the rear wheels is locked. Thus, the severity of a sideways skid depends on the duration between the later lockup of front wheels and the later lockup of rear wheels.

4.4.2.2. Effect of Initial Speed and Adhesion Coefficient on Directional Stability

Both initial speed and adhesion coefficient affect sideways skid behavior. At low speeds, the yaw angle is small even if the rear wheels are locked much earlier than the front wheels. At high speeds, however, the yaw angle may become dangerously large. When braking forces are applied only on rear wheels, the stopping distance on the dry road s_{dry} is shorter than that on the wet road s_{wet}, and typically $s_{dry} \approx 0.7s_{wet}$. The yaw angle on dry roads is also smaller than that on wet roads. However, the curves of the yaw angle versus time elapsed in rear-wheel skid lie actually on top of each other for both wet and dry roads. Hence, there is not much difference in sideways skid on wet and dry roads, except the extra braking time needed on wet roads.

4.4.2.3. Difference between Front- and Rear-Wheel Skids

When front wheels are locked and rear wheels are rolling, sideways skid would occur at the front axle if it is subject to lateral force. Thus, the velocity v_1 of the middle point of the front axle C_1 is at an angle α to the longitudinal direction, as illustrated in Figure 4.6a [4.1]. The velocity v_2 of the middle point of the rear axle C_2, however, remains in the longitudinal direction for the absence of skid there. As a result, the vehicle undergoes a curved motion, with the instant center of rotation O being located at the intersection of lines perpendicular to v_1 and v_2. The centrifugal force F_c produced at the centroid C is in the direction opposite to the skid, which cancels out the effect of skid and stabilizes the vehicle. However, since front wheels are locked, the vehicle cannot follow a curved path due to the loss of steering. Instead, the vehicle moves straight ahead along the tangential direction since no lateral force can be produced from the ground.

FIGURE 4.6 Vehicle motion in sideslip conditions. (a) Sideslip of the front axle and (b) sideslip of the rear axle.

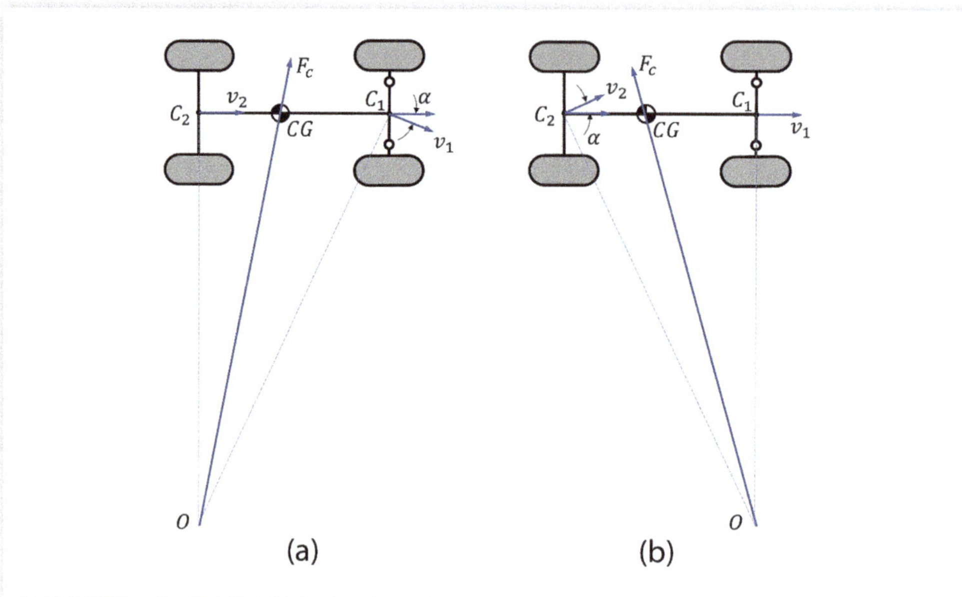

In contrast, when front wheels are rolling and rear wheels are locked, sideways skid would occur at the rear axle, as illustrated in Figure 4.6b [4.1]. Since the resultant centrifugal force is in the same direction as that of a skid, the sideways skid is intensified into a dangerous spinning motion and the vehicle becomes unstable.

4.4.2.4. Summary of Directional Stability

When only front wheels are locked or front wheels are locked earlier than rear wheels, the vehicle is in a stable condition. Despite the loss of steering, the vehicle travels almost straight ahead and decelerates.

When rear wheels are locked earlier than front wheels for a certain time, the vehicle can start skidding sideways under the action of even a minor lateral force when traveling above a certain speed. Also, the sideways skid of the rear axle worsens with the increased slippery road surface or extended stopping distance and braking time.

From the perspective of vehicle stability, the dangerous rear axle skid must be avoided first of all, by preventing the rear axle from lockup or lockup earlier than the front axle. Next, the steering capability of the vehicle can be retained by preventing the front axle or both axles from lockup. Finally, the directional stability can be best preserved by keeping both front and rear axles in a rolling condition without lockup.

4.5. Front and Rear Proportioning of a Braking Force

As discussed in the previous section, the proportioning of a braking force between front and rear axles is crucial because it affects the directional stability and utilization degree of the adhesion condition.

Given adequate braking force, there are three possible cases in terms of braking stability [4.1, 4.3]:

1. The front wheels are locked before the rear wheels.
2. The rear wheels are locked before the front wheels.
3. The front and rear wheels are locked simultaneously.

In case 1, the vehicle is stable, despite the loss of steering. Thus, the tire–ground adhesion is not fully utilized, as discussed later. In case 2, the vehicle is unstable since a sideways skid can take place at the rear axle, with a poor utilization of tire–ground adhesion. In case 3, the tire–ground adhesion is utilized better than that in the other two cases because a sideways skid of the rear axle can be avoided and front wheels only lose their steering at maximum braking effort.

4.5.1. Normal Reaction Forces from the Ground to Front and Rear Wheels

The forces acting on a vehicle in the braking process is illustrated in Figure 4.7 [4.3]. Neglecting air resistance and moments due to rolling resistance and rotating mass deceleration, we can obtain the normal reaction force from the ground to the front and rear wheels, respectively:

$$F_{z1} = mg\left(L_2 + zh_g\right)/L \tag{4.22}$$

$$F_{z2} = mg\left(L_1 - zh_g\right)/L \tag{4.23}$$

where

L is the wheelbase

z is the braking ratio

FIGURE 4.7 Force analysis in the braking process.

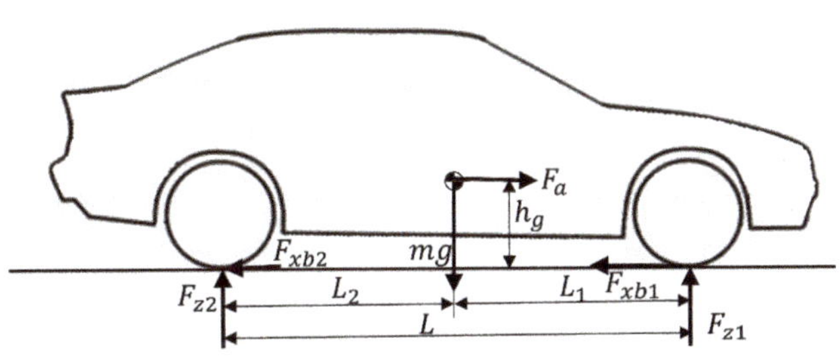

As a representation of the severity of braking, the braking ratio is defined as the ratio of vehicle deceleration a_b to the gravitational acceleration constant g, that is, $z = a_b/g$. When both front and rear wheels are locked, we have $a_b = fg$, and therefore,

$$F_{z1} = mg\left(L_2 + fh_g\right)/L \tag{4.24}$$

$$F_{z2} = mg\left(L_1 - fh_g\right)/L \tag{4.25}$$

Clearly, the normal reaction forces from the ground to the front and rear wheels vary linearly with the braking ratio and ground adhesion. In addition, if braked while a vehicle is moving forward, then F_{z1} increases and F_{z2} decreases.

4.5.2. Ideal Curve of Front and Rear Braking Force Proportioning

As mentioned earlier, the tire–ground adhesion can be best utilized when the front and rear wheels are locked at the same time. Based on this principle, an ideal front and rear wheel brake force proportioning curve can be obtained. To lock the front and rear wheels simultaneously, the front and rear braking forces must be equal to their corresponding adhesion forces individually, and their sum must be equal to the total adhesion force, that is,

$$\begin{cases} F_{f1} = fF_{z1} \\ F_{f2} = fF_{z2} \\ F_{f1} + F_{f2} = fmg \end{cases} \tag{4.26}$$

Substitution of Equations (4.24) and (4.25) into Equation (4.26) yields

$$\begin{cases} \dfrac{F_{f1}}{F_{f2}} = \dfrac{L_2 + fh_g}{L_1 - fh_g} \\ F_{f1} + F_{f2} = fmg \end{cases} \tag{4.27}$$

Thus, the relationship between rear and front braking forces is given by

$$F_{f2} = \frac{1}{2}\left(\frac{mg}{h_g}\sqrt{\frac{4h_g L}{mg}F_{f1} + L_2^2} - \frac{mgL_2}{h_g} - 2F_{f1}\right) \tag{4.28}$$

The ideal front and rear wheel brake force proportioning curve can be plotted based on Equation (4.28), which is often known as the I-curve, as illustrated in Figure 4.8 [4.4]. When both front and rear wheels are locked at the same time, the brake force is equal to the tire–ground adhesion force and ground braking force, that is, $F_{b1} = F_{f1} = F_{xb1}$ and $F_{b2} = F_{f2} = F_{xb2}$.

FIGURE 4.8 Ideal curve of front and rear braking force distribution.

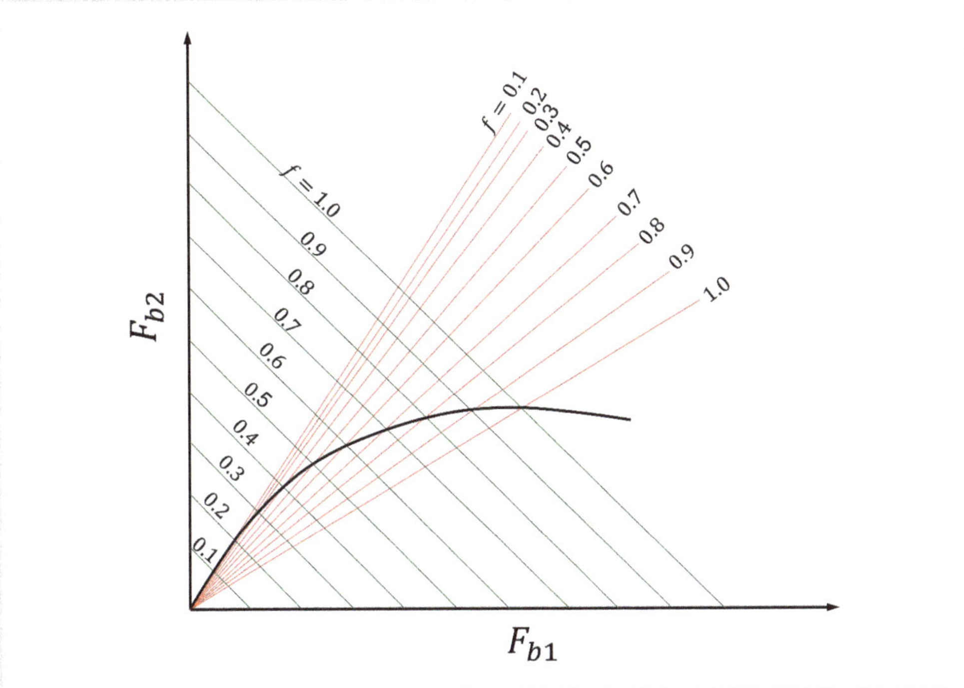

4.5.3. Fixed Proportioning of Front and Rear Brake Forces

The proportioning ratio of the front and rear brake forces β can be defined as

$$\beta = \frac{F_{b1}}{F_b} \tag{4.29}$$

where F_b is the total brake force, $F_b = F_{b1} + F_{b2}$. Apparently,

$$F_{b2} = \frac{1-\beta}{\beta} F_{b1} \tag{4.30}$$

If the proportioning ratio is fixed, then Equation (4.30) implies that the relationship between F_{b2} and F_{b1} is a straight line passing through the origin, as shown in Figure 4.9 [4.4]. Since a large number of vehicles have a fixed proportioning ratio, this line is often called the actual proportioning line of front and rear brake forces, or line β. For a given proportioning ratio, there is only one adhesion coefficient that allows front and rear wheels to lock simultaneously, which is right at the intersection of the I-curve and line β, that is, point A. The adhesion coefficient $f_{I\beta}$ at point A is given by

$$f_{I\beta} = \frac{L\beta - L_2}{h_g} \tag{4.31}$$

FIGURE 4.9 Actual distribution of front and rear braking forces.

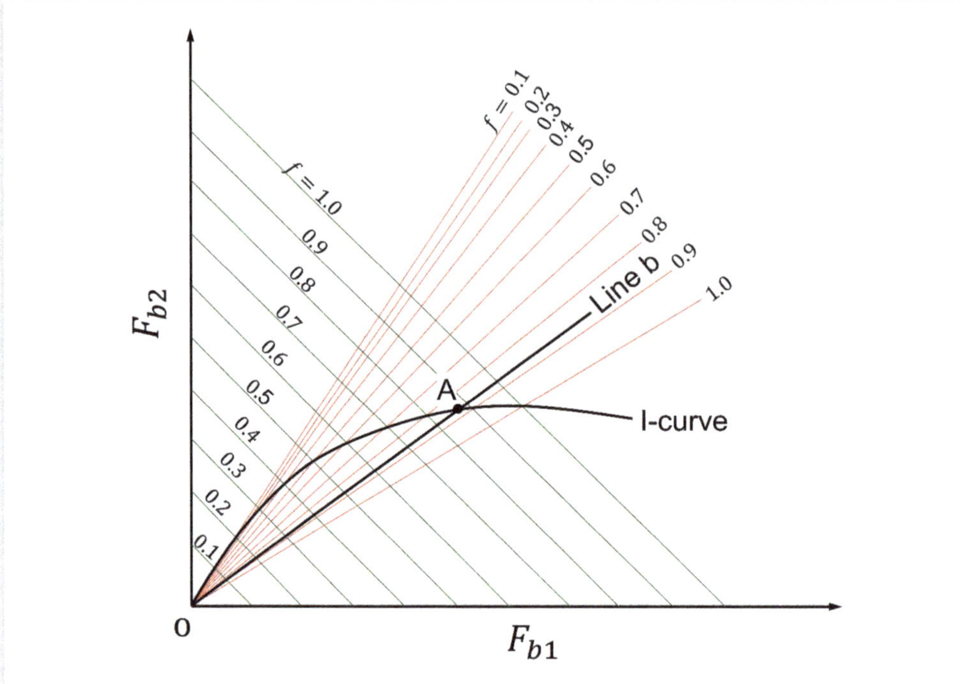

Depending on the vehicle configuration, the value of the adhesion coefficient $f_{I\beta}$ is an indicator of the braking performance of a vehicle. The corresponding braking deceleration at point A is termed critical deceleration.

4.5.4. Analysis of the Braking Process of EVs with Fixed Brake Proportioning

When the front wheels are locked and rear wheels are in free rolling, the ground braking force is equal to the tire-ground adhesion force for front wheels:

$$F_{xb1} = fF_{z1} = \frac{f}{L}\left(mgL_2 + F_{xb}h_g\right) = \frac{f}{L}\left(mgL_2 + F_{xb1}h_g + F_{xb2}h_g\right) \tag{4.32}$$

By rearranging Equation (4.32), we have

$$F_{xb2} = \frac{1}{h_g}\left(\frac{L - fh_g}{f}F_{xb1} - mgL_2\right) \tag{4.33}$$

By taking different values of f, a group of the so-called f-lines can be plotted, as illustrated in Figure 4.10 [4.1].

FIGURE 4.10 The f-lines and r-lines.

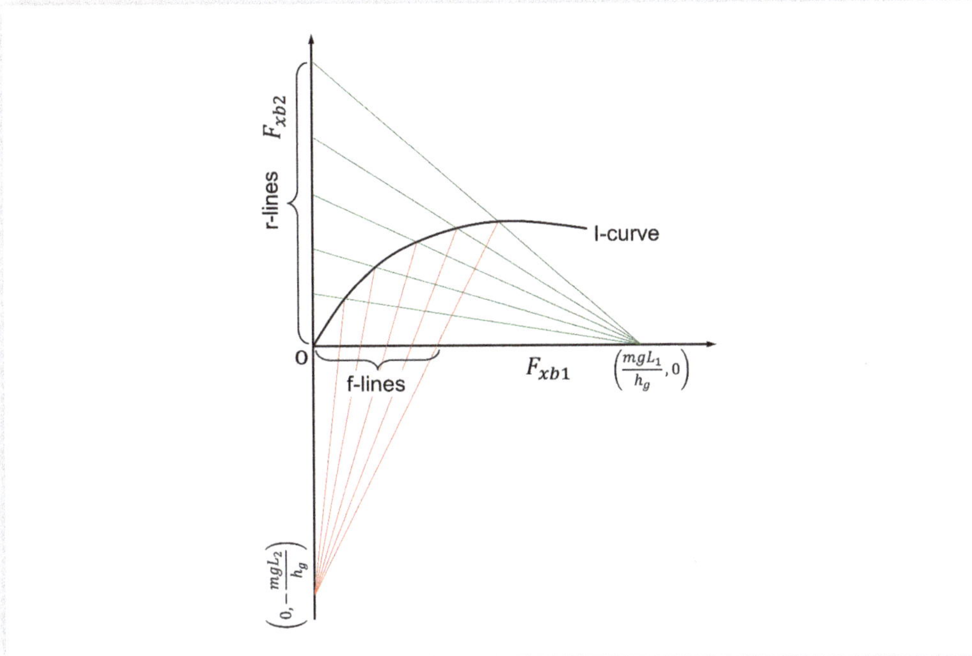

Similarly, when the rear wheels are locked and front wheels are in free rolling, the ground braking force is equal to the tire–ground adhesion force for rear wheels:

$$F_{xb2} = fF_{z2} = \frac{f}{L}\left(mgL_1 - F_{xb}h_g\right) = \frac{f}{L}\left(mgL_1 - F_{xb1}h_g - F_{xb2}h_g\right) \quad (4.34)$$

By rearranging Equation (4.34), we obtain

$$F_{xb2} = \frac{f}{L + fh_g}\left(-h_g F_{xb1} + mgL_1\right) \quad (4.35)$$

By taking different values of f, a group of the so-called r-lines can also be plotted, as illustrated in Figure 4.10.

Since the I-curve represents the situation of both front and rear wheels being locked, the f-lines above the I-curve and the r-lines below the I-curve are meaningless.

The braking process can be analyzed by means of the aforementioned I-curve, line β, f-lines, and r-lines in the following three cases according to the tire–ground adhesion coefficient f:

1. If $f < f_{I\beta}$, for instance, $f = 0.2$ and $f_{I\beta} = 0.47$, as illustrated in Figure 4.11 [4.4], the front and rear brake forces, F_{b1} and F_{b2}, rise along line β, so do the ground braking forces F_{xb1} and F_{xb2} since neither front nor rear wheels are locked, until point A is reached. The front wheels start to lock at point A, where line β intersects with the line of $f = 0.2$ among f-lines, with a braking ratio z of 0.16. With a further increase

in brake forces F_{b1} and F_{b2}, the ground braking forces F_{xb1} and Fx_{b2} will follow f-lines to increase until point B is reached, with F_{xb1} no longer equal to F_{b1} due to the lockup of front wheels. Correspondingly, since the rear wheels remain unlocked, the front and rear brake forces F_{b1} and F_{b2} still grow along line β until point B′ is reached, with the rear brake forces F_{b2} equal to the rear ground braking force F_{xb2}. Once point B is reached, both the front and rear wheels are locked, where the braking ratio peaks at 0.2.

Thus, when line β is under the I-curve, the front wheels are always locked before the rear wheels. As a result, the vehicle will lose its ability to steer, although it remains directional stable.

2. If $f = f_{I\beta}$, for instance, $f = f_{I\beta} = 0.47$, the brake forces F_{b1} and F_{b2} rise along line β, until both front and rear wheels are locked simultaneously at point C. The corresponding braking ratio peaks at $f_{I\beta}$, that is, 0.47, as shown in Figure 4.11.

   Clearly, although remaining directionally stable, the vehicle will lose its ability to steer at point C.

3. If $f > f_{I\beta}$, for instance, $f = 0.6$ and $f_{I\beta} = 0.47$, the brake forces F_{b1} and F_{b2} and ground braking forces F_{xb1} and Fx_{b2} rise along line β since both front and rear wheels remain unlocked until point D is reached. At point D, the rear wheels start to lock, with a braking ratio of 0.56. With a further increase in brake forces F_{b1} and F_{b2} along line β toward point E′, the ground braking forces F_{xb1} and Fx_{b2} will follow r-lines toward point E. The front brake force F_{b1} remains equal to the front ground braking force F_{xb1}, until points E′ and E are reached. At point E, r-lines intersect with the I-curve, where the front and rear wheels are all locked. The corresponding braking ratio peaks at 0.6.

FIGURE 4.11 Analysis of the braking process on roads of various adhesion coefficients.

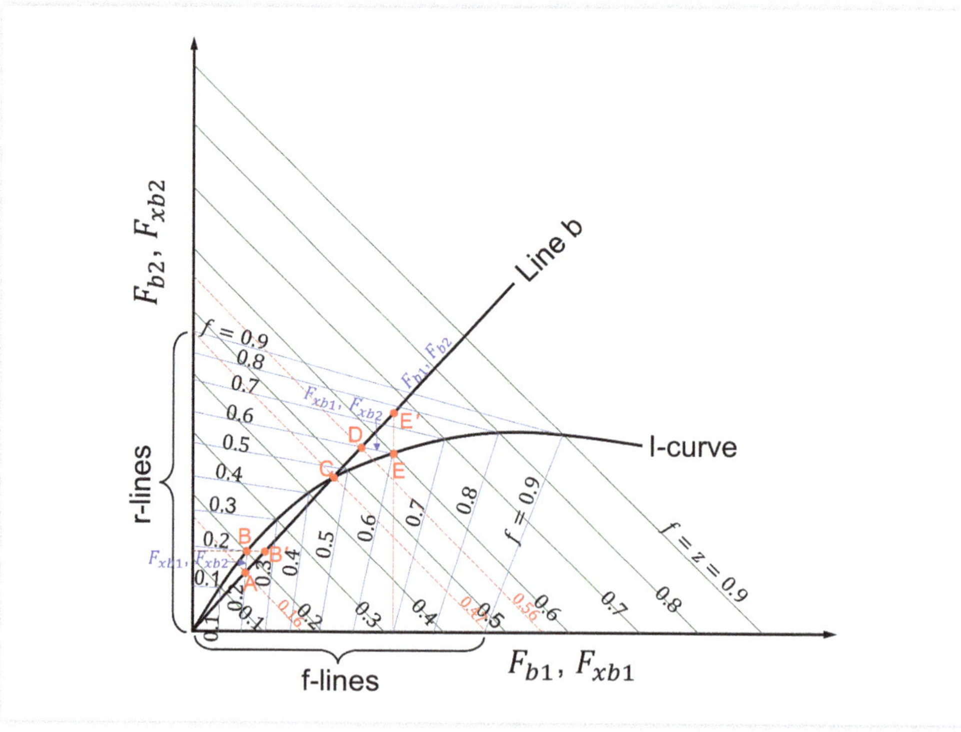

Therefore, when line β is above the I-curve, the rear wheels are always locked before the front wheels. As a result, the vehicle is prone to a sideways skid and loss of directional stability.

4.5.5. Adhesion Utilization and Braking Efficiency

Per the previous discussion, it is desired to set the maximum braking ratio close to the value that is just about to trigger the earliest lockup of any wheels, to prevent the vehicle from skid or loss of steering.

The braking ratio z is equal to the tire–ground adhesion coefficient $f_{I\beta}$ if the front and rear wheels are locked simultaneously in braking. For any other road with an adhesion coefficient $f \neq f_{I\beta}$, the braking ratio z must be smaller than the adhesion coefficient f to avoid skid or loss of steering. Thus, the adhesion can only be best utilized on the road of $f = f_{I\beta}$. In order to judge the rationality of brake force proportioning, adhesion utilization f_{ui} for the ith axle can be defined as

$$f_{ui} = \frac{F_{xbi}}{F_{zi}} \tag{4.36}$$

where

F_{xbi} is the ground braking force of the ith axle

F_{zi} is the normal reaction force of the ground to the ith axle under the same braking ratio z

Adhesion utilization improves as the difference between z and f_u reduces.

Based on Equation (4.36), the adhesion utilization for the front wheels is

$$f_{uf} = \frac{F_{xbf}}{F_{zf}} = \frac{\beta mgz}{\frac{mg}{L}\left(L_2 + zh_g\right)} = \frac{\beta zL}{L_2 + zh_g} \tag{4.37}$$

Similarly, adhesion utilization for the rear wheels is

$$f_{ur} = \frac{F_{xbr}}{F_{zr}} = \frac{(1-\beta)mgz}{\frac{mg}{L}\left(L_1 - zh_g\right)} = \frac{(1-\beta)zL}{L_1 - zh_g} \tag{4.38}$$

An example of the adhesion utilization curves for the front and rear axles of a typical vehicle with and without loads is illustrated in Figure 4.12 [4.1]. Optimal adhesion utilization can be found along the diagonal, where $f_u = z$. For empty vehicles, the f_{ur} curve lies completely above the diagonal, implying that the rear wheels are always locked earlier than the front wheels. Adhesion utilization much greater than the braking ratio also indicates poorly proportioned front and rear brake forces. For instance, in Figure 4.12, we have $f_{uf} = f_{ur} = 0.3$ for the loaded vehicle that is braked at the braking ratio $z = 0.3$, and the minimal tire–ground adhesion coefficient $f = 0.3$ to avoid lockup of any wheels. For any roads of $f < f_{I\beta}$, since the front wheels lock first, adhesion utilization should be selected according to f_{uf}, while f_{ur} curves are meaningless. In contrast, for any roads of $f > f_{I\beta}$, since the rear wheels lock first, adhesion utilization should be selected according to f_{ur}, while f_{uf} curves are meaningless. Adhesion utilization can also be selected by taking curves above the diagonal since the braking ratio cannot exceed the adhesion utilization, that is, $z \le f_{uf}$ or $z \le f_{ur}$.

FIGURE 4.12 Adhesion utilization versus braking ratio.

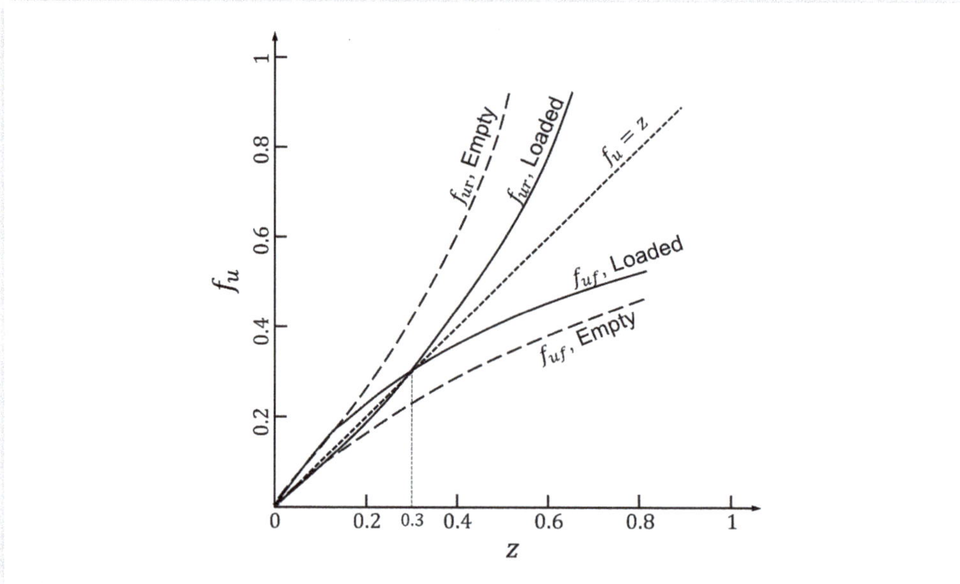

Adhesion utilization and rationality of the braking force proportion can also be expressed by braking efficiency, which is defined by the ratio of the maximum braking ratio without locking the axle to the tire–ground adhesion coefficient. For front axles, braking efficiency is

$$E_{bf} = \frac{z}{f_f} = \frac{L_2}{\beta L - f_f h_g} \tag{4.39}$$

For rear axles, braking efficiency is

$$E_{br} = \frac{z}{f_r} = \frac{L_1}{(1-\beta)L + f_r h_g} \tag{4.40}$$

As a function of the adhesion coefficient, braking efficiency is 100% when adhesion efficiency is equal to $f_{I\beta}$. For a loaded vehicle, braking efficiency drops on the road with an adhesion coefficient $f < f_{I\beta}$ as the front axle locks first, while efficiency falls on the road with an adhesion coefficient $f > f_{I\beta}$ as the rear axle locks first. For an empty vehicle, the braking efficiency is typically subject to rear wheel lockup only.

4.5.6. Requirements for Brake Proportioning

As discussed previously, line β should always lie below the I-curve to avoid the sideways skid of the rear axle. On the other hand, it is desired to set line β as close as possible to the I-curve to keep the front wheels away from lockup or loss of steering. From the adhesion utilization perspective, the utilization curve of front wheels should situate above both diagonal and utilization curves of rear wheels and as close as possible to the diagonal.

In addition, braking proportioning is subject to regional standards or regulations where the vehicle will operate. Two examples of such standards or regulations are discussed now.

4.5.6.1. Requirements of Federal Motor Vehicle Safety Standard (FMVSS)

For vehicles with a gross vehicle weight rating (GVWR) of 3500 kg or less, the FMVSS 135 of the US specifies requirements for service brakes and associated parking brake systems in order to ensure safe braking performance under normal and emergency driving conditions. The requirements for adhesion utilization are established for vehicles without any ABS. For all braking ratios between 0.15 and 0.80, each adhesion utilization curve for a rear axle shall be situated below a line defined by $z = 0.9f_u$, or $f_u < z/0.9$.

4.5.6.2. Requirements of ECE

Requirements for brake force proportioning are imposed by the ECE Safety Regulations on the two-axle vehicles.

For sedans with all braking ratios within $z = 0.2–0.8$, the adhesion utilization of the front axle should not only situate above that of the rear axle but also satisfy $f_u \le (z + 0.07)/0.85$. When braking ratios $z = 0.3–0.5$, under the condition that the adhesion utilization of the rear axle does not exceed $f_u = z + 0.05$, the adhesion utilization of the rear axle is allowed to exceed that of the front axle.

For trucks with all braking ratios $z = 0.15–0.3$, the adhesion utilization of the front axle should situate above that of the rear axle. When braking ratios $z = 0.2–0.8$, adhesion utilization should satisfy $f_u \le (z + 0.07)/0.85$. Or, when braking ratios $z = 0.15–0.3$, the adhesion utilization of each of the axles must situate between the straight line $f_u = z \pm 0.08$ and the optimal line $f_u = z$. When braking ratios $z \ge 0.3$, the adhesion utilization of the rear axle must satisfy $f_u \le (z - 0.0188)/0.74$.

In addition, for both sedans and trucks, the sequence for wheel lockup and the limit of adhesion utilization are specified. For instance, if the adhesion utilization of the front axle lies above that of the rear axle, the front wheels can be locked first to ensure the directional stability in braking. The adhesion utilization must satisfy $f_u \le (z + 0.07)/0.85$ and situate as close as to the optimal line $f_u = z$ to maximize adhesion utilization.

For FWD vehicles, with all braking ratios $z = 0.2–0.8$, the ECE Regulations also require that rear wheel adhesion utilization not fall below the limits defined by the following formula:

$$\begin{cases} f_{uf} = \dfrac{z+0.07}{0.85} \cdot \dfrac{mg}{L} \cdot \left(L_1 + zh_g\right) \\ f_{ur} = mgZ - f_{uf} \end{cases} \tag{4.41}$$

4.5.6.3. Variable Brake Proportioning

Today, most EVs are equipped with variable brake proportioning systems to bring the proportion close to the I-curve and to meet the requirements of regulations. In such a system, the brake forces to the front and rear axles can be adjusted with factors such as braking rate and load by means of proportional valves. In the design of braking proportioning curves, priority is given to braking stability, although stopping distance must be considered as well. The variable brake proportioning is typically realized by a deceleration sensing proportioning valve (DSPV). The brake proportioning curve is preferably designed to situate below the I-curve, leaving some room for vehicle stability [4.5].

4.5.7. ABS

ABS is a closed-loop control system that senses the occurrence of wheel lockup, releases the brake on the locked wheel instantly, and reapplies it as the wheel spins up again in braking. More-advanced ABSs can even release the brake before the wheel lockup, apply a brake force according to the adhesion condition, and alter the brake proportioning to front and rear wheels. ABSs not only reduce the stopping distance significantly on both dry and wet road surfaces, with a more prominent effect on wet road surfaces, but also allow the vehicle to follow the curved path in cornering.

A typical ABS consists of four-wheel speed sensors, a brake control module, a hydraulic control modulator, and a warning light. It has three hydraulic pipelines, allowing independent brake pressure control. The wheel speed is constantly monitored and processed by the control module in braking. A command signal is transmitted to the hydraulic control modulator to release the brake pressure if wheel lockup is determined imminent. With the wheel speeds up again, the brake is pressurized once more.

Two-position two-way valves are used in hydraulic control modulators to control brake pressure. The pressure increases as the inlet valve opens up, while the outlet valve shuts off. Conversely, the pressure drops as the inlet valve shuts off, while the outlet valve opens up. The pressure remains constant when both valves shut off.

The process of wheel lockup can be analyzed by using a simplified single wheel model, as shown in Figure 4.13 [4.5]. The kinetic equation of a braked wheel is

$$I_w \dot{\omega} = F_{xb} R - T_b \tag{4.42}$$

where

- I_w is the rotating inertia of the wheel
- $\dot{\omega}$ is the angular acceleration of the wheel
- R is the wheel radius
- T_b is the brake torque on the wheel
- F_{xb} is the ground braking force $F_{xb} = F_z \varphi_b$

FIGURE 4.13 A model for lockup analysis.

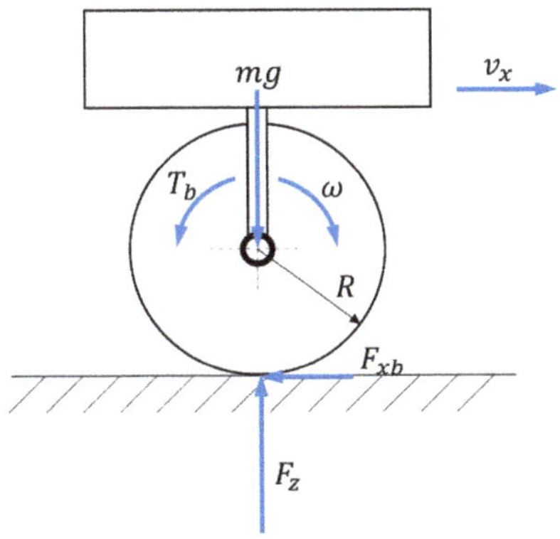

To simplify the problem, it is assumed that the wheel speed and load remain constant, and the pressure in such wheel cylinder p is a linear function of time in the lockup process. Thus, $F_z = mg$ and $p(t) = p_0 t$. The brake torque T_b, therefore, is given by

$$T_b = A_c K_{ef} r_b p(t) \tag{4.43}$$

where

A_c is the area of the wheel cylinder

K_{ef} is the brake efficiency factor, that is, friction force produced by unit force acting on the wheel cylinder

r_b is the equivalent radius of the friction force

Let $T_0 = p_0 A_c K_{ef} r_b$, then Equation (4.43) becomes

$$T_b = T_0 t \tag{4.44}$$

In addition, the adhesion coefficient can be assumed to be two linear functions of the slip ratio with different slopes:

$$\varphi_b = \begin{cases} \dfrac{s}{s_{bp}} \varphi_{bp} & 0 \le s \le s_{bp} \\ \varphi_{bs} + \dfrac{(1-s)(\varphi_{bp} - \varphi_{bs})}{1 - s_{bp}} & s_{bp} < s \le 1 \end{cases} \tag{4.45}$$

The slip ratio can be expressed by

$$s = \frac{v_x - \omega R}{v_x} = 1 - \frac{\omega}{\dfrac{v_x}{R}} = 1 - \frac{\omega}{\omega_0} \tag{4.46}$$

where the initial angular speed of the wheel $\omega_0 = \dfrac{v_x}{R}$.

For $0 \le s \le s_{bp}$, by substituting Equations (4.43) through (4.45) into Equation (4.42) and rearranging, we have

$$I_w \dot{\omega} + \frac{\varphi_{bp} mgR}{s_{bp}\omega_0}\omega = \frac{\varphi_{bp} mgR}{s_{bp}} - T_0 t \tag{4.47}$$

For $s_{bp} < s \le 1$, similarly, we have

$$I_w \dot{\omega} - \frac{\left(\varphi_{bp} - \varphi_{bs}\right) mgR}{\left(1 - s_{bp}\right)\omega_0}\omega = \varphi_{bs} mgR - T_0 t \tag{4.48}$$

Both first-order ordinary differential equations (4.48) and (4.49) can be solved using numerical methods, for example, the Runge–Kutta method, for parameters such as $\dot{\omega}$, ω, and s.

In a typical anti-lock process based on the logic threshold control algorithm [4.6], once the brake is applied, the longitudinal speed of the wheel center is taken as the initial reference speed of the vehicle v_{ref1} if the absolute value of angular deceleration of the wheel is smaller than the threshold $|\dot{\omega}_1|$, that is, $|\dot{\omega}| < |\dot{\omega}_1|$. Apparently, the reference speed of the vehicle $v_{ref} = v_{ref1} + \dot{\omega}Rt$. Let the threshold of the slip ratio be s_1, and the brake pressure p is held constant once $\dot{\omega}$ reaches $\dot{\omega}_1$ and $s < s_1$. When $s > s_1$, brake pressure p is reduced. Accordingly, $|\dot{\omega}|$ decreases until $|\dot{\omega}| = |\dot{\omega}_1|$, where the brake pressure stops declining and becomes constant. With continued acceleration of the wheel due to its rotating inertia, the acceleration threshold for judging the low-adhesion coefficient road surface $\dot{\omega}_2$ is passed. Once the acceleration threshold for the high-adhesion coefficient road surface $\dot{\omega}_3$ is reached, the brake pressure increases again. When the wheel acceleration drops back to the threshold $\dot{\omega}_3$, the brake pressure is kept constant. As the wheel acceleration further drops to $\dot{\omega}_2$, the brake pressure increases slowly to maximize the utilization of the peak adhesion coefficient. With the wheel deceleration dropping back to $\dot{\omega}_1$, a cycle is complete. Thereafter, this cycle will be repeated until the vehicle comes to a complete stop. Furthermore, modern ABSs allow the slip ratio to be held in the proximity of the optimal value in braking.

4.6. Braking Performance Tests

4.6.1. Road Tests

The brake performance of a vehicle is mainly evaluated by road tests. Parameters such as stopping distance, deceleration, and the time to stop are measured in the cold and hot tests, along with directional stability tests of braking during cornering and lane changing.

The tests are typically conducted on clean, flat, hard surface roads of a slope less than 1% and an adhesion coefficient of at least 0.72–0.75, with the wind speed lower than 5 m/s and air temperature in the range of 0–35°C. The vehicle should be preheated by running at 80–90% of its top speed for at least one hour before tests.

The initial speed, stopping distance, time to stop, and lateral offset are measured primarily by instruments such as fifth wheels, pressure sensors, and decelerometers.

For a vehicle equipped with an ABS, the adhesion utilization ratio λ_u should be measured by tests, which is defined as the ratio of the maximum braking rate to the adhesion coefficient when the ABS is in operation, that is, $\lambda_u = z/f$. The test results have to meet the requirement of $\lambda_u \geq 0.75$ on the roads of two different adhesion coefficients f_1 and f_2, with $f_1 \leq 0.3$ and $f_2 \approx 0.8$. Moreover, it is also required that in braking from the initial speed of 50 km/h, no lockup occurs on joint pavements from high adhesion f_h to low adhesion f_l, and vice versa, with $f_h \geq 0.5$ and $f_h/f_l > 2$. For bisectional roads where the left and right wheels run on different pavements, the same requirements apply. In addition, the steering angle must not exceed 120° in the first 2 sec and the total steering angle not exceed 240° when the steering wheel is used to correct the vehicle direction.

4.6.1.1. Cold Braking Tests

Cold braking tests are conducted with a brake temperature not exceeding 100°C. It is generally required that the brakes be burnished 200 times at a deceleration of 3.5 m/s^2 before the cold tests. The vehicle is first accelerated to a speed 3–5 km/h over the specified initial speed before braking and then coast down to the exact initial speed before applying emergent braking till a complete stop. If the vehicle passes through the 3.5 m limit in the test road width or the yaw angle varies larger than 8° during the test, the brake system of the test vehicle must be readjusted before the test continues.

4.6.1.2. Hot Braking Tests

Hot braking tests involve two stages, namely, brake heating and braking efficiency measurement.

Brake heating is often measured by repeated braking, in which the vehicle is accelerated to 80% of the top speed and then decelerated to 40% of the top speed at a deceleration of 3 m/s^2. A total of 15–20 times of braking is repeated with a time interval of 45–60 sec between each application of brakes. Brake heating can also be performed on a downhill road of a 7% slope and 1.7 km long at a constant speed of 40 km/h. The brakes are heated to 250–270°C for passenger vehicles, 140–150°C for mid-duty trucks, and 170–200°C for heavy-duty trucks.

Braking efficiency is measured before, after, and in the middle of brake heating a number of times to evaluate the thermal decay of the brakes.

Continuous braking on a long downhill road is another type of hot braking test, which is typically conducted on a road with a slope of 6–10% and a length of 7–10 km at a constant speed of 30 km/h. The braking efficiency is measured immediately thereafter.

4.6.1.3. Braking-in-a-Turn Tests

Braking-in-a-turn tests are carried out on a level, dry road. The tests start with the vehicle making a circular motion of a given radius until it reaches one of the following two stable conditions: first, 40–50 m in the turning radius, 5 ± 0.5 m/s^2 in lateral acceleration, or 51–57 km/h in vehicle speed, and second, 100 m in the turning radius, 4 ± 0.4 m/s^2 in lateral acceleration, or 72 km/h in vehicle speed. Once the stable condition is reached, the brake pedal is depressed quickly after releasing the acceleration pedal, while the angle of a steering wheel remains unchanged. With the deceleration being held constant in each test, a number of tests are carried out for different decelerations. Based on the data recorded in tests, the maximum yaw rate, variation in the yaw angle, and lateral offset of a path can be plotted against deceleration for evaluating the directional stability during braking in a turn.

Braking-in-a-turn tests are often conducted on a wet road as well because distinctive directional stability problems can develop when the adhesion coefficient is low.

4.6.2. Rig Tests

Indoor rig tests are generally used for measuring the frictional torque of brakes or for a quick checkup of the braking efficiency of a vehicle. These tests can be completed in a limited space and time.

There are two types of brake testers, namely, plate and roller, as described now.

4.6.2.1. Plate Brake Tester

A plate brake tester is typically made of four movable flat plates. The distance between the centerlines of the left and right plates is equal to the track, while the distance between the centerlines of the front and rear plates is equal to the wheelbase. The length of the plate is greater than the wheel diameter, typically about 1 m. The vehicle is driven up to the plates at a speed of 5–10 km/h in the test and then the brake pedal is depressed. The longitudinal brake force between each wheel and plate is measured individually by a force sensor. This type of tester can better reflect the load shift in braking and imitate the adhesion condition of the roads than their roller-type counterparts. However, the measurements of brake drum roundness and brake force variations with the pedal force on plate brake testers are not as easy as on roller brake testers.

4.6.2.2. Roller Test Stands

A roller brake tester generally has two pairs of rollers mounted nearly flush with the ground each on the left and right sides with the distance between their centerlines equal to the track of test vehicles. The test wheels are placed on between the rollers, and other wheels are supported on the ground. There are usually transverse groove patterns on the roller surface to keep the adhesion coefficient above 0.65. Additional loading devices are sometimes used to increase the loads on test wheels. Driven by the rollers, the test wheels spin in the test to imitate road driving conditions. When the brake pedal is depressed, the rollers apply a moment opposite to the brake moment, until the slip ratio between tires and rollers reaches a specified value (typically 20%) and then the rollers stop driving test wheels. The tangential friction force acting from rollers on test wheels can be measured with force sensors, which is actually the brake force of test wheels. In terms of test results, this type of stand is less sensitive to the adhesion coefficient than the plate type of stand.

4.7. Regenerative Braking

In the braking process of an EV, its kinetic energy can be converted or transferred and then stored in the form of electric energy, kinetic energy, elastic potential energy, and so forth, for later utilization. In most EVs, kinetic energy is converted back into electric energy and stored in batteries. For those EVs equipped with flywheel energy storage devices, a certain portion of the kinetic energy of the vehicle is transferred into the kinetic energy of the flywheel and stored there.

Regenerative braking has a significant effect on the energy economy of EVs. Theoretically, it is estimated that nearly 17% of energy savings can be achieved with regenerative braking in NEDC if the kinetic energy of the vehicle can be fully recovered.

4.7.1. Mechanical and Electrical Compound Braking

In a typical EV, the electric motor is used as a generator to convert the kinetic energy into the electric energy for storage in a battery. Thus, in addition to the mechanical brake force, there is an electric brake force provided by the electric motor, resulting in mechanical and electrical compound braking. In general, the total brake force F_b on the circumference of the tire is

$$F_b = F_{bm} + F_{be} = F_{bm1} + F_{bm2} + F_{be1} + F_{be2} \tag{4.49}$$

where

F_{bm} is the total mechanical brake force
F_{be} is the total electrical brake force
F_{bm1} and F_{bm2} are the mechanical brake forces on front and rear axles, respectively
F_{be1} and F_{be2} are the electrical brake forces on front and rear axles, respectively

Obviously, for an FWD EV, Equation (4.49) reduces to

$$F_b = F_{bm1} + F_{bm2} + F_{be1} \tag{4.50}$$

And for an RWD EV, we have

$$F_b = F_{bm1} + F_{bm2} + F_{be2} \tag{4.51}$$

The mechanical and electrical brake forces can be applied either simultaneously or in sequence, with the former being called parallel braking and the latter, serial braking, as shown in Figure 4.14a and 4.14b, respectively [4.7]. Compared to parallel braking, the pedal feel and energy recovery efficiency of the serial braking are considerably better because the priority can be given to regenerative braking and the hydraulic pressure is adjusted to accommodate the additional regenerative brake force.

FIGURE 4.14 Schematic of parallel braking and serial braking. (a) Parallel braking and (b) serial braking.

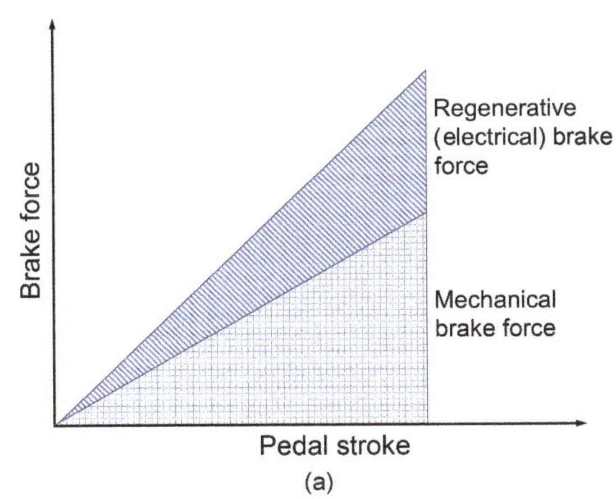

Courtesy of Automotive Engineering.

FIGURE 4.14 **(Continued)** Schematic of parallel braking and serial braking. (a) Parallel braking and (b) serial braking.

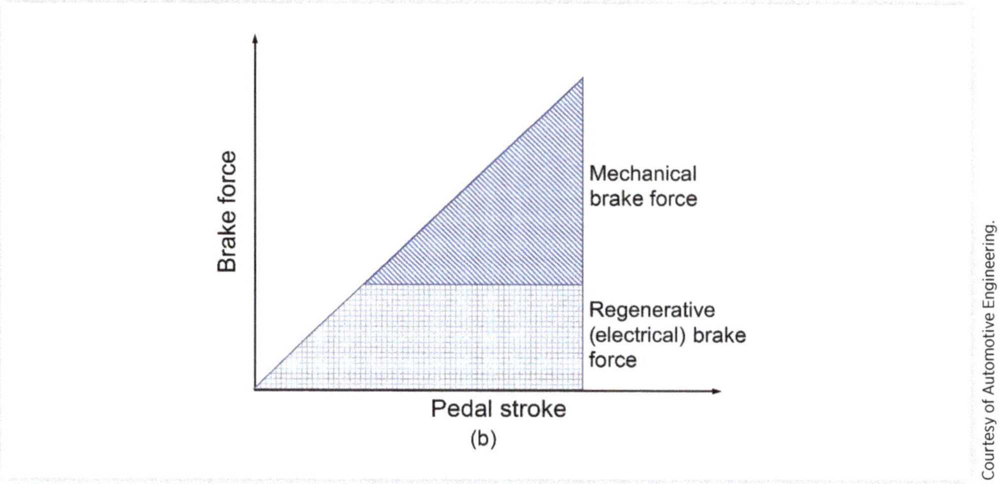

The SOC of a battery must be taken into consideration in the decision-making of compound braking. Regenerative braking is, in general, disabled when SOC is high (typically when SOC > 0.95), due to the difficulty in charging the battery. In these circumstances, only the mechanical brake force is applied.

The vehicle speed is also a factor that must be taken into account. At high speeds, it is recommended to apply a mechanical brake force only for its higher reliability to ensure safety.

Finally, the electric brake force is subject to the limits imposed by the characteristics of the electric motor. As illustrated in Figure 4.15, below a certain limit of electric motor speed, little energy can be recovered because of the minimal kinetic energy of the vehicle and the low efficiency of the electric motor. Thus, regenerative braking is deactivated when the vehicle speed is below a certain threshold, usually corresponding to an electric motor speed of 500 rpm. At higher electric motor speeds, on the other hand, the brake torque is restricted by the peak torque and peak power of the electric motor.

FIGURE 4.15 Effects of electric motor characteristics on regenerative braking.

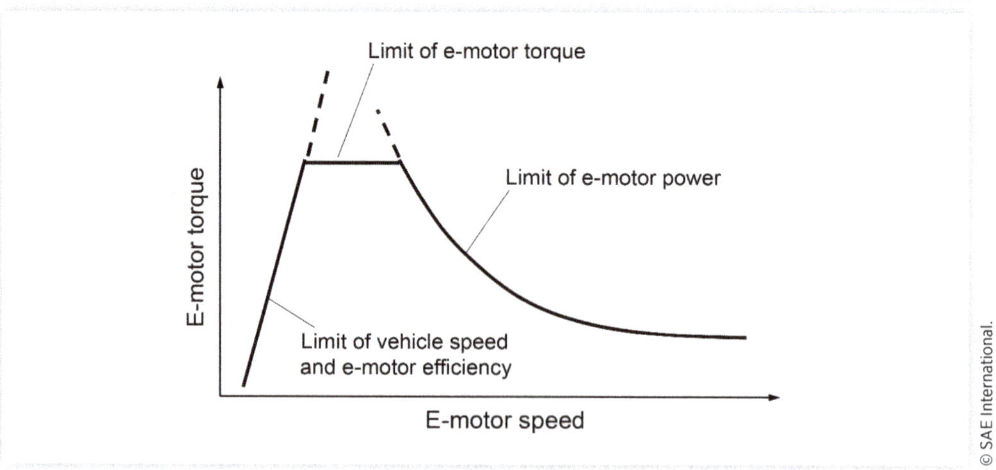

4.7.2. Brake Proportioning for Maximization of Regenerative Energy

When the braking ratio $z \geq 0.8$, it is generally agreed that braking efficiency and safety have higher priority than energy recovery from regenerative braking. Involving critical links such as vehicle control unit (VCU), controller area network (CAN), and MCU the reliability of regenerative braking is deemed, in general, not on par with that of mechanical braking systems. Thus, it is a normal practice to rely solely on a mechanical braking system for such emergent braking.

4.7.2.1. Brake Proportioning for FWD EVs

For FWD EVs, a serial braking strategy can be taken to maximize the regenerative energy recovery. By making the electric brake force on the front axle as large as possible, the total braking force proportioning lies in the area enclosed by the I-curve, horizontal axis, curves of ECE lower limit, and f-lines, as shown in Figure 4.16. As an example of conservative strategies, the brake force proportioning can follow the curve OABCD [4.8], assuming the electric motor torque is adequate. When the braking ratio is small, for example, $z \leq 0.22$ (where the curve of ECE lower limit intersects with the abscissa axis), the total brake force is purely electric from the front axle. For a braking ratio of $0.22 < z \leq 0.35$ (where the curve of ECE lower limit intersects with f-lines of $z = 0.5$), the total brake force is proportioned by following the curve of ECE lower limit. When $0.35 < z \leq 0.5$, the total brake force is proportioned by following the f-line. Finally, for $z > 0.5$, the total brake force proportioning follows the I-curve. By shifting line BC toward the right to f-lines higher than $z = 0.5$, the brake proportioning becomes more aggressive in favor of energy recovery, at a price of undermined braking reliability. As an example of more aggressive brake proportioning strategies, the f-line of $z = 0.9$ (line 3) is followed since the brake proportioning can be taken within the area defined by curve 1, line 2, and line 3, as illustrated in Figure 4.17 [4.9]. The lower limit per ECE Regulations is presented by a straight line for simplicity.

FIGURE 4.16 A conservative brake proportioning strategy for FWD EVs.

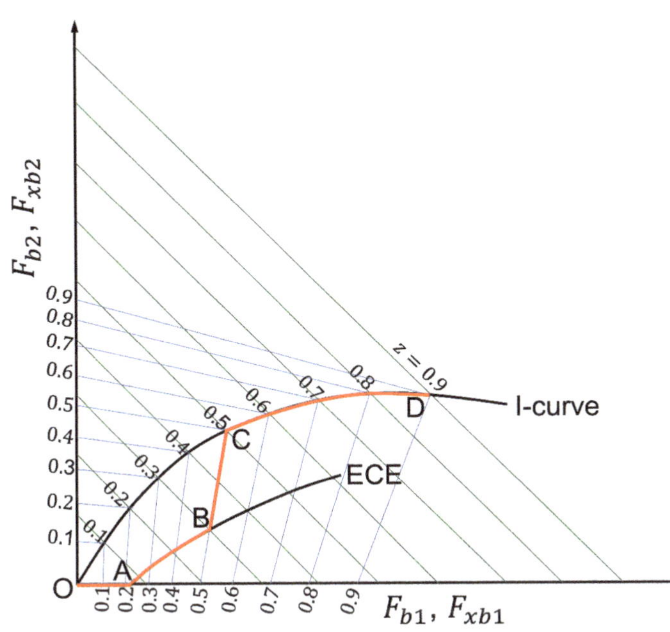

FIGURE 4.17 An aggressive brake proportioning strategy for FWD EVs.

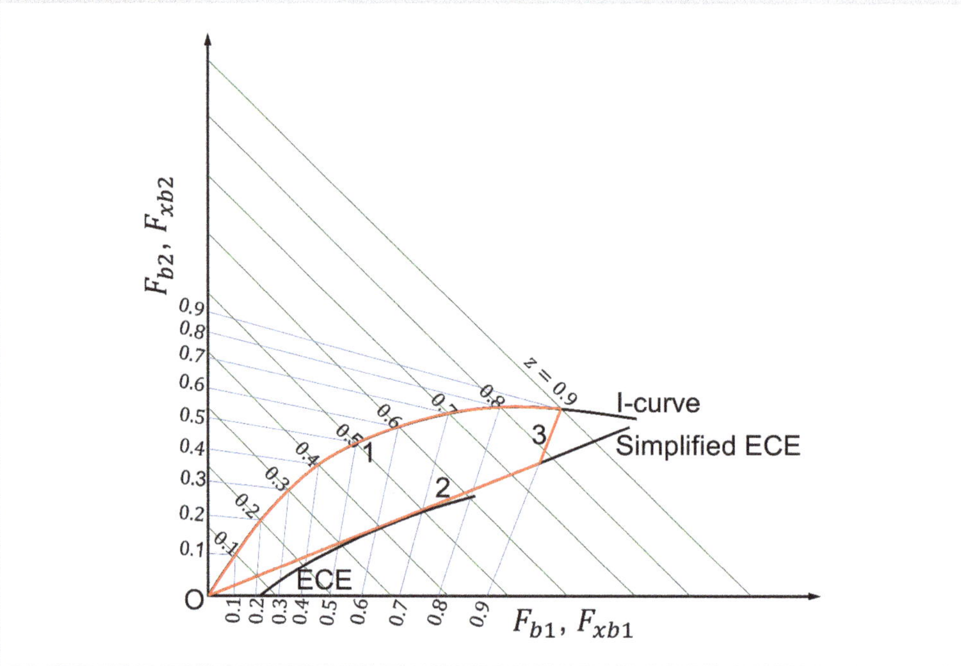

Courtesy of Automotive Engineering.

4.7.2.2. Brake Proportioning for RWD EVs

For RWD EVs equipped with variable brake proportioning control valves and ABSs, the mechanical brake proportioning strategy can be improved from aforementioned line β (i.e., line OA in Figure 4.18) to line OC of a smaller slope, which is tangential to the curve of the ECE requirement for a minimum rear-wheel brake force [4.10]. As a result, the rear-wheel regenerative brake force increases from the portion of the slash-shadowed area to that of the grid-shadowed area, without violating the ECE regulation or risking braking stability. Note that all these lines must pass the origin O due to their nature of parallel braking. However, any potential failure in the regenerative braking system can cause the braking efficiency of strategy OC to fall below that of strategy OA, resulting in a violation of ECE requirements [4.10].

In addition, a serial brake proportioning strategy can be taken to further improve the regenerative energy recovery without penalty in directional stability. It is generally reckoned acceptable to have rear wheels locked up first when the adhesion coefficient is smaller than 0.2 [4.10]. Thus, the mechanical brake force can follow line ODA passing point D, as shown in Figure 4.19, while the rear-wheel regenerative brake force can be selected within the slash-shadowed area. The regenerative energy can be maximized by using line OF as the mechanical brake force, leaving both slash- and grid-shadowed areas for the rear-wheel regenerative brake force. Note that the line OF is tangential to the curve of the ECE requirement for a minimum rear-wheel brake force. In case of any failure in the regenerative braking system, the braking efficiency of strategy OF would be lower than that of strategy ODA, while the ECE requirement would be violated [4.10].

FIGURE 4.18 A parallel braking-based brake proportioning strategy for rear-wheel-drive EVs.

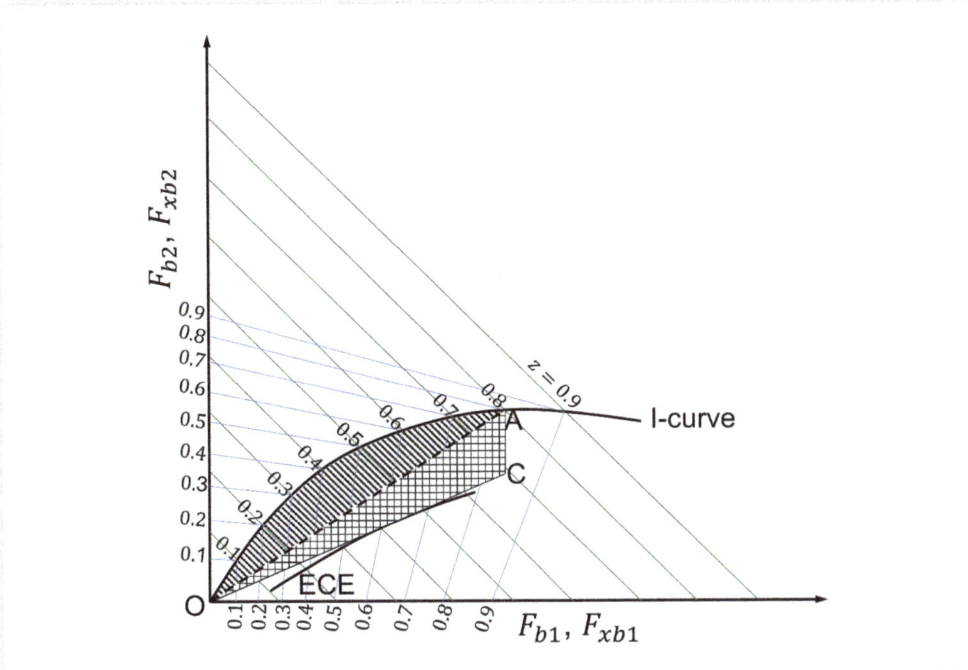

Courtesy of Automotive Engineering.

FIGURE 4.19 A series braking-based brake proportioning strategy for rear-wheel-drive EVs.

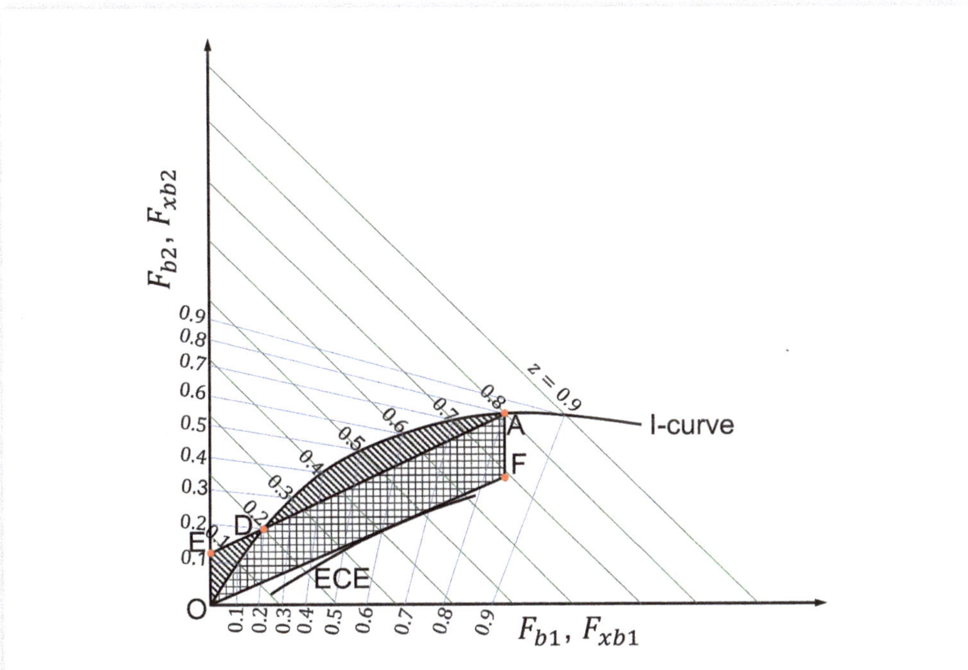

Courtesy of Automotive Engineering.

4.7.2.3. Brake Proportioning for All-wheel Drive EVs

The electric drive system of a typical all-wheel drive (AWD) EV consists of two or more electric motors, mounted on both front and rear axles. The following rules are generally followed in brake proportioning:

1. The brake proportioning for AWD EVs is generally selected between the I-curve and the ECE lower limits to ensure directional stability and regulatory compliance.
2. Because of the difficulty in charging the battery of a high SOC and the concerns about the reliability of electrical braking, regenerative braking is disabled when SOC is higher than 0.9 or when the vehicle speed exceeds an upper limit v_{max}, for example, 90 km/h.
3. In order to maximize the energy being recovered in the braking process, electric motors must operate at optimal points corresponding to their efficiency maps. The system efficiency of braking energy recovery is given by

$$\eta_b = \frac{T_{be1} i_1 \eta_1 (T_{be1}, \omega_{e1}) + T_{be2} i_2 \eta_2 (T_{be2}, \omega_{e2})}{T_{be1} i_1 \eta_{t1} + T_{be2} i_2 \eta_{t2}} \tag{4.52}$$

where

T is the output torque of the electric motor
ω is the output angular speed of the electric motor
i is the transmission ratio
η_t is the transmission efficiency
subscripts 1 and 2 denote front and rear axles, respectively

With the definition of the electric braking force proportioning,

$$\lambda = \frac{T_{be1} i_1 \eta_{t1}}{T_{be1} i_1 \eta_{t1} + T_{be2} i_2 \eta_{t2}} \tag{4.53}$$

Equation (4.52) can be transformed into

$$\eta_b = \lambda \eta_1 (T_{be1}, \omega_{e1}) + (1 - \lambda) \eta_2 (T_{be2}, \omega_{e2}) \tag{4.54}$$

Thus, the operating points of the electric motor for a given moment in regenerative braking can be obtained by solving the following constrained optimization problem [4.11]:

$$\begin{cases} \lambda^{opt} = \underset{\lambda}{\arg\max} \{\eta_b\} \\ 0 \le T_{be1} \le T_{be1max}(\omega_{e1}) \\ 0 \le T_{be2} \le T_{be2max}(\omega_{e2}) \\ T_{be1} i_1 \eta_{t1} + T_{be2} i_2 \eta_{t2} \le z_{emax} mgR \\ 0 < v \le v_{max} \\ F_{b2,i} \le T_{be2} i_2 \eta_{t2} \le F_{b2,ECE} \end{cases} \tag{4.55}$$

where the first and second inequalities represent the constraints imposed by electric motor characteristics, the third inequality represents the constraint by the maximum braking ratio allowed for electrical braking, the fourth inequality represents the constraint imposed by the maximum vehicle speed permissible for regenerative braking, and the fifth inequality represents the constraint imposed by the I-curve and ECE lower limit curve for brake force proportioning.

The problem described by Equation (4.55) can be solved by algorithms such as PSO. The details of PSO are omitted here.

References

[4.1]. Yu, Z., *Automobile Theory*, 3rd ed. (Beijing: China Machine Press, 2000), 1-36, ISBN:7-111-02076-6.

[4.2]. Harned, J., Johnston, L., and Scharpf, G., "Measurement of Tire Brake Force Characteristics as Related to Wheel Slip (Antilock) Control System Design," SAE Technical Paper 690214 (1969), doi:https://doi.org/10.4271/690214.

[4.3]. Gillespie, T., *Fundamentals of Vehicle Dynamics* (Warrendale: Society of Automotive Engineers, Inc., 1992), 21-118, ISBN-10:1560911999, ISBN-13:978-1560911999.

[4.4]. van Zanten, A. and Kost, F., "Brake-Based Assistance Functions," in Winner, H., Hakuli, S., Lotz, F., and Singer, C. (eds), *Handbook of Driver Assistance Systems* (Cham: Springer, 2015), 919-967, doi:10.1007/978-3-319-12352-3_40.

[4.5]. Limpert, R., *Brake Design and Safety* (Warrendale: Society of Automotive Engineers, Inc., 1992), ISBN-10:1560912618, ISBN-13:978-1560912613.

[4.6]. Al-mola, M., Mailah, M., Muhaimin, A., and Abdullah, M., "Intelligent Active Force Controller for an Anti-lock Brake System Application," in Mastorakis, N., Mladenov, V., and Bojkovic, Z. (eds), *Latest Advances in Systems Science and Computational Intelligence* (Sydney: World Scientific and Engineering Academy and Society, 2012), 210-215, ISBN:9781618040947.

[4.7]. Zhang, J., Lv, C., and Li, Y., "Status Quo and Prospect of Regenerative Braking Technology in Electric Cars," *Automotive Engineering* 36, no. 8 (2014): 911-919, doi:10.19562/j.chinasae.qcgc.2014.08.003.

[4.8]. Liu, L., Ji, F., Yang, S., and Xu, B., "Control Strategy for Electro-Mechanical Braking Based on Curves of ECE Regulations and Ideal Braking Force," *Journal of Beijing University of Aeronautics and Astronautics* 39, no. 1 (2013): 138-142.

[4.9]. Yang, Y., Zhao, H., and Zhu, M., "A Study on the Control Strategy for Maximum Energy Recovery by Regenerative Braking in Electric Vehicles," *Automotive Engineering* 35, no. 2 (2013): 105-110, doi:10.19562/j.chinasae.qcgc.2013.02.002.

[4.10]. Zhang, K., Xu, L., Hua, J., Li, J. et al., "A Comparative Study on Regenerative Braking System and Its Strategies for Rear-Wheel Drive Battery Electric Vehicles," *Automotive Engineering* 37, no. 2 (2015): 125-131, doi:10.19562/j.chinasae.qcgc.2015.02.001.

[4.11]. Tang, Y., "Research on Driving and Braking Control Strategy of Dual Motor Four-Wheel Drive Vehicle," Master of Engineering thesis, School of Transportation, Northeast Forestry University, Harbin, 2019.

5

Handling and Stability

Vehicle handling refers to the way a vehicle responds to the inputs of a driver and how it moves along the road. Vehicle stability indicates its capability of maintaining a stable operation against external interference. Vehicle handling and stability are critical factors that determine the ease of operation and driving safety.

5.1. Cornering Characteristics of Tires

The cornering characteristics of tires are mainly presented by how a sideslip angle relates to the cornering force and aligning torque, which fundamentally impacts vehicle handling and stability.

5.1.1. Coordinate System

Tire dynamics are established in the Cartesian coordinate system, as shown in Figure 5.1 [5.1]. The wheel plant is the central plane of the tire normal to its spin axis. The x-axis is the intersection of the wheel plane and ground, with the forward direction being positive. The z-axis is perpendicular to the ground plane, with the upward direction being positive. The origin O, that is, the center of tire–road contact, is at the intersection of the projected line of the spin axis on the ground and the x-axis. The y-axis in the ground plane passes through the origin O, constituting a right-handed coordinate system.

FIGURE 5.1 Tire coordinate system.

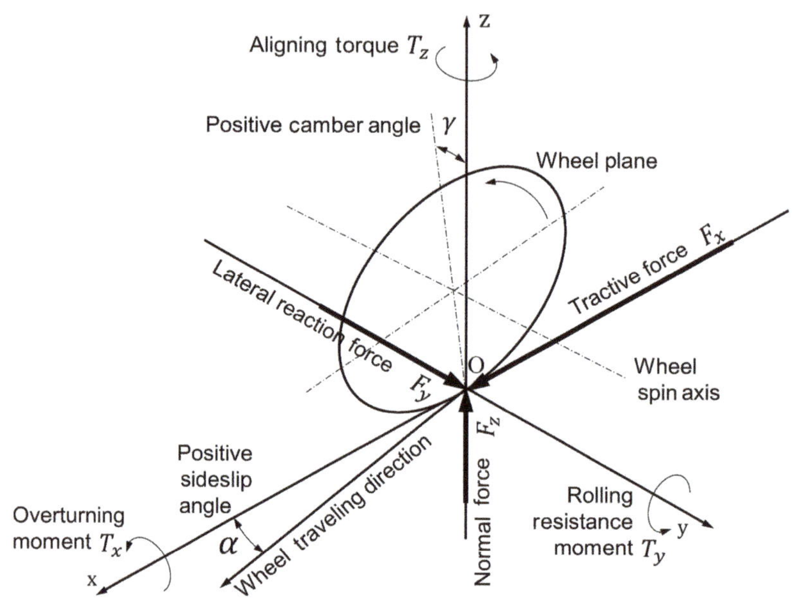

Taheri, Shahyar & Ab, & Sandu, Corina & Taheri, & Taheri, Saied. (2014). Development and Implementation of a Hybrid Soft Soil Tire Model (HSSTM).

The following terminology is often cited with the coordinate system, as defined previously:

1. Tractive force F_x: It is the longitudinal force acting on the tire from the ground. The tractive force $F_x > 0$ when the vehicle is accelerating, and $F_x < 0$ when the vehicle is braking.
2. Lateral reaction force F_y: It is the lateral reaction force acting on the tire from the ground, which is perpendicular to F_x and F_z and produced by a lateral force acting at the wheel center in the opposite direction.
3. Normal force F_z: It is the vertical force normal to the ground plane, which is actually equal to the wheel load. The normal force $F_z > 0$ when the force is upward.
4. Overturning moment T_x: The moment causes the tire to roll about the x-axis, which is also called the roll moment, tilting moment, or bank moment.
5. Rolling resistance moment T_y: The rolling resistance moment $T_y < 0$ when it tends to hinder the tire from rolling forward, which is sometimes called the pitch moment.
6. Aligning torque T_z: The moment tends to turn the tire on the z-axis, which is also called the yaw moment, bore torque, or self-aligning torque.
7. Sideslip angle α: It is the angle between the wheel traveling direction and the x-axis. The sideslip angle $\alpha > 0$ when the tire is slipping to the left.
8. Camber angle γ: It is the angle between the wheel plane and the z-axis. The camber angle $\gamma > 0$ when the tire leans to the right.

It must be noted that the coordinate system defined here is different from that adopted by SAE International. The z-axis is perpendicular to the tire print and downward in the SAE International convention.

5.1.2. Lateral Forces in Cornering

When a vehicle travels on a curved or sideway tilted road, or in a lateral wind, there is a lateral force F_y acting at the center of the wheel, which is transmitted to the tire by the rim. This force produces a reaction force $F_{y\alpha}$ acting from the ground on the tire. Known as the cornering force, the reaction force $F_{y\alpha}$ is actually the resultant lateral force from the distributed pressure at the tire–road interface, as illustrated in Figure 5.2 [5.2]. Under the action of lateral force F_y, a lateral velocity v_y component is developed due to the deformation of the tire, causing the wheel to travel in a direction off the wheel plane, that is, a phenomenon called sideslip. The angle between the wheel traveling direction and the wheel plane is the aforementioned sideslip angle α. Obviously, the sideslip angle α is related to both lateral force F_y and cornering force $F_{y\alpha}$.

FIGURE 5.2 Tire deformation in cornering.

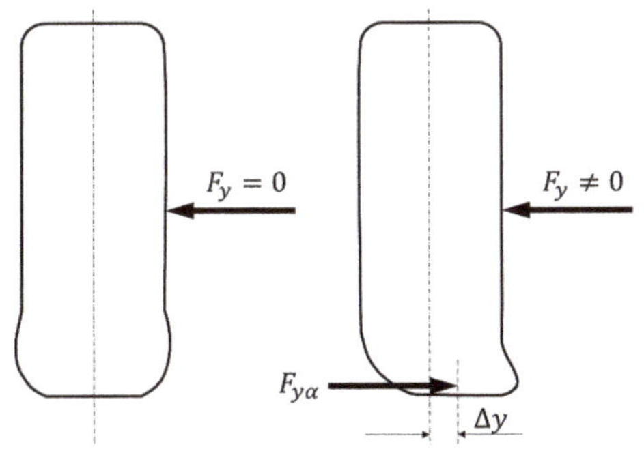

The relationship between the sideslip angle α and the cornering force $F_{y\alpha}$ is shown in Figure 5.3 [5.3]. When the sideslip angle $\alpha < 5°$, the cornering force $F_{y\alpha}$ increases almost linearly with the sideslip angle α. The slope at $\alpha = 0$ is referred to as cornering stiffness C_α, that is,

$$C_\alpha = \left.\frac{\partial F_{y\alpha}}{\partial \alpha}\right|_{\alpha=0} \tag{5.1}$$

FIGURE 5.3 Cornering characteristics of a tire.

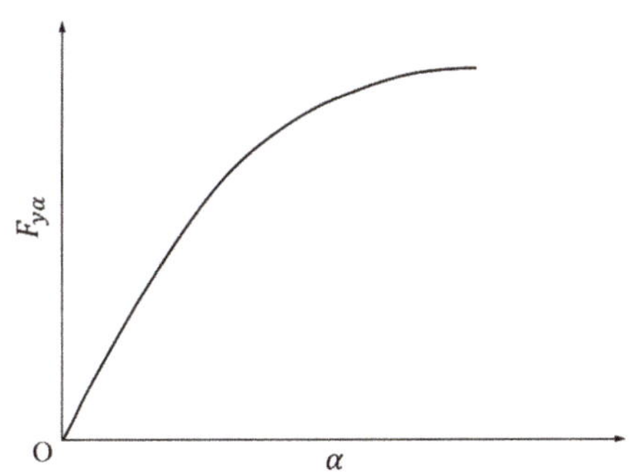

Per the coordinate convention defined in the last section, the positive sideslip angle α would produce a positive cornering force $F_{y\alpha}$, implying positive cornering stiffness C_{α}.

As a tire characteristic parameter, cornering stiffness significantly impacts the handling and stability of the vehicle. The typical cornering stiffness of passenger car tires ranges from 28,000 to 80,000 N/rad. When the cornering force $F_{y\alpha}$ is large, the curve of $F_{y\alpha}$ versus α levels off, and the sideslip angle α grows at a greater rate, with partial sideslip starting to develop at the tire–road interface. A complete sideslip finally occurs when the cornering force reaches the adhesion limit. The larger the maximum cornering force, the better the ultimate cornering performance of a vehicle.

5.1.3. Factors Affecting Cornering Behavior

The maximum cornering force depends on the adhesion condition of the tire–road interface, involving factors such as the structure, material, tread pattern, inflation pressure of the tire, the structure, material, wetness of the road surface, the camber angle of the wheel, and the vertical load.

5.1.3.1. Tire-Related Factors

Cornering stiffness is largely affected by the size, shape, and structural parameters of the tire. Compared to smaller-size, nonradial, or nylon-cord radial tires, larger-size, radial, or steel-cord tires generally exhibit higher cornering stiffness.

The flatness ratio of a tire can significantly affect its cornering stiffness [5.4]. The flatness ratio is the aspect ratio of the maximum cross-sectional width and height measured from the rim to the tire tread, expressed as a percentage. Tires of small flatness ratios ranging from 40 to 60% have been widely used in EVs to improve their cornering stiffness.

The vertical load alters frequently in driving, which has a great impact on cornering stiffness. With the increase in vertical load, cornering stiffness rises first, peaks, and then drops because of dramatically uneven contact at the tire–road interface under excessively large vertical loads.

Inflation pressure also has a considerable influence on cornering stiffness. With the rise of inflation pressure, cornering stiffness increases and then levels off when inflation pressure becomes excessively high.

Under the combined action of cornering and tractive forces, the cornering behavior is affected by the tractive force. For a given sideslip angle, the cornering force decreases with an increase in the tractive force, due to the changing elasticity in the lateral direction. Under a large tractive force, the cornering force reduces substantially, while adhesion approaches its upper limit. With most of the adhesion utilized by the tractive force, little adhesion is available for cornering force. Similar characteristics are observed for a tire under the action of a braking force. For a given tire under a certain vertical load, the cornering force can be plotted against a tractive force for a range of sideslip angles, as shown in Figure 5.4 [5.5]. The envelope of these curves, known as the friction ellipse or friction circle, represents the maximum resultant force achievable under a given adhesion condition.

FIGURE 5.4 Effect of tractive load on the cornering force.

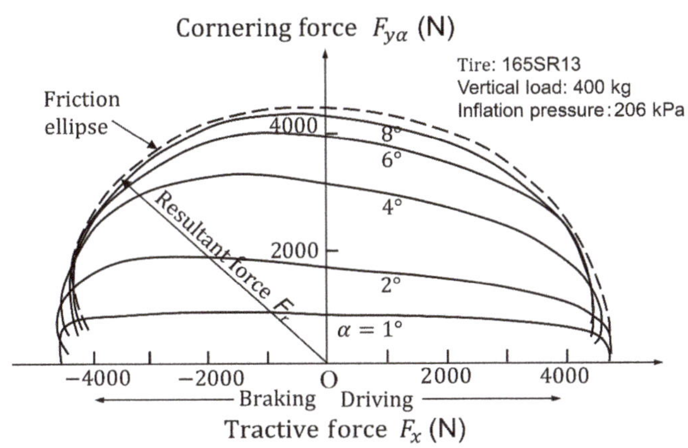

5.1.3.2. Road-Related Factors

The pavement, roughness, and wetness of a road have significant effects on cornering characteristics, in particular on the maximum cornering force.

When a vehicle operates on a water-covered road, the cornering force can diminish while a phenomenon called hydroplaning takes place. The cornering force may drop to zero, regardless of pavement roughness or tire pattern when the vehicle operates at high speeds in the water of a certain depth.

5.1.4. Aligning Torque

Aligning torque is the torque about the vertical axis produced by the lateral force at the tire–road interface, which helps turn the wheel back to the forward direction in cornering.

Under the action of a lateral force, the tire advances at an angle with respect to the wheel plane, which is known as the sideslip angle. With the tread elements coming into contact with the road surface, the tire deflects sideways with respect to the tread elements, with deflection larger in the rear than in the front. Assuming that the lateral reaction force of the road surface acting on tread elements is proportional to the deflection, a profile of a lateral force is developed on the tire print, as shown in Figure 5.5 [5.6]. Because of the asymmetry in the lateral force profile, the resultant lateral force F_y over the tire print leans toward the rear of the projected wheel spin axis on the tire print by a distance e, which is called a pneumatic trail. Clearly, the aligning torque $T_z = eF_y$.

FIGURE 5.5 Mechanism of aligning torque.

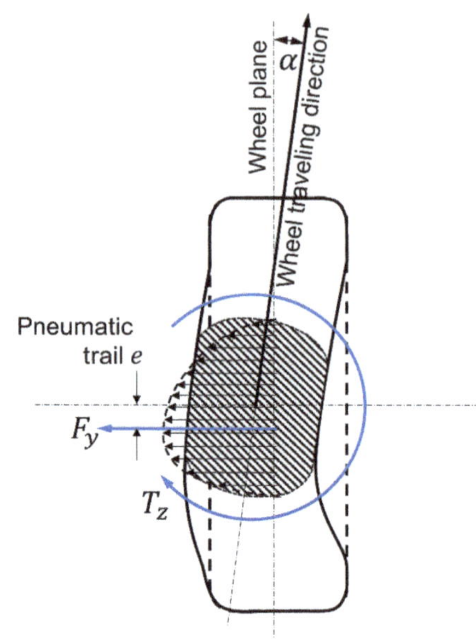

Courtesy of Elsevier.

Sideslip occurs when the lateral force acting on the tread elements is large enough to overcome the friction at the tire–road interface. Initiated from a localized area toward the rear portion of the tire print, sideslip grows in area with an increase in the lateral force, until it occurs throughout the tire print. On the contrary, the pneumatic trail drops with an increase in the lateral force, causing aligning torque to decrease at large sideslip angles. As the sideslip angle increases, the aligning torque rises at small sideslip angles, then peaks around 3–7° of sideslip angles, and finally becomes negative. It is also evident that the aligning torque rises as the vertical load increases.

With an increase in the tractive force, aligning torque increases at small tractive forces and then peaks before dropping to a small positive value. On the other hand, with an increase in the braking force, aligning torque gradually reduces to zero and then becomes negative.

The formation and structural parameters of tires affect the aligning torque as well. For a given sideslip angle, the aligning torques of large-size tires are greater than those of small-size tires, and the aligning torques of radial tires are greater than those of bias tires.

The aligning torque can be increased by reducing the inflation pressure of the tire since the pneumatic trail increases with the extended tire print in the longitudinal direction.

5.1.5. Behavior of a Tire with the Camber Angle

The camber angle γ is the inclination angle of the tire about the z-axis. As illustrated in Figure 5.6 [5.3], for the tire rolling about its axis OO′ at a positive camber angle, it is required that a lateral force $F_{y\gamma}$ should act at the wheel center. Meanwhile, a lateral force $F_{y\gamma}$ is produced by the road surface acting on the tire, which is often referred to as the camber thrust or camber force.

FIGURE 5.6 Camber angle and camber thrust.

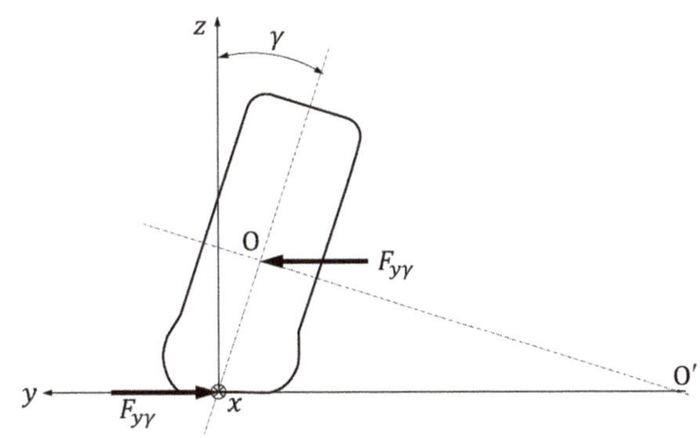

Camber thrust holds almost a linear relationship with the camber angle; therefore,

$$F_{y\gamma} = C_\gamma \gamma \tag{5.2}$$

where C_γ is camber stiffness. The strict definition of camber stiffness is

$$C_\gamma = \left. \frac{\partial F_{y\gamma}}{\partial \gamma} \right|_{\gamma=0} \tag{5.3}$$

Note that the camber thrust is actually negative for a positive camber according to the coordinate system defined in Section 5.1.1.

Under the combined action of the cornering force and camber thrust, the overall lateral force is

$$F_y = F_{y\alpha} + F_{y\gamma} \tag{5.4}$$

When both sideslip angle and camber angle are adequately small, Equation (5.4) becomes

$$F_y \approx \alpha C_\alpha + \gamma C_\gamma \tag{5.5}$$

The lateral force is usually linearly proportional to the sideslip angle at small sideslip angles and then levels off as the sideslip angle becomes large.

The camber angle produces aligning torque due to the contribution of the camber angle to the lateral force. The overall aligning torque can be approximated by a superposition of the sideslip and camber effects:

$$T_z \approx \alpha C_{T\alpha} + \gamma C_{T\gamma} \tag{5.6}$$

where $C_{T\alpha} \approx 0.013$ N·m/deg, $C_{T\gamma} \approx 0.0003$ N·m/deg for radial tires, and $C_{T\alpha} \approx 0.01$ N·m/deg, $C_{T\gamma} \approx 0.001$ N·m/deg for nonradial tires.

With an increase in the camber angle, the tire–road contact area shrinks, causing the lateral adhesion force to drop. Taking a camber angle of nearly zero degrees helps mitigate this effect on the cornering performance of a vehicle.

In summary, a positive sideslip force produces a negative sideslip angle and a negative aligning torque, while a positive camber thrust creates a negative camber angle and positive aligning torque, according to the coordinate convention discussed in Section 5.1.1.

5.2. Vehicle Response to the Input of the Front Wheel Angle

Since the input of the front wheel angle is typically a result of a steering operation, the way that the vehicle responds to such an input reveals the most fundamental aspects of handling and stability of the vehicle without accounting for the effects of the steering system.

A two-wheel planar model without roll motion is illustrated in Figure 5.7, which is often known as a bicycle model [5.3]. The vehicle motion is assumed to be restricted in the x-y plane, with the roll angle about the x-axis, the pitch angle about the y-axis, and the displacement along the z-axis ignored. Thus, the vehicle has two degrees of freedom (DOFs) remaining, that is, the lateral displacement along the y-axis and the yaw motion about the z-axis.

FIGURE 5.7 A two-wheel planar vehicle model.

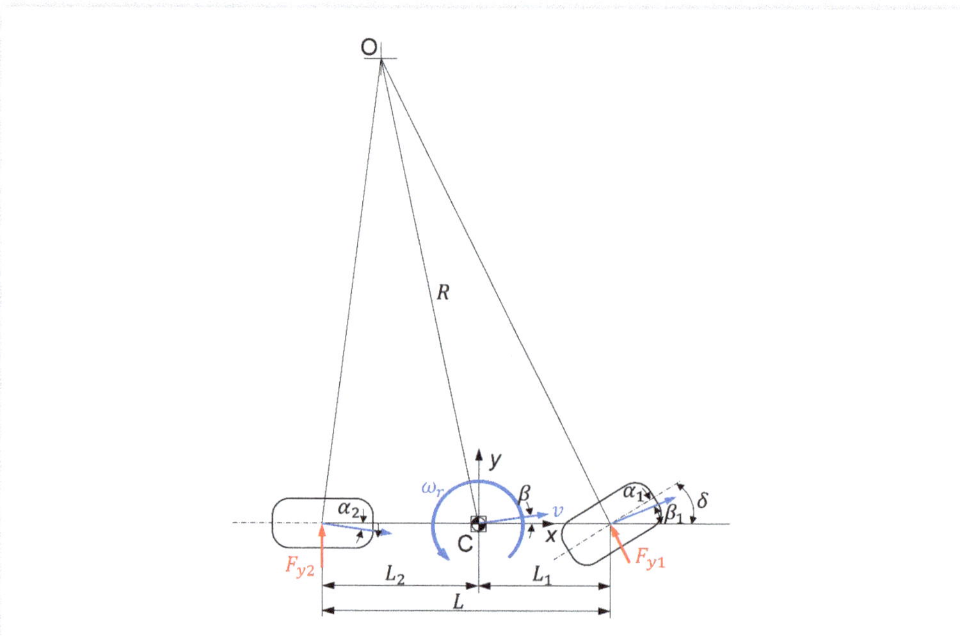

Within the coordinate system attached to the vehicle, the vehicle motion is governed by the following equations:

$$\begin{cases} \Sigma F_x = ma_x \\ \Sigma F_y = ma_y \\ \Sigma T_z = I_z \dot{\omega}_r \end{cases} \tag{5.7}$$

where

m is the vehicle mass

I_z is the vehicle mass moment of inertia about the z-axis

$\dot{\omega}_r$ is the angular acceleration about the center of rotation O

The forces acting on the two-wheel vehicle are given by

$$\begin{cases} \Sigma F_x = F_{x1} \cos\delta + F_{x2} - F_{y1} \sin\delta \\ \Sigma F_y = F_{y1} \cos\delta + F_{y2} + F_{x1} \sin\delta \\ \Sigma T_z = L_1 F_{y1} - L_2 F_{y2} \end{cases} \tag{5.8}$$

The translational acceleration components of the vehicle mass center are

$$\begin{cases} a_x = \dot{v}_x - v_y \omega_r \\ a_y = \dot{v}_y + v_x \omega_r \end{cases} \tag{5.9}$$

Substituting Equations (5.8) and (5.9) into Equation (5.7) yields

$$\begin{cases} F_{x1} \cos\delta + F_{x2} - F_{y1} \sin\delta = \mathrm{m}\left(\dot{v}_x - v_y \omega_r\right) \\ F_{y1} \cos\delta + F_{y2} + F_{x1} \sin\delta = m\left(\dot{v}_y + v_x \omega_r\right) \\ L_1 F_{y1} - L_2 F_{y2} = I_z \dot{\omega}_r \end{cases} \tag{5.10}$$

Assuming that the vehicle velocity along the x-axis is a constant, that is, $\dot{v}_x = 0$, the first equation in Equation (5.10) becomes independent. Thus, when both the steer angle δ and global sideslip angle at the centroid of mass, β, are small, we have

$$F_{y1} \approx C_{\alpha 1} \alpha_1 \tag{5.11}$$

$$F_{y2} \approx C_{\alpha 2} \alpha_2 \tag{5.12}$$

$$\beta \approx \frac{v_y}{v_x} \tag{5.13}$$

In addition, the global sideslip angle for the front wheel β_1 is

$$\beta_1 = \frac{v_y + \omega_r L_1}{v_x} = \beta + \frac{\omega_r L_1}{v_x} \tag{5.14}$$

The sideslip angles for the front and rear wheels are, respectively, given by

$$\alpha_1 = -(\delta - \beta_1) = \beta + \frac{\omega_r L_1}{v_x} - \delta \tag{5.15}$$

$$\alpha_2 = \frac{v_y - \omega_r L_2}{v_x} = \beta - \frac{\omega_r L_2}{v_x} \tag{5.16}$$

Substituting Equations (5.11) through (5.16) into the second and third equations in Equation (5.10), they can be reduced to

$$\begin{cases} \frac{\omega_r}{v_x}(C_{\alpha 1}L_1 - C_{\alpha 2}L_2) + (C_{\alpha 1} + C_{\alpha 2})\beta - C_{\alpha 1}\delta = m(\dot{v}_y + v_x\omega_r) \\ \frac{\omega_r}{v_x}(C_{\alpha 1}L_1^2 + C_{\alpha 2}L_2^2) + (C_{\alpha 1}L_1 - C_{\alpha 2}L_2)\beta - C_{\alpha 1}L_1\delta = I_z\dot{\omega}_r \end{cases} \tag{5.17}$$

The fundamental characteristics of the vehicle in responding to the steer angle input can be obtained by solving Equation (5.17). It is evident that one of the most significant parameters describing the response of a vehicle to a steering angle input is the angular velocity about the center of rotation ω_r, which is known as yaw velocity or yaw rate.

5.3. Steady-State Response to the Input of the Front Wheel Angle

The evaluation of steady-state response to the steer angle input involves a number of different parameters, among which yaw velocity gain is employed most often since it is a direct quantification of the system output with respect to the input.

5.3.1. Yaw Velocity Gain

Yaw velocity gain is defined as the ratio of the yaw rate to the steer angle at front wheels, that is, ω_r/δ, which is sometimes called the yaw rate response. As the vehicle moves at a constant speed, the steady-state response to the step input of a steer angle at front wheels is obviously a circular motion. Since the yaw rate is a constant in the steady-state condition, that is, $\dot{v}_y = 0$, $\dot{\omega}_r = 0$, Equation (5.17) becomes

$$\begin{cases} \frac{\omega_r}{v_x}(C_{\alpha 1}L_1 - C_{\alpha 2}L_2) + (C_{\alpha 1} + C_{\alpha 2})\beta - C_{\alpha 1}\delta = mv_x\omega_r \\ \frac{\omega_r}{v_x}(C_{\alpha 1}L_1^2 + C_{\alpha 2}L_2^2) + (C_{\alpha 1}L_1 - C_{\alpha 2}L_2)\beta - C_{\alpha 1}L_1\delta = 0 \end{cases} \tag{5.18}$$

The yaw velocity gain can be obtained by solving Equation (5.18),

$$\frac{\omega_r}{\delta}=\frac{v_x/L}{\frac{mg}{L}\left(\frac{L_1}{C_{\alpha 2}}-\frac{L_2}{C_{\alpha 1}}\right)\frac{v_x^2}{Lg}+1}=\frac{v_x/L}{\left(\frac{W_2}{C_{\alpha 2}}-\frac{W_1}{C_{\alpha 1}}\right)\frac{v_x^2}{Lg}+1}=\frac{v_x/L}{K\frac{v_x^2}{Lg}+1} \tag{5.19}$$

where $K=\frac{W_2}{C_{\alpha 2}}-\frac{W_1}{C_{\alpha 1}}$, known as the understeer gradient, in deg. The understeer gradient K is an important parameter characterizing the turning response properties of a vehicle. It dictates the magnitude and direction of the steer angle input required to achieve a given yaw rate output. With each term being the ratio of the loads on the axle to the cornering stiffness of tires, the understeer gradient reflects the difference in sideslip angles between the rear and front tires. Thus, the steady-state turning response can be classified into three categories in terms of the value of the understeer gradient K, as detailed here.

1. Neutral steer

   When $K = 0$, we have $\frac{\omega_r}{\delta}=v_x/L$. Therefore, $\delta=\frac{\omega_r L}{v_x}=\frac{L}{R}$. Known as the Ackerman angle, the angle δ is actually the steer angle required for the vehicle turning at low vehicle speeds with little sideslip at the front and rear tires. As illustrated in Figure 5.8 [5.3], yaw velocity gain is linearly dependent on longitudinal velocity, but the steer angle is a constant for making a turn of a constant radius. This implies that the sideslip angles resulting from lateral acceleration increase identically at the front and rear tires.

2. Understeer

   When $K > 0$, we have $\frac{\omega_r}{\delta}<v_x L$. The yaw velocity gain is, therefore, smaller than that in a neutral steer scenario. As the vehicle speed increases, the yaw velocity gain increases, then peaks, and finally drops at high speeds.

   As illustrated in Figure 5.8, the yaw velocity gain peaks at the vehicle speed v_{ch}, which is known as characteristic speed. By taking the derivative of Equation (5.19), it can be shown that when $\mathrm{d}\left(\frac{\omega_r}{\delta}\right)/\mathrm{d}v_x=0$, $v_{ch}=\sqrt{Lg/K}$ and the yaw velocity gain drops by half as compared to a neutral steer vehicle of the identical wheelbase L. At the characteristic speed, the steer angle becomes twice the Ackerman angle. Apparently, the characteristic speed v_{ch} reduces with increasing K as the understeer intensifies. For a given vehicle speed, the yaw velocity gain declines as K increases and the understeer intensifies. Since the sideslip angle caused by lateral acceleration at the front tires is larger than that at rear tires, the front wheels must have a greater steer angle to maintain the turning radius.

3. Oversteer

   When $K < 0$, we have $\frac{\omega_r}{\delta}>v_x L$. The yaw velocity gain is, therefore, larger than that in a neutral steer scenario, as shown in Figure 5.8. The yaw velocity gain increases monotonically with vehicle speed and approaches infinity at the vehicle speed v_{cr}, which is known as critical speed. Since K is negative in value, the oversteer intensifies with decreasing K. By solving $K\frac{v_x^2}{Lg}+1=0$, we obtain $v_{cr}=\sqrt{-Lg/K}$. The critical speed is wheelbase-dependent. A short wheelbase vehicle has a low critical speed for a certain level of oversteer, while the oversteer intensifies as the critical speed drops for a given wheelbase. Since the sideslip angle caused by lateral acceleration at the rear tires is larger than that at front tires, the front wheels must have a reduced steer angle to maintain the turning radius and must be zero at critical speed.

FIGURE 5.8 Yaw velocity gain as a function of vehicle speed.

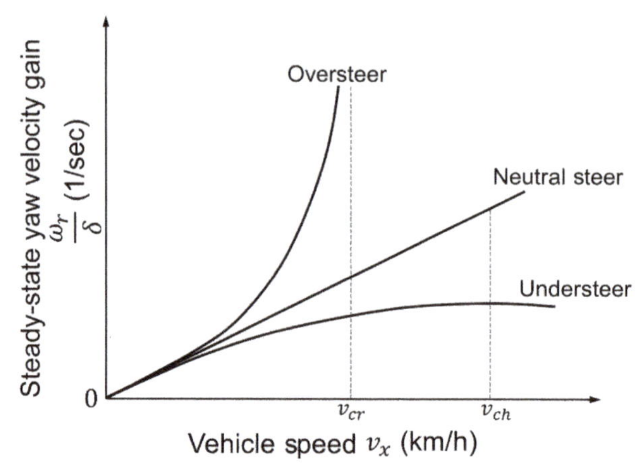

An oversteer vehicle becomes directionally unstable when the critical speed is reached, inasmuch as a huge yaw rate is produced by a tiny steer angle of front wheels. Consequently, dramatic turning, sideslip, or rollover can occur. Hence, all vehicles today are of understeer characteristics to an appropriate extent.

For modern sedans, the averaged value of the understeer gradient K typically ranges from 0.059 to 0.069, with lateral acceleration varying from 0.3 to 0.5g [5.4]; the steering wheel-based steady-state yaw velocity gain $\frac{\omega_r}{\delta_{sw}}$ ranges from 0.16 to 0.33 sec^{-1} when vehicle speed v_x = 22.35 m/s and lateral acceleration a_y = 0.4g [5.4]. Regulations have been established in some countries that require the test data of yaw velocity gain fall into a satisfactory region between upper and lower limits in vehicle evaluation tests.

5.3.2. Ratio of Turning Radii

The ratio of turning radii R/R_0 is an alternative parameter for evaluating steady-state response to a given steer angle input, where R is the turning radius at a given speed and lateral acceleration and R_0 is the turning radius at both speed and lateral acceleration of nearly zero with a negligible tire sideslip angle.

According to the previous discussion on neutral steer, it is evident that $R_0 = L/\delta$. By rearranging Equation (5.19), we have

$$R = \frac{v_x}{\omega_r} = \frac{\left(K\frac{v_x^2}{Lg}+1\right)L}{\delta} = \left(K\frac{v_x^2}{Lg}+1\right)R_0 \tag{5.20}$$

Thus,

$$\frac{R}{R_0} = K\frac{v_x^2}{Lg}+1 \tag{5.21}$$

When $K = 0$, $R/R_0 = 1$, the vehicle is in neutral steer with a constant turning radius $R = R_0$, which does not vary with vehicle speed; when $K > 0$, $R/R_0 > 1$, the vehicle has a turning radius always larger than R_0, which increases with the square of vehicle speed; and when $K < 0$, $R/R_0 < 1$, the vehicle has a turning radius always smaller than R_0, which decreases with increasing vehicle speed.

5.3.3. Static Margin

There exists a neutral steer point (NSP) on the longitudinal axis of a vehicle at which the action of a lateral force produces equal sideslip angles at front and rear wheels, which can be denoted by α. Thus, the lateral forces at the front and rear axles are $F_{y1} = C_{\alpha 1}\alpha$ and $F_{y2} = C_{\alpha 2}\alpha$, respectively. As shown in Figure 5.9 [5.4], by solving the moment equilibrium equation about the NSP C_n, its distance from the front axle L_1' can be obtained:

$$L_1' = \frac{F_{y2}L}{F_{y1} + F_{y2}} = \frac{C_{\alpha 1}L}{C_{\alpha 1} + C_{\alpha 2}} \tag{5.22}$$

FIGURE 5.9 NSP and center of mass.

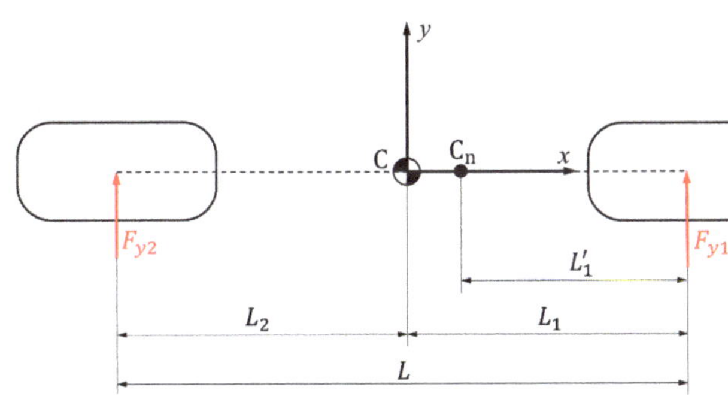

The static margin M_s is defined as the ratio of the distance between the NSP C_n and center of mass C to the wheelbase of the vehicle:

$$M_s = \frac{L_1' - L_1}{L} = \frac{C_{\alpha 1}}{C_{\alpha 1} + C_{\alpha 2}} - \frac{L_1}{L} \tag{5.23}$$

When the NSP coincides with the center of mass, $M_s = 0$. The vehicle has neutral steer behavior because identical sideslip angles are produced at the front and rear wheels by the lateral force acting on the center of mass.

When the NSP is located in the rear of the center of mass, $M_s > 0$. The vehicle is understeer as the sideslip angle produced at the front is greater than that at rear wheels.

When the NSP is located in the front of the center of mass, $M_s < 0$. The vehicle is oversteer since the sideslip angle produced at the front is smaller than that at rear wheels.

5.3.4. Difference in Sideslip Angles

The difference in absolute sideslip angles is also an indicator of the steady-state response of a vehicle to the steer angle input due to its relationship with the understeer gradient K. Per the coordinate convention defined in Section 5.1.1, a negative lateral force F_y produces a positive lateral acceleration a_y with a positive sideslip angle α, that is,

$$F_y = \frac{L}{L_1} F_{y2} = \frac{L}{L_2} F_{y1} = -ma_y \tag{5.24}$$

Thus,

$$K = \frac{W_2}{C_{\alpha 2}} - \frac{W_1}{C_{\alpha 1}} = g\left(\frac{mL_1}{LC_{\alpha 2}} - \frac{mL_2}{LC_{\alpha 1}}\right) = g\left(\frac{mL_1\alpha_2}{LF_{y2}} - \frac{mL_2\alpha_1}{LF_{y1}}\right) = \frac{g}{a_y}(\alpha_1 - \alpha_2) \tag{5.25}$$

It is clear that the difference in front and rear sideslip angles, $\alpha_1 - \alpha_2$, determines the sign of the understeer gradient K. When $\alpha_1 - \alpha_2 = 0$, the vehicle is neutral steer as $K = 0$; when $\alpha_1 - \alpha_2 > 0$, the vehicle is understeer as $K > 0$; and when $\alpha_1 - \alpha_2 < 0$, the vehicle is oversteer as $K < 0$.

Substituting $v_x = \omega_r R$ and $v_x^2 = a_y R$ into Equation (5.19) yields

$$\delta = \frac{Ka_y}{g} + \frac{L}{R} \tag{5.26}$$

Substituting Equation (5.25) into Equation (5.26), we obtain

$$\delta = \alpha_1 - \alpha_2 + \frac{L}{R} \tag{5.27}$$

or

$$R = \frac{L}{\delta - (\alpha_1 - \alpha_2)} \tag{5.28}$$

At very low vehicle speeds, we have $\alpha_1 \approx \alpha_2 \approx 0$ and $R = R_0 = L/\delta$; as vehicle speed increases, $R > R_0$ if $\alpha_1 - \alpha_2 > 0$, indicating that the vehicle is understeer with suppressed turning effort; on the contrary, $R < R_0$ if $\alpha_1 - \alpha_2 < 0$, implying that the vehicle is oversteer with enhanced turning effect.

When the vehicle acceleration exceeds 3–4 m/s², the sideslip angle difference $\alpha_1 - \alpha_2$ no longer holds a linear relationship with the lateral acceleration. Such a nonlinear sideslip behavior may cause both sideslip angle and yaw rate of the front or the rear wheel to vary drastically for some vehicles. As a result, the turning radii of such vehicles can increase or decrease sharply, instead of maintaining a circular motion.

5.4. Instantaneous Response to the Step Input of the Front Wheel Angle

5.4.1. Equation of Motion and Its Solution

The global sideslip angle β can be calculated by rearranging the second equation in Equation (5.17):

$$\beta = \frac{I_z\dot{\omega}_r + C_{\alpha 1}L_1\delta - \frac{\omega_r}{v_x}\left(C_{\alpha 1}L_1^2 + C_{\alpha 2}L_2^2\right)}{C_{\alpha 1}L_1 - C_{\alpha 2}L_2} \tag{5.29}$$

Since $\beta = v_x/v_y$, $\dot{\beta} = \dot{v}_x / v_y$, taking the derivative on both sides of Equation (5.29) and substituting into the first equation in Equation (5.17) yields the one-DOF differential equation for forced vibration:

$$\ddot{\omega}_r + 2\xi\omega_0\dot{\omega}_r + \omega_0^2\omega_r = C_1\dot{\delta} + C_0\delta \tag{5.30}$$

where

ξ is the damping ratio, $\xi = -\dfrac{m\left(C_{\alpha 1}L_1^2 + C_{\alpha 2}L_2^2\right) + I_z\left(C_{\alpha 1} + C_{\alpha 2}\right)}{2m\omega_0 v_x I_z}$

ω_0 is the natural circular eigenfrequency, $\omega_0 = \sqrt{\dfrac{mv_x^2\left(C_{\alpha 1}L_1 - C_{\alpha 2}L_2\right) + C_{\alpha 1}C_{\alpha 2}L^2}{mv_x^2 I_z}} = \dfrac{L}{v_x}\sqrt{\dfrac{C_{\alpha 1}C_{\alpha 2}}{mI_z}\left(\dfrac{Kv_x^2}{Lg} + 1\right)}$

the coefficients C_1 and C_0 are $C_1 = -\dfrac{C_{\alpha 1}L_1}{I_z}$ and $C_0 = \dfrac{C_{\alpha 1}C_{\alpha 2}L^2}{mv_x I_z}$

For a step input at the front wheels of a vehicle, the turning angle can be expressed as

$$\begin{cases} \delta = 0 & t < 0 \\ \delta \geq \delta_0 & t \geq 0 \\ \dot{\delta} = 0 & t > 0 \end{cases} \tag{5.31}$$

Thus, Equation (5.30) can be simplified further for $t > 0$

$$\ddot{\omega}_r + 2\xi\omega_0\dot{\omega}_r + \omega_0^2\omega_r = C_0\delta_0 \tag{5.32}$$

The general solution for the second-order constant coefficients inhomogeneous differential equation, that is, Equation (5.32), is the sum of a particular solution and a homogeneous solution, that is, $\omega_r = \omega_{rp} + \omega_{rh}$.

The particular solution is

$$\omega_{rp} = \frac{C_0\delta_0}{\omega_0^2} = \frac{v_x / L}{K\frac{v_x^2}{Lg} + 1}\delta_0 = \frac{\omega_r}{\delta}\delta_0 = \omega_{r0} \tag{5.33}$$

where ω_{r0} is the yaw rate in the steady-state condition.

The homogeneous solution is

$$\omega_{rh} = \begin{cases} Ce^{-\xi\omega_0 t}\sin\left(\omega_0\sqrt{1-\xi^2}\,t + \Omega\right) & \xi < 1 \\ (C_1 + C_2 t)e^{-\omega_0 t} & \xi = 1 \\ C_3 e^{\left(\sqrt{\xi^2-1}-\xi\right)\omega_0 t} + C_4 e^{\left(-\sqrt{\xi^2-1}-\xi\right)\omega_0 t} & \xi > 1 \end{cases} \tag{5.34}$$

where C, C_1, C_2, C_3, C_4, and Ω are integral constants, which can be determined by the initial condition of the motion.

When $\xi < 1$, the damping ratio is small. The yaw rate $\omega_r(t)$ is a sine function with reduced amplitude converging to the steady-state yaw rate ω_{r0}. When $\xi = 1$, the damping ratio is called the critical damping ratio. The yaw rate $\omega_r(t)$ increases monotonically and approaches ω_{r0}. When $\xi > 1$, the damping ratio is large. If the vehicle speed is lower than the critical speed v_{xc}, the yaw rate $\omega_r(t)$ increases monotonically and approaches ω_{r0}; if the vehicle speed is higher than the critical speed v_{xc}, the yaw rate $\omega_r(t)$ diverges and approaches infinity, implying an unstable operating condition.

All vehicles are generally of small damping ratio characteristics. The yaw rate for $\xi < 1$ is

$$\omega_r(t) = \frac{C_0\delta_0}{\omega_0^2} + Ce^{-\xi\omega_0 t}\sin\left(\omega_0\sqrt{1-\xi^2}\,t + \Omega\right) \tag{5.35}$$

Let $\omega_d = \omega_0\sqrt{1-\xi^2}$, then Equation (5.35) becomes

$$\omega_r(t) = \frac{C_0\delta_0}{\omega_0^2} + Ce^{-\xi\omega_0 t}\sin(\omega_d t + \Omega) \tag{5.36}$$

Given such initial conditions at $t = 0$ as $\omega_r = 0$, $v_y = 0$, $\delta = \delta_0$, Equation (5.17) yields $\dot{\omega}_r = -\frac{L_1 C_{\alpha 1}\delta_0}{I_z} = C_1\delta_0$, while the integral constants become

$$C = \frac{\omega_r}{\delta}\delta_0\sqrt{\frac{\left(\frac{mv_x L_1\omega_0}{LC_{\alpha 2}} + \xi\right)^2}{1-\xi^2} + 1} \tag{5.37}$$

$$\Omega = \arctan\left(\frac{\sqrt{1-\xi^2}}{\frac{mv_x L_1\omega_0}{LC_{\alpha 2}} + \xi}\right) \tag{5.38}$$

Thus, the instantaneous response of the vehicle to the step input of the front wheel angle can be derived from Equation (5.35):

$$\omega_r(t) = \frac{\omega_r}{\delta}\delta_0\left[1 + \left(\sqrt{\frac{\left(\frac{mv_x L_1\omega_0}{LC_{\alpha 2}} + \xi\right)^2}{1-\xi^2} + 1}\right) e^{-\xi\omega_0 t}\sin(\omega_d t + \Omega)\right] \tag{5.39}$$

It is clear that when $t = \infty$, $\omega_r(t) = \frac{\omega_r}{\delta}\delta_0 = \omega_{r0}$. Thus, when $0 < t < \infty$, $\omega_r(t)$ is a decay sine function that approaches the stead-state yaw rate ω_{r0}, as shown in Figure 5.10 [5.4], whereas the decay rate increases with the damping ratio.

US regulations require that the instantaneous response to a step input of a steer angle fall into the area between the specified lower and upper limits. In the evaluation test, the vehicle travels in a straight line at a speed of 40 or 110 km/h and then turns the steering wheel at an angular speed of no less than 500 deg/s, with a lateral acceleration of 0.4g in entering the steady-state condition.

Several parameters in Equation (5.39) are of significance to the instantaneous response of the vehicle, as discussed in the following section.

5.4.1.1. Rise Time t_r

Rise time t_r refers to the time required for the yaw rate to reach ω_{r0} for the first time following a step input in a steer angle, as illustrated in Figure 5.10. A small rise in time is generally desirable in that the vehicle responds quickly to the turning of the steering wheel.

FIGURE 5.10 Instantaneous response of a vehicle to a step input in a steer angle.

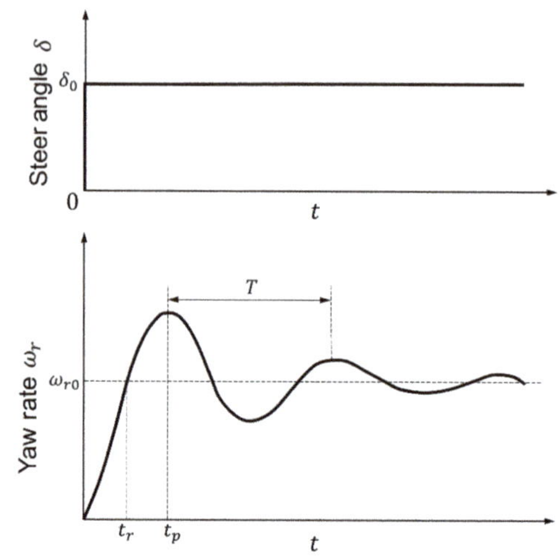

Rise time t_r can be obtained by letting $\omega_r(t = t_r) = \omega_{r0} = \frac{C_0\delta_0}{\omega_0^2}$ in Equation (5.39):

$$t_r = -\frac{\Omega}{\omega_d} = -\frac{\arctan\left(\dfrac{\sqrt{1-\xi^2}}{\dfrac{mv_x L_1\omega_0}{LC_{\alpha 2}}+\xi}\right)}{\omega_0\sqrt{1-\xi^2}} \tag{5.40}$$

5.4.1.2. Peak Time t_p

Peak time t_p is the time required for the yaw rate to reach the maximum for the first time following a step input in steer angle, as illustrated in Figure 5.10. Peak time t_p can be solved by taking the derivative of Equation (5.39) and letting $\frac{d\omega_r(t)}{dt} = 0$:

$$t_p = \frac{\arctan\left(\dfrac{\sqrt{1-\xi^2}}{\xi}\right)}{\omega_0\sqrt{1-\xi^2}} + t_r \tag{5.41}$$

The typical peak time for sedans typically ranges from 0.23 to 0.59 sec [5.4].

5.4.1.3. Circular Eigenfrequency ω_0 of Yaw Rate Fluctuation

A reasonably high eigenfrequency ω_0 is desirable to improve the instantaneous response of the vehicle. With $f_0 = \frac{\omega_0}{2\pi}$, the eigenfrequency f_0 typically ranges from 0.9 to 1.0 Hz for sedans.

5.4.1.4. Damping Ratio ξ

The damping ratio can be expressed by

$$\xi = -\frac{m\left(L_1^2 C_{\alpha 1} + L_2^2 C_{\alpha 2}\right) + I_z\left(C_{\alpha 1} + C_{\alpha 2}\right)}{2L\sqrt{mI_z C_{\alpha 1} C_{\alpha 2}\left(\frac{K}{Lg}v_x^2 + 1\right)}} \tag{5.42}$$

The damping ratio of sedans are typically within the range of 0.5–0.8 [5.4].

5.4.2. Stability Conditions of Instantaneous Response

As mentioned in the previous section, the solution to the two-DOF equation of motion of the vehicle, that is, Equation (5.32), is the sum of a particular solution $\frac{C_0\delta_0}{\omega_0^2}$ and a homogeneous solution. Since the particular solution $\frac{C_0\delta_0}{\omega_0^2}$ has a finite value, the system stability depends only on the homogeneous solution, which is discussed in the following two cases:

1. When $\xi \le 1$, the convergence of solution expressed in Equation (5.34) depends on the sign of $\xi\omega_0$; the solution is convergent if $\xi\omega_0 > 0$, and divergent if $\xi\omega_0 < 0$. Per Equation (5.31), we have

$$\xi\omega_0 = \frac{m\left(C_{\alpha 1}L_1^2 + C_{\alpha 2}L_2^2\right) + I_z\left(C_{\alpha 1} + C_{\alpha 2}\right)}{2mv_x I_z} \tag{5.43}$$

   Clearly, $\xi\omega_0$ is always positive. Thus, the homogeneous solution converges and approaches zero.

2. When $\xi > 1$, the convergence of solution to the homogeneous equation of Equation (5.32) requires the eigenvalue for its secular equation to be negative. In other words, the following equation must hold:

$$-\xi\omega_0 \pm \omega_0\sqrt{\xi^2 - 1} < 0 \tag{5.44}$$

   Since ξ is a positive real number, it is evident that Equation (5.44) is valid as long as ω_0 is a positive real number. From the coefficients for Equation (5.31), we have

$$\omega_0 = \frac{L}{v_x}\sqrt{\frac{C_{\alpha 1}C_{\alpha 2}}{mI_z}\left(\frac{Kv_x^2}{Lg} + 1\right)} \tag{5.45}$$

To ensure ω_0 is a real number and $\omega_0 > 0$, it is obvious that the following relationship must hold:

$$\frac{Kv_x^2}{Lg} + 1 > 0 \tag{5.46}$$

For a vehicle of $K \geq 0$, the condition in Equation (5.46) is satisfied at any given v_x, implying both understeer and neutral steer vehicles are stable. On the other hand, for an oversteer vehicle, $K < 0$, the condition Equation (5.46) is satisfied only when

$$v_x < \sqrt{-\frac{Lg}{K}} \tag{5.47}$$

Therefore, a critical speed can be defined as $v_{xc} = \sqrt{-\frac{Lg}{K}}$ for oversteer vehicles, similar to that in the steady-state scenario. The vehicle is stable if $v_x < v_{xc}$ and becomes unstable if $v_x \geq v_{xc}$.

5.5. Yaw Rate Frequency Response Characteristics

Applying Fourier transform to both sides of Equation (5.31) produces

$$-\omega^2\omega_r(\omega) + 2\omega_0\xi j\omega\omega_r(\omega) + \omega_0^2\omega_r(\omega) = C_1 j\omega\delta(\omega) + C_0\delta(\omega) \tag{5.48}$$

where

δ is the steer angle or system input
ω_r is the yaw rate or system output
j is the complex operator

By rearranging Equation (5.48), the system response function can be obtained:

$$H(j\omega) = \frac{\omega_r(\omega)}{\delta(\omega)} = H_1(\omega) + jH_2(\omega) \tag{5.49}$$

where

$$H_1(\omega) = \frac{2C_1\xi\omega_0\omega^2 + C_0\left(\omega_0^2 - \omega^2\right)}{\left(\omega_0^2 - \omega^2\right)^2 + 4\xi^2\omega_0^2\omega^2}$$

and

$$H_2(\omega) = \frac{C_1\omega\left(\omega_0^2 - \omega^2\right) - 2C_0\xi\omega_0\omega}{\left(\omega_0^2 - \omega^2\right)^2 + 4\xi^2\omega_0^2\omega^2}$$

Thus, the amplitude-frequency characteristics of the system is

$$A(\omega) = \sqrt{\left[H_1(\omega)\right]^2 + \left[H_2(\omega)\right]^2} \tag{5.50}$$

which reflects the distortion level of the yaw rate response of a vehicle to the input from the driver at various frequencies.

The phase-frequency characteristic of the system is

$$\Phi(\omega) = \arctan\frac{H_2(\omega)}{H_1(\omega)} \tag{5.51}$$

which quantifies the distortion level of the yaw rate response of a vehicle with respect to the steer angle. In the yaw rate frequency response characteristics of a typical sedan, the amplitude is nearly a constant in the low-frequency region and then increases with frequency. It peaks at a certain frequency at which the resonance occurs. The amplitude drops with a further rise in frequency. On the other hand, the hysteresis in the phase angle is minimal at low frequencies and increases with frequency.

It is generally desired to have a flat amplitude-frequency characteristics curve with a broad frequency band and a high resonance frequency. Thus, a small ratio A_r/A_0 and a high frequency f_r are advantageous for enhancing the yaw stability of a vehicle. As far as the phase-frequency characteristics are concerned, a small phase angle hysteresis would warrant a fast and nimble response. The hysteresis angles $\Phi_{0.1}$ and $\Phi_{0.6}$ are commonly utilized to evaluate the sensitivity of the vehicle response to relatively slow and fast turning of the steering wheel, respectively. A good engineering practice is to keep $\Phi_{0.1}$ nearly zero and make $\Phi_{0.6}$ as small as possible.

5.6. Suspension Effects on Handling Stability

The simplified analysis in the previous section has shown that the vehicle response to a step input in a steer angle is largely determined by the understeer gradient K, which is, in turn, closely related to the sideslip angles of front and rear wheels. In addition, the sideslip angles of front and rear wheels depend only on the centroid of the vehicle and the sideslip stiffness of the front and rear tires when the influences of the camber angle, load shift, and driving force are ignored.

However, in reality, all these factors contribute to the fundamentally tire elasticity-related sideslip angles, to some extent. For a vehicle in cornering, the load shifts in the lateral direction from the inside to outside wheels because the roll moment acting on the body and suspension is developed in the presence of lateral acceleration. In the presence of a camber angle, the roll moment also causes the camber angle to vary due to the motion and deflection of the suspension linkage. Furthermore, the driving force produces tangential reaction forces acting on wheels. Consequently, these factors can alter the cornering stiffness and camber-related lateral force of the tires, which causes the tire elasticity-related sideslip angles to vary.

In addition to the tire elasticity-related sideslip angle, the total sideslip angle should at least include the following:

1. Roll steer angle: Under the action of roll moment, this is the angle between the wheel plane and the kingpin axis produced by the motion of the guide rod and the steering rod at the front suspension, as well as by the motion of the guide rod at the rear suspension.
2. Compliance steer angle: Under the action of roll moment, this is the angle between the wheel plane and the kingpin axis produced by the compliance of the guide rod and the steering rod at the front suspension and by the compliance of the guide rod at the rear suspension.

Thus, the total sideslip angle is affected by the configuration and structural parameters of the suspension, steering, transmission, and driveline systems. The influence of suspension on the total sideslip angle is discussed in the following section.

5.6.1. Suspension-Related Cornering Behavior

For a vehicle in cornering, two instantaneous roll centers can be employed to characterize the lateral turning of the vehicle body with respect to the ground, with one for the front axle and the other for the rear axle. The roll center is defined as the point at which the lateral force is transferred from the axle to the sprung mass. It is actually the point about which the axle rolls when subjected to a roll moment. Figure 5.11 is an example of the roll center for an axle with a single wishbone independent suspension [5.4]. The line connecting the front and rear roll centers is called the roll axis, as illustrated in Figure 5.12.

FIGURE 5.11 The roll center for an axle with a single wishbone independent suspension.

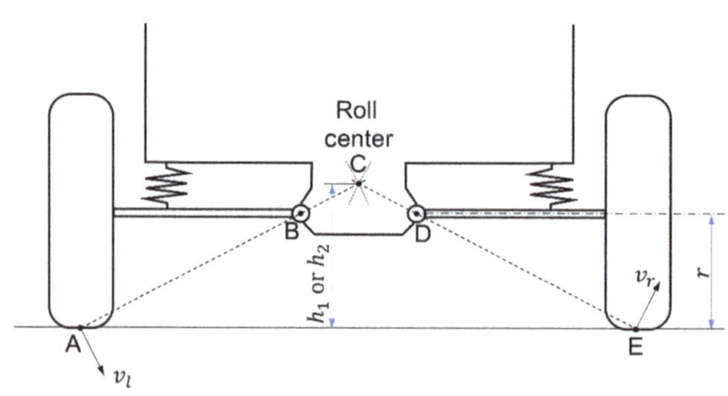

FIGURE 5.12 The roll axis and roll centers.

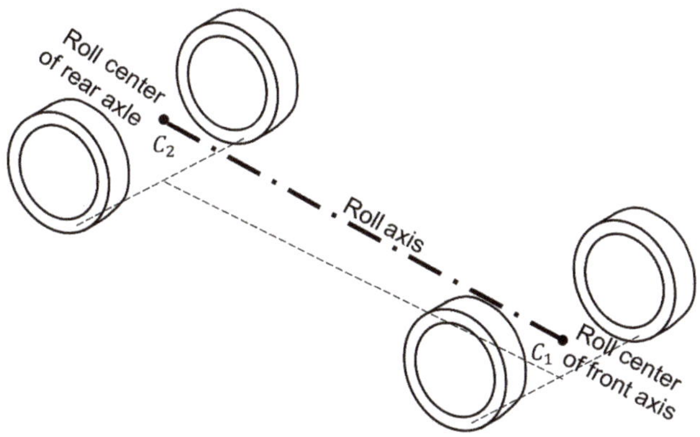

5.6.1.1. Roll Stiffness

A suspension of any type can be equivalent to two springs that are set apart laterally to develop a moment resisting the roll motion of the vehicle body with respect to the axle. Thus, roll stiffness K_Φ can be defined as the resisting moment T_r produced per unit roll angle between the body and the axle Φ_r:

$$K_\Phi = \frac{dT_r}{d\Phi_r} \tag{5.52}$$

For a nonindependent suspension in roll motion (Figure 5.13), it can be shown that roll stiffness is

$$K_\Phi = \frac{1}{2}K_s b^2 \tag{5.53}$$

where

K_s is the vertical rate of the left or right spring
b is the lateral distance between the springs

FIGURE 5.13 A nonindependent suspension in roll motion.

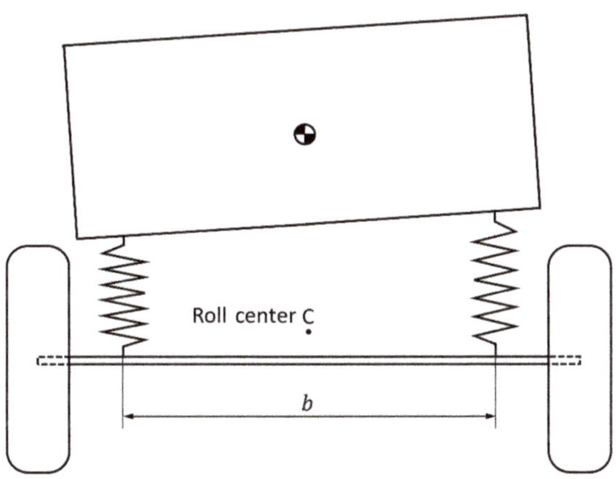

For an independent suspension in roll motion (Figure 5.14), roll stiffness is

$$K_{\Phi} = \frac{1}{2} K_s \left(\frac{t_r s_1}{t_1} \right)^2 \tag{5.54}$$

where
- t_r is the wheel track
- s_1 is the distance from the center of the spring to the hinge point of the crossarm
- t_1 is the length of the crossarm

Roll stiffness for a sedan typically ranges from 300 to 2500 N·m/deg for the front axle and from 180 to 1000 N·m/deg for the rear axle [5.7].

FIGURE 5.14 An independent suspension in roll motion.

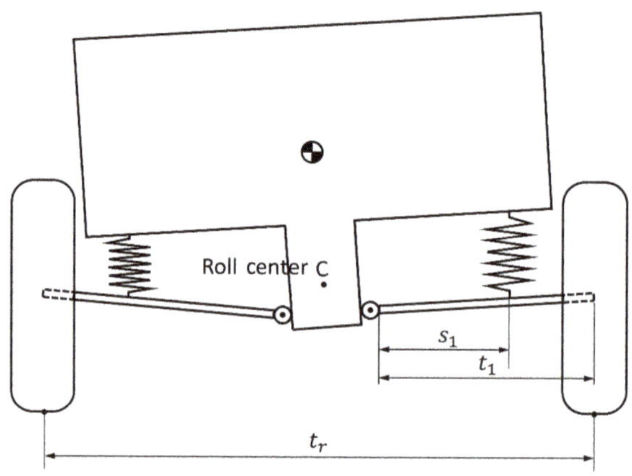

5.6.1.2. Roll Angle of the Body

For a vehicle in cornering, the body rolls about the roll axis due to the action of a lateral force. The roll angle of the body is a parameter that largely affects handling stability and ride smoothness since it is related to the yaw rate in steady-state and instantaneous response of the vehicle. An excessive roll angle of the body not only makes the driver feel unsafe and unstable but also makes passengers uncomfortable. On the other hand, a very small roll angle of the body during cornering may imply a high roll stiffness, which can worsen the smoothness when one side of the wheels goes over a pit hole or a bump.

The roll angle of the body Φ can be computed by

$$\Phi = \frac{\Sigma M_{\Phi}}{\Sigma K_{\Phi}} \tag{5.55}$$

where
- ΣM_{Φ} is the sum of roll moments acting on the body
- ΣK_{Φ} is the overall stiffness of the suspensions and stability bars

The sum of roll moments is

$$\sum M_\Phi = m_s \frac{V^2}{R} h + m_s g e - m_u \frac{V^2}{R}(h_1 - r) - m_u \frac{V^2}{R}(h_2 - r) \tag{5.56}$$

where

m_s is the sprung mass
e is the lateral offset of the center of mass due to roll moment (as shown in Figure 5.15)
m_u is the unsprung mass
h_1 and h_2 are the roll center heights at the front and the rear, respectively
V is the vehicle velocity
R is the radius of the turn
r is the tire radius

FIGURE 5.15 Forces and geometries for computing the roll angle of the body.

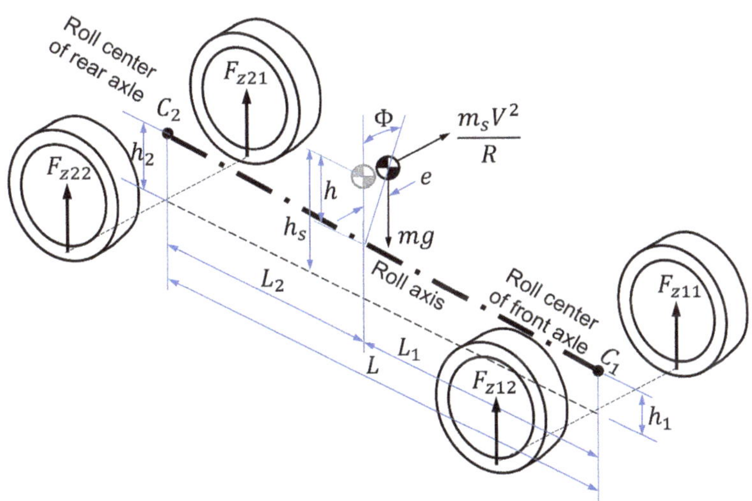

On the right-hand side of Equation (5.56), the first term is the roll moment produced by the centrifugal force acting on the sprung mass; the second term is the roll moment produced by the gravitational force of the sprung mass due to the lateral offset of the mass center; the third and fourth terms are the roll moments produced by a centrifugal force acting on the unsprung mass of the front and rear independent suspensions.

For a small roll angle Φ, $e \approx h\Phi$, Equation (5.56) becomes

$$\sum M_\Phi = \left[m_s h - m_u (h_1 + h_2 - 2r)\right]\frac{V^2}{R} + m_s g h \Phi \tag{5.57}$$

The overall stiffness of the suspensions and stability bars is

$$\Sigma K_{\Phi} = K_{\Phi 1} + K_{\Phi 2} + K_{\Phi 3} \tag{5.58}$$

where

$K_{\Phi 1}$ is the stiffness of front suspension
$K_{\Phi 2}$ is the stiffness of rear suspension
$K_{\Phi 3}$ is the stiffness of the stability bar

Substitution of Equations (5.56) and (5.58) into Equation (5.55) yields

$$\Phi = \frac{\left[m_s h - m_u \left(h_1 + h_2 - 2r\right)\right]\frac{V^2}{R} + m_s gh\Phi}{K_{\Phi 1} + K_{\Phi 2} + K_{\Phi 3}} \tag{5.59}$$

Solving Equation (5.59), we obtain the roll angle of the body:

$$\Phi = \frac{\left[m_s h - m_u \left(h_1 + h_2 - 2r\right)\right]\frac{V^2}{R}}{K_{\Phi 1} + K_{\Phi 2} + K_{\Phi 3} - m_s gh} \tag{5.60}$$

A roll rate can be defined as the roll angle of the body per lateral acceleration in terms of gravitational acceleration g:

$$r_{\Phi} = \frac{d\Phi}{da_y} = \frac{\left[m_s h - m_u \left(h_1 + h_2 - 2r\right)\right]g}{K_{\Phi 1} + K_{\Phi 2} + K_{\Phi 3} - m_s gh} \tag{5.61}$$

The roll rate of the body for a sedan typically ranges from 3 to 7 deg/g [5.3, 5.7].

5.6.1.3. Load Shift

On a level ground, the respective vertical loads on the front and rear axles of a vehicle in stationary or uniform linear motion are

$$F_{z1} = \frac{L_2}{L} mg \tag{5.62}$$

$$F_{z2} = \frac{L_1}{L} mg \tag{5.63}$$

Although a common practice is to make left and right wheels equally loaded, a small offset e is, however, assumed with the vertical load biased toward the left per conventions illustrated in Figure 5.15. Thus, the vertical loads on each of the four wheels are

$$F_{z11} = F_{z1}\left(\frac{1}{2} + \frac{e}{t_1}\right) = \frac{L_2}{L}\left(\frac{1}{2} + \frac{e}{t_1}\right) mg \tag{5.64}$$

$$F_{z12} = F_{z1}\left(\frac{1}{2} - \frac{e}{t_1}\right) = \frac{L_2}{L}\left(\frac{1}{2} - \frac{e}{t_1}\right)mg \tag{5.65}$$

$$F_{z21} = F_{z2}\left(\frac{1}{2} + \frac{e}{t_2}\right) = \frac{L_1}{L}\left(\frac{1}{2} + \frac{e}{t_2}\right)mg \tag{5.66}$$

$$F_{z21} = F_{z2}\left(\frac{1}{2} - \frac{e}{t_2}\right) = \frac{L_1}{L}\left(\frac{1}{2} - \frac{e}{t_2}\right)mg \tag{5.67}$$

where t_1 and t_2 are the front and rear tracks, respectively.

For a vehicle in cornering, lateral load shift causes unequal vertical loads on the left and right wheels, which impacts the sideslip characteristics of tires and consequently the vehicle response. In some circumstances, the effect is so significant that an understeer vehicle can become oversteer. The variations of the vertical loads on left and right wheels, respectively, are

$$\Delta F_{z11} = \frac{1}{t_1}\left(\frac{L_2}{L}h_1 m_s \frac{V^2}{R} + h_{u1} m_{u1}\frac{V^2}{R} + K_{\Phi 1}\Phi\right) \tag{5.68}$$

$$\Delta F_{z12} = -\Delta F_{z11} \tag{5.69}$$

$$\Delta F_{z21} = \frac{1}{t_1}\left(\frac{L_1}{L}h_2 m_s \frac{V^2}{R} + h_{u2} m_{u2}\frac{V^2}{R} + K_{\Phi 2}\Phi\right) \tag{5.70}$$

$$\Delta F_{z22} = -\Delta F_{z21} \tag{5.71}$$

where $K_{\Phi 1}$ and $K_{\Phi 2}$ are the roll stiffness of the front and rear suspensions, respectively, and h_{u1} and h_{u2} are the heights of front and rear unsprung mass centroids, respectively, which can be approximated by the corresponding wheel radii. On the right side of Equations (5.68) and (5.70), the first term represents the influence of centrifugal force acting on the sprung mass, the second term is the effect of centrifugal force acting on the unsprung mass, and the third term is the recovery moments acting on the body by the front and rear axles.

Apparently, the vertical loads on each of the four wheels for a vehicle in cornering are $F_{z11} + \Delta F_{z11}$, $F_{z12} + \Delta F_{z11}$, $F_{z21} + \Delta F_{z21}$, and $F_{z21} + \Delta F_{z22}$. The variations of the vertical loads are negative on inside wheels and tend to reduce the vertical loads there. On the contrary, the variations of the vertical loads are positive on outside wheels and tend to increase the vertical loads there.

5.6.1.4. Effect of Load Shift on Cornering Stiffness

Since the cornering stiffness of a tire is vertical load-dependent, the stiffness peaks at a certain vertical load, as shown in Figure 5.4. The rated vertical load is typical two-thirds of that at the maximum cornering stiffness.

For a vehicle subject to a lateral force, with the vertical load shift between left and right wheels ignored, the sideslip angle of each tire is

$$\alpha_0 = \frac{F_y}{2K_0} \tag{5.72}$$

where

F_y is the total lateral reaction force from the ground to the tire on an axle
K_0 is the cornering stiffness of each tire, as shown in Figure 5.16 [5.8]

In reality, the load shift is inevitable in cornering. As a result, the left and right cornering stiffnesses change to K_1 and K_2, respectively. If the averaged cornering stiffness of each tire is $K_0^{'} = \frac{K_1 + K_2}{2}$, the sideslip angle of each tire then becomes

$$\alpha = \frac{F_y}{K_1 + K_2} = \frac{F_y}{2K_0^{'}} \tag{5.73}$$

From Figure 5.16, it is obvious that the averaged cornering stiffness $K_0^{'}$ drops with the decrease in the vertical load difference between left and right wheels. Thus, the vehicle in cornering becomes increasingly understeer if a large load shifts toward the front axle, while becoming increasingly oversteer if a large load shifts toward the rear axle.

FIGURE 5.16 Effect of load shift on cornering stiffness of left and right tires.

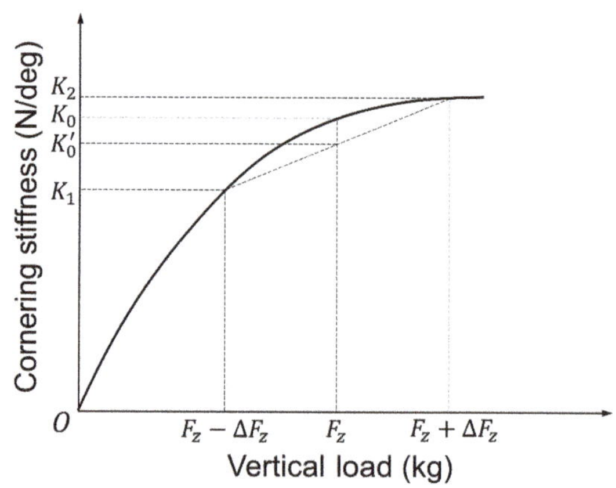

5.6.1.5. Camber Change in Cornering

Camber refers to the inclination of a wheel outward from the body, which produces a lateral force called camber thrust or camber force.

Since both body roll and suspension jounce contribute to camber change, the total camber change in cornering is given by

$$\gamma = \Phi + \gamma_b \tag{5.74}$$

where γ is the total camber change with respect to the ground, Φ is the roll angle of the vehicle, and γ_b is the camber angle of the wheel with respect to the body, as shown in Figure 5.17, with subscripts i and o denoting the inner and outer wheels, respectively.

FIGURE 5.17 Camber change during cornering.

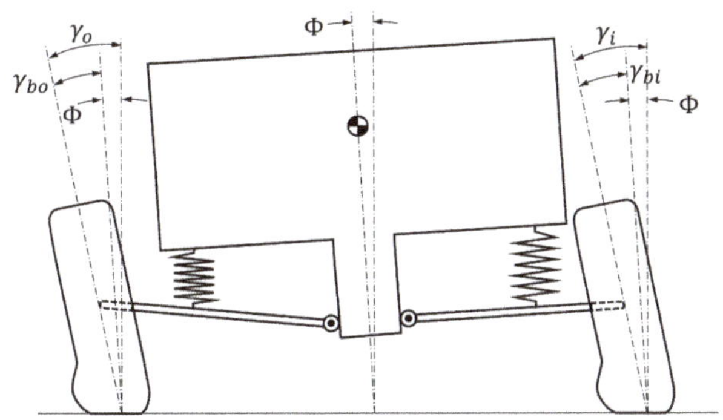

For a vehicle in cornering, the camber angle may lean toward the direction of lateral reaction force from the ground or its opposite direction, or remain unchanged. The variation of camber angle can result in changes in camber thrust and sideslip angle. According to Equation (5.4), the overall lateral reaction force acting on the tire from the ground is the sum of cornering force and camber thrust. Thus, the sideslip angle can be expressed as

$$\alpha = \frac{F_{y\alpha}}{C_\alpha} = \frac{F_y - F_{y\gamma}}{C_\alpha} = \frac{F_y}{C_\alpha} - \frac{C_\gamma}{C_\alpha}\gamma \tag{5.75}$$

According to Equation (5.75), when the camber on a wheel is in the same direction of lateral reaction force from the ground, the camber tends to reduce the sideslip angle; when the camber is in the opposite direction of lateral reaction force, in contrast, the camber tends to augment the sideslip angle. In addition, the increase in the camber angle deteriorates the lateral adhesion of a tire and, in turn, worsens the ultimate lateral acceleration of the vehicle. Thus, to maximize cornering performance, the suspension should be carefully designed to make camber angle appropriate so that the outside wheels are as perpendicular as possible to the ground in cornering as they take most of the vertical loads. Some types of suspensions, for example, single wishbone independent suspension, should be avoided since the camber may reverse its direction with respect to lateral reaction force when subjected to large lateral acceleration, resulting in augmented sideslip angle and worsened cornering performance.

The camber change produced by body roll can be described by roll camber, or camber gradient $\frac{\partial \gamma_b}{\partial \Phi}$, which is a function of wheelbase, roll angle, and suspension geometry. For a typical sedan, the front roll camber ranges from 0.61 to 0.88 deg/deg and the rear roll camber ranges from 0 to 0.85 deg/deg [5.6].

5.6.2. Roll Steer

Roll steer refers to the variation of steer angle produced by the front or rear wheels due to the rolling motion in cornering. For front axles, roll steer is the turning of front wheels about a kingpin produced by the kinematics of suspension linkage interacting with steering linkage for front axles, which is actually the suspension linkage interference in steering. For rear axles, roll steer is the turning of rear wheels about the vertical axis, which is simply a result of the kinematics of suspension linkage. Roll steer delays steer input as time is needed for sprung mass to roll.

To quantify the effect of the roll angle of the body on the roll steer angle, a roll steer coefficient ϵ can be defined as

$$\epsilon = \frac{\partial \delta}{\partial \Phi} \tag{5.76}$$

Clearly, a positive roll steer coefficient implies that the body roll causes wheels to steer to the same side as the body roll, opposite to the side that the vehicle is turning to. Thus, the positive roll steer on the front axle is actually understeer as it tends to steer the vehicle out of the turn. In contrast, the positive roll steer on the rear axle is oversteer.

For a typical sedan, the front roll steer varies from 0.2 deg/deg understeer to 0.1 deg/deg oversteer, and the rear roll steer varies from 0.13 deg/deg understeer to 0.06 deg/deg oversteer [5.6].

A nonindependent suspension can be kinematically equivalent to a trailing arm system. The inclination of the trailing arm is equal to that of the axle with respect to the longitudinal axis of the vehicle. For the solid rear axle shown in Figure 5.18, the initial inclination angle of the trailing arm is β, which is identical for both sides of the wheels. β is defined as positive when the orientation of the trailing arm is angled downward. For a vehicle in cornering, the trailing arm angle on the inside wheel turns downward, that is, $\beta_i > \beta$, while the inside wheel is pulled forward. Conversely, the trailing arm angle on the outside wheel turns upward, that is, $\beta_o < \beta$, while the outside wheel is pushed backward. As mentioned earlier, such a positive roll steer ($\beta > 0$) results in an oversteer effect on the solid rear axle. Apparently, negative roll steer ($\beta < 0$) produces an understeer effect, while no roll steer ($\beta = 0$) yields a neutral steer effect.

FIGURE 5.18 Nonindependent suspension equivalent to a trailing arm system. (a) Initial inclination angle and (b) inclination angle in cornering.

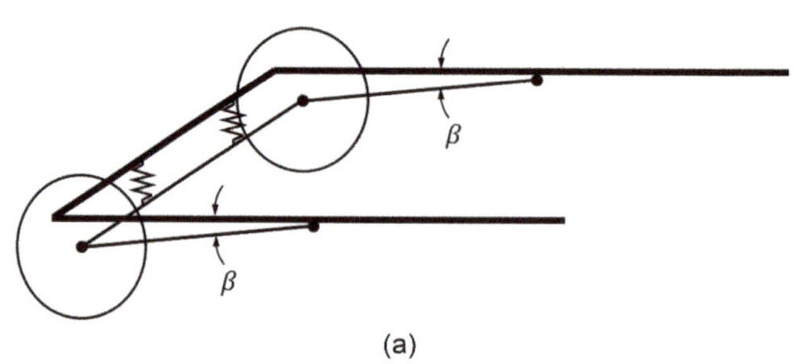

FIGURE 5.18 **(Continued)** Nonindependent suspension equivalent to a trailing arm system. (a) Initial inclination angle and (b) inclination angle in cornering.

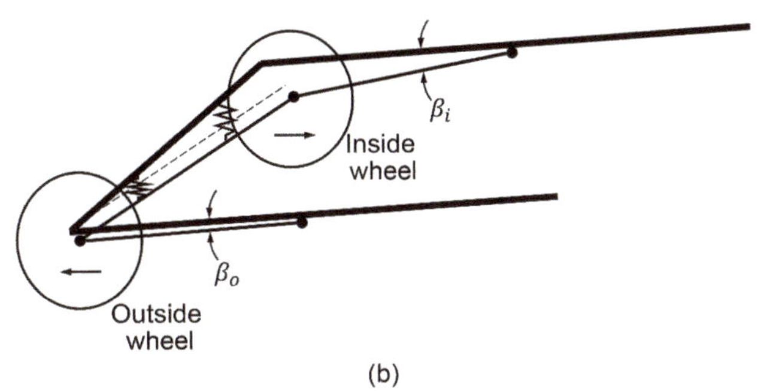

For independent suspensions, the roll moment causes the axle to turn about a vertical axis. From a kinematics perspective, the turning of either axles or wheels about the vertical axis is equivalent to the sideslip angle of tires in effect. The roll steer effect can be assessed by taking the kinematics of the suspension and its interaction with the steering mechanism into consideration.

5.6.3. Compliance Steer Related to Lateral Force

Compliance steer refers to the variation in the steer angle produced by the deflection of suspension linkage under the action of a lateral force in cornering. It is actually the rotation of the wheel about a yaw center or the kingpin. A compliance steer coefficient can be defined as the compliance steer angle per lateral force, that is, $\frac{\partial \delta}{\partial F_y}$. Most of the deflection comes from the rubber bushings used in suspension linkages for the purpose of NVH reduction.

For solid axles, understeer occurs if deflection causes the front yaw center to move forward or the rear yaw center to move backward, as illustrated in Figure 5.19a. On the contrary, oversteer occurs if deflection causes the front yaw center to move backward or the rear yaw center to move forward, as illustrated in Figure 5.19b. Thus, compliance steer is practically another effective means of making a vehicle of proper understeer. It is generally desired to produce an appropriate understeer angle on the outside wheels as they take most of the loads in cornering, with a compliance steer angle to reduce toe-in on front wheels and to augment toe-in on rear wheels.

For a typical sedan, the lateral force compliance steer ranges from 0.60 deg/kN understeer to 0.22 deg/kN oversteer for the front and from 0.02 deg/kN understeer to 0.22 deg/kN oversteer for the rear [5.7].

FIGURE 5.19 Compliance steer due to suspension deflection in cornering. (a) Deflection-resulted understeer and (b) deflection-resulted oversteer.

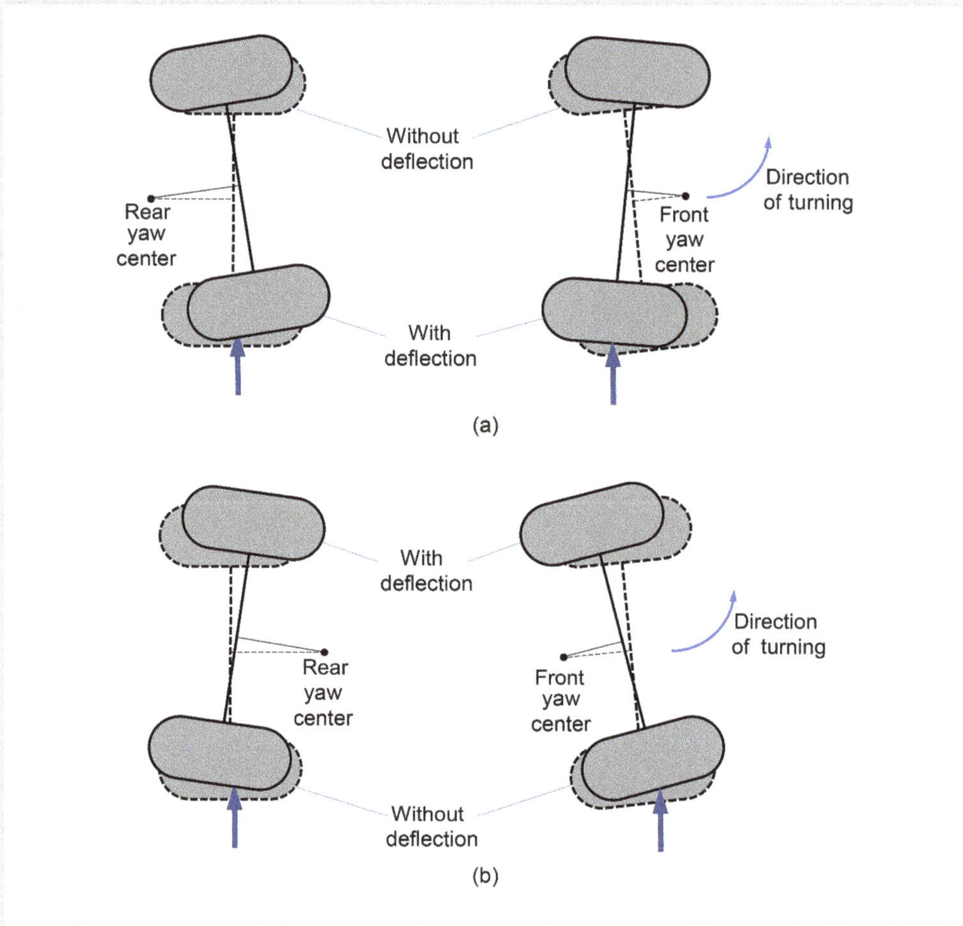

5.6.4. Compliance Steer Related to Aligning Torque

As discussed in Section 5.1.4, aligning torque is a fundamental characteristic of all tires. Aligning torque causes deflections in the suspensions and wheels, producing additional steer angles at the front and rear wheels. This compliance steer angle due to aligning torque can be quantified by the coefficient $\frac{\partial \delta}{\partial T_z}$, which represents the compliance steer angle per unit aligning torque. Given the aligning torque T_z, the compliance steer angle due to aligning torque, δ_a, is given by

$$\delta_a = T_z \frac{\partial \delta}{\partial T_z} \tag{5.77}$$

Due to the strong nonlinearity in the compliance steer angle associated with aligning torque, the coefficient $\frac{\partial \delta}{\partial T_z}$ for a tire oriented in the longitudinal direction can be dramatically greater than that at the angle to the longitudinal direction. For a typical sedan, the off-center aligning compliance steer ranges from 0.004 to 0.016 deg/(N·m) understeer for the front and from 0 to 0.0025 deg/(N·m) oversteer for the rear [5.7].

The direct influence of the aligning torque-associated compliance steer angle on the handling of a vehicle can be evaluated by the understeer gradient:

$$K_a = mg\frac{e}{L}\left(\frac{1}{C_{\alpha 1}} + \frac{1}{C_{\alpha 2}}\right) \tag{5.78}$$

Obviously, the aligning torque effect on handling is always positive since both $C_{\alpha 1}$ and $C_{\alpha 2}$ are positive. In other words, the aligning torque promotes understeer and stabilizes the vehicle at all times.

5.6.5. Compliance Camber Change

The deflection in independent suspension linkage under the action of a lateral force can result in the camber change of wheels, thus affecting the stability and instantaneous response of the vehicle. Such an influence can be quantified by the coefficient $\frac{\partial \gamma}{\partial F_y}$, which represents the compliance camber angle per unit lateral force.

For a typical sedan, the lateral force compliance camber of suspension ranges from 0.24 to 0.75 deg/kN understeer for the front and from 0.20 to 0.82 deg/kN oversteer for the rear [5.7].

5.7. Steering System Effect on Handling

The steering system imposes more influence on vehicle handling than any other system in the chassis.

5.7.1. Functionality of the Steering System

Traditionally, the function of the steering system is two-fold, namely, providing directional control of the vehicle and road feel to the driver. With the advent of autonomous driving vehicles, the necessity for the steering wheel and road feeling, however, vanishes.

The direction control of the steering system is made through the manipulation of the steerable wheel angle about the kingpin, either manually by the driver operating the steering wheel in traditional vehicles or automatically by electromechanical actuators in self-driving vehicles.

The state of vehicle motion is fed back to the driver via the steering system in the form of tactile, sight, hearing, and bodily perceptions, which is generally called road feeling. A good road feeling is essential in the evaluation of vehicle handling and stability for traditional vehicles.

As an important part of road feeling, the reaction force from the steering wheel to the hands of the driver plays a critical role in vehicle handling. The steering wheel force is often characterized in these typical operating conditions: during strong lateral acceleration, for example, 0.8*g*; in a circular motion at a fixed steering wheel angle; and during weak lateral acceleration with the steering wheel positioned at little angle off-center. Other than the steering wheel force, the aforementioned understeer gradient K and yaw rate frequency response are also of significant influence on vehicle handling.

Besides the steering ratio, friction, and efficiency of the steering system, the steering wheel force is also dependent on factors such as kingpin position, load on the tire, tire pressure, mechanical properties of the tire, ground adhesion, rotating inertia of steerable wheels, and overall vehicle dynamic characteristics. Among them, the geometric parameters related to the kingpin position have a remarkable influence on the steering wheel force, aligning torque, and straight-line driving performance, as illustrated in Figure 5.20 [5.9].

FIGURE 5.20 Kingpin position-related geometric parameters.

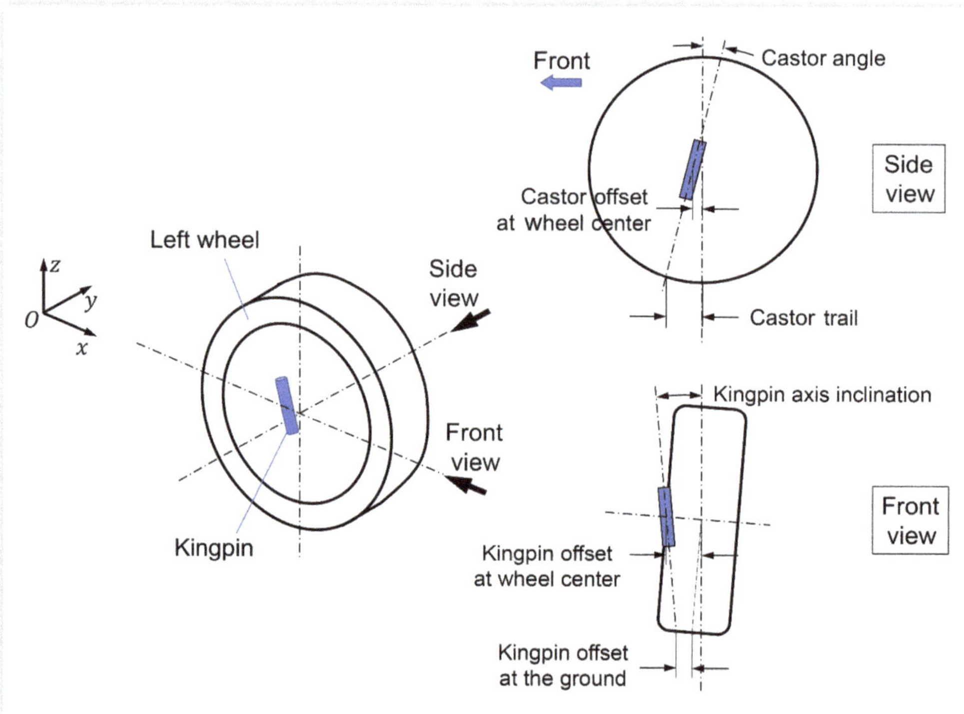

The requirements for steering system characteristics vary with driving conditions. When driving at low speeds with weak lateral acceleration, emphasis is put on the force, overall turning angle, and self-aligning performance of the steering wheel. In contrast, when driving at high speeds with weak lateral acceleration and little steering wheel turning angle, it is important to have proper yaw rate frequency characteristics, the ability to keep moving straight ahead, and self-aligning performance, along with good steering wheel force characteristics. Since large steering sensitivity is desirable at high speeds, the steering wheel-to-road wheel ratio must be kept moderate, which would cause difficulties in turning the steering wheel at low speeds. Although this problem can be remedied by using a power steering system, it is generally recommended to maintain the steering force above a certain level as the vehicle speed increases and to keep a proper rate of steering force change with the lateral acceleration. As far as the relationship between the steering wheel force and lateral acceleration is concerned, a linear proportion is considered the best, while a slight drop in the steering wheel force with increased lateral acceleration at high lateral acceleration is considered acceptable. The characteristics of a typical power steering system are illustrated in Figure 5.21. The system produces a relatively large assisting torque at low vehicle speeds, while the assisting torque diminishes as vehicle speed increases, preventing the steering wheel from becoming too light at high speeds.

FIGURE 5.21 Characteristics of a typical power steering system.

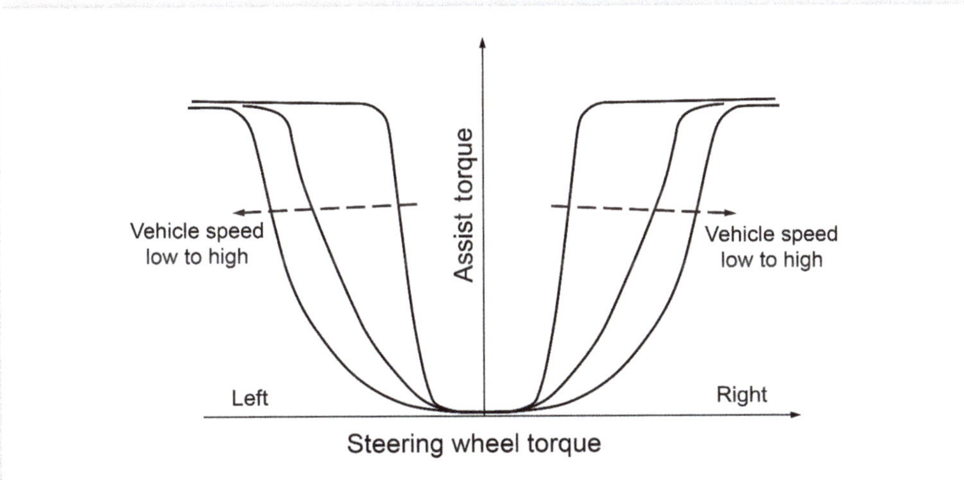

In the operation of the steering wheel, a certain amount of force is required to overcome the resistance produced by self-aligning torque and system damping. Meanwhile, the driver receives feedback about the tire–road interaction from the steering wheel. Thus, the criteria shown in Figure 5.22 are of particular importance [5.4].

FIGURE 5.22 Steering wheel force-related criteria for high-speed handling.

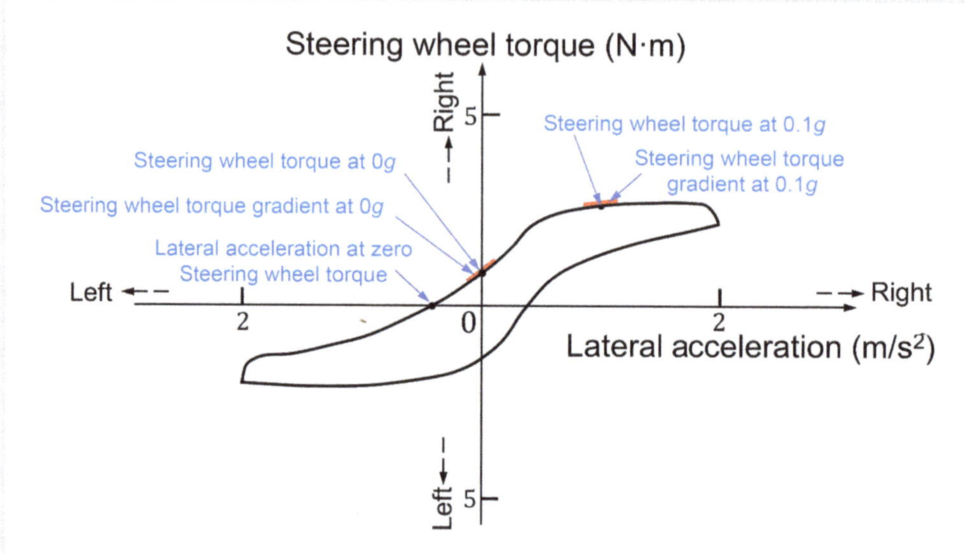

5.7.2. Fundamental Handling Characteristics

The most fundamental handling behavior of a vehicle is characterized by the stability of the vehicle in cornering, depending on whether the vehicle is understeer or oversteer. A steady-state circular test is often carried out to determine such behavior, in which the vehicle is first driven at the lowest stable speed along a circular path of a certain radius, for example, 15 m, and then accelerated slowly and evenly until the lateral acceleration reaches a certain value, for example, 6.5*g* or the maximum limited by vehicle instability or the output power of the electric drive system. Parameters such as the steering angle, turning radius, roll angle of the vehicle body, lateral acceleration, yaw rate, and longitudinal speed of the vehicle are acquired in the test and then processed to obtain the initial turning radius, lateral acceleration at zero steering angle, and the degree of understeer and roll angle at 0.2*g* lateral acceleration, of which the last three parameters are often quoted for the assessment of fundamental handling performance of the vehicle.

5.7.3. Steering Effort Characteristics

Steering effort is an important aspect of vehicle handling. A proper steering effort is necessary to ensure a clear road feeling in making a turn with the steering wheel. Neither an excessively heavy nor light steering effort is acceptable to the driver. To evaluate steering effort, tests are often performed by a driver steering the vehicle at a certain speed, for example, 10 m/s, along an 8-shaped path. Parameters such as the steering wheel torque, lateral acceleration, yaw rate, maximum steering torque, and longitudinal speed of vehicle are acquired in the test. The test data are then processed to find the maximum steering torque, maximum steering force, average steering force, steering work, average friction torque of the steering wheel, average friction force of the steering wheel, and so on, among which the maximum steering force and the average steering force are often used to quantify the steering effort.

In particular, the steering wheel torque at 0*g* in Figure 5.22 is a crucial parameter that indicates the variation in effort needed to turn the steering wheel off the center at various lateral accelerations. Its value is dependent on the Coulomb friction and viscous resistance within the steering system, as well as vehicle phase lag.

Additionally, the steering wheel torque at 0.1*g* in Figure 5.22 indicates the steering wheel effort needed for the vehicle just deviating from straight-line driving conditions. The steering wheel torque for a vehicle equipped with a power steering system is much smaller than that of the vehicle without such a system.

Finally, the steering wheel torque gradient at 0.1*g* in Figure 5.22 gives an indication of the road feeling for the vehicle just deviating from a straight-line driving condition, which is the rate of change in the steering wheel torque with respect to the lateral acceleration of the vehicle at 0.1*g* lateral acceleration. The steering wheel torque gradient for a vehicle equipped with a power steering system is much smaller than that of the vehicle without such a system.

5.7.4. On-Center Handling Characteristics

On-center handling characteristics are a critical indicator for evaluating the dynamic behavior of a road vehicle when the steering wheel is oriented near the center at both low and high speeds.

At low speeds, the so-called on-center handling tests are carried out while the vehicle is driven in a steady-state condition around a circular path of a certain radius, for example, 15 m, with the driver's hands off the steering wheel for a certain duration, for example, 4 sec, at a certain lateral acceleration, for example, 4*g*, on a level course. On the other hand, the high-speed on-center handling tests are conducted by turning the steering wheel to create a certain lateral acceleration, for example, 0.2*g* while the vehicle is driven straight at a certain speed, for example, 100 km/h or 70% of the nominal top speed. When the vehicle is stabilized, the steering wheel is then released for a certain duration, for example, 4 sec. With the torque and angle of the steering wheel, vehicle speed, and yaw rate acquired in the tests, the lateral acceleration is simply the product of the vehicle speed and yaw rate. Parameters such as time to stabilization, residual yaw rate, yaw rate overshoot, natural frequency of yaw mode, relative damping coefficient, and total variance of the yaw rate are acquired in these tests, among which the residual yaw rate and the total variance of yaw rate are often the most crucial ones for evaluating the on-center handling performance of the vehicle.

The evaluation of vehicle handling performance at high speeds involves linear, nonlinear, and on-center handling behaviors, as listed in Table 5.1 [5.10].

TABLE 5.1 Indices for high-speed handling evaluation.

Type of handling behavior	Indices	Unit
Linear	Understeer gradient	deg/g
	Steering sensitivity	g/100°
	Linear range of steering sensitivity	g
	Response time to lateral acceleration	sec
	Roll gradient	deg/g
	Roll damping	—
	Yaw rate damping	—
	Pitch gradient—acceleration and braking	deg/g
Nonlinear	Maximum lateral acceleration	g
	Stability in releasing throttle	g
	Road holding in single protrusion	deg
On-center handling	Minimum steering sensitivity	g/100°
	Ratio of steering sensitivity	—
	Steering lag	deg
	Lateral acceleration at zero steering torque	g
	Steering wheel torque gradient ratio	—
	Torque work sensitivity	g^2/100 N·m
	Static steering force	N
	Minimum turning diameter at outside of tire	m

Specifically, the lateral acceleration at zero steering wheel torque in Figure 5.22 is an indication of self-aligning performance, which is influenced by the system resistance and vehicle response lag as well. The best practice is to keep its value within certain limits as excessive effort is needed to turn the steering wheel back to the center or away from the center when its value is too large or too small.

Moreover, the steering wheel torque gradient at 0*g* in Figure 5.22 represents the road feeling, which is the rate of change in the steering wheel torque with respect to the lateral acceleration of the vehicle at zero lateral acceleration. Primarily affected by the geometric parameters of the kingpin and the overall gear ratio in the steering system, it is also influenced by the torsion bar stiffness, design of the revolving valve, and friction in the power steering system.

The typical values for aforementioned criteria are given in Table 5.2 [5.4].

TABLE 5.2 Typical values for steering-related handling criteria.

Vehicle size class	Large	Medium	Compact	Compact	Sports	Medium	Medium	Sports
Country of production	US	US	US	US	US	Other	Other	Other
Drive axle	Rear	Front	Front	Front	Rear	Rear	Front	Rear
Steering system	Power	Power	Power	Manual	Power	Power	Power	Manual
Lateral acceleration at zero steering wheel torque (*g*)	−0.110	−0.055	−0.063	−0.038	−0.054	−0.071	−0.045	−0.109
Steering wheel torque at 0*g* (N·m)	1.29	0.81	0.78	0.96	0.95	1.57	0.90	1.90
Steering wheel torque gradient at 0*g* (N·m/g)	8.8	16.5	13.0	27.5	18.5	25.1	20.6	20.6
Steering wheel torque at 0.1*g* (N·m)	1.90	2.19	1.83	3.94	2.27	3.46	2.43	4.09
Steering wheel torque gradient at 0.1*g* (N·m/g)	4.2	4.7	6.6	15.9	8.4	12.4	7.9	20.3

5.7.5. Movement Interference between the Steering System and Suspension during Body Roll

Poor kinematic coordination between the steering system and nonindependent suspension can result in motion interference as the vehicle body rolls, causing the front wheels to turn and impacting the handling in straight-line driving conditions. It is, therefore, desired to keep the interference as small as possible.

5.7.6. Compliance Steer Related to Steering System Stiffness

The theoretical steer angle at front wheels is equal to the ratio of the steering wheel angle and the overall gear ratio of the steering system. However, elastic deflection in the steering system occurs when the self-aligning torque acts on the steer wheels from the ground. This compliance steer angle is equal to the ratio of self-aligning torque and the angular stiffness of the steering system, involving components such as the steering gear, linkage, and mounting surface. Thus, the total steer angle is actually the difference in the theoretical steer angle and the compliance steer angle, if the friction between the steering system and the front suspension is ignored. For a given steering wheel angle, the compliance steer angle tends to increase understeer if steering system stiffness is low or decreases understeer if steering system stiffness is high.

Since the steering system actually produces a much larger compliance steer angle than suspension, excessive understeer can occur if the stiffness of the steering system is not sufficient. From the perspective of handling stability and good road feeling at high speeds, it is a common practice to make the stiffness of the steering system adequately high, particularly around the on-center position.

5.8. Drivetrain System Effect on Handling

The drivetrain system has a strong influence on the tangential reaction force from the ground, which consequently impacts the cornering characteristics and handling of a vehicle.

5.8.1. Mechanisms of the Tractive Force Effect on Cornering Characteristics

Since the tangential reaction force from the ground is produced by the tractive force, it affects the understeer gradient due to a few different mechanisms. First, the demand for an augmented tractive force in acceleration during cornering results in a reduction in the cornering force, as per Figure 5.7. Hence, to meet the requirement of the cornering force, the sideslip angle at the driving axle will have to grow inevitably, which tends to increase understeer for FWD vehicles. Second, load shifts from front to rear wheels during strong acceleration in a curved path cause the vertical load on the front axle to decrease and the vertical load on the rear axle to increase. This leads to a rise in the understeer gradient, thus increasing the understeer of the vehicle. Third, the rise in the self-aligning torque with the increase in the tractive force also tends to increase understeer. Fourth, the compliance steer angle produced by the driving torque at the front axle through half shafts tends to increase understeer. Assuming the rolling resistance and the torque produced by the normal reaction force are negligible, the driving torque about the kingpin axis for the front wheel is

$$T_{kp} = F_x r_s \cos\psi \cos\chi + T_d \sin(\psi + \zeta) = F_x \left[r_s \cos\psi \cos\chi + R\sin(\psi + \zeta) \right] \tag{5.79}$$

where

F_x is the tractive force
r_s is the kingpin offset at the ground (i.e., scrub radius)
ψ is the kingpin axis inclination
χ is the castor angle
T_d is the driving torque at the front axle
ζ is the angle between the half shaft and the y-axis
R is the tire radius, as illustrated in Figure 5.23

FIGURE 5.23 The relationship between driving torque and tractive force.

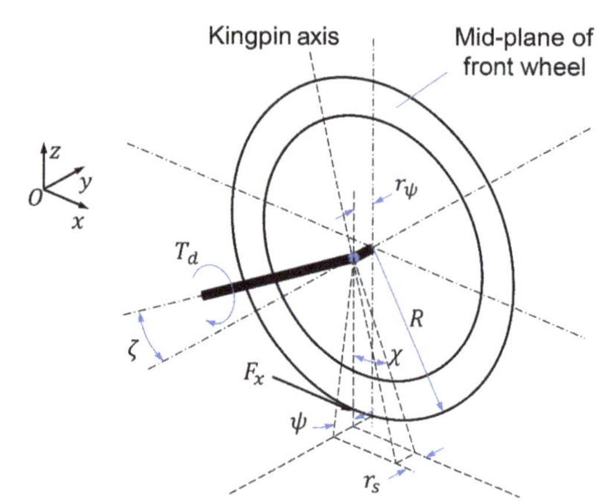

Since both ψ and χ are small, we have cos $\psi \approx$ cos $\chi \approx 1$, and Equation (5.79) can be simplified to

$$T_{kp} = F_x \left[r_s + R\sin(\psi + \zeta) \right] \tag{5.80}$$

In cornering, the body roll causes the angle ζ to increase at the inside wheel and to decrease at the outside wheel. It is obvious from Equation (5.80) that the T_{kp} changes in the same way correspondingly, producing a torque on the steering linkage. Thus, the deflection of the steering linkage results in a compliance steer angle that gives an understeer tendency at front wheels.

It is worth noting that, however, the effects of the three mechanisms increase oversteer in the process of regenerative braking. Substantial changes in steering characteristics of high-performance FWD vehicles can occur if the accelerator pedal is released suddenly in a strong acceleration. Consequently, the vehicle in cornering could abruptly turn further into the direction of the steering angle, which is a phenomenon called tuck-in. Not only with FWD vehicles, tuck-in also occurs with RWD vehicles due to the extra oversteer produced by the enlarged sideslip angle at the rear axle subjected to a braking force. The use of automatic transmissions, limited slip differentials, or suspensions capable of increasing understeer by compliance steer in braking can reduce or eliminate the influence of tuck-in.

5.8.2. Cornering Stability Control by Tractive and Braking Forces

The effect of tractive force on cornering behavior can be readily revealed by the variation in the yaw rate with longitudinal acceleration on the ground of low adhesion. Some studies have shown that with the FWD vehicle being understeer and the RWD vehicle oversteer, the AWD vehicle with equal torque split to front and rear axles can be understeer with little variation in the yaw rate. It is implied that the cornering characteristics can be altered by adjusting the torque split proportion to front and rear axles.

The effect of the braking torque split proportion to front and rear axles on the cornering characteristics of tires is quite similar to that of driving torque. In fact, handling and stability of a vehicle are easier to control by means of a braking force, rather than a tractive force, due to the convenience of adjusting the braking force applied at every wheel, as in the case of ABSs.

The control of a tangential reaction force generally falls into four categories, as elaborated now.

5.8.2.1. Stability Control through Tractive Force

The traction control system (TCS) is an example of directional stability control by adjusting the tractive force at each of the driving wheels. This method generally involves two aspects of control: first, applying braking torque to the wheel with less traction and redistributing the tractive force to the wheel with more traction, by making use of either mechanically or electronically controlled differentials, and second, limiting the total tractive force to the driving wheels by lowering the output torque of electric drive motors.

5.8.2.2. Stability Control through Braking Force

The ABS represents a typical directional stability control method that adjusts the braking force to the wheels to prevent them from locking up. This is implemented in two ways: reducing the braking force to the wheel that is about to lock up and then increasing the braking force to the wheel that is turning faster than others, which is an operation often repeated alternately in a continuous manner, and regulating the total braking force applied at all wheels to ensure a good slip ratio by lowering the hydraulic pressure in the brake system.

5.8.2.3. Stability Control through Front and Rear Tractive Force Proportion

The cornering characteristics of AWD vehicles can be altered by adjusting the tractive force proportion to the front and rear axles. As an implementation of this control methodology, the electronic traction system (ETS) makes the driving path more predictable by better keeping the vehicle close to a neutral steer. In principle, the ETS controls the yaw moment indirectly.

5.8.2.4. Stability Control through Inner and Outer Tractive Force Proportion

The directional stability of AWD vehicles can also be improved by controlling the tractive force proportion to the inner and outer wheels in cornering. For an ordinary vehicle equipped with mechanical differentials, the tractive forces acting on the inner and outer wheels from the same differential are always equal, with the combined force always in the longitudinal direction. As a more-advanced system, the direct yaw moment control system (DYCS) alters cornering characteristics under steady-state and extreme conditions by applying a yaw moment directly to the vehicle. The mechanism of such a control method is explained in the following text with a simplified two-dimensional model, as shown in Figure 5.24.

FIGURE 5.24 Mechanism of yaw moment control.

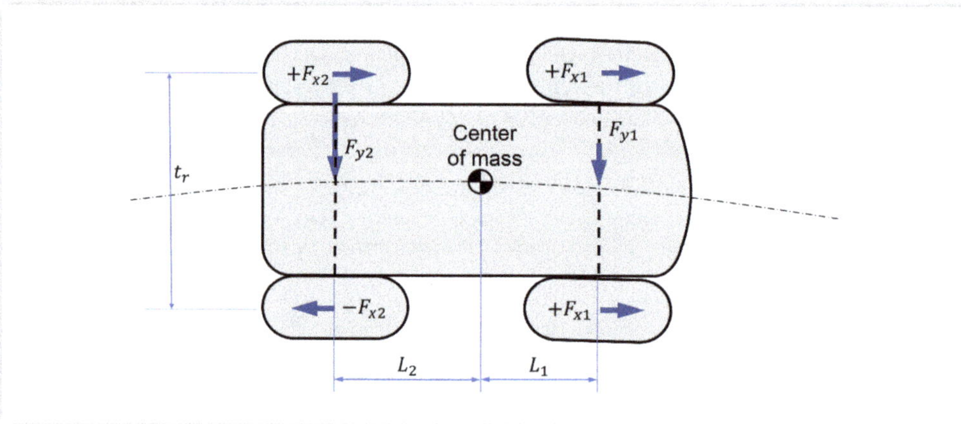

The directional stability of the vehicle in cornering is governed by the following equations of motion:

$$F_{y1} + F_{y2} = ma_y \tag{5.81}$$

$$F_{y1}L_1 - F_{y2}L_2 + F_{x2}t_r = I_z\dot{\omega}_r \tag{5.82}$$

where

F_{y1} and F_{y2} are the lateral forces at front and rear wheels, respectively
m is the vehicle mass
a_y is the lateral acceleration of the vehicle at the center of mass
L_1 and L_2 are the distances of front and rear axles to the center of mass, respectively
F_{x1} is the tractive force
F_{x2} is the additional tractive force to produce yaw moment by DYCS
t_r is the wheel track
I_z is the rotating inertia of the vehicle about the z-axis
$\dot{\omega}_r$ is the yaw acceleration

It is obvious from Equation (5.82) that for a given yaw acceleration, the additional yaw moment $F_{x2}t_r$ provided by DYCS helps reduce F_{y1} and increase F_{y2}. Consequently, the sideslip angle α_1 at the front wheel is smaller and the sideslip angle α_2 at the rear wheel is larger, producing a negative effect on understeer. Conversely, a yaw moment in the direction opposite to what is shown in Figure 5.24 would have a positive effect on understeer. For instance, in the case that the ground reaction forces at front wheels approach the adhesion, further acceleration can cause the vehicle in cornering to skid sideways and fail to follow the curved path. The yaw moment from DYCS, however, can help reduce F_{y1} and increase F_{y2}, thus preventing skid from occurring at front wheels and taking advantages of redundant adhesion at rear wheels. As a result, the acceleration performance and directional stability of the vehicle during cornering can be improved.

5.8.3. Electronic Control Systems for Handling Stability

The ABS, TCS, and electronic stability control (ESC) are electronic systems most frequently employed in EVs for handling stability control. The ABS improves the stability of a vehicle in the braking process by modulating the brake force, while the TCS prevents the vehicle from loss of traction in the acceleration process by controlling the brake force on the driven wheels. On the other hand, the ESC deals with the steering stability of a vehicle in cornering by applying drift-out and yaw moment control through the brakes and electric traction motors.

5.8.3.1. ABS

The ABS has been widely applied in automobiles since it was first put into mass production in the early 1970s.

A typical ABS consists of the brake master cylinder, a wheel speed sensor, a hydraulic modulator unit, and an electronic control unit (ECU), in addition to the components of a brake system such as the brake pedal, master cylinder, and wheel cylinder. During braking, the wheel speed is constantly monitored by the wheel speed sensor and ECU. If the wheel speed is found diminishing, then the wheel lockup is deemed imminent, and therefore, the ECU will control the solenoid to release the brake pressure, allowing the wheel speed to pick up. Once the wheel regains speed, the ECU will command the solenoid to increase the brake pressure again. On a road surface of low adhesion, this process often takes place repeatedly, resulting in cycling changes in wheel speed until the vehicle comes to a full stop.

In general, the ABS operates in terms of adhesion coefficient. When the slip ratio is small, the ABS does not operate. With a continued increase in the slip ratio, the ABS intervenes at the point where the slip ratio becomes excessive by releasing the brake pressure. As a result, the slip ratio starts to decline. Once the slip ratio drops back to an acceptable range, the ABS reapplies the brake pressure to stop the vehicle. When such a cyclic process is executed, the slip ratio can be maintained near the peak of the adhesion coefficient curve to maximize the braking efficiency.

5.8.3.2. TCS

The TCS comes into general use a little later than the ABS.

The TCS consists of essentially the same components as the ABS and is, therefore, often provided as a supplement to the ABS.

The TCS operates by monitoring the speed of each wheel continuously during vehicle acceleration. When one or more driven wheels are detected spinning significantly faster than the others (an indication of loss of traction), the ECU invokes the hydraulic valves to apply brake pressure to the wheels with lessened traction. As a result, more torque will be transferred to the wheel or axle of greater traction due to the mechanical action within the differential. This process often takes place in conjunction with the ECU, reducing the torque from the electric traction motor.

In addition to improving the stability of the vehicle in acceleration under slippery conditions, the TCS can also help cornering safely. When the driven wheels lose traction and skid sideways, understeer in FWD vehicles and oversteer in RWD vehicles occurs. The TCS can mitigate understeer or oversteer by limiting torque to the overdriven wheels.

5.8.3.3. ESC

The application of ESC in automobiles started in the 1990s. It is also referred to as vehicle dynamic control (VDC), electronic stability program (ESP), or dynamic stability control (DSC).

ESC consists of a steering angle sensor, a wheel speed sensor, a lateral acceleration sensor, a yaw rate sensor, and an ECU, whereas the ECU is often shared by different systems such as ABS or TCS. In more-sophisticated versions of ESC, a longitudinal acceleration sensor and a roll rate sensor may also be included to provide more-accurate data about the vehicle motion. A high-speed network, for example, CAN or Local Interconnect Network (LIN), is required for the ESC to communicate with controllers for different systems on the vehicle, such as AWD system or active suspension system.

The ESC constantly monitors the steering angle and vehicle direction in normal driving. Once a mismatch between the actual direction and the driver's intended direction is detected, the ESC intervenes by applying brakes to individual wheels asymmetrically in order to create a moment about the vertical axis. This moment reacts against the skid and brings the vehicle back in line with the driver's commanded direction, preventing the vehicle from probable loss of steering control. Furthermore, the ESC may reduce the traction motor torque to slow the vehicle down.

The ESC differs from the ABS or TCS mainly in that it incorporates yaw rate control. To ensure each wheel receives the correct brake force, the ESC incorporates a hydraulic modulator similar to that used in ABS or TCS. Besides a hydraulic pump, additional pressure boosters, for example, active vacuum brake boosters, precharge pumps, or pressure generator assemblies, are often equipped to meet the pressure gradient demands when a sharp increase in brake pressure becomes necessary.

For a vehicle equipped with both ESC and ABS, the ESC logic typically has a higher priority than that of ABS since ESC allows up to 50% slip ratio to stabilize the vehicle in comparison to up to 20% in the ABS.

For a vehicle equipped with both ESC and TCS, the TCS logic is usually assigned with higher priority as the higher brake pressure from the ESC pressure booster would damage the hydraulic system of TCS.

5.8.3.4. Front Axle Skid and Its Control by Yaw Moment and Brake Force

Front axle skid typically occurs when a ramp steering input is applied at front wheels. As a result, the front axle may fail to follow the curved path and skid sideways. In an investigation of the vehicle response to various ramp steering inputs at the initial speed of 70 km/h, the sideslip angle of the front wheel is found to increase with the rise in the steering angle and then levels off at a certain point when the lateral force stops growing. Meanwhile, the cornering radius cannot be reduced further, imposing an upper limit on the cornering ability of the vehicle. On the contrary, the sideslip angle at the rear wheel is smaller than that corresponding to the maximum cornering force, allowing the utilization of greater sideslip angles to produce a larger cornering force at the rear wheel [5.3]. A typical relationship between the sideslip angle at the front wheel and the cornering radius at various initial vehicle speeds for a ramp steering input indicates that the cornering radius declines with a front sideslip angle only when the latter is smaller than a certain threshold, that is,

$$\alpha_1 < \alpha_{1th} \tag{5.83}$$

In addition, the initial speed does not have much influence on the threshold sideslip angle α_{1th}.

To reduce the cornering radius, both external yaw moment and external braking force are effective. However, the vehicle may become unstable when an excessively large yaw moment is applied. In contrast, the application of an external braking force reduces the cornering radius without affecting the stability of the vehicle, and its effect increases with time as it causes the vehicle to decelerate.

5.8.3.5. Rear Axle Skid and Its Control by Yaw Moment and Brake Force

A rear axle skid can take place when a sinusoidal steering input is applied with a magnitude exceeding a certain threshold. As a result, the sideslip angles at the centroid, front, and rear wheels increase drastically, causing the vehicle to spin out and lose its stability. As illustrated in Figure 5.25, the stability criterion can be approximated as [5.11]

$$\left|C_1\beta + C_2\dot{\beta}\right| < 1 \tag{5.84}$$

where

β is the sideslip angle at the centroid
$\dot{\beta}$ is the sideslip angular velocity at the centroid
C_1 and C_2 are constants

FIGURE 5.25 Stability criterion for a vehicle subject to sinusoidal steering angle input.

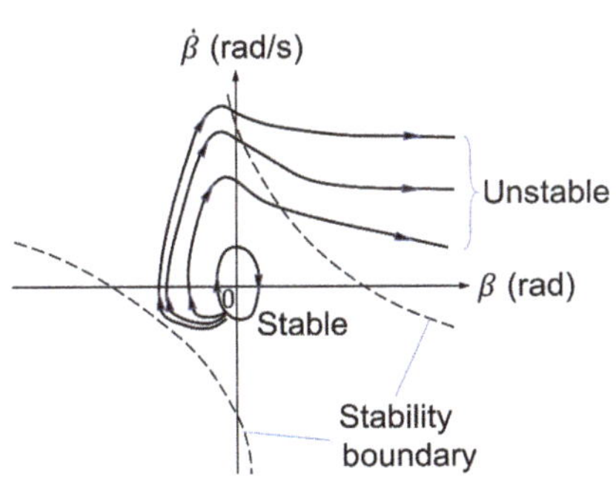

For a vehicle subject to a sinusoidal steering input, exerting an external yaw moment can largely reduce the maximum sideslip angle at the centroid, while exerting an external braking force has no such effect since the stability gain due to deceleration is canceled out by the inward yaw moment produced by a vertical load shift from the rear to the front during braking [5.3].

5.8.3.6. Control Strategies in Braking Force Application

The braking force can be applied either on individual wheels or on multiple wheels. Figure 5.26 shows the yaw moment produced by the braking force on each wheel [5.4], which is dependent on factors such as the magnitude of brake force, vertical load, the relationship between longitudinal and lateral forces defined by the friction ellipse, and the position of the wheel relative to the vehicle centroid.

FIGURE 5.26 External yaw moment produced by an external braking force exerting on each wheel.

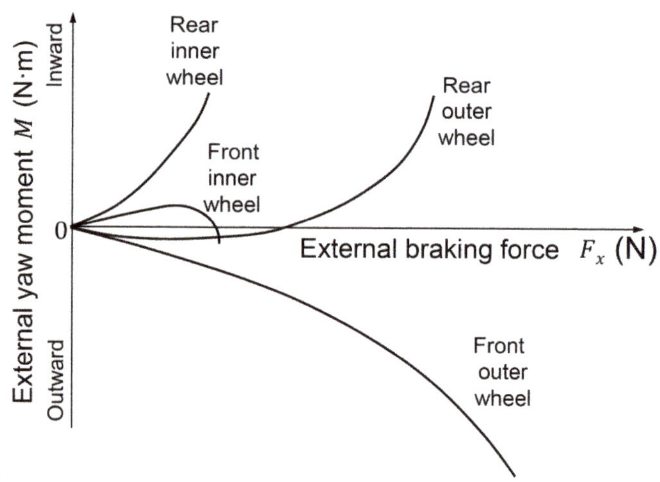

Reprinted with permission from Yu, Z.-S. (2009) Automotive Theory. China Machine Press, Beijing.

To prevent the skid from taking place at the rear axle, it is necessary to apply the braking force at the front outer wheel since the fact that the yaw moment is monotonically proportional to the braking force makes it easy to control.

To meet the desired stability control target in the nominal slip ratio, the braking force applied at the front outer wheel is recommended to follow the control strategy:

$$\beta + K\dot{\beta} = 0.1 + s/3 \quad (5.85)$$

On the other hand, the braking force applied at either one of inner wheels or the rear outer wheel seems to produce the inward yaw moment needed for correcting the sideslip at the front axle. However, it is difficult to implement the control by applying the braking force on only one wheel since the yaw moment either increases sharply or shifts its direction with an increase in the braking force. Thus, it is more effective to control the sideslip at the front axle through the application of appropriate braking forces at four wheels, instead of individual wheel.

5.9. Rollover

Rollover is defined as any maneuver in which a vehicle rotates 90° or more about its longitudinal axis so that its body makes contact with the ground, which is one of the most dangerous maneuvers that a vehicle can experience. One or more factors involving the defects in the vehicle structure, the operation of the driver, and the road condition can result in a rollover. There are, in general, two types of rollover in terms of its causes: maneuver-induced rollover, essentially due to excessive lateral acceleration that causes the centroid to fall outside of outer wheels, and tripped rollover, which is often a result of uneven or soft road surface or other obstructions. The fundamental mechanics involved in the process of the maneuver-induced rollover is explained in the following quasistatic analysis of a rigid vehicle.

5.9.1. Quasistatic Rollover Analysis of a Rigid Vehicle

In this analysis, the vehicle is assumed to be in steady-state cornering with the deflections of suspensions and tires ignored. Figure 5.27 illustrates the rollover-related forces acting on the rigid vehicle in the roll plane. If the transverse slope angle of the road φ is small, then $\sin\varphi \approx \varphi$, $\cos\varphi \approx 1$, and the equation of moment equilibrium about the center of contact for the outside tires is

$$ma_y h_g + F_{zi} t_r - mg\varphi h_g - \frac{1}{2} mgt_r = 0 \tag{5.86}$$

where

- m is the vehicle mass
- a_y is the lateral acceleration
- h_g is the centroid height
- F_{zi} is the total vertical reaction force acting on the front and rear tires from the ground
- t_r is the wheel track

FIGURE 5.27 Rollover-related forces acting on the rigid vehicle in the roll plane.

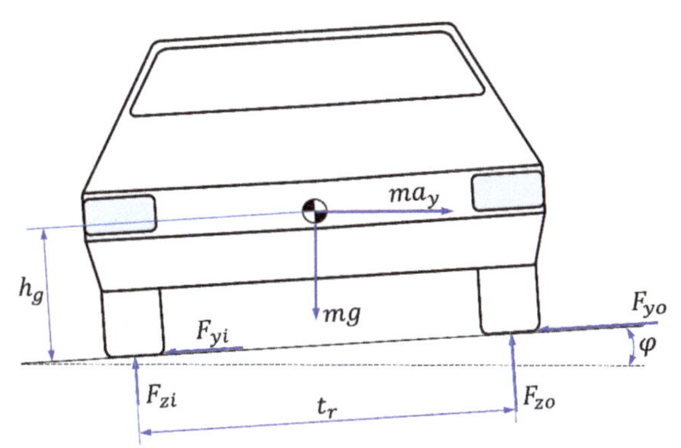

By rearranging Equation (5.86), we have

$$\frac{a_y}{g} = \varphi + \left(\frac{1}{2} - \frac{F_{zi}}{mg}\right)\frac{t_r}{h_g} \tag{5.87}$$

It is obvious from Equation (5.87) that the total vertical load on the inside tires $F_{zi} = mg/2$ if the cross-slope angle is selected as $\varphi = a_y/g$. This is exactly the purpose and method of cross-slope design in highways, where φ typically produces a lateral acceleration of 0–0.1g [5.10]. In addition, the total vertical load on the inside tires F_{zi} drops with the increase in lateral acceleration a_y. When F_{zi} reaches zero, the vehicle starts to roll over as it can no longer maintain equilibrium in the roll plane. The lateral acceleration at this point is referred to as the rollover threshold:

$$\frac{a_y}{g} = \varphi + \frac{t_r}{2h_g} \tag{5.88}$$

When the transverse slope φ is zero, the rollover threshold is $\frac{t_r}{2h_g}$, indicating that the resistance to rollover depends solely on the basic structural parameters t_r and h_g of the vehicle. It is worth noting that the actual threshold is smaller than the estimates obtained from Equation (5.88). The rollover threshold is quite different for various types of vehicles, as listed in Table 5.3 [5.4].

The adhesion coefficient of a good road surface can typically reach 0.8, implying the skid starts when lateral acceleration comes to 0.8g. It can be seen from the values in Table 5.3 that the medium and heavy trucks can spin out first and then roll over, while the cars and pickup trucks could start to skid sideways prior to rollover. However, the statistics of accidents indicate that it is not always the case for cars and pickup trucks, which calls for a more in-depth analysis of the rollover process.

TABLE 5.3 Typical rollover threshold values for various types of vehicles.

Type of vehicles	C.G. height of sprung mass (cm)	Wheel track (cm)	Rollover threshold (g)
Sports car	46-51	127-154	1.2-1.7
Compact car	51-58	127-154	1.1-1.5
Luxury car	51-61	154-165	1.2-1.6
Pickup truck	76-89	165-178	0.9-1.1
Passenger van	76-102	165-178	0.8-1.1
Medium truck	114-140	165-190	0.6-0.8
Heavy truck	154-216	178-183	0.4-0.6

5.9.2. Quasistatic Rollover Analysis of a Vehicle with Suspension

The physical model of a vehicle with suspension in steady-state cornering is shown in Figure 5.28 [5.3]. Taking the moment equilibrium about the center of contact for the outside tires yields

$$ma_y h_g + F_{zi} t_r + m_s g \Phi (h_g - h_r) - \frac{1}{2} m_s g t_r = 0 \tag{5.89}$$

where
- m_s is the sprung mass
- Φ is the roll angle of the vehicle
- h_r is the height of the roll center

FIGURE 5.28 Rollover-related forces acting on a suspended vehicle in the roll plane.

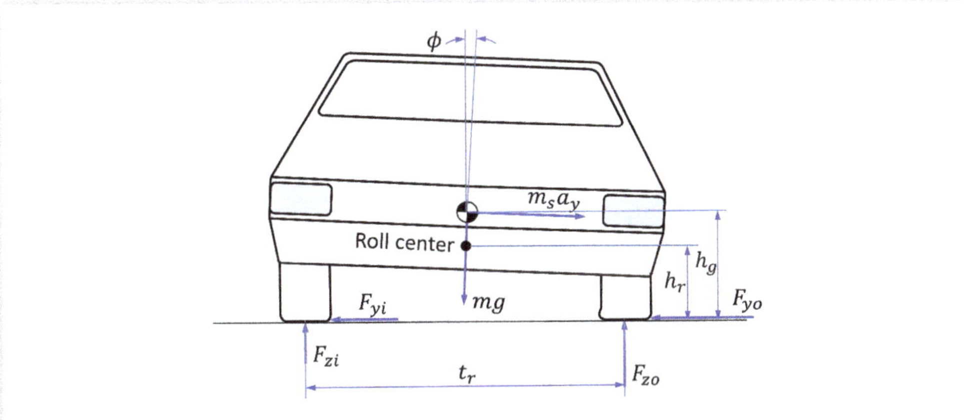

Since $\Phi = r_\Phi a_y$, the rollover threshold can be obtained by letting $F_{zi} = 0$ and rearranging Equation (5.89):

$$\frac{a_y}{g} = \frac{1}{\left[1 + \left(1 - h_r / h_g\right) r_\Phi\right]} \frac{t_r}{2h_g} \tag{5.90}$$

where r_Φ is the roll rate. Compared to the rigid vehicle, the deflection of suspension causes the roll center to fall below the centroid, thus $1 + \left(1 - \frac{h_r}{h_g}\right) r_\Phi > 1$. Besides, when the vehicle is subject to lateral forces, the deformation of outside tires makes the center of tire–road contact deviate toward the inside, causing the wheel track t_r to decline. It is apparent from Equation (5.90) that the rollover threshold becomes smaller than that of a rigid vehicle, that is, $\frac{t_r}{2h_g}$.

5.9.3. Transient Rollover Analysis

When the lateral acceleration varies quickly with time, the rollover behavior of a vehicle can be better represented with a transient rollover model. A simple model for transient rollover analysis in a roll plane can be created by adding a roll moment inertia I_s to the centroid of sprung mass, as shown in Figure 5.27. With the details of differential equations of motion given in the literature [5.9, 5.10, 5.11], their solution indicates that the vehicle response to the step input in lateral acceleration is similar to that of a damped single-DOF system. As illustrated in Figure 5.29 [5.3], the roll angle overshoots when it rises to the equilibrium level the first time and then oscillates before finally subsiding to the steady-state roll angle at the equilibrium position. Thus, the transient rollover threshold should be designed lower than that in the steady state as the inside wheels can be lifted off the ground at a lower lateral acceleration input in the transient state than that in the steady state. In fact, the rollover threshold under a step steering input is approximately 30% lower than $\frac{t_r}{2h_g}$ for cars and SUVs and about 50% lower for trucks.

FIGURE 5.29 Roll response to a step input in lateral acceleration.

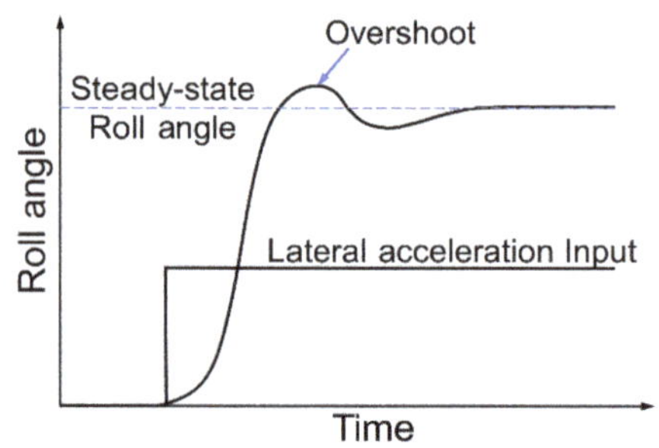

The extent of overshoot depends primarily on roll damping. The rollover threshold is the lowest without damping and rises with the increase in the damping ratio.

The vehicle response to a sinusoidal lateral acceleration input is dependent on the frequency of the input, as shown in Figure 5.30 [5.3]. The rollover thresholds at zero frequency approach the steady-state values predicted from the quasistatic vehicle model with suspension. With increasing frequency, the rollover thresholds decline until a minimum is reached near the roll resonant frequency and then rise quickly. For cars and SUVs, the roll resonant frequency is 1.5 Hz and higher. In such a frequency range, the amplitudes of steering input are generally small, producing a lateral deviation typically less than 0.25 m. Therefore, the chance of rollover due to roll resonance is slim for cars and SUVs. In contrast, the roll resonant frequency is lower than 1 Hz for heavy trucks. A typical lane-changing type of maneuver typically falls right into this frequency range, normally producing a lateral deviation of over 2 m to avoid obstacles or other vehicles on the road. As a result, roll resonance can be excited, making heavy trucks prone to rollover accidents.

FIGURE 5.30 Effect of sinusoidal steering input frequency on the rollover threshold.

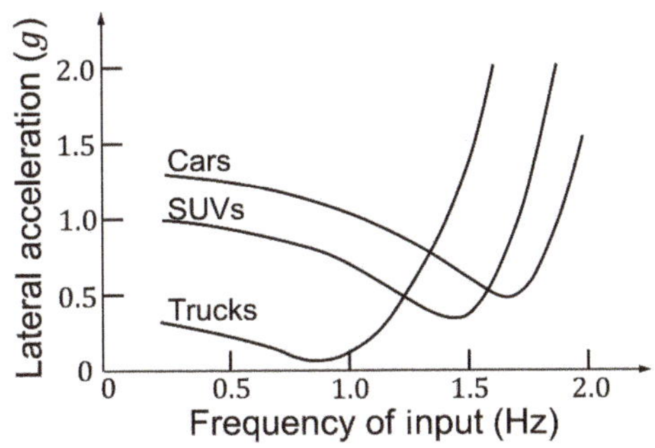

5.10. Vehicle Handling Tests

Vehicle handling tests are commonly carried out on specific test sites, before which the axle load proportion, inflation pressure, and tire tread condition should be examined.

The parameters measured in handling tests and the corresponding instruments are listed in Table 5.4.

TABLE 5.4 Parameters measured in handling tests and corresponding instruments.

Parameters	Instruments
Vehicle speed	Noncontact speedometer, fifth wheel
Time	Timing signal generator
Steering torque and angle	Measurement steering wheel with multiturn potentiometers
Lateral acceleration	Accelerometer
Yaw rate	Two-DOF angular velocity gyroscope
Yaw angle	Three-DOF directional gyroscope
Roll angle	Three-DOF vertical gyroscope

5.10.1. Low-Speed Steering Effort Tests

In the test, the vehicle is driven at a speed of 10 km/h by following a lemniscate curve defined by

$$r = d\sqrt{\cos 2\theta} \tag{5.91}$$

The minimum radius of the curvature, $R_{min} = d/3$, occurs at $\theta = 0$ and $\theta = \pi$. In setting up the lemniscate curve, R_{min} should be determined by taking 1.1 times the minimum turning radius of the test vehicle and rounding to the nearest integer.

With the steering torque and angle data recorded in the test, a torque angle curve can be plotted. The steering effort is generally evaluated by the maximum torque, force, or work at the steering wheel.

5.10.2. Steady-State Steering Characteristics Tests

These tests are conducted to measure the steady-state yaw response to a steering input when the vehicle reaches a stable operating condition.

The tests are carried out on a level field by following a circle on the ground with a diameter of 15 or 20 m. The vehicle speed and yaw rate v_0 and ω_{r0} are measured when the vehicle is driven at the lowest stable speed along the circle with a steering wheel angle of δ_{s0}. Since both the sideslip angles of tires and the centrifugal force are negligible at such a low speed, the turning radius, therefore, becomes

$$R_0 = \frac{v_0}{\omega_{r0}} \tag{5.92}$$

While keeping the steering wheel angle δ_{s0} constant, the vehicle is slowly accelerated until the lateral acceleration reaches 6.5 m/s^2, with a longitudinal acceleration kept below 0.25 m/s^2. With the vehicle speed v and yaw rate ω_r data measured in the tests, the turning radius and lateral acceleration can be obtained by $R = v/\omega_r$ and $a_y = v\omega_r$, respectively. The turning radius increases with speed for an understeer vehicle and decreases for an oversteer vehicle.

With the difference between the front and rear sideslip angles obtained by

$$a_1 - a_2 = 57.3\left(\frac{1}{R_0} - \frac{1}{R}\right)L \tag{5.93}$$

where

- R is the turning radius
- L is the wheelbase

At last, the value $(a_1 - a_2)$ versus a_y can be plotted.

5.10.3. Transient Yaw Response Tests

The tests are typically carried out on a level field with a step steering input. The vehicle is driven straight at the beginning and then the steering wheel is turned with an angular speed of 200–500 deg/s once the test speed is reached. The test speed is either 70% of the top speed or both 40 and 110 km/h. The angular displacement of both steering wheel varies with vehicle speed, but a steady circular motion with a certain lateral acceleration, for example, 1–3 m/s^2, is required. With both the steering angle and accelerator pedal kept unchanged, the vehicle enters a circular motion eventually. With the vehicle speed, time, steering angle, yaw rate, and lateral acceleration taken down in the test, the time trace of the ratio of yaw rate gain to steady-state yaw rate gain, that is, $(\omega_r/\delta)/(\omega_{r0}/\delta)$, can be plotted, from which parameters such as the response time, overshoot, and time constant can be obtained.

5.10.4. Self-Aligning Tests

The tests are conducted on a level field, with the vehicle moving along a circular path of a radius of 15 m. The steering wheel is released suddenly when the lateral acceleration reaches 4 m/s^2 by adjusting the vehicle speed. The front wheels return to the straight driving position under the action of self-aligning torque. Time t, vehicle speed v, steering angle δ_s, and yaw rate ω_r are recorded to produce the ω_r versus t curve.

For a vehicle with a top speed of over 100 km/h, the self-aligning tests at 70% of the top speed are required. The test starts with the vehicle moving straight and then the driver turns the steering wheel to produce a lateral acceleration of 2 m/s^2. The self-aligning test is followed by releasing the steering wheel suddenly. It is generally required that the steering wheel should return itself to the center position without an excessive residual yaw rate or swing left and right.

5.10.5. Pulse Steering Tests

The tests aim to determine the frequency response of the yaw rate. Ideally, a sinusoidal pulsed steering angle input is applied and the yaw rate is measured as the output, with the frequency characteristics obtained by Fourier transform of the data. Since it is difficult to apply a precise sinusoidal pulsed steering angle by a human driver, a simpler pulsed angle input is actually applied to the steering wheel instead. The vehicle is operated at a constant speed that is 70% of the top speed on a level field, with a pulse width of 0.3–0.5 sec. The magnitude of the pulse has to produce a lateral acceleration of 4 m/s^2. The vehicle should return to straightforward operation quickly after the swing in its direction during the pulsed steering input. The time, steering angle, vehicle speed, yaw rate, and lateral acceleration are recorded in the test and then processed with Fourier transform to obtain the frequency characteristics.

5.10.6. On-Center Handling Tests

See Section 5.7.4 for the description of such tests.

References

[5.1]. Taheri, S., Sandu, C., and Taheri, S., "Development and Implementation of a Hybrid Soft Soil Tire Model (HSSTM)," Paper Presented at in *the 18th International Conference of the ISTVS*, Seoul, Korea, September 2014.

[5.2]. Jazar, R., *Vehicle Dynamics: Theory and Application* (New York: Springer, 2008), 121, doi:10.1007/978-0-387-74244-1.

[5.3]. Gillespie, T., *Fundamentals of Vehicle Dynamics* (Warrendale: Society of Automotive Engineers, Inc., 1992), 21-118, ISBN-10:1560911999, ISBN-13:978-1560911999.

[5.4]. Yu, Z., *Automobile Theory*, 3rd ed. (Beijing: China Machine Press, 2000), 1-36, ISBN:7-111-02076-6.

[5.5]. Reimpell, J., Stoll, H., and Betzler, J., *The Automotive Chassis: Engineering Principles*, 2nd ed. (Oxford, UK: Butterworth-Heinemann, 2001), ISBN:0750650540.

[5.6]. Heisler, H., *Advanced Vehicle Technology*, 2nd ed. (Burlington: Elsevier, 2002), ISBN:0750651318.

[5.7]. Riede, P. Jr., Leffert, R., and Cobb, W., "Typical Vehicle Parameters for Dynamic Studies Revised for the 1980's," SAE Technical Paper 840561 (1984), doi:https://doi.org/10.4271/840561.

[5.8]. Bastow, D., Howard, G., and Whitehead, J., *Car Suspension and Handling*, 4th ed. (Warrendale: Society of Automotive Engineers, Inc., 2004), ISBN:0-7680-0872-7.

[5.9]. SAE International, "Vehicle Dynamics Terminology," SAE Standard J670, Revised June 2022.

[5.10]. Ryan, J., Fuja, S., and Schmid, H., "Objective Ride and Handling Goals for the 1997 Chevrolet Corvette," SAE Technical Paper 970091 (1997), doi:https://doi.org/10.4271/970091.

[5.11]. Koibuchi, K., Yamaoto, M., Fukada, Y., and Inagaki, S., "Vehicle Stability Control in Limit Cornering by Active Brake," SAE Technical Paper 960487 (1996), doi:https://doi.org/10.4271/960487.

6

Ride Comfort

The passengers riding an EV are subject to a wide spectrum of noise and vibration mainly coming from the electric motor, drivetrain, and tire–road and body–air interactions. The ride quality discussed in this chapter primarily covers the vibrations caused by tire–road interaction with frequencies ranging from 0.5 to 25 Hz.

As one of the important criteria for evaluating a vehicle, ride quality quantifies the impact of vibrations on passengers' comfort during its operation. It is generally required that such an impact be kept within a certain limit for passenger vehicles and the cargo be kept intact for trucks. The ride quality is usually evaluated by the subjective perception of passengers and supplemented by some objective engineering metrics, which are far from perfect yet reflect all aspects of the human perception of vibrations.

The ride quality is characterized by the dynamic system comprised of road, vehicle, and human involving excitation, transmission, and perception of the vibrations, as shown in Figure 6.1. The excitation sources such as the electric motor, drivetrain, wheels, tires, and road roughness are the inputs that act on the vehicle, which is essentially a dynamic system. The behavior of system response depends on the physical properties of the components on the transmission path, that is, the elasticity, damping ratio, mass, and inertia of these related structures. The low-frequency vibrations at the seat, cabin, and suspension are the system outputs that ultimately determine the passenger's perception of the vehicle, typically by tactile and visual sensations.

FIGURE 6.1 The dynamic system associated with ride quality.

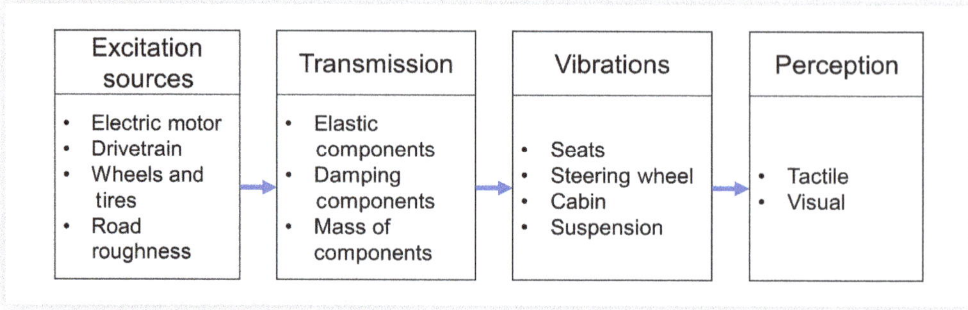

In order to improve the ride quality, the vehicle vibration responses to given excitation inputs must be kept within certain limits by controlling the output characteristics of the dynamic system.

6.1. Excitation Sources

The vibrations of an EV are mainly excited from either on-board or road–tire interaction-related sources. The on-board sources primarily consist of rotating components such as electric motors, drivetrain, and wheel and tire assemblies. The road–tire interaction is mostly related to the road roughness and vehicle speed.

6.1.1. Electric Motors

As the primary power source of an EV, the electric motor is one of the most important sources of vibration excitation as it rotates and delivers the torque to the drivetrain and wheels.

Although electric motors produce power in a continuous manner, their output torque is not constant in magnitude—there are always torque ripples due to the nonsinusoidal counter-electromotive forces and phase currents. In fact, the presence of stator cogging and the geometry of magnets make it difficult to produce a completely sinusoidal counter-electromotive force. It is also impossible to get rid of harmonic contents completely from the phase currents modulated by a state-of-the-art inverter.

A torque fluctuation rate λ_{tr} can be defined to quantify the severity of torque ripples:

$$\lambda_{tr} = \frac{T_{rmax} - T_{rmin}}{T_{e0}} \tag{6.1}$$

where

T_{rmax} and T_{rmin} are the maximum and minimum of the torque ripples, respectively
T_{e0} is the steady-state torque of the electric motor

6.1.1.1. Torque Ripples Related to Nonsinusoidal Magnetic Field

Due to the distortion in the excitation field of a permanent magnetic rotor or in the spatial distribution of stator windings of a practical three-phase electric motor, the magnetic field in an electric motor is always nonsinusoidal in the strict sense. Consequently, the output torque contains components of 6k times the fundamental frequency [6.1]:

$$T_r = 1.5 p_0 \psi_{d6n} \cos 6k\theta_e i_q, \quad k = 1,2,3,\ldots \tag{6.2}$$

where

p_0 is the number of pole pairs

ψ_{d6n} is the d-axis flux linkage of the stator associated with mutual inductance with permanent magnets

θ_e is the angle between the direct axis of the rotor and the axis of phase A stator winding

i_q is the stator the d-axis current

The torque fluctuation rate is [6.1]

$$\lambda_{tr} = \frac{2\psi_{d6n}}{\psi_f} \tag{6.3}$$

where ψ_f is the permanent magnet flux linkage.

The frequency of torque ripples is

$$f_{tr} = \frac{2n}{5} = 6f \tag{6.4}$$

where

n is the motor speed

f is the phase current frequency

6.1.1.2. Torque Ripples Related to Switching Frequency of Inverter

In an inverter, the DC power is inverted into AC by a chopper circuit using the pulse width modulation (PWM) method. Thus, the phase current fluctuates at the switching frequency, causing the torque to oscillate at the same frequency. Assuming a sinusoidal magnetic field, the torque ripples associated with the switching frequency is

$$T_r = E\Delta i = \left(V - iR - E\right)\frac{E}{L}\Delta T_{on} \tag{6.5}$$

where

E is the counter-electromotive force

Δi is the magnitude of AC

V is the voltage

R is the resistance

L is the inductance

ΔT_{on} is the time of on-state in a PWM cycle given by

$$\Delta T_{on} = \left(\frac{\sqrt{3}}{2}\cos\theta - \frac{1}{2}\sin\theta\right)\frac{\sqrt{3}V}{V_{DC}} T_s \tag{6.6}$$

where θ is the angle between the reference voltage vector and the d-axis.

The torque fluctuation rate is [6.1]

$$\lambda_{tr} = \left(\frac{V - E}{i} - R \right) \frac{1}{L} \Delta T_{on} \tag{6.7}$$

6.1.1.3. Torque Ripples Related to Current Sensor Errors

Two current sensors are needed for the feedback control of electric motors. The current sensor errors, mainly DC offset and unequal gain, can result in torque ripples.

The DC offset error refers to the DC component in current sensor signals. The torque ripple due to DC offset error can be expressed as

$$T_r = \left[\Delta_a \cos\left(\theta_e - \frac{2\pi}{3} \right) - \Delta_b \cos\theta_e \right] \frac{\sqrt{3}EI}{\omega_m} \tag{6.8}$$

where

Δ_a and Δ_b are the ratio of the DC component to the actual current value I in feedback current corresponding to phases A and B, respectively
θ_e is the angle between the d-axis and the stator winding axis of phase A
ω_m is the angular speed of the motor

In the worst scenario, both Δ_a and Δ_b are equal to the maximum value of DC offset error, that is, $\Delta_a = \Delta_b = \Delta_{max}$, and the torque fluctuation rate is, thus, given by [6.1]

$$\lambda_{tr} = 2\Delta_{max} \sin\left(\theta_e - \frac{2\pi}{3} \right) \tag{6.9}$$

On the other hand, the unequal gain error refers to the difference between gains imposed in the amplification process of the two current sensor signals.

The torque ripple due to unequal gain error is [6.1]

$$T_r = (k_a - k_b) \frac{\sqrt{3}EI}{2\omega_m} \sin\left(2\theta_e - \frac{2\pi}{3} \right) \tag{6.10}$$

where k_a and k_b are the ratio of measuring error to actual current value I in feedback current corresponding to phases A and B, respectively. The torque fluctuation rate is [6.1]

$$\lambda_{tr} = \frac{4}{\sqrt{3}} \frac{k_a - k_b}{2 + k_a + k_b} \tag{6.11}$$

Considering the worst scenario of $k_a = -k_b = k_{max}$, the torque fluctuation rate becomes [6.1]

$$\lambda_{tr} = \frac{2}{\sqrt{3}} k_{max} \tag{6.12}$$

6.1.1.4. Torque Ripples Related to the Resolver

The initial angle of the resolver must be calibrated so that the resolver's initial angle set in the motor controller software is consistent with that of the electric motor. If not, the output power of the electric motor and the torque control accuracy will be affected, which can result in vehicle vibrations. Termed judder or jitter, this type of vehicle vibration is often observed at the start-up of the vehicle, mainly in the longitudinal direction.

6.1.2. Drivetrain

The aforementioned torque variations from electric motors are superimposed and transmitted along the drivetrain, resulting in vehicle vibrations mainly by two mechanisms: first, torsional resonance due to the excitation frequency coinciding with the natural frequency of the system comprised of the rotor and other rotating components in the drivetrain; and second, the torsional vibrations excited on the electric motor housing and then transmitted to the vehicle through the powertrain mounts.

Due to the compliance in electric powertrain mounts, the torque variations can cause the system to vibrate in three translational and three rotational directions. Of all the vibration directions, the most significant one is in the pitch direction about the lateral axis of a transverse electric motor (or about the longitudinal axis of the electric motor mounted in the north–south direction) excited by drive torque fluctuations. These fluctuations occur at the natural frequency and subharmonics of the powertrain housing or the rotor–geartrain assembly. To isolate the excitation from the vehicle body, the resonance frequency of the powertrain mounts should be designed below the lowest excitation frequency, with their axes parallel to each other. Torque variations above the resonance frequency are absorbed by the rotating inertia of the rotor–geartrain assembly and, therefore, attenuated.

In addition, the drivetrain can excite ride vibrations by itself, primarily due to the mass imbalance and secondary couples of the driveshaft. For a multispeed drivetrain, the gear shift can be another source of ride vibrations.

6.1.2.1. Driveshaft-Related Excitation

The combination of any imperfections, such as geometric asymmetry, nonstraightness, elastic deflection, off-center support members, and clearance, can lead to a mass imbalance of a driveshaft.

The imbalance produces a force vector on the support members of the shaft in the plane perpendicular to the shaft axis that rotates at the same speed as the shaft, with the magnitude being

$$F_{im} = m_{im}\omega_s^2 r_{im} \tag{6.13}$$

where

m_{im} is the mass imbalance

ω_s is the shaft speed

r_{im} is the radial distance from mass imbalance to the axis of rotation

For FWD vehicles, such an excitation exerts additional forces on the half shafts and, thus, affects the vehicle motion in the longitudinal and vertical directions. For RWD vehicles, the excitation may cause the vehicle to vibrate in all three directions. Since the elastic deflection increases with the rotation speed, the geometric asymmetry grows. Thus, it is difficult to dynamically balance the shaft for all but a particular speed.

On the other hand, the secondary couple, or rocking torque, is produced when a universal joint is employed to connect two driveshafts at an angle. Acting in the planes of yokes, the secondary couple tends to bend the two shafts and rock them about their bearings. Thus, the bearings are cyclically loaded at the rate of two cycles per shaft revolution. The magnitude of rocking torque depends on the configuration of universal joints. For Cardan joints, the rocking torque of driving yoke is given by the following equation [6.2]:

$$M_{bi} = T_{in} \tan\theta \cos\beta \tag{6.14}$$

The rocking torque of the driven yoke is

$$M_{bo} = T_{in} \sin\theta \sin\beta \tag{6.15}$$

where

T_{in} is the input torque
θ is the joint angle
β is the orientation angle of the driving yoke from the position where it is normal to the joint angle plane

For Rzeppa joints, the rocking torques of driving and driven yokes are identical, that is,

$$M_{bi} = M_{bo} M_{bi} = M_{bo} = T_{in} \tan\frac{\theta}{2} \tag{6.16}$$

It is obvious that the rocking torques vary with the rotation of driveline components, imposing cyclic excitation forces on the vehicle. These forces can be transmitted as torsional vibrations along the driveline and superimposed on aforementioned excitation forces generated by electric motors. The torsional vibrations result in the variation of drive forces at the tire–ground interface and produce longitudinal vibrations in the vehicle. In addition, the excitation forces can also be transmitted into the interior of the vehicle via the support members of driveshafts, deteriorating the ride.

The use of appropriate types of joints or reduction in the joint angle can help minimize the excitation from secondary couples.

6.1.2.2. Gear Shift-Related Excitation

For EVs equipped with multispeed transmissions, the gear shift can result in the variation of tractive force, typically for the following reasons.

1. Torque interruption. Due to the nature of gear shift elements such as synchronizers or dog clutches, the torque interruption in some transmissions, for example, AMTs, is intrinsically inevitable. The output torque drops to zero during the period of torque interruption.
2. Torque management. In powershift transmissions (e.g., DCTs and ATs), electric motor torque is often reduced during gear shift to improve the shift quality and the longevity of the transmission and drivetrain as a whole. The output torque can drop over 30% during the period of torque interruption.
3. Torque hole and overshoot. The dynamic process of gear shift generally produces a torque hole and torque overshoot in the output torque. The width and depth of the torque hole, as well as the overshoot, depend on the control of on-going and off-going shift elements.

The gear shift-related excitation is typically quantified by the so-called shift jerk, which is defined as the rate of longitudinal acceleration resulting from torque variation during the shift, that is,

$$j = \frac{da}{dt} \tag{6.17}$$

where a is the longitudinal acceleration of the vehicle.

The shift jerk can be minimized by shortening the period of torque interruption, reducing the amount of torque management, or optimizing the pressure control in on-going and off-going elements.

6.1.3. Wheels and Tires

The manufacturing imperfections in wheel and tire assembly, for example, dimensional deviations, mass imbalance, and stiffness variations, can result in force and moment alterations at the tire–ground interface as it rolls.

6.1.3.1. Dimensional Deviation-Related Excitation

Dimensional deviations can be classified into radial runouts and lateral runouts.

The wheel and tire assembly can be modeled by an array of radial springs around its circumference, as shown in Figure 6.2 [6.3]. The free radial runout represents the dimensional nonuniformities with springs at their free lengths, while the loaded radial runout characterizes the rolling nonuniformities with springs compressed to their lengths under nominal load.

FIGURE 6.2 The model of wheel and tire assembly.

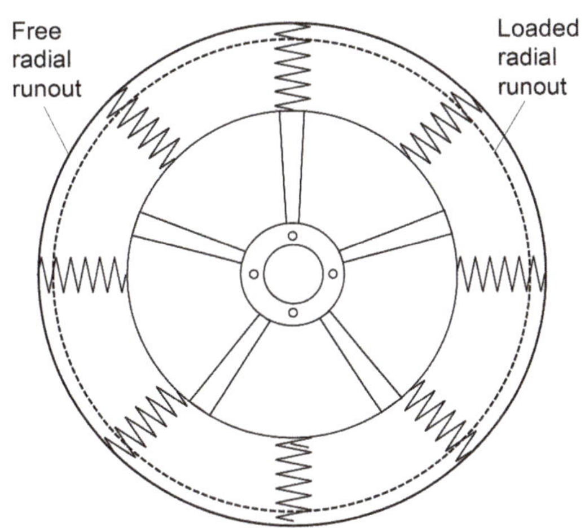

The excitation caused by radial runouts in a wheel and tire assembly can be presented by their major harmonics, which are equivalent to imperfect shape in effect, as shown in Figure 6.3 [6.3].

FIGURE 6.3 Major harmonics in excitation resulted from radial runouts in a wheel and tire assembly. 1st harmonic (eccentric), (b) 2nd harmonic (oval), (c) 3rd harmonic (triangular), and (d) 4th harmonic (square).

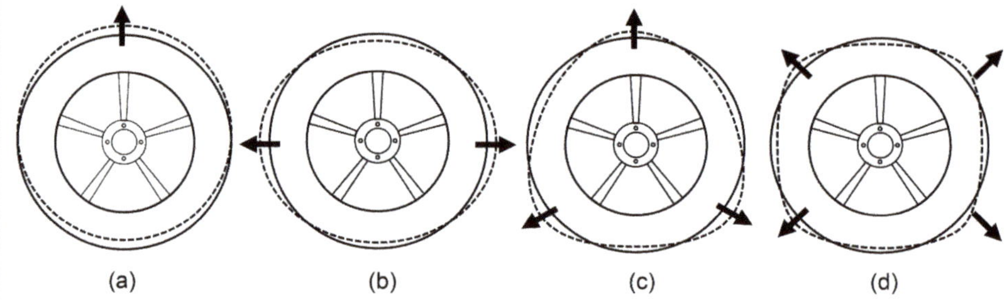

Specifically, the displacement produced by i^{th} of the harmonics can be obtained by [6.4]

$$z_i = \Delta R \sin\left(i\omega_w t\right) \tag{6.18}$$

where

ΔR is the difference between the maximum and minimum radii

ω_w is the angular speed of the assembly

The longitudinal and vertical excitation forces produced by the first harmonic radial runout, that is, eccentricity, are, respectively, given by [6.4]

$$F_x = em_w\omega_w^2 \cos\left(\omega_w t + \varphi\right) \tag{6.19}$$

$$F_z = em_w\omega_w^2 \sin\left(\omega_w t + \varphi\right) \tag{6.20}$$

where

e is the eccentricity

m_w is the mass of wheel and tire assembly

φ is the initial phase of eccentricity

On the other hand, the lateral runouts produced moments about both longitudinal and vertical axes are, respectively, given by [6.4]

$$M_x = I_{xx}\ddot{\beta}_x + I_{y'y'}\omega_w\dot{\beta}_z \tag{6.21}$$

$$M_z = I_{zz}\ddot{\beta}_z - I_{y'y'}\omega_w\dot{\beta}_x \tag{6.22}$$

where

I is the rotational inertia of wheel and tire assembly about the axis specified by the subscript

β is the deviation angle, as shown in Figure 6.4 [6.4]

FIGURE 6.4 The schematic of lateral runouts.

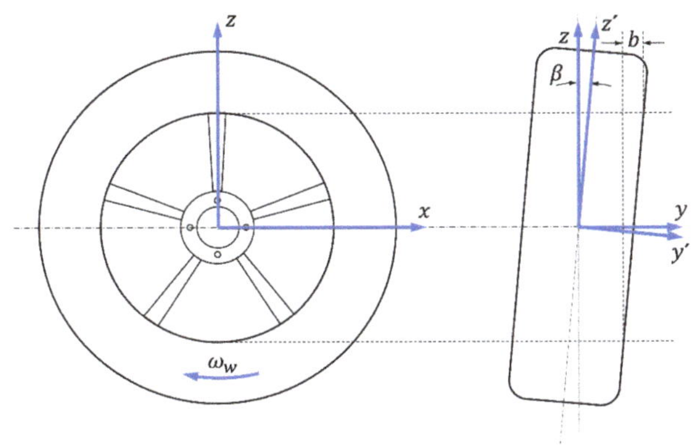

Assuming the deviation angle β is small, the deviation angle can be estimated by

$$\beta = \frac{b}{d_{rm}} \tag{6.23}$$

where

b is the lateral runout at the rim horn

d_{rm} is the diameter of the rim horn

6.1.3.2. Mass Imbalance-Related Excitation

Asymmetrical mass distribution about the axis of the tire rotation, often termed static imbalance, results in a rotating force in the radial direction, producing both radial and longitudinal excitation in the wheel plane.

The excitation forces in the radial and longitudinal directions are, respectively, expressed by

$$F_x = m_{im}\omega_w^2 r_{im} \cos(\omega_w t + \varphi) \tag{6.24}$$

$$F_z = m_{im}\omega_w^2 r_{im} \sin(\omega_w t + \varphi) \tag{6.25}$$

where

φ is the initial phase angle of imbalance

m_{im} is the imbalance mass

r_{im} is the radius of imbalance, as shown in Figure 6.5 [6.4]

FIGURE 6.5 Asymmetrical mass imbalance of a wheel and tire assembly.

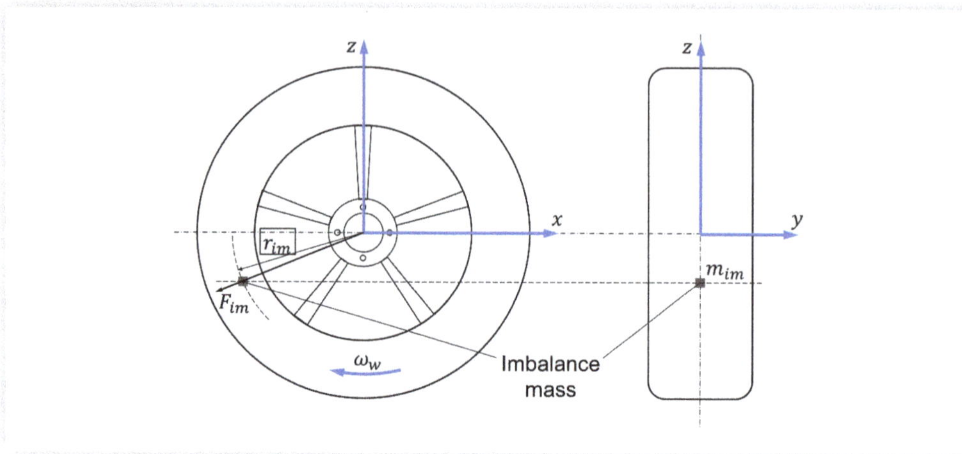

Courtesy of Condition Monitoring and Fault Diagnosis (CMFD 2014)At: Sharif University of Technology, Tehran, Iran.

On the other hand, asymmetrical mass distribution along the axis of rotation, termed dynamic imbalance, produces moments about the longitudinal and vertical axes. As illustrated in Figure 6.6 [6.4], if the pair of imbalance masses is out of phase by 180°, then the imbalance forces cancel each other out. The moments about the longitudinal and vertical axes are, respectively, given by

$$M_x = am_{im}\omega_w^2 r_{im}\cos(\omega_w t + \varphi) \tag{6.26}$$

$$M_z = am_{im}\omega_w^2 r_{im}\sin(\omega_w t + \varphi) \tag{6.27}$$

where a is the axial distance between the pair of imbalance masses.

FIGURE 6.6 Mass imbalance along the axis of rotation of a wheel and tire assembly.

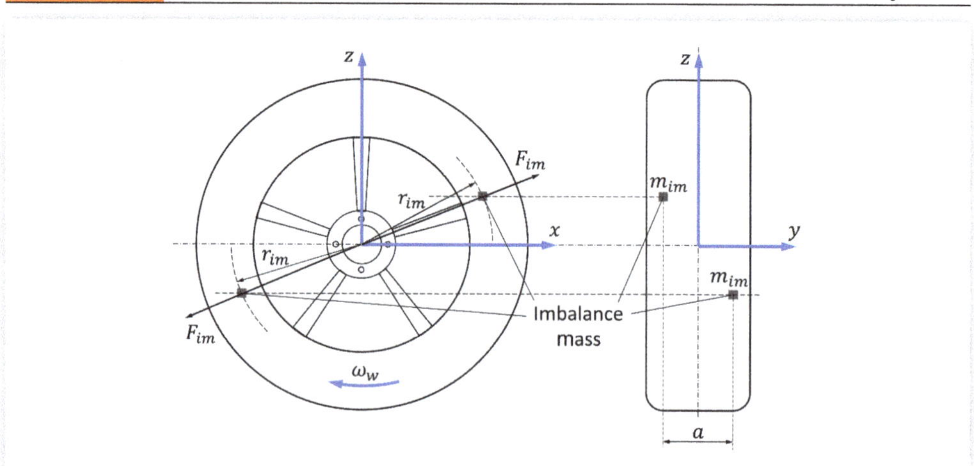

Courtesy of Condition Monitoring and Fault Diagnosis (CMFD 2014)At: Sharif University of Technology, Tehran, Iran.

6.1.3.3. Stiffness Variation-Related Excitation

The stiffness variations are mainly a result of tire construction, for example, the overlaps of a number of plies of steel or fabric material. Moreover, the manufacturing defects and permanent deformations in the wheel hub can also produce stiffness variations indirectly although their effects are substantially attenuated by the tire that mounts on the wheel hub.

The stiffness variations around the circumference of a tire can be represented by radial springs in the model illustrated in Figure 6.2. In general, the stiffness variations make the wheel and tire assembly displace vertically, producing a displacement similar to the effect of radial runouts.

Since the order of harmonics is dependent on the number of stiffness variations around the circumference, the displacement produced by i^{th} of the harmonics can be obtained by

$$z_i = \frac{F_z}{k_z}\sin(i\omega_w t) \tag{6.28}$$

where

F_z is the vertical load on the wheel and tire assembly
k_z is the vertical stiffness of the wheel and tire assembly

6.1.4. Road Roughness

The elevation of the road surface along the wheel tracks with respect to a datum plane, $\zeta(x)$, is called unevenness function or roughness profile, with x being the distance along the road. Due to the random nature of the elevation profile, the power spectral density (PSD) function is commonly used to represent its statistical characteristics. The PSD for unevenness of the road can be represented by

$$G_\zeta(\nu) = G_\zeta(\nu_0)\left(\frac{\nu}{\nu_0}\right)^{-w} \tag{6.29}$$

where

$G_\zeta(\nu)$ is the PSD amplitude in m^3
ν is the wavenumber (i.e., spatial frequency) in m^{-1}
$G_\zeta(\nu_0)$ is the reference PSD amplitude at wavenumber ν_0 in m^3 with $\nu_0 = 0.1\ m^{-1}$
w is the frequency index

As illustrated in Figure 6.7 [6.5], the PSD for a given road is a straight line if plotted in a bilogarithmic coordinates system. The unevenness of road surfaces can be classified into eight different classes, denoted "A" through "H" from excellent to very poor. The geometric mean of the PSD magnitude at reference wavenumber ν_0 for each of the eight classes with a frequency index $w = 2$ is listed in Table 6.1 according to ISO 8608:2016 [6.5], along with the geometric mean of the root-mean-square of the elevation ζ for the wavenumber ranging from 0.011 to 2.83 m^{-1}.

FIGURE 6.7 Road unevenness classification according to ISO 8608:2016.

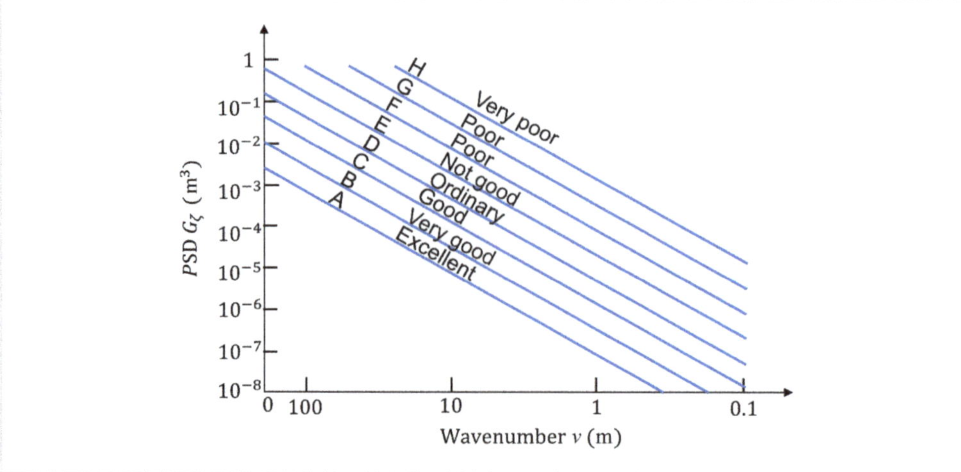

TABLE 6.1 Road roughness classification.

Class of road surface	$G_\zeta(\nu_0)$ at $\nu_0 = 0.1$ m^{-1} ($\times 10^{-6}$ m^3) Geometric mean	σ_ζ for 0.011 m$^{-1} < \nu <$ 2.83 m^{-1} ($\times 10^{-3}$ m^3) Geometric mean
A	16	3.81
B	64	7.61
C	256	15.23
D	1,024	30.45
E	4,096	60.90
F	16,384	121.80
G	65,536	243.61
H	262,144	487.22

When a vehicle moves along the road, the road roughness affects the vehicle as a vertical displacement input to the wheels that excites the ride vibrations. Hence, transformations of elevation profile into displacement, velocity, and acceleration as functions of time help understand the dynamics of the ride. Therefore, in addition to the PSD based on the elevation data, the first- and second-order derivatives of unevenness function $\zeta(x)$ with respect to x are also employed as supplementary descriptions of the statistical characteristics of the road unevenness:

$$G_{\dot{\zeta}}(\nu) = (2\pi\nu)^2 G_\zeta(\nu) \tag{6.30}$$

$$G_{\ddot{\zeta}}(\nu) = (2\pi\nu)^4 G_\zeta(\nu) \tag{6.31}$$

where $G_{\dot{\zeta}}(\nu)$ is in m and $G_{\ddot{\zeta}}(\nu)$ is in m^{-1}.

It can be shown that the spatial frequency-based PSD is related to temporal frequency-based PSD by

$$G_\zeta(f) = \frac{1}{V} G_\zeta(v) \tag{6.32}$$

where f is the temporal frequency corresponding to the vehicle speed V on a road with the unevenness of spatial frequency v. The temporal frequency f is related to the spatial frequency v by

$$f = V v \tag{6.33}$$

Substitution of Equations (6.32) and (6.33) into Equation (6.29) yields

$$G_\zeta(f) = \frac{1}{V} G_\zeta(v_0)\left(\frac{v}{v_0}\right)^{-w} = G_\zeta(v_0) v_0^w \frac{V^{w-1}}{f^w} \tag{6.34}$$

In addition, the temporal frequency-based unevenness velocity and acceleration can be obtained by substituting Equation (6.4) into Equations (6.2) and (6.3) and then reorganizing

$$G_{\dot{\zeta}}(f) = (2\pi f)^2 G_\zeta(f) = 4\pi^2 v_0^w \frac{V^{w-1}}{f^{w-2}} G_\zeta(v_0) \tag{6.35}$$

$$G_{\ddot{\zeta}}(f) = (2\pi f)^4 V G_\zeta(f) = 16\pi^4 v_0^w \frac{V^{w-1}}{f^{w-4}} G_\zeta(v_0) \tag{6.36}$$

It is apparent from Equations (6.34) through (6.36) that the temporal frequency-based unevenness functions $G_\zeta(f)$, $G_{\dot{\zeta}}(f)$, and $G_{\ddot{\zeta}}(f)$ are all proportional to the reference PSD amplitude $G_\zeta(v_0)$ and vehicle speed V.

The vertical displacement inputs to all the wheels must be taken into account in modeling the ride behavior of a vehicle. Thus, the auto-power spectrum of the roughness profile for a particular wheel can be used to evaluate the influence from the same wheel, while the cross-power spectra of the roughness profiles for other wheels reflect the influence from other wheels. The auto- and cross-power spectra for a four-wheel vehicle can be expressed by

$$G_{jk}(v) = \lim_{X\to\infty} \frac{1}{X} Z_j^*(v) Z_k(v) \tag{6.37}$$

where

$Z_j(v)$ and $Z_k(v)$ are the Fourier transform of the roughness profiles $\zeta_j(x)$ and $\zeta_k(x)$, respectively

$Z_j^*(v)$ and $Z_k^*(v)$ are the conjugate complex of $Z_j(v)$ and $Z_k(v)$, respectively

X is the interval along the wheel track x

subscripts j, k = 1, 2, 3, and 4 represent the four wheels

The Fourier transform of the roughness profile for each of the four wheels is

$$\left.\begin{aligned} F_1(v) &= F\left[\zeta_1(x)\right] = F\left[\zeta_l(x)\right] = Z_l(v) \\ F_2(v) &= F\left[\zeta_2(x)\right] = F\left[\zeta_l(x-L)\right] = e^{-i2\pi vx} Z_l(v) \\ F_3(v) &= F\left[\zeta_3(x)\right] = F\left[\zeta_r(x)\right] = Z_r(v) \\ F_4(v) &= F\left[\zeta_4(x)\right] = F\left[\zeta_r(x-L)\right] = e^{-i2\pi vx} Z_r(v) \end{aligned}\right\} \tag{6.38}$$

where $Z_l(v)$ and $Z_r(v)$ are the Fourier transform of the roughness profile along the tracks of left and right wheels $\zeta_l(x)$ and $\zeta_r(x)$, respectively. Let $G_{ll}(v)$, $G_{rr}(v)$, $G_{lr}(v)$, and $G_{rl}(v)$ denote the auto- and cross-power spectra of the roughness profiles along the tracks of left and right wheels, the auto- and cross-power spectra for a four-wheel vehicle can be expressed by substituting Equation (6.38) into Equation (6.37):

$$\left.\begin{aligned} G_{11}(v) &= G_{22}(v) = G_{ll}(v) \\ G_{33}(v) &= G_{44}(v) = G_{rr}(v) \\ G_{12}(v) &= G_{21}^*(v) = e^{-i2\pi vx} G_{ll}(v) \\ G_{34}(v) &= G_{43}^*(v) = e^{-i2\pi vx} G_{rr}(v) \\ G_{14}(v) &= G_{41}^*(v) = e^{-i2\pi vx} G_{lr}(v) \\ G_{32}(v) &= G_{23}^*(v) = e^{-i2\pi vx} G_{rl}(v) \\ G_{13}(v) &= G_{31}^*(v) = G_{lr}(v) \\ G_{42}(v) &= G_{24}^*(v) = G_{rl}(v) \end{aligned}\right\} \tag{6.39}$$

Since the cross-power spectral density is generally a complex number, the cross-power spectral density between the left and right wheels becomes

$$G_{lr}(v) = \left|G_{lr}(v)\right| e^{-i\varphi_{lr}(v)} \tag{6.40}$$

where $|G_{lr}(v)|$ is the amplitude of the cross-spectrum of $\zeta_l(x)$ and $\zeta_r(x)$ that represents the degree of linear correlation between the content at a frequency v of the left and right wheel tracks, and $\varphi_{lr}(v)$ is the phase spectrum of $\zeta_l(x)$ and $\zeta_r(x)$ that represents the averaged phase difference between the content at a frequency v of the left and right wheel tracks. In addition to the cross-power spectral density functions, the statistical characteristics between the unevenness of the two-wheel tracks can also be expressed by their coherence function

$$C_{lr}^2(v) = \frac{\left|G_{lr}(v)\right|^2}{G_{ll}(v) G_{rr}(v)} \tag{6.41}$$

The coherence function $C_{lr}^2(\nu)$ describes the degree of linear correlation between the content at a frequency ν of the left and right wheel tracks in the frequency domain. Both the amplitude ratio and phase difference of the content at a frequency ν in $\zeta_l(x)$ and $\zeta_r(x)$ vary randomly if $C_{lr}^2(\nu)=0$, while remain constant if $C_{lr}^2(\nu)=1$.

The differences in amplitude and phase between the unevenness of left and right wheel tracks cause the vehicle to vibrate about its roll angle. The PSD of roll angle displacement is related to the PSD of vertical displacement and the coherence function by

$$\frac{G_{\Phi}(\nu)}{G_z(\nu)}=\frac{2}{t_r^2}\left[1-C_{lr}^2(\nu)\right] \tag{6.42}$$

When the vehicle is moving at low speeds, the coherence function is small. It is obvious from Equation (6.42) that the input power from roll angle vibrations is quite significant in comparison with that from vertical vibrations. In contrast, the coherence function is large at high speeds and the input power from roll angle vibrations is much less than that from vertical vibrations.

The actual measurement data reveal that the averaged phase difference between the roughness profiles along the left and right tracks is approximately zero because the probability of the content at a frequency ν for either of the tracks to lead the other is the same as to lag. Therefore, when the roughness profiles along the left and right tracks, $\zeta_l(x)$ and $\zeta_r(x)$, are of the same statistical characteristics, that is, $G_{ll}(\nu) = G_{rr}(\nu) = G_\zeta(\nu)$, and the phase spectrum $\varphi_{lr}(\nu) = 0$, Equation (6.41) can be simplified to

$$G_{lr}(\nu)=G_{rl}(\nu)=C_{lr}(\nu)G_{\zeta}(\nu) \tag{6.43}$$

The spectral matrix of vertical displacement inputs into a four-wheel vehicle becomes

$$\left[G_{jk}(\nu)\right]=G_{\zeta}(\nu)\begin{bmatrix}1 & e^{-i2\pi\nu x} & C_{lr}(\nu) & e^{-i2\pi\nu x}C_{lr}(\nu)\\ e^{i2\pi\nu x} & 1 & e^{i2\pi\nu x}C_{lr}(\nu) & C_{lr}(\nu)\\ C_{lr}(\nu) & e^{-i2\pi\nu x}C_{lr}(\nu) & 1 & e^{i2\pi\nu x}\\ e^{i2\pi\nu x}C_{lr}(\nu) & C_{lr}(\nu) & e^{i2\pi\nu x} & 1\end{bmatrix} \tag{6.44}$$

6.2. Vehicle Response to Excitations

The vehicle response to the aforementioned excitations depends on the dynamic characteristics of the system. Such a dynamic model generally involves seven DOF, that is, the vertical, pitch, and roll motion of the sprung mass, as well as the vertical motion for each of the four wheels (unsprung mass), to represent the ride behavior of a vehicle.

6.2.1. Dynamic Model of a Vehicle

To further simplify the problem, a planar vehicle model can be adopted by assuming bilateral (or left/right) symmetry and ignoring tire damping, as illustrated in Figure 6.8 [6.6]. With the left and right wheels combined into one and the roll motion of sprung mass left out, the model is reduced to four DOF. The sprung mass m_2 can be broken down into the sprung mass at the front axle, centroid, and rear axle, that is, m_{2f}, m_{2c}, and m_{2r}, respectively, and connected with a rigid rod of negligible mass. Thus, to retain kinetic equivalence, the following three conditions must be satisfied, that is, the total mass, centroid position, and moment of inertia remaining the same.

$$m_{2f} + m_{2c} + m_{2r} = m_2 \tag{6.45}$$

$$m_{2f} L_1 = m_{2r} L_2 \tag{6.46}$$

$$I_y = m_2 r_y^2 = m_{2f} L_1^2 + m_{2r} L_2^2 \tag{6.47}$$

where r_y is the radius of sprung mass spin about the y-axis. The sprung mass at the front axle, centroid, and rear axle can be obtained by solving Equations (6.45) through (6.47):

$$m_{2f} = m_2 \frac{r_y^2}{LL_1} \tag{6.48}$$

$$m_{2r} = m_2 \frac{r_y^2}{LL_2} \tag{6.49}$$

$$m_{2c} = m_2 \left(1 - \frac{r_y^2}{L_1 L_2}\right) \tag{6.50}$$

where L is the wheelbase.

FIGURE 6.8 A planar vehicle model.

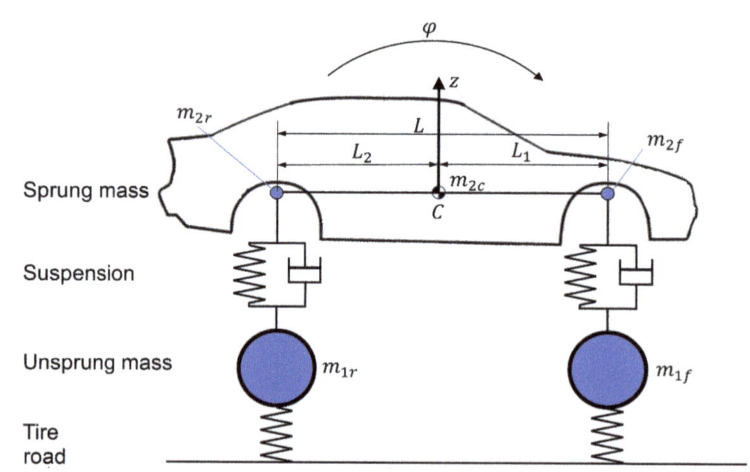

It is apparent from Equation (6.50) that $m_{2c} = 0$ when $\frac{r_y^2}{L_1 L_2} = 1$. In fact, the sprung mass distribution coefficients, $\frac{r_y^2}{L_1 L_2}$, of most vehicles range from 0.8 to 1.2. When $\frac{r_y^2}{L_1 L_2} = 1$, the vertical motion of each axle mass due to road roughness is independent from the other axle. Thus, the model of four DOF shown in Figure 6.8 can be further reduced to the quarter vehicle model of two DOF, as illustrated in Figure 6.9 [6.3].

FIGURE 6.9 A quarter vehicle model.

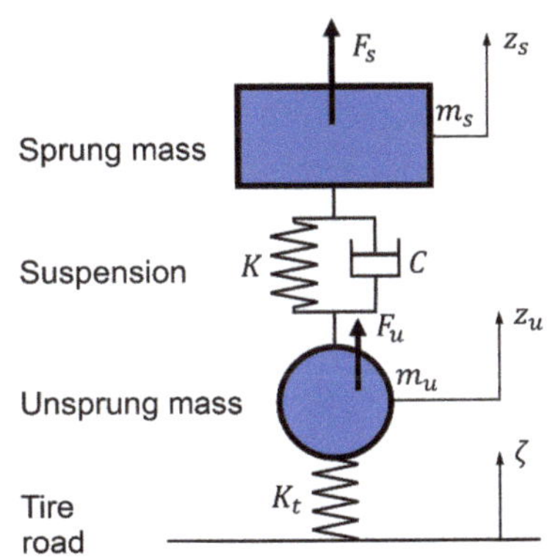

The equations of motion governing this system can be obtained by applying Newton's second law to the sprung and unsprung masses, respectively:

$$\begin{cases} m_s\ddot{z}_s + C(\dot{z}_s - \dot{z}_u) + K(z_s - z_u) = F_s \\ m_u\ddot{z} + C(\dot{z}_u - \dot{z}_s) + K(z_u - z_s) + K_t(z_u - \zeta) = F_u \end{cases} \quad (6.51)$$

where

- m_s is sprung mass
- z_s is sprung mass displacement
- C is suspension damping coefficient
- z_u is unsprung mass displacement
- K is suspension stiffness
- F_s is the force on sprung mass
- m_u is unsprung mass
- K_t is tire stiffness
- ζ is road displacement
- F_u is force on unsprung mass

The second-order ordinary differential equations, that is, Equation (6.51), can be solved by application of classical methods such as Laplace transforms, with details omitted here. The sprung mass motion in response to force on the axle, force on the sprung mass, and road displacement inputs are of particular interest in engineering practice. Let $C_2 = C/m_s$, $K_1 = K_t/m_s$, $K_2 = K/m_s$, $\chi = m_u/m_s$, and j be the complex operator, and the amplitude ratios for these characteristics are as follows [6.4]:

$$\frac{\ddot{z}_s}{F_s/m_s} = \frac{\left[\chi\omega^4 - (K_1 + K_2)\omega^2\right] + \mathrm{j}\left(C_2\omega^3\right)}{\left[\chi\omega^4 - (K_1 + K_2\chi + K_2)\omega^2 + K_1K_2\right] + \mathrm{j}\left[K_1C_2\omega - (1+\chi)C_2\omega^3\right]} \tag{6.52}$$

$$\frac{\ddot{z}_s}{F_u/m_s} = \frac{\left(K_2\omega^2\right) + \mathrm{j}\left(C_2\omega^3\right)}{\left[\chi\omega^4 - (K_1 + K_2\chi + K_2)\omega^2 + K_1K_2\right] + \mathrm{j}\left[K_1C_2\omega - (1+\chi)C_2\omega^3\right]} \tag{6.53}$$

$$\frac{\ddot{z}_s}{\ddot{\zeta}} = \frac{\left(K_1K_2\right) + \mathrm{j}\left(K_1C_2\omega\right)}{\left[\chi\omega^4 - (K_1 + K_2\chi + K_2)\omega^2 + K_1K_2\right] + \mathrm{j}\left[K_1C_2\omega - (1+\chi)C_2\omega^3\right]} \tag{6.54}$$

6.2.2. Fundamental Behavior of Sprung Mass

As the vibration frequency of sprung mass is usually much lower than the natural frequency of unsprung mass (typically 10–15 Hz), the tire deflection is trivial. Thus, the mass and stiffness of the tire can be neglected and the road displacement can be assumed to input directly at the unsprung mass (refer Figure 6.9), that is, $m_u = \zeta$. The first equation of Equation (6.51) then becomes

$$m_s\ddot{z}_s + C\left(\dot{z}_s - \dot{\zeta}\right) + K\left(z_s - \zeta\right) = 0 \tag{6.55}$$

The solution to Equation (6.55) is

$$z_s = Ae^{-nt}\sin\left(\omega_{0d}t + \varphi\right) \tag{6.56}$$

where

A is the amplitude
$n = \sqrt{C/(2m_s)}$
ω_{0d} is the damped natural frequency of sprung mass, $\omega_{0d} = \sqrt{\omega_0^2 - n^2}$
ω_0 is the undamped natural frequency of sprung mass, $\omega_0 = \sqrt{K/m_s}$
φ is the initial phase angle

A damping ratio ξ can be defined by

$$\xi = \frac{n}{\omega_0} = \frac{C}{2\sqrt{m_sK}} \tag{6.57}$$

Thus,

$$\omega_{0d} = \omega_0 \sqrt{1-\xi^2} \tag{6.58}$$

Apparently, the damped natural frequency of sprung mass ω_{0d} drops as damping ratio ξ. The damping ratio ξ of a vehicle suspension generally falls in the range from 0.2 to 0.4. As ξ is around 0.25 for most modern vehicles, it can often be approximated that $\omega_{0d} \approx \omega_0$ in engineering practice.

In addition, it is indicated by Equation (6.56) that the amplitude of the sprung mass vibration decays with time. The amplitude ratio of any two successive peaks is given by

$$\frac{A_1}{A_2} = \frac{Ae^{-nt}}{Ae^{-n(t+T)}} = e^{nT} \tag{6.59}$$

A logarithmic decrement δ can be defined as

$$\delta = \ln\frac{A_1}{A_2} = nT = \xi\omega_0 \frac{2\pi}{\omega_0\sqrt{1-\xi^2}} = \frac{2\pi\xi}{\sqrt{1-\xi^2}} \tag{6.60}$$

where T is the period of the oscillation and

$$T = \frac{2\pi}{\omega_{0d}} = \frac{2\pi}{\omega_0\sqrt{1-\xi^2}} \tag{6.61}$$

With the measured attenuation data, logarithmic decrement δ is known. The damping ratio ξ can then be obtained by

$$\xi = \sqrt{\frac{\delta}{\delta + 4\pi^2}} \tag{6.62}$$

6.2.3. Frequency Response to Road Excitation

By taking the Fourier transform on Equation (6.55), the frequency response function of the system to the road excitation can be expressed by

$$H(j\omega) = \frac{K + j\omega C}{\left(K - m_s\omega^2\right) + j\omega C} \tag{6.63}$$

By defining the frequency ratio $\lambda = \omega/\omega_0$ and substituting the damping ratio as of Equation (6.57), Equation (6.63) becomes

$$H(j\omega) = \frac{1 + j(2\lambda\xi)}{(1-\lambda^2) + j(2\lambda\xi)} \tag{6.64}$$

The corresponding amplitude–frequency characteristics are the module of Equation (6.64):

$$|H(j\omega)| = \left|\frac{z_s}{\zeta}\right| = \sqrt{\frac{1+(2\lambda\xi)^2}{(1-\lambda^2)^2 + (2\lambda\xi)^2}} \tag{6.65}$$

The effect of frequency ratio on the amplitude–frequency characteristics at various damping ratios is shown in Figure 6.10 [6.3]. It can be seen that $\log\left|\frac{z_s}{\zeta}\right|$ approaches the asymptotes of slope 0, ∞, −1, and −2 as $\log\lambda \ll 0$, $\log\lambda = 0$, and $\log\lambda \gg 0$ without damping ($\xi = 0$), and $\lambda \gg 1$ with damping ($\xi = 0.5$), respectively. In particular, at low frequencies ($\log\lambda \le -0.12$), the damping ratio has little influence on the system response, with $\left|\frac{z_s}{\zeta}\right|$ only barely larger than zero; at resonance frequencies ($-0.12 \le \log\lambda \le 0.15$), the resonance occurs where $\left|\frac{z_s}{\zeta}\right|$ peaks, and the peak value is substantially attenuated with the increase in the damping ratio; at high frequencies ($\log\lambda \ge 0.15$), $\left|\frac{z_s}{\zeta}\right|$ is less than 1; therefore, the road excitation input is attenuated, and the damping effect declines as the damping ratio rises.

FIGURE 6.10 Amplitude-frequency characteristics of the sprung mass response to road excitation.

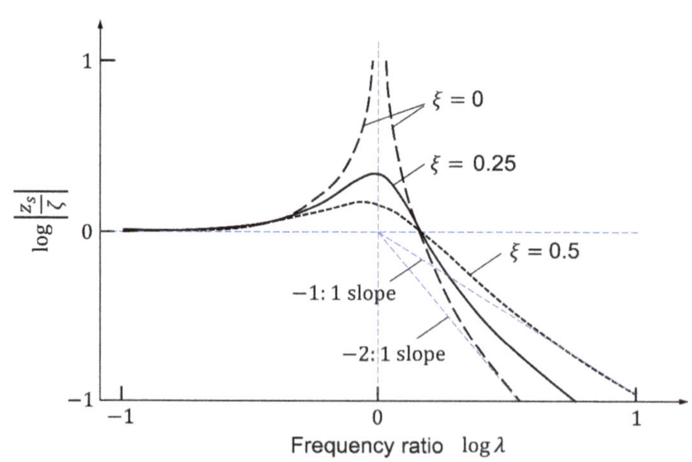

At the resonance frequency, $\omega = \omega_0$ and $\lambda = 1$; thus,

$$\left|\frac{z_s}{\zeta}\right|_{\omega=\omega_0} = \sqrt{\frac{1}{4\xi^2}+1} \tag{6.66}$$

6.2.4. Statistical Characteristics of System Response

The system response to the random road roughness input can be characterized by a statistical method to determine the probability of a given level response.

In general, under the assumption that the vehicle system is linear, the PSD of system response $G_r(f)$ is related to the PSD of road input $G_\zeta(f)$ by

$$G_r(f) = \left|H(f)\right|^2 G_\zeta(f) \tag{6.67}$$

where r stands for any of the response quantities $\ddot{z}_s$, h_d, or F_{zd}, and $|H(f)|$ is the module of system response function $H(f)$ for the system response r to the road input ζ.

In fact, as the probability for response quantities $\ddot{z}_s$, h_d, or F_{zd} being positive is the same as being negative, the mean values of these quantities are nearly zero. Their variances are, therefore, identical to mean-square values, that is,

$$\sigma_r^2 = \int_0^\infty G_r(f)\mathrm{d}f = \int_0^\infty \left|H(f)\right|^2 G_\zeta(f)\mathrm{d}f \tag{6.68}$$

where σ_r is the standard deviation.

When the road roughness and vehicle speed are known, the effect of suspension parameters can be evaluated and optimized by Equations (6.67) and (6.68).

Due to the stochastic nature of the road excitation, the ride comfort requirements are often put forward by the probability distributions of response quantities. A great deal of measurement data suggests that both the road displacement input and the system response will follow normal distribution with zero mean. For a response quantity r, its probability distribution of response quantities depends only on standard deviation σ_r. Let P represent probability, and the probability of the absolute value of the response quantity amplitude exceeding the threshold value $\lambda\sigma_r$, that is, $P = P(|r| > \lambda\sigma_r)$, is given by

$$P\left(|r| > \lambda\sigma_r\right) = \frac{1}{\sigma_r\sqrt{2\pi}}\int_{\lambda\sigma_r}^{\infty} \mathrm{e}^{-\frac{x^2}{2\sigma_r^2}}\mathrm{d}r \tag{6.69}$$

Some typical values are listed in Table 6.2 [6.6].

TABLE 6.2 Typical probability values for normally distributed response data.

λ	1	2	2.58	3	3.29
$P(\|r\| > \lambda\sigma_r)$	31.7%	4.6%	1%	0.3%	0.1%

Reprinted with permisssion from Yu, Z.-S. (2009) Automotive Theory. China Machine Press, Beijing.

Since it is often assumed that the frequency index $w = 2$ for most of the road surfaces, by taking Equations (6.34) and (6.67) into account, the PSD of sprung mass response expressed in Equation (6.38) becomes

$$G_r(f) = |H(f)|^2 G_\varsigma(f) = 4\pi^2 v_0^2 V G_\varsigma(v_0)|H(f)|^2 \tag{6.70}$$

Apparently, the PSD of sprung mass response $G_r(f)$ is independent on the frequency:

$$G_\varsigma(f) = \frac{1}{V} G_\varsigma(v_0)\left(\frac{v}{v_0}\right)^{-w} = G_\varsigma(v_0) v_0^w \frac{V^{w-1}}{f^w} \tag{6.71}$$

In addition, the temporal frequency-based unevenness velocity and acceleration can be obtained by substituting Equation (6.69) into Equations (6.70) and (6.71) and reorganizing

$$G_{\dot\varsigma}(f) = (2\pi f)^2 G_\varsigma(f) = 4\pi^2 v_0^w \frac{V^{w-1}}{f^{w-2}} G_\varsigma(v_0) \tag{6.72}$$

$$G_{\ddot\varsigma}(f) = (2\pi f)^4 V G_\varsigma(f) = 16\pi^4 v_0^w \frac{V^{w-1}}{f^{w-4}} G_\varsigma(v_0) \tag{6.73}$$

In the absence of external forces, road displacement, and damping, the system equations, that is, Equations (6.69) and (6.70), can be reduced to the following:

$$\begin{cases} m_s \ddot{z}_s + K(z_s - z_u) = 0 \\ m_u \ddot{z}_u + K(z_u - z_s) + K_t z_u = 0 \end{cases} \tag{6.74}$$

In the dual-mass system, each mass has its own natural frequency if the other mass does not move. The natural frequencies for the sprung mass and unsprung mass are, respectively,

$$\omega_0 = \sqrt{K / m_s} \tag{6.75}$$

$$\omega_{u0} = \sqrt{(K + K_t) / m_u} \tag{6.76}$$

The main frequencies of the dual-mass system are given by the characteristic equation as follows:

$$\omega^4 - \left(\omega_{u0}^2 + \omega_0^2\right)\omega^2 + \omega_{u0}^2\omega_0^2 - \frac{\omega_0^2 K}{m_u} = 0 \tag{6.77}$$

By solving Equation (6.79), we can obtain two main frequencies of the dual-mass system:

$$\omega_{1or2}^2 = \frac{\omega_{u0}^2 + \omega_0^2}{2} \mp \sqrt{\frac{\left(\omega_{u0}^2 + \omega_0^2\right)^2}{4} - \frac{KK_t}{m_u m_s}} \tag{6.78}$$

For a typical vehicle, since $m_u \ll m_s$, and $K \ll K_t$, it can be found out from Equation (6.80) that $\omega_1 \approx \omega_0$, $\omega_2 \approx \omega_{u0}$, and $\omega_1 < \omega_0 < \omega_{u0} < \omega_2$.

Moreover, the amplitude ratios of these two dominant modes are as follows:

The first-order dominant mode is

$$\left(\frac{z_{u0}}{z_{s0}}\right)_1 = \frac{\omega_0^2 - \omega_1^2}{\omega_0^2} \tag{6.79}$$

and the second-order dominant mode is

$$\left(\frac{z_{u0}}{z_{s0}}\right)_2 = \frac{\omega_0^2 - \omega_2^2}{\omega_0^2} \tag{6.80}$$

As illustrated in Figure 6.11 [6.6], a low-frequency resonance is excited as the first-order dominant mode when the excitation frequency approaches ω_1.

FIGURE 6.11 Two dominant modes of a dual-mass system of two DOF.

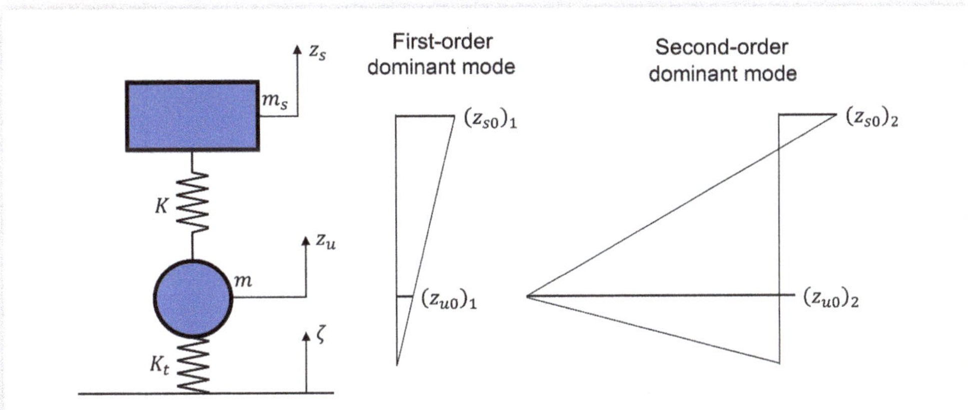

Since the amplitude of the sprung mass is much greater than that of the unsprung mass, it is therefore referred to as body-type vibrations. On the contrary, a high-frequency resonance is excited as the second-order dominant mode when the excitation frequency is in the vicinity of ω_2. As a result, the amplitude of the unsprung mass is much greater than that of the sprung mass. By assuming the sprung mass as stationary and the presence of damping, the high-frequency behavior of unsprung mass can be evaluated by simplifying Equation (6.51) into the following:

$$m_u \ddot{z}_u + C\dot{z}_u + (K + K_t)z_u - K_t\zeta = 0 \tag{6.81}$$

For simply harmonic vibrations, the frequency response function of the tire displacement with respect to road displacement is

$$\frac{z_u}{\zeta} = \frac{K_t}{K + K_t - m_u\omega^2 + j\omega C} \tag{6.82}$$

Substitution of the natural frequency in Equation (6.76) and the damping ratio of the unsprung mass, $\xi_u = C / \sqrt{4m_u(K + K_t)}$, into Equation (6.82) yields

$$\frac{z_u}{\zeta} = \frac{K_t / (K + K_t)}{1 - (\omega / \omega_u)^2 + j(2\xi_u \omega / \omega_u)} \tag{6.83}$$

When high-frequency resonance occurs at $\omega = \omega_u$, it can be shown that the vertical acceleration of the unsprung mass is

$$\left| \frac{\ddot{z}_u}{\zeta} \right|_{\omega = \omega_u} = \frac{\omega_u K_t / (K + K_t)}{2\xi_u} \tag{6.84}$$

The acceleration of unsprung mass at high-frequency resonance can be lowered by the reduction in tire stiffness K_t. On the other hand, a decrease in unsprung mass m_u does not make much difference in the acceleration of unsprung mass as this would elevate both natural frequency ω_{u0} and damping ratio ξ_u. Nevertheless, the dynamic load on the unsprung mass, $m_u \ddot{z}_u$, drops, leading to a decline in the relative dynamic load $F_d/m_u g$.

6.2.4.1. Transfer Characteristics of the Dual-Mass System

Let $S_1 = j\omega C + K$, $S_2 = -\omega^2 m_s + K + j\omega C$, and $S_3 = -\omega^2 m_u + K + K_t + j\omega C$, and the frequency response functions of the sprung and unsprung masses for the dual-mass system defined by Equations (6.51) and (6.52) are, respectively,

$$\frac{z_s}{z_u} = \frac{S_1}{S_2} \tag{6.85}$$

and

$$\frac{z_u}{\zeta} = \frac{S_2 K_t}{S_2 H_3 - S_1^2} \tag{6.86}$$

By defining the stiffness ratio $r_k = K_t/K$ and the damping ratio $\xi = C/\sqrt{4m_u K}$, the amplitude–frequency characteristics of the sprung and unsprung masses become, respectively [6.6],

$$\left|\frac{z_s}{z_u}\right| = \sqrt{\frac{1+4\xi^2\lambda^2}{\left(1-\lambda^2\right)^2+4\xi^2\lambda^2}} \tag{6.87}$$

$$\left|\frac{z_u}{\zeta}\right| = r_k\sqrt{\frac{\left(1-\lambda^2\right)^2+4\xi^2\lambda^2}{\Delta}} \tag{6.88}$$

where

λ is the frequency ratio, $\lambda = \omega/\omega_0$

$$\Delta = \left\{\left[1-\left(\frac{\omega}{\omega_0}\right)^2\right]\left[1+r_k-\chi\left(\frac{\omega}{\omega_0}\right)^2\right]-1\right\}^2 + 4\xi^2\left(\frac{\omega}{\omega_0}\right)^2\left[r_k-(\chi+1)\left(\frac{\omega}{\omega_0}\right)^2\right]^2$$

The typical amplitude–frequency characteristics given by Equations (6.87) and (6.88) are plotted in Figure 6.12 [6.6], with $\gamma = 9$ and $\chi = 0.1$. When the excitation frequency is low, that is, $f \le f_0$, both $\left|\frac{z_s}{z_u}\right|$ and $\left|\frac{z_u}{\zeta}\right|$ approach 1; when the excitation frequency is high, that is, $f \ge f_u$, $\left|\frac{z_s}{z_u}\right|$ approaches the asymptote of –1:1 slope, while $\left|\frac{z_u}{\zeta}\right|$ approaches that of –2:1 slope. A low-frequency resonance occurs at $f = f_0$ for the sprung mass, where $\left|\frac{z_s}{z_u}\right|$ peaks. When the frequency becomes higher than $\sqrt{2}f_0$, the suspension starts to attenuate the resonance, with $\left|\frac{z_s}{z_u}\right|$ dropping below 1. Moreover, a high-frequency resonance occurs at $f = f_u$ for the unsprung mass, where $\left|\frac{z_u}{\zeta}\right|$ peaks. When the frequency exceeds f_u, the resonance is attenuated by the tire, with $\left|\frac{z_u}{\zeta}\right|$ dropping below 1.

FIGURE 6.12 Typical amplitude-frequency characteristics. (a) Sprung mass and (b) unsprung mass.

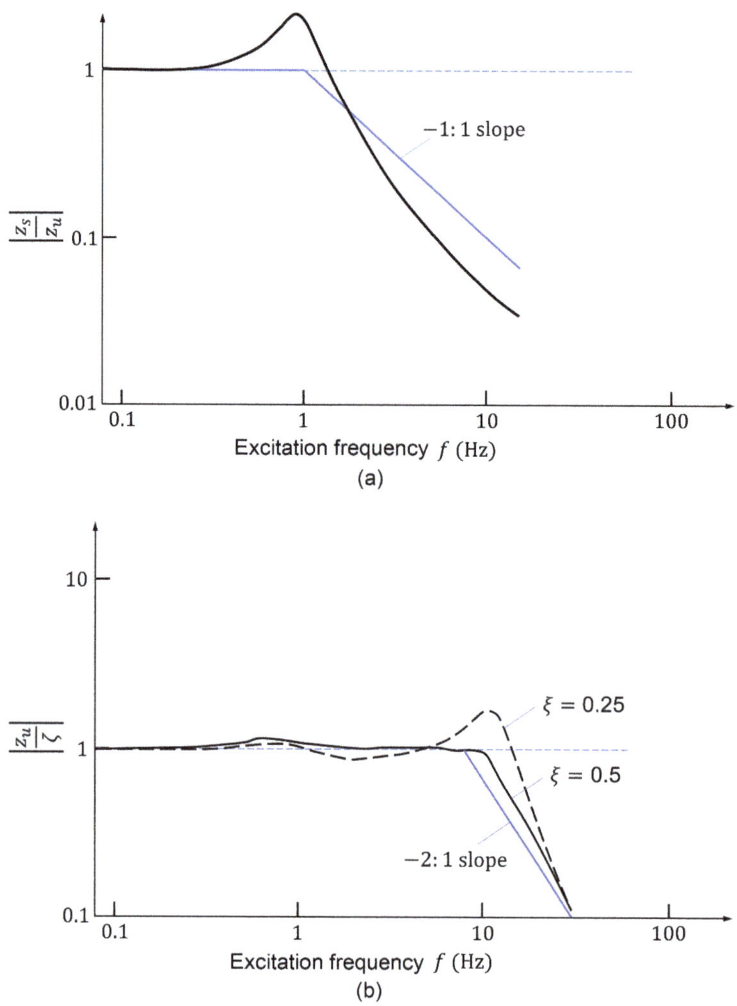

The frequency response function of sprung mass with respect to road displacement can be obtained by combining Equations (6.85) and (6.86):

$$\frac{z_s}{\zeta} = \frac{z_s}{z_u}\frac{z_u}{\zeta} = \frac{S_1}{S_2}\frac{S_2 K_t}{S_2 S_3 - S_1^2} = \frac{S_1 K_t}{S_2 H_3 - S_1^2} \tag{6.89}$$

The corresponding amplitude–frequency characteristics can be derived by applying Equations (6.87) and (6.88):

$$\left|\frac{z_s}{\zeta}\right| = \left|\frac{z_s}{z_u}\right|\left|\frac{z_u}{\zeta}\right| = r_k\sqrt{\frac{1+4\xi^2\lambda^2}{\Delta}} \tag{6.90}$$

The $\left|\frac{z_s}{\zeta}\right|$ value is actually the superimposition of $\left|\frac{z_s}{z_u}\right|$ and $\left|\frac{z_u}{\zeta}\right|$ in terms of the respective slope of their asymptotes, with peaks at the natural frequencies of sprung and unsprung masses, that is, f_0 and f_u. The road excitation is attenuated by the suspension at $f \geq f_0$ and then further attenuated by the tire at $f \geq f_u$.

6.2.4.2. Amplitude–Frequency Characteristics of Sprung Mass Acceleration

The vehicle body (or sprung mass) acceleration is an important quantity in evaluating ride comfort. Its amplitude–frequency characteristics can be determined by Equation (6.90):

$$\left|\frac{\ddot{z}_s}{\dot{\zeta}}\right| = \omega\left|\frac{z_s}{\zeta}\right| = \omega r_k\sqrt{\frac{1+4\xi^2\lambda^2}{\Delta}} \tag{6.91}$$

The typical amplitude–frequency characteristics feature two resonance peaks. An increase in damping ratio helps attenuate the resonance around the natural frequency of the suspension f_0.

6.2.4.3. Amplitude–Frequency Characteristics of Suspension Deflection

The dynamic deflection of the suspension spring is closely related to its travel and setup of the travel limiter. If not designed appropriately, the ride comfort can be deteriorated by an increase in the chances of hitting the limiter. The frequency response function of the dynamic deflection of the suspension spring can be derived by substituting Equations (6.86) and (6.89) into its definition:

$$\frac{h_d}{\zeta} = \frac{Z-z}{\zeta} = \frac{Z}{\zeta} - \frac{z}{\zeta} = \frac{(S_1 - S_2)K_t}{S_2S_3 - S_1^2} \tag{6.92}$$

where h_d is the dynamic deflection of the suspension spring.

Thus, the amplitude–frequency characteristic of suspension deflection becomes

$$\left|\frac{h_d}{\dot{\zeta}}\right| = \frac{r_k\lambda^2}{\omega\sqrt{\Delta}} \tag{6.93}$$

Similarly, a typical amplitude–frequency characteristic of suspension deflection features two resonance peaks.

6.2.4.4. Amplitude–Frequency Characteristics of Dynamic Load on Wheels

The dynamic load on a wheel can affect the adhesion between the tire and the ground, undermining driving safety. Since the static load on a wheel is $F_{zs} = (m_u + m_s)g = (1 + \chi)m_s g$ and the dynamic load on a wheel is $F_{zd} = (z_u - \zeta)K_t$, the frequency response function of the relative dynamic load on a wheel is

$$\frac{F_{zd}}{F_{zs}\zeta} = \frac{z_u - \zeta}{\zeta} \frac{K_t}{(1+\chi)m_s g} = \left(\frac{S_2 K_t}{S_2 S_3 - S_1^2} - 1\right)\frac{\chi K_t}{(1+\chi)m_u g} \tag{6.94}$$

The amplitude–frequency characteristics of relative dynamic load on the wheel are

$$\left|\frac{F_{zd}}{F_{zs}\zeta}\right| = \frac{r_k \omega}{g}\sqrt{\frac{\left(\frac{x\lambda^2}{1+\chi} - 1\right)^2 + 4\xi^2\chi^2}{\Delta}} \tag{6.95}$$

The typical amplitude–frequency characteristics of relative dynamic load on the wheel feature two resonance peaks that are largely affected by the damping ratio ξ.

6.2.4.5. Statistical Characteristics of System Response to Road Excitation

Based on the amplitude–frequency characteristics expressed in Equations (6.93), (6.95), (6.97) and the road speed power spectrum depicted in Equation (6.35), the statistical characteristics of system response to the random input from road excitations can be obtained in the form of mean-square because the mean values of these responses are all zero. Thus, the mean-square of body acceleration is

$$\sigma_{\ddot{z}_s}^2 = \int_0^\infty \left|\frac{\ddot{z}_s}{\dot{\zeta}}\right|^2 G_{\dot{\zeta}}(f)\mathrm{d}f = 4\pi^2 v_0^w V^{w-1} G_\zeta(v_0)\int_0^\infty f^{2-w}\left|\frac{\ddot{z}_s}{\dot{\zeta}}\right|^2 \mathrm{d}f \tag{6.96}$$

Similarly, the mean-square of dynamic deflection of suspension spring is

$$\sigma_{h_d}^2 = \int_0^\infty \left|\frac{h_d}{\dot{\zeta}}\right|^2 G_{\dot{\zeta}}(f)\mathrm{d}f = 4\pi^2 v_0^w V^{w-1} G_\zeta(v_0)\int_0^\infty f^{2-w}\left|\frac{h_d}{\dot{\zeta}}\right|^2 \mathrm{d}f \tag{6.97}$$

and the mean-square of relative dynamic load on the wheel is

$$\sigma_{\frac{F_{zd}}{F_{zs}}}^2 = \int_0^\infty \left|\frac{F_{zd}}{F_{zs}\dot{\zeta}}\right|^2 G_{\dot{\zeta}}(f)\mathrm{d}f = 4\pi^2 v_0^w V^{w-1} G_\zeta(v_0)\int_0^\infty f^{2-w}\left|\frac{F_{zd}}{F_{zs}\dot{\zeta}}\right|^2 \mathrm{d}f \tag{6.98}$$

Equations (6.96) through (6.98) can be solved only numerically most of the time.

6.3. Effects of System Parameters on Vibration Response

The effects of sprung mass natural frequency, mass ratio, stiffness ratio, and damping ratio on the vibration response of the dual-mass system are elaborated here.

The effect of sprung mass natural frequency on vibration response can be revealed by the variations of mean-square roots $\sigma_{\ddot{z}_s}$, σ_{h_d}, and $\sigma_{\frac{F_{zd}}{F_{zs}}}$ with sprung mass natural frequency. The effects of sprung mass natural frequency on vibration responses are significant, which are often presented by the curves of vertical acceleration of sprung mass, dynamic deflection of suspension spring, and relative dynamic load on the wheel as functions of the excitation frequency under various sprung mass natural frequencies at given χ, r_k, and ξ.

In contrast, the variations of mean-square roots $\sigma_{\ddot{z}_s}$, σ_{h_d}, and $\sigma_{\frac{F_{zd}}{F_{zs}}}$ with the mass ratio indicate that the effect of the mass ratio on vibration response, often reflected by the curves of vertical acceleration of sprung mass, dynamic deflection of suspension spring, and relative dynamic load on the wheel as functions of the excitation frequency under various mass ratios at given f_0, r_k, and ξ, is insignificant. Thus, the increase in the mass ratio χ (or decrease in unsprung mass) affects primarily driving safety.

Moreover, the variations of mean-square roots $\sigma_{\ddot{z}_s}$, σ_{h_d}, and $\sigma_{\frac{F_{zd}}{F_{zs}}}$ with the stiffness ratio indicate that the effect of the stiffness ratio on vibration response is also trivial, as the curves of vertical acceleration of sprung mass, dynamic deflection of suspension spring, and relative dynamic load on the wheel under various stiffness ratios versus relative dynamic load on the wheel at given f_0, χ, and ξ nearly lie on top of each other for relatively low excitation frequencies. Thus, the decrease in stiffness ratio r_k (e.g., reduction in tire inflation pressure) helps improve the adhesion at the tire–ground interface.

At last, the variations of mean-square roots $\sigma_{\ddot{z}_s}$, σ_{h_d}, and $\sigma_{\frac{F_{zd}}{F_{zs}}}$ with the damping ratio indicate that the damping ratio has quite substantial influences on the vertical acceleration of sprung mass, dynamic deflection of suspension spring, and the relative dynamic load on the wheel. Therefore, the stiffness ratio ξ should be selected appropriately to compromise between ride comfort and driving safety.

6.4. Human Perception of Vibrations

The influence of vibrations on the human body is generally dependent on factors such as the frequency, intensity, direction, and duration of these vibrations. This subjective perception is also quite personal as it is substantially affected by the physical and psychological conditions of the passenger traveling in a vehicle.

The influence of these prolonged random vibrations on the human body is normally evaluated with the human model in seated, standing, and recumbent positions. Due to the multi-input, multiaxis nature of vibrations, three coordinate systems are employed to describe the vibration inputs from the seat, seat back, and feet. Thus, the translational and rotational vibrations can be taken into account by using the frequency weighting method. Different weighting curves can be selected to reflect the frequency dependency of the human perception.

In general, the human body is most sensitive to the translational vibration inputs in the x-, y-, and z-axes from the seat surface. When it comes to the effect of vibrations on health, the weight factors along the x- and y-axes are greater than that along the z-axis.

For human bodies, the internal organs go into resonance at frequencies ranging from 4 to 8 Hz, while the spine system is mostly vulnerable under frequencies ranging from 8 to 12.5 Hz. Thus, the frequency weighting factor along the z-axis vertical to the seat surface is most sensitive for frequencies ranging from 4 to 12.5 Hz. On the other hand, human bodies are more sensitive to horizontal vibrations than to vertical vibrations. Hence, the frequency weighting factors along the x- and y-axes for the seat surface are usually greater than that along the z-axis. Besides, in view of up to 3 Hz of the natural frequency of the single DOF system of the human body and seat as well as 1.2–2 Hz of the natural frequency of the vehicle body, it is generally recommended that the parameters of the human body and seat system be selected so that its natural frequency is at least 3 Hz or higher in order to avoid their overlap in the frequency band. A proper damping ratio should also be designed to ensure the vibration frequency ranging from 4 to 12 Hz is attenuated.

6.5. Evaluation of Ride Quality

The effect of vibrations on the comfort and health of passengers are generally evaluated using different methods that are dependent on the severity of vibrations.

Defined as the modulus of the ratio of the maximum instantaneous peak value of the frequency-weighted acceleration signal to its root-mean-square (RMS) value, the crest factor is often employed to indicate the severity of vibrations although it is not necessarily the case. The basic evaluation method is used for the crest factor less than 9; otherwise, the alternative evaluation method is used instead.

6.5.1. Basic Evaluation Method

In the basic evaluation method, the weighted RMS acceleration a_w is expressed as

$$a_w = \sqrt{\frac{1}{T}\int_0^T a_w^2(t)\,dt} \tag{6.99}$$

where

$a_w(t)$ is the weighted acceleration of translational or rotational vibrations as a function of time, in m/s^2 or rad/s^2

T is the time duration of vibrations under evaluation (typically $T = 120$ sec)

The numerical values of the weighting factors can be found by referring to relevant standards, for example, ISO2631-1.

For evaluating the comfort of a seated person, the following vibration total value a_v is recommended:

$$a_v = \sqrt{\left(1.4a_{wx}\right)^2 + \left(1.4a_{wy}\right)^2 + a_{wz}^2} \tag{6.100}$$

where a_{wx}, a_{wy}, and a_{wz} are the weighted RMS accelerations relative to the orthogonal axes x, y, and z, respectively.

In addition, a weighted vibration level L_w is sometimes used as an alternative:

$$L_w = 20\log\frac{a_w}{a_0} \tag{6.101}$$

where a_0 is the reference RMS acceleration, $a_0 = 10^{-6}$ m/s^2.

The human perception is listed against different levels of vibration severity in terms of a_w and L_w in Table 6.3 [6.7].

TABLE 6.3 Human perception versus a_w and L_w.

Human perception	Weighted RMS acceleration a_w (m/s^2)	Weighted vibration level L_w (dB)
Not uncomfortable	<0.315	110
A little uncomfortable	0.315-0.63	110–116
Fairly uncomfortable	0.5-1.0	114–120
Uncomfortable	0.8-1.6	118–124
Very uncomfortable	1.25-2.5	112–128
Extremely uncomfortable	>2.0	126

6.5.2. Alternative Evaluation Method

As the alternative evaluation method, the fourth power vibration dose method is more sensitive to peaks than the basic evaluation method. The fourth power vibration dose value (VDV) is defined as

$$\mathrm{VDV} = \sqrt[4]{\int_0^T a_w^4(t)\,\mathrm{d}t} \tag{6.102}$$

where

$a_w(t)$ is the instantaneous frequency-weighted acceleration of translational or rotational vibrations as a function of time, in m/s^2 or rad/s^2

T is the time duration of vibrations under evaluation (typically T = 120 sec)

VDV is in m/s$^{1.75}$ or rad/s$^{1.75}$

6.5.3. Sound Quality Evaluation

In general, the sound quality in the passenger cabin can be evaluated using subjective and objective approaches.

In the subjective approach, the annoyance rating is often assessed by a group of professional human assessors according to the criteria listed in Table 6.4 [6.8]. Ranging from 1 to 10, the annoyance ratings can be processed statistically to come up with the mean and standard deviation.

TABLE 6.4 Annoyance rating criteria of subjective sound quality evaluation.

Annoyance rating	Description
1	Unacceptable to all
2	Unacceptable to most customers
3	Some customers cannot accept
4	Most customers complain
5	Few customers accept, most customers complain
6	Some customers accept, some customers complain
7	Most customers accept, few customers complain
8	Strict customers accept, very few customers complain
9	Very good, most customers like it
10	Very pleasant

Courtesy of Noise and Vibration Control.

The objective approach, on the other hand, involves the use of psychoacoustic metrics for quantifying monotonic noise called the tone-to-noise ratio (TNR) and total tone-to-noise ratio (total TNR). The TNR is defined as [6.9]

$$\text{TNR} = 10\log_{10}\left(\frac{P_t}{P_n}\right) \tag{6.103}$$

where P_t is the sound pressure level of the tone in Pa and P_n is the sound pressure level of the broadband noise in Pa with the sound pressure level of the tone subtracted. Thus [6.9],

$$P_n = \frac{\Delta f_m}{\Delta f_m - \Delta f_t}\left(P_m - P_t\right) \tag{6.104}$$

where

Δf_m is the bandwidth of the broadband noise
Δf_t is the bandwidth of the tone
P_m is the sound pressure level of the masking broadband noise in Pa

The total TNR is defined as [6.9]

$$\text{Total TNR} = 10\log_{10}\left(\sum_{i=1}^{k}\text{TNR}_i\right) \tag{6.105}$$

where k is the number of discrete tones. Compared to the TNR that is capable of characterizing the whine noise of a single order, the total TNR is more suitable for the presence of the whine noise of multiple orders at the same time.

Besides the TNR and total TNR, other comprehensive approaches are available that take additional psychoacoustic objective parameters into account, such as Zwicker loudness, sharpness, roughness, waviness, and AI index (voice intelligibility). A regression model can be established in the form of linear, quadratic, cubic, logarithmic, or exponential functions, and the model accuracy can be evaluated and improved by comparing the results with those of a subjective approach with Spearman's correlation analysis [6.9]. A neural network model is utilized sometimes as an alternative method as well.

References

[6.1]. Liang, R., "An Investigation on the Torsional Vibrations of an Electric Vehicle Powertrain," Master's thesis, School of Automotive Studies, Tongji University, Shanghai, 2008.

[6.2]. Oh, S. and Woscek, J., "Analysis of Rzeppa and Cardan Joints in Monorail Drive Train System," *International Journal of Mechanical Engineering and Robotics Research* 4, no. 1 (2015): 1-11, doi:10.18178/ijmerr.

[6.3]. Gillespie, T., *Fundamentals of Vehicle Dynamics* (Warrendale: Society of Automotive Engineers, Inc., 1992), 21-118, ISBN-10:1560911999, ISBN-13:978-1560911999.

[6.4]. Sina, N. and Nasiri, S., "Excitation Behavior of Tire and Wheel Assembly Faults in Shape of Non-uniformity in a Vehicle," Paper Presented at in *8th Condition Monitoring & Fault Diagnosis Conference*, Sharif University of Technology, Iran, February 2014.

[6.5]. ISO, "Mechanical Vibration—Road Surface Profiles—Reporting of Measured Data," ISO Standard 8608:2016, 2016.

[6.6]. Yu, Z., *Automobile Theory*, 3rd ed. (Beijing: China Machine Press, 2000), 1-36, ISBN:7-111-02076-6.

[6.7]. ISO, "Mechanical Vibration and Shock—Evaluation of Human Exposure to Whole-Body Vibration—Part 1: General Requirements," ISO Standard 2631-1:1997, 1997.

[6.8]. Diao, K., Wang, X., and Wang, W., "Subjective and Objective Evaluation Model of Sound Quality of Drive Unit Tonal Noise of Electric Vehicles," *Noise and Vibration Control* 41, no. 3 (2021): 187-191, doi:10.3969/j.issn.1006-1355.2021.03.032.

[6.9]. Wang, Y., Gu, G., and Chen, D., "Analysis of Sound Quality and Electromagnetic Noise for Electric Vehicle," *Automobile Applied Technology* 4 (2018): 3-6, doi:10.16638/j.cnki.1671-7988.2018.04.001.

7

Trafficability

Trafficability of an EV refers to its ability to pass through poor roads or areas without roads (e.g., soft or uneven roads) and overcome various obstacles such as slopes, steps, bushes, moats, and water, at a sufficiently high average speed. In general, trafficability can be classified as ground bearing capacity-related or ground geometry-related. Mainly depending on the physical properties of the ground and structural and geometric parameters of a vehicle, the trafficability is also closely related to the dynamic performance, maneuverability, ride, stability, and visibility of the vehicle.

7.1. Indices and Geometric Parameters for Trafficability Evaluation

7.1.1. Indices for Trafficability Evaluation Related to Ground Bearing Capacity

The ground bearing capacity-related trafficability is commonly evaluated by three indices, that is, traction coefficient, traction efficiency, and energy utilization index.

7.1.1.1. Traction Coefficient

The traction coefficient C_{tr} is defined as the traction force for towing per unit vehicle weight:

$$C_{tr} = \frac{F_t}{mg} \tag{7.1}$$

where

F_t is the tractive effort for towing
m is the vehicle mass

C_{tr} is an indication of the ability of a vehicle to accelerate, climb a slope, and tow other vehicles on soft ground.

7.1.1.2. Traction Efficiency

Traction efficiency η_{tr} is defined as the ratio of output power to input power of the traction wheels:

$$\eta_{tr} = \frac{F_t V}{T_t \omega} = \frac{F_t R(1-s)}{T_t} \tag{7.2}$$

where

V is the vehicle speed
T_t is the input torque to driving wheels
ω is the angular velocity of driving wheels
R is the dynamic radius of driving wheels
s is the slip ratio

η_{tr} measures the energy loss in power transfer at the driving wheels associated with two physical processes: first, the heat generation due to friction between cord and rubber layers in tires; and second, the compaction and deformation of soil under tires.

7.1.1.3. Energy Utilization Index

The energy utilization index I_{eu} is defined as the output power per unit energy consumption:

$$I_{eu} = \frac{F_t V}{P_b} \tag{7.3}$$

where P_b is the power consumption of the battery.

7.1.2. Structural and Geometric Parameters Related to Trafficability

The trafficability of a vehicle deteriorates with a decrease in its ground clearance. In extreme circumstances, the vehicle can get trapped in some terrain as the clearance between the vehicle and the ground becomes inadequate, when the front end, rear end, or underbody of the vehicle touches the ground.

The structural and geometric parameters for evaluating the trafficability of a vehicle include minimum ground clearance h_{min}, approach angle γ_1, departure angle γ_2, ramp angle β, and minimum turning radius r_{min}, as illustrated in Figure 7.1.

FIGURE 7.1 Structural and geometric parameters for trafficability assessment of a vehicle.

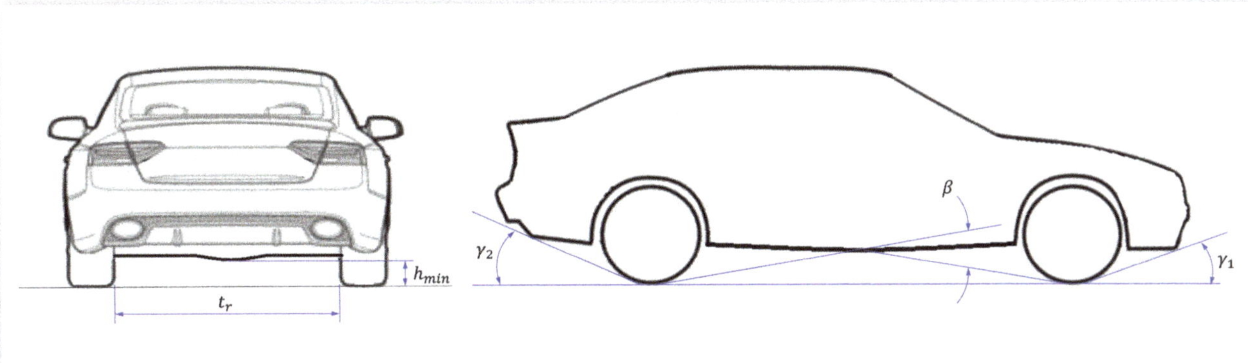

7.1.2.1. Minimum Ground Clearance h_{min}

The minimum ground clearance h_{min} is defined as the distance between the supporting plane and the lowest point within $0.8t_r$ around the central area of the underbody when the vehicle is fully loaded and stationary. It is an indication of the ability of a vehicle to pass through the bumps on the ground without hitting them. The trafficability improves with increasing h_{min} as the likelihood of the bottom of the vehicle colliding with the ground reduces.

7.1.2.2. Approach Angle γ_1

The approach angle γ_1 is defined as the angle between the ground plane and the line that connects to the protruding point of the front end and is tangential to the front wheels when the vehicle is fully loaded and stationary. The trafficability improves with increasing γ_1 as the likelihood of the frond end colliding with the ground reduces.

7.1.2.3. Departure Angle γ_2

The departure angle γ_2 is defined as the angle between the ground plane and the line that connects to the protruding point of the rear end and is tangential to the rear wheels when the vehicle is fully loaded and stationary. The trafficability improves with increasing γ_2 as the likelihood of the rear end colliding with the ground reduces.

7.1.2.4. Ramp Angle β

The ramp angle β is defined as the acute angle formed by the two planes intersecting at the lowest portion of the underbody, each tangential to the peripheral of front and rear tires and both perpendicular to the longitudinal symmetric plane of the vehicle when the vehicle is fully loaded and stationary. It represents the outline size of a vehicle passing through obstacles such as hills and arch bridges without hitting them. The trafficability improves with increasing β as the likelihood of the central underbody colliding with the ground reduces.

7.1.2.5. Minimum Turning Radius r_{min}

The minimum turning radius r_{min} is defined as the radius of the trajectory circle of the center plane of outside turning wheels on the supporting plane when the vehicle is driven at the minimum stable speed with the steering wheel turned to its extreme angle. It reflects the ability of the vehicle to pass through narrow band or go around insurmountable obstacles. The trafficability improves with decreasing r_{min} as the maneuverability becomes better.

7.1.2.6. Turning Path Circle

The turning path refers to the projected outlines of the vehicle on the supporting plane when the vehicle is driven at the minimum stable speed with the steering wheel turned to its extreme angle. The smallest circle that keeps the entire turning path inside is referred to as the outer circle of the turning path, while the largest circle that keeps the entire turning path outside is referred to as the inner circle of the turning path. The difference in the radii of outer and inner circles is the width of the space that the vehicle takes to make a turn. The maneuverability and trafficability are enhanced as the difference in radii decreases.

The typical structural and geometric parameters of some vehicles are listed in Table 7.1 [7.1].

TABLE 7.1 Typical structural and geometric parameters of some vehicles.

Type of EVs	h_{min} (m)	γ_1 (°)	γ_2 (°)	r_{min} (m)
4×2 Sedan	0.12-0.20	20-30	15-22	3.5-6.5
4×4 Sedan/SUV	0.21-0.37	45-50	35-40	5.0-7.5
4×2 Truck	0.25-0.30	25-60	25-45	4.0-7.0
4×4 or 6×6 Truck	0.26-0.35	45-60	35-45	5.5-10.5
6×4 or 4×2 Bus	0.22-0.37	10-40	6-20	7.0-11.0

Reprinted with permisssion from Yu, Z.-S. (2009) Automotive Theory. China Machine Press, Beijing.

7.2. Physical Properties of Soft Ground

Sand, snow, soil, and swamp are several examples of soft ground that a vehicle can drive onto. In these circumstances, the drive wheels exert a force that is tangential to the ground. With the shear deformation occurring in the soft ground, a shear force is developed as the propulsion force that drives the vehicle. Meanwhile, the loads on the wheels compact the sand, snow, or soil, resulting in tracks that produce resistance to the vehicle. The drawbar pull, or the net traction force on soft ground, is actually the difference between the propulsion and resistance forces.

7.2.1. Relationship between Shear Stress and Shear Deformation in Soft Ground

In general, there are two types of soft ground, that is, clayey or frictional. Clay and wet snow are examples of clayey ground, while sand and frozen grainy snow are instances of frictional ground. Their shear characteristics differ from each other completely.

For the clayey ground, the maximum shear force depends only on the cohesion of the soil or wet snow and the area of the tire–ground interface, with no influence from the vertical load on the tires:

$$F_x = AC \tag{7.4}$$

where

A is the area of the tire–ground interface

C is the cohesion coefficient of the snow or wet snow

As far as the frictional ground is concerned, no shear force can be produced when the sand or frozen grainy snow is loose. When compacted, however, friction occurs between particles, eliminating the relative motion. As a result, the friction force is produced at the interface between the tire and stationary sand or frozen grainy snow that is compacted by the normal force. Proportional to the vertical load on the tire, the maximum traction force can be expressed per Coulomb's law of friction:

$$F_x = F_z \tan\varphi \tag{7.5}$$

where

F_z is the vertical load on the tire

φ is the friction angle

In reality, most of the soft ground is neither purely clayey nor frictional. Instead, the soft ground often behaves as a mixture of clayey and frictional grounds. The maximum shear force can therefore be expressed as

$$F_x = AC + F_z \tan\varphi \tag{7.6}$$

Divided by A on both sides of Equation (7.6), we obtain

$$\tau_{max} = C + \sigma \tan\varphi \tag{7.7}$$

where

τ_{max} is the maximum shear stress

σ is the normal stress on the shear interface

Both C and φ can be obtained by means of shear strength measuring instruments.

For plastic ground such as wet clay, loose sand, dry snow, and most of the disturbed soil, the shear stress increases with the shear deformation as the latter increases from zero and then gradually approaches the maximum shear stress τ_{max}. The relationship can be expressed by

$$\tau = \tau_{max}\left(1 - e^{-\frac{\delta_s}{E_s}}\right) = \left(C + \sigma \tan\varphi\right)\left(1 - e^{-\frac{\delta_s}{E_s}}\right) \tag{7.8}$$

where

E_s is the shear deformation modulus of the plastic ground

δ_s is the shear deformation

The slope of the $\tau - \delta_s$ curve at the origin can be obtained by taking a derivative of Equation (7.8):

$$\left.\frac{d\tau}{d\delta_s}\right|_{\delta_s=0} = \left.\frac{\tau_{max}}{\delta_s} e^{-\frac{\delta_s}{E_s}}\right|_{\delta_s=0} = \frac{\tau_{max}}{\delta_s} \tag{7.9}$$

Apparently, E_s is equal to the distance from the ordinate axis to the intersection of the line tangential at the origin with the extension line of the level portion. As a measure of the deformation under the action of maximum shear stress, E_s varies from approximately 2.5 for loose sand to 0.6 for compacted clay, depending on the firmness of the ground.

For brittle ground, for example, compacted sand, silt, frozen snow, and undisturbed brittle soil, the shear stress increases with the shear deformation as the latter increases from zero until it peaks; then, it decreases gradually to a residual stress τ_r. The relationship between the shear stress and corresponding deformation is given by

$$\tau = \left(C + \sigma \tan\varphi\right)\frac{\lambda}{\lambda_{max}} \tag{7.10}$$

where

$\lambda = e^{\left(-K_2+\sqrt{K_2^2-1}\right)K_1\delta_s} - e^{\left(-K_2-\sqrt{K_2^2-1}\right)K_1\delta_s}$

$\lambda_{max} = \max(\lambda)$

K_1, K_2 are constants

In engineering practice, the peak in the $\tau - \delta_s$ curve can be neglected most of the time for its insignificant effect on the traction force, and Equation (7.8) can be used in the shear stress calculation, instead of Equation (7.9).

7.2.2. Relationship between Subsidence and Vertical Loads

By simplifying the tire–ground interface into a plane, the pressure on the unit area of the plane, p, can be expressed as

$$p = kz^n \tag{7.11}$$

where z is the subsidence, n is the subsidence index, and $k = \frac{k_c}{b} + k_\varphi$, in which k_c is the cohesion modulus, k_φ is the friction modulus, and b is the minor axis of the tire mark. The values of k_c, k_φ, and n can be measured by instruments on homogeneous samples of soil, sand, clay, or snow. The field measurements are carried out with a plate larger than 5–10 cm in width at a downward speed of 2.5–5 cm/s to minimize the inhomogeneity of the soil, sand, clay, or snow.

Typically, curves of subsidence as a function of pressure for various soft grounds measured with the same plate are measured with homogeneous pressure in the absence of horizontal forces. In reality, the horizontal forces can hardly be eliminated when the vehicle is traveling. It is observed that the shear deformation resulting from horizontal forces would increase the subsidence. This additional subsidence is referred to as sliding subsidence. The characteristic parameters of some common types of soft ground are listed in Table 7.2 [7.2].

TABLE 7.2 Characteristic parameters of common types of soft ground.

Type of soft ground	Humidity (%)	n	k_c (kN/m^{n+1})	k_φ (kN/m^{n+2})	C (kPa)	φ (°)
Dry sand (Land Locomotion Lab.)	0	1.1	0.95	1528.43	1.04	28
Sandy loam (Land Locomotion Lab.)	15	0.7	5.27	1515.04	1.72	29
	22	0.2	2.56	43.12	1.38	38
Sandy loam Michigan (Strong, Buchele)	11	0.9	52.53	1127.97	4.83	20
	23	0.4	11.42	808.96	9.65	35
Sandy loam (Hanamoto)	26	0.3	2.79	141.11	13.79	22
	32	0.5	0.77	51.91	5.17	11
Clayey soil (Thailand)	38	0.5	13.19	692.15	4.14	13
	55	0.7	16.03	1262.53	2.07	10
Heavy clay (WES)	25	0.13	12.70	1555.95	68.95	34
	40	0.11	1.84	103.27	20.69	6
Lean clay (WES)	22	0.2	16.43	1724.69	68.95	20
	32	0.15	1.52	119.61	13.79	11
Snow (Harrison)	–	1.6	4.37	196.72	1.03	19.7
	–	1.6	2.49	245.90	0.62	23.2

Note: The names in parentheses are the organizations that conducted the tests.

7.2.3. Effect of the Density of Semifluid Mud and Snow on Trafficability

Simplified hydrodynamic solutions can be applied to solve the resistance that the vehicle is subjected to when traveling in semifluid mud or snow. The resistance is dependent not only on vehicle speed and immersion area but also on the density of the mud, ρ, and the drag coefficient C_D, with the resistance proportional to both ρ and C_D.

The trafficability of snow can be analyzed using the stress–strain method discussed earlier. In addition, the density and depth of snow need to be taken into account. If the snow depth is less than the ground clearance of the vehicle, then the snow of any density cannot block the vehicle; if the snow depth is greater than 150% of the ground clearance of the vehicle, then light-duty vehicles can travel on snow of density larger than 350 kg/m^3, while heavy-duty vehicles can travel on snow of density larger than 500 kg/m^3. The previous discussion applies to dry snow only because of the greater density and lower strength of wet snow.

7.3. Drawbar Pull of EVs

The drawbar pull is the difference between the traction force and the resistance force. Thus, the traction force and the resistance force must be determined first to obtain the drawbar pull.

7.3.1. Traction Force Acting on the Vehicle from Soft Ground

The traction force that propels the vehicle on soft ground is dependent on the shear characteristics of the soft ground. When the traction force is produced, the shear deformation occurs in the ground, causing the traction wheel to slip backward relative to the ground. The relationship between such a slip motion and the traction force is of significance as it affects both the average speed and energy consumption of the vehicle.

For a rigid traction wheel, the slip velocity of a point on the tread of the tire v_s is a function of angle θ and slip ratio s_r:

$$v_s = \omega R\left[1-(1-s_r)\cos\theta\right] \tag{7.12}$$

where

ω is the angular speed of the wheel

R is the radius of the wheel, as shown in Figure 7.2 [7.2]

FIGURE 7.2 Shear deformation of a rigid wheel at the ground-tread interface.

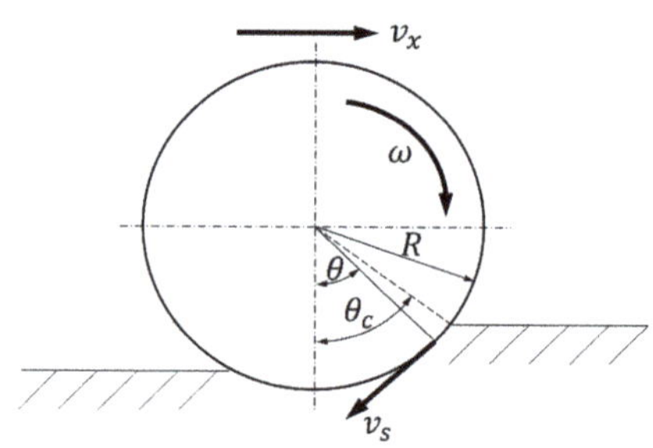

The shear deformation at the ground–tread interface is

$$\delta_s = \int_0^{\theta_c} v_s dt \tag{7.13}$$

Substitution of Equation (7.12) into Equation (7.13) yields

$$\delta_s = R\left[(\theta_c-\theta)-(1-s_r)(\sin\theta_c-\sin\theta)\right] \tag{7.14}$$

where θ_c is the angle corresponding to the portion of the tire that contacts the ground. Combining Equation (7.8) with Equation (7.14), the distribution of shear stress becomes

$$\tau(\theta) = \left(C+\sigma(\theta)\tan\varphi\right)\left(1-e^{-\frac{R}{E_s}\left[(\theta_c-\theta)-(1-s_r)(\sin\theta_c-\sin\theta)\right]}\right) \tag{7.15}$$

where the normal stress $\sigma(\theta)$ can be determined by Equation (7.11).

The traction force can be given by the integration of shear stress about the angle between the wheel–ground interface

$$F_x = \int_0^{\theta_c} bR\tau(\theta)\cos\theta \mathrm{d}\theta \tag{7.16}$$

7.3.2. Resistance on the Vehicle Traveling on Soft Ground

When a vehicle travels on the soft ground, the tire compacts and pushes the soil, sand, snow, or swamp. The resulting reaction force acts as a resistance to the vehicle, retarding its motion. Such a resistance force can be divided into compaction resistance and bulldozing resistance due to the different mechanisms involved. Besides, the elastic deformation of a pneumatic tire results in additional resistance due to hysteresis losses. Hysteresis is a characteristic of a deformable material such that the energy absorbed in the process of deformation is greater than the energy released in the process of recovery.

7.3.2.1. Soft Ground Resistance to a Rigid Rolling Wheel

For adequately soft ground, the rolling of a rubber tire on the ground can be approximated as that of a rigid tire tread. As illustrated in Figure 7.3 [7.1], assumed to be completely radial, the reaction force acting on the rigid driven wheel from the soft ground is $\sigma = p = kz^n$, according to Equation (7.11). Thus, the horizontal component of the reaction force is actually the compaction resistance F_{cx}:

$$F_{cx} = b\int_0^{\theta_c} \sigma R\sin\theta \mathrm{d}\theta \tag{7.17}$$

FIGURE 7.3 Soft ground resistance on a rigid driven wheel.

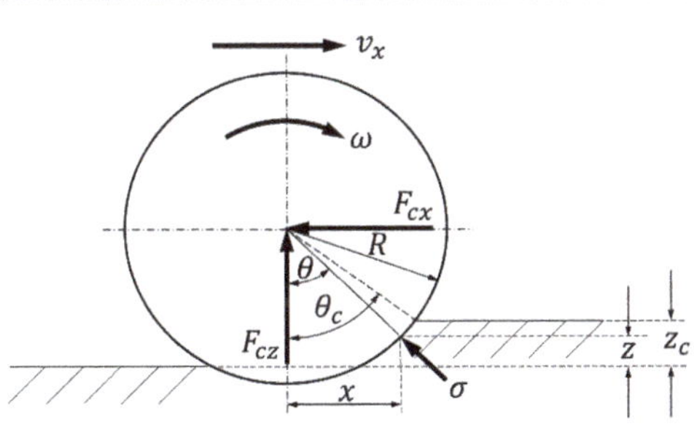

and the vertical component of the reaction force F_{cz} actually bears the vertical load on the wheel,

$$F_{cz} = b\int_0^{\theta_c} \sigma R\cos\theta \mathrm{d}\theta \tag{7.18}$$

Since $\sigma R \sin \theta \mathrm{d}\theta = p\mathrm{d}z$ and $\sigma R \cos \theta \mathrm{d}\theta = p\mathrm{d}z$, Equation (7.17) becomes

$$F_{cx} = b\int_0^{z_c} \left(\frac{k_c}{b} + k_\varphi\right) z^n \mathrm{d}z = \left(k_c + bk_\varphi\right)\frac{z_c^{n+1}}{n+1} \tag{7.19}$$

and Equation (7.18) becomes

$$F_{cz} = b\int_0^{R\sin\theta_c} p\mathrm{d}x = \int_0^{R\sin\theta_c} \left(\frac{k_c}{b} + k_\varphi\right) z^n \mathrm{d}x \tag{7.20}$$

By examining the geometry, we notice

$$x^2 = R^2 - \left[R^2 - (z_c - z)\right]^2 = 2R(z_c - z) - (z_c - z)^2 \tag{7.21}$$

When the subsidence is small, $(z_c - z)^2$ is negligible, $x^2 \approx 2R(z_c - z)$, and $\mathrm{d}x = -\frac{R\mathrm{d}z}{x} = -\sqrt{\frac{R}{2(z_c - z)}}\mathrm{d}z$. Let $\zeta^2 = z_c - z$, $\mathrm{d}\zeta = -\frac{\mathrm{d}z}{2\zeta}$, and the Equation (7.20) becomes

$$F_{cz} = \left(k_c + bk_\varphi\right)\sqrt{2R}\int_0^{\sqrt{z_c}} \left(z_c - \zeta^2\right)^n \mathrm{d}\zeta \tag{7.22}$$

By substituting $(z_c - \zeta^2)^n$ with the first two terms in its expansion, z_c can be solved from Equation (7.22):

$$z_c = \left[\frac{3F_{cz}}{\left(k_c + bk_\varphi\right)(3-n)\sqrt{2R}}\right]^{\frac{2}{2n+1}} \tag{7.23}$$

By substituting Equation (7.23) into Equation (7.19), the compaction resistance is obtained:

$$F_{cx} = \frac{1}{(n+1)\left(k_c + bk_\varphi\right)^{\frac{1}{2n+1}}(3-n)^{\frac{2n+2}{2n+1}}}\left(\frac{3F_{cz}}{\sqrt{2R}}\right)^{\frac{2n+2}{2n+1}} \tag{7.24}$$

From Equation (7.24), it can be inferred that increasing the wheel radius is a more effective way to reduction of compaction resistance than increasing the tread width. Equation (7.24) is valid for any rigid wheels in homogeneous soil, sand, snow, or swamp with medium subsidence. Its accuracy improves with increasing wheel radius or decreasing subsidence. Only when the wheel radius is smaller than 0.25 m, the accuracy starts to drop considerably.

Because the slip is not taken into account in the derivation of Equation (7.24), it is more accurate for clayey soil than for dry sand with significant slip and subsidence.

On a soft and loose ground, the wheel bulldozes the ground to form a bulge wave in the front, producing a bulldozing resistance F_{bx}:

$$F_{bx} = b\left(Cz_c K_c + 0.5 z_c^2 \rho g K_r\right) \tag{7.25}$$

where

$K_c = (N_c - \tan\varphi)\cos^2\varphi$

$K_r = \left(\dfrac{2N_r}{\tan\varphi} + 1\right)\cos^2\varphi$

N_c and N_r are the load-carrying capacity coefficients of the ground

φ is the friction angle

C is the cohesion coefficient

ρ is the density of the sand, soil, or snow

It can be noticed from Equation (7.25) that F_{bx} is proportional to b and the resistance is greater on a narrower tire of larger radius than on a wider tire of a smaller radius.

For an extremely soft and loose ground, the bulldozing resistance F_{bx} can be assessed by

$$F_{bx} = b\left(0.67 Cz_c K_c' + 0.5 z_c^2 \rho g K_r'\right) \tag{7.26}$$

where $K_c' = \left(N_c' - \tan\varphi'\right)\cos^2\varphi'$, $K_r' = \left(\dfrac{2N_r'}{\tan\varphi'} + 1\right)\cos^2\varphi'$, N_c' and N_r' are the load-carrying capacity coefficients of the ground with local shear failures, and $\tan\varphi' = \dfrac{2}{3}\tan\varphi$.

For mud ground, the shape immersed in the mud is of significance to the resistance, where bulldozing resistance overtakes compaction resistance as the major concern. If there is a hard bottom beneath the mud, the bulldozing resistance is given by

$$F_{bx} = \frac{1}{2}\rho C_D A v_x^2 \tag{7.27}$$

where ρ is the density of the mud, A is the projected area of the immersed portion of the tire in the plane perpendicular to the direction of travel, C_D is the drag coefficient that is a function of Reynolds number $R_e = \rho h v_x/\mu$, in which h is the depth of immersion, and μ is the dynamic viscosity of mud.

7.3.2.2. Soft Ground Resistance to a Pneumatic Tire

The rolling behavior of a tire is dependent on its inflation pressure and the firmness of the soil. On very soft and loose soil, when the sum of the inflation pressure p_i and the pressure p_c produced by the stiffness of the carcass is greater than the bearing pressure p_g at the lowest point of the tire circumference that the ground acts on it, then the pneumatic tire would roll just like a rigid wheel. In contrast, when the soil is quite firm, the tire tread touching the ground is flattened. In order to calculate the compaction resistance, it must be determined first whether the tire rolls as a rigid or elastic one.

In practice, p_c can be determined by experiments in which the tire with an inflation pressure p_i is placed on a hard, smooth, and level ground under vertical load F_z. The ground pressure p_g can be calculated by $p_g = F_z/A_g$, where A_g is the area of the tire mark. The ground pressure can be plotted as a function of inflation pressure for various vertical loads. Apparently, $p_c = p_g - p_i$ when p_i is known.

The minimum value of $p_i + p_c$ required to maintain the pneumatic tire in a circular shape as it rolls on the soft ground is referred to as the critical pressure p_{cr}.

When $p_i + p_c \geq p_{cr}$, the tire remains circular and rolls like a rigid wheel. The compaction resistance acting from ground on the tire can be estimated using Equation (7.24). The bearing pressure at the lowest point of the tire circumference is given by using Equations (7.11) and (7.23):

$$p_g = \left(\frac{k_c}{b} + k_\varphi\right) z_c^n = \left(\frac{k_c}{b} + k_\varphi\right)^{\frac{1}{2n+1}} \left(\frac{3F_{cz}}{(3-n)b\sqrt{2R}}\right)^{\frac{2n}{2n+1}} \tag{7.28}$$

Obviously, $p_g = p_{cr}$ in this circumstance.

When $p_i + p_c < p_{cr}$, the part of the tire that comes into contact with the ground becomes flattened, and therefore, $p_g = p_i + p_c$. The subsidence is given by

$$z_c = \left(\frac{p_i + p_c}{k_c/b + k_\varphi}\right)^{\frac{1}{n}} \tag{7.29}$$

The compaction resistance on the rolling elastic tire can be obtained by substituting Equation (7.29) into Equation (7.19):

$$F_{cx} = \left(k_c + bk_\varphi\right)\frac{z_c^{n+1}}{n+1} = \frac{\left[b\left(p_i + p_c\right)\right]^{\frac{n+1}{n}}}{(n+1)\left(k_c + bk_\varphi\right)^{\frac{1}{n}}} \tag{7.30}$$

A hysteresis loss is resulted by the deformation of a pneumatic tire as it rolls. The elastic resistance force associated with the process can be measured in experiments in which the tire of various inflation pressures rolls on a smooth road surface. With the rolling resistance force F_{ex} acquired in the experiments, an elastic resistance force per unit load can be approximated by

$$f_{ex} = \frac{F_{ex}}{F_z} \tag{7.31}$$

An empirical formula can be used to estimate f_{ex}:

$$f_{ex} = \frac{v_x}{p_i^{\kappa}} \tag{7.32}$$

where κ is the empirical index. Thus, the elastic resistance force becomes

$$F_{ex} = F_z \frac{v_x}{p_i^{\kappa}} \tag{7.33}$$

On the other hand, the bulldozing resistance F_{bx} on the pneumatic tire can be estimated by Equation (7.25) or Equation (7.26).

7.3.3. Calculation of the Drawbar Pull

The drawbar pull can be obtained by subtracting resistance from the traction force, that is,

$$F_{dp} = F_x - F_r \tag{7.34}$$

The drawbar pull reflects the strength reserve of the soft ground, which is the net propulsion force for vehicle acceleration, grading, overcoming the resistance of rough roads, or pulling other vehicles. As a function of the slip ratio, the drawbar pull is often used to evaluate various electric powertrain systems. It is obvious that the system provides greater drawbar pull, which is advantageous.

7.4. Tractive Trafficability

In the previous discussion, the effect of shear stress at the tire–ground interface on the resistance of soil is not taken into account. To calculate tractive trafficability, the shape of the interface and the stress distribution on the interface must be determined first, and then, the force equivalence equation is solved to obtain the corresponding parameters of tractive trafficability.

For simplicity, the tire–ground interface is assumed to be a plane connecting to a circular arc section. The shear stress and pressure distributions can be evaluated using Equations (7.8) and (7.11), respectively. By neglecting the elastic and bulldozing resistances on the tire, the forces acting on an elastic tire rolling on a soft and loose ground are illustrated in Figure 7.4 [7.1].

FIGURE 7.4 Forces acting on the traction wheel rolling on a soft ground.

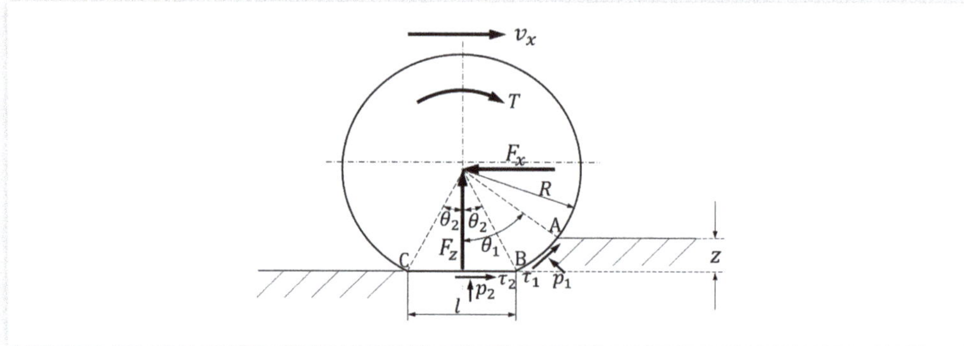

Based on force equilibrium in the vertical direction, the vertical force acting on the wheel from the ground is

$$F_z = bR\int_{\theta_2}^{\theta_1}(p_1\cos\theta + \tau_1\sin\theta)\,d\theta + 2bRp_2\sin\theta_2 \tag{7.35}$$

Similarly, the horizontal force acting on the wheel from the ground is

$$\begin{aligned} F_x &= bR\int_{\theta_2}^{\theta_1}(\tau_1\cos\theta - p_1\sin\theta)\,d\theta + b\int_0^l \tau_2\,dx \\ &= bR\int_{\theta_2}^{\theta_1}(\tau_1\cos\theta - p_1\sin\theta)\,d\theta + bR\cos\theta_2\int_{-\theta_2}^{\theta_2}\tau_2\left(1+\tan^2\theta\right)d\theta \end{aligned} \tag{7.36}$$

The torque acting on the wheel is

$$T = bR^2\int_{\theta_2}^{\theta_1}\tau_1\,d\theta + bR\cos\theta_2\int_{-\theta_2}^{\theta_2}\tau_2 d\theta = bR^2\int_{\theta_2}^{\theta_1}\tau_1\,d\theta + bR^2\cos^2\theta_2\int_{-\theta_2}^{\theta_2}\tau_2\left(1+\tan^2\theta\right)d\theta \tag{7.37}$$

where

p_1 and τ_1 are the normal and shear stresses on the circular arc section of the tire–ground interface, respectively

p_2 and τ_2 are the normal and shear stresses on the planar section of the interface, respectively

l is the length of the interface in the wheel plane, with $l = 2R\sin\theta_2$

b is the tire width

The subsidence of any point on the circular arc section of the tire–ground interface is given by

$$z = R(\cos\theta - \cos\theta_1) \tag{7.38}$$

By substituting Equation (7.38) into Equation (7.11), the normal stress on the circular arc section can be obtained

$$p_1 = \left(\frac{k_c}{b} + k_\varphi\right) R^n \left(\cos\theta - \cos\theta_1\right)^n \tag{7.39}$$

The normal stress on the planar section of the interface is

$$p_2 = p_g = p_c + \epsilon p_i \tag{7.40}$$

where

p_g is the bearing pressure at the tire–ground interface
p_c is the pressure produced by the stiffness of the carcass
p_i is the inflation pressure
ϵ is the empirical factor

Obviously, $p_1 = p_2$ at point B; thus,

$$\cos\theta_2 = \cos\theta_1 - \frac{\left(\frac{p_g}{k}\right)^{\frac{1}{n}}}{R} \tag{7.41}$$

The shear stresses can be derived from Equation (7.8):

$$\tau_1 = \left(C + p_1 \tan\varphi\right)\left(1 - e^{-\frac{\delta_{s1}}{E_s}}\right) \tag{7.42}$$

$$\tau_2 = \left(C + p_2 \tan\varphi\right)\left(1 - e^{-\frac{\delta_{s2}}{E_s}}\right) \tag{7.43}$$

The shear deformations δ_{s1} and δ_{s2} can be obtained from the following equations:

$$\delta_{s1}(\theta) = R\left[(\theta_1 - \theta) - (1 - s_r)(\sin\theta_1 - \sin\theta)\right] \tag{7.44}$$

$$\begin{aligned}\delta_{s2}(\theta) = \delta_{s1}(\theta) + s_r x = R\left[(\theta_1 - \theta) - (1 - s_r)(\sin\theta_1 - \sin\theta)\right]\\ + s_r R\cos\theta_2\left(\tan\theta_2 - \tan\theta_1\right)\end{aligned} \tag{7.45}$$

To solve these equations, θ_1 and θ_2 can be obtained first by iteration of Equations (7.35) and (7.38) through Equation (7.45), and then, F_x and T can be solved by substituting θ_1 and θ_2 into Equations (7.36) and (7.37). Finally, the traction coefficient and traction efficiency can be estimated by Equations (7.1) and (7.2), respectively.

7.5. Conditions for Obstacle Clearance

The conditions for clearing the road obstacles in the front of the vehicle and below the underbody between the wheelbase in the longitudinal direction of the vehicle are discussed here.

7.5.1. Obstacle Clearance of the Front End

Illustrated in Figure 7.5 is a vehicle with a front overhang L_f moving into a ditch of depth h [7.2]. For simplicity, it is assumed that the bottom of the vehicle's front end is on the plane that passes all front and rear wheels. The condition for the bottom of the front end to clear the bottom of the ditch is

$$L_f < \frac{R}{\sin(\alpha + \beta_1)} \tag{7.46}$$

where

α is the inclination angle of the longitudinal axis of the vehicle
β_1 is the slope angle at the bottom of the ditch

By examining the geometry in Figure 7.5, it is obvious that

$$h_1 = h_2 + R = h - h_3 - h_4 + R \tag{7.47}$$

Because of $h_1 = L\sin\alpha, h_3 = \dfrac{R\sin\alpha}{\sin(\alpha + \beta_1)}$, and $h_4 = \dfrac{R\cos\alpha\tan\beta_1}{\sin(\alpha + \beta_1)}$, Equation (7.47) becomes

$$L\sin\alpha = h - \frac{R\sin\alpha}{\sin(\alpha + \beta_1)} - \frac{R\cos\alpha\tan\beta_1}{\sin(\alpha + \beta_1)} + R \tag{7.48}$$

FIGURE 7.5 Geometry for analysis of obstacle clearance of the vehicle's front end.

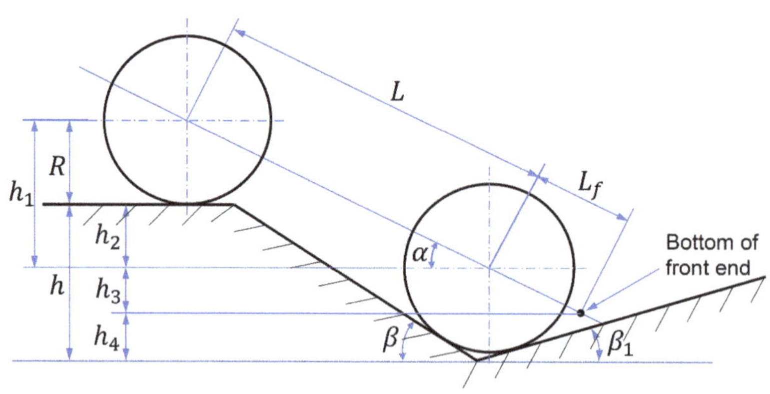

Rearranging Equation (7.48), we obtain

$$C_1 \sin\alpha \cos\alpha + C_2 \sin^2\alpha - C_3 \cos\alpha - C_4 \sin\alpha = 0 \quad (7.49)$$

where

$C_1 = L \sin \beta_1$

$C_2 = L \cos \beta_1$

$C_3 = (h + R) \sin \beta_1 - R \tan \beta_1$

$C_4 = (h + R) \cos \beta_1 - R$

After solving Equation (7.49) for α $(0 < \alpha < \beta)$, we can calculate the front overhang L_f for the bottom of the front end to clear obstacles using Equation (7.46).

7.5.2. Obstacle Clearance of the Underbody

To examine the obstacle clearance of the vehicle underbody, the planes tangential to the front and rear wheels are assumed to be the upper envelope of the obstacle, as shown in Figure 7.6 [7.2]. When the vehicle moves with respect to the obstacle, all the vertexes of possible obstacles, that is, the intersection of the two planes, form a cylinder of radius R_{ob}. As shown in the side view (Figure 7.6), the positions of the intersection points A and B are dependent on angle α_0, which is actually the extreme position for the front or rear wheels just about rolling over the vertex of the obstacle. The angle ACB is referred to as the ramp angle, which is an indication of how high a hump that the vehicle can negotiate without scraping the underbody or becoming high-centered. Obviously, the condition for the underbody to clear the obstacle is that the distance from the top of the obstacle (point C) to the bottom of the underbody is greater than zero, $h > 0$, that is,

$$h_m - R + \left(R + R_{ob}\right)\sin\alpha_0 - R_{ob} > 0 \quad (7.50)$$

where h_m is the ground clearance at the center plane between the front and rear wheels.

FIGURE 7.6 Geometry for analysis of obstacle clearance of the vehicle underbody.

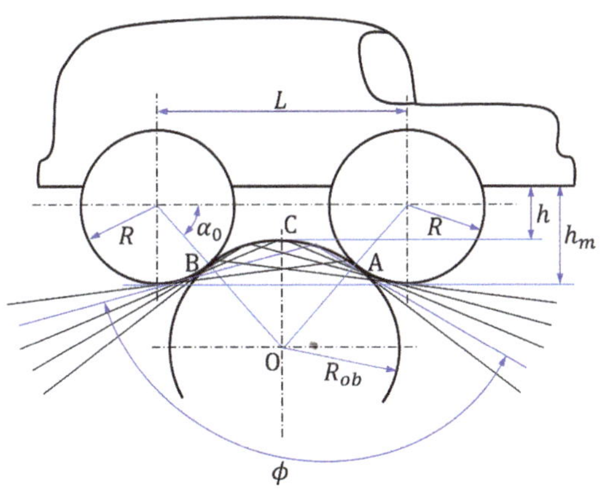

Since we have $2(R + R_{ob}) \cos \alpha_0 = L$, the condition Equation (7.50) becomes

$$h_m > R + R_{ob} - \sqrt{(R + R_{ob})^2 - \frac{L^2}{4}} \tag{7.51}$$

It can be proved that

$$R + R_{ob} = \frac{RL^2(\cos^2\phi - \cos\phi)}{2L^2 \sin^2\phi - 2R^2(1+\cos\phi)^2} + \frac{1}{2}\sqrt{\left[\frac{RL^2(\cos^2\phi - \cos\phi)}{L^2\sin^2\phi - R^2(1+\cos\phi)^2}\right]^2 + \frac{L^4}{L^2\sin^2\phi - R^2(1+\cos\phi)^2}} \tag{7.52}$$

Thus, the minimum ground clearance h_m required for the underbody to clear obstacles can be estimated by substituting Equation (7.52) into the right side of Equation (7.51).

7.6. Overcoming Terrain Steps and Trenches

For off-road driving, it is required that the vehicle be able to overcome terrain steps or trenches. Due to the slow speed of crossing these obstacles, the static equilibrium equations can be utilized to analyze the process.

7.6.1. FWD Vehicle Overcoming Steps

An FWD vehicle crossing steps on hard ground is assumed in this discussion, as shown in Figure 7.7.

FIGURE 7.7 An FWD vehicle rolling over a step.

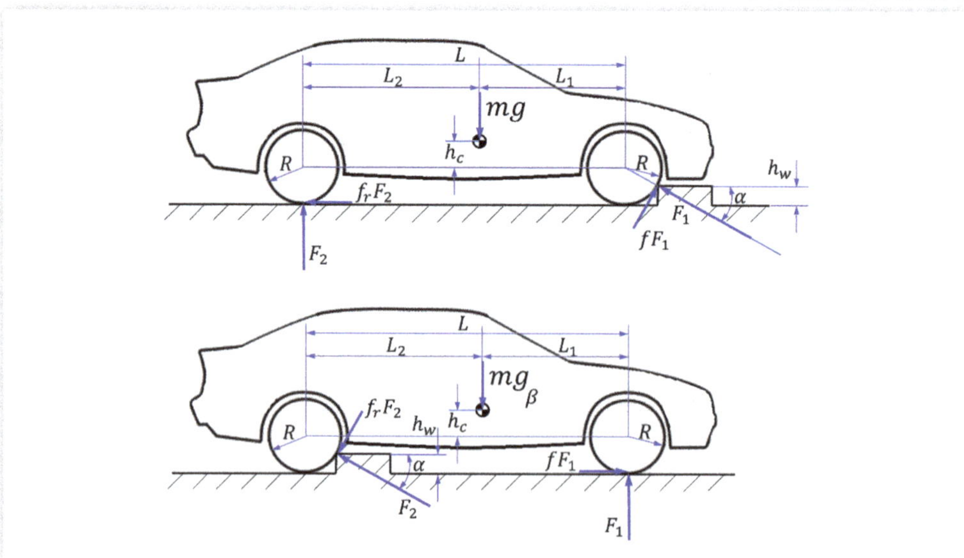

When the front wheel, that is, the drive wheel, is about rolling up the step, the force and moment equilibrium equations can be written as

$$\begin{cases} fF_1 \sin\alpha - F_1 \cos\alpha - f_r F_2 = 0 \\ fF_1 \cos\alpha + F_1 \sin\alpha + F_2 - mg = 0 \\ fF_1 R + mgL_1 - f_r F_2 R - F_2 L = 0 \end{cases} \tag{7.53}$$

where
F_1 is the reaction force acting on the front wheel from the step
F_2 is the load on the rear wheel
f is the adhesion coefficient
f_r is the rolling friction coefficient

Rearranging Equation (5.73) yields

$$\left(\frac{f+f_r}{f}\frac{L_1}{L} - \frac{f_r R}{L} - 1\right)\sin\alpha - \left(\frac{1}{f}\frac{L_1}{L} - \frac{f_r}{f}\frac{R}{L} - \frac{1}{f} - f_r\right)\cos\alpha + \frac{f_r R}{L} = 0 \tag{7.54}$$

By neglecting the rolling friction on hard ground, that is, letting $f_r = 0$, the Equation (7.54) is reduced to

$$f\sin\alpha - \cos\alpha = 0 \tag{7.55}$$

Since $\sin\alpha = \frac{R-h_w}{R} = 1 - \frac{h_w}{R}$, we can obtain the following from Equation (7.55):

$$\left(\frac{h_w}{R}\right)_f = 1 - \frac{1}{\sqrt{1+f^2}} \tag{7.56}$$

where the ratio $\left(\frac{h_w}{R}\right)_f$ represents the step height that the front wheel is able to overcome. It is obvious that the ability of an FWD vehicle to overcome steps at its front wheels is independent of any structural parameters of the vehicle.

On the other side, when the rear wheel (driven wheel) is about rolling over a step, the equations become

$$\begin{cases} fF_1 - f_r F_2 \sin\alpha - F_2 \cos\alpha = 0 \\ F_1 + F_2 \sin\alpha - f_r F_2 \cos\alpha - mg = 0 \\ F_1 L + fF_1 R - f_r F_2 R - mgL_2 = 0 \end{cases} \tag{7.57}$$

Rearranging Equation (7.57) yields

$$\left(\frac{f_r - f}{f}\frac{L_1}{L} + \frac{f_r R}{L} - 1\right)\sin\alpha + \left(\frac{1 - ff_r}{f}\frac{L_1}{L} + \frac{R}{L} + f_r\right)\cos\alpha - \frac{f_r R}{L} = 0 \tag{7.58}$$

Similarly, by assuming $f_r = 0$, we obtain

$$f(L - L_1)\sin\alpha - (L_1 + fR)\cos\alpha = 0 \tag{7.59}$$

Again, since $\sin\alpha = 1 - \frac{h_w}{R}$, we have

$$\left(\frac{h_w}{R}\right)_r = 1 - \frac{L_1 + fR}{\sqrt{(fL - fL_1)^2 + (L_1 + fR)^2}} = 1 - \frac{1}{\sqrt{1 + \left(\frac{f - fL_1/L}{L_1/L + fR/L}\right)^2}} \tag{7.60}$$

where the ratio $\left(\frac{h_w}{R}\right)_r$ is the step height that the rear wheel is able to overcome as a driven wheel. For a vehicle with equal load distribution on the front and rear axles, that is, $L_1 = L_2 = 0.5L$, it is obvious that $\left(\frac{h_w}{R}\right)_r < \left(\frac{h_w}{R}\right)_f$. Thus, the ability of such a vehicle to overcome steps is limited by the rear wheel.

7.6.2. RWD Vehicle Overcoming Steps

An RWD vehicle crossing steps on hard ground is assumed in this discussion, as shown in Figure 7.8.

FIGURE 7.8 An RWD vehicle rolling over a step.

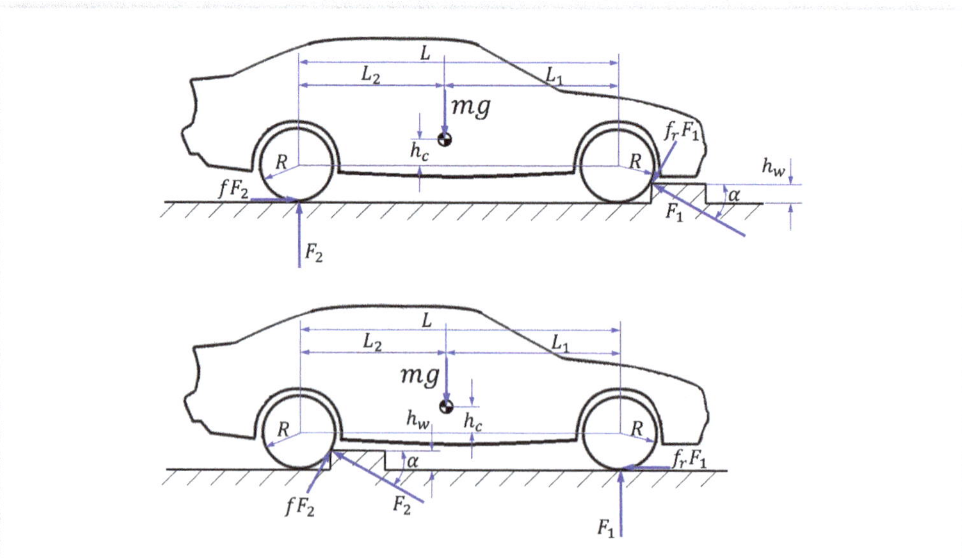

When the front wheel, that is, the driven wheel, is about rolling up the step, the force and moment equilibrium equations can be written as

$$\begin{cases} fF_2 - f_r F_1 \sin\alpha - F_1 \cos\alpha = 0 \\ F_2 + F_1 \sin\alpha - f_r F_1 \cos\alpha - mg = 0 \\ f_r F_1 R + F_2 L - fF_2 R - mgL_1 = 0 \end{cases} \tag{7.61}$$

Rearranging Equation (7.61) yields

$$\left(\frac{f+f_r}{f}\frac{L_1}{L}+\frac{fR}{L}-\frac{f_r}{f}\right)\sin\alpha+\left(\frac{1-ff_r}{f}\frac{L_1}{L}+\frac{R}{L}-\frac{1}{f}\right)\cos\alpha-\frac{fR}{L}=0 \tag{7.62}$$

By assuming $f_r = 0$, Equation (7.62) becomes

$$\left(\frac{L_1}{L}+\frac{fR}{L}\right)\sin\alpha+\left(\frac{1}{f}\frac{L_1}{L}+\frac{R}{L}-\frac{1}{f}\right)\cos\alpha-\frac{fR}{L}=0 \tag{7.63}$$

Substituting $\sin\alpha = 1-\frac{h_w}{R}$ yields the approximate solution of Equation (7.63):

$$\left(\frac{h_w}{R}\right)_f = 1-\frac{L-L_1-fR}{\sqrt{(fL_1)^2+(L-L_1-fR)^2}} = 1-\frac{1}{\sqrt{1+\left(\frac{fL_1/L}{1-L_1/L-fR/L}\right)^2}} \tag{7.64}$$

It is clear from Equation (7.64) that $\left(\frac{h_w}{R}\right)_f$ increases with a rise in either L_1/L or R/L. That is, an RWD vehicle with high load at the rear axle or of large-size wheels has advantages in overcoming steps at its front wheels.

When the rear wheel (drive wheel) is about rolling over a step, the equations become

$$\begin{cases} fF_2 \sin\alpha - F_2 \cos\alpha - f_r F_1 = 0 \\ F_1 + F_2 \sin\alpha + fF_2 \cos\alpha - mg = 0 \\ F_1 L + fF_2 R - f_r F_1 R - mgL_2 = 0 \end{cases} \tag{7.65}$$

Rearranging Equation (7.65) yields

$$\left(\frac{f+f_r}{f}\frac{L_1}{L}-f_r\frac{R}{L}-\frac{f_r}{f}\right)\sin\alpha+\left(\frac{ff_r-1}{f}\frac{L_1}{L}+\frac{f_r}{f}\frac{R}{L}-f_r\right)\cos\alpha+\frac{f_rR}{L}=0 \tag{7.66}$$

With $f_r = 0$, Equation (7.66) can be simplified into

$$f\sin\alpha-\cos\alpha=0 \tag{7.67}$$

By substituting $\sin\alpha=1-\frac{h_w}{R}$, we obtain the approximate solution of Equation (7.67):

$$\left(\frac{h_w}{R}\right)_r=1-\frac{1}{\sqrt{1+f^2}} \tag{7.68}$$

Apparently, the ability of an RWD vehicle to overcome steps at its rear wheels is not affected by vehicle parameters.

7.6.3. AWD Vehicle Overcoming Steps

An AWD vehicle crossing steps on hard ground is assumed in this discussion, as shown in Figure 7.9.

FIGURE 7.9 An AWD vehicle rolling over a step.

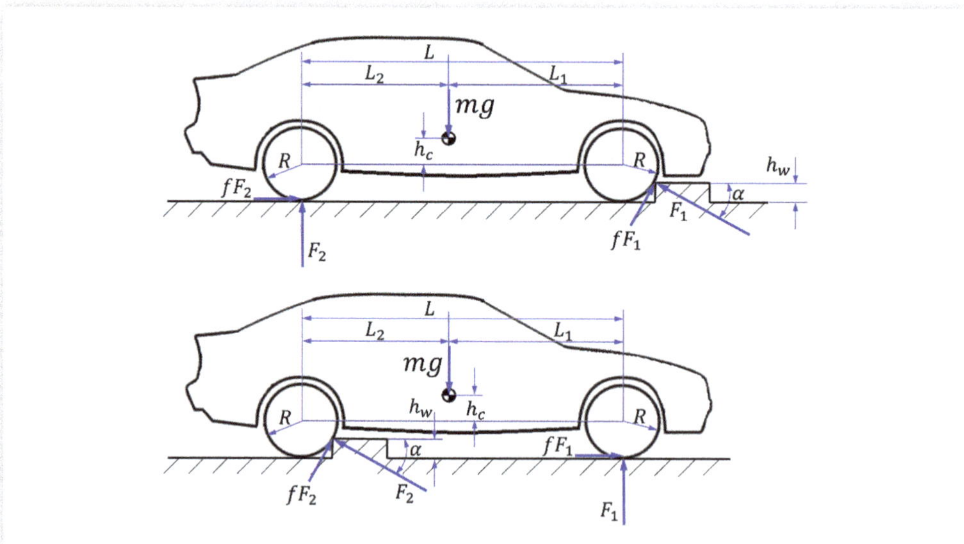

When the front wheel is about rolling up the step, the force and moment equilibrium equations can be written as

$$\begin{cases} fF_1 \sin\alpha + fF_2 - F_1 \cos\alpha = 0 \\ F_1 \sin\alpha + fF_1 \cos\alpha + F_2 - mg = 0 \\ F_2 L - fF_1 R - fF_2 R - mgL_1 = 0 \end{cases} \tag{7.69}$$

Rearranging Equation (7.69) yields

$$\left(1 - \frac{fR}{L}\right)\sin\alpha + \left(\frac{1+f^2}{f}\frac{L_1}{L} + \frac{R}{L} - \frac{1}{f}\right)\cos\alpha + \frac{fR}{L} = 0 \tag{7.70}$$

Substituting $\sin\alpha = 1 - \frac{h_w}{R}$, and letting $A = 1 - \frac{fR}{L}$ and $B = \frac{1+f^2}{f}\frac{L_1}{L} + \frac{R}{L} - \frac{1}{f}$, then $\left(\frac{h_w}{R}\right)_f$ is obtained by solving the following equation:

$$\left(A^2 + B^2\right)\left(\frac{h_w}{R}\right)_f^2 - 2\left(B^2 + A\right)\left(\frac{h_w}{R}\right)_f + 1 = 0 \tag{7.71}$$

The solution of Equation (7.71) indicates that the $\left(\frac{h_w}{R}\right)_f$ rises with decreasing L/R and increasing L_1/L. Hence, enlarging the size of wheels, shortening the wheelbase, or shifting the load toward the rear axle helps the front wheel of a vehicle overcome steps. When L_1/L is adequately large, the vehicle can overcome a step that is taller than the wheel radius.

When the rear wheel is about rolling over a step, the equations become

$$\begin{cases} fF_1 + fF_2 \sin\alpha - F_2 \cos\alpha = 0 \\ F_1 + fF_2 \cos\alpha + F_2 \sin\alpha - mg = 0 \\ F_1 L + fF_1 R + fF_2 R - mgL_2 = 0 \end{cases} \tag{7.72}$$

Rearranging Equation (7.72) yields

$$\left(\frac{2L_1}{L} + \frac{fR}{L} - 1\right)\sin\alpha + \left(\frac{f^2 - 1}{f}\frac{L_1}{L} - \frac{R}{L} - f\right)\cos\alpha + \frac{fR}{L} = 0 \tag{7.73}$$

Noting that $\sin\alpha = 1 - \frac{h_w}{R}$, and letting $A = \frac{2L_1}{L} + \frac{fR}{L} - 1$, $B = \frac{f^2 - 1}{f}\frac{L_1}{L} - \frac{R}{L} - f$, and $C = \frac{fR}{L}$, we can then solve the following equation for $\left(\frac{h_w}{R}\right)_r$:

$$\left(A^2 + B^2\right)\left(\frac{h_w}{R}\right)_r^2 + 2\left(2AC - A^2 - B^2\right)\left(\frac{h_w}{R}\right)_r + \left(A - C\right)^2 = 0 \tag{7.74}$$

The solution of Equation (7.74) reveals that $\left(\frac{h_w}{R}\right)_r$ increases with increasing L/R or decreasing L_1/L. In other words, vehicles with a long wheelbase or high load at the front axle are advantageous in overcoming steps at their rear wheels.

For FWD or RWD vehicles, their step obstacle capability depends only on the adhesion coefficient at the drive wheel. For AWD vehicles, however, their step obstacle capability is significantly higher than that of FWD or RWD vehicles, although it varies in a range for its dependency on vehicle parameters. In the design of an AWD vehicle, variable L_1/L can be adjusted to improve the step obstacle capability by balancing h_w/R values for steps at its front and rear wheels.

7.6.4. Evaluation of Trench Obstacle Capability

To evaluate the trench obstacle capability of a vehicle, the following equation can be used to convert the aforementioned step obstacle capability h_w/R to the trench obstacle capability:

$$\frac{b_t}{R} = 2\sqrt{2\frac{h_w}{R} - \left(\frac{h_w}{R}\right)^2} \tag{7.75}$$

where b_t is the trench width. Thus, b_t/R can be calculated when h_w/R is known.

7.7. Testing of Trafficability

7.7.1. Brief Descriptions of Trafficability Tests

The testing of trafficability generally involves the measurements of geometric parameters, the drawbar pull performance, and the obstacle-surmounting capability of the vehicle.

The trafficability-related geometric parameters are measured when the vehicle is fully loaded. These parameters include minimum ground clearance, approach angle, departure angle, ramp angle, and minimum turning radius.

The physical properties of the test ground and the geometry of obstacles are normally measured prior to the tests. The density, adhesion coefficient, cohesion coefficient, shear deformation modulus, subsidence index, cohesion modulus, friction modulus, and drag coefficient of the soil are instances of the physical properties, while the geometry of obstacles may include the slope, vertical height of the vertical obstacle, trench width, depth of snow, water, or mud.

The off-road drawbar pull performance testing of a vehicle is carried out on typical poor roads such as sand, soft soil, snow, grassland, mud, swamp, and paddy field, with parameters such as slip ratio, soil resistance, drive torque on traction wheel, contact patch area, and ground pressure of tires, as well as drawbar pull being measured.

The vehicle's capability of surmounting various obstacles such as steps, trenches, steep slopes, side slopes, ridges, bushes, and rivers is examined in the tests as well.

7.7.2. Measurements of Soil Parameters

The aforementioned physical properties of soils are typically measured by using a bevameter. Primarily functioning the same way, most of the bevameters consist of a torque motor, loading cylinders, shear annulus, penetration plates, sensors, pressure gauges, amplifiers, data acquisition devices, and so on.

The soil subsidence z and the corresponding pressure under the penetration plates are acquired for the calculation of pressure bearing characteristics. On the other hand, the shear characteristics are measured by means of the shear annulus.

For the given angular displacement of the shear annulus, the total torque acting on it is the sum of torques on the side and bottom surfaces, which can be expressed in terms of shear stresses on these surfaces

$$T = T_s + T_b = 2\pi h\lambda\left(R_o^2 + R_i^2\right)\tau_s + \frac{2\pi}{3}\left(R_o^3 - R_i^3\right)\tau_b = \frac{2\pi}{3}\left(R_o^3 - R_i^3\right)\tau_{eq} \tag{7.76}$$

where T is the total torque acting on the shear annulus; T_s and T_b are the torques on the side and bottom surfaces, respectively; τ_s and τ_b are the shear stresses on the side and bottom surfaces, respectively; R_o and R_i are the outer and inner radii of the shear annulus, respectively; h is the vane height of the shear annulus; τ_{eq} is the equivalent shear stress, with $\tau_{eq} = \tau_b + qh\tau_s$; $q = 3\left(R_o^2 + R_i^2\right)/\left(R_o^3 - R_i^3\right)$; and λ is a factor, with $\lambda = 1$ when the soil subsidence of the shear annulus $z \le h/2$, and $\lambda = 2\sin^2(\varphi/2 + \pi/4)$ when $z > h/2$.

When the shear annulus rotates, the average shear displacement on the bottom is

$$\delta = \frac{\pi\left(R_o + R_i\right)\theta}{360} \tag{7.77}$$

where θ is the rotation angle in deg.

Per Coulomb's law, the variation of maximum shear stress τ_{max} with the normal stress p is given by

$$\tau_{max} = C + p\tan\varphi \tag{7.78}$$

where

C is the cohesion coefficient of soil

φ is the internal friction angle of soil

Since the maximum shear stress can be measured for various normal stresses, the cohesion coefficient and the internal friction angle of the soil can be obtained by

Equation (7.78). To eliminate the effect of shear action on the side of the shear annulus, the following equations can be employed:

$$C = \frac{C'}{1 + q\lambda} \tag{7.79}$$

$$\varphi = \varphi' \tag{7.80}$$

where

C' is the cohesion coefficient of soil with the effect of shear action on the side of the shear annulus

φ' is the internal friction angle of soil with the effect of shear action on the side of the shear annulus

References

[7.1]. Yu, Z., *Automobile Theory*, 3rd ed. (Beijing: China Machine Press, 2000), 1-36, ISBN:7-111-02076-6.

[7.2]. Bekker, M., *Introduction to Terrain-Vehicle Systems* (Ann Arbor: The University of Michigan Press, 1969), ISBN-10:0472041444, ISBN-13:978-0472041442.

8

Vehicle-Level Design

The overall design of an EV involves the determination of its body style, major vehicle-level parameters and dimensions, powertrain configuration, and the drawing of the general layout sketch and the checkup of movement interference.

8.1. Categorization and Classification of EVs

Electric cars, buses, and trucks are the most common types of EVs. Each of these types of EVs can be classified by various schemes developed by government and organizations.

8.1.1. Classification of Electric Cars

Electric cars are typically classified by their wheelbase instead of engine displacement as for their internal combustion engine (ICE)-powered counterparts. Table 8.1 gives a comparison of electric car classifications in Europe and the US. Note that there is currently no unified definition of electric car classifications across the world yet, and the numbers in the table are intended for reference only.

TABLE 8.1 Electric car classification in Europe and the US.

European Commission						
Car segment	Approximate overall length (m)	Description	Reference wheelbase (m)	US*	Combined interior and cargo volume* (L)	Examples
—	<2.7	Quadricycle (microcar)	Any	Two-seaters	Any	Wuling Air EV E230
A	2.7–3.7	City car	<2410	Minicompact	<2405	Chery QQ Ice Cream Fiat 500e
B	3.7–4.2	Small car	2410–2800	Subcompact	2405–2830	Chevy Bolt Nissan Leaf
C	4.2–4.5	Medium car	2800–3100	Compact	2830–3110	Tesla Model 3
D	4.5–4.8	Large car	3100–3400	Mid-size	3115–3395	Mercedes-Benz EQS
E	4.8–5.0	Executive car	≥3400	Large	≥3400	Tesla Model S
F	>5.0	Luxury car		Full-size luxury		Lucid Air
J		Sport utility car		SUV and off-road vehicle		Audi Q4 e-tron
M		Multipurpose car		Multipurpose car		Chrysler Pacifica Electric
S		Sport coupe		Sport coupe		Tesla Roadster

* US Environmental Protection Agency (EPA)

8.1.2. Classification of Electric Buses

Passenger-carrying electric buses can be classified by the number of axles, namely, two-axle buses and three-axle buses. However, in some countries, they are classified by their overall length or the number of seats, as listed in Table 8.2.

TABLE 8.2 Examples of electric bus classification by length or number of seats.

Class of electric bus	Small bus	Medium bus	Large bus
Overall length L_a (m)	$L_a \le 6$	$6 \le L_a \le 9$	$L_a > 9$
Number of seats	≤9	10-18	≥19

8.1.3. Classification of Trucks

Trucks can be classified by the number of axles with the number of trailers.

8.2. Body Style

The body styles vary substantially for different categories of EVs. The basic body styles of cars, buses, and trucks are summarized as follows.

8.2.1. Body Styles of Cars

Similar to traditional ICE-powered cars, the most common body styles of cars include sedan, coupe, station wagon, convertible, sports car, multipurpose vehicle (MPV) or minivan, and SUV.

8.2.1.1. Sedan

A sedan, or saloon, typically features the so-called three-box configuration as the three compartments are all separated from one another. Housing the e-powertrain and other accessories in most designs, the front compartment may alternatively provide accommodation for luggage instead of e-powertrain in some cases, thanks to the flexibility in packaging the increasingly compact e-powertrain systems. In general, the passenger compartment in the middle has four doors, with the cargo compartment, or trunk, in the back.

There are several different body styles in terms of shapes from the roof to the rear of the sedans.

1. Notchback
   As illustrated in Figure 8.1a, the roof is connected to the rear body by lines of distinct angles with approximately horizontal extension from the bottom of the rear window to the back end.
2. Fastback
   As illustrated in Figure 8.1b, the roof is connected to the tail by a curve of a nearly single slope, with the rear window serving as the cover of the cargo compartment. With a streamlined roofline, fastback sedans generally have a low aerodynamic drag and a relatively large cargo space.

FIGURE 8.1 Body styles of cars. (a) A notchback sedan, (b) a fastback sedan, (c) a hatchback sedan, (d) a coupe, (e) a station wagon, (f) a convertible, (g) a sports car, (h) an MPV, and (i) a SUV.

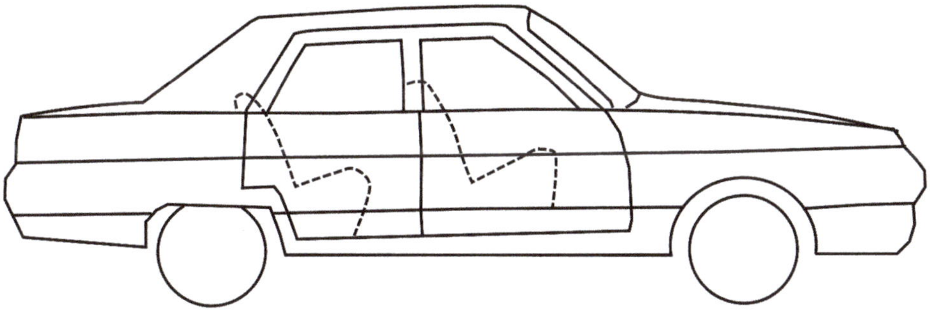

(a)

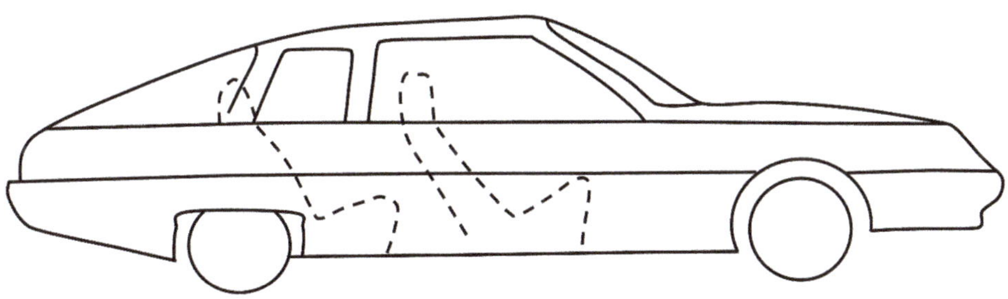

(b)

FIGURE 8.1 **(Continued)** Body styles of cars. (a) A notchback sedan, (b) a fastback sedan, (c) a hatchback sedan, (d) a coupe, (e) a station wagon, (f) a convertible, (g) a sports car, (h) an MPV, and (i) a SUV.

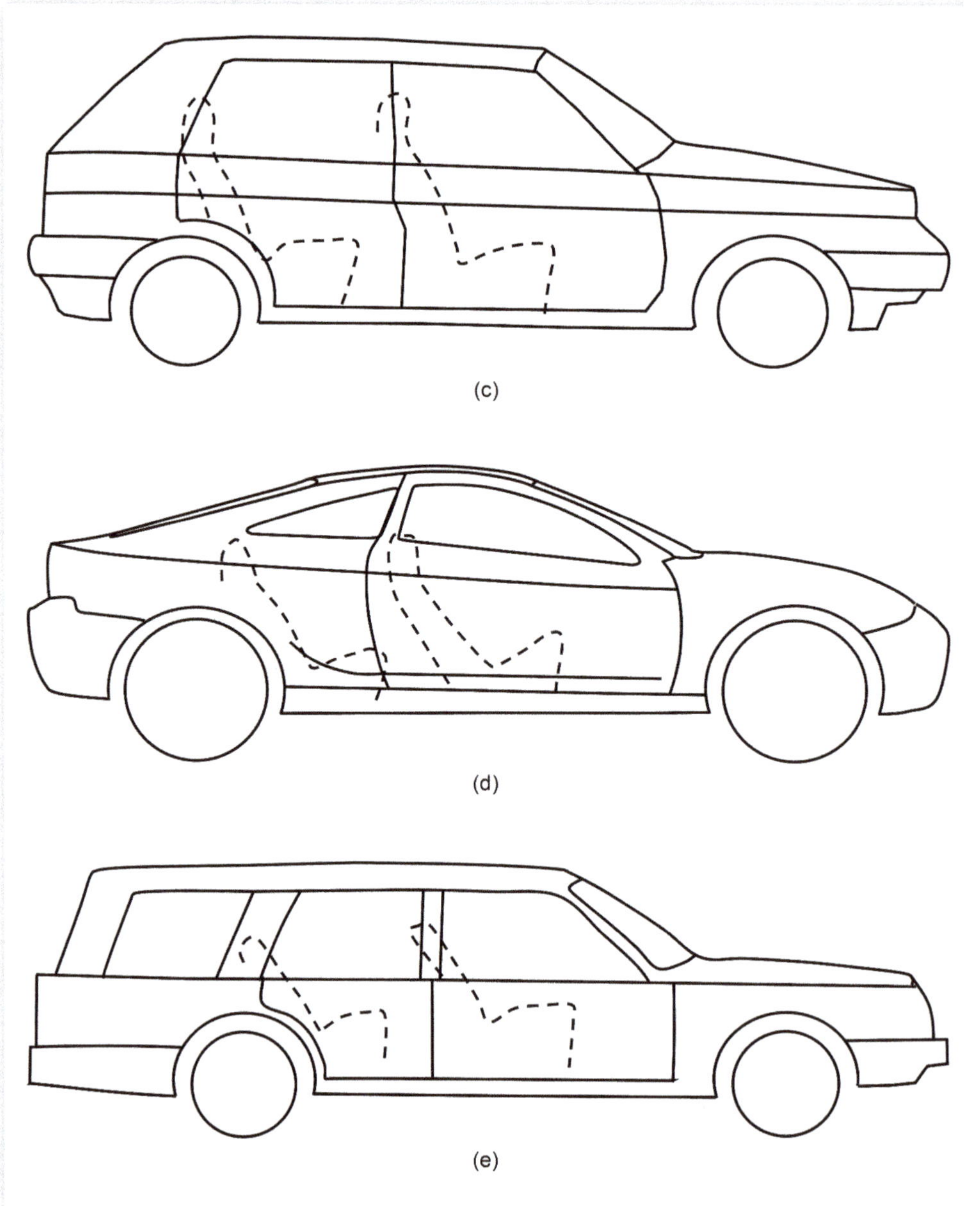

FIGURE 8.1 **(Continued)** Body styles of cars. (a) A notchback sedan, (b) a fastback sedan, (c) a hatchback sedan, (d) a coupe, (e) a station wagon, (f) a convertible, (g) a sports car, (h) an MPV, and (i) a SUV.

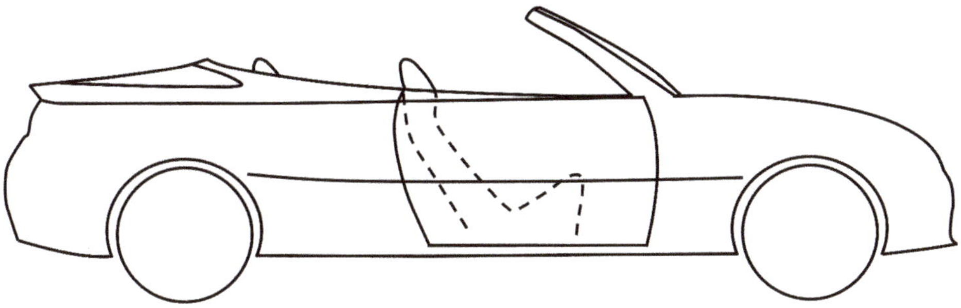

(f)

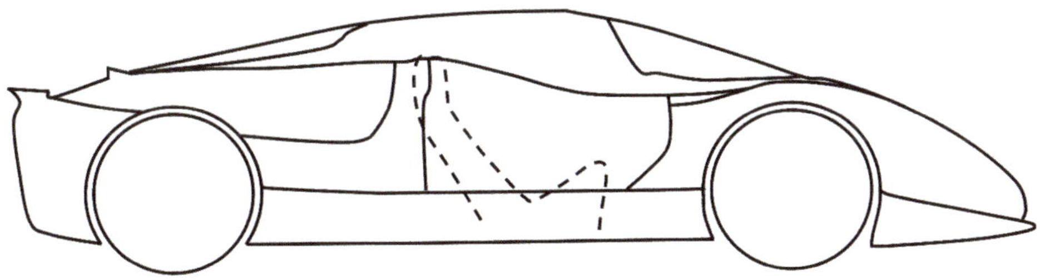

(g)

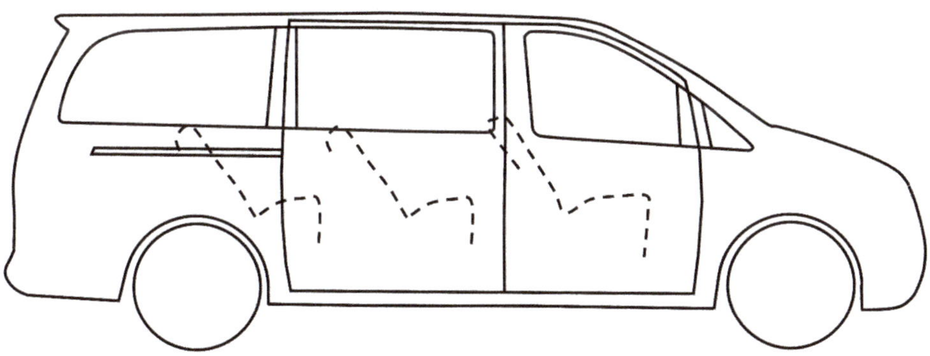

(h)

FIGURE 8.1 **(Continued)** Body styles of cars. (a) A notchback sedan, (b) a fastback sedan, (c) a hatchback sedan, (d) a coupe, (e) a station wagon, (f) a convertible, (g) a sports car, (h) an MPV, and (i) a SUV.

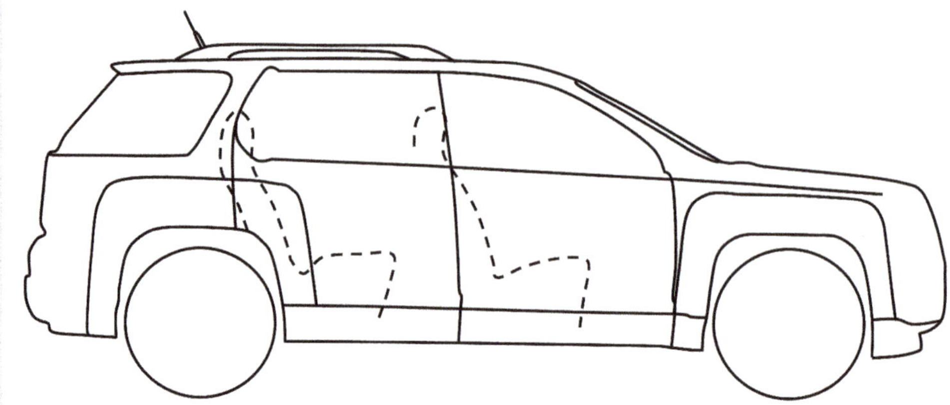

(i)

3. Hatchback

   As illustrated in Figure 8.1c, the roof is longer than that of the notchback, with the rear window and the cover of the cargo compartment combined into a single rear door that swings upward. Most hatchback sedans have a relatively small cargo space.

8.2.1.2. Coupe

Generally, with two doors for passenger egress, a coupe features a sloping or truncated rear roofline, as shown in Figure 8.1d. A coupe is generally sportier and more compact than its sedan counterpart.

8.2.1.3. Station Wagon

With the roofline extended rearward, the notchback sedan becomes a station wagon, which features a shared passenger and cargo volume and a fifth door (or liftgate) at the back, as illustrated in Figure 8.1e. Thus, a station wagon has actually a two-box configuration, with one or two rows of removable or folding seats behind the driver that can be flexibly reconfigured to prioritize either passenger or cargo volume.

8.2.1.4. Convertible

The roof can be retracted or restored by mechanisms such as a foldable framework with a soft top made from cloth or other fabric, retractable hardtops, or detachable hardtops. The weakness of this body style (Figure 8.1f), sometimes also called cabriolet, is usually a reduction in structural rigidity, along with limited cargo space.

8.2.1.5. Sports Car

The roofline is ultimately streamlined for maximization of dynamic performance and handling, as illustrated in Figure 8.1g. Often with limited interior space, sports cars typically have two seats although some have two additional smaller seats in the back. Sports cars feature either RWD with a rear-mounted e-powertrain or AWD with separate e-powertrains at the front and rear axles.

8.2.1.6. MPV/Minivan

MPVs or minivans usually feature a tall roof, a low and fully flat floor, high seating, a sliding door for rear passengers, and a rear liftgate, as shown in Figure 8.1h. With a one-box or two-box configuration, MPVs or minivans are often based on car platforms with unibody construction and reconfigurable or fold-flat seats in the second and third rows for flexibility in transporting up to seven passengers or cargo. Most MPVs or minivans use FWD or AWD layout.

8.2.1.7. SUV

With a roofline similar to a station wagon, SUVs generally feature a raised ground clearance and AWD capability to combine elements from off-road vehicles to that of road-going vehicles, as shown in Figure 8.1i. Mostly of a two-box unibody design, SUVs can be either two-door or four-door body style, with no difference from a crossover.

8.2.2. Body Styles of Buses

There are two major body styles for buses in terms of their outline, namely, flat-nose and dog-nose, as shown in Figure 8.2a and 8.2b. Dog-nose design is rarely used nowadays except for school buses due to its extra safety in front-end collision, while most buses adopt a flat-nose design to maximize the volume of the passenger compartment. To reduce aerodynamic drag, high-speed intercity buses often have a slanting windshield at the front. On the other hand, short-distance transit buses often feature a low and flat floor, few seats, and multiple wide doors, allowing easy access to a large number of passengers.

FIGURE 8.2 Body styles of buses. (a) Flat-nose, (b) dog-nose, and (c) articulated.

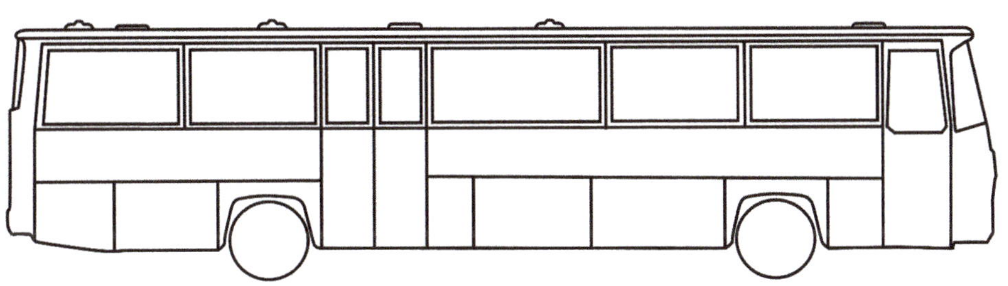

(a)

FIGURE 8.2 **(Continued)** Body styles of buses. (a) Flat-nose, (b) dog-nose, and (c) articulated.

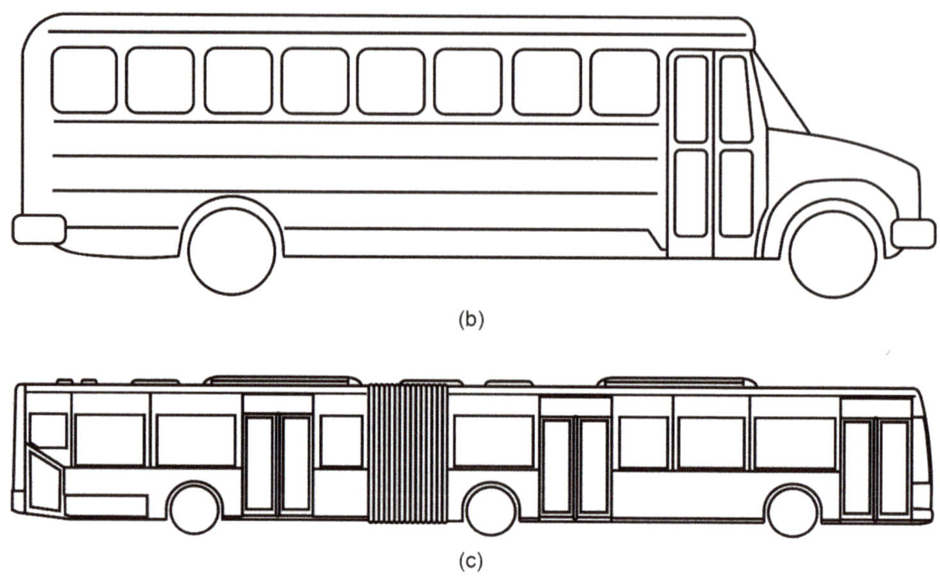

Mostly with one or two doors, buses typically can be single-decker, double-decker, or split-level in terms of their internal configuration. The double-decker design increases the number of seats more than the single-decker at a cost of a reduction in lateral stability, while the split-level is a mixture of the other two as an intermediate solution.

With the addition of one or more axles, articulated buses consist of two or more rigid sections linked by articulations that pivot relative to each other, allowing an increased length, axle load, and passenger capacity, as illustrated in Figure 8.2c. The articulation is usually covered by bellows, with a plate on the floor that enables passengers to move freely between these sections. Articulated buses may have three or more articulations and doors, depending on the number of sections.

8.2.3. Body Styles of Trucks

The body styles of trucks can generally be categorized into long-nose, short-nose, and flat-nose in terms of the shape of their front end and roofline, as shown in Figure 8.3. Long-nose trucks give better protection to the people in passenger cabs than flat-nose trucks in the case of front collision, while the latter provides more space behind the cab for cargo containers than the former at the cost of reduced collision safety. The short-nose trucks come in between the other two as a compromise in cargo-carrying capacity and front collision safety. The characteristics of these body styles are compared in Table 8.3.

FIGURE 8.3 Body styles of trucks. (a) Long-nose, (b) short-nose, (c) flat-nose, and (d) pickup.

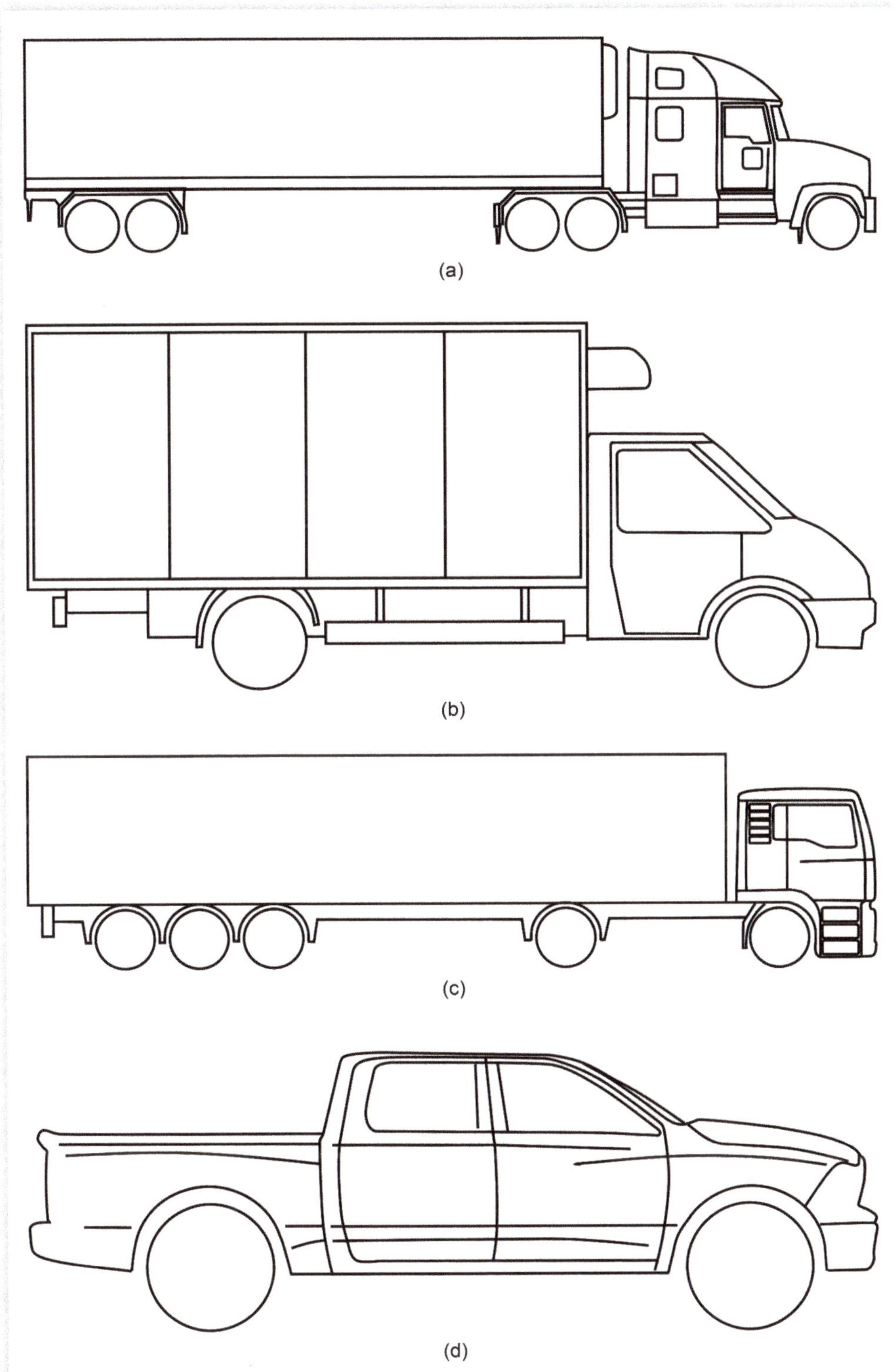

TABLE 8.3 Characteristics of commercial truck body styles.

Features	Long-nose	Short-nose	Flat-nose
Curb weight	Heavy	Medium	Light
Total length and wheelbase	Long	Medium	Short
Mobility	Poor	Medium	Good
Driver's field of vision	Poor	Medium	Good
Cargo space	Small	Medium	Large
Safety in front collision	Good	Medium	Poor
Aerodynamic drag	Small	Medium	Large
Access to front compartment	Easy	Difficult	Easy
Driving mechanism	Simple	Simple	Complex
Get in and out of the cab	Convenient	Inconvenient	Difficult

From an aerodynamic perspective, long-nose trucks can be better streamlined than short-nose trucks, while flat-nose trucks are certainly the worst. Thus, long-nose design is more advantageous for long-distance high-speed trucks for the purpose of freight transportation in view of the energy economy. Trucks usually have either an open cargo bed or an enclosed cargo compartment in the rear. The aerodynamic drag can be reduced by putting a cover on the open cargo bed and by letting the top of the cargo compartment flush with the roof of the passenger cab.

A pickup truck is the only truck designed for noncommercial usage featuring a passenger cab with two or four doors in the front and an open cargo bed in the rear, as shown in Figure 8.3d. A long-nose front-end design is most frequently adopted for the reduction in aerodynamic drag.

As far as the position of the electric drive system is concerned, trucks are mostly rear-mount rear-wheel drive (RR) or AWD as it produces good grip in the presence of cargo loading. However, front-mount rear-wheel drive (RR) and mid-mount RWD layouts are occasionally used for cost-saving or driveline-shortening purposes as well.

8.3. Electric Drive System Layout

The layout of electric drive systems in an EV is more flexible than in a traditional ICE-powered vehicle, mainly because of the compactness of modern electric drive systems.

8.3.1. Number of Axles and Axle Load

Although passenger vehicles usually have two axles, large commercial vehicles, for example, trucks and buses, can use three or more axles since the load on each axle is typically limited by regulations to prevent the road surface from overloading. In addition, the load capacity of tires is also a factor in the consideration of axle load. The cost of a vehicle, obviously, increases with the number of axles due to the complicity involved in adding more axles, particularly driving axles, to the vehicle. Nevertheless, distributed electric drive systems, for example, wheel-side motors or in-wheel motors, can potentially simplify or completely eliminate the driveline system of an EV.

The axle load is defined as the vertical load carried by each of the axles from front to rear. Note that the axle load should not be split to each side of the vehicle even if the two electric drive systems mounted side-by-side for the same axle are not physically connected to each other.

8.3.2. Basic Principles of Electric Drive System Layout

The number of drive wheels may or may not equal to the total number of wheels in an EV. Thus, this fundamental characteristic of a vehicle is generally presented in the form of "number of drive wheels × total number of wheels." For example, 4×2 stands for a four-wheel vehicle with two drive wheels.

The performance of a vehicle can be improved by increasing the number of drive wheels, but it often results in reduced energy efficiency due to imperfect coordination in allocating power to four or more drive wheels.

The electric drive system layout refers to positioning electric drive systems in the chassis and designating drive wheels, both of which have effects on various functions and performances of an EV.

8.3.2.1. Electric Drive System Layout of Electric Cars

Electric cars generally take one of the three layouts: front-mount front-wheel drive (FF), RR, and AWD. Among these layouts, FF is currently used in most electric cars, particularly in Class C and below, primarily because of the convenience of packaging the electric drive system in the front compartment and the understeer characteristics due to increased front axle load. On the other hand, RR is more suitable for sports cars that emphasize acceleration and gradeability performances. The pros and cons of these layouts are compared in Table 8.4.

TABLE 8.4 Comparison of electric drive system layouts of electric cars.

Features	FF	RR	AWD
Axle load	Large at front	Large at rear	Balanced
Steer characteristics	Understeer	Oversteer	Neutral
Obstacle crossing ability	Good	Poor	Excellent
Gradeability	Poor	Good	Excellent
Operating mechanism*	Simple	Complex	Complex
CV joint	Required	Not needed	Required at front
Lifespan of tire	Poor	Good	Fair
Trunk volume	Large	Small	Fair
Cost of frontal collision Repair	High	Low	High

* Mechanical (none shift-by-wire) type only.

8.3.2.2. Electric Drive System Layout of Electric Buses

Mid-mount rear-wheel drive (MR) and RR are the electric drive system layouts often employed in electric buses. The reason for FF being abandoned in electric buses is that the front axle load may become insufficient to provide traction when the bus travels up a slope or most passengers occupy the rear portion of the bus. On the other hand, articulated buses usually have an electric drive system in each section. Two electric drive systems can be utilized side-by-side at the rear axle as an alternative to a single system configuration for reducing the total weight of the electric powertrain. The pros and cons of these layouts are compared in Table 8.5.

TABLE 8.5 Comparison of electric drive system layouts of electric buses.

Features	Single e-motor MR	Single e-motor RR	Double e-motor RR
Axle load	Balanced	Large at rear	Large at rear
Floor	High	Low	Low
Effect on passenger compartment and seat layout	Large	Small	Small
Under-floor luggage space	Small	Fair	Fair
Access to electric drive system for repair	Difficult	Fair	Fair
Operating mechanism*	Complex	Complex	Complex
Comfort at rear	Fair	Good due to increased sprung mass	Good due to increased sprung mass

* Mechanical (none shift-by-wire) type only.

8.3.2.3. Electric Drive System Layout of Electric Trucks

Electric trucks commonly utilize the RWD layout, that is, MR or RR. Similar to the situation in electric buses, the FWD is not used in electric trucks because the load share at the front axle reduces when the truck is loaded. For heavy-duty trucks, two or more drive axles in the rear can be used where distributed electric drive systems, such as in-wheel motors or wheel-side motors, can become a competitive solution to simplify the driveline system. The pros and cons of these layouts are compared in Table 8.6.

TABLE 8.6 Comparison of electric drive system layouts of electric trucks.

Features	Single e-motor MR	Single e-motor RR	Double e-motor RR
Axle load	Balanced	Large at rear	Large at rear
Cargo bed floor	High	Low	Low
Access to electric drive system for repair	Difficult	Fair	Fair
Operating mechanism*	Complex	Complex	Complex

* Mechanical (none shift-by-wire) type only.

8.4. Selection of Overall Design Parameters

8.4.1. Dimensional Parameters

In addition to the overall length, width, and height, the dimensional parameters of importance to the design of an EV include wheelbase, wheel track, front overhang, and rear overhang, along with the dimensions for body features, for example, length of nose and size of cargo box.

8.4.1.1. Overall Dimension of the Vehicle

The overall length, width, and height are the most fundamental dimensional parameters in the design of an EV, which are normally subject to restrictions imposed by road vehicle regulations.

For a car, the overall length L_a is the sum of the wheelbase L, front overhang L_f, and rear overhang L_r. A coefficient C can be defined by $C = L/L_a$, with C statistically falling in the range from 0.56 to 0.65. The overall width of a car is related to the internal width and the door thickness on both sides, and it must be at least 1.41 m to accommodate three passengers in its back seat. The overall height of a car is the sum of its ground clearance h_c, the thickness of the floor and lower parts h_f, the height of the interior h_i, and the height of the roof structure h_r.

8.4.1.2. Wheelbase

The wheelbase L has an influence on a number of design parameters. For example, the overall length, curb weight, minimum turning radius, and the minimum radius of terrain that the vehicle is able to pass all grow with increasing wheelbase L. In addition, the wheelbase L affects the allocation of axle loads as well. With a decrease in wheelbase, the maneuverability of a vehicle improves, but the braking distance and handling can be compromised due to excessive load shift between the front and rear axles in braking or hill climbing, and the ride comfort can be impaired by intensified pitch vibrations of the vehicle body.

In general, the wheelbase is longer for a car of higher class and for a bus or truck of higher occupant or cargo-carrying capacity. The value of the wheelbase for an EV can be selected by referring to Table 8.7.

TABLE 8.7 Wheelbases of various classes of EVs.

Type of vehicle	Class	Wheelbase L (m)
Car	Minicompact (A)	1.96-2.46
	Subcompact (B)	2.41-2.60
	Compact (C)	2.60-2.70
	Mid-size (D)	2.70-2.88
	Full-size (E)	2.75-3.00
	Full-size luxury (F)	3.00-3.21
	SUV and off-road vehicle (J)	2.70-3.00
	Multipurpose car (M)	2.70-2.85
	Sport coupe (S)	2.35-2.65
Bus	City bus	3.0-7.2
	Intercity bus	3.3-7.9
Truck	Pickup	2.8-3.9
	Light-duty	2.3-3.4
	Medium-duty	3.4-5.6
	Heavy-duty	3.8-6.6

8.4.1.3. Wheel Tracks

As wheel tracks increase, both the interior width and the roll stiffness of the vehicle rise, at the price of augmenting the curb weight, the overall width, and the minimum turning radius though. Hence, an excessive front or rear wheel track is apparently not acceptable due to the upper limit of the overall width.

The front wheel track should be selected to accommodate the electric drive system (if a centralized FF vehicle), front frame, and suspension, as well as front wheels, with adequate clearance among steering linkage, frame, and wheels and for the front wheels to turn.

The selection of rear wheel track must leave sufficient clearance among longitudinal beams, rear suspension, and tires, with enough space to cover everything including the width of rear suspension, the width of tires, and the distance between longitudinal beams.

The value of the wheel tracks for an EV can be selected by referring to Table 8.8.

TABLE 8.8 Wheel tracks of various classes of EVs.

Type of vehicle	Class	Wheel track B (m)
Car	Minicompact (A)	1.29-1.31
	Subcompact (B)	1.42-1.58
	Compact (C)	1.51-1.61
	Mid-size (D)	1.55-1.61
	Full-size (E)	1.58-1.69
	Full-size luxury (F)	1.66-1.68
	SUV and off-road vehicle (J)	1.55-1.71
	Multipurpose car (M)	1.49-1.61
	Sport coupe (S)	1.46-1.80
Bus	City bus	1.74-2.13
	Intercity bus	1.74-2.10
Truck	Pickup	1.65-1.90
	Light-duty	1.30-1.57
	Medium-duty	1.60-1.85
	Heavy-duty	1.83-2.07

8.4.1.4. Front and Rear Overhangs

A short overhang at the front or rear can result in an increase in the approach or departure angle, which improves the trafficability of the vehicle and the field of vision of the driver. If the front overhang is too short, however, the front collision safety worsens and packaging the front bumper and other underhood components may become difficult.

For a car, the selection of rear overhang must take the trunk size into account. The rear overhang of a bus must not exceed a certain upper limit to prevent the center of mass from falling behind the rear wheel as passengers are crowed in the back. The selection of rear overhang for a track should take the size of the cargo box, wheelbase, and axle load allocation into consideration.

The value of the front and rear overhangs for an EV can be selected by referring to Table 8.9.

TABLE 8.9 Front and rear overhangs of various classes of EVs.

Type of vehicle	Class	Front overhang L_f (m)	Rear overhang L_r (m)
Car	Minicompact (A)	0.48–0.54	0.42–0.46
	Subcompact (B)	0.82–0.86	0.70–0.75
	Compact (C)	0.85–0.96	0.78–0.89
	Mid-size (D)	0.66–0.86	0.78–1.02
	Full-size (E)	0.90–0.98	0.95–1.15
	Full-size luxury (F)	0.90–0.94	1.00–1.15
	SUV and off-road vehicle (J)	0.84–1.08	0.79–1.21
	Multipurpose car (M)	0.91–0.99	0.85–0.93
	Sport coupe (S)	0.85–0.89	0.70–0.75
Bus	City bus	1.2–3.5	0.8–3.7
	Intercity bus	1.1–2.7	1.4–3.5
Truck	Pickup	0.8–1.0	1.1–1.3
	Light-duty	1.1–1.3	1.2–1.5
	Medium-duty	1.1–1.4	1.2–2.2
	Heavy-duty	1.3–1.6	0.7–2.4

8.4.1.5. Other Dimensional Parameters of a Truck

The length of a truck head refers to the distance from the front end of the front bumper to the rear of the cab, which is obviously affected by the length of the nose. The appearance, habitability, and sometimes accessibility to the electric drive system should all be considered in selecting a truck head size. The truck head size generally ranges from 2.5 to 3.0 m for long-nose trucks and from 1.4 to 1.5 m for flat-nose trucks.

The cargo box should be sized to a specific type of cargo to meet the nominal tonnage. The internal width should be selected as large as possible in complying with regulations and standards in order to shorten the length and the height of side boards. Note that for high-speed trucks the aerodynamic drag increases as the width of the cargo box because of the augmented frontal area. Shortening the length of the cargo box is usually preferred as it helps reduce the curb weight. The recommended height of side boards ranges from 0.45 to 0.65 m as it has an influence on the height of the center of mass and the convenience in loading and unloading cargos.

8.4.2. Mass-Related Parameters

The mass-related parameters, including curb weight, carrying capacity, carrying capacity per unit curb weight, and axle load distribution, have significant effects on many characteristics of an EV.

8.4.2.1. Curb Weight, GVW, and GCVW

Curb weight is the total mass of a vehicle with standard equipment and all necessary operating consumables such as lubricant, brake fluid, coolant, and A/C refrigerant, while not loaded with either passengers or cargo.

Curb weight impacts the energy economy, safety, and cost of the vehicle. The ultimate design goal, therefore, is always to meet the requirements on carrying capacity, safety, and NVH with the minimum curb weight. Lightweight design is particularly important for EVs due to the limited on-board energy storage capacity that the state-of-the-art battery technology can possibly render. The curb weight can be reduced by using lightweight materials or optimizing the topology and geometry of a vehicle and its components.

In vehicle design, the curb weight can be estimated by adding up the weights of all subsystems. The average curb weight per passenger for various cars and buses is listed in Table 8.10.

TABLE 8.10 Average curb weight per passenger of various electric cars and buses.

Type of vehicle	Class	Averaged curb weight per passenger (kg/passenger)
Car	Minicompact (A)	178–258
	Subcompact (B)	230–382
	Compact (C)	302–409
	Mid-size (D)	322–476
	Full-size (E)	365–444
	Full-size luxury (F)	469–478
	SUV and off-road vehicle (J)	351–513
	Multipurpose car (M)	281–412
	Sport coupe (S)	326–980
Bus*	Small bus	115–300
	Medium bus	80–275
	Large bus	108–412

* Standing passengers included for city buses.

On the other hand, the GVW refers to the maximum operating weight of a vehicle including the chassis, body, electric drive system, fluids, accessories, driver, passengers, and cargo, but excluding that of any trailers. Thus, GVW is actually the combined weight of the curb weight and the weight of passengers and cargo.

Finally, the gross combined vehicle weight (GCVW) refers to the maximum allowable weight of both the loaded tow vehicle and the loaded trailer that the tow vehicle can handle safely. Hence, GCVW is the combined weight of GVW and the loaded trailer.

8.4.2.2. Carrying Capacity

For a car or a bus, the carrying capacity refers to the number of passengers it is legally allowed to carry.

The carrying capacity for a car is quantified by the total number of seats including the driver seat. The carrying capacity typically ranges from two to four for a car of minicompact or below and four to seven seats for a car of subcompact or above.

The carrying capacity for a bus is quantified by the total number of passengers who are either seated or standing. The carrying capacity of a long-distance bus or a touring bus is usually equal to the total number of seats. The number of standing passengers can be estimated as 8–10 passengers/m^2 of floor area.

For a truck, the carrying capacity refers to the payload that is legally allowed to carry on a hard, well-paved road, which can decrease by 15–25% on a gravel road. For an off-road vehicle, the carrying capacity is specified as that on a dirt road.

8.4.2.3. Payload per Unit Curb Weight

The payload per unit curb weight reflects the lightweight design level of a vehicle. Advanced structural design and manufacturing processes are key to the ever-increasing payload per unit curb weight. The payload per unit curb weight of trucks is listed in Table 8.11.

TABLE 8.11 Payload per unit curb weight of electric trucks.

Type of vehicle	Class	Payload per unit curb weight
Truck	Pickup	0.14–1.06
	Light-duty	0.36–1.20
	Medium-duty	1.20–1.41
	Heavy-duty	0.97–1.86

8.4.2.4. Axle Load Distribution

Often expressed in percentile with respect to the total vertical loading, axle load distribution refers to the vertical loading of each axle acting on the supporting plane when the vehicle, either unloaded or fully loaded, is in stationary condition.

Axle load distribution can influence the performance and tire lifespan of a vehicle. For all tires to wear evenly with almost the same lifespan, it is desired to make axle load distribution as uniform as possible. However, the dynamic performance and trafficability of a vehicle can be improved by leaning axle load distribution toward the drive axle, while the steering stability can be enhanced by increasing axle load distribution on the steering axle.

The vehicle system layout has a remarkable influence on axle load distribution. The typical axle load distribution of cars and trucks is listed in Table 8.12.

TABLE 8.12 Axle load distribution of electric cars and trucks.*

Type of vehicle		Front axle (%)	Rear axle (%)
Car	FF	46-58	42-54
	RR	46-50	50-54
	AWD	34-51	48-66
Truck	Pickup	43-50	50-57
	Light-duty	37-70	30-63
	Medium-duty	32-60	40-65
	Heavy-duty	36-61	39-64

* With no occupants or cargo loads.

8.4.3. Performance-Related Parameters

8.4.3.1. Dynamic Parameters

1. Acceleration time
   Acceleration time refers to the time taken to accelerate the vehicle to the specified speed with the maximum acceleration from a standstill. The most quoted is the acceleration time of 0–100 km/h, but sometimes 0–60 km/h or 0–60 mph is quoted instead. Nowadays, the 0–100 km/h acceleration time of an electric car typically is well below 10 sec.
2. Gradeability
   Gradeability refers to the maximum grade that the vehicle is able to overcome when it is fully loaded on a well-paved road, which is usually expressed in percentile. In general, the gradeability is required to be greater than 30% for most road vehicles and greater than 60% for off-road vehicles.
3. Top speed
   The top speed refers to the maximum velocity that a vehicle is able to reach continuously for a certain period of time, typically for at least 30 sec. Most electric cars have a top speed ranging from 150 to 200 km/h, while some high-end models are capable of reaching 250 km/h or higher.
4. Specific power and specific torque
   Specific power is defined as the ratio of the maximum combined power of all the electric drive systems in a vehicle to GVW. As an indication of the dynamic performance, the specific power of a car is typically larger than that of a bus or a truck, which tends to grow with increasing GVW, as shown in Figure 3.1. In addition, the specific power of electric passenger vehicles has been increasing rapidly in the past decade, driven by the de facto fierce competition in dynamic performance, as indicated in Figure 8.4. Note that road vehicles are often subject to a certain minimum specific power requirement to prevent them from blocking traffic due to their poor dynamic performance.

FIGURE 8.4 Specific power of some electric passenger vehicles in the past decade.

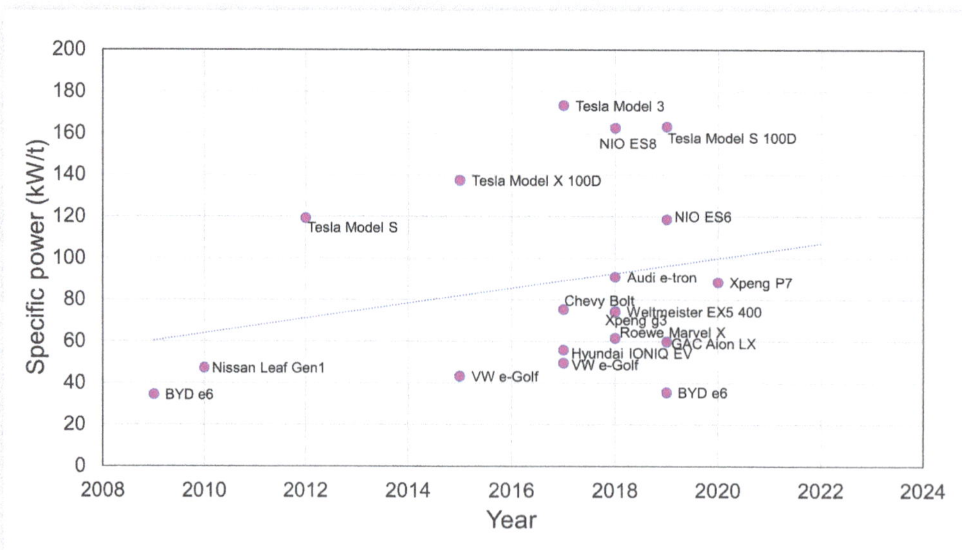

On the other hand, specific torque refers to the maximum combined torque of all the electric drive systems in a vehicle that can produce GVW. It is generally used as an indication of the towing capacity of a vehicle.

8.4.3.2. Energy Economy Parameters

The energy consumption per 100 km for a vehicle operating at a constant speed or at variable speeds specified by a particular driving cycle on a flat, well-paved road is employed to evaluate the energy economy. The most frequently used driving cycles are NEDC and WLTC, as mentioned in Chapter 2.

The energy economy of small vehicles is generally better than that of large vehicles. Today, the energy economy of EVs continues to improve, with only 10–13 kWh/100 km for some electric passenger car models, as shown in Figure 8.5.

FIGURE 8.5 Trend of energy consumption of EVs.

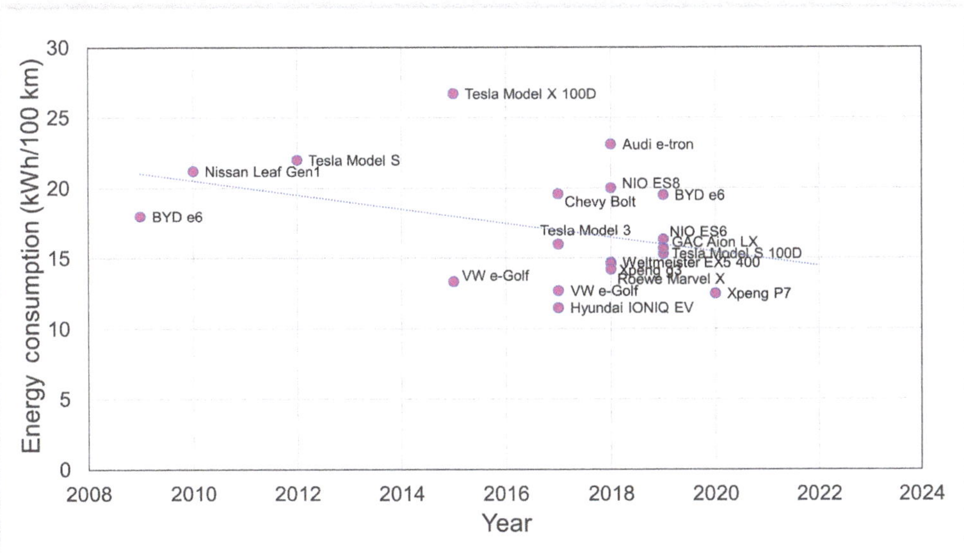

8.4.3.3. Trafficability and Handling Parameters

1. Trafficability-related geometric parameters
   The minimum turning radius is dependent on the maximum turning angle of steering wheels, the wheelbase, the wheel track, etc. To improve the maneuverability of a vehicle, a small value of the minimum turning radius is preferred. The selection of the minimum turning radius is also subject to the minimum turning radius values specified by applicable regulations or standards.

   The minimum turning radii of various types of vehicles are listed in Table 8.13.

TABLE 8.13 Minimum turning radii of various classes of EVs.

Type of vehicle	Class	Minimum turning radius r_{min} (m)
Car	Minicompact (A)	3.5–4.3
	Subcompact (B)	4.3–5.4
	Compact (C)	5.2–5.8
	Mid-size (D)	5.7–6.0
	Full-size (E)	5.1–6.2
	Full-size luxury (F)	5.4–6.2
	SUV and off-road vehicle (J)	5.1–6.2
	Multipurpose car (M)	5.3–6.1
	Sport coupe (S)	5.1–6.0
Bus	Mini	5.0–6.5
	Medium	6.5–8.5
	Large	8.5–15.0
Truck	Pickup	5.4–7.3
	Light-duty	4.9–8.0
	Medium-duty	8.0–12.0
	Heavy-duty	12.0–1.0

Other trafficability-related geometric parameters, for example, the minimum ground clearance, approach angle, departure angle, and ramp angle, need to be determined as well. These parameters can be selected from the values listed in Table 7.1.

2. Handling parameters
The parameters related to vehicle handling include the body roll angle, steering characteristic parameter, and pitch angle during braking, which need to be determined.

When the vehicle is in a circular motion at a constant speed with a centripetal acceleration of 0.4*g*, the body roll angle is preferably within 3°, and it should never exceed 7° in extreme cases.

When the vehicle is in a circular motion at a constant speed with a centripetal acceleration of 0.4*g*, the difference in front and rear wheel sideslip angles is typically used as a steering characteristic parameter. The vehicle can be made reasonably understeer by keeping such a parameter within 1–3°, in order to ensure good handling stability.

The pitch angle during braking at 0.4*g* deceleration should not exceed 1.5° to warrant good ride comfort.

3. Braking performance-related parameters
The braking distance and average braking deceleration are the two primary parameters related to the braking performance of a vehicle, which need to be determined according to the applicable regulations or standards.

Note that the requirements for the braking distance and average braking deceleration are different for the service brake system and the emergency brake system, with typically more stringent requirements for the former than for the latter. The emergency brake system needs to function when the service brake system is in partial failure.

4. Ride comfort-related parameters

   For ride comfort, parameters related to NVH, ambient air, spaciousness, and convenience of operation are usually considered.

   The NVH-related parameters include the frequency and vertical acceleration of vibrations and interior noise.

   The ambient air-related parameters include air temperature, humidity, and air quality which are dependent on the performance of A/C, heaters, ventilators, and air purification systems.

   The parameters related to spaciousness include the interior volume and size of doors and passages.

   The convenience of operation should be addressed by parameters that define the proper layout of pedals, handles, buttons, switches, and other elements involved in the vehicle operation to follow ergonomic principles.

8.5. Selection of Battery System

8.5.1. Battery Type Selection

There are a number of factors to consider when selecting the type of battery cell for powering an EV. Traditionally, lead acid and Ni-MH batteries were used for the traction of EVs, but they are obsolete nowadays due to their relatively low specific energy density. Today, most EVs are driven by Li-ion batteries. A comparison of key specifications of some of the most common types of batteries is given in Table 8.14.

In selecting the appropriate battery type for an EV, the most important requirements are safety, energy density, and cost, which are all related to the type of chemistry. It is well-known that the P-O bond in the lithium iron phosphate crystal is stable and difficult to decompose. Even at high temperature or overcharge, it will not collapse and heat up like lithium cobalt oxide or form strong oxidizing substances. Therefore, it is generally deemed that lithium ferrophosphate (LFP) batteries are the best choice if safety is the first priority. However, due to the intensive competition on the range of EVs, Nickel Manganese Cobalt (NMC) batteries are increasingly utilized because they have an edge over LFP batteries in terms of energy density in spite of their slightly higher costs for containing precious metals such as nickel, manganese, and cobalt. Nevertheless, the chemistry of NMC batteries, particularly with high nickel content, is more active in nature than that of LFP batteries. Hence, a properly designed BMS becomes a necessity to ensure the safe operation of the battery system.

With the rapid development of battery technology, the existing types of batteries will surely be replaced by new types, which may have completely different chemistry. Some examples of promising battery types include high nickel ternary lithium ion with silicon carbon cathode, solid lithium, and air metal batteries.

TABLE 8.14 Key specifications of most common types of batteries.

	Cell chemistry						
				Li-ion			
	Lead acid	Nickel cadmium	Nickel metal hydride	Lithium cobalt oxide	Lithium manganese oxide	Lithium iron phosphate	Lithium nickel manganese cobalt oxide
Specifications	SLA/VRLA	NiCd	NiMH	$LiCoO_2$/LCO	$LiMn_2O_4$/ LMO	$LiFePO_4$/LFP	$LiNiMnCoO_2$/ NMC
Anode	Lead	Cadmium	Metal hydride	Graphite	Graphite	Graphite	Graphite
Cathode	Lead dioxide	Nickel oxide hydroxide	Nickel oxide hydroxide	Lithium cobalt oxide	Lithium manganese oxide	Lithium iron phosphate	Lithium nickel manganese cobalt oxide
Energy density by mass (Wh/kg)	30-50	45-80	60-120	150-200	100-150	90-210	150-220
Energy density by volume (Wh/L)	60-75	100	400	560	420	300-350	580
Specific power (W/kg)	180	150-200	250-1000			200-1200	
Nominal cell voltage (V)	2.0	1.2	1.2	3.6	3.8	3.3	3.7
Life cycle	200-300	1000-2000	500-800	500-1000	500-1000	1000-2000 4000-12000	1000-2000
Maximum discharge rate*	5C	10C-20C	5C	1C-3C	1C-30C	30C	1C-2C
Maximum charge rate*	1.5C	0.5C-1C	1C	0.7C-1C	0.7C-3C	0.7C-1C	0.7C-1C
Charge efficiency (%)	50-92	70-90	70-85	90	90	90	90
Thermal runaway temperature (°C)	120	300	—	165-190	210-240	220-240	160-230
Pros	Highly stable and reliable; low cost; high safety	Tolerance to extreme temperatures	High specific power	High specific energy	High specific power; safer than cobalt	Safest Li-ion batteries	High specific energy and power
Cons	Poor energy density	Memory effect	Energy density lower than Li-ion; poor charging efficiency	Limited specific power cobalt is expensive	Mediocre specific energy	Mediocre specific energy; high self-discharge rate	Thermal stability not the best

* These are the maximum values that do not significantly affect the battery life.

8.5.2. Selection of Battery System Parameters

The battery capacity is a measure of the charge stored by the battery, which represents the maximum amount of energy that can be extracted from the battery under certain specific conditions. Thus, the total capacity of a battery can be expressed by

$$E_{tot} = V \cdot C \tag{8.1}$$

where

V is the voltage of the battery
C is the charge capacity of the battery
E_{tot} is the total capacity of a battery typically in Wh (watthours)

The charge capacity C, also referred to as ampere-hour capacity, is a crucial parameter of batteries. Both the discharge rate and the charge rate of a battery are commonly presented in terms of a ratio with respect to C. For instance, a discharge rate of $0.5C$ indicates that the battery discharges at half of its charge capacity, and it would take 2 h to deplete the battery. On the other hand, the discharge rate can also be noted by the time taken for the discharge, with C_x standing for x hours of discharge. For example, with a charge capacity of 60 Ah, a battery system can provide a discharge current of 30 A for 2 h, namely, at a discharge rate of $C_2 = 60/2 = 30$ A. The capacity of a traction battery used in EVs is typically quoted for 5 h of discharge, that is, at a discharge rate of C_5 or $0.2C$. Note that, for a given battery system, E_{tot} is not a constant. In fact, depending on the discharge rate, E_{tot} reduces if the battery system is drained quickly. Therefore, E_{tot} should be quoted at a specific discharge rate, that is, ampere hour rate C_x.

First, the selection of battery system capacity must meet the total energy requirement of the EV, which depends mainly on vehicle design variables such as the range of the EV, the efficiency of the electric drive system, the mass of the vehicle, the aerodynamic drag coefficient, the frontal area, and rolling resistance. Therefore, the sum of electric energy stored in all battery packs of an EV must be no less than the battery energy consumption given in Equation (2.3):

$$\sum_{i=1}^{n} E_{tot,i} \geq E_b \tag{8.2}$$

where

$E_{tot,i}$ is the electric energy stored in the ith battery pack of totally n packs
E_b is the battery energy consumption

The volume of each battery pack can be estimated by its total energy capacity and the volumetric energy density of the available battery pack:

$$V_{b,i} = \frac{E_{tot,i}}{E_{v,i}} \tag{8.3}$$

where

$V_{b,i}$ is the volume of each battery pack
$E_{v,i}$ is the volumetric energy density of the battery pack

The mass of each battery pack can be estimated by its total energy capacity and the specific energy of the available battery pack,

$$m_{b,i} = \frac{E_{tot,i}}{E_{m,i}} \tag{8.4}$$

where

$m_{b,i}$ is the mass of each battery pack
$E_{m,i}$ is the specific energy of each battery pack

Second, the selection of battery system capacity must meet the total power requirement of the EV, which is primarily related to vehicle design variables such as the maximum speeds on various road conditions (particularly the slope of the road), the mass of the vehicle, the aerodynamic drag coefficient, the frontal area, and rolling resistance. The maximum charge or discharge power of a battery system is dependent on its specific power, namely, the amount of power per unit mass of the battery. Thus, the sum of electric power supplied by all battery packs of an EV must be no less than the battery power given in Equation (2.4):

$$\sum_{i=1}^{n} m_{b,i} P_{m,i} \geq |P_b| \tag{8.5}$$

where

$P_{m,i}$ is the specific power of each battery pack, which is assumed to be a positive value without considering the power flow direction

P_b is the battery power

Per the convention given in Equation (2.4), a positive P_b corresponds to discharging the battery and a negative P_b corresponds to charging the battery.

As a matter of fact, batteries of excellent specific energy normally do not have a superior specific power, and vice versa. Today, batteries of high specific energy are typically utilized for EV applications as range maximization is given higher priority than vehicle performances at high speeds or during regenerative braking. The determination of the maximum charge or discharge rate must comply with the recommended values from battery pack manufacturers, which is usually less than 6°C for state-of-the-art lithium-ion batteries.

An extra redundancy in the maximum power may be needed to protect the battery system from being overloaded in extreme operating conditions, for example, overtaking at high speeds or regenerative braking at relatively hard stops.

8.5.3. Packing of the Battery System

Packing is a critical design step to integrate the battery cells into an assembly, which is often referred to as a battery pack. The specifications of the battery system, such as safety, weight, lifespan, ruggedness, ingress protection, cyclability, and cost, are strongly influenced by the battery pack design. For lithium-ion batteries, it is essential to integrate the BMS into the pack to ensure the safety, reliability, and durability of the system.

In general, the packing of a battery system involves a couple of different levels of integration. At the most fundamental level, a number of electrochemical cells, each consisting of an anode and a cathode separated by an insulating separator and embedded in an electrolyte, are interconnected in series or parallel or both to form a module. In such a module, each individual cell may come with voltage, current, and temperature monitoring sensors, as a part of the BMS. At the next level, a group of modules is connected to form a battery pack, along with the sensors and electrical circuits for signal acquisition and processing, which constitute the entire BMS. In addition, the battery pack also contains a thermal management system comprising air or liquid cooling medium and channels, PCM with corresponding structures, or their combinations. To bear the vertical load due to the mass of a battery system and prevent the battery system from external damage, the battery pack is typically made of materials of high strength with a stiffener. A high ingress protection code, typically IP67 or higher, must be complied in designing the seals, connectors, and wiring for battery packs. To further simplify the structure and improve the energy density of a battery pack, cells can be integrated to form a pack directly without assembling them into modules in the first place.

Lightweight is one of the most important considerations in battery pack design. A packing efficiency can be used to evaluate the lightweight level of a battery pack design, which is defined as the ratio of the specific energy of a battery pack to that of all cells as a whole, that is,

$$\varepsilon_p = \frac{E_p}{\sum_{i=1}^{n} E_{ci} / n} = \frac{m_p}{\sum_{i=1}^{n} m_{ci}} \tag{8.6}$$

where

E_p and m_p are the specific energy and mass of the battery pack, respectively

E_{ci} and m_{ci} are the specific energy and mass of the ith battery cell among a total number of n cells, respectively

At present, the packing efficiency lies between 0.6 and 0.7 for an electric passenger vehicle and above 0.8 for an electric bus.

Cylindrical cells, prismatic cells, and pouch cells are the most common types of cells that are used in EVs today. Cylindrical cells are very mature in technology and come with standardized sizes or aspect ratios. For example, 18650 cells refer to the cells of 18 mm in diameter and 65 mm in length. Hardly deformed, they are mechanically stable because their cylindrical structure can withstand internal pressure to a certain extent when thermal runaway occurs internally. The gap between cells also facilitates the design of cooling channels around each cell. With little cell-to-cell variation in the physical (particularly electrochemical) properties and high energy density of cylindrical cells, it is easy to produce a pack of high energy density and good consistency without much needs for battery equalization. However, the limited capacity of each cylindrical cell makes it necessary to connect a large number of cells in parallel and then in series to meet the voltage and capacity requirements, thus calling for a complicated BMS for monitoring and controlling the cells and modules.

On the other hand, prismatic cells are a relatively new format with no standard size or aspect ratio established yet to date. Due to the ease in packaging these relatively large-capacity cells, prismatic cells are more widely used in EV applications today than cylindrical cells. However, these cells are prone to deform during thermal runaway, which requires a more rugged pack cover to protect the occupants of the vehicle. Besides, the lack of gap between cells makes it difficult to design cooling channels around each cell to meet the cooling requirements. To overcome the cell-to-cell variation in electrochemical properties, a passive or active battery equalization design may become a necessity to ensure the full utilization of pack capacity and designed lifespan.

Pouch cells are soft and flexible ones that usually have most of their components enclosed in an aluminum-coated polymer film, with only two tabs remaining outside as anode and cathode connectors. In case of a failure, pouch cells would only blow up and crack, with no risk of explosion. With a small internal resistance and high volumetric energy density, they are also light and easy to be made into custom sizes. Nevertheless, the accumulation of gases generated inside each pouch cell during every charge and discharge cycle causes it to swell. The need for a cooling plate in between pouch cells undermines their advantage in energy density.

The comparison of the packing efficiencies of cylindrical, prismatic, and pouch cells is listed in Table 8.15.

TABLE 8.15 Packing efficiencies of major types of battery cells.

	Packing efficiency		
	Cylindrical cell	Prismatic cell	Pouch cell
Module-level	85	89	85
Pack-level	65	70	60

For EVs operating in an extremely cold environment, a sophisticated thermal management system, involving both heating and cooling of the battery pack, may become essential. In this case, a heater, typically using the electrical energy from traction batteries, can be added to the system on top of a radiator to provide both cooling and heating capabilities.

Presently, there is a trend to integrate the controller of a BMS into the VCU or into one of the domain controllers, in a never-ending pursuit of system hardware simplification and cost reduction.

8.6. Selection of Electric Drive Systems

8.6.1. Electric Motor Selection

A few different types of electric motors considered to be suitable for application in EVs as the prime mover include DC brushed (BDC) motors, brushless DC (BLDC) motors, permanent magnet synchronous motors (PMSMs), induction motors (IMs), and switched reluctance motors (SRMs).

8.6.1.1. BDC Motors

The principle of a BDC motor is illustrated in Figure 8.6. Although a BDC motor is able to produce high torque at low speeds with a simple controller, its operation requires that brushes slide over the mechanical commutator continuously, coming in contact with different segments of the commutator. As a result, the brushes and commutators are prone to wear. Besides, the friction between brushes and commutators puts a limit on the maximum motor speed, making it difficult to downsize the motor. Consequently, BDC motors generally have a poor power density.

FIGURE 8.6 Principle of BDC motors.

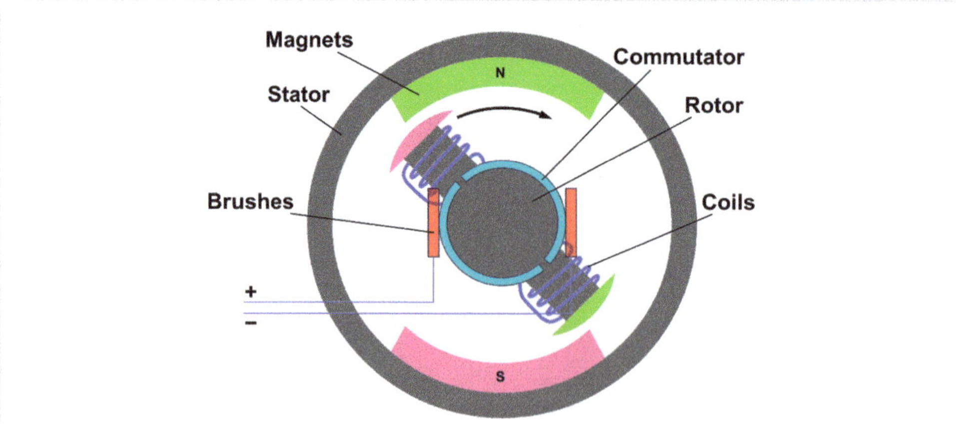

8.6.1.2. BLDC Motors

The principle of a BLDC motor is illustrated in Figure 8.7. It is a synchronous motor using a DC electric power supply. The permanent magnet rotor follows the rotating magnetic field produced by switching the DC to the windings using an electronic closed-loop controller. Hence, the BLDC is typically driven by either a square wave or a trapezoidal-shaped current. The speed and torque of the motor are controlled by adjusting the phase and amplitude of the DC. BLDC motors not only require less maintenance but also have higher efficiency than BDC motors because an electronic commutation circuit is used instead of mechanical commutators. Capable of operating at high speeds, BLDC motors generally have a remarkably higher power density than BDC motors. However, the output torque can fluctuate slightly when a BLDC motor operates at low speeds.

FIGURE 8.7 Principle of BLDC motors.

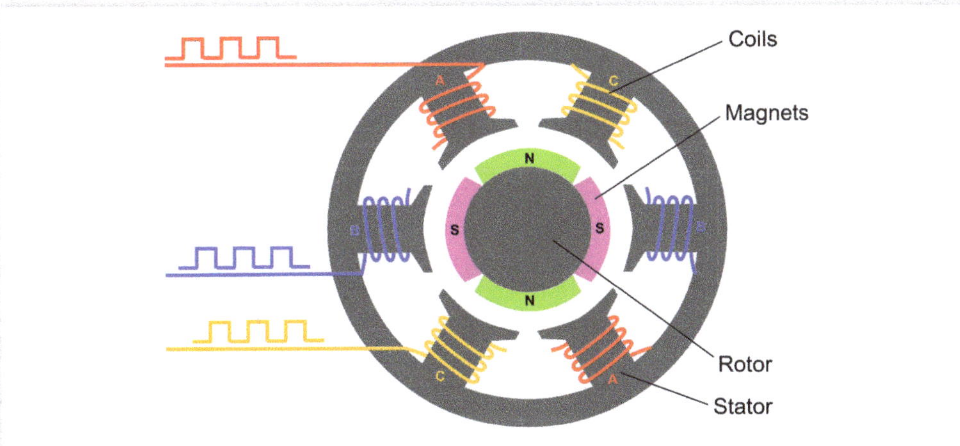

8.6.1.3. PMSMs

The principle of a PMSM is illustrated in Figure 8.8. A PMSM is a synchronous motor, in which a constant magnetic field is produced by permanent magnets embedded in the rotor, while a rotating magnetic field is created by stator windings that connect to a three-phase AC electric power supply. The rotor poles interlock with the rotating magnetic field at a synchronous speed. Driven by sinusoidal waveforms, a PMSM produces a torque with significantly less ripple than a BLDC motor. PMSMs in EV applications are mainly closed-loop controlled with variable-frequency drive (VFD) control methods such as vector control (field-oriented control, i.e., FOC) or direct torque control (DTC). Capable of maintaining a full torque at low speeds, PMSMs usually have high efficiency and power density, with a broad range of operating speeds. However, when the permanent magnet material is subjected to high temperature, vibration, and overload current, its magnetic permeability may decrease, or a demagnetization phenomenon may occur, which may reduce the performance of PMSMs. As accurate position feedback is necessary for torque generation, PMSMs are typically more expensive than BLDC motors due to their more complex construction and the use of costly neodymium-based permanent magnets.

FIGURE 8.8 Principle of PMSMs.

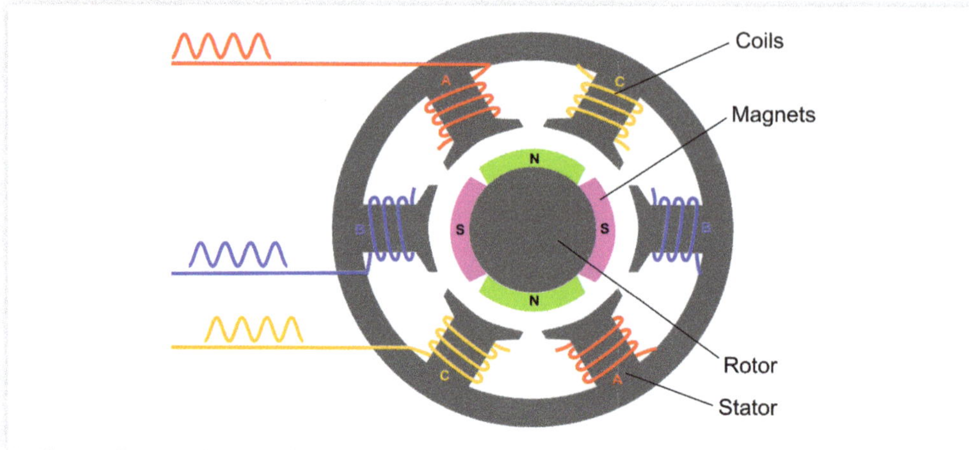

8.6.1.4. IMs

The principle of an IM is illustrated in Figure 8.9. An IM produces a torque by the rotor in which the required electric current is obtained by electromagnetic induction from the magnetic field of the stator windings. For rotor currents to be induced, the rotor speed must be lower than that of the rotating magnetic field of the stator, with the difference being called slip. As an asynchronous motor, the rotor speed of an IM drops under load, and the slip increases to an extent so that sufficient torque is created to turn the load. IMs are typically controlled by the vector control method, which allows to control the torque and speed independently. With a simple and strong structure, IMs are not only capable of producing torque with little ripple and noise but are also highly reliable and less expensive than PMSMs. However, IMs are generally less efficient than PMSMs due to the energy consumed for excitation of the magnetic field, along with a lower power density.

FIGURE 8.9 Principle of IMs.

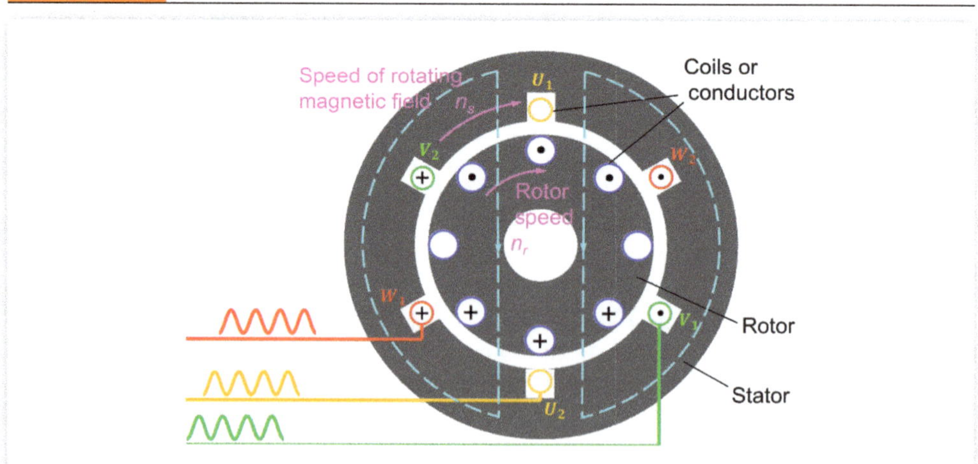

8.6.1.5. SRMs

SRMs operate by reluctance torque that is produced by delivering power to the windings in the stator rather than the rotor. Made of soft magnetic material, the magnetic reluctance of a solid salient-pole rotor, typically made of silicon steel laminates, produces a force that tends to align the rotor pole with the nearest stator pole. The windings of successive stator poles are switched in sequence by a controller so that the magnetic field of the stator always leads the rotor pole to keep the rotor rotating while the angle of the rotor is acquired by a position sensor. The required sequential pulses are given to the power circuitry by the controller with precise phase activation timing. Not only are SRMs efficient, but they are also reliable since the topological structure of their inverters prevents them from short-circuit faults. In spite of their simple, robust, low-cost, and high-speed capable construction, SRMs are not widely used in EVs yet because of drawbacks such as torque ripple, noise, and vibrations. The principle of SRMs is illustrated in Figure 8.10.

FIGURE 8.10 Principle of SRMs.

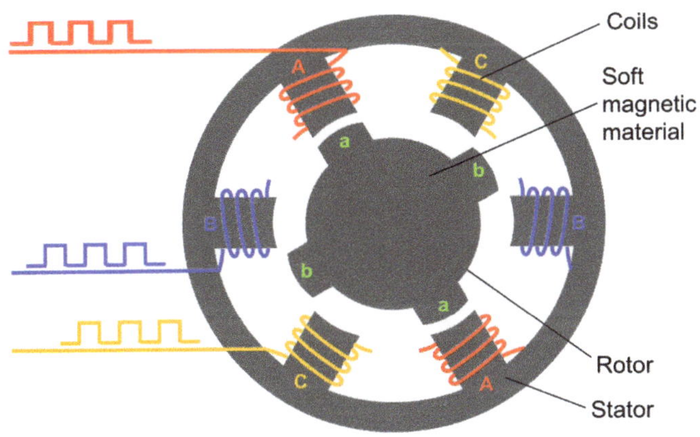

The state-of-the-art characteristics of aforementioned types of electric motors are compared in Table 8.16.

In view of the distinct characteristics of these electric motors, the application of two different types of motors in the same vehicle becomes an alternative electric drive system solution. For example, the electric drive efficiency and driving range can be improved by using a PMSM mainly for acceleration at low speeds and an IM primarily for cruising at high speeds.

TABLE 8.16 Comparison of characteristics of different types of electric motors.

Characteristics	BDC motor	BLDC motor	PMSM	IM	SRM
Overload capability (%)	200	300	300	300-500	300-500
Peak efficiency (%)	85-89	95-97	95-97	94-95	90
Full-load efficiency (%)	80-87	85-97	85-97	90-92	78-86
Power factor	—	0.90-0.93	0.90-0.93	0.82-0.85	0.60-0.65
Proportion of constant torque region	1:2-1:3	1:2.25	1:2.25	1:5	1:3
Speed range (rpm)	4000-6000	4000-10000	4000-20000	12000-20000	>15000
Reliability	++	++++	++++	+++++	++++
Robustness	+	+++	+++	++++	+++++
Size	+	+++++	+++++	+++	+++++
Power density	+	+++++	+++++	+++	++++
Controllability	+++++	+++	+++	+++	+++
Complexity of controller	+++++	+++	+	+	+++
Cost	+++++	+++	+	++++	+++
Torque ripple	+++++	++	+++++	+++++	+
Noise and vibration	+	+++	+++++	+++++	++

Note: + = poor, ++ = fair, +++ = good, ++++ = very good, and +++++ = excellent.

8.6.2. Selection of Electric Drive System Parameters

Based on the overall vehicle parameters described in Section 8.4, the electric drive system parameters, for example, rated power and torque, peak power and torque, maximum speed, extreme speed, number of gears, and system voltage, must be determined accordingly.

8.6.2.1. Rated Power and Torque

Since the efficiency and power factor of an electric motor are usually high in the low-torque region around base speeds, it is generally desired that the electric drive efficiency at the cruise speed can be maximized, while the vehicle can operate constantly at the top speed. The normal practice is to select the electric motor so that it operates around its base speed when the vehicle travels at the cruise speed and operates at its maximum speed when the vehicle travels at the top speed. For an electric motor, however, its base speed at rated power, n_{rb}, may not equal to its base speed at peak power, n_{pb}, as illustrated in Figure 8.11, although it is often assumed to be identical in engineering practice for simplicity; that is, $n_{pb} = n_{rb} = n_b$.

FIGURE 8.11 Selection of rated and peak characteristics.

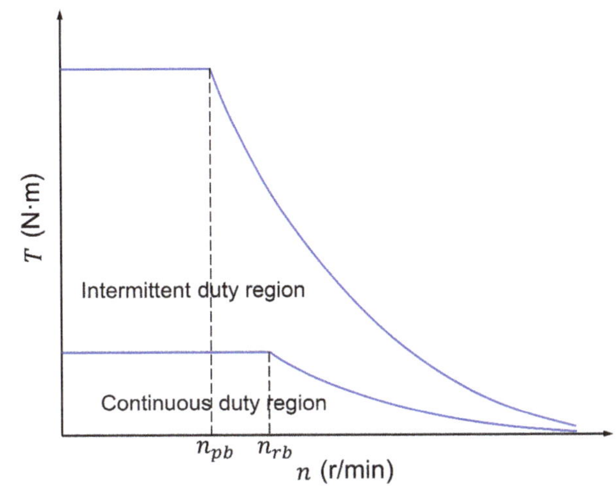

Based on the design principles discussed earlier, the rated power must satisfy the maximum power required for the EV to operate continuously for a long time (e.g., 30 min) at its nominal top speed on a level road. That is,

$$P_{rated} \geq \left(m_a gf + \frac{C_d A V_{max}^2}{21.15} \right) \frac{V_{max}}{3600\eta_T} \tag{8.6}$$

where

P_{rated} is the rated power of the electric motor
m_a is the GVW of the vehicle
f is the rolling friction coefficient
C_d is the drag coefficient of the vehicle
A is the frontal area
V_{max} is the top speed of the vehicle
η_T is the transmission efficiency

To match the base speed of the motor with the cruise speed of the vehicle, which is most frequently operated at

$$n_b = \frac{i_g i_{fd} V_n}{0.377 R_d} \tag{8.7}$$

where

n_b is the base speed of the motor
i_g is the transmission ratio
i_{fd} is the final drive ratio
V_n is the cruise speed of the vehicle
R_d is the dynamic radius of wheels

The rated torque can be calculated by

$$T_{rated} = \frac{9554.3P_{rated}}{n_{rb}} \tag{8.8}$$

where

T_{rated} is the rated torque
P_{rated} is the rated power
n_{rb} is the base speed corresponding to rated power

8.6.2.2. Peak Power and Torque

The peak power must satisfy the maximum power required for the EV to operate continuously for at least 30 sec to 3 min. The maximum power demands may occur under several different operating conditions: first, accelerating with a fully depressed accelerator pedal; second, fast overtaking on the highway; third, traveling at short-term top speed (generally for 30 sec to 3 min); and fourth, climbing a road of large slope at high speeds, as elaborated as follows.

1. Peak power and torque demands from fully depressed pedal acceleration
   On a level road, the 0–100 km/h acceleration time t_{100} can be calculated by

$$t_{100} = \frac{1}{3.6}\int_0^{100} \frac{\delta m_a}{\frac{3600\eta_T P_{max}(V)}{V} - \left(m_a g f + \frac{C_d A V^2}{21.15}\right)} dV \tag{8.9}$$

where

δ is the rotating mass conversion factor
V is the vehicle speed
P_{max} is the maximum power of the electric motor

Obviously, P_{max} is a function of vehicle speed V:

$$P_{max}(V) = \begin{cases} \frac{5 i_g i_{fd} T_{peak}}{18 R_d} & V \le V_{pb} \\ P_{peak} & V > V_{pb} \end{cases} \tag{8.10}$$

where

T_{peak} and P_{peak} are the peak torque and peak power of the electric motor, respectively
V_{pb} is the vehicle speed corresponding to the base speed at peak power

Note that the transmission ratio i_g varies with vehicle speed if a multispeed transmission is used.

Thus, the peak power $P_{1,peak}$ and peak torque $T_{1,peak}$ required for fully depressed pedal acceleration can be determined by using Equations (8.9) and (8.10).

2. Peak power demand from fast overtaking on the highway
   The peak power required for fast overtaking on the highway $P_{2,peak}$ can be estimated by assuming a vehicle traveling at a speed of 10% higher than the speed limit on a level highway:

$$P_{2,peak} = \left(m_a g f + \frac{C_d A V_{lmt}^2}{21.15} \right) \frac{1.1 V_{lmt}}{3600 \eta_T} \tag{8.11}$$

   where V_{lmt} is the speed limit on the highway.
3. Peak power demand for traveling at short-term top speed
   The peak power required for the vehicle to travel at its short-term top speed on a level road is given by

$$P_{3,peak} = \left(m_a g f + \frac{C_d A V_{st}^2}{21.15} \right) \frac{V_{st}}{3600 \eta_T} \tag{8.12}$$

   where V_{st} is the short-term top speed, which can typically be maintained for 30 sec–3 min.
4. Peak power demand for climbing a road of large slope at high speeds
   The peak power required for the vehicle to climb a road of large slope at high speeds $P_{4,peak}$ can be evaluated by

$$P_{4,peak} = \left(m_a g f \cos\alpha + \frac{C_d A V_{grade}^2}{21.15} + m_a g \sin\alpha \right) \frac{V_{grade}}{3600 \eta_T} \tag{8.13}$$

   where

   α is the road slope

   V_{grade} is the vehicle speed on the slope
5. Peak torque demand for overcoming the maximum slope
   The peak torque required for overcoming the maximum slope $T_{5,peak}$ as specified by the gradeability of the vehicle can be calculated by neglecting the aerodynamic drag:

$$T_{5,peak} = \left(m_a g f \cos\alpha + m_a g \sin\alpha \right) \frac{R_d}{i_g i_{fd} \eta_T} \tag{8.14}$$

Finally, the peak power must satisfy the maximum power required in driving conditions 1 through 4 as discussed earlier:

$$P_{peak} = \max\left(P_{1,peak}, P_{2,peak}, P_{3,peak}, P_{4,peak} \right) \tag{8.15}$$

With the accelerator pedal fully depressed, the peak power needs to be maintained for at least 30 sec–3 min.

The peak torque must satisfy the maximum torque required in driving conditions 1 and 5:

$$T_{peak} = \max\left(T_{1,peak}, T_{5,peak}\right) \tag{8.16}$$

With the accelerator pedal fully depressed, the peak torque needs to be maintained for at least 30 sec–3 min.

8.6.2.3. Maximum Motor Speed

The maximum motor speed must satisfy not only the required maximum motor speed at each gear shift but also the maximum speed corresponding to the top speed operation in the top gear.

8.6.2.4. Extreme Motor Speed

The extreme motor speed is the instantaneous speed at which the motor can operate safely for a very short period of time, which should be at least 5% higher than the aforementioned maximum motor speed. In no circumstances can the motor speed exceed its extreme speed, even in the presence of a system failure.

8.6.2.5. System Voltage

Many factors must be taken into consideration in selecting the system voltage for an EV. First of all, the availability of battery cell modules and packs is a major concern, along with the convenience, interchangeability, and versatility of the battery system. Besides, the available power electronics and accessories such as DC-DC converter, A/C, and heater should be taken into account as well. Finally, the system voltage can be selected by evaluating the gains and penalties in the efficiency, weight, and cost of these subsystems and components.

Currently, there is a trend to elevate the system voltage in the EV industry, mainly driven by the need for superfast charging. By doing so, the electric current in power electronic circuitry can also be reduced for a given power requirement, which helps in turn to bring down not only the capacity and discharge rate of the battery pack but also the diameter and size of electric components such as wire harnesses, connectors, and switches. However, the increase in system voltage raises the requirements for the voltage level of power electronics of the motor controller and the insulation design of the motor components, bringing the costs up.

The selection of system voltage should also comply with applicable regulations and standards. Although most EVs currently have a system voltage ranging from 300 to 600 V, it might not take long before a system voltage of 800 V or higher becomes the norm.

8.6.3. Electric Drive System Mounts

In general, the electric drive system mounts play three roles: first, serving as supporting members to balance the gravitational force acting on the electric drive system; second, providing the reaction torque needed to balance the electric drive torque for holding the electric drive system in place within allowable displacement; and third, isolating the vibrations of electric drive system and preventing them from transferring into the passenger compartment.

In fact, the vibrations of an electric drive system stem mainly from its torque ripple, especially during operating conditions such as creeping, acceleration, deceleration, and braking. Since these vibrations are insignificant most of the time and the electric drive torque is typically very high, the emphasis in selecting electric drive system mounts is, therefore, usually placed on providing the reaction torque rather than isolating vibrations.

Since most electric drive systems are not left-right symmetric, and the half shafts are often tilted at considerable angles with respect to the centerline of the motor shaft, a three-point centroid principle is often adopted in the layout of electric drive system mounts. That is, the centroid of the electric drive system must fall in the triangle formed by connecting the three mounting points in the top view. In the meanwhile, the three mounting points should be positioned so that the center of the triangle is as close to the centroid of the electric drive system as possible.

Although the electric drive system mounts can be chosen from rubber, air spring, hydraulic, semiactive, and active types, most EVs employ rubber mounts for their simplicity, reliability, and high stiffness in the form of cylindrical bushing. The characteristics of these mounts can be generally represented by their dynamic stiffness and the loss angle. Dynamic stiffness refers to the deformation under dynamic loading or excitation force. The loss angle is defined as the phase angle between the applied displacement signal and measured force signal in a dynamic stiffness test as a measure of damping amount in an elastomer.

A rubber mount is typically comprised of sheet metal and rubber pieces that are bound together in the form of compression, shear, or compound. The compression type is capable of bearing relatively large loads in the vertical direction and a small excitation force in the lateral direction. The shear type can generally be used when the excitation force is large and the vertical load is light. The compound type is suitable for the scenario when both the vertical load and excitation force are large, while the compression shear ratio can be made adjustable to fit the needs of various load and excitation conditions. Neither dynamic stiffness nor the loss angle of a rubber mount varies significantly with excitation frequency.

As far as the layout of electric drive system mounts, the traditional torque roll axis (TRA) methodology is not appropriate for EVs, due to the excessive mount forces and change in system attitude [8.1]. To minimize the mount forces and system attitude change, it is generally recommended that the elastic center of each mount be placed as far away from TRA as possible for maximizing the torsion resistance of the mounting system. Consequently, the reduced mount forces allow the use of low-stiffness mounts, which helps improve their vibration isolation rate. In addition, the mounts should be so positioned that the forces are as evenly distributed as possible.

In addition, over 80% of the decoupling rate between the frequency of e-motor excitation force and the rigid modal frequency is desired in pitch direction [8.1]. It is also recommended that the vibration isolation rate between 400 and 8000 Hz be greater than 30 dB, which can be attained by lowering the static stiffness of mounts and confining the dynamic-to-static stiffness ratio [8.1]. It is preferable to design the mount brackets as compact as possible with a modal frequency higher than 1000 Hz [8.1].

8.7. Selection of Suspensions

8.7.1. Front Suspension

Today, almost all electric passenger vehicles, including electric pickups, employ independent front suspension to facilitate the mounting of the e-powertrain and prevent the steering wheel from shimmying.

Capable of meeting basic chassis performance requirements, MacPherson struts are used in many entry-level electric passenger vehicles as front suspension for their simplicity, compactness, and low cost. However, the relatively higher mounting point of their shock absorbers makes it more difficult to fit large-size tires than double wishbone suspensions or multilink suspensions.

Since electric drive systems are relatively smaller in size than their internal combustion counterparts, the addition of an upper swing arm becomes appropriate to the front suspension of EVs to improve their lateral support capacity and steering performance. Hence, double wishbone suspensions and their derivative, that is, multilink suspensions, are alternatives to consider, especially for high-end electric passenger vehicles. However, they are both more complicated in structure, more difficult to develop, and more expensive than MacPherson struts.

Besides, large-size tires are often needed for EVs to meet the minimum ground clearance and half-shaft angle requirements when a big final drive gear is used, or to keep a reasonable tire-to-body height ratio in styling simply for a more appealing look. Due to the increased mass of battery system, large-size tires may also become necessary to accommodate large-size brake discs and calipers to meet the braking performance requirement.

8.7.2. Rear Suspension

As far as the rear suspension is concerned, either nonindependent suspension or independent suspension can be utilized in practice. When making a selection of rear suspensions, factors such as the influence on the packaging of the rear e-powertrain and the restrictions on the width and length of the battery pack must be taken into account. As a general rule of thumb, multilink rear suspension is recommended for most electric passenger vehicles, although torsional beam rear suspension is sometimes used as an alternative for its low cost and high load-carrying capacity. In multilink suspension, three to five links are typically used for controlling various DOFs of rear wheels. Ideally, five control links are needed to limit five out of the total six DOFs, as a suspension of three or four control links usually does not offer full control over the behavior of wheels. If it is assumed that all control links feature pivot supports at both ends, then a three-link suspension allows free vertical, longitudinal, and rotational movement; a four-link suspension allows free vertical and rotational movement; and the five-link suspension allows free vertical movement only. As a variant of multilink suspension, H-arm suspension is often adopted in EVs as well for its high stiffness, less degree of rotation, little influence on the battery pack size, and low development effort owning to well-decoupled functionality among its control arms. With the H-arm typically arranged as the lower control arm, an extra horizontal control arm is normally added to the H-arm suspension to make the system of a single DOF. Nevertheless, in addition to the shortcomings mentioned in the last section, torsional beam rear suspension also imposes a significant restriction on the width and length of the battery pack and on the rear-mounted e-powertrain for RWD or full-wheel drive vehicles. Hence, it is generally not recommended for EVs of high performance in handling and ride comfort nor for off-road applications. If it is used for any reason, careful tuning of suspension design parameters is necessary to ensure satisfactory results in handling and ride comfort performances of the vehicle. On the other hand, leaf spring suspensions are sometimes used in electric commercial vehicles, such as buses and trucks.

8.8. Selection of Tires

The load index and speed rating are the most important criteria in selecting tires for an EV. Besides, low-noise and rolling resistance tires should be selected whenever possible, along with considerations on characteristics such as adhesion, wear, mass, interchangeability, ease of replacement, and cost.

Compared to the tires on traditional vehicles, the tires on EVs generally require higher load capacity, better torque fluctuation tolerance, lower rolling resistance, and noise, due to the extra mass of the battery system, fast torque response of the e-motor, and demands for driving range and compartment quietness.

In layout design, the maximum stationary load $F_{z,\, max}$ on each tire must be made close to the nominal load of the tire F_{nom}, to maximize the utilization of its load capacity. The ratio $F_{z,max}/F_{nom}$ should be kept within the range of 90–100% to prevent tires from overloading. A value near its lower bound is typically selected for light-duty EVs because they operate frequently at high speeds and under large dynamic loading.

Among various types of tires in terms of carcass structure, the radial tires are the best choice for most road vehicles due to their LRR, minimal rise in temperature, and good adhesion and cushioning performance, while bias tires only find their applications in construction vehicles most of the time.

There are different types of tire tread groove patterns designed for various types of road surfaces, namely, longitudinal, transversal, combined, and off-road groove patterns [8.2, 8.3]. The suitable pavement types of these groove patterns are listed in Table 8.17.

TABLE 8.17 Suitable pavement types for various tire tread groove patterns.

Tire tread groove pattern	Suitable pavement type
Longitudinal	Good asphalt or cement pavement
Transversal	Earth-rock pavement
Combined	Uncertain pavement
Off-road	Poor pavement or no road

Low-profile tires often find their applications in high-performance EVs for boosting the handling at cornering. However, low-profile tires result in a firmer ride over bumps than normal-profile tires, deteriorating ride comfort.

8.9. Overall Vehicle Layout Design

The coordinate system for the overall vehicle layout design follows the convention defined in Figure 8.12 [8.4]. The overall vehicle layout design involves producing the layout drawing and checking whether the structure and dimensions of each selected subsystem meet the requirements of vehicle-level parameters and dimensions so that a reasonable overall layout plan can be worked out.

FIGURE 8.12 The coordinate system for overall vehicle layout design.

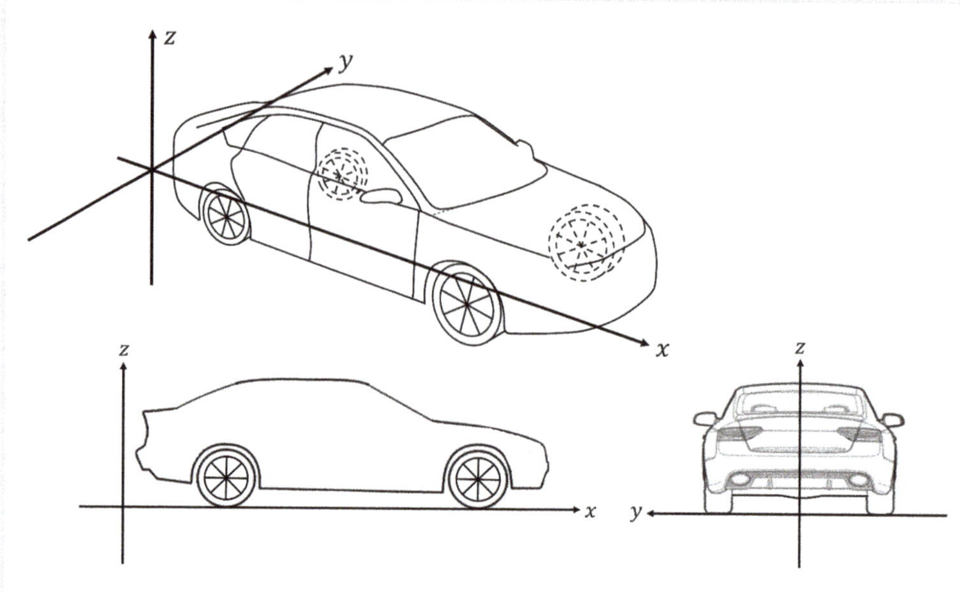

8.9.1. Battery System Layout

Dependent primarily on the dynamic performance, range, and environmental temperature, the battery system capacity of an EV can be determined based on the method elaborated in Section 8.5.2.

The layout of the battery system in an EV can significantly affect the tire load, braking performance, suspension system, and vehicle attitude. Hence, the layout of the battery system must meet the requirements in aspects such as heat dissipation, collision safety, trafficability, axle load allocation, CG, and handling stability.

The battery system is typically integrated into a vehicle between the front and rear axles to balance out the vertical load due to its substantial mass. Mostly mounted to the frame of the vehicle underneath the floor, either inside or outside of the passenger compartment, the battery system can also be arranged under the rear seats and even under the front seats in rare cases. Although most battery systems are mounted to the vehicle as a single pack or a couple of separate packs, a modern design approach is to integrate modules or cells directly into a fully enclosed battery compartment in the chassis.

In laying out the battery system, it is first necessary to determine the cooling method of the battery pack, as it has a considerable impact on the required clearance around the pack. In addition, other factors such as the floor structure of the vehicle body, the chassis parts, the electrical components, the internal and external trim parts, and whether a spare tire is needed and where it is placed influence the battery system layout as well.

The general rules in the layout design of battery systems are as follows:

1. The ground clearance under the battery pack must be not only greater than the minimum ground clearance of the vehicle at full loads but also greater than the required clearance at the maximum dynamic deflection of suspension, with no portion of the battery pack lower than the surrounding underbody structure of the vehicle.

2. The number of mounting bolts around the pack must be sufficient to connect it to the underbody of the vehicle, in order to improve its strength and vibration modes.
3. A protective fender, typically made of steel and bolted to the bottom of the battery pack, should be considered to prevent it from road hazards.
4. Collision safety structures on each side of the battery pack, particularly crumple zones for lateral collisions, must be carefully designed to prevent battery cells from catching fire and exploding due to extrusion and deformation caused by collision, rolling, or falling during an accident. Load shift paths during the vehicle crash should be designed properly, ideally by topology optimization of material distribution. Other than the longitudinal beam on the left and right sides of the battery pack, the longitudinal impact loads can also be transferred by making use of the battery pack structure, for example, the longitudinal beams of the battery pack. Nevertheless, the positional deviation between the lateral beams of the battery pack and the vehicle should be minimized to reduce the pressure on the battery cells and modules in the case of a side collision.
5. The stress and fatigue life of the load-carrying structure of the battery pack must be examined carefully to withstand the static and dynamic loads, particularly due to the long-term vibrations and impacts from the road–tire–suspension interaction.
6. The cooling structure for the battery system must satisfy the worst operating condition with maximum heat generation in high ambient temperature with air, liquid, or phase-change material (PCM) cooling.
7. The axle load distribution and the CG height of the EV must be checked to ensure proper drivability and ride comfort.
8. Arranging the battery pack in the luggage compartment should be avoided wherever possible in order to maintain a proper luggage space.

8.9.2. Powertrain System Layout

The layout of the electric powertrain depends on the number and location of the power sources. Unlike traditional ICE vehicles, EVs often have more than one power source, leading to great flexibility in their layout.

Generally speaking, the powertrain system layout of EVs can be divided into two types: centralized and distributed.

A centralized powertrain system layout typically features one or two traction motors, each of which mates with a transmission, and the power is transmitted to the wheels on both sides through the differential. Illustrated in Figure 8.13a through 8.13f is the layout of several typical centralized powertrain system layouts, including single-motor front-mount front-wheel drive (FF), single-motor mid-mounted RWD (MR), single-motor rear-mounted rear-wheel drive (RR), dual-motor AWD, dual-motor front-mounted front-wheel drive (FF), and dual-motor rear-mounted rear-wheel drive (RR), respectively.

FIGURE 8.13 Centralized powertrain system layout.

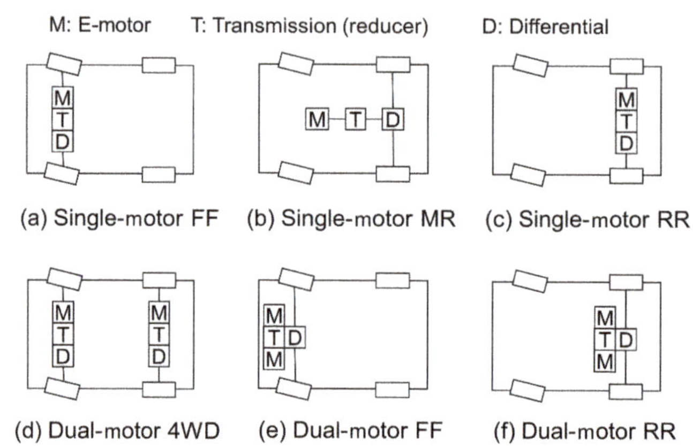

Distributed powertrain system layouts can be divided into two types: in-wheel motors (or hub motors) and wheel-side motors. The hub motor is directly installed within the wheel, with the outer-rotor type of hub motors driving the wheel directly, and the inner-rotor hub motors generally driving the wheel through a gear reduction mechanism. The wheel-side motor mostly drives the wheel through a reducer and a drive shaft. Whether in-wheel motors or wheel-side motors are used, they are usually arranged in pairs on both sides of the vehicle symmetrically with respect to the vertical longitudinal zero plane. Shown in Figure 8.14a through 8.14i are several typical distributed powertrain system layouts, including outer-rotor hub motor FWD, outer-rotor hub motor four-wheel drive, outer-rotor hub motor RWD, inner-rotor hub motor FWD, inner-rotor hub motor four-wheel drive, inner-rotor hub motor RWD, side-wheel motor FWD, side-wheel motor four-wheel drive, and side-wheel motor RWD, respectively.

FIGURE 8.14 Distributed powertrain system layout.

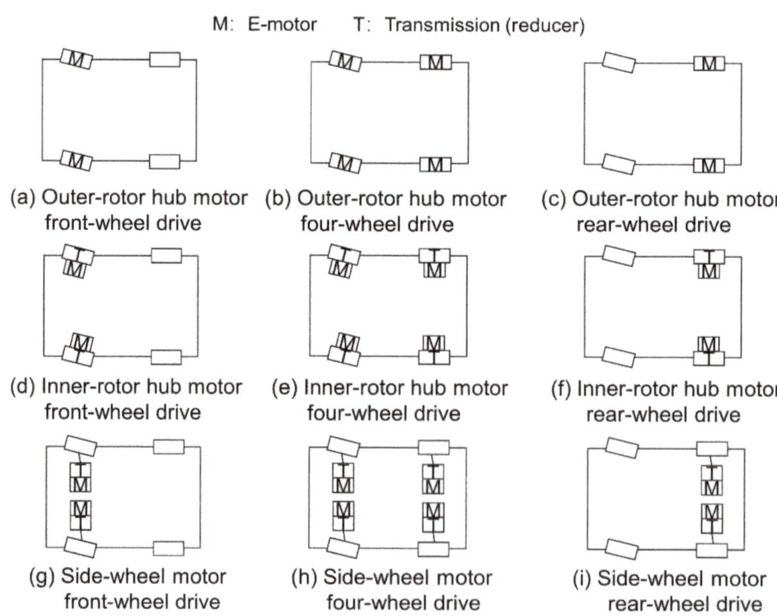

In addition, there are also three-motor four-wheel-drive powertrain layouts that combine the centralized and distributed arrangements. Illustrated in Figure 8.15a and 8.15b are typical examples of such layouts, including side-wheel motor FWD with single-motor RWD, and single-motor FWD with side-wheel motor RWD, respectively.

FIGURE 8.15 Three-motor four-wheel-drive powertrain layouts.

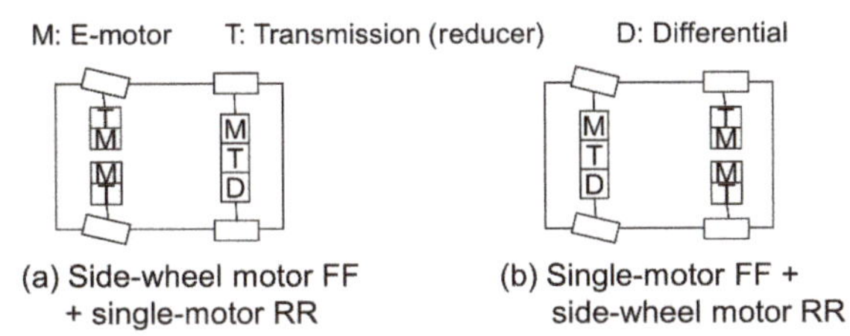

The ground clearance and the driver's field of vision are the most important aspects to consider in the layout of the electric drive system in terms of its vertical position. The distance from the lowest point of the electric drive system, typically the bottom of the oil pan, to the ground in fully loaded condition must meet the minimum ground clearance requirement. On the other hand, it is preferred to position the electric drive system lower to prevent the height of the front compartment cover from affecting the driver's field of vision in an FF configuration, or to prevent the height of the rear floor and back seat from affecting the interior space in a RR configuration. Besides, lowering the electric drive system helps bring down the mass centroid as well, which improves the stability of the vehicle. A 25 mm clearance below the front compartment cover is recommended to avoid interfering with the top of the electric drive system and other components mounted in the compartment in shutting off the cover. The vertical position of the electric drive system is labeled by the distance from the ground to the point where the centerline of the electric motor output shaft intersects with the front face of its cover. In addition, the vertical positions of the radiator and cooling fan can be determined accordingly by following the similar principle as described earlier, with their centerlines typically aligned with each other.

In the lateral direction, a layout with the outputs of the electric drive system centered about the longitudinal plane of the vehicle is always preferred not only because the left and right drive shafts and mounting brackets can be made identical but also because the load on the suspension can be made more uniform in the lateral direction. Unsymmetric layout is often used, however, in engineering practice for various reasons, mostly because of other constraints such as ground clearance or clearance to other components, or simply because the electric drive system is laterally unsymmetric itself in terms of its center of mass and output planes on the right and left. When an unsymmetric left-right layout becomes inevitable, a rule of thumb is to keep the angle between the centerline of differential output and the centerline of half shaft within 25° and the angle between the centerline of half shaft and the centerline of the wheel within 10° to avoid excessive wear of the half-shaft splines, as the wheel plane is aligned with the longitudinal direction.

The layout of the front-rear position of the electric drive system must take load distribution and usable passenger or cargo space into account. The front-rear position of the electric drive system is usually specified by the horizontal distance from the centerline of the front wheel to the point where the centerline of the electric motor output shaft intersects with the front face of its cover. The radiator and cooling fan are generally set at least 25 mm apart in the absence of a fan shield, or at least 50 mm apart if a fan shield is used in between to better guide the airflow.

8.9.3. Steering System Layout

The steering wheel must be positioned at a proper angle between the wheel plane and the horizontal plane so that the driver can make turns comfortably with ample clearance around it, especially to the windshield. Moreover, the position of the steering wheel should not block the driver's view of the instrument panel with the blind spot in front of the vehicle minimized.

When the steering gear connects to the steering wheel directly with a rigid shaft, the projected angle in the horizontal plane between the steering shaft and the vehicle centerline should not exceed 5° to prevent the foot manipulation from being affected.

When the steering gear connects to the steering wheel with one or two universal joints and a steering shaft, the steering gear can be mounted on the vehicle frame with its axis at an angle with the centerline of the steering wheel. The steering shaft and steering column are typically not in the same plane to prevent injury caused by backward deformation of steering wheel in a front collision.

The steering gear should be positioned near the instantaneous center of leaf springs to avoid motion interference between the steering mechanism and suspension, when a leaf spring suspension is used in the front, as described in Sections 11.2.6 and 11.3.4.

The angle between the Pitman arm and the drag link is preferably 90° to maximize the transmission efficiency. A right angle between the steering arm and the drag link is also desirable for the same reason.

8.9.4. Layout of Suspensions

For an EV, the layout of suspensions must fit the package space required for passengers, the electric drive system, and the battery system. The additional stiffness and mass due to the presence of a battery system often call for a strong suspension restraint and high load-carrying capacity.

The shock absorber is preferably positioned vertically to maximize its effective stroke for damping out vibrations, although it can be oblique when there is not enough space to do so.

For leaf spring suspensions, the leaf springs in the front are typically arranged under the longitudinal beams to maximize the turning angle of steering wheels, while the leaf springs in the rear are arranged outside the frame to improve the roll stiffness. In addition, the understeer characteristics of a vehicle can be enhanced by making the front leaf springs high at its front end and low at the rear end, and the rear leaf springs low at its front end and high at the rear end. Besides, it is helpful to alleviate the impact load from the road surface by using the fixed hinge at the front and movable ear at the back.

In general, although the compact size of the electric drive system gives quite a flexibility in selecting suspensions, the suspension design is complicated by its entangled relationship with vehicle topology.

8.9.5. Brake System Layout

The layout of the brake pedal should follow the rules given in the next section. A checkup on motion interference is needed to make sure that no interference occurs when wheels bounce off the ground.

The brake lines should be laid out in a neat and reliable manner, with a minimum distance of 5 mm between parallel lines and 20 mm between crosslines. A transitional hose should be utilized if there is relative motion between two points on the same line.

8.9.6. Layout of Operator Pedals

The accelerator pedal and brake pedal should be so positioned that the driver can easily switch between them by only changing his or her ankle angle. The accelerator pedal is typically arranged a bit lower than the brake pedal, with at least 60 mm of distance between them. The pedal force must be designed appropriately to allow the driver to operate with ease. For instance, the accelerator pedal force should be designed light enough so that it can be depressed effortlessly for a long time. The brake pedal should be placed closer to the driver than the accelerator pedal for easy operation, particularly when great force is applied in emergent braking.

The pedal force is significantly affected by the pedal position relative to the seat in both longitudinal and vertical directions [8.2, 8.3]. The dimensions related to pedal position are given in Figure 8.16, with the recommended values listed in Table 8.18.

FIGURE 8.16 Dimensions related to the pedal position.

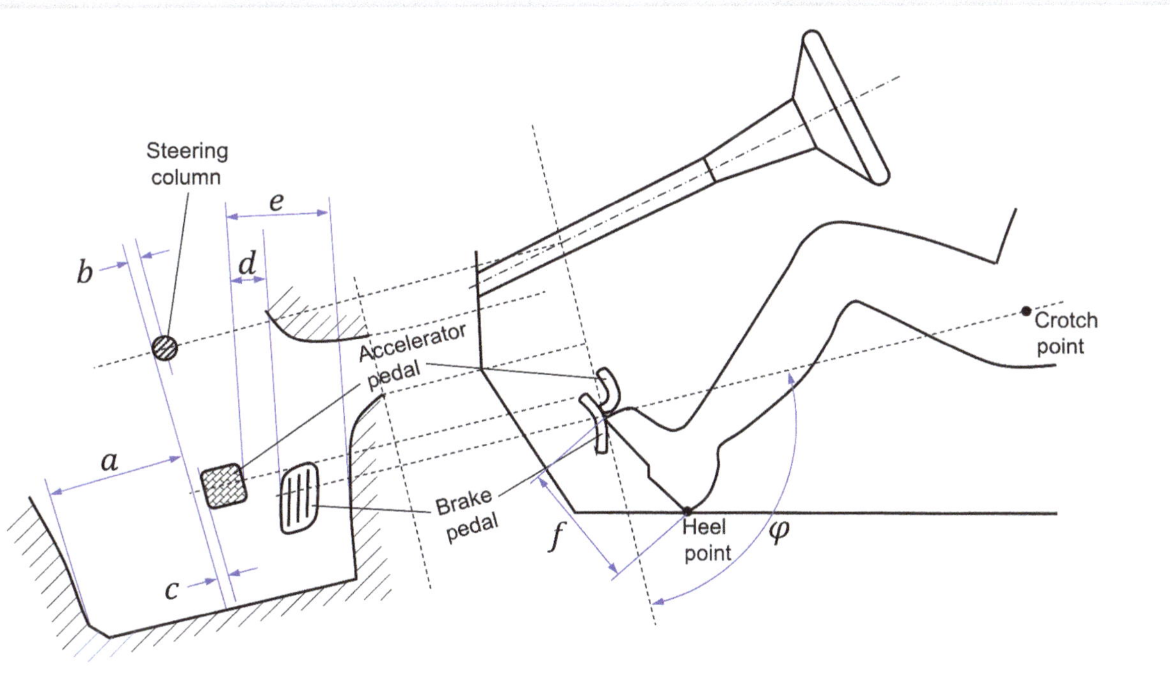

TABLE 8.18 Recommended dimensions for pedal positioning.

Dimension	Recommended value
a	255–265 mm
b	0–50 mm
c	25–35 mm
d	65–75 mm
e	165–175 mm
f	195–205 mm
φ	80–100°

8.9.7. Layout of Luggage Space and the Spare Tire

The layout of luggage space is more flexible in EVs than in their ICE-powered counterparts. Other than the space in the rear portion of the vehicle, additional luggage space can be made available in the front compartment as well. The trunk volume typically ranges from 0.4 to 0.9 m^3 for a medium- to large-size sedan.

Placed in the bottom of the trunk for most passenger vehicles, the spare tire may also be mounted on the interior sidewall or exterior rear wall of the luggage compartment for some SUVs.

8.9.8. Vehicle Interior Layout

Safety and comfort are the primary concerns when it comes to the interior layout of the passenger compartment.

The human body size in the region where the vehicle is intended to be used must be considered to determine its interior dimensions. Per SAE SP142, the human body measurements are comprised of sixteen dimensions from *A* to *P*, along with an additional ten dimensions related to passenger cab layout, as shown in Figure 8.17 [8.5]. Such statistical distributions of anthropometric data are usually expressed in the form of mean and standard deviation.

FIGURE 8.17 Human body measurements and dimensions related to the passenger cab layout.

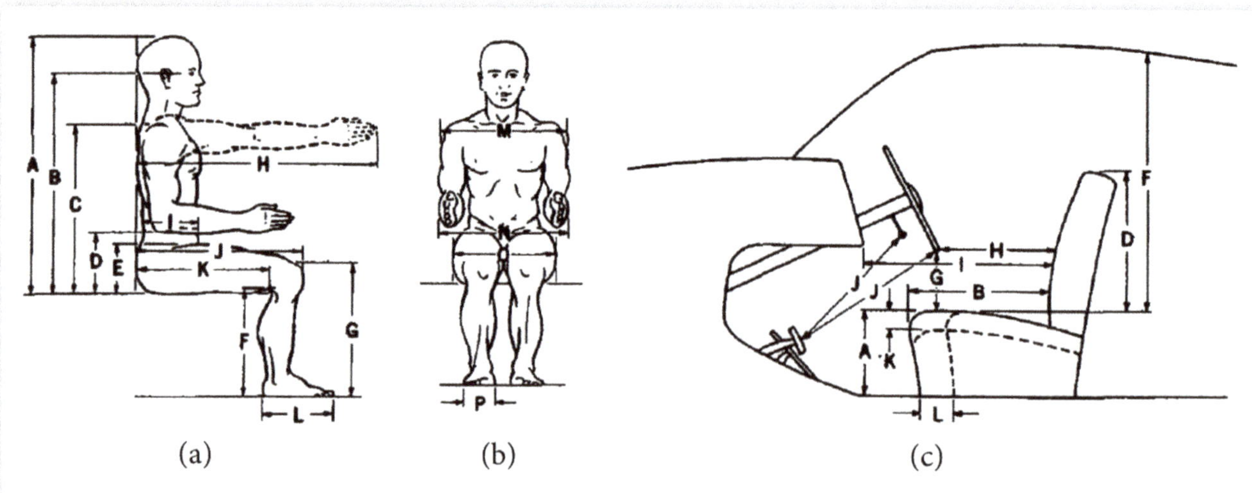

Traditionally, human body models, typically of the sizes of 10th, 50th, and 95th percentiles, are used for interior layout design, representing short, medium, and tall body shapes, respectively. The major joints of human body are modeled with hinges, allowing the angles between body parts to change, with angle rulers showing the readings. These models are made of plexiglass or plywood, in general, at the scales of 1:5, 1:2, or 1:1. The 50th percentile body model is employed to determine the basic size of the interior components, for example, seats and surrounding space, while the models of 10th and 95th percentiles are used to establish the seat adjustment.

The layout design is mainly carried out in the lateral view of the vehicle body by laying the human body model on it, as illustrated in Figure 8.18. At the beginning of the interior layout, a seating reference point, that is, R-point, is defined as the theoretical hip position of an occupant when the seat is set in the rearmost and lowermost seating position. After the prototype vehicle is built, the actual hip position relative to the seat, typically measured with a seated dummy, is referred to as the H-point. Both the R-point and the H-point specify the pivot points between the torso and the thigh when the occupant is seated. The layout design is usually deemed acceptable if the H-point is within a distance of 15 mm lateral and 10 mm vertical from the R-point, whereas the backrest angle is within 3° from the design value.

FIGURE 8.18 The seating reference point and accelerator heel point.

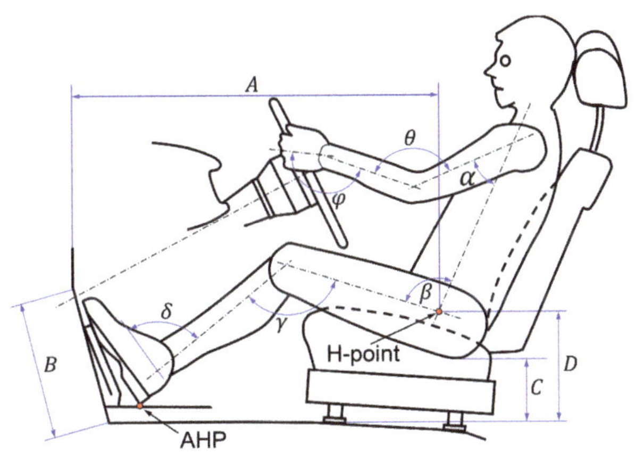

In designing the H-point with respect to the seat, the compression amount of the seat must be taken into account. With the accelerator heel point, AHP, placed on the vehicle floor, the height from the H-point to the heel point AHP can be adjusted to make the sitting angles α, β, γ, and δ, along with dimensions A, B, and C fall into recommended ranges. The recommended ranges for these design parameters for an electric sedan are given in Table 8.19.

Nowadays, digital human models (DHMs) are increasingly used in the layout design as an alternative to traditional physical models for the reduction in development time and cost.

TABLE 8.19 Recommended values for design parameters related to layout design in the lateral view of an electric sedan.

Design parameters	α	β	γ	δ	θ	φ
Recommended values (°)	10-45	95-115	95-145	87-110	80-120	170-190
Design parameters	***A***	***B***	***C***	***D***		
Recommended values (mm)	≥698	≥306	≥100	250-405		

8.9.9. Sizing of Vehicle Body

With the seat adjusted to the averaged position in both longitudinal and vertical directions, a point F at a distance of S from the H-point can be drawn on the line that is tilted at an angle ϕ from the vertical line on the lateral view of the layout, as shown in Figure 8.19, which is the highest point of the head of the 50th percentile driver. The interior roof point F_1 can be drawn that is at a distance S_1 above point F, corresponding to the clearance required above the head. Then, the exterior roof point above the driver, point F_2, can be determined by adding another distance S_2 above point F_1, which includes the thicknesses of the interior mask, isolation layer, and body panels. Finally, the highest point of the roof on the mid-plane cross section, point F_3, takes an additional distance S_3 above point F_2. The recommended values of S_1, S_2, and S_3 for a sedan are given in Table 8.20 [8.3].

FIGURE 8.19 Dimensions related to the sizing of a vehicle body.

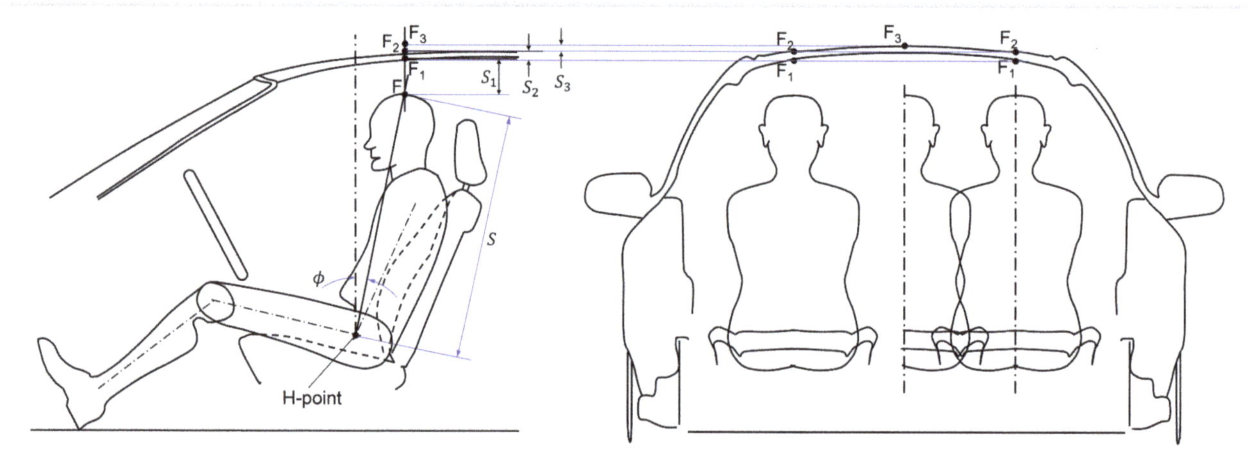

TABLE 8.20 Recommended values of S_1, S_2, and S_3 for a sedan.

Body layout parameters	ϕ (°)	S_1 (mm)	S_2 (mm)	S_3 (mm)
Recommended values	8-10	100-135	15-25	20-40

Courtesy of Wangyu Wang.

Similarly, the highest point of the roof for the rear seats can be established. Then, the outline in the longitudinal direction can be obtained by connecting the highest points of the roof at the front and rear.

The outline in the lateral direction can be drawn by connecting the highest point of the roof on the mid-place, F_3, to the highest points of the roof above the left and right seats, F_2.

Furthermore, the inclination of the side wall of the vehicle body must be taken into consideration as it affects the convenience of getting in and out of the vehicle. For example, the upper body of a passenger has to bend about 30° to get in a sedan when the side wall is vertical, versus only 0–10° when the upper edge of the door indents about 100–150 mm with respect to the lower sill edge. However, an excessive indentation can impact a number of design elements such as the proportion of the upper and the lower body, the vehicle appearance, space inside the door needed for the glass lift, the shoulder clearance from the glass, and elbow clearance from the internal surface of the door. It is generally desired that the shoulder clearance from the glass be greater than 100 mm and the elbow clearance from the internal surface of the door be greater than 70 mm.

8.9.10. Mounting Positions of Seat Belts and Airbags

The seat belts and airbags protect the occupants from being injured in a secondary collision.

Mainly used to restrain the occupants in their seats in the event of a front collision, seat belts can substantially reduce the human body acceleration at the moment of the crash. There are two-point, three-point, and four-point seat belts available. The two-point seat belts are used mostly for passenger seats of long-distance buses and the central seats of trucks where only the lower bodies of the occupants are restrained. The three-point seat belts are currently employed in the front seats of most vehicles and all rear seats in passenger vehicles, while four-point seat belts are applied only in racing cars.

Due to their safe nature, the design of seat belt anchorages must comply with applicable standards and regulations in the region where the vehicle is used. As far as the anchorages for a three-point seat belt, the upper end of the upper torso restraint shall be located within the acceptable range, while a line from the belt contact point to the H-point shall form an angle with the horizontal in order to determine the anchorage for the pelvic belt. Refer to those standards and regulations for details.

Airbags are an auxiliary occupant restraint system that provides additional soft cushioning and restraint on top of the seat belts during a collision. Consisting of the airbag cushion, a flexible fabric bag, an inflation module, and an impact sensor, the airbag inflates extremely quickly during the collision to reduce injuries between the occupant and the interior of the vehicle or even between the pedestrian and the exterior of the vehicle. A luxury vehicle can equip up to ten airbags in a variety of configurations such as driver, passenger, side-curtain, B- and C-pillar-mounted side-impact, seat- and door-mounted, knee bolster, inflatable seat belt, and pedestrian airbag modules.

The airbags are typically mounted in the steering wheel or in the dashboard at the front of the occupant.

8.10. Motion Correctness Check

Motion interference may occur in the parts or components of the vehicle where relative motion takes place, causing the vehicle to malfunction. Motion in the wrong direction can also produce serious consequences such as injuries, deaths, property damages, and waste of development time and resources.

To prevent motion in the wrong direction and to minimize motion interference, a motion correctness check must be carried out as the final step of the overall design.

To prevent the vehicle from moving in the wrong direction, a motion correctness check must be carried out on the electric drive system to ensure that the vehicle moves forward when the electric motor rotates in the same direction as what is designed. In addition, a motion correctness check must be conducted on the steering system mechanisms to make sure that the vehicle turns in the same direction as the steering wheel.

In order to minimize motion interference, motion correctness checks must be performed wherever relative motions occur, for example, the suspension bounce with respect to the drivetrain, the wheel bounce with respect to the wheel well, and the clearance between the front wheel and the wheel well at the extreme turning angle.

References

[8.1]. Kang, Q., Gu, P., and Zuo, S., "A Research on Design Principles for Electric Drive Assembly Mounts of Pure Electric Vehicles," *Automotive Engineering* 41, no. 11 (2019): 1235-1242, doi:10.19562/j.chinasae.qcgc.2019.011.002.

[8.2]. Liu, W., *Automobile Design* (Beijing: Tsinghua Press, 2001), ISBN:9787302045298.

[8.3]. Wang, W., *Automobile Design*, 3rd ed. (Beijing: China Machin Press, 2000), ISBN:7111076133.

[8.4]. ISO, "Road Vehicles—Vehicle Dynamics and Road-Holding Ability—Vocabulary," ISO Standard 8855:2011, 2011.

[8.5]. SAE International, "Human Body Measurements and Vehicle Dimensions," SAE SP142, 1955.

9

Battery System Design

The traction battery system is an energy storage device that not only supplies electric energy to the electric powertrain system to propel the vehicle but also absorbs the electric energy produced in a regenerative braking process. It must ensure the vehicle to reach the specified driving range and dynamic performance in specified ambient conditions. In addition, it must have a sturdy structure to minimize the damage to the internal battery cells in any collision, without catching fire or exploding in case of puncture, extrusion, and fall.

In general, a battery system for EVs must satisfy the following fundamental requirements:

1. Adequate power supply to the electric propulsion system to drive the vehicle under any driving conditions within a specified temperature range.
2. Structural integrity under specified collision conditions, with the maximum dynamic loading of 50*g* longitudinal acceleration in frontal collisions.
3. Structure of high strength and stiffness to prevent deformation, fracture, fatigue, vibration, and noise caused by static and dynamic loading, particularly in global structural resonance due to the vehicle interaction with the road and airflow.
4. High electrical safety by constantly monitoring the status of high-voltage circuitry and cutting off the high-voltage output automatically when a failure occurs, with functions such as insulation protection, leakage prevention, error-proof cable connectors, safety switches for emergency power off, and high-voltage labels.
5. Good electromagnetic compatibility (EMC) to prevent electromagnetic radiation from interfering with CAN signals, with electromagnetic radiation disturbance and immunity levels complying with applicable standards and regulations.

6. Lightweight and compact structure for battery capacity and driving range maximization, as well as for easy packaging onto the vehicle.
7. High reliability and durability, particularly for the connectors of interchangeable battery packs.
8. High ingress protection rating to prevent moisture, water, or dust from getting into the battery pack, typically no lower than IP67.
9. Good manufacturability and maintainability.

A typical battery back consists of battery cells or modules, electrical wiring, a battery thermal management system (BTMS), and a BMS. A battery cell is the smallest unit of a battery, while a battery module comprises a number of battery cells that are organized in series and parallel to form a subsystem. A BTMS consists primarily of mechanical structures and components for maintaining a uniform battery temperature in the desired range to prevent the low temperature, high temperature, and nonuniformity temperature distribution from affecting the battery performance. On the other hand, a BMS is mainly a set of electronic devices that generally include sensors, wires, fuses, and electrical circuit boards for monitoring and controlling the battery. The functionality of a BMS is to protect the battery by operating it within the operating margin of safety, monitoring its state, calculating and reporting secondary data, balancing cells, and controlling the operation of BTMS.

With the rapid development of battery and its system integration technology, there is a clear trend toward further structural simplification and an increased level of integration into the chassis of EVs in pursuit of safety, compactness, reliability, and weight reduction.

9.1. Types of Battery Cells

There are generally three types of battery cells, namely, cylindrical, prismatic, and pouch. These types of cells are compared in Table 9.1 in terms of a variety of metrics. Since each of the three types comes with advantages and drawbacks, they are all widely employed in the production of EVs today.

TABLE 9.1 Comparison of three types of cells.

	Battery cell type		
Metrics	**Cylindrical**	**Prismatic**	**Pouch**
Cell casing material	Aluminum or steel	Aluminum or steel	Aluminum plastic film
Fabrication process	Rolled	Laminated or rolled	Laminated or rolled
Energy density	Medium	Medium	High
Cell-to-pack mass ratio	Low	High	Medium
Cell-to-cell variation	Low	Medium	Medium
Changing and discharging rate	Medium	High	Medium
Safety	Medium	Medium	High
Productivity	High	Medium	Medium
Degree of standardization	High	Medium	Low
Strength of cell case	High	High	Low
Cost	Medium	Medium	High
Requirement of fabrication process	Low	Low	High
Suitability for irregular space	High	Low	Low

9.2. Battery Pack Architecture

The battery pack architecture depends heavily on the battery cell type and the packaging space on the vehicle [9.1].

The cylindrical cells are typically arranged vertically in the pack, as shown in Figure 9.1a. In contrast, prismatic and pouch cells can be arranged in various ways, as shown in Figure 9.1b through 9.1d, depending on the size of cells and the space available in the vehicle. In some vehicles, prismatic and pouch cells can be arranged using a T-shaped architecture (Figure 9.2a) to utilize the space between the left and right seats and under the rear seats, with battery cells or modules standing vertically on their bottom as shown in Figure 9.1b. However, this can result in reduced interior space for passenger vehicles, particularly the headroom for passengers in rear seats. Therefore, floor architecture is a much more popular choice for most electric passenger vehicles, in which battery cells or modules are arranged in the slab format under the vehicle floor as shown in Figure 9.2b. These battery cells or modules are placed either face down or side down, as shown in Figure 9.1c and 9.1d, respectively.

FIGURE 9.1 Typical orientations of battery cells in a battery pack. (a) Cylindrical cells in vertical arrangement, (b) prismatic or pouch cells in vertical arrangement, (c) prismatic or pouch cells in face- down arrangement, and (d) prismatic or pouch cells in side-down arrangement.

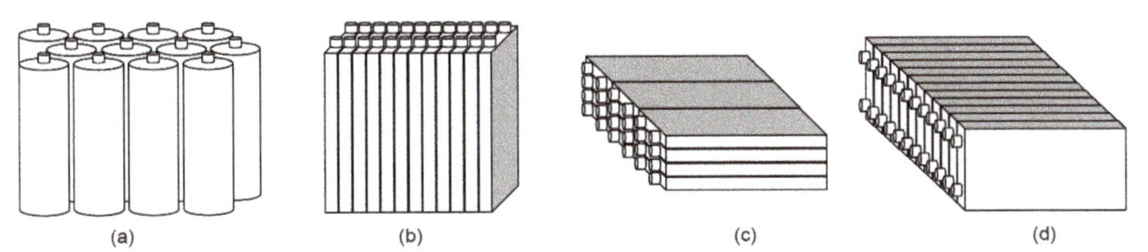

FIGURE 9.2 Typical shapes of battery packs in electric passenger vehicles. (a) T-shaped format and (b) slab format.

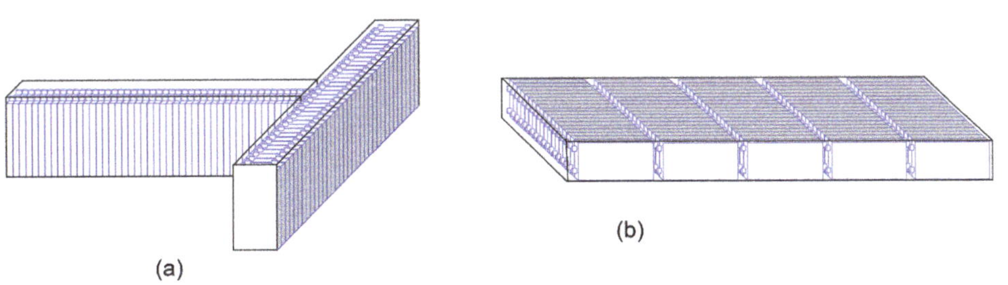

In addition, a clearance is needed between adjacent battery cells to accommodate their expansion that may occur when they are in operation. Sometimes, these clearances are foam-filled to provide extra cushion and insulation, thus preventing neighboring battery modules from direct contact and conductive heat transfer. In a more complex design practices, thermal conductive shims may be placed between adjacent battery cells for connecting to the liquid cooling structure as a part of the active BTMS.

To reduce the complexity of the battery system, large-capacity battery cells are preferable. Nevertheless, the consistency among battery cells generally worsens with increasing the charge capacity of batteries. At present, the charge capacity ranges from 3 to 5 Ah for cylindrical batteries, and from 50 to 600 Ah for prismatic and pouch batteries, for EV applications.

For large vehicles such as buses and trucks, multiple battery packs, typically of identical specifications, are used instead of a single one for ease of maintenance, as well as for the reduction in the thermal runaway scale and replacement cost.

9.3. Electrical Topology of Battery Packs

The electrical topology, namely, the interconnection of battery cells or modules is determined in this step. At its fundamental level, the battery cells can be either connected in series or parallel, and the electric parameters of the corresponding circuit are elaborated next.

9.3.1. Battery Cells Connected in Series

When the battery cells are connected in series, the positive terminal of each cell is connected to the negative terminal of the next, as shown in Figure 9.3a. Thus, the total electromotive force, E, is given by

$$E = \sum_{i=1}^{n} E_i \tag{9.1}$$

where

n is the total number of battery cells

E_i is the electromotive force (EMF) of the ith battery cell

FIGURE 9.3 Ways of connecting battery cells. (a) Serial connection and (b) parallel connection.

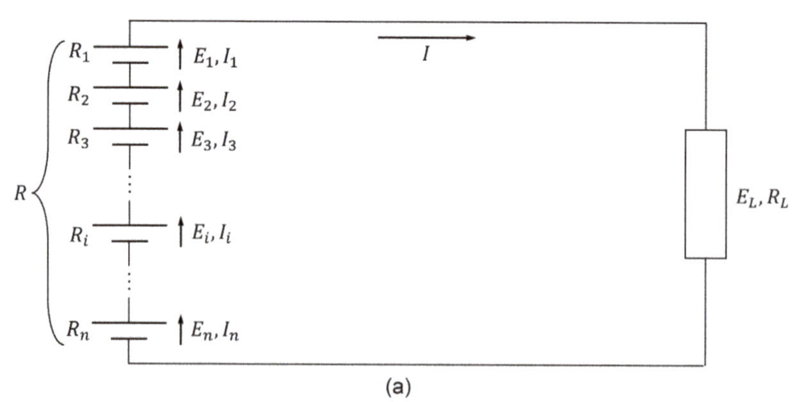

FIGURE 9.3 **(Continued)** Ways of connecting battery cells. (a) Serial connection and (b) parallel connection.

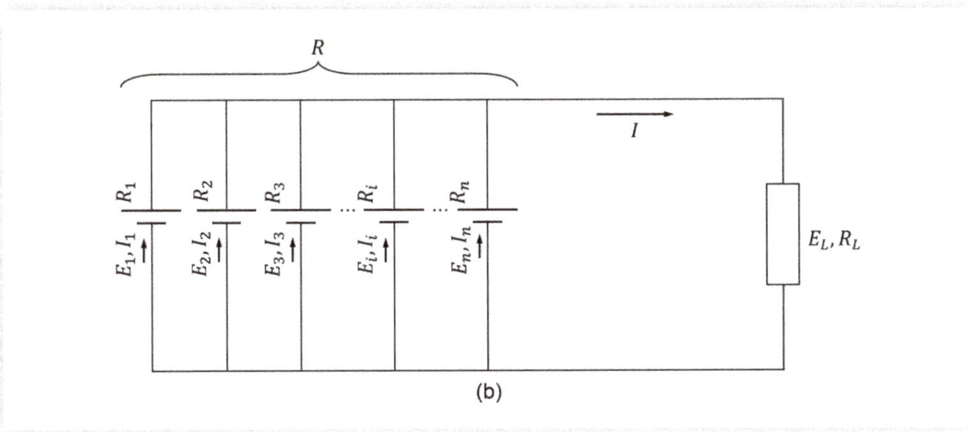

The total internal resistance of such a group of battery cells in serial connection, R, is given by

$$R = \sum_{i=1}^{n} R_i \tag{9.2}$$

where R_i is the internal resistance of the ith battery cell. The current, I, in the closed circuit consisting of these battery cells and a resistive load R_l can be obtained by

$$I = \frac{E}{R_l + R} = \frac{\sum_{i=1}^{n} E_i}{R_l + \sum_{i=1}^{n} R_i} \tag{9.3}$$

Note that the serial connection of battery cells increases the overall voltage of the battery system but does not increase the capacity. Consequently, a low system current is generated, meaning thinner wiring can be used.

9.3.2. Battery Cells Connected in Parallel

When the battery cells are connected in parallel, all the positive terminals are connected together and all the negative terminals are connected together, as shown in Figure 9.3b. Thus, the total electromotive force, E, is given by

$$E = E_1 = E_2 = \cdots = E_n \tag{9.4}$$

where the electromotive forces of the 1st, 2nd, …, n^{th} battery cells are all assumed equal. The total internal resistance of such a group of battery cells in series connection, R, is given by

$$R = \frac{R_1}{n} \tag{9.5}$$

where R_1 is the internal resistance of the 1st battery cell, and it is assumed that the internal resistances of the 1st, 2nd, …, n^{th} battery cells are all equal. The current in a closed circuit consisting of these battery cells and a resistive load R_l can be obtained by

$$I = I_1 + I_2 + \cdots + I_n = \frac{E}{R_l + \frac{R_1}{n}} \tag{9.6}$$

Note that the parallel connection of battery cells can increase the ampere-hour capacity of the battery system but does not elevate its voltage. In addition, if one of the cells malfunctions, the remaining cells in the system can still provide power. However, thicker wiring is needed to operate with the larger current.

9.3.3. Battery Cells Connected in Series–Parallel

Battery cells can be connected in series–parallel to increase both their EMF and amp-hour capacity. Basically, as shown in Figure 9.4a and 9.4b, respectively, their interconnection can be divided into two configurations: first, parallel first and then in series, or second, in series first and then parallel.

FIGURE 9.4 Basic configurations of series–parallel wired battery cells. (a) Parallel first and then in series and (b) series first and then parallel.

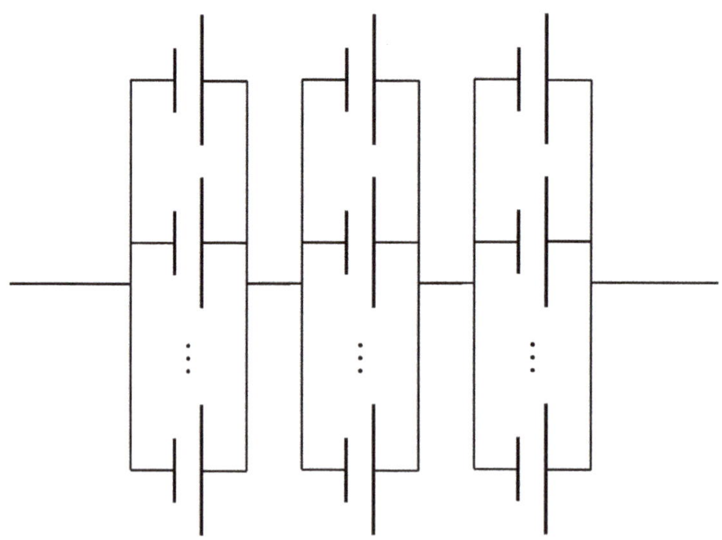

(a)

FIGURE 9.4 (Continued) Basic configurations of series-parallel wired battery cells. (a) Parallel first and then in series and (b) series first and then parallel.

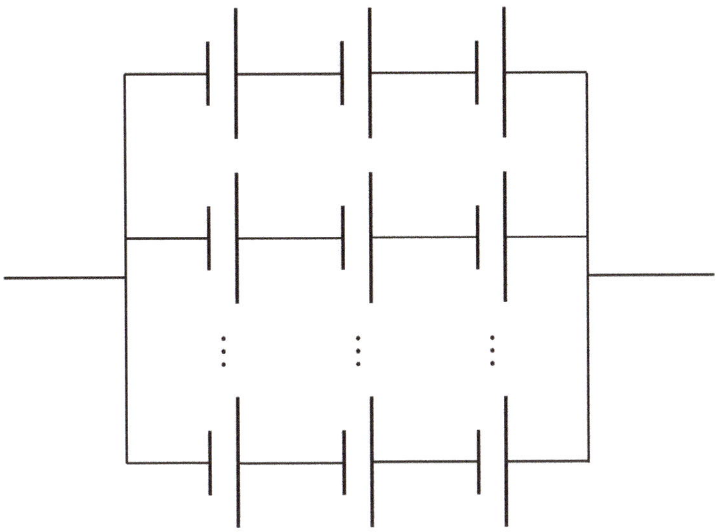

(b)

When the cells are wired in parallel first and then in series, the cells wired in parallel can be treated as a single cell, which can be monitored using a sensor. As a result, the wiring for monitoring the BMS can be simplified, and its cost can be reduced accordingly. However, when the battery cells are of relatively large capacity, this kind of configuration may cause nonuniformity in cell current and difficulty in increasing the overcurrent capability due to an excessive number of terminals in parallel connection and the associated large current. Therefore, this type of configuration is only suitable for a slow charging system with low power requirements.

On the other hand, when the cells are wired in series first and then in parallel, the system would have a strong overcurrent capability and a uniform current distribution among circuit paths. Hence, this configuration can satisfy the needs of a battery system where fast charging and high power are essential. However, as each circuit path needs to be independently monitored, such a configuration would incur relatively high costs in BMS hardware.

9.3.4. Guidelines for the Design of Pack Topology

For a given number of battery cells, the total available energy remains the same regardless of their interconnection being series or parallel. However, to meet the requirement for a high electromotive force, more battery cells or modules must be connected in series, which deteriorates the reliability of the system per Cannikin Law. Thus, parallel connections should be used wherever possible to improve the reliability of the system or module. Typically, a battery module is made of a number of battery cells that are first connected in series and then in parallel, as shown in Figure 9.5a. Then, the battery system is formed by connecting these modules in series to attain the specified electromotive force, as illustrated in Figure 9.5b. As a result, the number of battery cells and modules can be determined.

FIGURE 9.5 Electrical topology of battery packs. (a) Configuration of a battery module and (b) configuration of a battery system.

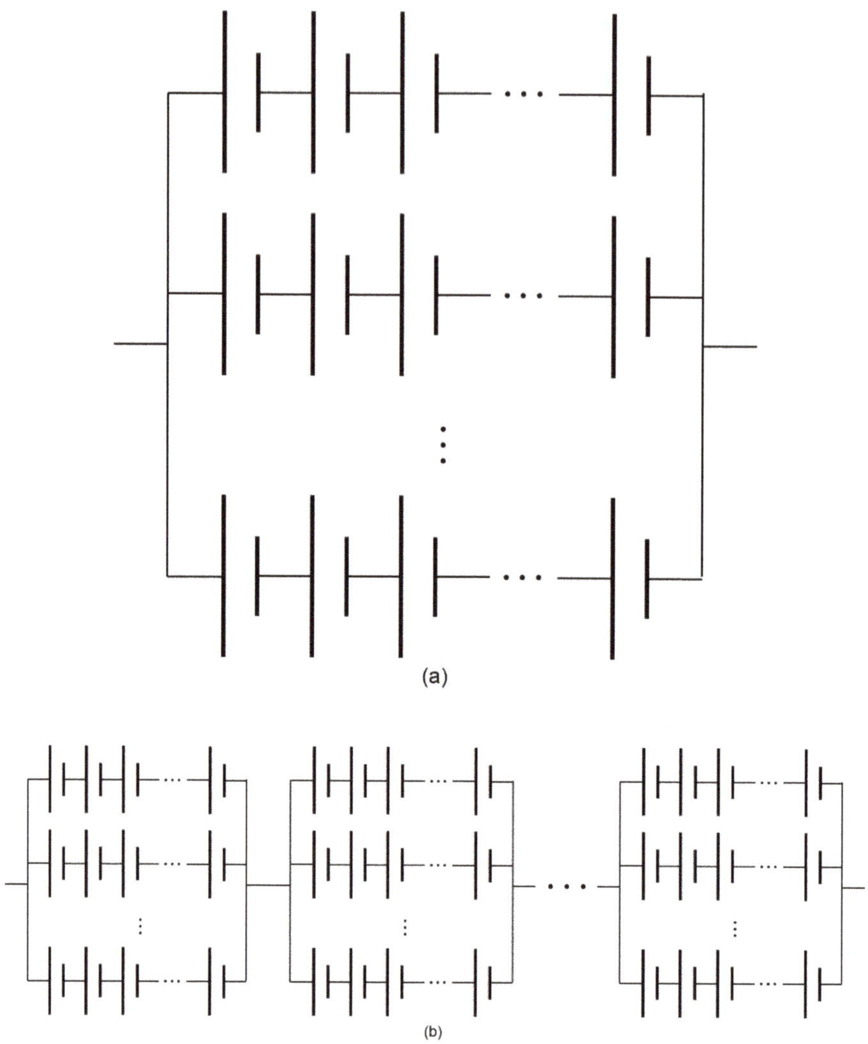

The maximum number of battery cells that can be wired in series varies from battery to battery, and it is usually a good engineering practice to follow the recommendation of the battery manufacturer. Besides, the cells connected in series must have the same voltage and capacity rating, in order to prevent possible damage to the cells. In contrast, there is no limit to the number of battery cells being wired in parallel. Nevertheless, care must be taken on proper system fusing to prevent catastrophic consequences of the high current available in case of accidental short circuiting. The cells connected in parallel must be identical in voltage and preferably the same capacity as well.

9.4. Design of Battery Packs

The overall design parameters related to a battery pack, such as the total energy capacity, volume, mass, and system voltage, are determined using the method elaborated in Sections 8.5.2 and 8.6.2, and then, other design parameters of the battery pack can be determined and selected accordingly.

9.4.1. Structural Design of Battery Packs

First of all, in view of the substantial mass of a typical battery pack, its structural design must meet the strength and stiffness requirements. Although mostly featuring a boxy shape, cross members, stiffener, and fillet should be used wherever needed to reduce the local stresses and deformations to the allowable level. Other than steel or aluminum alloy, other novel materials, such as glass fiber-reinforced composites, sheet molding compounds, and carbon fiber-reinforced composites, have increasingly found their applications in battery packs to improve the specific energy of the battery packs. However, anticorrosion treatment is often necessary to protect the steel-made casing from mechanical failures under long-term operation in a broad range of temperatures.

Besides, the action of impact loading on the battery pack during a collision must be taken into account as well. No structural failure is allowed under the maximum dynamic load corresponding to 50*g* of longitudinal acceleration in a frontal collision. In order to maintain the structural integrity of the battery pack during a crash, various vehicle structures, for example, longitudinal beams, transverse cross members, side sills, doors, and floor panels, can be used as barriers against head-on, rear-end, side, and bottom collisions. It is particularly important to protect the bottom of battery packs from the impacts of road hazards by using structural reinforcement wherever needed.

Next, the battery pack is continuously subjected to low-frequency vibrations, particularly in the vertical direction, when the vehicle is driven on a road. More severe vibrations, or shocks, are produced when the vehicle travels on uneven surfaces, for example, potholes, bumpers, or bridge abutments. The structure of the battery pack and the electrical components, for example, terminal connectors, contactors, bus bars, and circuit boards, must be designed properly to prevent them from fatigue failures or loss of electrical continuity. To absorb vibrations and avoid the vertical movements of battery modules or cells, damping pads of various shapes may be used between them or against frame structures. Tensioning bolts and retainer frames can also help apply a compression force on the top surface of the battery pack.

Additionally, uniform compression load around all the mating surfaces of a battery pack is essential to ensure an effective, long-lasting seal. The unevenness, surface quality, and deformation must also be controlled within certain limits as well. High-temperature-resistant sealants, for example, solid silica gel, are recommended.

Furthermore, at least one pressure relief valve is recommended to prevent injury and damage in case that thermal runaway occurs. Cross members can be employed to separate the battery pack into multiple compartments, while mounting flanges, brackets, or spacers can be used to hold the battery modules in place. Flame retardant materials can also be utilized inside and outside of a battery pack to reduce its propagation speed. A manual service disconnect (MSD), preferably positioned in the proximity of the driver, is often equipped for quickly isolating the high-voltage battery system in case of emergency or during maintenance.

Moreover, a proper clearance distance must be maintained to warrant the high-voltage safety of a battery pack. Clearance distance refers to the shortest distance in air between two conductive parts, which shall be dimensioned to withstand the required impulse withstand voltage. For equipment directly connected to the low-voltage mains, the required impulse withstand voltage is the rated impulse voltage. Larger clearances may be required due to mechanical influences such as vibration or applied forces. On the other hand, creepage distance means the shortest distance along the surface of a solid insulating material between two conductive parts, which is determined based on the long-term RMS value of the working voltage across it. A general rule of thumb is to keep the clearance equal to the shortest creepage distance in the design of battery packs.

Finally, once the geometrical dimensions are determined, the load distribution must be checked to make sure it falls into the acceptable range. The checkup can be carried out by means of regular CAD software.

9.4.1.1. Integration of Battery Modules

Nowadays, a common practice in the integration of the battery pack is assembling a number of battery cells into a battery module first and then grouping several battery modules into a battery pack.

The following considerations should guide the design of a battery module. First, battery cells should be held in place in a compact volume using cell holders, with thermal management channels for their heat-up and cooling. Second, the assembly of battery modules requires moderate tightness and sufficient strength of each structural component to prevent deformation or damage due to internal and external forces of the battery. Third, the connection, preferably a flexible one, should be reliable with the conductive distance between the battery cells as short as possible. Fourth, the conductive capacity of each conductive connection part should meet the maximum overcurrent capacity of the electrical equipment. Fifth, insulation protection measures, such as insulation gaps and creepage distances, must be taken for the serial and parallel high-voltage connections.

Note that standards or regulations may apply to the design of battery modules.

The constructions of battery modules for cylindrical, prismatic, and pouch cells are slightly different, as elaborated next.

1. Module design for cylindrical cells

   The battery module for cylindrical cells is primarily comprised of such components as honeycomb-type cell holders, busbars, insulating plates or films, a bottom plate, and a top cover, as shown in Figure 9.6. The honeycomb-type cell holders, commonly made of plastics, are employed to retain battery cells in designated positions and ensure that no excessive displacement occurs under vibration and shock conditions. Therefore, they should meet such fundamental design requirements as high strength and stiffness, high temperature resistance, and lightweight. The battery module is usually designed into a relatively symmetric structure as much as possible to achieve uniform current density distribution and temperature field. Particular attention must be paid to the wires at the places of the module inlet and outlet. Simulations of the temperature field of the battery module are commonly used to optimize the shape of the busbar in this regard.

   Thermal conductive adhesives, for example, thermal silica, are often used to better conduct the heat from cells to the cooling structure where the heat is further dissipated using liquid cooling and PCM cooling mechanisms, as discussed in the next section.

A good design practice is the usage of a fuse for connecting each cell to the module. In case that excessive temperature is detected, the fuse on the faulty cell blows automatically to disconnect it from the module.

Moreover, a control electronic board (not shown in Figure 9.6) is usually mounted within the module on the top or on the side for collecting the signals from voltage and temperature sensors.

FIGURE 9.6 Battery module for cylindrical cells.

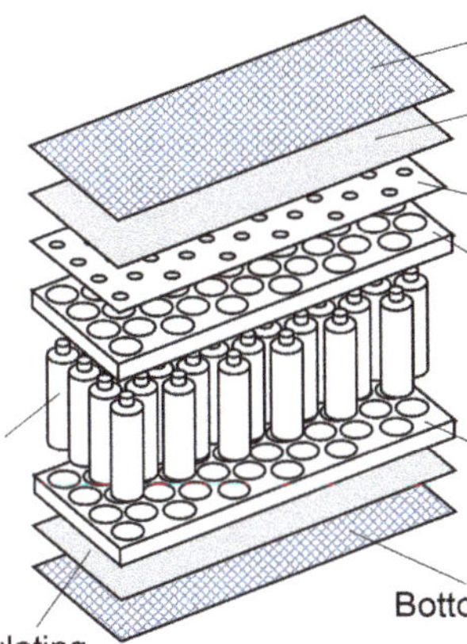

2. Module design for prismatic cells
   A typical battery module for prismatic cells mainly consists of busbars, wire harness boards, insulating plates or films, and plates, retainers, and covers for forming the external shell, as illustrated in Figure 9.7. Specifically, wire harness boards are used to isolate the cells from the low-voltage sampling lines. As an important structural component for safety, their design must be optimized to ensure functionality and reliability with reduced weight and cost. The external plates and covers are typically made of steel or aluminum, with the latter being used increasingly for weight reduction purposes. The cells are fastened in place usually by bolts or pressure strips.

   Due to the flat side walls, prismatic cell modules generally have poor resistance to internal pressure and are thus notorious for their side swelling problem, particularly for large-capacity prismatic cells. It is estimated that they can gain up to 1% in volume after 50 charge cycles due to the generation of gases during charges and discharges. This problem can cause an increase in internal resistance, partial depletion of the electrolyte, or even shell rupture, which may seriously affect the safety and cycle life of the cell. The use of the reinforced shell structure or pressure relief mechanism may help alleviate the problem.

   To standardize the battery modules, a series of module dimensions are applicable in some countries or regions. Make sure that these standards, if mandatory, comply with the battery module design.

   Finally, two safety devices are often equipped in a battery module for prismatic cells. One is an overcharge safety device (OSD), which is actually a metal membrane used together with the fuse typically connected to the positive terminal. When overcharged, the pressure inside the battery rises, causing the OSD to trigger an internal short circuit and generate an instantaneous large current. Subsequently, the fuse blows, cutting off the internal current loop of the battery. The other is a nail safety device (NSD), which is a metal layer, for example, copper foil, added to the outer surface of the coil. When acupuncture occurs, the local large current generated at the acupuncture position quickly reduces the current per unit area through the large-area copper sheet, which can prevent the local overheating at the acupuncture position and slow down the thermal runaway of the battery.

FIGURE 9.7 Battery module for prismatic cells.

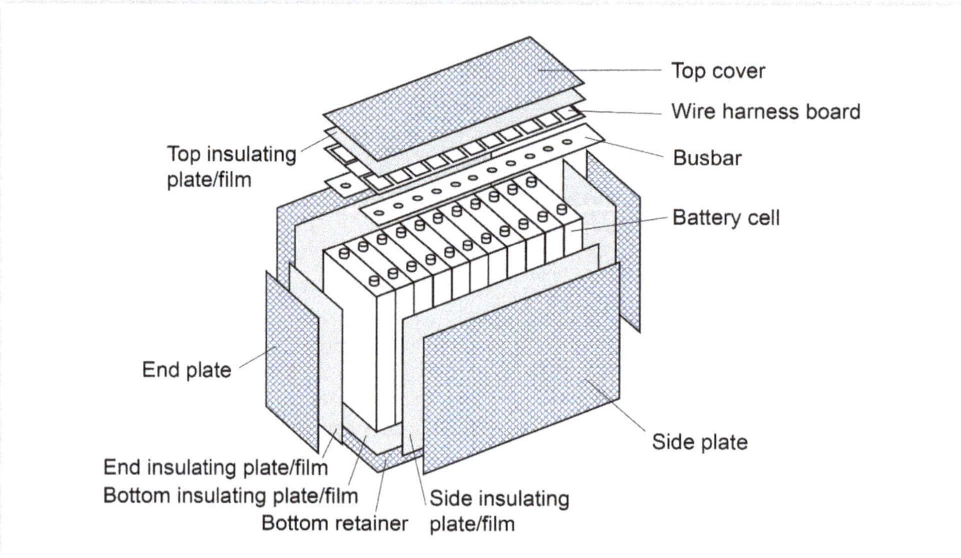

3. Module design for pouch cells
   Although there are various designs for the battery module of pouch cells, the module is generally composed of heat sink plates, foam spacers, busbars, wire harness boards, end plates, and a top cover, as shown in Figure 9.8. The heat sink plates are sandwiched between the pouch cells to dissipate the heat generated during charging and discharging processes. The foam spacers are used to separate the pouch cells and absorb their expansion while keeping a reasonable pressure on the components within the module. The cell expansion can be a result of heat generation caused by the internal resistance in the charging or discharging process, or gas production due to electrolyte decomposition or electrochemical reactions with impurities, for example, water, inside the cell. To hold the cells, spacers, and heat sink plates in place and prevent any relative motions between them, adhesives are often applied at contact surfaces, along with tensioning straps or long bolts applied on the end plates. The end plates are sometimes designed into a cartridge to better retain the pouch cells in between. In some designs, springs are used to create opposite pressure on the pouches to prevent them from gaining too much volume. The cell swelling can result in extra force on the end plates or cartridge and additional pressure on the surfaces of pouch cells that can in turn affect the electric performance of these cells. Thus, the coupled influences among the cell swelling, surface pressure, and their electric performance should

be carefully calculated with experiments and simulations to make a compromise between the two extreme scenarios, that is, constant pressure and constant displacement. Optimized design parameters, such as the thickness of end plates, the size of long bolts, the cooling effect of the heat sink plate, and the position, width, and thickness of the tensioning straps, can be obtained by controlling the maximum relative cell deformation during the life cycle of the battery module within a certain allowable percentage. For weight reduction, end plates can be made of aluminum or high-strength steel instead of regular low-carbon steel, while busbars can be made of aluminum instead of copper.

FIGURE 9.8 Battery module for pouch cells.

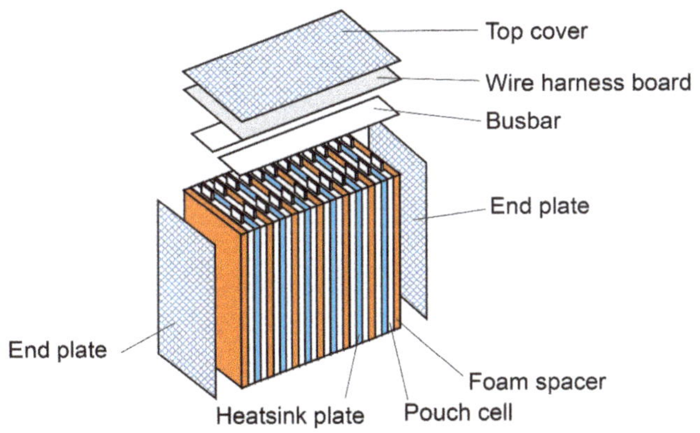

As far as the heat sink plate design is concerned, liquid cooling and PCM cooling are widely used today and sometimes supplemented with thermal conductive adhesives that help transfer the heat within the pouch cells to the external surface of the module and then dissipate into the environment. These measures can reduce the propagation of thermal runaway to a certain extent.

The sensors for collecting voltage and temperature, the electronic circuits or devices for passive or active cell equalization or their combination, and relays for on–off control of cells are often integrated into wire harness boards or module controllers, typically in the form of flexible printed circuit boards (FPCBs). It would be ideal to monitor the internal resistance of each cell. The signals from the module controller are then conveyed to the main controller of the BMS via CAN bus.

It is critical to maintain a low resistance in the busbar and a uniform contact resistance between the cell and busbar for power consumption minimization and correct detection of cell voltage. Before welding the tabs to busbars with processes such as ultrasonic bonding, the surfaces to be joined must be perfectly cleaned to guarantee the weld strength and the quality of electric contact preferably using laser cleaning. Or, laser welding can be used to replace ultrasonic bonding after the laser cleaning to combine the cleaning and welding into a single operation. To ensure a safe operation, fuses are utilized to disconnect the faulty cell from the module if its temperature is abnormally high, and firewalls are also employed between battery modules to retard the propagation of thermal runaways. In addition, fire-resistant materials should also be considered to prevent the structural collapse of the battery module.

9.4.1.2. Trends in Battery System Integration

Rapid advancements in the energy density of battery systems have brought about revolutions to the traditional ways of battery system integration, among which large-module approach, cell-to-pack (CTP), cell-to-chassis (CTC), and cell-to-body (CTB) appear to be the most promising ones. These technologies are continuously pushing the packing efficiency of the battery system beyond the limits, which has currently reached 84% or so.

1. Large-module approach
   In this approach, large-size battery modules are used instead of small modules, typically by removing the side plates from the previous small modules and connecting them with strips to enlarge the module. As a result, the number of modules could be reduced by as much as an order of magnitude.

   In general, the large-module approach offers advantages such as weight reduction, space utilization enhancement, a decrease in the number of parts and components, assembly process simplification, and cost-savings. By using the large-module approach, it is estimated that volume utilization can be improved by 15–20% and the number of parts and components reduced by 15–20%, with a 30% increase in productivity, in comparison with the regular-module approach. However, its drawback is that the repair cost for a faulty cell may be considerable if a large module, instead of a small module, needs to be replaced.
2. CTP technology
   The CTP technology aims at the direct integration of battery cells into a pack without the intermediate step of modules, enhancing the energy density and reducing the costs of the battery packs.

   Similar to the large-module approach, the CTP technology is deemed to be beneficial in the enhancement of the energy density of the battery pack and, thus, the driving range. The number of parts and components can be reduced dramatically. Nevertheless, it may also cause an increase in the repair cost if the entire battery pack, instead of a module, needs to be replaced. Additionally, due to the lack of module-level protection, the design issues, for example, cell expansion, deformation, cooling, and prevention of thermal runaway, all must be dealt at the pack level, making the engineering decisions more intricate than before. Also, the requirement on cell-to-cell uniformity becomes inevitably more stringent than that in the previous approach of using the module as the intermediate step of integration.
3. CTC technology
   The CTC technology refers to the integration of battery cells directly into the chassis of the vehicle, without the intermediate steps of modules or packs. Consequently, the number of parts and components could be reduced by 20% from that of the large-module approach or the CTP technology, with a 15% cost reduction in structural parts and components. Such a high degree of integration and modulization can improve the stiffness of the vehicle by 25% or so as well.

   The CTC technology is apparently more suitable for vehicles of a body-on-frame construction, especially in skateboard chassis, where the battery cell integration is decoupled to the vehicle body design. As a result, the vehicle body and chassis can be designed separately, simplifying the development process.

   Although the CTC technology offers the aforementioned advantages, it also brings challenges to the design in the following aspects: first, a high requirement for cell-to-cell uniformity in voltage, capacity, internal resistance, and so on; second, keeping serviceability and repair cost of the battery system within a reasonable range;

third, the base plate needs to meet the elevated requirements as more important structural parts in strength, stiffness, lightweight, air-tightness, shockproof, and so forth, with particularly robust battery protection at the bottom; fourth, design of battery swap mechanisms; and fifth, structural design for thermal runaway protection.

4. CTB technology
   The CTB technology is a new way of integrating battery cells into the vehicle body directly. The essence of this technology is utilizing the top cover of the battery pack concurrently with the floor of a unibody vehicle, as shown in Figure 9.9. Due to the sandwich structure of the battery pack, the stiffness of the vehicle body and collision safety (both frontal and lateral) can be improved significantly.

   However, care must be taken on the sealing performance and durability to prevent the ingress of water, mud, or dirt from outside into the battery pack and the leakage of harmful gas from the battery pack into the passenger compartment.

FIGURE 9.9 Schematic of CTB technology.

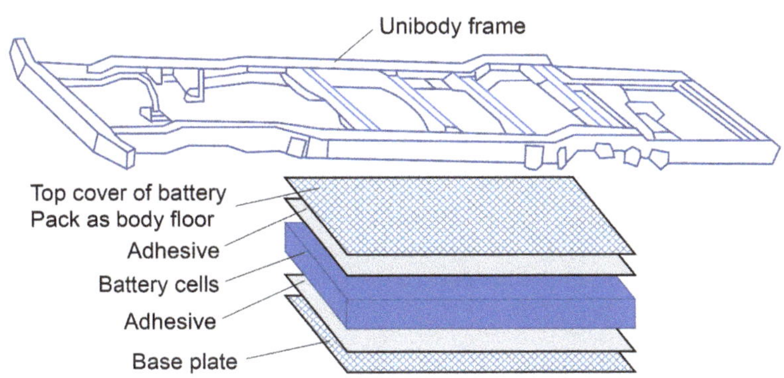

9.4.2. Design of a BTMS

To warrant the safety during the charge and discharge of the battery and prevent it from thermal runaway, the heat generated inside the battery must be dissipated effectively. In addition, a uniform temperature distribution in the battery pack is desired to maximize its performance and lifespan. Besides, to ensure EVs are operable with decent performances even under extreme temperatures, the traction battery must meet certain performance requirements within the temperature range from −40 to 60°C. Thus, a good design of BTMS is essential from all these perspectives.

At present, the basic BTMSs for EVs can be classified into types such as air-cooled, liquid-cooled, PCM, HP, or thermoelectric coolers (TECs).

Air-cooled battery systems are still used in some EVs due to their simplicity, lightweight, and low cost. Compared to natural convection, most air-cooled BTMSs feature forced convection because it can reduce the thermal gradient in the battery pack [9.2]. The cooling air comes from the exterior of the vehicle, from the cabin, or from an auxiliary heat exchanger, but it must be preconditioned to meet the cleanliness and humidity requirements. The airflow rate must be controlled properly as excessive airflow causes energy consumption to increase without much effect on the temperature level and its uniformity within the battery pack. Besides, the airflow inlet and outlet must be arranged at proper heights above the ground to prevent water from entering the battery pack when the vehicle operates in deep water.

Liquid-cooled battery systems are applied in most EVs today, especially in high-end models, for their effectiveness, compactness, and low energy consumption. The energy consumption of this type of BTMSs is higher than that of the air-cooled BTMSs though, due to significantly more power needed for fluid circulation if forced convection is implemented. Other than ordinary coolant media such as water, water/glycol mixture, or mineral oil, some novel coolants have been proposed and studied, including nano-emulsions and liquid metal. Cooling plates are the most common implementation for prismatic and pouch batteries, as shown in Figure 9.10a [9.2], while wavy channels with liquid coolant can be adopted as one of the practical solutions for cylindrical batteries, as shown in Figure 9.10b [9.2].

FIGURE 9.10 Liquid-cooling structures for batteries of different shapes. (a) For prismatic and pouch cells and (b) for cylindrical cells.

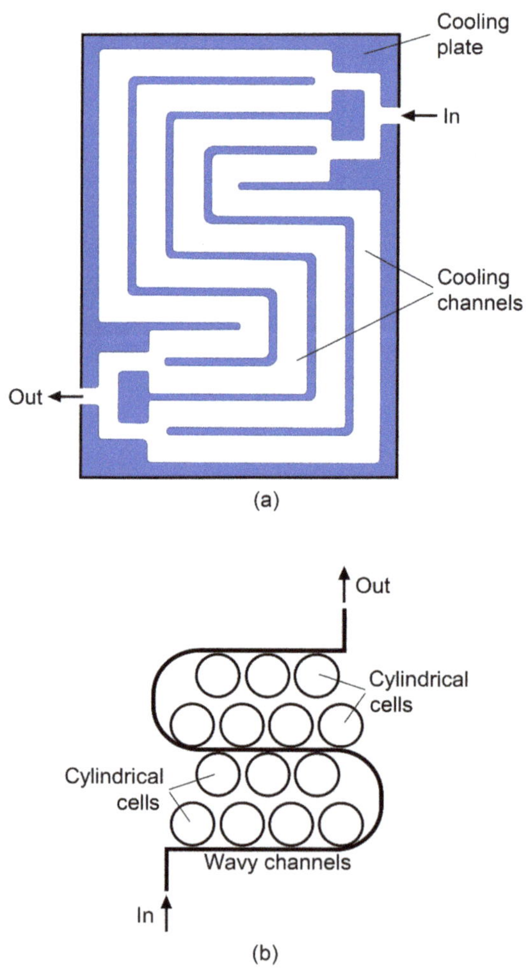

By using refrigerant as a coolant medium, a better cooling effect can be achieved, with greatly reduced temperature differences between batteries due to the transition phase of the refrigerant at a nearly constant temperature [9.2].

PCM-based BTMSs are a passive solution particularly suitable for cylindrical batteries as such a medium can easily reach the unused spaces between batteries, greatly improving the temperature uniformity. With high latent heat capacity, PCMs are capable of keeping the battery temperature within a small range around their melting point. However, measures must be taken to avoid thermal instability, volume increase, and leakage due to the complete melting of PCM and to improve the heat dissipation caused by its low thermal conductivity [9.2].

Composite phase change materials (CPCM) can be applied to BTMSs as an alternative solution to heating up the batteries by means of a wire preheating system, to improve the thermal conductivity, or to alleviate the leakage and volume-changing issues [9.2].

As another passive solution, HP-based BTMSs offer an enhancement in heat dissipation and temperature uniformity owing to the extremely high thermal conductivity of HPs. Generally compact and lightweight, with no power consumption, no maintenance needed, and a long life cycle, they are often used together with forced air or liquid cooling for better cooling effect in the condensation section [9.2].

TEC is a relatively new technology that is often coupled with other methods to improve the efficiency of existing BTMSs. By converting voltage to temperature difference, TECs can be used for both heating and cooling purposes.

The characteristics of these types of BTMSs are compared in Table 9.2 [9.2].

In order to improve the heat transfer process and the controllability of the system operation, the active and passive methods described earlier can be combined to form a hybrid BTMS. The most promising combinations are as follows: first, PCM with air or liquid; second, HP with air or liquid; third, PCM and HP with air or liquid; and fourth, TECs with another basic method [9.2].

TABLE 9.2 Comparison of BTMS characteristics.

		Active				Passive		
Characteristics		Air-cooled	Liquid-cooled (direct)	Liquid-cooled (indirect)	Air-cooled	Liquid-cooled (direct)	CPCM	Air-cooled
Cooling capacity		Very small	High	Medium	High	Very small	Small	Medium
Temperature distribution		Uneven	Uneven	Uneven	Uneven	Even	Even	Uneven
Energy consumption		Low	High	Medium	High	None	None	None
Weight		Light	Heavy	Medium	Medium	Heavy	Heavy	Light
Size		Large	Medium	Compact	Compact	Large	Large	Compact
Complexity		Simple	Medium	Complex	Complex	Medium	Medium	Simple
Costs		Low	Medium	Medium	High	Low	Medium	High
Possibility of heating		No	No	No	Yes	No	No	No
Reliability		High	Medium	Medium	Low	Very low	Low	High
Adaptability	Cylindrical	Easy	Moderate	Difficult	Difficult	Easy	Easy	Difficult
	Prismatic	Moderate	Difficult	Easy	Easy	Easy	Easy	Easy
	Pouch	Moderate	Difficult	Easy	Easy	Moderate	Moderate	Easy

Courtesy of Applied Sciences.

9.4.3. Design of a BMS

A BMS is a distributed electronic device that controls the operation of the battery system and protects it by constantly evaluating several important criteria such as SOC, state of health (SOH), state of power (SOP), and state of safety (SOS), based on the acquired data from sensors. SOC is an indicator of the charge level of the battery, while SOH indicates the remaining capacity of the battery as a percentile of the original capacity. SOP is the amount of power available for a defined time interval given the current power usage, temperature, and other conditions. SOS represents the condition when the battery is dangerous for use in vehicles, and it can be estimated by means of thermal runaway, current, voltage, SOC, and SOH.

A BMS is typical of the following functions:

1. Monitoring the state of the battery system, such as voltage, temperature, current, and coolant flow of the BTMS.
2. Processing the data from the sensors of the monitoring portion of the system to come up with a number of parameters such as voltage, current, internal impedance, charge, energy, operating time, and number of cycles.
3. Communicating internally with the hardware operating at a cell level and externally with high-level controllers, for example, VCU, using serial, bus, and wireless types of communications.
4. Protecting the batteries by keeping them operating within their safe operating zone in terms of current, voltage, and temperature, with ground fault or leakage current detection, as well as by preventing them from excessive inrush current to load capacitors in connecting to load circuit with a precharge system.
5. Equalizing the voltage or SOC of all the battery cells to maximize the capacity and prevent localized undercharging or overcharging.

9.4.3.1. Topology of BMS Controllers

There are three types of topologies in terms of BMS controllers: centralized, distributed, and modular.

The centralized topology has a single controller connecting to battery cells through multiple wires, as illustrated in Figure 9.11. It is compact, reliable, and economical and is generally used in little electric cars with relatively small battery capacity, low system voltage, and few battery cells. Data acquisition, processing, and control are handled by a single controller board, with relays controlled by the VCU. The controller board can be divided into high- and low-voltage areas. The high-voltage area of the hardware is responsible for acquiring cell voltages and the total system voltage, as well as monitoring the insulation resistance. The low-voltage area includes circuits for power supply, CPU, CAN communication, and control circuits. The fact that all the wiring harnesses are routed directly to the controller causes the probability of signal interference to increase, although some are inevitable.

The distributed topology features a cell supervising circuit (CSC) installed on each cell, with a single communication cable between the controller and CSCs, as shown in Figure 9.12. Although the most expensive among the three types, it offers the simplest installation and cleanest assembly, thus particularly suitable for large battery systems. It typically consists of a master controller and several slave controllers. The slave controllers are responsible for acquiring the voltage and temperature data and are often installed near the modules that it monitors, while the master controller not only processes the data from the slave controllers to evaluate the various states of the battery but also controls the BTMS, relays, diagnosis, communication, and the charge and discharge of batteries. A high-voltage slave controller is sometimes utilized to monitor the total current and voltage of the battery pack and report the hardware-integrated data to the master controller for SOC and SOH calculations, with precharge and insulation detection functionality as well.

FIGURE 9.11 A BMS controller of centralized topology.

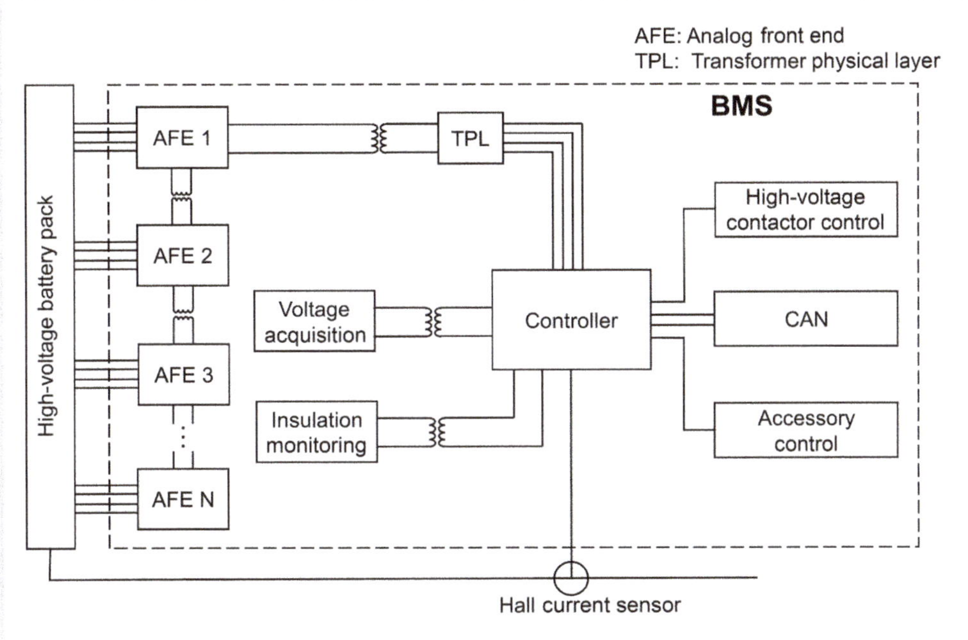

FIGURE 9.12 A BMS controller of distributed topology.

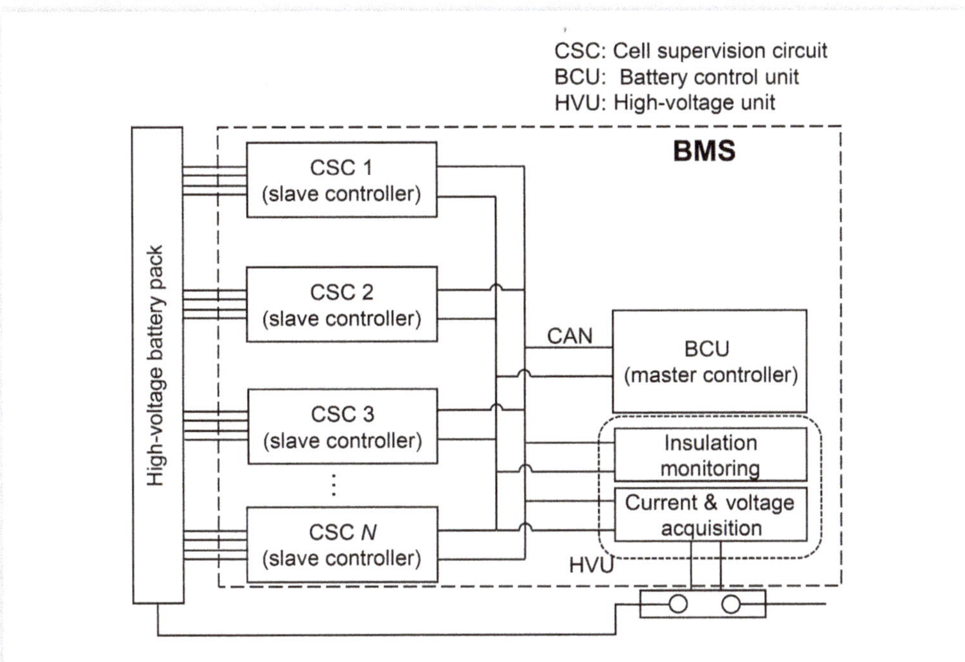

The modular topology uses a few controllers, each handling a certain number of cells, with communication between the controllers. As a compromise of the centralized and distributed topologies, the modular topology is mainly applied to battery packs of irregular arrangement of modules. Compared with the distributed topology, the modular topology usually comes up with fewer controllers and components, but the cost of the system may not go down much.

9.4.3.2. Cell Equalization

Cell equalization refers to equalizing the cell SOC in a battery system to a certain level within an acceptable range of difference, which is critical to prevent the battery pack from performance degradation and premature failures. Cell equalization methods can be generally classified into passive and active.

The passive cell equalization method equalizes the cell voltages to the lowest cell voltage by dissipating the extra energy of higher-voltage cells typically using switching shunt resistors for lithium-ion batteries [9.3]. Hence, such a method is less energy-efficient due to losses at resistors during equalization operation and, therefore, not appropriate for high-power applications. Illustrated in Figure 9.13 is a schematic of passive cell equalization circuits [9.4].

FIGURE 9.13 A schematic of passive cell equalization circuit.

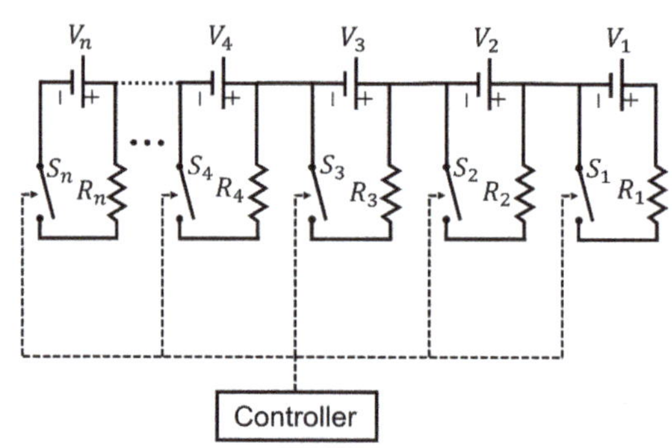

On the other hand, the active cell equalization method equalizes the cell voltages by transferring energy from higher-voltage cells to low-voltage cells using capacitors, inductors, or power electronic converters [9.4]. Capable of equalizing cells in both charging and discharging operations, this method features less energy and time consumption than the passive method, thus suitable for high-power applications. However, active cell equalization methods have not been applied broadly to date due to their higher system complexity and cost. An active cell equalization circuit with a single capacitor is illustrated in Figure 9.14 [9.4].

FIGURE 9.14 A schematic of an active cell equalization circuit.

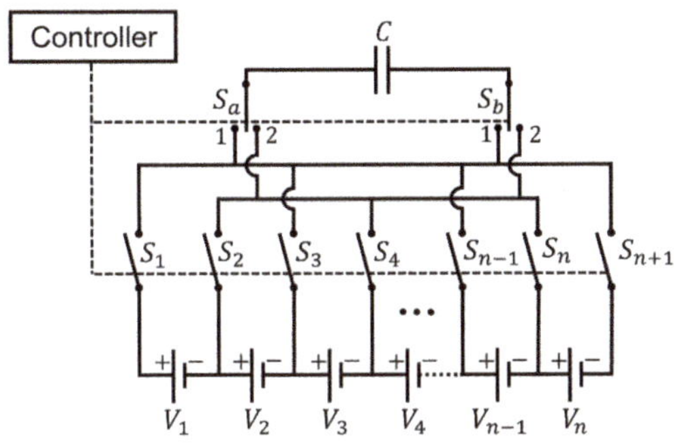

9.4.3.3. State Estimation of Batteries

1. Estimation of battery SOC
   The SOC of a battery at a given time t can be defined by

$$SOC(t) = \frac{Q(t)}{Q_n} \tag{9.7}$$

where
  $Q(t)$ is the battery capacity at the time t
  Q_n is the nominal capacity

Numerous methods have been proposed to estimate the SOC in real time to date, which can generally be classified into the four categories as follows.

The first category is the direct measurement method, which mainly includes the open circuit voltage (OCV) method, terminal voltage method, impedance method, and electrochemical impedance spectroscopy (EIS) method. These methods are rarely used alone for estimating SOC for various reasons. For instance, it is difficult to estimate SOC using the OCV method for iron phosphate lithium-ion batteries due to their flat OCV versus SOC curves. The excessive error in SOC estimation because of the sudden drop in the terminal voltage at the end of discharge prevents it from being applied as a stand-alone method. In addition, the impedance method is hardly used to estimate SOC due to the insensitivity of internal resistance change to SOC. Furthermore, the EIS method also faces challenges in the generation of various AC frequencies and detection of responses in real time in the presence of other voltage and currents.

The second category is book-keeping estimation methods. Using battery discharging current data as the input, this category includes the Coulomb counting method and the modified Coulomb counting method. For the Coulomb counting method, the SOC at the time t is estimated by the following equation:

$$SOC(t) = SOC(t - \Delta t) + \frac{I(t)}{Q_n}\Delta t \tag{9.8}$$

where

$SOC(t - \Delta t)$ is the estimated SOC value in the last time step
$I(t)$ is the discharge current
Δt is the time step

On the other hand, the modified Coulomb counting method can significantly improve the accuracy of SOC estimation by using the corrected current $I_c(t)$ instead of $I(t)$:

$$SOC(t) = SOC(t - \Delta t) + \frac{I_c(t)}{Q_n}\Delta t \tag{9.9}$$

$$I_c(t) = k_2 I^2(t) + k_1\, I(t) + k_0 \tag{9.10}$$

where k_2, k_1, and k_0 are constants from experimental data. Nevertheless, both methods require correct battery capacity and initial SOC, which is not always known. Besides, these methods are not capable of correcting the estimated SOC by themselves once the value goes wrong. The sensing errors can cause the errors in estimated SOC to accumulate over time [9.5].

The third category is adaptive system methods. Various adaptive systems for SOC estimation have been developed, such as backpropagation (BP), radial basis function (RBF), and fuzzy neural networks, along with fuzzy neural logic, support vector machine (SVM), and Kalman filter. BP neural network is the most popular SOC estimation method based on artificial neural networks for advantages such as nonlinear mapping, self-organization, and self-learning. The SOC is estimated using the recent history of voltage, current, and ambient temperature. The RBF neural network can be used to estimate the SOC based on the input data such as the terminal voltage, discharging current, and battery temperature. It is proven that a good agreement between estimated SOC and experimental data can be achieved, with acceptable computing speed as well [9.6]. The fuzzy neural network-based soft computing approach can be used for SOC estimation through a fusion with B-spline membership functions and a reduced-form genetic algorithm [9.6]. The drawback of this method is that it relies on the battery data from the manufacturer, which are not available most of the time. The fuzzy logic method is a powerful tool for modeling nonlinear and complex systems. The estimated SOC can be kept error-free from the time-dependent variation by compensating it with a learning system built on fuzzy logic models. Or, the fuzzy logic models can be used to analyze the data obtained by other methods, for example, EIS or Coulomb counting. However, this method usually requires considerable computing and storage resources. SVM is more robust than a least-square estimation methodology as it is less sensitive to small changes. The drawbacks of the Coulomb counting SOC estimation method can be eliminated with SVM, resulting in improved accuracy [9.6]. The Kalman filter, extended Kalman filter (EKF), and unscented Kalman filter (UKF) have been extensively studied for SOC estimation. In general, these methods can improve the robustness of the SOC estimation algorithm by eliminating the sampling noise and reducing the dependency of book-keeping methodology on the initial SOC. Nevertheless, the parameters in the Kalman filter model can vary with time, and corrections are, therefore, needed constantly. It may also be difficult to complete the computation within a sampling period due to the great amount of computation involved. In addition, rounding and truncation errors can accumulate due to the limited computer word length.

The fourth category is hybrid methods, which improve the accuracy of SOC estimation and robustness of the algorithm by combining two or more individual methods discussed earlier. Coulomb counting and terminal voltage methods can be combined to substantially increase the accuracy of SOC estimation. For example, a simple maximum capacity adaptation algorithm can be introduced to take the aging effect into account [9.6]. Moreover, Coulomb counting and Kalman filter methods can also be combined to largely reduce the error in SOC estimation. For instance, the Kalman filter method can be used to converge the approximate initial value to its real value, which is then employed to estimate the SOC for the extended operating time using the Coulomb counting method [9.5].

2. Estimation of battery SOH

At present, there is no consensus on the definition of SOH. It is sometimes defined as the ratio of the maximum battery charge to its nominal capacity in percentile, namely,

$$SOH(t) = \frac{Q_{max}(t)}{Q_n} \tag{9.11}$$

where Q_{max} is the maximum battery charge at the time t. Obviously, Q_{max} is a function of time t due to battery aging.

The most intuitive behavior of battery aging is the decline in its capacity and the rise in its internal impedance. With the SOH of a new battery being 100%, the battery is generally considered not acceptable when SOH is 80% or below.

The battery SOH can be estimated by methods based on the analysis of experimental data or system modeling.

One type of the SOH estimation method is based on experimental data analysis. These methods can be classified into direct measurement methods and indirect analysis methods. With the direct measurement methods, the experimental data such as current, voltage, and temperature are acquired and analyzed to obtain the characteristic parameters that reflect the decay in battery capacity. In the indirect analysis methods, other parameters such as voltage trajectories, incremental capacity, and differential voltage are used to analyze the battery SOH.

The other type of the SOH estimation method is based on system modeling. These methods can be classified into self-adaptive methods and data-driven methods. Examples of self-adaptive methods are joint estimation, synergy estimation, and fusion methods. Data-driven methods include empirical regression, optimization, and sample entropy methods.

3. Estimation of battery SOP
The SOP of a battery is defined as the ratio of peak power to nominal power, where the peak power stands for the maximum power that a battery is capable of providing or taking in discharge or charge for a certain duration of time persistently.

    The four methods for estimating the battery SOP are as follows.

    The first method involves hybrid pulse power characterization (HPPC). Based on the upper and lower cutoff voltage of the battery, HPPC can be used to calculate the instantaneous peak current and power, but it is not suitable for estimating the continuous power for a certain duration of time. It is widely applied due to its simplicity in the algorithm. However, the Rint model used by this method is not capable of capturing the dynamic voltage characteristics of the battery. As a result, HPPC tends to overpredict the SOC, which may lead to overcharge or overdischarge of the battery.

    The second method is based on voltage constraints, which overcomes the shortcomings of the HPPC method and takes into account the change in the OCV of the battery during the peak power prediction period, thus yielding a more reliable result. Nevertheless, due to the static nature of the Rint model, the influence of dynamic polarization and hysteresis of the battery on peak current is not taken into account.

    The third method is based on SOC constraints, which estimate the peak current of a battery based on the constraints in the maximum and minimum allowable SOC. This method is often used in combination with other methods because it tends to overestimate the peak current when the allowable SOC range is large.

    The fourth is the multiconstrained dynamic (MCD) method. In the MCD method, the dynamic voltage of a battery is characterized by the Thevenin model with the RC network. As a result, the SOP can be estimated using the MCD method in real time by incorporating multiple constraint variables such as terminal voltage, current, and SOC. At the same time, the influence of dynamic response characteristics such as electrochemical kinetics, thermodynamics, and hysteresis effects on the SOP prediction results is considered as well.

4. Estimation of battery SOS
Battery SOS represents the condition when the battery is dangerous for use in vehicles, and it can be estimated by several means, such as its thermal runaway, current, voltage, SOC, and SOH. However, there is no consensus on the definition of battery SOS to date.

    In one of the attempted definitions, the battery SOS is defined as a probability function that can be calculated as the product distribution of several individual subfunctions [9.7]:

$$SOS(x) = \prod_{i=1}^{n} f_i(x_i) \tag{9.12}$$

where x is all types of state and control variables that describe the behavior of the battery, for example, voltage, temperature, charging and discharging currents, and internal impedance, at a given time t, and $f_i(x_i)$ is the ith of the total n safety subfunctions that is actually inversely proportional to battery abuse [9.7]:

$$f_i(x_i) = \frac{1}{\left(\frac{1}{\zeta} - 1\right)\left(\frac{x_i - x_{i,100}}{x_{i,\zeta} - x_{i,100}}\right)^2 + 1} \tag{9.13}$$

where ζ is a lower value of each subfunction that guarantees the safe behavior of the battery. A nonexhaustive list of these safety subfunctions may include temperature, current, voltage, SOC, and SOH, which all influence the SOS in one way or another [9.7].

In this approach [9.7], the SOS of 1.0 indicates that the battery is completely safe, while $SOS = \zeta$ can be a warning as one of the subfunctions may have reached the lower value ζ. If $SOS = \zeta^n$, then the SOS has reached a lower bound of acceptable safety, as all the subfunctions have reached the lower value ζ. Obviously, the battery is unsafe if $SOS < \zeta^n$.

9.4.3.4. EMC Design Principles

The electromagnetic interference (EMI) from electric motors, DC/AC inverter circuits, AC/DC rectifier circuits, and DC/DC converter circuits may influence the normal operation of BMS. For example, errors may occur in the acquired voltage or current data.

Both low- and high-voltage systems can result in EMI. For example, inductive devices such as switches, relays, or DC motors in a 12 V system can produce a transient voltage of up to 100 V in amplitude for a duration of approximately 1 ms in on/off operations. The transient voltage is coupled to other electronic systems by conduction through a shared power supply. In addition, the main chips, clock circuits, trigger circuits, data lines, and signal lines of various controllers can generate EMI in the frequency band of 150 kHz–2.5 GHz during their operation. Other electrical devices, for example, brushed DC motors, may generate electrical sparks during their operation, resulting in a wide spectrum of electromagnetic radiation noise. On the other hand, during the operation of a high-voltage system, the high-speed switching of power electronics, for example, IGBTs or MOSFETs, of the motor controller and the DC converter can produce highly transient voltage and current in the circuit. This strong EMI can be conducted or radiated to BMS, affecting its normal operation [9.8-9.9].

In order to minimize the influence of EMI on the normal operation of BMS, particular measures can be taken in the design of its power supply, data acquisition circuit, PCB layout and wiring, and software.

1. EMC design of BMS power supply
   The BMS power cord typically connects in parallel with the 12 V lead-acid battery, the low-voltage output terminal of DC/DC, and the low-voltage power supply terminal of the motor controller, in sharing the power supply with other on-board electrical equipment as well. To prevent the low-frequency transients, high-frequency interference, and common-mode interference generated by DC/DC and other electrical equipment from being coupled into the BMS through the power supply, the following measures can usually be considered: first, adding a transient voltage suppressor (TVS) to suppress transient interference in power input and provide electrostatic discharge (ESD) protection; second, using high-current magnetic beads to suppress high-frequency interference in power input and suppress the transmission of high-frequency interference from inside BMS to outside; third, building a common-mode filter to remove the common-mode noise and harmonic interference in the power input; and

fourth, constructing an LC filter circuit to filter out differential-mode interference in the power input [9.8-9.9].

Besides, the stability and anti-interference ability of the 5 V power supply in the BMS directly affect the accuracy and stability of signal acquisition. For example, the use of an LC π-type differential-mode filter circuit can filter out the differential-mode interference via the power line and suppress the differential-mode interference that may be conducted from inside the BMS to outside to a certain extent [9.8-9.9].

Moreover, the analog power supply can provide a stable power supply for the BMS analog acquisition and operational amplifier circuits. For example, in a single-ended flyback switching regulator circuit with positive and negative output voltages, the selection of base frequency for operation needs to avoid the sensitive frequency bands for conduction and radiation immunity [9.8-9.9].

2. EMC design of sensitive signal acquisition circuit
   The battery current is acquired as an mV-level weak signal. For example, when Mn-Cu precision shunt resistor type of sensors are used, the signal amplitude is small and prone to interference, causing inaccuracy in the collected current data. The common-mode choke inductors and capacitors can be considered at the BMS input to perform common-mode filtering on the collected signals [9.8-9.9].
3. EMC design of interface circuit
   Serial magnetic beads and parallel decoupling capacitors are recommended for each pin to filter the conduction coupling of external high-frequency interference. The choice of magnetic beads and capacitors should not only effectively filter high-frequency interference but also take the level change speed and current magnitude of the pin signal into consideration. Different magnetic beads and bypass capacitors should be selected based on the capacitance value, packaging form, and parasitic parameters [9.8-9.9].
4. EMC considerations in PCB design
   Generally, in a four-layer BMS circuit board, the middle two layers are the power layer and the ground layer, and both top and bottom layers are the signal layers. The power layer is divided into 5 V digital power supply, as well as 12 and 15 V analog power supplies. The ground layer is separated according to the function, with individual ground for each of the analog circuits, digital circuits, and high-current power output circuits. It is crucial to reduce the trace length of the signal acquisition circuit. The shielding effect can be improved by making a BMS with an aluminum shell, copper-clad around the PCB, and good grounding. Also, EMC simulation software can be used to model the PCB for optimization [9.8-9.9].
5. Software filtering for EMC
   In addition to hardware-related ECM measures, common software filtering methods such as first-order lag filtering can also be used to effectively eliminate instantaneous pulse interference and random interference in the process of data acquisition, so as to make the signal smoother and prevent instantaneous abnormal data [9.8-9.9].

References

[9.1]. Arora, S., and Kapoor, A., "Mechanical Design and Packaging of Battery Packs for Electric Vehicles," in Pistoia, G., Liaw, B., Arora, S., and Kapoor, A. (eds), *Behaviour of Lithium-Ion Batteries in Electric Vehicles* (Cham: Springer, 2018), 175-200, doi:10.1007/978-3-319-69950-9_8.

[9.2]. Buidin, T. and Mariasiu, F., "Battery Thermal Management Systems: Current Status and Design Approach of Cooling Technologies," *Energies* 14, no. 16 (2021): 4879, doi:10.3390/en14164879.

[9.3]. Hemavathi, S., "Overview of Cell Balancing Methods for Li-Ion Battery Technology," *Energy Storage* 3, no. 2 (2021): 1-12, doi:10.1002/est2.203.

[9.4]. Moore, S. and Schneider, P., "A Review of Cell Equalization Methods for Lithium Ion and Lithium Polymer Battery Systems," SAE Technical Paper 2001-01-0959 (2001), doi:https://doi.org/10.4271/2001-01-0959.

[9.5]. Tang, X., Mao, X., Lin, J., and Koch, B., "Li-Ion Battery Parameter Estimation for State of Charge," Paper Presented at in *American Control Conference*, San Francisco, June 29–July 01, 2011, doi:10.1109/ACC.2011.5990963.

[9.6]. Chang, W., "The State of Charge Estimating Methods for Battery: A Review," *International Scholarly Research Notices* 2013 (2013): Article ID 953792, doi:10.1155/2013/953792.

[9.7]. Castillo, E., Niedermeier, F., and Jossen, A., "Calculation of the State of Safety (SOS) for Lithium Ion Batteries," *Journal of Power Sources* 324 (2016): 509-520, doi:10.1016/j.jpowsour.2016.05.068.

[9.8]. Bai, Y., "Electromagnetic Compatibility and Electromagnetic Compatibility Design," *Radio Engineering* 38, no. 11 (2008): 34-36, doi:10.3969/j.issn.1003-3106.2008.11.012.

[9.9]. Li, X., Wang, L., He, J., and Wang, Z., "Research of Electromagnetic Compatibility Optimization Technology of Battery Management System," *Advanced Technology of Electrical Engineering and Energy* 33, no. 3 (2014): 39-34, doi:10.3969/j.issn.1003-3076.2014.03.008.

10

Electric Drive System Design

The electric drive system for EVs, sometimes also called e-powertrain, is the mechanism that converts electric power to mechanical power for driving the vehicle. Such an electric drive system, traditionally, consists of an electric motor and its mating drivetrain, which delivers power to the drive wheels. Alternatively, in an EV equipped with distributed drive systems, electric motors can be mounted in the hub of the wheels to drive the vehicle directly with no need for a drivetrain at all, as a bit of futuristic rendition.

The electric drive system must satisfy the torque and power requirements for all specified operating conditions of the vehicle. Meanwhile, it is desired that the electric drive system works in its operable region of high system efficiencies. The fundamental requirements of an EV on its electric drive system are as follows:

1. Power, torque, and speed ranges must meet the dynamic performance requirements of the vehicle.
2. The system must operate at optimal operating points to maximize its energy efficiency and, thus, improve the energy economy of the vehicle.
3. The electric motor must allow the four-quadrant operation, with bidirectional energy conversion and transmission capability between electrical energy and mechanical energy, as well as between the battery and the wheel.

4. The park gear (P), reverse gear (R), neutral gear (N), and drive gear (D) in the driving mode selector helps in securing the vehicle in place, backing up the vehicle, decoupling wheels from the electric motor, or driving the vehicle in the forward direction, respectively.
5. In the drive gear (D), a proper torque is needed to provide creeping functionality for the vehicle to start, without rolling back on a slope.
6. In addition to the mechanism behind the park gear (P), a second park gear is highly recommended to provide an extra set of safety mechanisms to secure the vehicle in place.
7. High efficiency, high reliability, low noise, compactness, lightweight, good manufacturability, ease of operation and maintenance, and low costs.

10.1. Electric Drive System Configurations

The basic electric drive system configurations for EVs can generally be divided into layshaft and coaxial types in terms of the position of the axle shafts of transmission relative to the shaft of the electric motor. The layshaft type of configuration typically features a layshaft gearing system that consists of an input shaft (i.e., motor shaft), an intermediate shaft (also called the output shaft), and a differential, which splits the power to the axle shafts for driving the wheels, as shown in Figure 10.1a [10.1]. The coaxial type of configuration, however, can either adopt a layshaft or planetary gearing system, with a hollow motor shaft allowing one of the axle shafts to go through in the middle, as shown in Figure 10.1b and 10.1c, respectively.

FIGURE 10.1 Basic electric drive system configurations for EVs. (a) Layshaft, (b) coaxial layshaft, and (c) coaxial planetary.

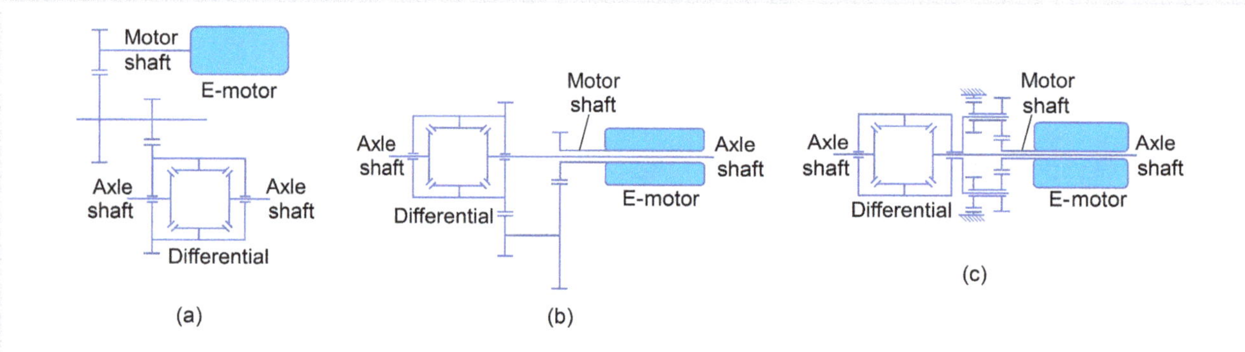

Reproduced from, Yuan, Y., Electric Vehicle Transmissions: Structure and Control (Beijing: China Machine Press, 2020), ISBN: 978-7-111-66809-1, with the permission of China Machine Press.

The electric drive systems of the layshaft configuration are simple and reliable, with low manufacturing costs. In contrast, those of the coaxial configuration are more compact and lightweight, with a high manufacturing cost except in the case of mass production.

10.2. Design of Electric Motors

The design of electric motors involves a multidisciplinary and systematic approach, which is beyond the scope of this book. The interested reader is referred to relevant textbooks or design manuals for a complete knowledge set of the subject. The discussion in the following subsections is not intended as a comprehensive introduction to this broad and profound knowledge system. Rather, it serves mostly as fundamentals and guidelines for making compromises in the engineering design of electric motors.

10.2.1. Topology of Electric Motors

Based on the structural topology, a rotary electric motor consists of at least a stator and a rotor, which can be arranged one inside another or side by side in a coaxial manner, as shown in Figure 10.2. The stator is the stationary part of the motor, and the rotor is the moving part that rotates about its central axis, with an air gap between them. According to the orientation of the magnetic flux, these two types of motors are also called radial flux motors and axial flux motors, respectively.

FIGURE 10.2 Topology of electric motors. (a) Radial flux and (b) axial flux.

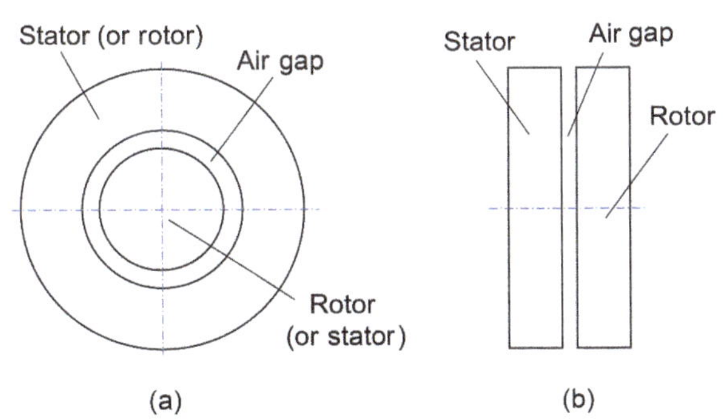

In radial flux motors, the flux is generated perpendicular to the axis of rotation. A typical stator contains a circular support known as a yoke, which is outfitted with "teeth" containing electromagnetic coils. These teeth function as alternating magnetic poles. The magnetic poles of the rotor interact with the alternating magnetic flux of the wound stator teeth, producing the torque.

In axial flux motors, on the contrary, the flux is generated parallel to the axis of rotation because of the way the electromagnetic coils are wound. Axial flux motors are typically of higher power density than radial flux motors for several reasons. First, the magnets in axial flux motors are located further away from the central axis, resulting in greater torque due to the larger lever. Second, the flux path is shorter in axial flux motors and is, therefore, an inherently more efficient topology in the electromagnetic sense. Third, there is no coil overhang at all in axial flux motors with 100% of the winding being fully active, in comparison to a large part of coil overhang, that is, inactive windings, in radial flux motors. Fourth, the cooling is better in axial flux motors because the windings can be in direct contact with the motor casing, while steel in the stator of radial flux motors is not a very good heat conductor. Special structural arrangements are, however, often needed to eliminate the significant magnetic pull between the rotor and the stator in axial flux motors. In addition, high surface flatness and gap uniformity need to be held over the large air gap area, resulting in stringent requirements for the manufacturing and assembly accuracy of rotors and stators. Axial flux motors have not been widely used in EVs yet.

A detailed comparison of the radial flux motors and axial flux motors is given in Table 10.1.

TABLE 10.1 Comparison of radial flux and axial flux motors.

Characteristics	Radial flux motors	Axial flux motors
Torque density (N·m/kg)	3–8	15–24
Power density (kW/kg)	1.3–4.6	3–7
Cooling	Mediocre	Excellent
High speed capability	Excellent	Mediocre
Compactness	Smaller radial dimension	Smaller axial dimension
Weight	Mediocre	Excellent
Manufacturability	Good	Poor
Cost	Average	High

10.2.2. Basic Design Parameters and Equations

An electric motor is fundamentally a machine in which the conversion takes place between electrical and mechanical energy. The power directly related to the main parameters of an AC motor is referred to as apparent power, denoted as S:

$$S = mEI \tag{10.1}$$

where

m is the number of phases of armature winding
E is the induction EMF per phase in V
I is the current per phase in A

The apparent power is the combination of active power and reactive power:

$$S = \sqrt{\left(P^2 + Q^2\right)} \tag{10.2}$$

where P is the active power, also called the true power, in W (watt), which is the real electric power consumption of the electric motor, and Q is the reactive power in volt-ampere reactive (VAR), which is the power merely absorbed and returned by the load due to the reactive properties of loads such as inductors and capacitors.

10.2.2.1. Radial Flux Motors

The armature diameter and length are the main dimensions of a radial flux motor. For an AC motor, the armature diameter and length refer to the inner diameter and the length of its stator, respectively.

The EMF per phase in Equation (10.1) is given by [10.2, 10.3]

$$E = 4K_f K_w f N \Phi \tag{10.3}$$

where K_f and K_w are the form factor and winding factor of the electric motor, respectively (for a sinusoidal air gap magnetic flux field, $K_f = 1.11$, and the fundamental winding factor is often substituted for K_w); f is the EMF frequency in Hz; N is the number of turns in series for each phase of the winding; and Φ is the air gap flux per pole in Wb.

Per its definition, we know that

$$f = \frac{pn}{60} \tag{10.4}$$

where

p is the number of pole pairs

n is motor speed in rpm

Besides, the air gap flux per pole can be presented as

$$\Phi = a_p B_g \tau l_i \tag{10.5}$$

where

a_p is the equivalent polar arc coefficient

B_g is the maximum flux density over the air gap in T

l_i is the effective axial length of the armature in m

τ is the pole pitch in m

Here, τ can be calculated by

$$\tau = \frac{\pi D}{2p} \tag{10.6}$$

where D is the armature diameter. Combining Equations (10.4) through (10.6) into Equation (10.3) yields

$$E = \frac{1}{15} K_f K_w p n N a_p B_g \tau l_i \tag{10.7}$$

with

$$a_p = \frac{B_{gav}}{B_g} \tag{10.8}$$

where B_{gav} is the average flux density over the air gap in T. Moreover, the current per phase in Equation (10.1) can be obtained by

$$I = \frac{\pi D a_c}{2mN} \tag{10.9}$$

where a_c is the specific electric loading, defined as the total winding current per meter of armature periphery at the air gap.

Combining Equations (10.6) through (10.9) into Equation (10.1), we obtain [10.2, 10.3]

$$S = \frac{\pi^2}{60} K_f K_w a_p n a_c B_g D^2 l_a \tag{10.10}$$

Note that l_a is an approximate for l_i, which is the total length of the iron core of armature, as shown in Figure 10.3. Not only valid for AC motors, Equation (10.10) is also applicable for DC motors by letting $K_f = 1$ and $K_w = 1$.

FIGURE 10.3 Key dimensions of the iron core of a radial flux AC motor.

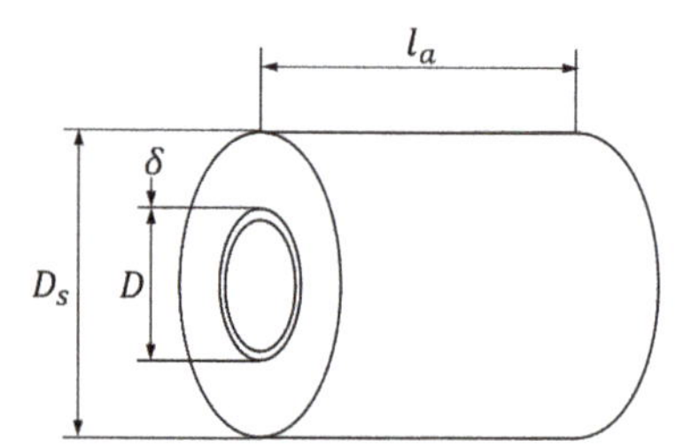

It is obvious from Equation (10.10) that the apparent power mainly depends on the motor speed n, the specific electromagnetic loading $a_c B_g$, and the characteristic quantity of rotor volume, $D^2 l_a$.

Since the volume of the rotor is proportional to $D^2 l_a$, a material utilization factor, K_A, can be defined to evaluate the torque level produced per unit volume of rotor:

$$K_A = \frac{S}{D^2 l_a n} = \frac{\pi}{30} \frac{T}{D^2 l_a} = \frac{\pi}{60} \frac{F}{D l_a} = \frac{\pi^2}{60} \sigma_t = \frac{\pi^2}{60} a_p K_f K_w a_c B_g \tag{10.11}$$

where

l_a is the full length of the armature core
n is the motor speed
S is the apparent power supplied to the load of the electric motor in volt-ampere (VA)
T is the electromagnetic torque corresponding to the apparent power
F is the electromagnetic force in the tangential direction
σ_t is the mean tangential electromagnetic stress on the rotor surface

The output power of an AC motor, P_{out}, is related to the apparent power by

$$P_{out} = \frac{SU\eta\cos\varphi}{E} = \frac{S\eta\cos\varphi}{K_E} \tag{10.12}$$

where

U is the phase voltage
η is the motor efficiency
φ is the power factor angle
K_E is the phase EMF coefficient of the stator winding, with $K_E = E/U$

Finally, based on the motor volume $V_m = \pi D_s^2 l_a / 4$, the output power per unit volume of the motor can be obtained to evaluate the power density of the motor by using Equation (10.12):

$$\frac{P_{out}}{V_m} = \frac{4\eta\cos\varphi}{\pi K_E}\frac{K_A}{K_D^2}n \tag{10.13}$$

where K_D is the outer-to-inner diameter ratio of the stator iron core, that is, $K_D = D_s/D$. Clearly, the power density of the motor is proportional to the material utilization factor and the motor speed and is inversely proportional to the square of the outer-to-inner diameter ratio of the stator iron core.

10.2.2.2. Axial Flux Motors

The axial flux motors usually have one or two air gaps. The presence of an axial electromagnetic force is a major drawback of single-air gap axial flux motors (Figure 10.4a). Therefore, most of the axial flux motors are dual-air gap type, as shown in Figure 10.4b, in which not only the axial electromagnetic forces cancel out each other but also the power density is improved. However, only the basic design parameters and equations for single-air gap axial flux motors are elaborated here for simplicity.

FIGURE 10.4 Schematic of axial flux motors. (a) Single-air gap and (b) dual-air gap.

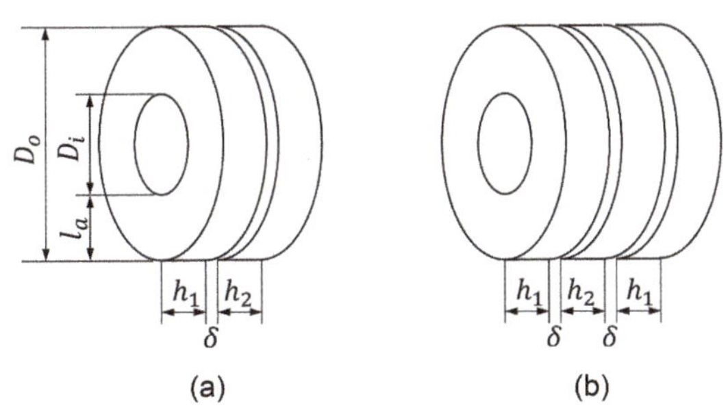

By ignoring the edge effects of magnetic fields around the planar air gap, the air gap flux per pole, Φ, is given by [10.2, 10.3]

$$\Phi = a_p l_a \tau_m B_g = \frac{\pi a_p}{8p}\left(D_o^2 - D_i^2\right)B_g \tag{10.14}$$

where

l_a is the radial length of the iron core of the armature
τ_m is the mean polar distance measured around the circumference of the armature at its mean diameter D_m
D_o is the outer diameter of the iron core of the armature
D_i is the inner diameter of the iron core of the armature

Substitution of Equation (10.14) into Equation (10.3) yields the EMF per phase:

$$E = \frac{\pi}{120} K_f K_w n N a_p B_g \left(D_o^2 - D_i^2\right) \tag{10.15}$$

The armature current is given by

$$I = \frac{\pi D_m a_{cm}}{2mN} = \frac{\pi\left(D_i + D_o\right)a_{cm}}{4mN} \tag{10.16}$$

where a_{cm} is the mean specific electric loading.

Noting that $l_a = (D_o - D_i)/2$ and $D_m = (D_i + D_o)/2$, we can obtain the following equation by substituting Equations (10.15) and (10.16) into Equation (10.1):

$$S = \frac{\pi^2}{60} K_f K_w a_p n a_{cm} B_g D_m^2 l_a \tag{10.17}$$

It is obvious that Equation (10.17) is quite similar to its radial flux motor counterpart in Equation (10.10). Since the armature diameter of an axial flux motor is not a constant with no relationship to the number of conductors in the slots, its specific electric loading varies with the armature diameter. Hence, the mean values, D_m and a_{cm}, are used in Equation (10.17) instead.

The material utilization factor is given by

$$K_A = \frac{S}{D_m^2 l_a n} = \frac{\pi^2}{60} K_f K_w a_p a_{cm} B_g \tag{10.18}$$

The electromagnetic torque is

$$T = \frac{S}{2\pi n/60} = \frac{30}{\pi} K_A D_m^2 l_a \tag{10.19}$$

The mean tangential electromagnetic stress on the rotor surface is

$$\sigma_t = \frac{F}{\pi D_m l_a} = \frac{2T}{\pi D_m^2 l_a} = \frac{60}{\pi^2} K_A \tag{10.20}$$

The output power per unit volume of the single-air gap axial flux motor is

$$\frac{P_{out}}{V_m} = \frac{S\eta\cos\varphi}{K_E \frac{\pi}{4}\left(D_o^2 - D_i^2\right)\left(h_1 + h_2\right)} = \frac{\eta\cos\varphi}{\pi K_E} \frac{D_m}{h_1 + h_2} K_A n \tag{10.21}$$

For the dual-air gap axial flux motor, as shown in Figure 10.4b, the output power per unit volume is given by

$$\frac{P_{out}}{V_m} = \frac{2S\eta\cos\varphi}{K_E \frac{\pi}{4}\left(D_o^2 - D_i^2\right)\left(h_1 + h_1 + h_2\right)} = \frac{2\eta\cos\varphi}{\pi K_E} \frac{D_m}{2h_1 + h_2} K_A n \tag{10.22}$$

By examining Equations (10.21) and (10.22), it is apparent that the output power density of an axial flux motor is proportional to the material utilization factor, motor speed, and mean armature diameter and inversely proportional to the axial length of the iron core of the armature. In addition, the output power density of the dual-air gap axial flux motor is roughly one-third higher than that of the single-air gap one if the armatures and rotors are identical in axial length [10.2].

10.2.3. Sizing of Electric Motors

As mentioned in Section 8.6.1, several different types of electric motors can be selected for the traction application of EVs. The PMSM, however, is undoubtedly the most popular choice for today's EVs. Thus, the structural design of electric motors will be elaborated here by focusing particularly on PMSMs.

As the first step of electromagnetic parameter design, sizing of the electric motor mainly involves the determination of its speed, electromagnetic loading, and aspect ratio.

10.2.3.1. Selection of Motor Speed

As discussed in the last section, the power density of both radial flux and axial flux motors is directly proportional to the motor speed. It is, therefore, of significance to beef up the motor speed in pursuit of maximal power density for gaining advantages in the size, cost, and ease of packaging. As a matter of fact, the maximum speed of traction motors has been increasing in the past decade from around 10,000 to over 20,000 rpm in production EVs.

In addition, the undergoing research and development projects are focusing on traction motors of speeds as high as 50,000 to 100,000 rpm. It is quite likely that the motor speed will keep going up in the foreseeable future.

Nevertheless, numerous structural and material problems need to be solved in the design of motors of extremely high speeds, particularly those related to bearings, rotor stress, and deformation, as well as cooling.

The selection of motor speed must also take the transmission or speed reduction mechanism into account since high-speed drivetrains are prone to issues in lubrication, noise, vibrations, etc.

10.2.3.2. Selection of Specific Electromagnetic Loading

Based on Equation (10.11), the material utilization factor K_A is proportional to the specific electromagnetic loading a_cB_g and the tangential stress of the rotor surface material. Thus, selecting a large specific electromagnetic loading a_cB_g or a rotor surface material of high tangential strength helps boost the material utilization and the power density of the motor.

However, copper losses (also called ohmic losses) rise with the specific electric loading a_c, while iron losses (also called core losses) grow with specific magnetic loading B_g as it saturates the core, both leading to excessive heat generation and thus making it necessary to use an upgraded thermal insulation or cooling mechanism [10.2]. Hence, both a_c and B_g need to be optimized in order to minimize the size and volume of the motor as well as the mass of the windings and the iron core, without incurring much penalty in copper and iron losses. The effects of specific electric and magnetic loadings on the size of an electric motor are discussed as follows:

1. Effects of specific electric loading a_c

   Current density J is defined as the current per unit area normal to the current of the armature conductor, namely,

$$J = \frac{2mNI}{\pi D t_{eq}} \tag{10.23}$$

where t_{eq} is the equivalent thickness of all the conductors on the armature surface calculated by assuming that the conductors between slots are evenly distributed around the armature surface [10.4]. By combining to Equation (10.9), Equation (10.23) can be rewritten as [10.2, 10.3]:

$$t_{eq} = \frac{a_c}{J} \tag{10.24}$$

The mass of copper in the slots of the stator winding is then given by

$$m_{css} = \pi D l_a t_{eq} \rho_c = \pi D l_a \rho_c \frac{a_c}{J} \tag{10.25}$$

where ρ_c is the density of copper winding. It is clear from Equation (10.25) that the mass of copper is proportional to t_{eq} and a_c and inversely proportional to J. In other words, the copper usage can be reduced by selecting a small specific electric loading a_c or a large current density J. Additionally, the mass of copper in the armature winding can be computed by

$$m_{cs} = \pi D l_a t_{eq} \rho_c \frac{l_c}{l_a} = \pi D l_c t_{eq} \rho_c \tag{10.26}$$

where l_c is the average half-turn length of winding.

For the iron core of a rotor, mostly made of neodymium iron boron (NdFeB), the quantity of iron material usage is dependent only on D^2l_a. In addition, the mass of the iron core of the stator, m_{is}, is not related to the aspect ratio either, which is given by

$$m_{is} = \frac{\pi}{4} D^2 l_a \rho_i \left[1 + 2\left(\frac{h_s}{D}\right) + 2\left(\frac{h_y}{D}\right)\right]^2 \tag{10.27}$$

where

ρ_i is the density of iron

h_s and h_y are the slot height and yoke height of the stator, respectively

On the other hand, the copper losses of the effective portion of windings, that is, the windings in the slots, Q_c, can be calculated by

$$Q_c = 2mNR_{ef} I^2 = 2mNl_{ef} \rho_{ce} IJ \tag{10.28}$$

where

R_{ef} is the electric resistance of the effective portion of each conductor

l_{ef} is the length of the effective portion of each conductor

ρ_{ce} is the resistivity of the conductor material

The copper losses of the effective portion of windings per unit armature surface area, q_c, are, thus, given by [10.2, 10.3]

$$q_c = \frac{Q_c}{\pi Dl} = \rho_{ce} a_c J \tag{10.29}$$

Furthermore, the per unit reactance value of AC windings is related to the specific electromagnetic loading by

$$X_{pu} = K \frac{a_c}{B_g} \tag{10.30}$$

where K is a scale factor that can be assumed a constant.

It is implied by Equation (10.30) that the selection of a_c affects the per unit reactance proportionally, resulting in a change in the operating characteristics of the motor. For a synchronous AC motor, a large a_c may not only raise the voltage change rate but also reduce the short-circuit current and short-circuit ratio, worsening both static and dynamic stability; for an AC IM, a large a_c can result in reduced starting torque, starting torque, and maximum torque.

For a given specific magnetic loading B_g, the increase in specific electric loading a_c helps gaining the following benefits: the size and volume of the motor shrink, with a reduction in the mass of the iron core material, and iron losses drop. Nevertheless, it also incurs two weaknesses: the amount of copper usage increases, as indicated by Equation (10.25), and copper losses per unit armature surface area surges, as indicated by Equation (10.29), leading to a further rise in winding temperature. In practice, the selection of specific electric loading a_c is made so that the product a_cJ does not exceed a certain limit due to its dependency on the heat generation and temperature rise of the motor [10.2, 10.3].

2. Effects of specific magnetic loading B_g
   For a given specific electric loading a_c, the increase in specific magnetic loading B_g brings the advantage of reducing the size and volume of the motor and saving the iron core material. However, this can incur some drawbacks: an increase in iron losses, due to the fact that the iron losses per unit volume of iron core are proportional to the square of flux density in the iron core; increases in both air gap magnetic potential drop and degree of magnetic saturation, causing a rise in the amount of copper used for excitation winding and excitation loss in a synchronous AC motor, or a deterioration in power factor due to elevated excitation current in an AC IM; and a change in the starting and operating characteristics of AC motors because of the reduction in the per unit reactance value, according to Equation (10.30).

10.2.3.3. Sizing of Radial Flux AC Motors

In practice, once the motor speed n is selected, the material utilization factor K_A can be determined according to its empirical relationship to the quantity S/n, typically in the form of $K_A = f(S/n)$. The characteristic quantity of rotor volume, $D^2 l_a$, is then obtained by using Equation (10.11). Next, the aspect ratio l_a/D can be chosen by following the guidelines:

1. Selection of aspect ratio
   For a radial flux AC motor, its aspect ratio refers to the ratio of the length to the diameter of its armature, namely, l_a/D, as shown in Figure 10.3. The selection of the aspect ratio for traction motors is primarily made according to the minimization of material usage.
   As indicated in the last section, neither the quantity of the iron material used in the rotor core nor that used in the stator core is dependent on the aspect ratio l_a/D. However, the mass of copper is affected by the aspect ratio l_a/D, according to Equation (10.26). The average half-turn length of winding l_c in Equation (10.26) can be expressed as $l_c = l_a + K_e D$, where K_e is a scale factor and $K_e \approx 2/p$. If both $D^2 l_a$ and t_{eq} are kept constant, then $D^2 l_a = D^3 \dfrac{l_a}{D} = const$ and, therefore, $D \propto \left(\dfrac{l_a}{D}\right)^{-1/3}$. It can be derived from Equation (10.26) that [10.2]

$$m_{cs} \propto \left(\frac{l_a}{D}\right)^{\frac{1}{3}} + K_e \left(\frac{l_a}{D}\right)^{-\frac{2}{3}} = f\left(\frac{l_a}{D}\right) \tag{10.31}$$

   Clearly, the mass of copper used in armature winding is a function of the aspect ratio l_a/D. The curves of the function $f\left(\dfrac{l_a}{D}\right)$ versus $\dfrac{l_a}{D}$ for $p = 1, 2, 3$, and 4 are shown in Figure 10.5. In general, the function $f\left(\dfrac{l_a}{D}\right)$ and the copper usage drop sharply as the aspect ratio l_a/D increases when $\dfrac{l_a}{D} < 1$ due to the relatively large amount of copper usage at the ends, with only mild rises when $\dfrac{l_a}{D} > 2$ though. The optimal aspect ratio l_a/D is given by [10.2]

FIGURE 10.5 Function $f\left(\frac{l_a}{D}\right)$ versus $\frac{l_a}{D}$ for p = 1, 2, 3, and 4.

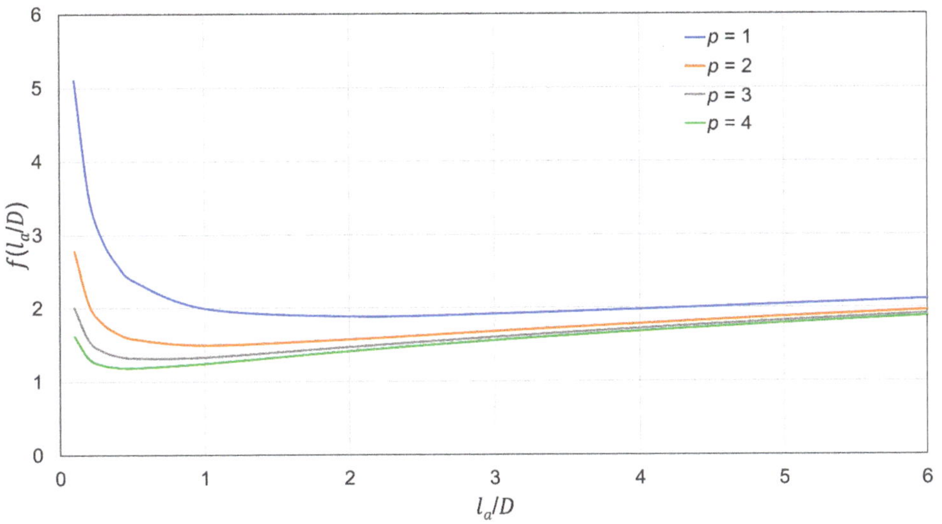

$$\frac{l_a}{D} = \frac{\pi}{2} p^{-2/3} \tag{10.32}$$

for an AC IM of $p > 1$, and

$$\frac{l_a}{D} = \frac{\pi}{4} p^{-1/2} \tag{10.33}$$

for a synchronous AC motor of $p > 1$.

Although the per-unit values of end leakage reactance, total leakage reactance, and electric resistance can be reduced by enlarging the aspect ratio, slim-type motors with the aspect ratio $\frac{l_a}{D} > 1$ are rarely used in traction applications due to the concerns about difficulties in electric powertrain packaging brought by their excessive axial dimension. The gain in the dynamic response of the motor owing to the reduction in rotor moment of inertia does not usually justify the use of slim motors either.

2. Selection of air gap length

The upper limit of air gap length is chosen by taking the armature reaction, reliability, and manufacturability into consideration, while the lower limit is determined by the excitation loss and power factor.

For an AC IM, the per-unit excitation current is given by

$$i_{pu} \propto \frac{g}{\tau} \frac{B_g}{a_c} \tag{10.34}$$

Thus, the power factor declines as the air gap length or the number of pole pairs increases. The air gap length for a radial flux AC IM is generally taken by

$$g = \begin{cases} 0.1 + 0.225P^{1/3} & p = 1 \\ 0.1 + 0.145P^{1/3} & p > 1 \end{cases} \tag{10.35}$$

where g is in mm and P is in kW.

For radial flux AC synchronous motor, the air gap length can be computed by

$$g = k_g \tau \frac{a_c}{B_g} \tag{10.36}$$

where g is in mm, τ is in m, a_c is in A/m, B_g is in T, and k_g is a constant that can be selected according to the pole and air gap design from the following values:

$$k_g = \begin{cases} 2.5 \times 10^{-4} & \text{Non-salient pole} \\ 7.0 \times 10^{-4} & \text{Salient pole, Constant air gap} \\ 4.5 \times 10^{-4} & \text{Salient pole, Sinusoidal air gap} \end{cases}$$

3. Determination of the armature diameter
   Finally, after the aspect ratio l_a/D is chosen by following the aforementioned guidelines, the armature diameter D can be determined by

$$D = \left[\frac{nK_A (l_a / D)}{S} \right]^{1/3} \tag{10.37}$$

which is derived from Equation (10.11).

10.2.3.4. Sizing of Axial Flux AC Motors

The steps of sizing axial flux AC motors are generally similar to those of sizing radial flux motors, as elaborated here:

1. Selection of the outer-to-inner diameter ratio
   For an axial flux AC motor introduced in Sections 10.2.1 and 10.2.2, the ratio of the radial length to the mean diameter of its armature core l_a/D_m is a major dimension ratio. Hence, we have

$$\frac{l_a}{D_m} = \frac{(D_o - D_i)/2}{(D_o + D_i)/2} = \frac{D_o / D_i - 1}{D_o / D_i + 1} \tag{10.38}$$

D_o/D_i is a more intuitive dimension ratio than l_a/D_m in the design. The apparent power S in Equation (10.17) can be expressed using the specific electric loading at the inner diameter, a_{ci}:

$$S = \frac{\pi^2}{240} K_f K_w a_i n a_{ci} B_g \left(D_o^2 - D_i^2\right) D_i \tag{10.39}$$

For given a_{ci} and B_g, the inner diameter D_i can be optimized to achieve the maximized apparent power S. By taking a derivative of the right-hand side of Equation (10.39) with respect to D_i, it can be found that the apparent power S reaches the maximum when $D_i = D_o / \sqrt{3}$, or $D_o / D_i = \sqrt{3}$.

In practice, as an excessively large ratio D_o/D_i may cause it difficult to accommodate the end windings around the inner periphery, the ratio D_o/D_i is, therefore, selected in the following range [10.2]:

$$D_o / D_i = \begin{cases} 1.4 - 1.7 & \text{when } S \le 1\,\text{kW} \\ 1.7 - 2.0 & \text{when } S > 1\,\text{kW} \end{cases} \tag{10.40}$$

Once the ratio D_o/D_i is chosen, the inner diameter D_i can be solved by Equation (10.39), with the motor speed and specific electromagnetic loading selected earlier.

2. Optimization of flux path size
For an axial flux motor, its volume of the iron core is given by [10.2]

$$V_i = \frac{\pi}{4}\left(D_o^2 - D_i^2\right)\left(h_s + h_y\right) \tag{10.41}$$

where h_s and h_y are the slot height and yoke height, respectively. The slot height is

$$h_s = \frac{1}{1 - \frac{l_a}{l_i}\frac{B_g}{B_{ta}}} \frac{a_{cm}}{f_u J} \tag{10.42}$$

where

l_i is the length of the iron core
B_{ta} is the apparent tooth flux density
f_u is the slot filling factor
a_{cm} is the specific electromagnetic loading at mean diameter

The yoke height is

$$h_y = \frac{\pi}{4}\frac{l_a}{l_i}\frac{D_m a_i}{p}\frac{B_g}{B_y} \tag{10.43}$$

where B_y is the yoke flux density.

Based on Equation (10.20), we can derive the tangential electromagnetic force per unit iron core volume [10.2]:

$$\frac{F}{V_i} = \frac{\sigma_t}{h_s + h_y} \tag{10.44}$$

Substituting Equations (10.38) and (10.43) into Equation (10.44) and taking derivative with respect to B_g, the optimal value of B_g that yields the maximum of $\frac{F}{V_i}$ can be obtained:

$$B_{g,opt} = \frac{l_i}{2l_a} B_{ta} \tag{10.45}$$

The optimal tooth width at the mean diameter is given by

$$\left(\frac{b_m}{t_m}\right)_{opt} = 0.5 \tag{10.46}$$

The optimal slot height and yoke height are given, respectively, by [10.2]

$$h_{s,opt} = \frac{2a_{cm}}{f_u J} \tag{10.47}$$

$$h_{y,opt} = \frac{\pi}{8} \frac{D_m a_i}{p} \frac{B_{ta}}{B_y} = \frac{\tau a_i}{4} \frac{B_{ta}}{B_y} \tag{10.48}$$

Accordingly, when $B_{ta} \approx B_y$, the volume of the iron core reaches the minimum:

$$V_{i,min} = \pi D_m l_a \left(\frac{2a_{cm}}{f_u J} + \frac{\pi D_m a_i}{8p} \right) \tag{10.49}$$

10.2.4. Design of a Magnetic Circuit

The design of a magnetic circuit involves the calculation of the magnetic potential difference along a given closed magnetic loop, the no-load-induced EMF, and other no-load characteristics.

The calculation of the magnetic potential difference is based on Ampere's circuital law, that is, the line integral of the magnetic field along any closed loop equals the total current flowing through the loop:

$$\oint_l \boldsymbol{H} \cdot \mathrm{d}\boldsymbol{l} = \sum_i F_i = I_{tot} \tag{10.50}$$

where

$\boldsymbol{H}$ is the magnetic field intensity
$\boldsymbol{l}$ is the path of the closed loop that usually goes through the centerline of magnetic poles
F_i is the magnitude of the excitation MMF for the ith pair of poles
I_{tot} is the total current through the loop

In order to simplify the calculation, the magnetic field in each portion of the motor along the closed loop can be divided into corresponding equivalent sections, as shown in Figure 10.6. Thus, Equation (10.50) can be rewritten as

$$\sum_{j=1}^{n} H_j l_j = F_p \tag{10.51}$$

where

H_j is the average magnetic field intensity in the jth section of the path
l_j is the length of the jth section of the path
F_p is the excitation MMF of each pole pair

FIGURE 10.6 Magnetic circuit.

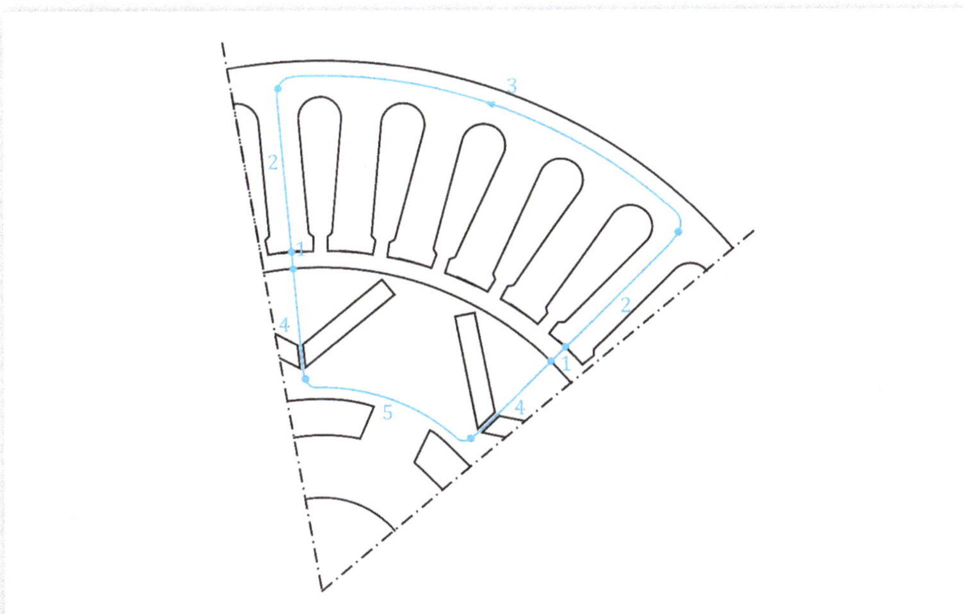

The sections of the path commonly include the air gap, stator teeth or poles, rotor teeth or poles, stator yokes, and rotor yokes. Among these sections of the magnetic path, the magnetic potential drop in the air gap is generally the largest, ranging from 60 to 85% of the total.

10.2.4.1. Magnetic Potential Drop in the Air Gap

In general, only the magnetic potential drop over the air gap along the pole centerline is calculated. Due to its small size, the flux density along the air gap length can be assumed constant. The magnetic potential drop of each pole F_g, thus, given by

$$F_g = C_s H_g g \tag{10.52}$$

where

C_s is the Carter coefficient that compensates the slot effect
H_g is the maximum magnetic intensity at the air gap
g is the minimum air gap along the pole centerline

The maximum magnetic intensity at the air gap is

$$H_g = \frac{B_g}{\mu_0} \tag{10.53}$$

where

B_g is the maximum flux density at the air gap
μ_0 is the air permeability

According to Equations (10.5) and (10.53), Equation (10.52) becomes

$$F_g = \frac{C_s B_g g}{\mu_0} = \frac{C_s \Phi g}{\mu_0 a_p \tau l_i} \tag{10.54}$$

For an AC motor, the flux per pole Φ can be obtained by rewriting Equation (10.3):

$$\Phi = \frac{E}{4K_f K_w f N} \tag{10.55}$$

It is clear from Equation (10.54) that the Carter coefficient C_s, equivalent polar arc coefficient a_p, and effective axial length of the armature l_i are needed for the calculation of the magnetic potential drop of each pole F_g. The methods for calculating these quantities are discussed next.

1. Carter coefficient C_s
   Assuming that there are slots and teeth on the surface of the rotor iron core and the inner surface of the stator is smooth, the presence of the slot opening would cause magnetic resistance at the air gap to increase. As a result, flux at the slot opening decreases, leading to a drop in air gap flux. To keep the flux per pole Φ unchanged,

the maximum flux density at the tooth top must be raised to B_g' from B_g in the absence of slots. Therefore, the Carter coefficient C_s is defined as [10.3]

$$C_s = \frac{B_g'}{B_g} \tag{10.56}$$

In routine engineering practice, the slot opening width b_s is larger than the air gap length g most of the time, then the Carter coefficient C_s can be calculated using the following empirical formulas [10.3]:

$$C_s = \frac{(b_s + 5g)t}{(b_s + 5g)t - b_s^2} \tag{10.57}$$

for open slots, where t is the tooth pitch, and [10.3]

$$C_s = \frac{(0.75b_s + 4.4g)t}{(0.75b_s + 4.4g)t - b_s^2} \tag{10.58}$$

for semiclosed slots.

When there are slots on both rotor and stator, the Carter coefficient C_s is dependent on their slot position relative to each other. When the centerline of the rotor slot is lined up with that of the stator slot, we have $C_s < C_{ss}C_{sr}$, where C_{ss} is the Carter coefficients for the slotted stator and the nonslotted rotor, and C_{sr} is the Carter coefficients for the slotted rotor and the nonslotted stator; when the centerline of the rotor tooth aligns to that of the stator slot, we have $C_s > C_{ss}C_{sr}$; when the rotor spins, an average can be taken between the above situations, that is, $C_s \approx C_{ss}C_{sr}$ [10.3].

2. Equivalent polar arc coefficient a_p
For an arbitrary air gap flux density distribution $B = B(x)$, the flux per pole can be expressed by

$$\Phi = l_i \int_0^{\tau} B(x)\mathrm{d}x \tag{10.59}$$

By comparing Equation (10.59) to Equation (10.5), the following relationship can be obtained:

$$a_p = \frac{\frac{1}{\tau}\int_0^{\tau} B(x)\mathrm{d}x}{B_g} = \frac{B_{gav}}{B_g} \tag{10.60}$$

where the average of air gap flux $B_{gav} = \frac{1}{\tau}\int_0^{\tau} B(x)\mathrm{d}x$. Note that Equation (10.60) is the same as Equation (10.8). Particularly, when the air gap flux density $B(x)$ has a sinusoidal distribution, the polar arc coefficient $a_p = 2/\pi \approx 0.637$.

However, for most AC IMs of $p \geq 2$, their excitation MMF has a top-hat distribution instead of a sinusoidal distribution due to the magnetic saturation in the tiny air gap, as illustrated in Figure 10.7 [10.3]. The polar arc coefficient depends on the degree of saturation of the magnetic circuit, especially at the stator teeth and rotor teeth. The saturation coefficient K_s can be defined as

$$K_s = \frac{F_{st} + F_{rt} + F_g}{F_g} \tag{10.61}$$

where

F_{st} is the magnetic potential drop at the stator tooth
F_{rt} is the magnetic potential drop at the rotor tooth

FIGURE 10.7 Air gap flux density distribution along the periphery of an armature.

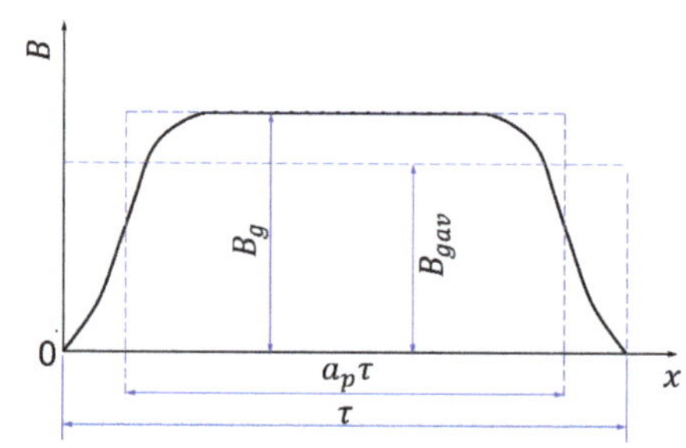

Courtesy of Suei Chen.

As far as salient pole synchronous motors are concerned, concentrated windings are adopted for field excitation, with a rectangular distribution of excitation MMF. The air gap flux density $B(x)$ is generally made into a sinusoidal distribution, that is,

$$B(x) = B_g^{'} \sin\frac{\pi x}{\tau} = \mu_0 H(x) = \mu_0 \frac{F_g}{g(x)} \tag{10.62}$$

where $B_g^{'}$ is the fundamental air gap flux amplitude, as shown in Figure 10.8 [10.3], with

$$B_g^{'} \approx B_g = \mu_0 \frac{F_g}{g} \tag{10.63}$$

FIGURE 10.8 Distribution of the air gap flux density of a salient pole synchronous motor.

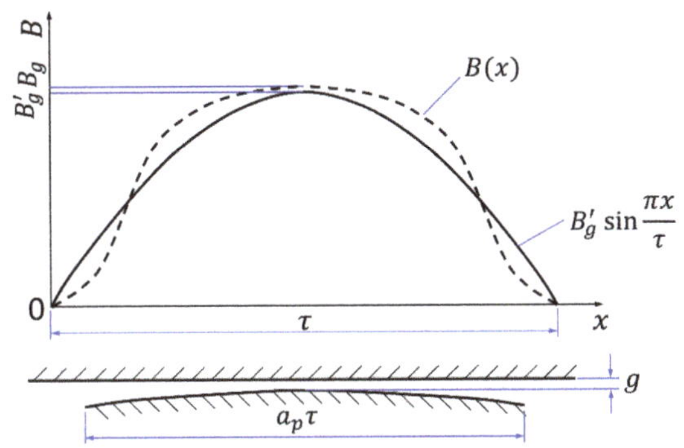

Courtesy of Suei Chen.

Substitution of Equation (10.63) into Equation (10.62) yields

$$g(x) \approx \frac{g}{\sin\frac{\pi x}{\tau}} \tag{10.64}$$

Clearly, Equation (10.64) requires that the air gap length should follow a cosecant distribution, which is difficult to implement. In practice, most of the salient pole synchronous motors for traction application utilize a uniform air gap for easy manufacturing, noise elimination, cost reduction, and other reasons. The equivalent polar arc coefficient falls generally in the range of 0.55–0.75; for small-size motors of not many poles, a value near the lower bound is often preferred to make adequate space for accommodating field windings between the poles. Let $\hat{b}_p$ be the polar arc length, that is, the angular distance spanned by the permanent magnet, then a polar arc coefficient can be defined as $\hat{b}_p / \tau$.

3. Effective axial length of the armature l_i

The axial length of the armature for calculating the maximum air gap flux density is actually longer than the full length of the iron core because of the so-called end effects, meaning a small portion of the flux passes through the air gap from the ends of the stator and rotor. The effective axial length of the armature l_i can generally be approximated by

$$l_i = l_a + 2g \tag{10.65}$$

where l_a is the axial length of the armature iron core.

10.2.4.2. Magnetic Potential Drop through Teeth

The magnetic potential drop through the teeth for each pole can be obtained by

$$F_t = H_t l_t \tag{10.66}$$

where

H_t is the magnetic field intensity passing through the teeth

l_t is the effective flux path length passing through the teeth

When the flux density through tooth $B_t < 1.8$ T, the degree of saturation through the magnetic material is not high. The magnetic resistance through the teeth is much lower than that through the slots because the permeability of a magnetic material is much higher than that of a winding material, for example, copper or insulation materials. As a result, almost all the flux entering the iron core surface from the air gap within a tooth pitch should pass through the tooth, rather than the slot [10.3]. For rectangular slots of which the side walls of a tooth are not parallel, as shown in Figure 10.9a, the magnetic potential drop through such a tooth can be calculated approximately by the following empirical formula [10.3]:

$$F_t = \frac{(H_{tr} + H_{tm} + H_{tt})l_t}{6} \tag{10.67}$$

where H_{tr}, H_{tm}, and H_{tt} are the magnetic field intensity at the tooth root, in the middle, and at the tooth top, respectively. For the tooth with nearly parallel side walls, such as the pear-shaped slots shown in Figure 10.9b [10.3], the magnetic potential drop through a tooth can be computed using

$$F_t = H_{t1/3} l_t \tag{10.68}$$

where $H_{t1/3}$ is the magnetic field intensity at one-third of the tooth height from the narrowest portion of the tooth.

FIGURE 10.9 Geometry of slots. (a) Rectangular slots and (b) pear-shaped slots.

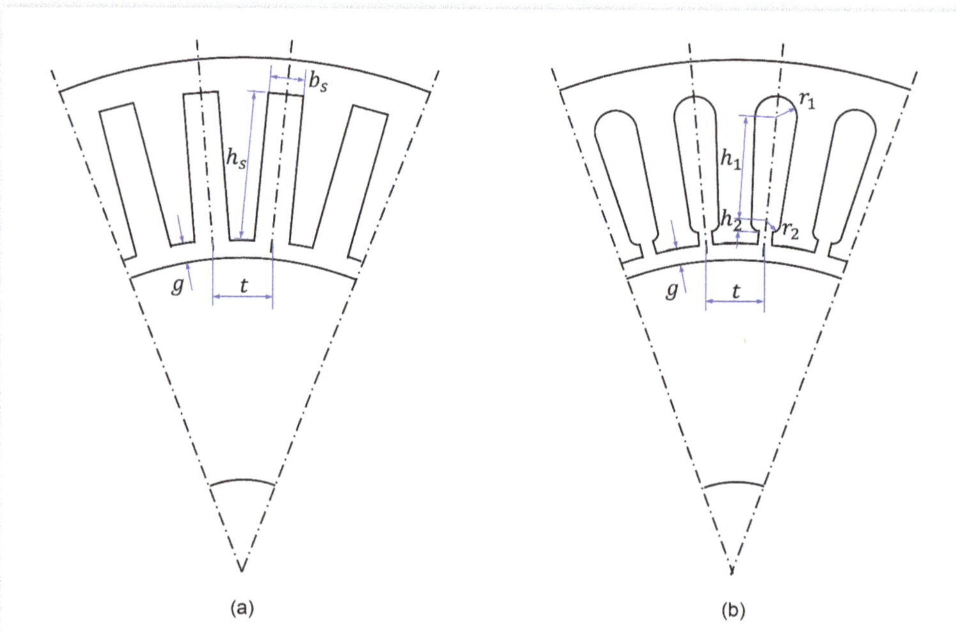

When the flux density through the tooth $B_t > 1.8$ T, the magnetic path through the tooth is quite saturated, resulting in the magnetic resistance through the slots close to that through the teeth. The magnetic field intensity through a tooth can be obtained by

$$H_t = \frac{B_{ta} - B_t}{\mu_0 k_s} \tag{10.69}$$

where B_{ta} is the apparent flux density, which is the flux density when all the flux with a tooth pitch enters the tooth, and k_s is the slot factor. For rectangular slots, the slot factor is

$$k_s = \frac{b_s l_i}{b_{t1/3} l_a K_i} \tag{10.70}$$

where

b_s is the slot width

$b_{t1/3}$ is the tooth width at one-third of the tooth height from the narrowest portion of the tooth

K_i is the stacking factor of the iron core

For pear-shaped slots, the slot factor is

$$k_s = \frac{(r_1 + r_2) l_i}{b_t l_a K_i} \tag{10.71}$$

where r_1 and r_2 are the radii of the slots, as illustrated in Figure 10.9b. The apparent flux density B_{ta} is given by

$$B_{ta} = \frac{l_i t B_g}{l_a b_t K_i} \tag{10.72}$$

The effective flux path length passing through the tooth in Equation (10.68) can be calculated according to the type of slots. For pear-shaped slots, $l_t = h_1 + h_2 + r_1/3$; for open slots, $l_t = h_s$; and for semiclosed slots, $l_t = h_s - h_o$, where h_o is the height of the slot opening.

To calculate the polar arc coefficient a_p, an estimate of the saturation coefficient K_s can be selected from the range 1.15–1.45 first and then the maximum air gap flux B_g can be obtained using Equation (10.5). Next, after F_{st}, F_{rt}, and F_g are calculated using Equation (10.51), an updated value of K_s can be computed using Equation (10.61). The aforementioned steps can be repeated until the updated value of K_s is close enough to the last value of K_s.

10.2.4.3. Magnetic Potential Drop through Yoke

The magnetic potential drop along the full length of the yoke flux path for each pole of an AC IM can be obtained by

$$F_y = H_{yav} l_y = \int_0^{l_y} \boldsymbol{H} \cdot d\boldsymbol{l} \tag{10.73}$$

where

H_{yav} is the average magnetic field intensity along the full length of the yoke flux path
l_y is the effective length of the yoke flux path per pole, with

$$l_y = \frac{\pi D_{yav}}{4p} \tag{10.74}$$

where D_{yav} is the average diameter of the yoke between two adjacent teeth, as illustrated in Figure 10.10.

FIGURE 10.10 Yoke flux path of an AC IM.

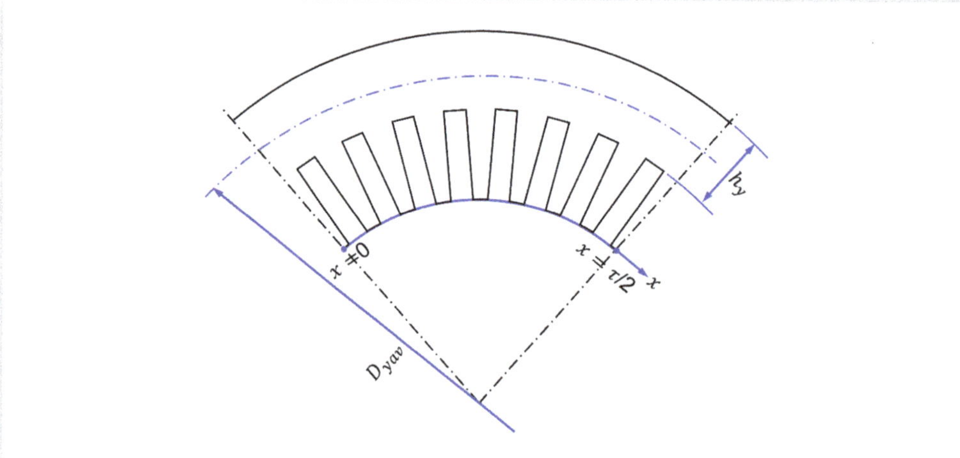

10.2.4.4. Pole Flux Leakage

It is desired that all the flux produced by the pole passes through the air gap, but the leakage is practically inevitable because there is always some flux that take the closed path through interpolar space without passing through the air gap. Hence, a pole flux leakage factor, denoted σ, can be defined as

$$\sigma = \frac{\Phi + \Phi_l}{\Phi} \tag{10.75}$$

where

Φ is the flux per pole passing through the air gap
Φ_l is the flux leakage per pole that does not pass through the air gap

For salient pole synchronous motors, it is generally recommended to keep the pole flux leakage factor σ under 1.35. An exccessive pole flux leakage factor reduces the power density of the motor and adversely influences its operating characteristics.

10.2.4.5. Excitation Current and No-load Characteristics

For AC IMs, the induced EMF does not vary significantly from no-load to rated load. Hence, it is only necessary to estimate the excitation current in no-load and rated load conditions [10.3].

The induced EMF per phase, E_{Nph}, under rated load can be estimated by

$$E_{Nph} \approx k_{EU} U_{Nph} \tag{10.76}$$

where

k_{EU} is the ratio of induced EMF to terminal voltage under rated load
U_{Nph} is the rated phase voltage of the winding

The ratio k_{EU} can be selected from the range of 0.85–0.96 for most of the AC IMs for traction applications. The ratio k_{EU} should be verified on completion of magnetic circuit computation, and iteration may be needed before the error between its updated value and last value is adequately small, for example, 0.5% [10.3].

The induced EMF per phase, E_{0ph}, under a no-load condition can be estimated by

$$E_{0ph} \approx U_{Nph} - I_0 X_{a1} \tag{10.77}$$

where

I_0 is the no-load current
X_{a1} is the leakage reactance per phase of the winding

The no-load excitation current for a salient synchronous motor with a concentrated winding is given by [10.3]

$$I_0 = \frac{F_p}{N_p} \tag{10.78}$$

where N_p is the number of turns in excitation winding per pole and F_p is the excitation MMF of each pole pair that can be obtained by Equation (10.51). The no-load excitation current for an AC multiphase motor with distributed winding can be obtained by [10.3]

$$I_0 = \frac{2pF_p}{0.9mNK_w} \tag{10.79}$$

10.2.5. Electromagnetic Design

Electrical resistance and reactance are electrical parameters that largely affect the performance and costs of electric motors. It is, therefore, of crucial importance to select these parameters properly in the design of motor windings. In addition, different types of winding are also elaborated briefly.

10.2.5.1. Calculation of Electrical Resistance

The AC resistance of a winding is always higher than the DC resistance because of the skin effect and proximity effect. Thus, the AC resistance of a winding can be expressed by

$$R_a = K_{AD} R \tag{10.80}$$

where

R_a is the AC resistance
R is the DC resistance
K_{AD} is the AC-to-DC resistance ratio

The phase resistance of the stator winding of an AC IM, R_s, is given by [10.3]

$$R_s = \frac{2Nl_c}{A_c Z_p} K_{AD} \rho_b \tag{10.81}$$

where

N is the number of turns in serial connection
l_c is the average length of the half-turn winding
A_c is the cross-sectional area of the winding conductor
Z_p is the number of branches in the parallel connection of the phase winding
ρ_b is the resistivity of the winding conductor at the base operating temperature

Note that Equation (10.81) is applicable for the phase resistance of armature (i.e., stator) windings of synchronous AC motors as well.

On the other hand, the phase resistance of rotor winding of an AC wound rotor IM, R_r, can also be calculated using a formula similar to Equation (10.81) by letting $K_{AD} = 1$ due to the negligible skin effect caused by the very low frequency of winding current during normal operation. In addition, the phase resistance of rotor winding can be converted into its stator winding by introducing a factor K_{rs}:

$$K_{rs} = \frac{m_s}{m_r}\left(\frac{N_s K_{ws}}{N_r K_{wr}}\right)^2 \tag{10.82}$$

where

m_s and m_r are the number of phases for the stator and rotor, respectively
N_s and N_r are the number of turns in serial connection for the stator and rotor, respectively
K_{ws} and K_{wr} are the fundamental winding factors of the stator and rotor, respectively

For an AC IMs with a squirrel-cage type rotor, the phase resistance of rotor winding, R_r, can be calculated by

$$R_r = R_b + \frac{R_r}{2\left(\sin\frac{\pi p}{Z_{rs}}\right)^2} \approx \left(\frac{l_b}{A_b} + \frac{Z_{rs} D_r}{2\pi p^2 A_r}\right)\rho_b \tag{10.83}$$

where

R_b is the electrical resistance of conductor bars
R_r is the electrical resistance of the end ring between adjacent conductor bars
Z_{rs} is the number of rotor slots
l_b is the length of a conductor bar
A_b is the cross-sectional area of a conductor bar
D_r is the average diameter of an end ring
A_r is the cross-sectional area of an end ring

The phase resistance of a squirrel-cage rotor winding can be converted to its stator winding by

$$K_{rs} = \frac{4m_s}{Z_r}(N_s K_{ws})^2 \qquad (10.84)$$

Again, $K_{AD} = 1$ during normal operation; however, K_{AD} becomes much larger due to the high frequency of rotor current during start-up.

10.2.5.2. Calculation of Electrical Reactance

In general, electrical reactance in windings can be divided into magnetizing reactance and leakage reactance. Magnetizing reactance corresponds to the fundamental magnetic field, while leakage reactance partially corresponds to the harmonic magnetic field. The per-unit value of electrical reactance is commonly used, allowing a convenient comparison of motors of different specifications such as rated power, speed, and voltage.

Per-unit magnetizing reactance, X_{mpu}, can be expressed as

$$X_{mpu} = \frac{I_N X_m}{U_N} \qquad (10.85)$$

where

I_N is the rated phase current
U_N is the rated phase voltage
X_m is the magnetizing reactance

The per-unit leakage reactance, X_{lpu}, is

$$X_{lpu} = \frac{I_N X_l}{U_N} \qquad (10.86)$$

where X_l is the leakage reactance.

The fundamental equations for the calculation of magnetizing reactance and leakage reactance are mentioned as follows:

1. Magnetizing reactance
   For a multiphase AC IM, the magnetizing reactance per phase armature winding can be calculated by

$$X_m = \frac{4}{\pi}\mu_0 f m \frac{l_i (N K_{ws})^2 \tau}{p g_{ef}} \qquad (10.87)$$

where

f is the frequency
g_{ef} is the effective air gap length
μ_0 is the air permeability, with $\mu_0 = 0.4\pi \times 10^{-6}$ H/m

For a nonsalient pole synchronous motor, the per-unit magnetizing reactance per phase is given by

$$X_{mpu} = C_m \frac{a_c}{B_{gs}} \tag{10.88}$$

where a_c is the electric loading, B_{gs} is the fundamental air gap flux amplitude corresponding to induced EMF that is equal to the rated voltage, and C_m is the coefficient, with

$$C_m = \frac{\sqrt{2}}{\pi} \frac{\mu_0 K_{ws} \tau}{g_{ef}} \tag{10.89}$$

Since $\frac{a_c}{B_{gs}} \propto N^2 l_i$, it is clear from Equations (10.88) and (10.89) that for a given τ/g_{ef}, the increase in electric loading a_c or the reduction in fundamental air gap flux B_{gs} would require the number of turns N to rise, causing unfavorable enlargement in the size of the motor. If necessary, the ratio a_c/B_{gs} should be adjusted to a proper value in the design to keep the magnetizing reactance-related parameters within reasonable ranges.

For a salient pole synchronous motor, its magnetizing reactance can be decomposed into direct-axis armature reaction reactance X_{ad} and quadrature-axis armature reaction reactance X_{aq}:

$$X_{ad} = C_{ad} X_m \tag{10.90}$$

$$X_{aq} = C_{aq} X_m \tag{10.91}$$

where C_{ad} is the ratio of the direct-axis field fundamental amplitude to its maximum of the armature and C_{aq} is the quadrature-axis field fundamental amplitude to its maximum of the armature, both of which are related to the equivalent polar arc coefficient a_p. The per-unit direct-axis armature reaction reactance is given by

$$X_{adpu} = \frac{C_{ad} F_a}{C_f F_N} = \frac{C_{ad} C_m}{C_f} \frac{a_c}{B_{gs}} \tag{10.92}$$

where F_a is the fundamental armature reaction MMF, F_N is the air gap MMF required by producing a no-load EMF equal to the rated voltage, and C_f is the direct-axis armature reaction coefficient, that is, the ratio of the fundamental amplitude of the rotor magnetic field to its maximum. Similarly, the per-unit quadrature-axis armature reaction reactance is

$$X_{aqpu} = \frac{C_{aq} F_a}{C_f F_N} = \frac{C_{aq} C_m}{C_f} \frac{a_c}{B_{gs}} \tag{10.93}$$

For a nonsalient pole synchronous motor, its magnetizing reactance is equal to the armature reaction reactance due to the almost equal magnetic resistance of its direct and quadrature axes, that is,

$$X_m = X_a \tag{10.94}$$

The per-unit magnetizing reactance per phase of a nonsalient pole synchronous motor can be estimated by

$$X_{mdpu} = X_{mqpu} \approx \frac{F_a}{F_N} \tag{10.95}$$

2. Leakage reactance
   Depending on the magnetic links produced at different locations of the winding, the leakage reactance of winding is generally the sum of the slot leakage, tooth-top leakage, end-winding leakage, and harmonic leakage. The leakage reactance of per phase winding, X_l, is given by

$$X_l = 4\pi\mu_0 f l_i \frac{N^2}{pq}\Sigma\lambda \tag{10.96}$$

with

$$\Sigma\lambda = \lambda_s + \lambda_t + \lambda_e + \lambda_h \tag{10.97}$$

where λ_s, λ_t, λ_e, and λ_h are the specific permeances of slot leakage, tooth-top leakage, end-winding leakage, and harmonic leakage, respectively.

With the derivation process omitted, the per-unit leakage reactance in Equation (10.86) can be expressed by

$$X_{lpu} = \frac{\sqrt{2}\pi\mu_0 \Sigma\lambda}{K_{ws} mq} \frac{a_c}{B_{gs}} \tag{10.98}$$

where a_c is the electric loading when the armature current $I = I_N$ and B_{gs} is the amplitude of air gap fundamental magnetic flux density corresponding to the induced EMF equal to the rated phase voltage. Similar to the magnetizing reactance, the leakage reactance is also proportional to the ratio a_c/B_{gs} for given $\Sigma\lambda$ and q, according to Equation (10.98). If necessary, small adjustments can be made on q or on the design parameters related to $\Sigma\lambda$ to match the leakage reactance to its preset value; for a large adjustment, however, a change in a_c or B_{gs} or one of the major dimensions of the motor should be considered [10.3].

The slot leakage, tooth-top leakage, end-winding leakage, and harmonic leakage will be elaborated in detail here.

First, for single-layer full-pitch windings, by making the assumptions that the core reluctance is negligible, the current is evenly distributed on the cross section of the conductor, and leakage lines of force in the slot are parallel to the bottom of the slot, the slot leakage per phase, X_{ls}, is given by [10.3]

$$X_{ls} = 4\pi\mu_0 f l_i \lambda_s \frac{N^2}{pq} \tag{10.99}$$

where

q is the number of slots per pole per phase
λ_s is the specific permeance of slot leakage

In fact, λ_s is dependent on the geometry of the slot, as illustrated in Figure 10.11 [10.3]. λ_s can be calculated by

$$\lambda_s = \frac{h_0}{b_0} + 0.62 \tag{10.100}$$

for round semiclosed slots (Figure 10.11a);

$$\lambda_s = \frac{h_0}{b_s} + \frac{h_1}{3b_s} \tag{10.101}$$

for rectangular open slots, as shown in Figure 10.11b;

$$\lambda_s = \frac{h_0}{b_0} + \frac{h_2}{b_s} + \frac{h_3}{3b_s} + 0.785 \tag{10.102}$$

for bottle-shaped semiclosed slots, as shown in Figure 10.11c; and

$$\lambda_s = \frac{h_0}{b_0} + \frac{2h_1}{b_0 + b_{s1}} + \frac{h_2}{b_{s1}} + \frac{2h_3}{3(b_{s1} + b_{s2})} \tag{10.103}$$

for beaker-shaped semiclosed slots, as shown in Figure 10.11d.

FIGURE 10.11 Single-layer winding and dimensions of various slots. (a) Round semiclosed, (b) rectangular open, (c) bottle-shaped semiclosed, and (d) beaker-shaped semiclosed.

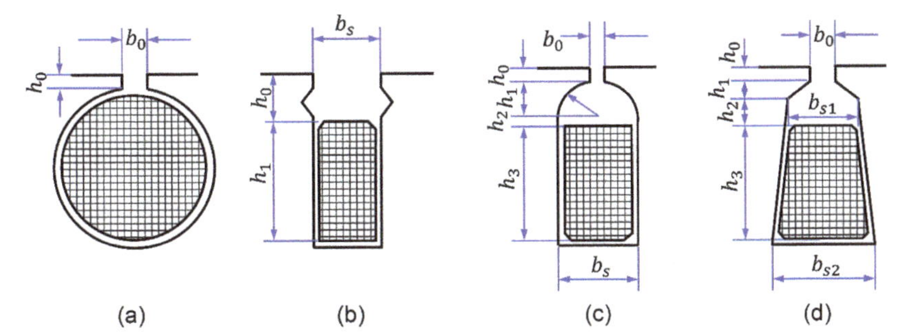

Courtesy of Suei Chen.

For double-layer full-pitch windings, by assuming a rectangular open slot with two coil sides on the upper and lower layers in it, as shown in Figure 10.12 [10.3], the slot leakage per phase, X_{ls}, is given by the same formula as Equation (10.98), with the specific permeance of slot leakage, λ_s, given by [10.3]

$$\lambda_s = \frac{1}{4}\left(\lambda_a + \lambda_b + 2\lambda_{ab}\right) \tag{10.104}$$

FIGURE 10.12 Double-layer full-pitch coil sides in a rectangular open slot.

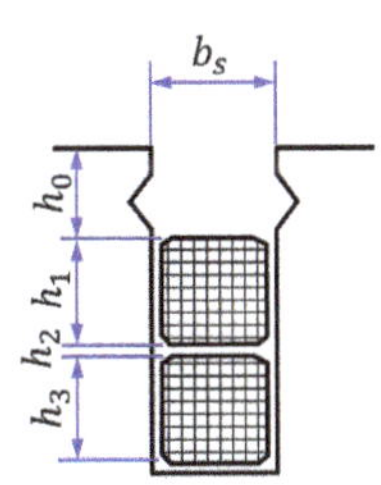

Courtesy of Suei Chen.

where λ_a and λ_b are the specific permeances corresponding to the self-inductances of upper and coil sides, respectively:

$$\lambda_a = \frac{h_0}{b_s} + \frac{h_1}{3b_s} \tag{10.105}$$

$$\lambda_b = \frac{h_0 + h_1 + h_2}{b_s} + \frac{h_3}{3b_s} \tag{10.106}$$

and λ_{ab} is the specific permeances corresponding to the mutual inductance of upper and coil sides:

$$\lambda_{ab} = \frac{h_0}{b_s} + \frac{h_1}{2b_s} \tag{10.107}$$

For double-layer short-pitch windings, the specific permeance of slot leakage, λ_s, is given by the same formula as Equation (10.99), with the specific permeance of slot leakage, λ_s, given by [10.3]

$$\lambda_s = \frac{1}{4}\left[\lambda_a + \lambda_b + (3\beta - 1)\lambda_{ab}\right] \tag{10.108}$$

where β is the ratio of the coil span to the pole pitch and $\beta < 1$ when the winding is short-pitched.

Second, the tooth-top leakage is caused by flux loops formed by some magnetic lines of force entering a tooth-top from another tooth-top, instead of passing through the air gap in the radial direction completely, which is a phenomenon mostly occurring in synchronous motors due to their relatively large air gaps.

For nonsalient synchronous motors, due to their uniform air gaps, the specific permeance of tooth-top leakage, λ_t, can be calculated by

$$\lambda_t = 0.0796\frac{g}{b_0} - 0.25\frac{b_0}{g}(1 - \sigma) + 0.2284 \tag{10.109}$$

where

$$\sigma = \frac{2}{\pi}\left[\tan^{-1}\frac{b_0}{2g} - \frac{g}{b_0}\left(1 + \frac{b_0^2}{4g^2}\right)\right] \tag{10.110}$$

For AC IMs, the tooth-top leakage is generally negligible due to their relatively small air gaps.

Third, the end-winding leakage per phase, X_{le}, is given by [10.3]

$$X_{le} = 4\pi\mu_0 f l_i \lambda_e \frac{N^2}{pq} \tag{10.111}$$

where λ_e is the specific permeance of end-winding leakage and

$$\lambda_e = 0.67(l_e - 0.64\tau)\frac{q}{l_i} \tag{10.112}$$

for ungrouped single-layer concentric windings, with l_e being the end length of half-turn coil;

$$\lambda_e = 0.47\left(l_e - 0.64\tau\right)\frac{q}{l_i} \tag{10.113}$$

for grouped single-layer concentric windings;

$$\lambda_e = 0.2 l_e \frac{q}{l_i} \tag{10.114}$$

for single-layer chain windings;

$$\lambda_e = 0.57\tau\left(\frac{3\beta - 1}{2}\right)\frac{q}{l_i} \tag{10.115}$$

for double-layer stacked windings [10.4]; and

$$\lambda_e = 0.57\tau\frac{q}{l_i} \tag{10.116}$$

for double-layer wave windings.

Although the aforementioned formulas are empirical ones that are derived with approximations and simplifications, they are valid for wound rotor windings as well. On the other hand, for squirrel cage windings, the specific permeance of end ring leakage is given by

$$\lambda_e = \frac{0.2523 Z_{rp}}{l_i}\left(\frac{l_e}{1.13} + \frac{D_r}{2p}\right) \tag{10.117}$$

where Z_{rp} is the number of rotor slots per pole, D_r is the diameter of the end ring, and l_e is the sum of the lengths of conductor bars protruding from the iron core on both ends, with $l_e = 0$ if the end rings cling to the iron core.

Fourth, although most of these harmonic waves are linked to the flux of rotor windings, they hardly produce any effective torque. It is, therefore, the EMF at fundamental frequency induced by the magnetic field of each harmonic wave that is deemed as leakage reactance drop, and the associated reactance is referred to as harmonic leakage reactance [10.3].

When multiphase symmetric currents are applied in the windings of a multiphase symmetric AC motor, a rotating magnetic field is produced in the air gap with the rotating speed of its νth order of harmonic waves given by

$$n_\nu = \frac{n_s}{\nu} \tag{10.118}$$

where n_s is the rotating speed of fundamental wave relative to the stator. The following analysis is made by making four assumptions: the permeability of ferromagnetic materials $\mu = \infty$; the currents in the coil sides of each slot concentrate on the centerline of the slot; the air gap is uniform and small, and the air gap field harmonics have radial components only, with the influence of the slot opening on each harmonic field being approximately calculated by the air gap coefficient; and the weakening effect of the current induced by each harmonic magnetic field in the other winding is negligible. Thus, the harmonic leakage corresponding to the νth order of harmonic waves can be expressed by

$$X_{lh\nu} = 4\mu_0 f l_i \frac{m\tau}{p g_{ef}} \left(\frac{N K_{w\nu}}{\nu} \right)^2 \tag{10.119}$$

where $K_{w\nu}$ is the winding factor corresponding to the νth order of harmonic waves. Hence, the total harmonic leakage corresponding to all orders of harmonic waves is

$$X_{lh} = 4\pi\mu_0 f l_i \lambda_h \frac{N^2}{pq} \tag{10.120}$$

with

$$\lambda_h = \frac{m}{\pi^2} \frac{q\tau}{g_{ef}} \Sigma\kappa \tag{10.121}$$

where

$$\Sigma\kappa = \Sigma \left(\frac{K_{w\nu}}{\nu} \right)^2 \tag{10.122}$$

Note that Equation (10.120) is valid for the stator windings of both AC IMs and synchronous motors, as well as for the rotor windings of wound rotor IMs.

10.2.5.3. Influence of the Skin Effect on Stator Winding Design

The skin effect refers to a tendency for AC to flow mostly near the outer surface of an electric conductor, such as metal wire, which is caused by the changing magnetic field resulting from the AC. As a result, the effective cross section of the conductor is reduced, and thus, the effective resistance is increased. When a single wire rod is placed within a slot, the current density is the highest at the slot opening and decreases gradually toward the bottom. Due to the skin effect, the electrical resistance of the wire rod increases since its AC resistance is larger than its DC resistance, while the slot leakage reactance decreases. Thus, a coefficient of AC resistance increase and a coefficient of leakage reactance decrease are often utilized to evaluate the influence of the skin effect on electrical resistance and reactance, respectively.

To minimize the energy loss caused by eddy currents, multiple thin wire rods are often utilized instead of a single large one, usually with a reduced thickness of less than 3–5 mm along the slot depth. If these thin wire rods are welded together at the ends of winding, a circulating current may develop between parallel strands due to the difference in the flux leakage-induced EMF among these rods. In order to prevent circulating current losses, the parallel thin wires in the slot are often woven into two rows for transposition, as shown in Figure 10.13 [10.3], making each thin wire take turns in all different positions in the slot. Except those made of such a single turn wire rod, a coil can be made of multiple turns of conductors connected in series as well, with each conductor often composed of several wires connected in parallel that are transposed by twisting their ends for the reduction of circulating current losses, as shown in Figure 10.14 [10.3].

FIGURE 10.13 Woven wire rods for transposition.

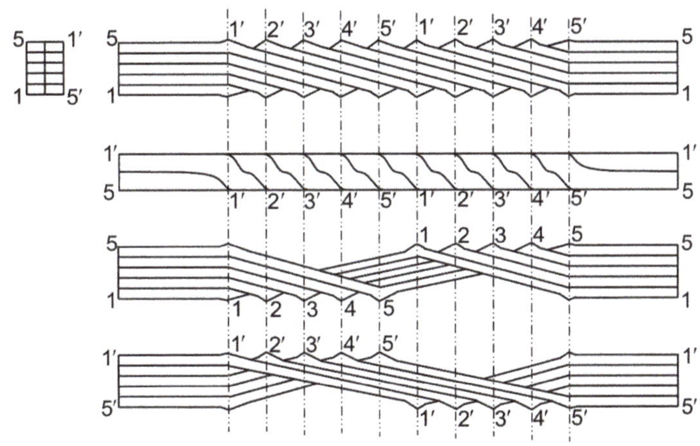

Courtesy of Suei Chen.

FIGURE 10.14 Twisted-end transposition for whole-coil windings.

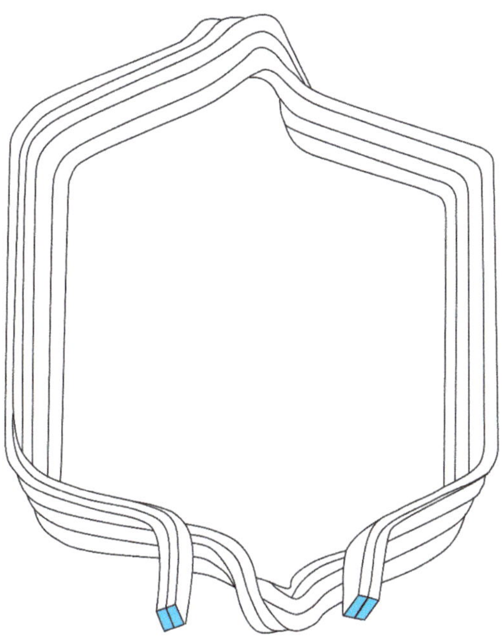

Courtesy of Suei Chen.

A simplified stator winding of an AC IM is discussed here to clarify the influence of the skin effect. Three assumptions are made: at first, the magnetic resistance of the iron core is negligible; next, the slot is infinitely deep with all edge effects ignored; and finally, the magnetic field vectors in the slot are parallel to the bottom of the slot everywhere.

If there are n pieces of wire of rectangular cross section stacked in a slot with only one piece along the width of the slot, as shown in Figure 10.15a, then the coefficient of AC resistance increase in the entire wire rods in the slot, K_r, is given by [10.3]

$$K_r = \frac{1}{n}\sum_{i=1}^{n} K_{ri} = \zeta(\theta) + \frac{n^2-1}{3}\varsigma(\theta) \tag{10.123}$$

where K_{ri} is the coefficient of AC resistance increase in the ith piece of wire,

$$\zeta(\theta) = \theta\frac{\sinh 2\theta + \sin 2\theta}{\cosh 2\theta - \cos 2\theta} \tag{10.124}$$

$$\varsigma(\theta) = 2\theta\frac{\sinh \theta - \sin \theta}{\cosh \theta + \cos \theta} \tag{10.125}$$

and

$$\theta = ah \tag{10.126}$$

with h being the thickness of the rectangular wire rod and

$$a = \sqrt{\frac{\omega\mu_0 b}{2\rho b_s}} = \sqrt{\frac{\pi f \mu_0 b}{\rho b_s}} \tag{10.127}$$

FIGURE 10.15 Simplified stator winding. (a) One piece along the x-axis and (b) m pieces along the x-axis.

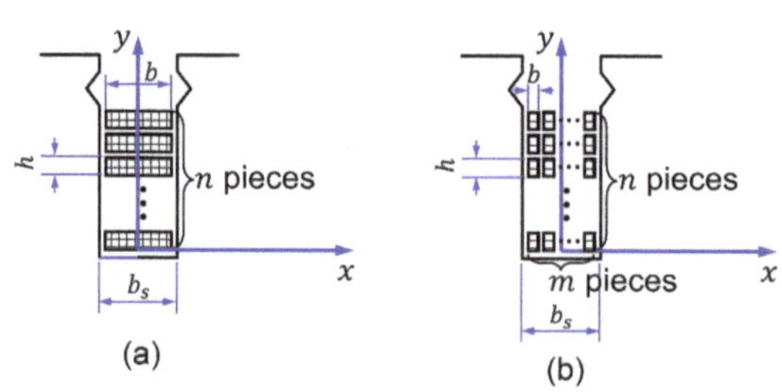

The coefficient of leakage reactance decrease, K_x, is given by

$$K_x = \frac{1}{n^2}\zeta'(\theta) + \frac{n^2-1}{n^2}\varsigma'(\theta) \tag{10.128}$$

where

$$\zeta'(\theta) = \frac{3}{2\theta}\frac{\sinh 2\theta - \sin 2\theta}{\cosh 2\theta - \cos 2\theta} \tag{10.129}$$

$$\varsigma'(\theta) = \frac{1}{\theta}\frac{\sinh\theta + \sin\theta}{\cosh\theta + \cos\theta} \tag{10.130}$$

In order to limit the eddy current loss and facilitate the weaving process, the thickness of the rectangular wire rod is usually quite small, so that $0 < \theta < 1$. Thus, Equation (10.123) can be simplified to

$$K_r = 1 + \frac{n^2 - 0.2}{9}\theta^4 \tag{10.131}$$

Similarly, Equation (10.128) becomes

$$K_x \approx 1 - \frac{1}{30}\theta^4 \tag{10.132}$$

which indicates that K_x is quite close to 1. Thus, the influence of the skin effect on the leakage reactance is often neglected for the stator winding of an AC motor.

When there are m pieces of wire along the width of the slot, as shown in Figure 10.15b, the coefficient of AC resistance increase is given by

$$K_r = \left(\frac{\pi\mu_0 fmnb}{3\rho b_s}\right)^2 h^4 \tag{10.133}$$

For a double-layer bar winding with only one turn per coil and each coil side composed of multiple stranded parallel conductors, transposition of conductors in slots is usually required. The coefficient of AC resistance increase for the bar conductors in slots can be calculated by Equation (10.133). However, the influence of the skin effect on end winding is not negligible. The coefficient of AC resistance increase can be estimated using Equation (10.133) as well, but n and b_s should be substituted by $n/2$ and b_{seq}, where b_{seq} is the equivalent width of end winding,

$$b_{seq} = b_e + 1.2h_e \tag{10.134}$$

where

b_e is the total width of each pole phase group at the end of the winding
h_e is the total height of end winding

With the coefficient of AC resistance increase at the end winding K_{re} taken into account, the total coefficient of AC resistance increase, K_{rtot}, can be computed by

$$K_{rtot} = \frac{K_r l_a + K_{re} l_e}{l_a + l_e} \tag{10.135}$$

where

l_a is the total length of the iron core
l_e is the half-turn length of end winding

For a double-layer coil winding of which each coil has multiple turns and each turn consists of several wires wound in parallel, the influence of the skin effect at the ends of winding is negligible. The wires in the slots are usually not transposed. When each turn of the coil is stacked along the direction of the slot height and the ends are welded together, it results in circulating current losses. In the case of $\theta \le 1$, the coefficient of AC resistance increase can be calculated by

$$K_r = \frac{760}{N_{ss}^2} h_c^4 \left(\frac{mbf\varepsilon}{b_s} \right)^2 + 1 \tag{10.136}$$

where

N_{ss} is the number of conductors connected in series per slot
h_c is the total height of copper wire in the slot
$\varepsilon = l_a/(l_a + l_e)$

Transposition can be made to end winding in order to reduce the circulating current losses.

In view of the fact that electrical resistance of a winding is inversely proportional to the thickness of the wire rod h and, meanwhile, the coefficient of resistance increase K_r is proportional to h^4 per Equation (10.131), there exists a critical thickness h_{cr} that yields the minimal AC resistance, with

$$h_{cr} = \frac{1.32}{a\sqrt{n}} \tag{10.137}$$

It must be ensured in the winding design that the thickness of the wire rod does not exceed the critical thickness h_{cr}, which corresponds to a coefficient of resistance increase $K_r \approx 1.33$ [10.3] for the wire rods in the slots.

10.2.5.4. Influence of the Skin Effect on Rotor Winding Design

For an AC IM with squirrel cage windings, the frequency of rotor current during start-up is much higher than that during normal operation. As a result, the rotor cage conductor bars are subjected to a prominent influence of the skin effect.

When rectangular slots are used, the coefficient of AC resistance increase is given by

$$K_r = \theta \frac{\sinh 2\theta + \sin 2\theta}{\cosh 2\theta - \cos 2\theta} \tag{10.138}$$

where $\theta = ah$, with h being the height of rectangular conductors on the cage. The coefficient of leakage reactance decrease is

$$K_x = \frac{3}{2\theta} \frac{\sinh 2\theta - \sin 2\theta}{\cosh 2\theta - \cos 2\theta} \tag{10.139}$$

10.2.5.5. Effect of Magnetic Saturation on Winding Design

Magnetic resistance in the iron core is often ignored in the calculation of electrical resistance and reactance, as discussed earlier. However, magnetic resistance always exists, particularly when a certain degree of magnetic saturation takes place somewhere in the magnetic circuit or leakage path of the motor. When magnetic saturation is taken into account, the magnetic reactance and armature reaction reactance will be less than what the aforementioned equations predict. A saturation factor can be utilized to reflect such an effect on the magnetizing reactance. When the iron core of an AC IM is magnetic saturated, its magnetizing reactance per phase armature winding, X_{ms}, is given by

$$X_{ms} = \frac{X_m}{K_s} \tag{10.140}$$

where X_m is the magnetizing reactance per phase armature winding when the iron core is unsaturated, as given by Equation (10.87), and K_s is the saturation factor.

For a salient pole synchronous motor, the armature reaction reactance of the direct axis and the quadrature axis are given, respectively, by

$$X_{ads} = k_d X_{ad} \tag{10.141}$$

$$X_{aqs} = k_q X_{aq} \tag{10.142}$$

where k_d and k_q are the factors related to the degree of magnetic saturation along the direct and quadrature axes, respectively. Besides, k_d and k_q are also dependent on the maximum-to-effective air gap length ratio g_{max}/g_{ef}, as illustrated in Figure 10.16. The saturation factor along the magnetic circuit is given by

$$K_s = \frac{F_g + F_t + F_y}{F_g} \tag{10.143}$$

where F_g, F_t, and F_y are the magnetic potential drops at the air gap, teeth, and yokes, respectively.

FIGURE 10.16 k_d and k_q versus K_s for various g_{max}/g_{ef}

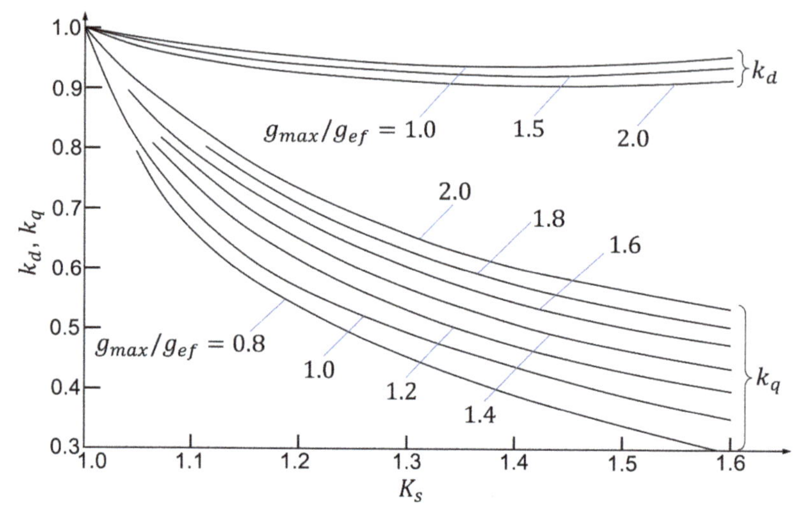

Courtesy of Suei Chen.

10.2.5.6. Types of Winding

Many different types of winding have been proposed, with each having its unique characteristics. For AC electric motors, their windings can generally be classified into two types, namely, lap winding and wave winding. Regardless of the winding type, the following design effects are generally desired: the waveforms of EMF and MMF are as close to sinusoidal as possible and the corresponding harmonic contents are kept as small as possible; the EMF and MMF are symmetrical, and the resistance and reactance are balanced; the winding utilization is high so that a larger fundamental EMF and MMF are obtained for a given amount of copper usage for the winding, while the amount of copper usage and the copper consumption is minimized for given EMF and MMF; and reliable insulation, high mechanical strength, good heat dissipation, and easy to manufacture.

In the lap winding, conductors are joined in such a manner that their parallel paths and poles are equal in number. In other words, each pole has a parallel path of conductors associated with it; in contrast, in the wave winding, the conductors are connected to two parallel paths irrespective of the number of poles, as shown in Figure 10.17 [10.5]. Each coil, made of either a single turn or multiple turns, has two coil sides and two ends. The two sides can effectively cut magnetic lines and induce an electromotive force, while the two ends simply function to connect successive conductors together in series. The winding pitch y is defined as the distance between two successive conductors that are directly connected together. In comparison to the pole pitch τ, a full pitch coil refers to the case if $y = \tau$; thus, the coil is short-pitched if $y < \tau$ or long-pitched if $y > \tau$.

FIGURE 10.17 Two basic types of winding. (a) Lap winding and (b) wave winding.

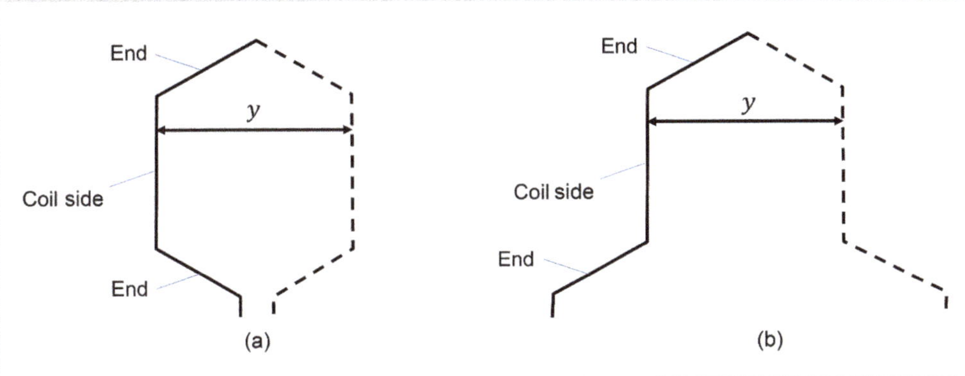

There can be single-layer or double-layer winding placed in a slot. In general, double-layer winding is more advantageous than single-layer winding in EMF waveform, harmonics, copper material savings, cost, ease of manufacture, and leakage reactance, despite increased need for insulation material. An example of the double-layer 3-phase lap winding of a 4-pole and 36-slot AC motor is taken herein to elaborate the design principle. The slot electrical potential star vectogram for this motor is illustrated in Figure 10.18 [10.5], which is actually the same as the electrical potential star vectogram for the upper-layer coil side. As an important design step, phase division refers to assigning the conductors in each slot to different phases of winding. It is generally desired to maximize the electrical potential of each phase and make the electrical potentials symmetrical among the three phases. To make the slot electrical potentials identical among three phases, the number of slots per phase per pole for all the three phases must be equal. In this example, since $Z = 36$, $m = 3$, $p = 2$, the number of slots per phase per pole $q = Z/(2mp) = 36/(2 \times 3 \times 2) = 3$. Thus, the result of phase division is listed in Table 10.2.

FIGURE 10.18 Slot electrical potential star vectogram.

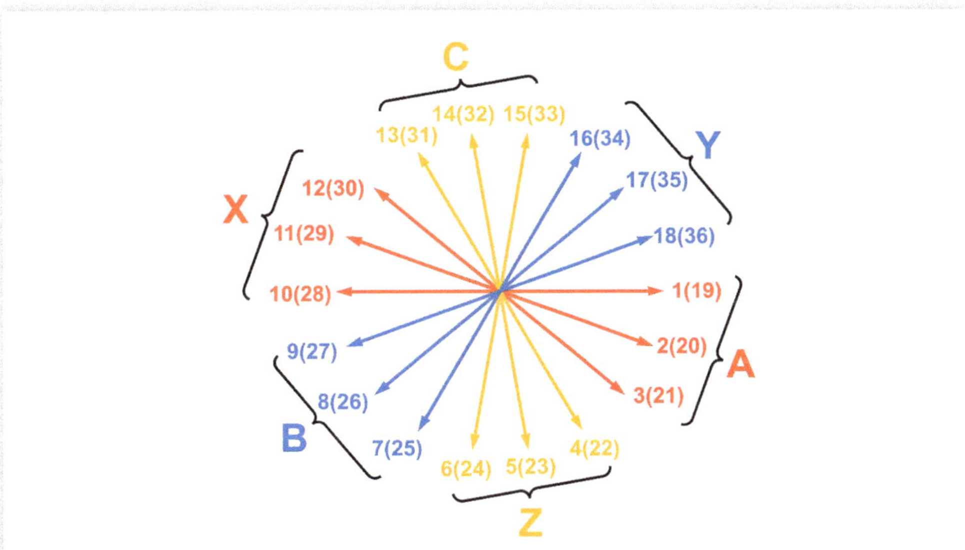

TABLE 10.2 Result of phase division for the 3-phase, 4-pole, 32-slot winding.

	Slot no. per phase					
Pole pair	**A**	**Z**	**B**	**X**	**C**	**Y**
1	1,2,3	4,5,6	7,8,9	10,11,12	13,14,15	16,17,18
2	10,20,21	22,23,24	25,26,27	28,29,30	31,32,33	34,35,36

In this phase division scheme, the surface of the armature under each pole is divided into three equal parts, with each phase occupying one equal part, called a phase belt. Here, since the width of each part is equivalent to a 60° electrical angle, it is, therefore, a 60° phase belt. Thus, a winding expansion diagram can be drawn, as shown in Figure 10.19 [10.5], in which the armature is cut axially from the center of the teeth and spread out on a plane, with the magnetic poles on the top. Note that only the coils for phase A are displayed for clarity. Let Z_p denote the number of branches in parallel connection of phase winding, then the total number of turns per parallel branch is given by $N_p = 2p \times q \times N_c/Z_p$, where N_c is the number of turns per coil. Obviously, the total number of turns per parallel branch N_p decreases with an increase in the number of branches in parallel connection Z_p, leading to a drop in the voltage that the winding can withstand and a rise in corresponding current.

FIGURE 10.19 Winding expansion diagram for the 3-phase, 4-pole, 32-slot lap winding, with only the coils for phase A displayed.

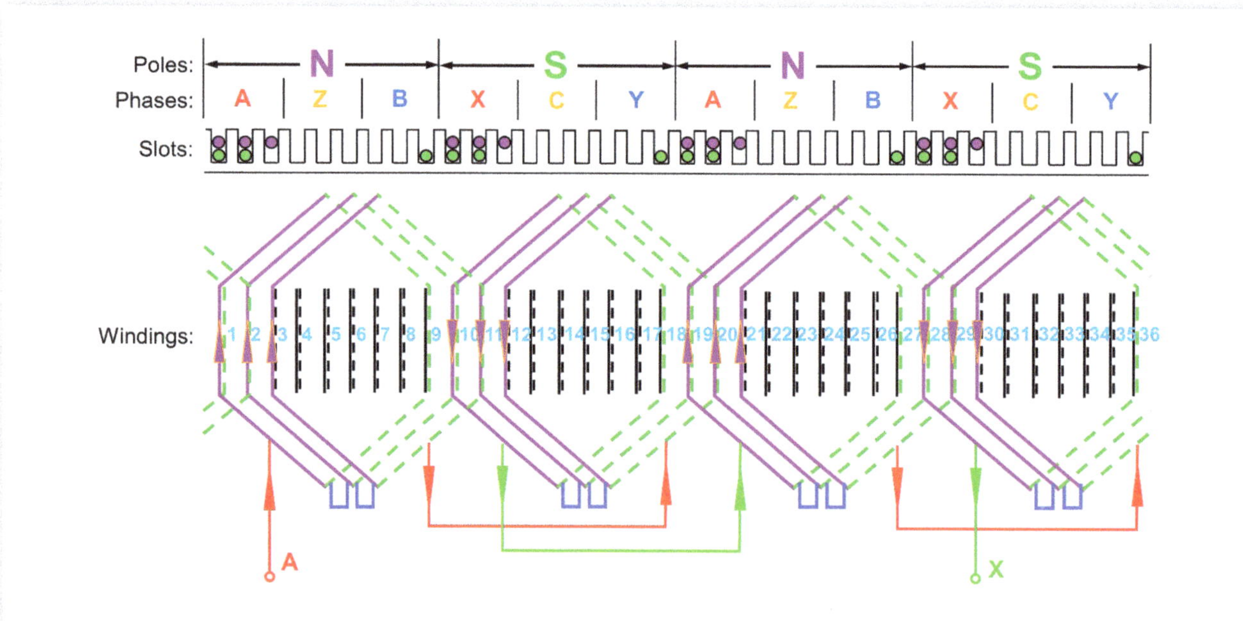

Courtesy of Lee Han.

10.2.6. Losses, Temperature Rise, and Cooling

To design a highly efficient electric motor, the various losses in it must be controlled to their minima since the efficiency of an electric motor in the driving mode under rated load can be expressed by

$$\eta_m = \left(1 - \frac{\sum P_L}{P_{Ni}}\right) \times 100\% \tag{10.144}$$

where

$\sum P_L$ is the sum of all the losses in the motor under rated power
P_{Ni} is the input power to the motor under rated power

When the motor operates in the electric generation mode, its efficiency can be calculated by

$$\eta_g = \left(1 - \frac{\sum P_L}{P_{No} + \sum P_L}\right) \times 100\% \tag{10.145}$$

where P_{No} is the output power of the motor under rated power.

The losses in electric motors can be classified into no-load losses and load losses. The no-load losses come from three sources: iron losses produced by the varying magnetic field in the iron core and permanent magnets; no-load stray losses in the iron core related to the flux variation, which is caused by the interaction of the slots in the stator and the rotor with the iron core or teeth on the other side of the air gap; and mechanical losses involving ventilation and bearings. These no-load losses can be measured in no-load tests, which usually change very little with the load increasing from zero to the rated value. On the other hand, load losses include the electrical losses resulting from electric current passing through winding conductors, and the load stray losses in the windings, iron core, and structural parts mainly due to the leakage field produced by the electric current in the stator or rotor. The load losses can be measured by short-circuit tests in synchronous motors [10.3].

10.2.6.1. Losses in Electric Motors

The losses in electric motors are elaborated, and the formulas for their calculation are presented in this section.

1. Iron losses

   These losses are actually the hysteresis and eddy current losses caused by either alternating magnetization or rotating magnetization in the iron core and permanent magnets.

   Hysteresis-related losses p_h per unit mass of the iron core can be estimated by

$$p_h = \sigma_h f B^a \approx \sigma_h f B^2 \tag{10.146}$$

   where

   σ_h is a constant depending on material property, as listed in Table 10.3 [10.3]
   f is the frequency of alternating magnetization
   B is the amplitude of flux density
   a is an exponent typically in a range of 1.6 – 2.2, which can be approximated as 2

TABLE 10.3 Material constants σ_h and σ_e.

Type of steel sheet	Thickness (mm)	$p_{10/50}$	σ_h W/(kg·Hz·T²)	σ_e W/(kg·Hz²·T²)
Low silicon content	0.5	2.80	0.045	0.00022
Medium silicon content	0.5	2.20	0.036	0.00016
High silicon content	0.5	2.00	0.029	0.00022

Courtesy of Suei Chen.

The eddy current losses p_e per unit mass of the iron core is calculated by

$$p_e = \sigma_e \left(fB \right)^2 \tag{10.147}$$

with

$$\sigma_e = \frac{\pi^2 \delta_i^2}{6\rho\rho_i} \tag{10.148}$$

where

ρ and ρ_i are the resistivity and density of the silicon steel sheet, respectively
δ_i is the thickness of silicon steel sheet
σ_e is a material constant related to its properties and thickness, as listed in Table 10.3

Note that in Equation (10.147), the reaction of eddy current on magnetic field is ignored. Such a reaction becomes prominent when the frequency is high or the sheet is thick since the flux is no longer uniformly distributed in the cross section of the sheet. As a result, the hysteresis losses would rise and the eddy current losses would drop.

Combining Equations (10.146) and (10.147), a specific iron loss can be obtained by

$$p_{he} = \sigma_h fB^2 + \sigma_e \left(fB \right)^2 \tag{10.149}$$

For the convenience of calculation, the specific iron loss is often expressed as

$$p_{he} \approx p_{10/50} B^2 \left(\frac{f}{50} \right)^{1.3} \tag{10.150}$$

where $p_{10/50}$ is the iron loss per unit mass when $B = 1$ T and $f = 50$ Hz, as listed in Table 10.2.

Thus, the total iron loss is given by

$$P_i = k_i p_{he} m_i \tag{10.151}$$

where k_i is an empirical factor that accounts for the increased losses due to the manufacturing imperfections, nonuniformity of flux density, nonsinusoidal variation in flux density with time, the difference between alternating magnetization and rotating magnetization of the iron core, and so forth; m_i is the mass of the silicon steel sheet subjected to the influence of alternating magnetization and rotating magnetization.

When the yoke iron loss for induction or synchronous motors is calculated using Equation (10.151), the specific iron loss p_{he} can be estimated using Equation (10.150), with B being the maximum flux density in the stator or rotor yoke; the empirical factor $k_i = 1.5$ for rated power $P_N < 100$ kVA and $k_i = 1.3$ for rated power $P_N \geq 100$ kVA; and m_i is the total mass of all yokes.

Similarly, when the iron loss in teeth for induction or synchronous motors is calculated using Equation (10.151), the specific iron loss p_{he} can be estimated using Equation (10.150) with B being the average flux density on the length of the tooth flux path; the empirical factor $k_i = 1.8$ for AC IMs, $k_i = 2.0$ for rated power $P_N < 100$ kVA, and $k_i = 1.7$ for rated power $P_N \geq 100$ kVA for synchronous motors; and m_i is the total mass of all teeth.

For PMSMs, the permanent magnets in the rotor spin together with the stator magnetic field at the synchronous speed. However, due to the cogging effect, nonsinusoidal distribution of winding MMF, and harmonic currents in the winding, eddy currents can occur in the permanent magnets. Although eddy current-related losses are generally less than iron losses in the iron core or winding losses, it cannot be ignored in many cases due to the poor heat dissipation condition around the rotor. The eddy current loss of permanent magnets, p_{pm}, can be estimated by

$$p_{pm} = \int_V \frac{J^2}{\sigma} \mathrm{d}V \tag{10.152}$$

where

J is the eddy current density

σ is the electrical conductivity of permanent magnets

V is the volume of permanent magnets

2. No-load stray losses

No-load stray losses in the iron core are caused by the harmonic flux in the air gap because of either the nonuniformity air gap permeance due to the presence of slots in the iron core or the harmonics in the spatial distribution of no-load field MMF. Consequently, eddy current is induced on the magnetic pole surface due to its relative motion with respect to the harmonic field. In addition, eddy current and hysteresis are also resulted in the teeth by their relative motion with respect to the harmonic field.

For synchronous motors, the eddy current loss per unit area of the magnetic pole surface, p_{ns}, is calculated by

$$p_{ns} = 0.5k_0 (nZ)^{1.5} (tB_0)^2 \tag{10.153}$$

where p_{ns} is in W/m^2, n is the armature speed relative to the magnetic pole in rpm, Z is the number of slots in the armature, t is the tooth pitch in m, B_0 is the amplitude of flux density in T, and k_0 is a coefficient with a value dependent on the sheet thickness of laminated magnets:

$$k_0 = \begin{cases} 4.6 & \text{for } 1.0\,\text{mm thick sheet} \\ 6.0 & \text{for } 1.5\,\text{mm thick sheet}. \\ 8.6 & \text{for } 2.0\,\text{mm thick sheet} \end{cases}$$

Note that Equation (10.153) is valid if the flux density is of a sinusoidal distribution within the range of pole pitch.

The total eddy current loss can be obtained by

$$P_{ns} = p_{ns} A_p \tag{10.154}$$

where A_p is the total area of the magnetic pole surface.

For AC IMs, both the iron cores of the stator and rotor are slotted and made of laminated magnets. The eddy current loss per unit rotor surface caused by an air gap permeance tooth harmonic field due to the presence of stator slots can be computed using Equation (10.153), with

$$B_0 = \beta C_s B_g \tag{10.155}$$

where C_s is the Carter coefficient, B_g is the air gap flux density, and β is a coefficient that is a function of b_0/g, with b_0 being the width of slot opening. Additionally, the coefficient k_0 is mainly dependent on the silicon content of laminated magnets and processing quality of the iron core:

$$k_0 = \begin{cases} 1.5 & \text{for low silicon content} \\ 3-5 & \text{for low silicon content, high quality} \\ 0.7 & \text{for high silicon content} \\ 1.5-3 & \text{for high silicon content, high quality} \end{cases}.$$

In AC IMs, the eddy current and hysteresis losses resulted in the stator teeth by their relative motion with respect to the harmonic field, p_{ps}, can be estimated by

$$p_{ps} \approx 0.07 m_{ts} \left(n Z_r B_{ps}\right)^2 \times 10^{-6} \tag{10.156}$$

where

- p_{ps} is the losses in the stator teeth in W
- n is the rotor speed in rpm
- m_{ts} is the mass of stator teeth in kg
- Z_r is the number of rotor teeth
- B_{ps} is the amplitude of flux density pulsation in stator teeth in T

Similarly, the eddy current and hysteresis losses in the rotor teeth, p_{pr}, can be estimated by

$$p_{pr} \approx 0.07 m_{tr} \left(n Z_s B_{pr}\right)^2 \times 10^{-6} \tag{10.157}$$

where

- p_{pr} is the losses in the rotor teeth in W
- n is the rotor speed in rpm
- m_{tr} is the mass of rotor teeth in kg
- Z_s is the number of stator teeth
- B_{pr} is the amplitude of flux density pulsation in rotor teeth in T

3. Mechanical losses

   Frictional losses in a roller bearing, p_b, is given by

$$p_b = 0.15\frac{Fv}{d}\times 10^{-5} \tag{10.158}$$

   where

   p_b is in W
   F is the load acting the bearing in N
   v is the circumferential velocity of the center of the roller in m/s
   d is the roller diameter in m

   For air-cooled motors, ventilation losses, p_v, can be estimated by

$$p_v = 1.75 q_v v_f^2 \tag{10.159}$$

   where

   p_v is in W
   q_v is the volumetric flow rate in m^3/s
   v_f is the circumferential velocity at the outer diameter of the cooling fan in m/s

4. Electrical losses

   Per Joule–Lenz law, electrical losses in the winds of an AC motor, p_{el}, is given by

$$p_{el} = mI^2R \tag{10.160}$$

   where

   m is the number of phases in the winding
   I is the phase current
   R is the DC resistance in each phase winding

   Note that Equation (10.160) is based on the assumption that all phase currents and resistances are equal.

5. Load stray losses

   For salient pole synchronous motors, load stray losses include the following four contributors:

   First, the additional loss in the stator winding during short circuit due to leakage field should be calculated. The fact that AC resistance in the stator winding is larger than DC resistance due to the skin effect leads to an additional loss of

$$p_{sk} = \left(K_{rtot} - 1\right) p_{el} \tag{10.161}$$

   where

   K_{rtot} is the total coefficient of AC resistance increase
   p_{el} is the DC electrical loss obtained from Equation (10.160)

Second, the additional losses in the metal parts near end winding during short circuit due to leakage field can be estimated using an empirical formula:

$$p_{mp} = 16.1 D_{si} f^2 \left(\frac{\tau A_s}{10^5} \right)^{2.5} \tag{10.162}$$

where

D_{si} is the inner diameter of the stator iron core in m
τ is the pole pitch in m
A_s is the current per unit length of the stator winding in A/m

Third, the sum of the losses on the surface of the rotor pole due to all field harmonics of the stator winding for a three-phase motor is given by

$$p_{fh} = \frac{2.1}{\sqrt[3]{q}} \left(\frac{k_\beta X_{adpu}}{C_s - 1} \right)^2 P_{ns} \tag{10.163}$$

where q is the number of slots per pole per phase; k_β is a coefficient that depends on the short pitch factor of winding, K_p, as listed in Table 10.4 [10.3]; X_{adpu} is the direct-axis armature reactive reactance; C_s is the Carter coefficient; and P_{ns} is the total eddy current loss per Equation (10.154).

TABLE 10.4 Relationship between k_β and K_p.

K_p	0.75	0.80	0.85	0.88	0.90	0.92	0.94	0.96	0.98	1.0
k_β	0.045	0.052	0.055	0.050	0.045	0.035	0.02	0.01	0.02	0.055

Courtesy of Suei Chen.

Fourth, if the air gap is uniform, the additional losses caused by eddy currents on the surface of magnetic poles by stator tooth harmonic potential can be estimated by the following empirical formula:

$$p_{th} = 1.2 \left[\frac{p X_{adpu}}{Z(C_s - 1)} \right]^2 P_{ns} \tag{10.164}$$

where

p is the number of pole pairs
Z is the number of slots in the stator
P_{ns} is the total eddy current loss per Equation (10.154)

On the other hand, for AC IMs, it is usually assumed that the load stray losses are 0.5% of the operating power.

10.2.6.2. Temperature Rise in Electric Motors

All the losses discussed in the last section are eventually dissipated into heat, that is, thermal energy, which is a highly disordered form of energy. Heat generation causes the temperature of all parts in an electric motor to rise, creating a temperature differential between the parts and their surrounding medium. It is critical to ensure that temperatures in all parts with heat generation do not exceed the acceptable upper limits.

1. Calculation of steady-state temperature rise
   In electric motor design, the temperature rises of the stator winding and iron core, as well as permanent magnets in the rotor, are of utmost importance. Due to its steady-state nature, the highest temperature rise of a particular component always has a certain relationship with the average temperature rise. Hence, in many cases, only the average temperature rise is calculated for evaluation.

   The average temperature rise can be estimated using the thermal circuit method. An example of an interior permanent magnet (IPM) type of the water-cooled synchronous motor is taken here to give a general idea about this method. As shown in Figure 10.20, the IPM has four heat generation sources, namely, stator winding, stator iron core, rotor iron core, and the permanent magnets within it. The heat generated in permanent magnets within the rotor, P_{rm}, is dissipated either into the air in the housing through the ends of magnets or to the rotor iron core through their interfaces with slots. The heat generated in rotor iron core, P_{rc}, is transferred either into the air within the housing through the ends or into the air gap through its outer surface and then subsequently into the stator iron core. The heat generated in the stator winding, P_{sw}, is dissipated either into the air in the housing through the end windings or to the stator iron core, then housing, and finally the ambient air. The heat generated in the stator iron core, P_{sc}, is transferred to the air in the housing through its ends or to the housing and subsequently to the ambient air. To simplify the problem, the following assumptions are made: the thermal conductivities of windings, iron cores, and motor housing are infinite, and all of the heat generated within the enclosure of motor housing is dissipated into the coolant and ambient air. Thus, the thermal circuit diagram of such a motor is shown in Figure 10.21, and the following equations can be written accordingly:

FIGURE 10.20 Schematic of an IPM.

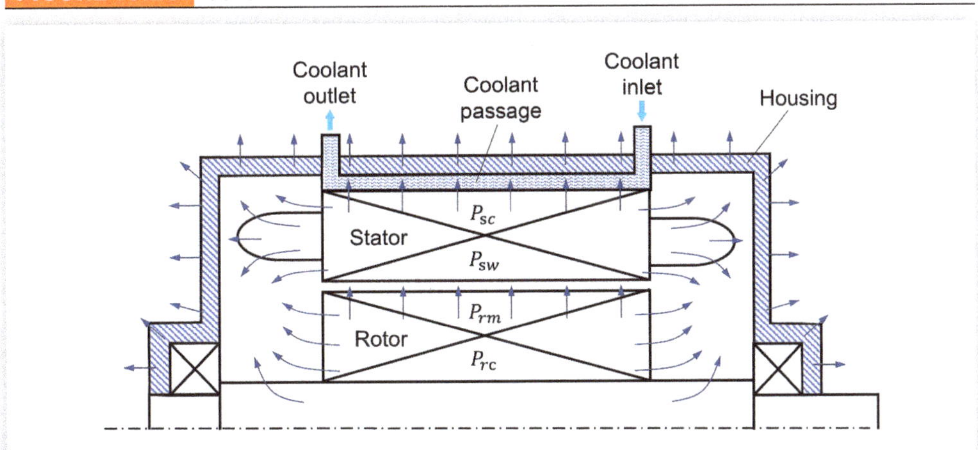

FIGURE 10.21 Thermal circuit diagram of an IPM.

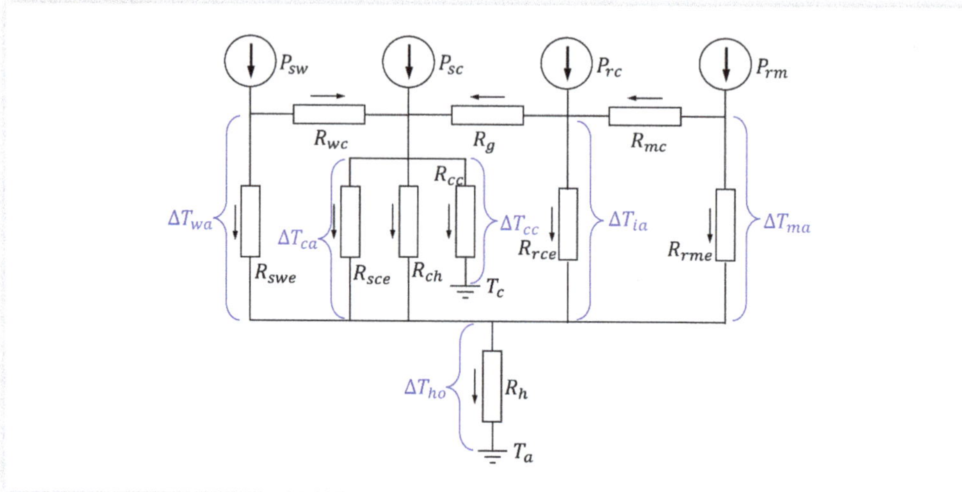

$$\begin{cases} \dfrac{\Delta T_{wa}}{R_{swe}} + \dfrac{\Delta T_{wa} - \Delta T_{ca}}{R_{wc}} = P_{sw} \\ \dfrac{\Delta T_{cc}}{R_{cc}} + \dfrac{\Delta T_{ca}}{\dfrac{1}{R_{sce}} + \dfrac{1}{R_{ch}}} - \dfrac{\Delta T_{wa} - \Delta T_{ca}}{R_{wc}} - \dfrac{\Delta T_{ia} - \Delta T_{ca}}{R_{g}} = P_{rc} \\ \dfrac{\Delta T_{ia}}{R_{rce}} + \dfrac{\Delta T_{ia} - \Delta T_{ca}}{R_{g}} - \dfrac{\Delta T_{ma} - \Delta T_{ia}}{R_{mc}} = P_{rc} \\ \dfrac{\Delta T_{ma}}{R_{rma}} + \dfrac{\Delta T_{ma} - \Delta T_{ia}}{R_{mc}} = P_{rm} \\ T_a + \Delta T_{ho} = T_c + \Delta T_{cc} - \Delta T_{ca} \end{cases} \quad (10.165)$$

where T_a and T_c are the ambient air temperature and the air temperature in the motor housing, respectively. The thermal resistances in Equation (10.165) can be calculated as follows.

Thermal resistance between end winding and air, R_{swe}, is given by

$$R_{swe} = \frac{\delta_{ei}}{k_{ei} A_{we}} + \frac{1}{h_{ei} A_{we}} \quad (10.166)$$

where

- δ_{ei} is the insulation thickness of the end winding
- k_{ei} is the thermal conductivity of end-winding insulation
- h_{ei} is the heat transfer coefficient of the end-winding surface
- A_{we} is the end-winding surface area for heat dissipation

The thermal resistances between the ends of the iron core and air in the housing, R_{sce} and R_{rce}, can be estimated using an equation similar to Equation (10.166).

Thermal resistance between the iron core and motor housing, R_{ch}, is calculated by

$$R_{ch} = \frac{\delta_{ch}}{k_{air} A_{ch}} \tag{10.167}$$

where

δ_{ch} is the gap between the iron core and motor housing
k_{air} is the thermal conductivity of air
A_{ch} is the heat transfer area between the iron core and motor housing

It is often approximated that $\delta_{ch} = 0.075$ mm and $k_{air} = 0.025$ W/m · °C.

Thermal resistance between the iron core and coolant passage, R_{cc}, is estimated by

$$R_{cc} = \frac{1}{h_{cc} A_{cc}} \tag{10.168}$$

where h_{cc} is the heat transfer coefficient between the iron core and coolant, with the thermal resistance between the iron core and inner surface of coolant passage wall ignored, and A_{cc} is the heat transfer surface area between the iron core and coolant.

Thermal resistance between the permanent magnets and air in the housing, R_{rme}, is estimated by

$$R_{rme} = \frac{1}{h_{ma} A_{ma}} \tag{10.169}$$

where

h_{ma} is the heat transfer coefficient between the magnets and air in the housing
A_{ma} is the corresponding heat transfer area

Thermal resistance between the motor housing and ambient air, R_h, is given by

$$R_h = \frac{1}{h_{ho} A_{ho}} \tag{10.170}$$

where

h_{ho} is the heat transfer coefficient between the exterior surface of the motor housing
A_{ho} is the exterior heat transfer area of the motor housing

Thermal resistance between the stator winding and iron core, R_{wc}, is estimated by

$$R_{wc} = \frac{\delta_{wc}}{k_{wc} A_{wc}} \tag{10.171}$$

where

δ_{wc} is the total thickness of the insulation layer between the winding and iron core
k_{wc} is the composite thermal conductivity
A_{wc} is the interface area between insulation and the iron core

By assuming that there are n insulation layers without any air in between, the composite thermal conductivity is given by

$$k_{wc} = \frac{\sum_{i=1}^{n} \delta_i}{\sum_{i=1}^{n} \frac{\delta_i}{k_i}} \tag{10.172}$$

where δ_i and k_i are the thickness and thermal conductivity of the ith layer of insulation, respectively.

Thermal resistance across the air gap between the stator and rotor, R_g, is calculated by

$$R_g = \frac{g}{k_{air} N_u A_{si}} \tag{10.173}$$

where

- g is the air gap length
- A_{si} is the inner surface area of the stator iron core
- N_u is the Nusselt number at the air gap

N_u can be obtained using the following empirical equation [10.3]:

$$N_u = \begin{cases} 0.23\left(\frac{D_{si}}{2g}\right)^{-0.25}\left(\frac{gV_r}{v}\right)^{0.5} & \text{when } 400 < \frac{gV_r}{v} < 10{,}000 \\ 0.0.36\left(\frac{D_{si}}{2g}\right)^{-0.25}\left(\frac{gV_r}{v}\right)^{0.7} & \text{when } 10{,}000 < \frac{gV_r}{v} < 50{,}000 \end{cases} \tag{10.174}$$

where D_{si} is the inner diameter of the stator iron core, V_r is the peripheral velocity of the rotor, and v is the dynamic viscosity of air, with $v = 0.0157 \times 10^{-3}$ m²/s at 20°C under atmospheric pressure.

Thermal resistance between the permanent magnets and rotor iron core, R_{mc}, is estimated by

$$R_{mc} = \frac{\delta_{mc}}{k_{mc} A_{mc}} \tag{10.175}$$

where

- δ_{mc} is the thickness of the insulation layer between the magnets and the iron core
- k_{mc} is the thermal conductivity of the insulation material
- A_{mc} is the total area between the interface of magnets and the iron core

With all the thermal resistances in Equation (10.165) obtained using Equations (10.166) through (10.175), five unknows ΔT_{wa}, ΔT_{ca}, ΔT_{cc}, ΔT_{ia}, and ΔT_{ma} can be found by simultaneously solving the group of five equations in Equation (10.165).

It is worth noting that three-dimensional heat conjugated computational fluid dynamics (CFD) models are frequently used in engineering practice for more precise results on the temperature rise prediction.

2. Calculation of transient temperature rise
To prevent excessive temperature rise in critical components, for example, iron cores or windings, during operating conditions such as short-time overtaking at high speed with peak power, the transient temperature rise often needs to be estimated in the design of electric motors. When the transient process is completed in minutes, only the temperature rise in winding is of interest since the temperature rise in the iron core is much slower. Thus, the winding can be simplified as a single isothermal heating element. Nevertheless, when the transient process takes tens of minutes, the temperature rise in both winding and iron core should be considered. The winding and iron core are often treated as separate isothermal heating elements that influence each other.

When the winding is assumed to be a single isothermal heating element, its temperature rise ΔT as a function of time t can be expressed by [10.3]

$$\Delta T = \Delta T_{\infty}\left(1 - e^{-\frac{t}{T}}\right) \tag{10.176}$$

where
ΔT_{∞} is the steady-state temperature rise
T is the time constant

Furthermore, we have

$$\Delta T_{\infty} = \frac{p}{hA} \tag{10.177}$$

and

$$T = \frac{cm}{hA} \tag{10.178}$$

where
p is the loss in the winding
h is the heat transfer coefficient between the winding surface and the surrounding medium
A is the winding surface area exposed to the surrounding medium
c is the specific heat capacity of the winding material
m is the winding mass

Note that the time constant T is dependent on the specific heat capacity c and cooling capacity hA and has nothing to do with the loss p, which typically lies in the range of 10–150 min. Moreover, when the process is over, the winding starts to cool down, with the temperature drop being

$$\Delta T = \Delta T_0 e^{-\frac{t}{T_c}} \tag{10.179}$$

where

ΔT_0 is the initial temperature rise

T_c is the time constant of the cooling process

In general, T_c is about 2–5 times longer than the time constant of the heating process T [10.3].

When the winding and iron core are each assumed to be an isothermal heating element, the temperature rise in the winding ΔT_w as a function of time t can be expressed by [10.3]:

$$\Delta T_w = K_{w1}\left(1 - e^{-\frac{t}{T_1}}\right) + K_{w2}\left(1 - e^{-\frac{t}{T_2}}\right) \tag{10.180}$$

where

$$K_{w1} = \frac{T_1 \Delta T_{w\infty} - \frac{p_w}{m_w c_w} T_1 T_2}{T_1 - T_2} \tag{10.181}$$

$$K_{w2} = \frac{\frac{p_w}{m_w c_w} T_1 T_2 - T_2 \Delta T_{w\infty}}{T_1 - T_2} \tag{10.182}$$

and

$$T_1 = \frac{1}{\frac{E}{2} - \frac{\sqrt{E^2 - 4F}}{2}} \tag{10.183}$$

$$T_2 = \frac{1}{\frac{E}{2} + \frac{\sqrt{E^2 - 4F}}{2}} \tag{10.184}$$

with

$$E = \frac{m_w c_w \left(h_{wc} A_{wc} + h_c A_c\right) + m_c c_c \left(h_w A_w + h_{wc} A_{wc}\right)}{m_w c_w m_c c_c} \tag{10.185}$$

$$F = \frac{\left(h_{wc} A_{wc} + h_w A_w\right)\left(h_c A_c + h_{wc} A_{wc}\right) - h_{wc}^2 A_{wc}^2}{m_w c_w m_c c_c} \tag{10.186}$$

Note that all the quantities in Equations (10.180) through (10.186) follow the same nomenclature as Equations 10.176 through (10.178), with the subscripts w and c denoting the winding and the iron core, respectively, and subscript wc denoting the interface between the winding and the iron core. Similarly, the temperature rise in the iron core, ΔT_c, as a function of time t can be expressed by [10.3]:

$$\Delta T_c = K_{c1}\left(1 - e^{-\frac{t}{T_1}}\right) + K_{c2}\left(1 - e^{-\frac{t}{T_2}}\right) \tag{10.187}$$

where

$$K_{c1} = \frac{T_1 \Delta T_{c\infty} - \frac{p_c}{m_c c_c} T_1 T_2}{T_1 - T_2} \tag{10.188}$$

$$K_{c2} = \frac{\frac{p_c}{m_c c_c} T_1 T_2 - T_2 \Delta T_{c\infty}}{T_1 - T_2} \tag{10.189}$$

To further simplify the calculation, the mass of winding insulation can be assumed to be 10–20% of winding mass, and the structural parts between the iron core and shaft can be approximated as 10–20% of iron core mass [10.3].

10.2.6.3. Cooling

Although traction motors can be either air-cooled or liquid-cooled, most of them are nowadays liquid-cooled though to maximize their power density.

Air-cooled motors are sometimes used in small EVs for their simplicity and low cost. Based on natural convection or forced convection, the air is guided into the interior or only blew through the exterior of the motor in the radial, axial, or mixed direction. The force convection air cooling requires a fan to move the air, which can be circulated either in an open-loop or a closed-loop manner.

Liquid-cooled traction motors are mostly either water-cooled or oil-cooled. Water cooling is essentially an indirect method as a passage is needed to separate the water from the iron core for adequate insulation. In contrast, oil cooling takes the advantage of good insulation characteristics of oil, allowing direct contact between the oil and cooled parts, for example, windings, to maximize the cooling effect. Obviously, a water pump or an oil pump is necessary to circulate the water or oil in the motor.

In cooling system design, if the surface area for heat rejection, A, is known, the cooling flow rate of the fluid, q, required for dissipating a certain amount of heat, Q, can be estimated by

$$q = \frac{Q}{hA} \tag{10.190}$$

where h is the heat transfer coefficient. The heat transfer coefficient h depends mainly on the characteristic length of the part-fluid system, fluid properties and velocity, and so on, which can be calculated using appropriate formulas in heat transfer textbooks.

The fluid passage may consist of various sections, such as straight, bend, sudden expansion, sudden contraction, and orifice, incurring pressure drops. As a result, the power of a fan or pump required to drive the fluid through the cooling passage, P, can be estimated by

$$P = \sum \Delta p \cdot q \tag{10.191}$$

where $\sum \Delta p$ is the total pressure drop through the fluid passage.

Based on the cooling flow rate q, head pressure $\sum \Delta p$, and power P, an appropriate fan or pump can be designed or selected.

10.2.7. Design Guidelines for PMSMs

In view of their wide application in EVs, the design guidelines for PMSMs are specifically elaborated in this section.

10.2.7.1. Design of PMSM Rotors

The rotor of a PMSM typically comprises a rotor core, a number of permanent magnets, a shaft, and two bearings. The rotor core is often made of laminated silicon steel sheets. As per the mounting position of the permanent magnets onto the rotor core, the rotors can be classified into surface permanent magnet (SPM) and IPM, as shown in Figures 10.22 and 10.23, respectively.

FIGURE 10.22 SPM type of rotors. (a) Mounted on the surface and (b) inserted into notches.

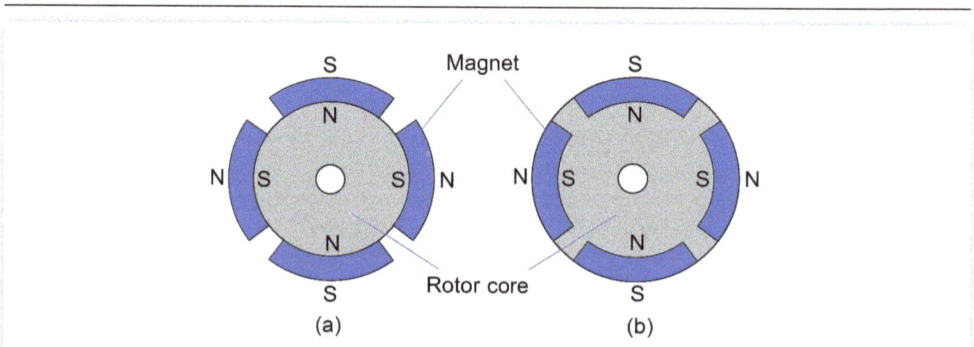

FIGURE 10.23 IPM type of rotors. (a) Radial, (b) tangential, (c) U-shaped combination, and (d) V-shaped radial.

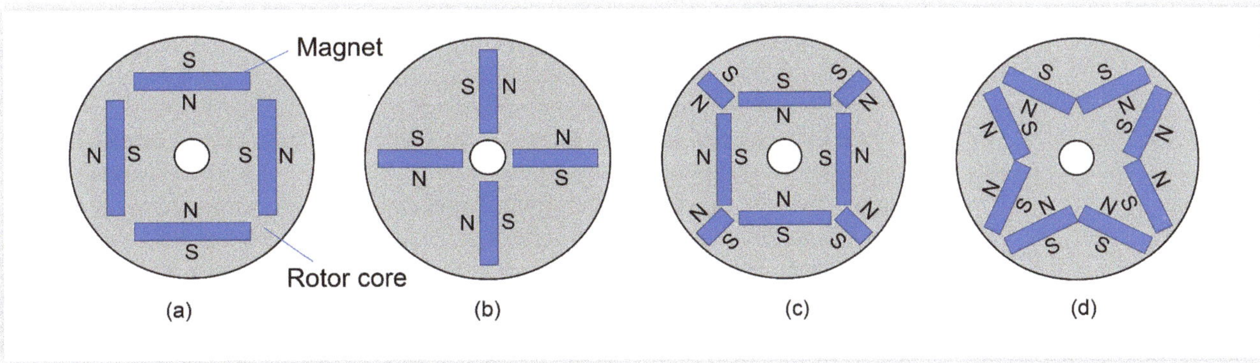

For the SPM type of rotors, magnets can be either mounted on the outer surface of the rotor (Figure 10.22a) or inserted into the notches flush with the outer diameter (Figure 10.22b). In general, there are no saliency effect and reluctance torque for this type of rotors, and the equivalent air gaps of the alternating and direct-axis magnetic circuits are large. Thus, the armature response is relatively small, with a poor field weakening ability and a small operation range for the constant power field weakening. Since the permanent magnets are directly exposed to the magnetic field in the air gap, they are prone to demagnetize. Therefore, this type of rotors is rarely used in electric motors for EV applications, in spite of easy manufacturing process and low cost.

On the other hand, for the IPM type of rotors, the design of the magnetic path usually takes one of such formats as radial arrangement (Figure 10.23a), tangential arrangement (Figure 10.23b), a U-shaped combination of radial and tangential arrangements (Figure 10.23c), or a V-shaped radial arrangement (Figure 10.23d). The pros and cons of these arrangements are discussed as follows. Most of the PMSMs used in traction applications today are IPM type as the saliency of the field poles produces a reluctance torque, which helps improve the power density of the motor. In contrast, the SPM type of motor does not produce such a reluctance torque.

The guidelines are elaborated here for several important design aspects:

1. Magnetic path design of the IPM type of rotors

   First, as far as the radial magnetic path is concerned, the magnetic path arrangement with a small effective air gap is generally in favor of field weakening. As a result, the constant power operating range of the motor can be expanded toward high speeds. In addition, this arrangement can produce reluctance torque, helping to improve the power density and overload capacity of the motor. Due to the small magnetic leakage coefficient, no isolation measures are needed, and the pole arc coefficient is easy to control. The rotors are of high strength, and the deformation of permanent magnets is negligible.

   Second, regarding the tangential magnetic path, two adjacent magnetic poles are connected in parallel to provide magnetic flux under one pole pitch, resulting in a larger magnetic flux per pole. As the number of pole pairs of the motor increases, magnetic flux enhancement becomes more prominent. The ratio of the reluctance torque to the total electromagnetic torque can reach up to 40%. However, since the magnetic leakage coefficient is usually quite large, isolation measures are often required.

Third, the U-shaped combination of radial and tangential magnetic paths combines the advantages of radial and tangential arrangements. Nevertheless, the structure and manufacturing process are complex, and the cost is high.

Fourth, in comparison to the radial magnetic path design, the arrangement with the V-shaped radial magnetic path can reduce iron losses, which increases the output torque. Besides, the mechanical strength and rotor stiffness can be improved further. For a given output torque, the weight of permanent magnets is also the lowest among the formats of magnetic paths mentioned earlier, thus helping in minimizing the material costs. However, the V-shaped radial magnetic path may yield a higher distortion rate in the waveform of no-load back EMF than other formats of magnetic paths, leading to excessive torque ripple. This drawback can be remedied by using a nonuniform circumferential air gap, as elaborated in the next subsection.

2. Design of the air gap structure on the rotor surface
   When a uniform circumferential air gap is used, there is a considerable amount of harmonic content in the waveform of air gap magnetic flux density, particularly the harmonics of third, fifth, and seventh orders, causing torque ripple, vibration, and iron losses. Several measures can be taken to resolve this problem: optimizing the shape of the permanent magnets, controlling the magnetizing energy for the permanent magnets so that the magnetic flux density varies sinusoidally, utilizing Halbach array arrangement of permanent magnets, and adopting a nonuniform circumferential air gap. Among these measures, the last one is usually the easiest to implement with a relatively low cost. Based on the principle of the eccentric magnet pole, the surface of the magnet pole can be analyzed to come up with the optimal shape for effective suppression of the harmonics in the magnetic flux density within the air gap.
3. Selection of air gap length
   The air gap length largely affects the performance, ease of assembly, manufacturing cost, and reliability of an electric motor. From the perspective of the electromagnetic performance of a PMSM, as the air gap length becomes larger, the power factor drops, while the speed range of field weakening shrinks, and the efficiency and torque density decline. However, the harmonic component of the air gap magnetic field decreases, and the motor is less prone to vibration and noise, with a reduction in stray losses. On the contrary, if the air gap length is too small, it would become difficult to assemble the motor and ensure coaxiality during its operation, which may lead to the failure of stator rubbing against the rotor, compromising the reliability of the motor. Hence, in the selection of air gap length, its influences on the vibration, noise, magnetic flux density, stray loss, assembly process, and production cost of the motor should be considered comprehensively.

   A parametric analysis can be performed to help make such a selection. The examples of the influences of air gap length on the quadrature and direct axis inductances, as well as saliency, are shown in Figure 10.24. To downsize the motor and enhance its efficiency, it is generally preferred to enlarge the proportion of reluctance torque in the total torque by increasing saliency, namely, selecting a small air gap length. Note that, however, an excessively high portion of reluctance torque in the total torque may result in large torque ripples and high stresses in the rotor structure.

FIGURE 10.24 Quadrature and direct axis inductance and saliency versus air gap length.

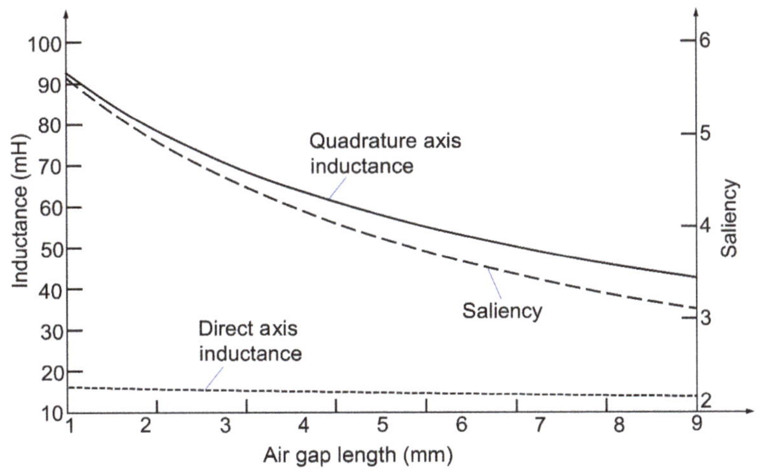

4. Slots for weight reduction
   When the distance from the bottom of the magnet slots to the hub of the rotor is large, additional slots should be considered to reduce the weight of the rotor, as illustrated in Figure 10.25. However, the presence of these slots should not affect the structural strength of the rotor or the magnetic path. The size and shape of these weight reduction slots should be optimized by means of simulation tools. Particular attention should be paid to stresses around the corners of these slots to keep them under the allowable level.

FIGURE 10.25 Rotor slots for weight reduction.

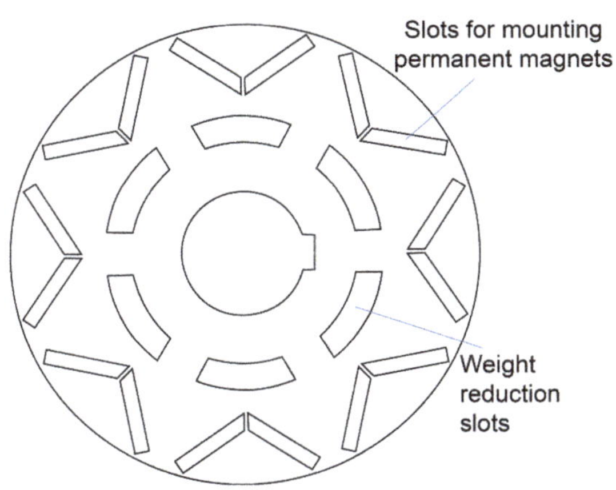

5. Selection of permanent magnetic materials
   Permanent magnets used in electric motors are often made of hard magnetic materials such as NdFeB, ferrite, alnico alloys (AlNiCo) and samarium cobalt (SmCo).
   The physical properties of these materials are listed in Table 10.5.

TABLE 10.5 Physical properties of permanent magnetic materials.

Permanent magnetic materials	Residual magnetic induction B_r (T)	Intrinsic coercivity H_{cJ} (kA/m)	Magnetic induction coercivity H_{cB} (kA/m)	Max. magnetic energy product $(BH)_{max}$ (kJ/m³)	Temperature coefficient $\alpha(B_r)$ (%/K)	Curie temperature T_c (°C)
NdFeB	1.1-1.52	955-2388	796-1115	239-446	-0.01	310-420
Ferrite	0.2-0.44	159-358	143-318	8-36	-0.18 to -0.09	450-465
AlNiCo	0.8-1.4	40-159	38-155	8-88	-0.03 to -0.02	800-860
Samarium cobalt	0.85-1.14	796-2388	637-844	127-255	-0.05 to -0.03	710-850

NdFeB is the most frequently used permanent magnetic material in today's electric motor design due primarily to its excellent magnetic properties. A major shortcoming of NdFeB is that the magnetic properties decay dramatically at high temperatures due to its relatively low Curie temperature and high temperature coefficient. Besides, the high content of iron and neodymium in NdFeB makes it easy to rust. Thus, a coating is often used to protect it from oxidization. Nevertheless, the use of such a rare earth element as neodymium also raises concerns about strategic resource availability and dependency.

The ferrite permanent magnetic material comes in two formulae: barium ferrite ($BaFe_{12}O_{19}$) and strontium ferrite ($SrFe_{12}O_{19}$). Strontium ferrite is more suitable for electric motor applications due to its slightly higher intrinsic coercivity than barium ferrite. With an almost linear demagnetization curve, low cost, and no need for rare earth and precious metal contents, the ferrite is an inferior substitute for NdFeB because its magnetic properties are not as good and are subject to relatively large influence of ambient temperature. Measures may be needed to prevent its demagnetization at low temperatures. In the current trend toward making magnets with no or little rare earth elements, ferrite permanent magnetic materials may meet the requirements if designed properly.

AlNiCo magnets can be produced by the casting or powder sintering process. AlNiCo magnets have high magnetic flux density, high resistance to corrosion, and excellent temperature stability. Casted AlNiCo is commonly used in PMSMs due to its slightly higher magnetic properties than sintered AlNiCo. With better mechanical characteristics than casted AlNiCo, sintered AlNiCo can be readily made into the desired shape.

SmCo has very good magnetic properties. Its demagnetization curve is close to a straight line and coincides with its recovery line. In addition to the very strong anti-demagnetization ability, SmCo also features a low temperature coefficient, high Curie temperature, and excellent magnetic stability, which make it an ideal choice for high-temperature applications. However, SmCo is typically expensive as it contains strategic metal cobalt and very rare metal samarium. As a result, it is normally employed only for special-purpose electric motors of high performance and exceptional reliability.

Finally, it is worth noting that the selection of permanent magnetic materials is also subject to applicable standards or regulations.

6. Selection of iron core materials
To channel and enhance the flux in electric motors, iron cores in both rotor and stator must be made of soft magnetic materials that are easily magnetized and demagnetized. The most popular soft magnetic material today is cold-rolled silicon steel (also called electrical steel) laminates, with a silicon content from 0.8 to 4.8%, which offer higher magnetic induction and lower iron loss than their hot-rolled counterparts. Either grain-oriented or non-grain-oriented, these silicon steel laminates typically come with a smooth surface, uniform thickness, high stacking coefficient, and good punching performance, with a thickness between. In addition, non-grain-oriented silicon steel laminates, usually of 0.2–0.35 mm thickness, outperform grain-oriented ones in reducing iron losses at high frequencies, and such an effect becomes prominent with the decrease in their thickness [10.6].

Moreover, amorphous alloys, for example, FeSiBP, are rising as a strong competitor to silicon steel for their great potential in further reducing iron loss, mainly caused by eddy currents, by as much as 80–95%, especially at high frequencies [10.7]. These Fe–Si–B amorphous alloy ribbons, produced in a rapid solidification process called melt-spun, are replacing silicon steel laminates for their superior magnetic properties, such as high resistivity (around 130 μΩ·cm, more than double that of silicon steel) and extremely small thickness (only 0.025 mm or so), primarily due to the lack of intrinsic magneto-crystalline anisotropy [10.8, 10.9]. The use of amorphous alloys for making iron cores also allows a significant improvement in the specific power density of the motor, as both pole counts and frequency can be raised, with the alternating frequency of magnetic field up to 1000 Hz achievable [10.10]. Nevertheless, amorphous alloy ribbons are thin, brittle, and hard, making it difficult and costly to form them using highly efficient processes, for example, stamping. Besides, amorphous alloy ribbons are normally not pressed tightly since their magnetic properties are highly sensitive to stresses. As a result, the iron cores made of amorphous alloy ribbons usually do not have a high stacking factor, making them susceptible to AC-induced noise and vibrations.

The basic properties of amorphous alloys and silicon steel are compared in Table 10.6.

TABLE 10.6 Basic properties of amorphous alloys and silicon steel.

Soft magnetic materials	Coercivity (A/m)	Hysteresis loss coefficient	Residual magnetic induction (T)	Max. relative permeability	Resistivity (μΩ · cm)	Density (kg/m)	Thickness (μm)
Silicon steel laminate	<30	177–179	2.03–2.4	4×10^4	45–52	7650–7700	350–500
Amorphous alloy ribbon	<4	1.4	1.56–1.63	>25×10^4	120–130	7180–7330	25(±4–5)

10.2.7.2. Design of PMSM Stators

The stator of a PMSM mainly comprises an iron core, a number of windings, a motor housing, and two end caps. Made of silicon steel laminates or amorphous alloy ribbons, the iron core has a number of evenly spaced stator teeth around the inner circumference of a circular-shaped yoke, as illustrated in Figure 10.26 [10.5]. The windings are embedded in the cogging between adjacent teeth.

FIGURE 10.26 Schematic of the iron core of a PMSM.

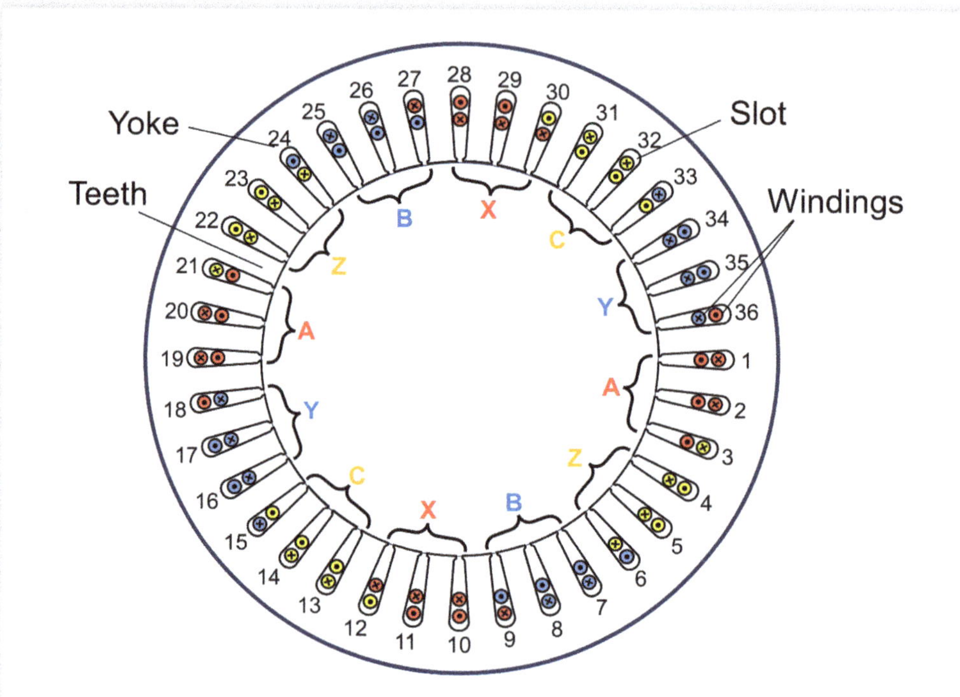

Courtesy of Lee Han.

1. Design of windings
   In general, stator windings can be made by concentrated and distributed methods, which can influence key design parameters such as the end-winding length, perfection of the sinusoidal waveform of the magnetic potential, and saliency. Hence, the winding methods largely affect the amount of copper used in the stator, length of the electric motor, torque ripples, noise and vibrations, and increase in reluctance torque.

   As far as the windings are concerned, round-wire distributed windings are used in most traditional PMSMs, as shown in Figure 10.27a [10.11]. However, hairpin windings are becoming the main stream design preference, as shown in Figure 10.27b.

FIGURE 10.27 Comparison of round wire with hairpin wire windings. (a) Round-wire winding and (b) hairpin winding.

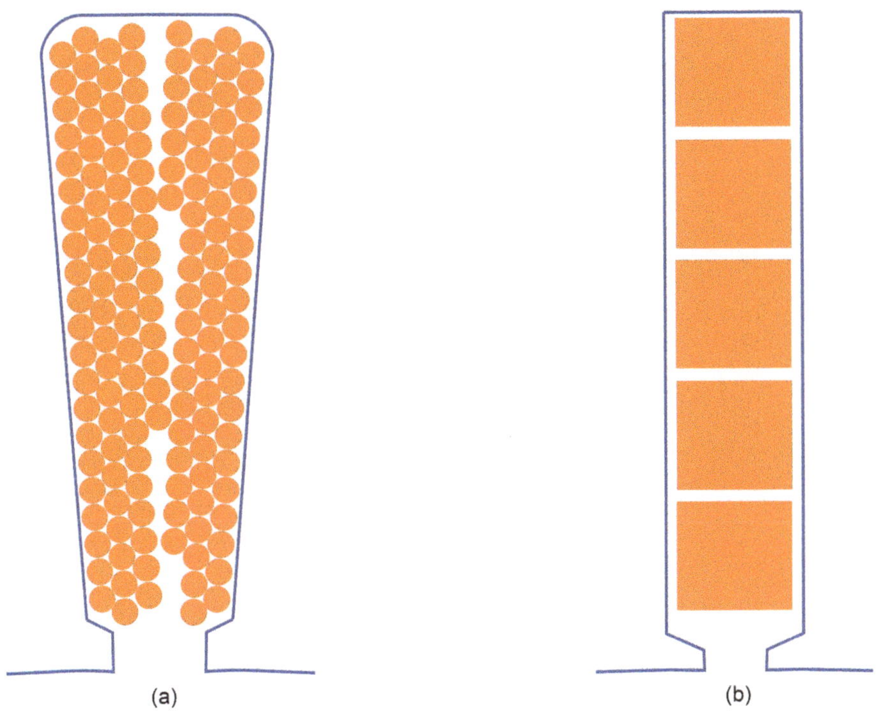

Courtesy of IEEE.

In general, hairpin windings offer the following advantages over the round-wire windings: 20–30% higher copper fill factor in the slot, which improves the power density or reduces the amount of copper for the PMSM of a given rated power, reduced copper loss due to increased cross-sectional area of the conductor, better cooling characteristics as a result of enhanced heat conduction for increased contact area between hairpin wires, elevated stiffness of armature, which helps suppress electromagnetic noise, and fifth, the end-winding lengths can be made shorter to save the copper and cut the cost.

However, the following difficulties must be overcome in the design of hairpin windings: thicker hairpin wires can cause the skin effect, increasing the loss, especially when the motor operates at high speeds; both a proper deformation allowance in design and strict quality control in the manufacturing process are needed for hairpin wires to prevent their bending and rebounding from damaging the insulation layer; high-precision, expensive equipment is necessary for mass production; and it is not easy to serialize the design by simply adjusting the length and number of turns.

In addition, the number of turns per slot has been increased from four to eight so far and very likely to exceed ten turns per slot in the near future, in order to further reduce the eddy current loss in hairpin windings at high rotating speeds. The augmented number of turns per slot can also help slow down the increase in AC resistance with the rise of rotating speed.

2. Design of the cooling structure
   Most of the PMSMs for traction applications are traditionally water-cooled, actually using an ethylene glycol water (50/50 volumetrically) coolant. Water cooling is actually an indirect cooling methodology as the fluid does not come into direct contact with the windings and iron core where the heat is generated.

   There are different topologies of cooling circuits for the stators of PMSMs, such as spiral, single U-shaped, bifurcated U-shaped, and axial [10.12], as illustrated in Figure 10.28 [10.13].

FIGURE 10.28 Different topologies of cooling circuits.

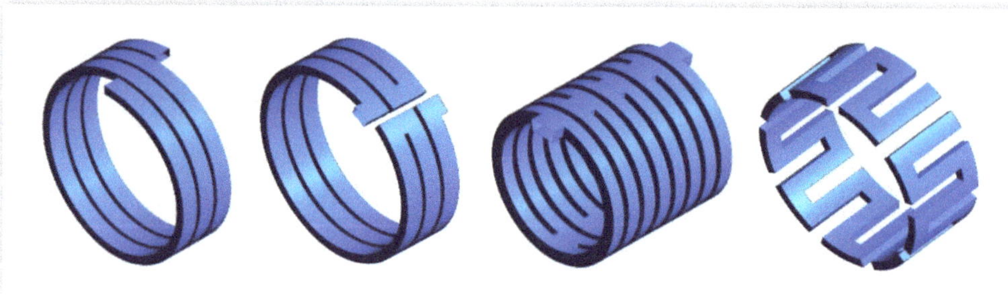

Courtesy of Marco S. D. Legarra.

Among these cooling circuit topologies, the axial topology generally has the largest inlet-to-outlet pressure drop, whereas the spiral topology has the smallest, with U-shaped in between. Therefore, although the spiral topology is employed in some occasions, U-shaped topologies are more frequently adopted for a balanced cooling effect and pressure drop. The number of cooling circuits must be selected carefully to maximize the cooling effect and minimize the flow resistance and power consumption of the pump. The optimal number of cooling circuits can be found by means of thermo-fluid simulation, which typically ranges from three to six.

On the other hand, oil-cooled PMSMs have been developed rapidly in recent years. It is widely reckoned to be a major trend in the cooling technology of electric motors because of the heat transfer enhancement caused by direct contact between the oil and windings as well as the iron core.

Although oil cooling can be divided into direct and indirect in terms of the way that the cooling oil comes in contact with the stator yoke, oil cooling usually refers to direct oil cooling most of the time as it apparently offers a better cooling effect. Moreover, direct oil cooling can be realized either by immersing the rotor and stator in the oil or by spraying oil to the winding ends and places where cooling is needed.

Furthermore, a traction PMSM is often highly integrated with its transmission, namely both sharing the same housing and common cooling media (or lubricant) in engineering practice. Compared to their water-cooled counterparts, the advantages of oil-cooled PMSMs are given as follows: High winding temperature often occurs at both ends of water-cooled windings where the cooling water cannot reach. With the oil cooling approach, oil can be delivered readily to high-temperature spots to bring down the local temperature; with a higher boiling point and lower solidifying point than water and ethylene glycol mixture, cooling oil offers a broader operating temperature range; the cooling oil has better insulation characteristics than water and ethylene glycol mixture, leading to improved electrical safety; with the maximum speed of electric motors increasing to 20,000 rpm level and beyond, direct oil cooling would become a vital solution to heat dissipation from an ever-decreasing packaging space of the PMSM stator.

Figure 10.29 illustrates the schematic of oil flow paths in an oil-cooled electric drive system with the e-motor and transmission integrated.

FIGURE 10.29 Schematic of oil flow paths in an oil-cooled electric drive system.

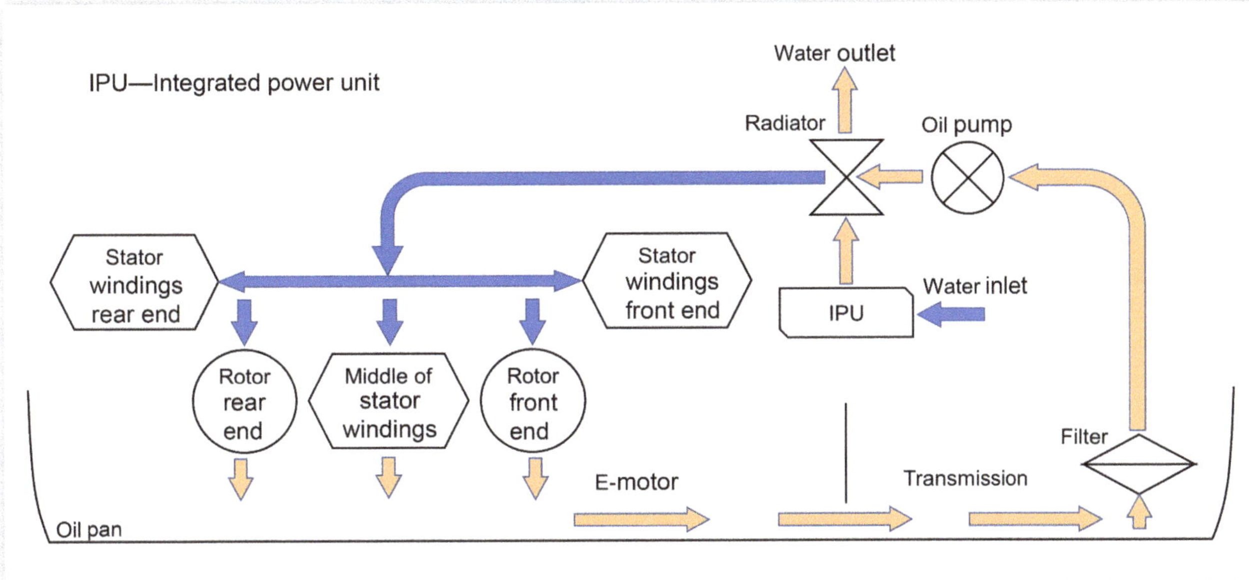

In general, there are several direct oil-cooling methods available for selection: oil-drenching cooling, oil-spray cooling, and oil fling-off cooling.

As for the oil-drenching cooling method, the cooling oil is delivered by an oil pump to locations such as stator core, end windings, both end of the rotor, and transmission, through a network of tubes with a number of holes along the way, where the oil drips down from above to those components. Although the oil-drenching cooling method offers the best cooling effect among these cooling methods, a large quantity of oil is normally needed.

With the oil-spray cooling method, the cooling oil is delivered to a few nozzles through a circuit and then sprayed to high-temperature spots such as end windings. Its cooling effect can be greatly affected by gravity, the rotating speed of rotor, and the design and mounting position of the nozzles. If not designed properly, this cooling method is prone to an uneven cooling effect.

The oil fling-off cooling method can result in a uniform cooling effect. However, its cooling effect may worsen at high rotating speeds when the cooling oil atomizes.

It is worth noting that these cooling methods can be applied in combination to enhance the cooling effect, at a price of increased system complexity though.

3. Selection of the aspect ratio
The aspect ratio of a PMSM motor mainly affects the moment of inertia of the rotor and, thus, the torque response time and the amount of copper needed in the stator winding. When the length-to-diameter aspect ratio increases, for a given rotor volume, the moment of inertia of the rotor drops and more copper is needed in the stator winding. As a result, the electric motor would respond to torque requests quicker.

It is generally preferred to select an aspect ratio so that the outer diameter of the rotor is roughly 1–2 times its core length.

4. Selection of the pole count
   With an increased number of pole pairs, both the amount of iron used in the yoke of the stator core and the volume of the electric motor decline for a given slot and pole ratio. However, such effects level off when the number of pole pairs exceed a certain number due to the relative increase in the volume of the insulation material.
   In addition, the frequency of input current increases as the number of pole pairs rises as well, causing iron losses to surge.
   Hence, in general, a relatively small number of pole pairs are preferred, especially for high-speed electric motors.
5. Selection of the slot-to-pole ratio
   As the slot-to-pole ratio increases, the distortion in the sinusoidal waveform of the MMF of the winding diminishes, and both torque ripples and iron losses decrease; however, the outer diameter of the stator increases due to the increase in the volume of slot insulation.
   On the other hand, the stator tooth width declines with the increase in the slot-to-pole ratio. If the stator teeth are too narrow, their poor mechanical strength may cause them to break in the vibrations caused by torque ripples or radial electromagnetic forces. With the rise in the slot-to-pole ratio, the manufacturing cost of the stator increases, while the winding becomes difficult to wind.
6. Selection of electrical loading
   For a given thermal loading, the outer diameter of stator declines first and then rises with the increasing electrical loading; meanwhile, both rotor volume and its moment of inertia drop, with an increase in the amount of copper used in the stator. Therefore, appropriate electrical loading is essential to balance the amount of copper and iron used in the motor for maximizing its power density.
7. Selection of slot opening width
   The slot openings affect the effective air gap between the stator and the permanent magnet, which, in turn, influence the air gap permeability and leakage inductance. As a result, the air gap magnetic flux density features a sawtooth-shaped waveform, thereby generating cogging torque and contributing to the torque ripple and noise during operation.
   With the decrease in slot opening width, leakage inductance, air gap magnetic flux density, and the saliency rise, thus improving the torque density of the electric motor. However, if the slot openings are too narrow, then the wire embedding process would encounter difficulties. It is generally recommended that slot openings should be kept small and only enough for wire embedding.
8. Selection of the slot type
   There are three slot types commonly employed for the stators: open, semiclosed, and closed, as shown in Figure 10.30. In order to make the assembly of windings easy, most of the stators use the semiclosed type of slots in traction motor applications, which can be one of the three subtypes: pear-shaped, tapered side with a flat bottom, or parallel side with a flat bottom, as illustrated in Figure 10.31.

FIGURE 10.30 Common types of stator slots. (a) Open, (b) semiclosed, and (c) closed.

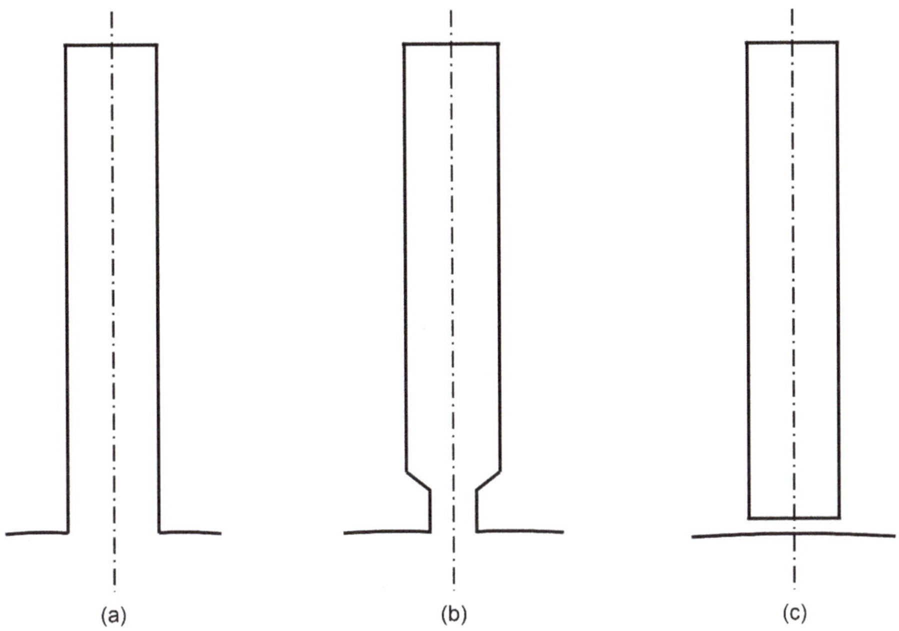

FIGURE 10.31 Types of stator slots for traction motor applications. (a) Pear-shaped, (b) tapered side with a flat bottom, and (c) parallel side with a flat bottom.

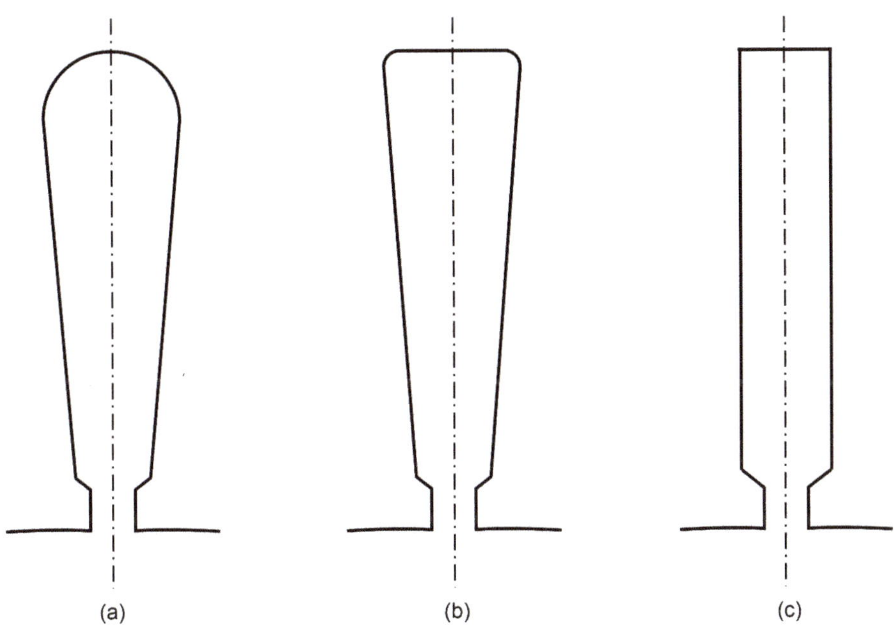

Pear-shaped slots are often utilized with random round wires for motors of small to medium power ratings, yielding mediocre power and torque densities. In this type of slots, the insulating paper fits well with the bottom of the semicircular groove, which enhances the slot filling factor.

Compared to the pear-shaped slots, the tapered-side slots with flat bottom enlarge the effective slot area and reduce the yoke thickness, thus usually suitable for the stators of large power ratings with a margin in yoke or tooth magnetic loads to further improve slot filling factor.

Parallel-side slots with a flat bottom are particularly suitable for form windings and hairpin windings to maximize the slot filling factor and reduce the current density.

10.3. Design of Electric Motor Controllers

An electric motor controller primarily consists of three parts: a low-voltage main control circuit board, high- and low-voltage driver boards, and a high-voltage main circuit. They are usually arranged into a three-layer structure with the high-voltage main circuit located closest to the electric motor for easy connection of three-phase power cables. The low-voltage control board is usually arranged farthest from the motor to minimize the EMI from the motor.

The circuit schematic of a typical electric motor controller is illustrated in Figure 10.32.

FIGURE 10.32 Circuit schematic of an electric motor controller.

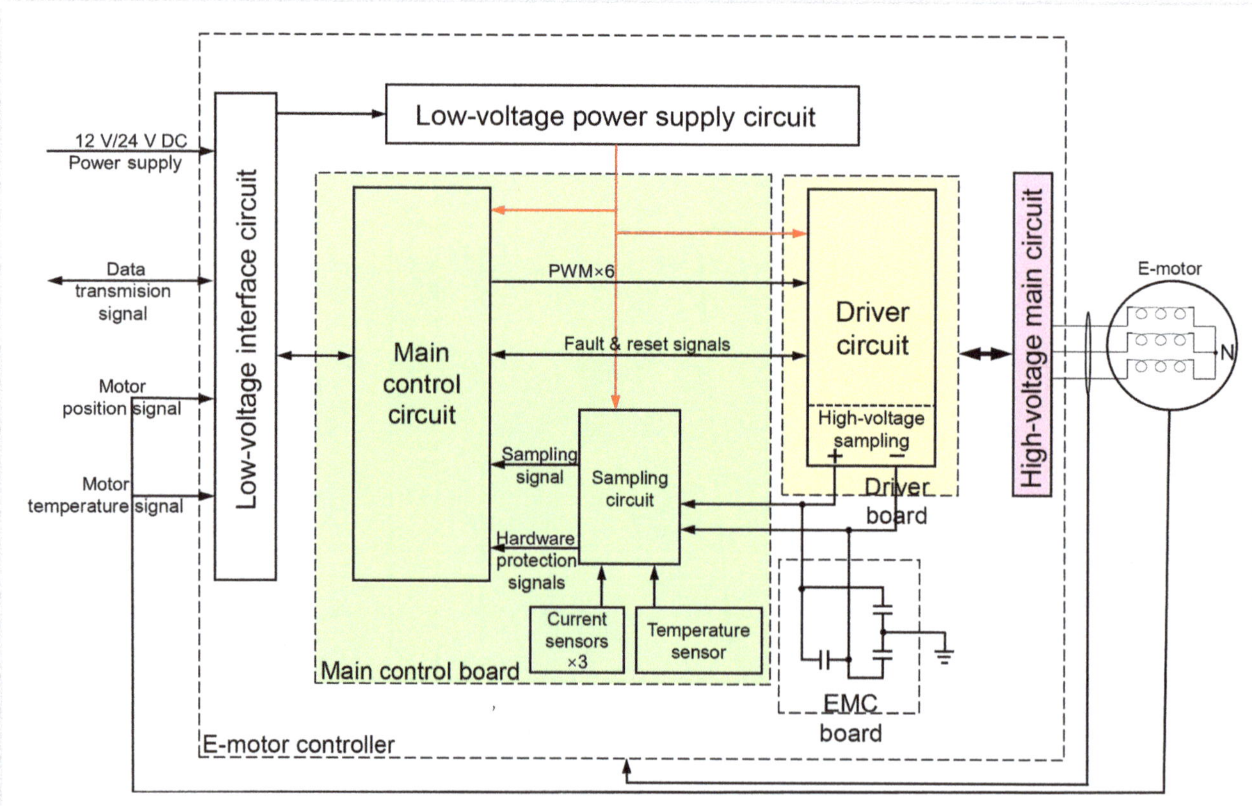

Automotive-grade electric chips, components, and devices are generally utilized to ensure reliability, durability, and environmental adaptability, with the allowable range of operating temperature of −40 to 125°C.

In addition, to reduce electromagnetic interference, some rules and principles need to be followed in the design of these circuit boards: separating digital circuits and analog circuits as much as possible; shrinking signal loops; and placing capacitors and spacing signal lines properly.

10.3.1. Design of Main Control Boards

The basic function of the main control board is to provide switching signals for power electronics, for example, IGBT or SiC MOSFET, to complete the DC-to-AC conversion. A pulse-width modulation (PWM) module is, therefore, needed to generate the PWM waveform. In addition, the design of the main control board also involves modules such as power supply, analog quantity acquisition, fault handling, communication, relay, and speed detection.

10.3.1.1. Power Supply Module

With a 12 or 24 V DC input from the on-board battery, the power supply module produces DC power supplies at various voltage levels, for example, 1.9, 3.3, 5, and 15 V, for processor and communication chips, as well as sensors. It is necessary to know the voltage, current, isolation option, and power-on sequence required by each chip on the board to design the power supply.

10.3.1.2. Analog Quantity Acquisition Module

In general, the analog signal acquisition involves current, voltage, and temperature, among which the current signal requires high accuracy as it is closely related to the real-time torque control.

The three-phase AC current is typically acquired with Hall-effect current sensors.

The DC bus voltage can be detected using Hall-effect current sensors or partial voltage sensors, with the latter being a more economical option. A separate DC bus voltage acquisition board is sometimes preferred to improve the high-voltage safety.

Motor temperature is often acquired by platinum thermal resistance sensors, which characterize a constant relationship between resistance value and temperature. The IGBT temperature is normally detected by negative-temperature coefficient (NTC) sensors for its low cost and simple supporting circuit.

All the signals from aforementioned sensors must be filtered with capacitors. The capacitor values should be kept small to avoid a large phase shift. An operational amplifier is often needed to adjust the voltage range of these signals to fit the thresholds of the master chip. Clamping diodes can also be used before the signals enter the master chip for this purpose. It is recommended the power supply and ground of the analog signal circuit should be separated from other power supply and ground on the board with magnetic beads to reduce the interference of the digital switch signal to the analog signal.

10.3.1.3. Fault Handling Module

Fault signals that may occur on the board include overvoltage, overcurrent, undervoltage, overtemperature, IGBT failure, and resolver decoding chip failure. Current and voltage signals processed by the analog operational amplifier circuit should be sent to the comparator circuit for comparison while being transmitted to the main control chip. The reference voltage of the comparator circuit is set by the resistor divider, so that the hardware fault signal can be generated when overcurrent or overvoltage occurs, which is more timely and reliable than the software-generated fault signal. Many control boards use a complex programmable logic device (CPLD) or a reset–set (RS) trigger to synthesize the fault signal for blocking the PWM output timely when a serious fault occurs. Often output by the open collector, the fault signal generated by the driver board should be raised up to an appropriate level with a resistor before being input to the CPLD or other circuits.

10.3.1.4. Communication Module

The control board needs to communicate with other controllers, for example, the VCU, to track the torque or speed command and transmit the fault signal. The control board is usually equipped with a communication chip based on the communicating protocols used in the vehicle, for example, CAN chip in most of the cases. Generally, it is recommended that the power supply of the communication chip be isolated for improved reliability. It is also desired to leave a place for the terminal matching resistance to eliminate the signal reflection in the communication cable when the transmission distance is long in the design of the CAN communication circuit.

10.3.1.5. Relay Module

Usually at least three relay circuits are required on the control board to control the precharge contactor, main contactor, and cooling fan or coolant pump. Because the contactor normally has a high rated voltage or a large control current, the I/O port of the main control chip cannot directly control the contactor. Thus, optocoupler circuits or solid-state relays are used for relaying.

10.3.1.6. Speed Detection Module

Rotor speed is generally detected using a resolver, a Hall-effect sensor, or an encoder [10.14].

Resolvers produce analog sine or cosine signals. They are capable of detecting a rotating speed ranging from 0 to 30,000 rpm at a resolution of 0.1 electrical degree within the temperature range from −40 to 220°C. Resolvers are highly reliable, robust, highly accurate, and of absolute positioning capability. However, they are expensive, relatively bulky, and troublesome to use as both a resolver to digital converter and a complex control algorithm are required. They are widely used today for their all-around capability in detecting very high speeds in a broad temperature range, good vibration resistance, and long lifespan [10.14]. A typical resolver consists of a rotor and two stators, which are positioned at 90° to each other, with its rotor fixed on the rotor of the electric motor. A sinusoidal excitation signal U_{ref} is applied to the excitation coil. When the rotor rotates, induced voltage signals U_{sin} and U_{cos} are generated in the two coils. According to the phase and amplitude of the induced voltage signal, the angle and speed of the rotor can be calculated. Such a calculation can be done by a special resolver decoding chip at a price of added hardware cost or by decoding algorithm in the control software.

Hall-effect sensors can be either analog or digital. Analog ones are suitable for the speed range from 0 to 55,000 rpm at the resolution of 1.0 electrical degree within the temperature range from −40 to 100°C. Despite of their low power consumption, small size, and long lifespan, they demand a readout circuit placed close to the signal transmission circuit due to their low signal level, which makes them prone to the influence of temperature and radiation. As a result, temperature compensation is often required, along with additional hardware and filters, and the output might not be sinusoidal for all motors. In contrast, digital Hall-effect sensors can operate at a speed ranging from 100 to 100,000 rpm at the resolution of 60 electrical degree under a temperature ranging from −40 to 150°C. They are cheap, simple, robust, and durable, but rarely utilized due to their low position resolution, particularly at a speed lower than 100 rpm [10.14].

Encoders can be divided into photoelectric and capacitive types. Photoelectric encoders can detect a speed from 0 to 10,000 rpm at the resolution of 0.001 electric degree under a temperature ranging from −40 to 85°C. Although their digital outputs make them less susceptible to interference, they are sensitive to water and condensation, and their inner parts are not as durable as other parts, which shortens their lifetime to a certain extent. On the other hand, capacitive encoders are only good for detecting a speed from 0 to 8000 rpm at the resolution of 0.2 electric degree under a temperature ranging from −40 to 125°C. By avoiding exposure to water and condensation, they are usually considered to be more reliable and robust in structure than their photoelectric counterparts. Nevertheless, the electronic parts mounted in the motor reduce their durability and lifetime in another way [10.14].

10.3.1.7. PWM Module

The PWM signal of the main control chip cannot be directly input to the drive board because the drive board usually has certain requirements for the level of the PWM signal. Hence, a voltage conversion may be needed for the PWM signal. Besides, there is no need to isolate the PWM signal because it is generally isolated on the drive board. If there is a CPLD on the control board to process the fault signal and PWM signal, the PWM signal output from the main control chip should pass through the CPLD before output, so that the CPLD can quickly block the PWM when a fault occurs to avoid damage to the driver board and power electronics. In some control board designs, the PWM signal may pass through a phase inverter, for example, Schmitt trigger inverter; impose an enable signal for output control; or be modulated to make the edges of the PWM signal steep. Special attention must be paid to whether the PWM level takes effect at a high or low level. A three-phase motor drive generally requires six PWM signals. However, if the motor controller has a braking function, it is best to add one more PWM signal for chopper control.

10.3.2. Design of Driver Boards

The driver board has two basic functions: isolating the low-voltage main control board from the high-voltage power electronic switching circuit and driving and protecting the power electronic switches such as IGBTs or SiC MOSFETs, to make them work properly. A gate driver is typically used to serve these purposes, which is essentially a power amplifier that takes a low-power input from the main control board and produces a high-current drive input for the gate of high-power switching devices such as IGBTs or SiC MOSFETs.

In essence, a gate driver consists of a level shifter in combination with an amplifier. An external push–pull output circuit is generally used to boost the maximum output current of the driver IC. An example of the basic push–pull output circuit with field effect tubes (FETs), for example, IGBTs or SiC MOSFETs, is shown in Figure 10.33. When the control signal V_i is at a high level, V_1 is turned on and the FETs corresponding to the output V_o is turned on; when the control signal V_i is at a low level, V_2 is turned on and the switches corresponding to the output V_o is turned off.

FIGURE 10.33 A basic push-pull output circuit with FETs.

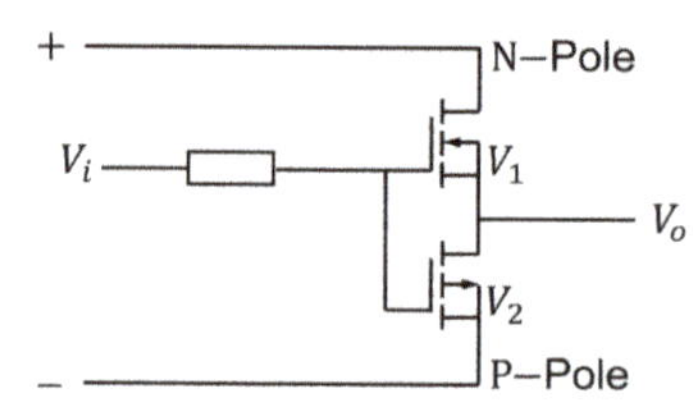

The gate driver circuit typically takes a full bridge topology, as shown in Figure 10.34. Referred to as an H-bridge, it may consist of two P-type FETs (Q_1 and Q_2) along with two N-type FETs (Q_3 and Q_4). By controlling the on and off of these switches, the operation of the electric motor can be fully controlled. For example, when Q_2 and Q_3 are on and Q_1 and Q_4 are off, the electric motor rotates in the direction indicated by the arrow; on the contrary, when Q_1 and Q_4 are on and Q_2 and Q_3 are off, the motor rotates in the opposite direction as indicated by the arrow.

FIGURE 10.34 A gate driver circuit of the full bridge topology.

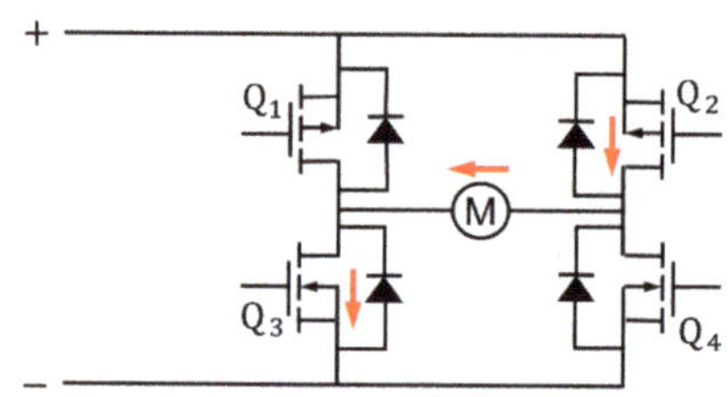

The gate drive circuit is often integrated with opto-isolators or magnetic ring pulse transformers to isolate the low-voltage control board from the high-voltage main circuit and protect the former from being damaged if any failure occurs in the latter. In spite of their small sizes, opto-isolators fall short in response, with a delay of more than 500 ns in general. Besides, an isolated auxiliary power supply is also needed for the output side. In contrast, transformers provide steep rising and falling edges with almost no delay. Furthermore, coreless transformers (CLTs) are becoming novel solutions to electrical isolation in recent years, as they offer the following advantages over the opto-isolators and pulse transformers: CLTs have high immunity to EMI and transients, and there is usually no degradation during their service life.

Currently, there are many specific gate driver modules available on the market, which usually have such well-around functions as over-current soft shutdown, high-speed isolation, under-voltage lockout, and fault signal output. Thus, these off-the-shelf integrated modules can be used to shorten the product development cycle by taking advantage of their protective functionality and high reliability, with no need for debugging.

In order to reduce the EMI of the gate driver circuit on other on-board controllers and the communication network, a grounding point can be added to the driver board to drain the interference signals.

10.3.3. Design of High-Voltage Main Circuits

A high-voltage main circuit refers to a power electronic module that produces three-phase AC current to drive the electric motor.

In general, power semiconductor devices such as IGBTs and SiC MOSFETs are packaged into a power electronic module mainly in three different ways, namely, single-sided cooling, double-sided cooling, and triple-or-above-sided cooling [10.15].

10.3.3.1. Single-Sided Cooling

This is a traditional approach to packaging power electronic modules. One side of the semiconductor devices is used for cooling and the other side for electrical connections. A typical structure of single-sided cooling power modules is shown in Figure 10.35. Heat generated within the semiconductor device can be conducted into the heat sink through a direct bonded copper (DBC) substrate and a baseplate. The thermal performance of this packaging methodology is intrinsically limited, which may lead to difficulties in further enhancing the power density of the system.

FIGURE 10.35 A typical structure of single-sided cooling power modules.

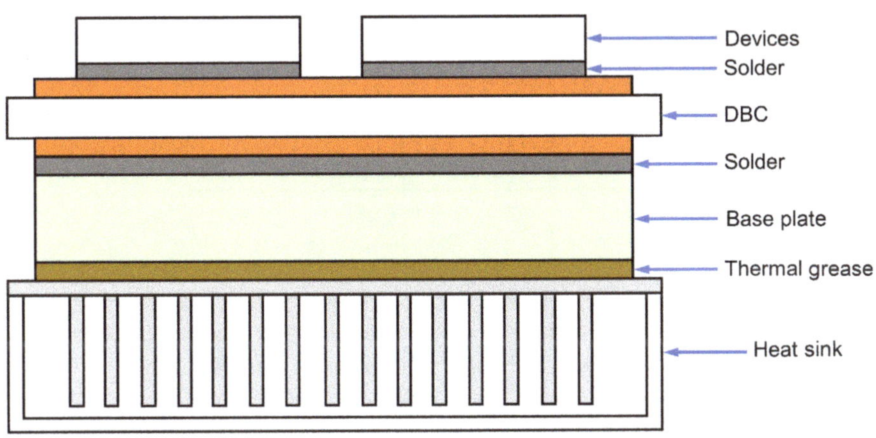

10.3.3.2. Double-Sided Cooling

In this architecture, the power semiconductor device is sandwiched between two DBC substrates with heat sinks on both sides of the device, with each of the substrates providing electrical connections as well as a path for heat transfer, as depicted in Figure 10.36. In addition, the wire bond-less interconnections are used to replace the classical wire bond. The baseplate can also be eliminated to lower the thermal resistance between the junction and case. As a result, the heat removal potential can be approximately doubled in comparison to the single-sided cooling design, with increased current handling capacity for the same size of package. The improved cooling capacity is deemed adequate to offset the removed thermal capacity of the baseplate as far as the thermal time constant of the module is concerned during short-time overloading condition [10.16]. The inner side of the heat sink is typically a pin-fin structure, allowing effective heat transfer with the coolant, while the top side of the heat sink acts as a second heat transfer path into the air. The shape, size, and density of the pins are needed to be optimized to maintain the junction temperature within limits under all operating conditions. Solder or a thermal interface material (TIM) can be applied on both sides of the module to improve the thermal contact.

FIGURE 10.36 A typical structure of double-sided cooling power modules.

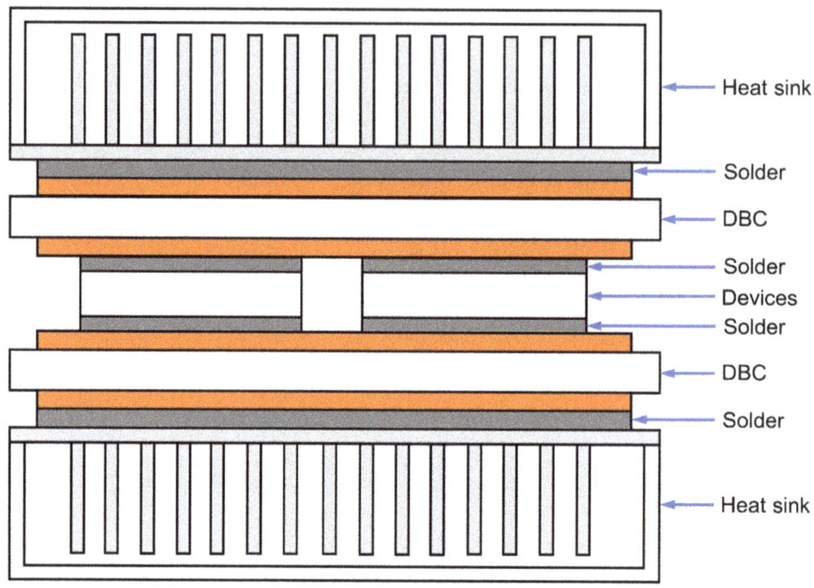

10.3.3.3. Triple-or-above-Sided Cooling

In this new architecture, independent structural components are used to provide cooling through three or more sides of power chips or by immersing them in coolant. As an example of such implementation, diodes and power chips are soldered between two copper lead frames using copper spaces for height adjustment. With the copper lead frames functioning as circuit interconnects, the power module has all the external electrical connections going out from the top side, with no DBC used. The power electronics bonded by lead frames are then placed between two aluminum pin-fin heat sinks with isolation sheets encapsulated together into a popsicle-shaped power module. Assembled in parallel, the power modules can serve as coolant dividers when they are immersed into the coolant channel in the main chassis and become structural elements of the cooling system. As a result, this architecture enables not only double-sided cooling by connecting the power electronics to lead frames on both sides but also all-sided cooling, except the top side at the power module level [10.15].

10.3.4. Design Guidelines for SiC Motor Controllers

SiC power modules are increasingly utilized in traction applications today. In order to unleash their full potential, attention must be paid to some specific aspects of the circuitry design involving SiC MOSFET modules, as elaborated in the following section.

10.3.4.1. Short-Circuit Protection

Compared to IGBT modules, SiC modules require a drive circuit of faster response for short-circuit protection. In addition, the increase in the DC bus voltage from 300–400 to 800 V or higher puts forward stricter requirements on the creepage and clearance distances. The secondary side drive and protection circuit for SiC modules mainly include gate drive circuit, active clamp protection circuit, gate clamp protection circuit, short-circuit protection circuit based on desaturation, and Miller effect protection circuit.

In general, there are two types of short-circuit faults. The first type is shoot-through of the bridge arm due to hardware failure or software fault. Due to the small inductance of the short-circuit loop (typically of a nanohenry order), the current rises quickly, making it difficult to protect the power module from this type of faults. This type of fault can be further classified into hard switching fault (HSF) and fault under load (FUL). When either an HSF or an FUL occurs, the SiC module can be burned out in a short time by the large amount of energy that it is subject to. The second type of short-circuit fault is interphase short circuit or short circuit to ground [10.17]. This type of fault features relatively large inductance (typically of microhenry order), making it easier to detect and shut off the faulty circuit than the first type of faults.

Hence, it is crucial to design appropriate circuits for fast and reliable detection of short-circuit faults and shutdown of the faulty portion of the circuit. The methods for short-circuit detection mainly include desaturation detection, parasitic inductance voltage detection, current sensing, shunt sensing, mirror current sensing, and gate charge sensing [10.17], as explained in the following paragraph.

Due to its simplicity and low cost, the desaturation detection method has been broadly used for short-circuit protection of IGBTs. When the SiC MOSFET is turned on and the voltage rises above the threshold, the comparator flips and sends a fault signal to turn off the device; when the SiC MOSFET is turned off, the transistor is turned on to pull down the voltage to a low level, and the detection circuit is shielded [10.17]. On PWM jumping to a high level, it is necessary to reserve enough dead time to prevent false triggering of the detection circuit before the SiC MOSFET is fully turned on. In medium- and high-power applications, the breakover voltage drop of SiC MOSFETs is relatively high. However, the high bus voltage requires multiple diodes to be connected in series to improve the reverse breakdown withstand voltage, which causes the potential to rise and is likely to reach the threshold, which, in turn, causes false triggering of the detection circuit. In addition, the drain-source voltage oscillation at the turn-on moment of SiC MOSFETs also increases the risk of false triggering of the detection circuit [10.17].

In contrast to the generally recognized short-circuit withstand time of about 10 μs in the industry for IGBTs, such a consensus has yet to be reached for SiC MOSFETs. Ranging from 1 to 6 μs for the state of the art, the short-circuit withstand time for SiC MOSFETs is considerably shorter than that of IGBTs, which raises the bar for the response time of the desaturation detection circuit to an unprecedented level.

Compared with the desaturation detection method, none of short-circuit detection methods are perfect either. Their advantages and disadvantages are listed in Table 10.7 [10.17].

TABLE 10.7 Short-circuit detection methods for SIC MOSFET.

Short-circuit detection methods		Advantage	Disadvantage
Desaturation detection	Diode sensing [10.18, 10.19, 10.20]	Simple and low cost	With blind spots and prone to false triggers
Parasitic inductance voltage detection	di/dt detection [10.21]	No blind spot and reliable	Prone to false triggering for HSF
	Current evaluation [10.22]	No blind spot and reliable	High peak current for FUL
	Two-stage RC-type current evaluation [10.23]	No blind spot and reliable	High cost
	RCD-type current evaluation [10.24]	No blind spot and reliable	High peak current for FUL
Current sensing	Hall device [10.25]	Convenient and no blind spot	Low accuracy
	PCB-type Rogowski coil [10.26]	Highly accurate	Complex circuit
Shunt sensing	Coaxial splitter [10.14,10.27]	Highly accurate	Large loss and high cost
	Nonlinear element [10.28]	Variable protection threshold	High cost and inconvenient installation
Gate charge detection [10.29, 10.30]		Quick response for HSF	Complex circuit and not applicable for FUL

10.3.4.2. Low-Stray Inductance Packaging

Today, most of the SiC MOSFET devices are still packaged using the traditional method for IGBTs, as shown in Figure 10.37. The bottom of the SiC MOSFET chip is soldered to the substrate through a direct bonding copper (DBC) liner, and its top is connected to the bus terminal through metal wire bonding. The whole package is then sealed using epoxy resin and silica gel. Despite its maturity and low cost, such a packaging method yields a large commutation circuit area. Thus, a relatively large stray inductance occurs, leading to serious voltage overshoot during the fast-switching process of SiC devices and other issues such as increased loss and electromagnetic interference. The metal wire bonding method, component pins, and the layout of chips are the key factors that cause the large commutation circuit area [10.18].

FIGURE 10.37 Typical structure of the traditional packaging method.

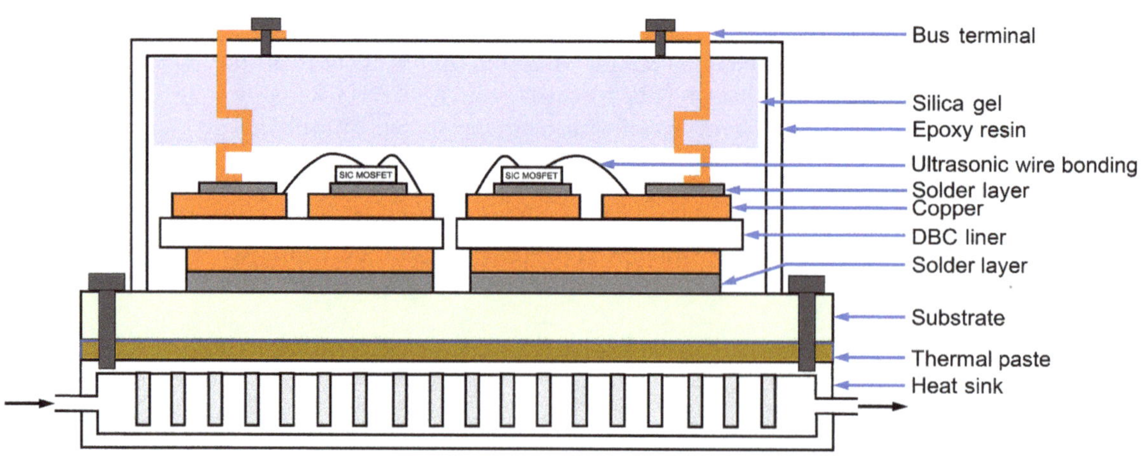

Hence, the following packaging methods are recommended to reduce the stray inductance in the SiC MOSFET devices:

1. For the metal wire bonding package structure, parasitic inductance mainly comes from the bonding wire, and its parasitic inductance can be approximately expressed as [10.31]:

$$L_\sigma = \frac{\mu_0 l}{2\pi}\left(\ln\frac{4l}{d} - 1.29\right) \tag{10.192}$$

where
l is the length of the bonding wire
d is the diameter of the bonding wire

Apparently, parasitic inductance can be reduced by cutting short the length of the bonding wire and enlarging its diameter, with a more prominent effect from the former.

2. The parasitic capacitance of the chip power circuit is primarily determined by the parasitic capacitance of the DBC ceramic layer, which can be expressed as [10.32, 10.33]:

$$C_\sigma = \varepsilon_0 \varepsilon_r \frac{(w_1 + 2a_1)^2}{h_3} \tag{10.193}$$

where
ε_0 is the vacuum permittivity
ε_r is the relative permittivity of ceramics
w_1 is the width of the power chip
a_1 is the width of exposed DBC ceramic layer beside the power chip
h_3 is the thickness of the DBC ceramic layer

Parasitic capacitance can, therefore, be lowered by increasing the thickness of the DBC ceramic layer and reducing the width of the exposed DBC ceramic layer besides the power chip.

3. Stray inductance can also be reduced by decreasing the cross-sectional area of the current flow path in the commutation circuit of SiC MOSFET [10.32]. For instance, the freewheeling diode of the upper and lower half-bridge SiC MOSFETs can be exchanged for the purpose.

Various packaging methods for SiC MOSFET modules in discrete and module formats are summarized in Table 10.8 [10.18]. The interested reader is referred to the cited references for further details of these methods.

TABLE 10.8 Typical structures of SiC power devices.

SiC MOSFET device	Metal wire bonding	Packaging method	Power level	Stray inductance (nH)
Discrete	Yes	0247, TO220, TO263, etc. [10.34, 10.35]	650-1700 V/5-100 A	10-20
	No	Flip patch [10.36]	650-1200 V/15-35 A	<5
Module	Yes	Traditional	1200-1700 V/20-500 A	20-30
		DBC+PCB hybrid [10.37-10.39]	1200-1700 V/20-100 A	<5
	No	SKiN [10.40-10.43]	1200 V/400 A	<1.5
		DLB, Cu-clip, SiPLIT [10.44-10.46]	650-1200 V/100-300 A	<5
		Double-sided welding (sintering) [10.47-10.49]	650 V-10 kV/50-300 A	<5
		Crimp [10.50-10.52]	1200 V/100-200 A	<10
		Three-dimensional [10.53-10.57]	1200 V/80 A	<1

Courtesy of CSEE (Chinese Society for Elec. Eng.).

10.3.4.3. Suppression of Bridge Arm Crosstalk

During the operation of SiC MOSFET devices, due to their fast-switching speed, a relatively large dv/dt is produced. Consequently, a crosstalk may occur between the upper and lower tubes of the same bridge arm, resulting in a possible false action due to its coupling to the gate through Miller capacitance. This problem worsens with the rise in switching speed. The bridge arm crosstalk can be suppressed by following the following design guidelines.

1. Connect an external capacitor in parallel with the gate source of the SiC MOSFET tube or increase the driving resistance to reduce dv/dt. This method, however, can cause a reduction in the switching speed of the device at the same time, and a drop in system efficiency due to the increase in switching loss [10.58].
2. When the power tube is turned off, a negative bias voltage can be added to the gate source, so as to superimpose on the positive crosstalk voltage generated during switching off to reduce the crosstalk voltage amplitude. Nevertheless, this method aggravates the negative crosstalk, which may damage the device. In addition, the need for a negative bias power supply increases the hardware cost [10.59].
3. An active clamping circuit composed of an auxiliary power tube and an auxiliary capacitor forms a low-impedance loop during the switching process, which can effectively suppress the crosstalk problem of the bridge arm without reducing the high switching speed of SiC MOSFET. The method requires an additional control signal, which increases the complexity of hardware and software design [10.60].
4. Dedicated isolation driver chips with integrated Miller clamps for SiC power devices are available from some SiC MOSFET suppliers, which are typically powered by dual-power supplies of positive and negative voltages. With a crosstalk suppression function similar to the aforementioned active clamping circuit, the crosstalk suppression parameters of the Miller clamp can be configured according to the specific needs.

10.3.4.4. Improvement of EMC

There are, in general, three essential elements in an EMI problem, namely, source, coupling path, and receiver. A source generates emission energy, and a coupling path transfers the emission energy to a receiver, which is sensitive to the radiated emission.

Usually, the EMI in a SiC MOSFET motor controller may come from several different sources:

1. During the turn-off process, the voltage on the SiC MOSFET rises at a rate of dv/dt, producing an additional voltage spike due to the presence of loop inductance.
2. During the turn-on process, the current through the SiC MOSFET surges at a rate of di/dt, resulting in an extra current spike due to the presence of loop inductance.
3. The recovery surge of freewheeling diodes.
4. The high-frequency components of the PWM waveform.

The EMI emission is usually transferred through three types of paths: electromagnetic coupling, or EM coupling, through which the radiate energy in the form of electromagnetic waves is generated via loop antennas (e.g., current loops) or rod antennas (e.g., leading wires), resulting in radiated interference, or the external electromagnetic radiation can also be received through these antennas, interfering with the internal circuit boards in the same way; inductive coupling, or H-coupling, through which the current and EMF are induced in one loop due to mutual inductance made by changing magnetic flux in the other loop; and capacitive coupling or E-coupling, through which the EMI is generated by the electric field coupling formed due to the capacitance between the wires and, thus, the cross-connection between one circuit and the other.

Electronic modules in a SiC MOSFET motor controller that are mostly sensitive to the EMI include CAN communication modules, reset modules, and sampling modules of temperature, position, phase current, bus current, voltage, and so on.

The design guidelines for improving the EMC of a SiC MOSFET motor controller are as follows:

1. In circuit board layout design, wiring should be neat and orderly to avoid redundancy, preferably along the bottom plate without flying lines. Ensure the integrity of the signal return path, without splitting or cutting it with any gaps in the loop ground plane or by changing the layers of the wiring reference plane, etc. In addition, the spacing between guard traces must have a minimum overlap of twice the trace width.
2. In selecting the type of shield for wires, combination shields, formed of two or more shields combined in the same wire, have the highest shielding effectiveness. In contrast, braided shields are better than aluminum foil in shielding low-frequency radiations, but they behave poorly at high frequencies due to the presence of braid gaps.
3. In making selection of diodes, the ones specially designed for automotive applications are preferred, ideally of high softness with neither spikes nor noise. Chips with low electromagnetic emission, high EMI immunity, and small input capacitance are recommended. Active chips with good EMC performance should be selected whenever possible, preferably with closely arranged, centrally distributed pins of power supply and ground. A small input capacitor helps reduce the magnitude of the current spikes that occur when switching logic states, thereby reducing magnetic field emissions and return currents, as well as allowing for

better matching to the transmission lines. The use of decoupling capacitors, typically integrated on-chip or in-package, can be an effective method to improve the EMC performance.

4. As far as the harness design associated with twisted pair wires, the current should form a loop through the twisted pair with a minimized loop area, with the signal return path laid as close as possible to the signal output path. In addition, to reduce the H-coupling, the negative wire of the twisted pair cannot be grounded at the load end, as this would not only increase the loop area but also weaken the ability of the twisted pair.
5. Keep the connector away from high-speed circuits, as the latter can easily generate potential differences of a few millivolts or larger, and the wires connected to the circuit board is an efficient antenna. If they are placed too close to each other, currents can be induced in the connected wires, resulting in excessive radiated emissions.
6. Suppress the time duration of high-speed signal rising and falling edge. An excessively long duration causes the low-frequency contents of the signal to rise dramatically, resulting in signal integrity and overheating problems with increased electromagnetic emissions due to the large amount of energy in the low-frequency band of digital signals.

10.3.4.5. Current Sharing among Parallel Paths

Limited by the manufacturing process and cost, a single state-of-the-art SiC MOSFET often falls short of meeting the current rating needs of traction applications for EVs. Hence, it is a common practice to employ parallelism, which would, however, inevitably result in uneven distribution of parasitic parameters and serious imbalance in dynamic and static currents among parallel current paths, especially at high switching frequencies. These unequal losses and stresses in voltage and current can not only make it necessary to derate the operation of SiC MOSFET power module but also increase its failure rate and shorten its lifespan. It is, therefore, desired to achieve an equal current sharing among these parallel paths. A variety of approaches have been proposed to make current sharing as uniform as possible among parallel paths, as summarized in Figure 10.38 [10.61].

FIGURE 10.38 Classification of current sharing approaches.

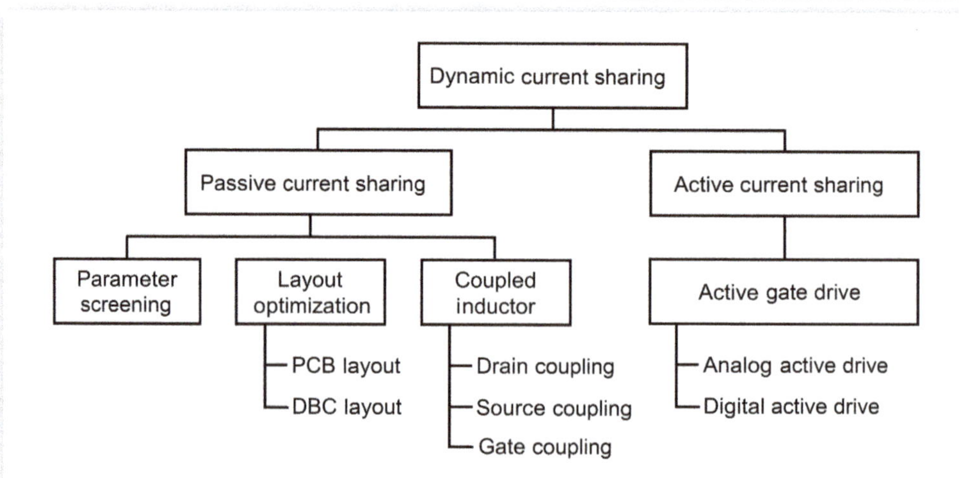

Courtesy of Wang, X.

1. Steady-state current sharing
The current sharing under steady state is relatively simple to deal with as only two main factors are involved, namely, the on-resistance of the SiC MOSFET and the parasitic parameters of the power loop, including parasitic resistance and parasitic inductance in the wiring. To avoid current sharing issues under steady state, it is recommended that parallel connection of MOSFETs be made with good on-resistance matching, and the wiring layout be optimized to reduce the effect of parasitic parameters and change the path of load current so that the flow through the parallel branches is as symmetrically as possible. In addition, the positive temperature coefficient (PTC) characteristics of on-resistance of SiC MOSFET give it equal current sharing ability to some extent.

In contrast, the dynamic current sharing is quite complex since it involves a great number of factors. The solutions to dynamic current sharing problems can be generally divided into passive and active ones.

The passive current sharing methods are elaborated next:

Parameter screening method: The SiC MOSFETs of consistent parameters are selected by means of a screening process for parallel connection, preferably from the same batch of wafer production. The parameters considered for screening include threshold voltage, transconductance, or transfer characteristics. The advantage of using transfer characteristics is that the difference between the transfer characteristics of two SiC MOSFETs hardly changes with temperature, and thus, the screening process can be made only at room temperature [10.62]. However, since the parametric characteristics must be obtained as a prerequisite for the application of this approach, all SiC MOSFETs must be tested and selected in advance before they are connected in parallel, which limits the use of this approach in mass production.

Layout optimization method: For upper transistors, the current sharing characteristics of the half-bridge parallel connection are better than those of the bare chip parallel connection, as their common source inductance can be kept consistent when the half bridges are connected in parallel. Nevertheless, this is not true for lower transistors. The position of the power port can be changed to improve the current sharing characteristics of lower transistors. For instance, the power port of two transistors connected in parallel can be placed in the middle, instead of at the beginning or end of the parallel branch, to ensure that their common source inductances are the same. However, this approach is effective for a small number of devices in parallel connection only and not applicable for a large number or odd number of devices in parallel connection [10.63].

Another methodology is to add decoupling capacitors on the left and right sides of the module, so that the current path in the transient state is more symmetrical than that of a single capacitor layout. Similar to the last approach, this methodology is not suitable for either a large number or odd number of devices in parallel connection. Or, a decoupling capacitor can be placed next to each parallel branch to provide switching energy, thereby ensuring the consistency of the common source inductance. At the same time, the parallel branches can be decoupled through the slots on the printed circuit board (PCB), with the current exchange among branches suppressed. However, this approach is not favorable to modularization due to the introduction of high-voltage capacitors. Meanwhile, there is no theoretical equation available for the calculation of the slot position and size, except simulation methods. The uniformity of dynamic current sharing can also be improved by changing the length and position of the DBC wires for each parallel branch within the module. However, as the number of parallel connections increases, it is necessary to introduce longer bonding wires to match the common source inductance of each branch, which can cause increased switching loss and voltage overshoot at both ends of the device during switching transients. On the other hand, the design of the drive circuit can be optimized to enhance the dynamic current sharing. Instead of using a single power amplifier for all parallel devices, the utilization of a power amplifier for each parallel device is recommended. As far as the influence of power module auxiliary source on parallel current sharing, the internal devices of the module with auxiliary sources offer better current sharing effect than the Kelvin source of discrete devices at the same switching frequency, although they essentially serve the same purpose of reducing common-source inductance and increasing switching speed. In addition, when power devices with Kelvin sources are connected in parallel, a small resistor, for example, as small as 100 mΩ, can be connected in series with the Kelvin source to suppress the circulating current.

Coupled inductor method: As an external passive device, the inductor can be placed at different positions to form drain coupling, source coupling, and gate coupling. Drain coupling puts a differential mode choke coil on the drain of a parallel device. When the differential mode current flows through, the coupling inductance of the branch generates a back EMF to reduce the differential mode current without affecting the common mode current. However, since the drain coupling needs to suppress a large differential mode power current, the inductor has a large volume. When expanding to parallel connection of multiple devices, although the adjacent tube coupling method is adopted, a large number of inductances are also required at the same time. In addition, due to the presence of leakage inductance, the voltage overshoot is worsened when the device is turned off, increasing the losses and the risk of device breakdown. Source coupling is different from drain coupling in principle. The increase in the power source inductances of parallel branches simultaneously can inhibit the uneven current sharing. The compensation effect enhances as the source inductance rises, but a larger source inductance can cause serious overshoot of the turn-off voltage. In order to solve such a contradiction, a coupling method can be adopted so that the differential mode current is inhibited with a large equivalent source inductance when the differential mode current flows through the power loop, and no large voltage overshoot is generated with a small equivalent source inductance when the common mode current flows through the power loop. This approach does not require a large volume of inductors since the power loop is affected by adjusting the drive loop through coupled inductors, but the disadvantage of inductor volume increasing with the rise of the coupled inductance remains. Gate coupling is achieved by connecting common-mode choke coils in series in the drive loop of each parallel device. This scheme helps suppress the uneven current caused by inconsistent parasitic inductances of the power loop on the parallel branch, and the suppression effect becomes prominent as the inductance value added to the common mode choke increases. However, only limited applications are found for this scheme as it cannot solve the dynamic unevenness caused by the difference in the parameters of the device itself, and the leakage inductance of the choke coil added to the drive circuit makes it prone to drive oscillations.

2. Dynamic current sharing

The active current sharing methods refer primarily to active gate drive methods, which can improve both switching characteristics and parallel current sharing characteristics by adding active devices to form a feedback circuit. A common active current sharing circuit consists of three units, namely, detection unit, control unit, and drive unit. The detection unit functions to collect uneven current information, the control unit processes the information and generates a corresponding correction signal, and the drive unit changes the drive signal of the corresponding parallel SiC MOSFET after receiving it to achieve current equalization. According to whether a digital controller is used, the active current sharing scheme can be divided into analog active drive and digital active drive, as discussed in the following paragraphs.

Analog active drive: In one of the analog active gate drive approaches, dynamic current sharing can be achieved by adjusting the drive resistance of SiC MOSFETs connected in parallel. Equivalent drive resistance in the drive unit consists of a fixed resistance connected in parallel with a low-power MOSFET, and its on-resistance can be changed by controlling the gate-source voltage of the low-power MOSFET. The control signal is derived from the integration of the unbalanced current information detected by the three-winding differential current sensor connected in series with the drains of the two parallel transistors. The output control signal increases with the nonuniformity in current sharing, causing the on-resistance of the adjustable MOSFET and, thus, the equivalent drive resistance to drop. As a result, the rate of current rise in the SiC MOSFET with a slow turn-on process is accelerated, and it finally matches with that of the other MOSFET in parallel connection. However, this approach increases the cost since it requires a high-bandwidth differential-mode current sensor and an operational amplifier. In another approach, the gate-source voltage of each parallel MOSFET is made to follow the first one through closed-loop control and finally ensures the consistency of the gate-source voltage of each parallel MOSFET. Nevertheless, the problem of uneven current caused by the inconsistent threshold voltage remains unsolved.

Digital active drive: With a current sensor connected in series to each parallel branch, the collected current information is transmitted to the digital controller, which controls the switching process by adjusting the delay in the driving voltage of each SiC MOSFET. This approach can be further improved by collecting the current information of each SiC MOSFET in parallel connection through its source inductance voltage integration, transmitting it to a CPLD, and making the current rising and falling time points as well as the slopes consistent across all parallel branches by controlling the delay and amplitude of the driving voltage. Note that, however, both approaches require multicycle adjustment, thus increasing the complicity and cost of the control circuit.

10.3.4.6. Thermal Management

Junction temperature is the actual operating temperature of semiconductors in electronic equipment. In operation, it is usually higher than the package case temperature. Different from IGBTs, SiC MOSFETs are capable of operating at a much higher junction temperature, typically 175°C or higher. Although this poses challenges to the packaging material design for SiC MOSFET power modules, it indeed offers possibilities to simplify their thermal management system. The fact that the thermal conductivity of silicon carbide is one order of magnitude higher than that of silicon (1490 versus 150 W/m·K) further helps achieve this design goal.

Owing to the high power density of the SiC MOSFET power modules, the design of their thermal management system faces unique engineering problems, particularly in high-temperature-resistant packaging, cooling technology, system modeling, junction temperature extraction, and thermal management strategies.

1. High-temperature-resistant packaging
In the design of high-temperature-resistant packaging, cooling capacity is an important consideration. The following design recommendations would help improve the reliability of high-temperature-resistant packaging.
First, copper or aluminum ribbons can be used, instead of wires, for chip front connection to enhance thermal conductivity [10.18].
Second, sintered silver connection is a better alternative than tin sheet or solder paste for its much higher thermal conductivity (approximately 200 versus 50 W/m·K), low sintering temperature, and high melting point [10.18].
Third, for ceramic substrates and metal baseplates, materials with high thermal conductivity and a thermal expansion coefficient close to that of SiC (around 3.7 ppm/K) are preferred. The thermal conductivity, thermal expansion coefficient, and flexural strength of several ceramic materials for substrates are listed in Table 10.9 [10.18]. Among these materials, AlN is currently an ideal substrate material because of its high thermal conductivity and proper thermal expansion coefficient, as well as reasonable cost. Thermal conductivity, the thermal expansion coefficient, and density of a few candidate materials are listed in Table 10.10 [10.18]. Among them, Cu-based alloys such as Cu/Mo, Cu/W, and Cu/C are superior due to their high thermal conductivity and appropriate thermal expansion coefficient, despite their high density and cost. Today, continuing efforts are being made in quest for optimal materials for substrates and baseplates.

TABLE 10.9 Properties of insulation materials for substrates.

Material	Thermal conductivity (W/m·K)	Thermal expansion coefficient (ppm/K)	Flexural strength (MPa)
Al_2O_3 (99%)	33	7.2	345
AlN	150–180	4.6	360
BeO	270	7.0	250
Si_3N_4	70	3.0	932

Courtesy of CSEE (Chinese Society for Elec. Eng.).

TABLE 10.10 Properties of common materials for baseplates.

Material	Thermal conductivity (W/m·K)	Thermal expansion coefficient (ppm/K)	Density (g/cm³)
Cu	398	17.8	8.96
Al	218	23	2.7
Cu/Mo	162–275	6.8–13	9.3–10
Cu/W	130–205	5.6–9	14.8–17
Cu/C	200	4	—
AlSiC	170–200	6.5–13.8	2.97–3.04

Courtesy of CSEE (Chinese Society for Elec. Eng.).

2. Cooling technology
   The SiC MOSFET chip or module is usually cooled by attaching it to a heat sink, through which heat is further dissipated by air cooling or liquid cooling.

   Microchannel cooling can be an effective way to enhance the heat transfer from the chip or module to the baseplate, by either making microchannels directly on the substrate of the chip or integrating microchannels into the thick metal layer under the chip [10.18, 10.64]. This technology allows the cooling medium to act directly on the chip without being affected by the thermal resistances from the multilayer structure of the module.

   In addition, phase change technologies such as HPs or sprays can be a good alternative to the single-phase air cooling or water cooling for improving heat dissipation due to their higher thermal conductivity and efficiency [10.18, 10.64].

   A simple comparison of the heat transfer coefficients among state-of-the-art cooling methods is illustrated in Figure 10.39 [10.64].

FIGURE 10.39 Comparison of various cooling technologies.

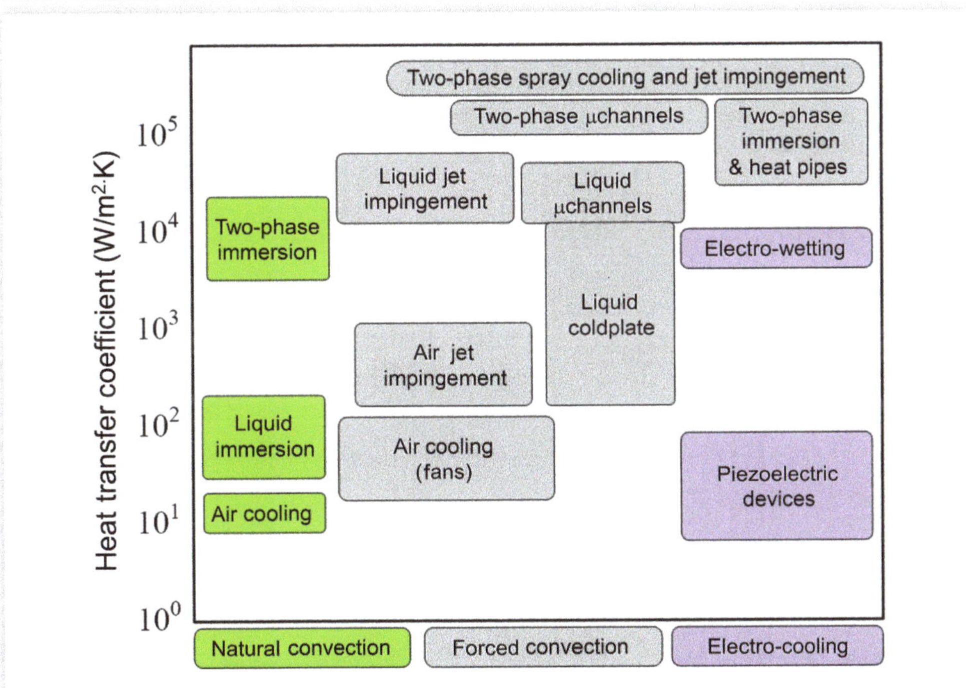

3. System modeling and junction temperature extraction
   Real-time estimation of junction temperature of SiC MOSFETs in a power module is of crucial importance to ensure their healthy operation and prolonged lifespan. For this purpose, a thermal network model is necessary to take the losses of power devices and thermal characteristics into account. In general, there are two types of thermal network model, namely, Cauer model and Foster model.

   The Cauer model is a physical model in which the thermal characteristics of each layer of heat dissipation material is represented correspondingly by a group of thermal resistance and heat capacity (RC) parameters in the actual physical model, as shown in Figure 10.40 [10.65]. The drawback of the Cauer model, however, is that heat capacity parameters are all relative to the reference point of ground. Once the reference point changes, thermal capacity parameters need to be re-evaluated and measured.

FIGURE 10.40 Schematic of a Cauer model.

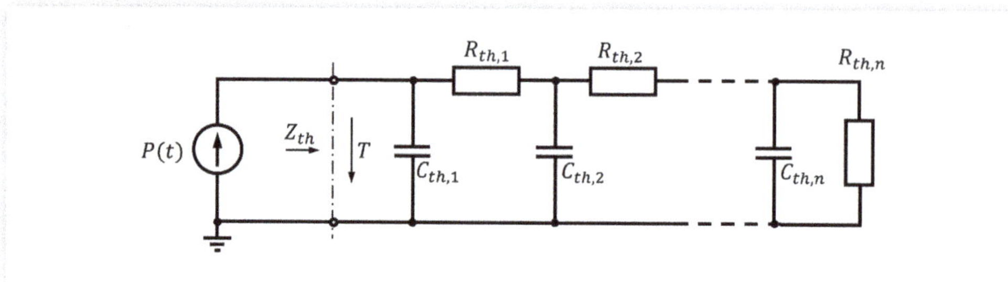

Courtesy of Shanghai Automobile.

The Foster model, on the other hand, is a local thermal network model in which the RC nodes are no longer in one-on-one correspondence with thermally conductive materials. Rather, the network nodes do not have any physical meaning, and only the temperature differences between junction temperature and sensor measurement point are needed, as illustrated in Figure 10.41 [10.65].

FIGURE 10.41 Schematic of a Foster model.

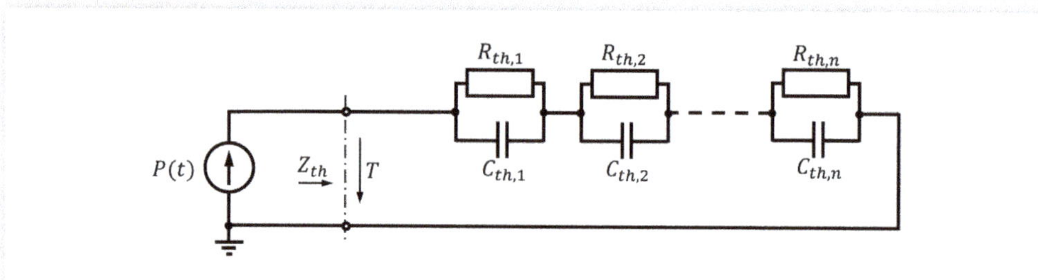

Courtesy of Shanghai Automobile.

In order to solve the junction temperature using the aforementioned thermal network model, the power loss of SiC MOSFET is required. In general, the power loss of SiC MOSFET, P_{loss}, is composed of conduction loss P_{cond} and switching loss P_{sw}, namely,

$$P_{loss} = P_{cond} + P_{sw} \tag{10.194}$$

The conduction loss P_{cond} can be calculated by

$$P_{cond} = V_{ds} I_d \tag{10.195}$$

where

V_{ds} is the turn-on voltage
I_d is the turn-on current

Both V_{ds} and I_d can be measured experimentally after the turn-on ringing is over. The switching loss P_{sw} is given by

$$P_{sw} = E_{sw} f_{sw} \tag{10.196}$$

where
E_{sw} is the switching energy
f_{sw} is the switching frequency

Switching energy E_{sw} is the sum of the turn-on switching energy E_{on} and turn-off switching energy E_{off}, that is,

$$E_{sw} = E_{on} + E_{off} \tag{10.197}$$

Turn-on switching energy E_{on} and turn-off switching energy E_{off} can usually be extracted by measuring the transient response of SiC MOSFET using a double pulse tester (DPT). Based on the experimental data obtained in such experiments at the given switching frequency f_{sw}, the switching energy can be obtained by the following polynomial regression equation:

$$E_{sw} = p_0 + p_1 I_d + p_2 T_j + p_3 I_d^2 + p_4 I_d T_j \tag{10.198}$$

where
p_0, p_1, p_3, and p_4 are polynomial coefficients
T_j is the junction temperature

As an alternative approach, the junction temperature of SiC MOSFETs in a power module can also be extracted in real time by means of monitoring temperature-sensitive electrical parameters (TSEPs), such as the on-state voltage drop in the off state of body diode when the SiC MOSFET has an intrinsic body diode. Since this type of approach is still under investigation, the interested readers are referred to the cited references for a detailed description of these approaches [10.66, 10.67, 10.68, 10.69, 10.70].

4. Thermal management strategies
Once the junction temperature of SiC MOSFET is extracted, it can be actively managed to prevent it from exceeding the predetermined upper limit and fluctuation level. Since SiC MOSFETs are capable of operating at a junction temperature much higher than silicon-based IGBTs, the target of thermal management often minimizes the fluctuation of junction temperature, instead of reducing junction temperature itself. The corresponding thermal management strategies are detailed as follows [10.71]:

   - **Dead time adjustment** [10.71]: The dead time refers to the duration when both the upper and lower switches on the same arm of the H-bridge are turned off simultaneously and the current flows through the body diode, which is required in the commutation process involving the two switches. When the load of the inverter declines at a certain moment of operation, the load current must be reduced to adapt to this change in the operating condition, causing the junction temperature to drop. To prevent the thermal stress brought by this situation, the dead time of the upper and lower switches on the same arm can be extended to increase the heat generation in the body diode so as to compensate for the potential drop in junction temperature.

- **Gate voltage adjustment** [10.71]: The gate voltage is one of the parameters that affect the switching characteristics of SiC MOSFETs. As the gate drive voltage increases within a certain range, both switching loss and conduction loss become lower and the turn-on speed becomes faster. Therefore, the gate drive voltage can be adjusted to maintain the junction temperature at a previous level, if needed. For instance, when the drain current decreases, the gate drive voltage can be reduced to keep the switching loss nearly constant, thus minimizing the influence on junction temperature.
- **Switching frequency adjustment** [10.71]: According to Equation (10.196), switching frequency has a significant effect on the switching loss of SiC MOSFETs. The junction temperature is related to the power loss of SiC MOSFETs by

$$T_j = T_c + P_{loss} R_{th} \tag{10.199}$$

where

T_c is the temperature of the motor controller housing

R_{th} is the total thermal impedance from the SiC MOSFET chip to the motor controller housing

It is apparent from Equations (10.196) and (10.199) that the junction temperature can be regulated by adjusting the switching frequency.

- **Combined methodology** [10.71]: When a single thermal management approach mentioned is not adequate to maintain the junction temperature within a certain limit, these approaches can be combined to reach the target more effectively.

10.4. Design of Drivetrains

Common drivetrains for EV applications can be categorized into fixed speed ratio reducers and multispeed transmissions. Either reducers or transmissions function to change the torque and speed of the traction motor to fit the operating condition of the vehicle. Many factors must be taken into consideration in making a selection between a single-speed reducer or a multispeed transmission for a given vehicle, such as market positioning, dynamic performance, energy economy, and costs.

Either layshaft or planetary gears can be used to achieve speed reduction and torque multiplication, at a fixed speed ratio for a reducer, or at stepped ratios or even continuously variable speed ratio for a transmission. However, traditional transmissions for internal combustion-powered vehicles, particularly manual transmissions (MTs), automatic transmissions (ATs), and continuously variable transmissions (CVTs), are generally not suitable for EVs. The main reasons for their obsolescence are the excessive number of gears, inconvenience in operation, compromised shift quality for MTs, and poor mechanical efficiency for ATs and CVTs due to the presence of a torque converter or a high-pressure oil pump.

Thus, multispeed stepped transmissions such as AMTs, DCTs, and planetary gear power shift transmissions (PSTs) are considered appropriate for EV applications, primarily for their potential to achieve high mechanical efficiencies and dynamic performance.

Most of the AMTs for EVs typically feature a layshaft configuration, using synchronizers for gear shifting without a clutch. The gear shifting process of such a clutchless AMT is completed in an automated manner by actuating a shift fork controlled by a transmission control unit (TCU). Although two- to four-speed AMTs have been widely applied for electric commercial vehicles for their simplicity, high efficiency, and low cost, they are rarely used in passenger EVs for their intrinsic shortcoming of power interruption during the gear shifting process. The schematic of a typical two-speed AMT is given in Figure 10.42 [10.1].

FIGURE 10.42 Schematic of a typical two-speed AMT.

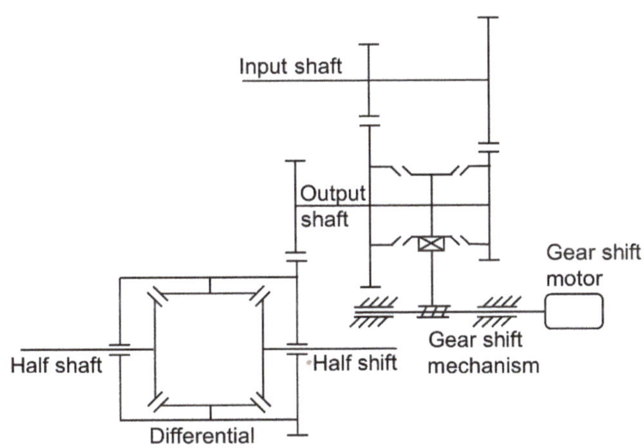

Of a layshaft gearing system, DCTs for EV applications typically feature two- to four-speed with dual-clutches, either wet or dry, that allow power shift without interruption during the shift process. The dual-clutches can be configured either on the input shaft or on the output shaft, as shown in Figure 10.43 [10.1]. Controlled by a TCU, the target gear can be preselected using the pertaining synchronizer and the gear shift is executed by engaging the on-coming clutch and disengaging off-going clutch. In a simpler configuration, synchronizers can be eliminated, whereas the shift can be made by switching clutches only. As a result, the shift time can be shortened and the cost reduced.

FIGURE 10.43 Schematic of typical two-speed DCTs. (a) Dual-clutches on the input shaft and (b) dual-clutches on the output shaft.

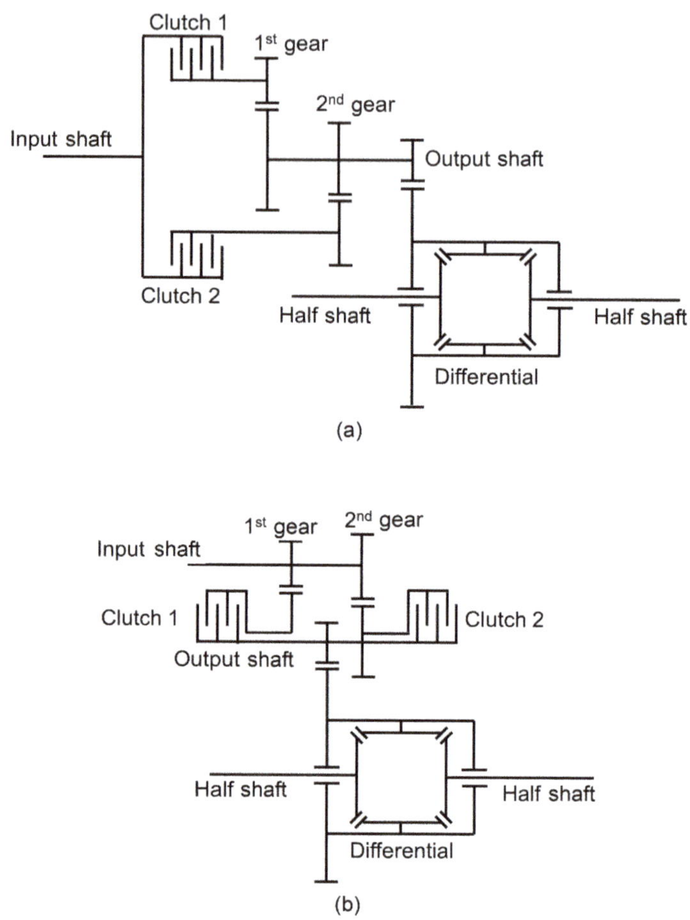

As for PSTs, one or more planetary gearsets are used with clutches or brakes for gear shift without power interruption. For a given input torque, a PST is generally more compact than an AMT or a DCT, particularly in a coaxial configuration. No physical gear selection is needed for the gear shifting process, with the shifting principle similar to an AT. The schematic diagrams of a non-coaxial and a coaxial two-speed PSTs are provided in Figure 10.44 [10.1].

Other than the multispeed stepped transmissions mentioned earlier, a novel type of electric variable transmission (EVT) is proposed for battery EV applications [10.1, 10.72, 10.73]. The speed ratio of an EVT is continuously variable within a certain range using electromagnetic and mechanical means or their combination, with no shift jerk at all. It is an electric drive system with highly integrated mechanical and electrical components to meet the compactness requirements.

FIGURE 10.44 Schematic of typical two-speed PSTs. (a) Non-coaxial and (b) coaxial.

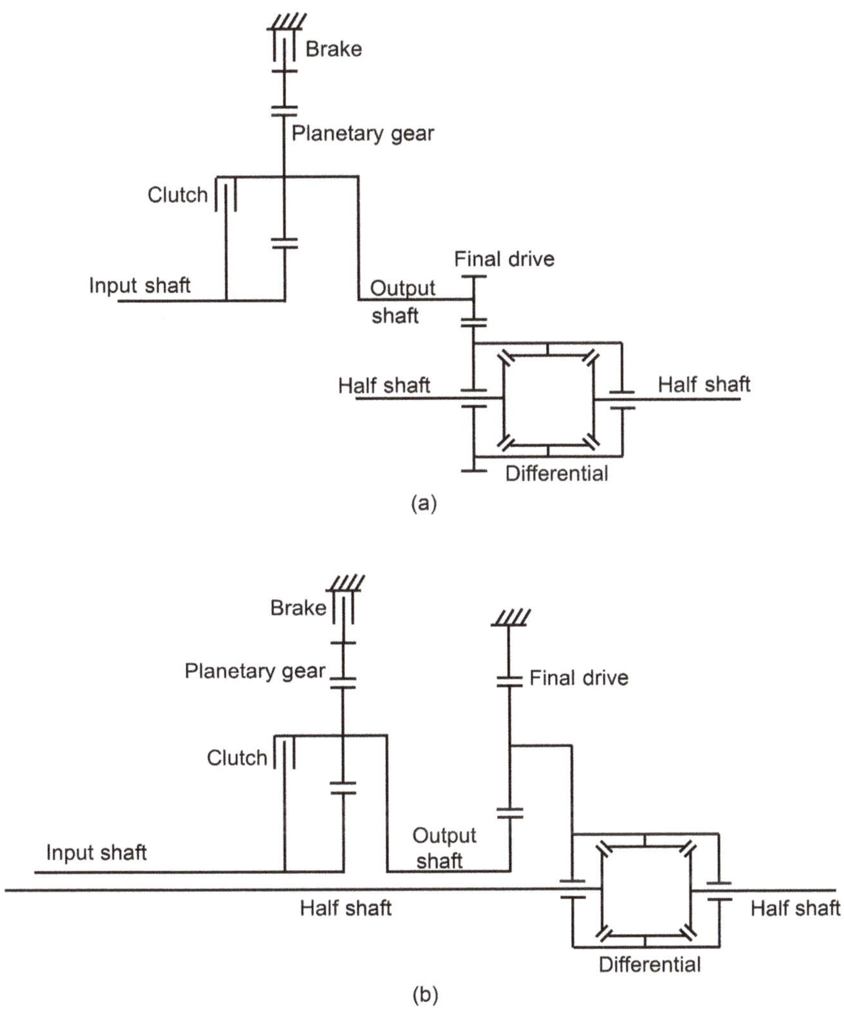

A typical EVT electric drive system consists of a dual-rotor motor, a planetary gearset, a final drive, and a differential, as shown in Figure 10.45 [10.1, 10.72, 10.73]. Coupling the torques from the two rotors with the planetary gearset, it provides a continuously variable optimal transmission ratio within a certain range to drive the vehicle according to the requirements of operating conditions and intention of the driver. The inner and outer rotors are arranged concentrically with the stator, with the inner rotor connected with the sun gear, the outer rotor connected with the ring gear, and the planet carrier connected with the driving wheels through a differential. In fact, there are a variety of architectures in terms of the system topology, that is, the connectivity between the rotors and the elements of the planetary set. In addition, the stator can be arranged not only between the inner and outer rotors but also outside the outer rotor. The use of a magnetic insulating material in the middle of the stator can decouple the magnetic field acting on the inner and outer rotors; otherwise, magnetic fields acting on the inner and outer rotors would become complicated. Three-phase windings are distributed on the stator, and permanent magnets are installed on the inner and outer rotors.

FIGURE 10.45 Schematic of an EVT.

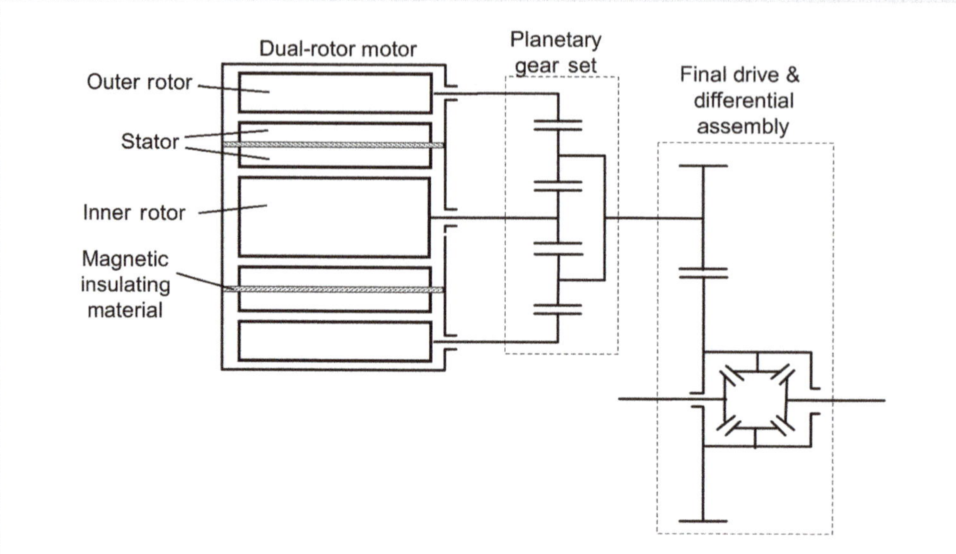

Obviously, the mechanical power of the inner and outer rotors is input to the sun gear and the ring gear, respectively, and coupled by the planet carrier to drive the vehicle through the differential. Under the condition of regenerative braking, the mechanical power of the vehicle is split in the planetary gearset through the differential and converted into electrical power by the motor system composed of the two rotors and the stator.

The main advantages of such an EVT electric drive system are as follows:

1. The operating points of the inner and outer motor subsystems can be selected to maximize either the overall efficiency of the system or the output power, which helps improve the driving range, acceleration, and gradeability of the vehicle.
2. The speed ratio of the system can be continuously switched to prioritize dynamic performance or energy economy at any time, without any shift jerk of stepped transmissions.
3. It can easily realize some difficult operating conditions for regular electric drive systems. For example, hill-hold, that is, the driver holds the vehicle stationary on a slope only by depressing the accelerator pedal. A regular e-motor can only be operated in hill-hold condition for a short period of time as it must operate in the stall condition. In contrast, the EVT has no problem at all offering such a function, as none of the inner and outer motor subsystems has to operate in the stall condition.
4. If the sun gear and the planet carrier are connected to the inner and outer rotors, respectively, and the ring gear is used as output, the EVT electric drive system can offer additional functions that regular electric drive systems do not have, such as speed augmentation and positive speed reversing. The output speed of the EVT output shaft can be higher than the speeds of both rotors. The engineering value of this functionality is that the maximum speed of the motor can be lowered in exchange for improvements in reliability, lifespan, and quietness. The working principle of positive speed reversing indicates that the vehicle can be driven in the reverse direction with both inner and outer rotors rotating in the positive direction, allowing the use of two-quadrant motor subsystems to achieve all-around vehicle operating conditions.

10.4.1. General Layout Design

In the general layout design, the overall structure and dimensional parameters of the main drivetrain components such as gears, shafts, bearings, shifting actuators, final drives, differentials, and housings are determined. In the early stage of the design, the dimensional parameters of these components are estimated and a corresponding layout plan is worked out. Then, a general layout sketch is made, which is helpful in evaluating its packaging in the vehicle. In the detailed design stage, with the three-dimensional assembly model established, the dimensions and structure of the drivetrain are precisely defined, and its packaging in the vehicle is further verified.

Most layshaft drivetrains for EV applications are of a two-shaft configuration [10.1]. They usually have an input shaft, an output shaft, and a final drive, with the pinion of the final drive arranged on the output shaft.

For planetary gear drivetrains, a simple planetary gearset is commonly used. Sometimes, however, the planetary gearset may work with a layshaft gearset to increase the reduction ratio, while the input shaft is in parallel with the output shaft. In fact, a combination of layshaft and planetary gearsets are often used in the same drivetrain for various reasons such as center distance makeup, cost reduction, and speed ratio flexibility.

In order to reduce the radial dimension of a reducer or a transmission, a coaxial arrangement is often used, in which the centerline of the input shaft coincides with that of the half shafts. In some circumstances, a coaxial reducer arrangement using a layshaft configuration is also utilized.

The general layout design of the drivetrain is also affected by its mounting location in the vehicle. The axial dimension is critical for an FF drivetrain in order to fit it into the space between driving wheels, while the radial dimension is crucial for an MR or RR drivetrain to ensure proper ground clearance and floor height.

The basic concepts and principles of laying out shafts, bearings, synchronizers, and gearsets are as follows [10.1]:

1. Adding or taking out a gearset on the transmission path will cause the rotation direction of the output shaft to reverse.
2. Arrange the gearset subject to high torque as close to the bearing as possible to reduce the deflection of the shaft under loads.
3. Minimize the distance between the bearings as much as possible to reduce the deflection of the shaft under loads and to avoid the tooth surface contact pattern deviating toward one side, as the consequent reduction in effective contact width can cause an increase in contact stress.
4. The shaft distance is determined by the center distance of the gearset corresponding to the maximum transmission ratio.
5. The synchronizer can be arranged on the input shaft, intermediate shaft, or output shaft, but it is mostly on the output shaft because the gear on the input shaft is often too small to accommodate the synchronizer.
6. Taking an example shown in Figure 10.46 [10.1], the total length of the synchronizer assembly, s_1, is usually provided by its supplier. The widths of the gears, b_1, b_2, and b_3 are the outcome of the gear design, while a gap of 1.5–2 mm is generally recommended for the axial gaps, c_1, c_2, and c_3, between gears as well as between the gear and the synchronizer. The center distance, A, between the input and output shaft can be determined by the method described in the following section.

FIGURE 10.46 An example of general layout design.

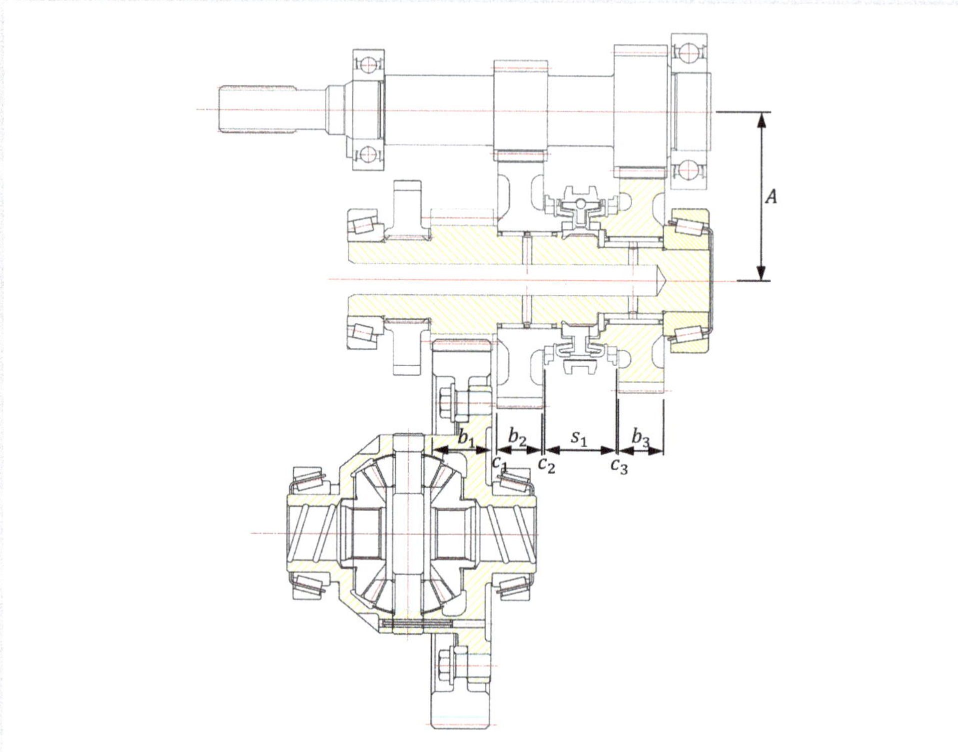

10.4.2. Determination of Basic Characteristic Parameters

For layshaft drivetrains, the center distance is the most important basic characteristic parameter, which significantly influences the overall size of the system. Generally speaking, the greater the center distance, the larger the size, volume, weight, and cost of the drivetrain.

The center distance is directly related to the torque transmitted by the system. The larger the torque transmitted, the greater the center distance. In the general layout design, the center distance needs to be determined first. An estimation of the center distance is given by the following empirical formula [10.30]:

$$A = K_A \left(T_{imax}\right)^{1/3} \tag{10.200}$$

where

K_A is the center distance factor
T_{imax} is the maximum input torque

The center distance factor K_A can be selected from Table 10.11 [10.1].

TABLE 10.11 Center distance factors for different types of EVs.

Type of EVs	Center distance factor K_A
Passenger	0.255
Commercial	0.278

Reproduced from, Yuan, Y., Electric Vehicle Transmissions: Structure and Control (Beijing: China Machine Press, 2020), ISBN: 978-7-111-66809-1, with the permission of China Machine Press.

For planetary gear drivetrains, the pitch diameter of the ring gear is a basic characteristic parameter, which determines the radial dimension of the transmission, to a large extent. The pitch diameter of the ring gear can be preliminarily estimated based on its statistical relationship with the maximum input torque.

Finally, in engineering practice, the estimated center distance A is usually rounded to an integer for easy measurement.

10.4.3. Design of Layshaft Gearsets

The gearsets for the electric drivetrain must be compact and lightweight. However, due to the very few forward speeds, all the gearsets are subject to a high ratio of load cycles, which poses great challenges to their fatigue life. Obviously, the lifespan of all gearsets in a reducer must not be shorter than that of the entire vehicle.

In addition, the noise and vibration issues in an electric drivetrain are often more prominent than its internal combustion-powered counterpart due to the lack of the masking effect of engine noise. The noise and vibration issues are further worthened by the higher rotating speeds of a traction motor than an ICE. All of these factors put forward more stringent requirements on the NVH design of the gearsets for electric drivetrains.

In short, the gearset design for an EV drivetrain generally requires compactness, light weight, low noise, long lifespan, and low cost. In order to achieve these design goals, nonstandard helical gears with profile modification are widely used. The meshing of helical gears sets in at a point, then the contact line crosses the tooth face diagonally, and finally the meshing ends at a point as well. This meshing process reduces the load change rate from beginning to end, minimizing the dynamic torque fluctuations simultaneously.

Due to very limited number of gearsets in an electric drivetrain, it commonly has an ample axial space for the layout of gearsets. Therefore, the design principle for these gearsets has changed from prioritizing bending strength to NVH performance. Taking into account the significant impact of the gear contact ratio on its NVH characteristics, on the basis of the prior-determined transmission ratio and the maximum input torque (see Chapter 3 for details), it is recommended that the layshaft gearsets for EV drivetrains be designed by following the steps that follow.

10.4.3.1. Initial Selection of the Center Distance

The initial center distance can be estimated using Equation (10.1) as discussed in the last section.

10.4.3.2. Calculate the Pitch Diameters

Based on the gear ratio i and the initial center distance A, the pitch diameter of the pinion d_1 is given by

$$d_1 = \frac{2A}{1+i} \tag{10.201}$$

The pitch diameter of the gear d_2 is

$$d_2 = id_1 \tag{10.202}$$

10.4.3.3. Initial Selection of Tooth Width, Number of Teeth, and Pressure Angle

The tooth width of the pinion, b_1, can be estimated using its pitch diameter by

$$b_1 = 0.65d_1 = b_2 \tag{10.203}$$

where it is assumed that the tooth width of the pinion is identical to that of the gear, b_2.

The choice of the number of teeth affects the strength and contact ratio of the gearset. In general, as the number of pinion teeth increases, its normal modulus decreases. As a result, its bending fatigue strength drops and its tooth height shortens accordingly. Nevertheless, the contact ratio grows, which improves the smoothness and quietness of gear meshing.

The initial selection of the number of teeth on the pinion, z_1, depends on whether it is standard or not. The minimum number of teeth without undercutting is 17 for a standard pinion gear; thus, $z_1 \geq 17$ is generally selected. For modified pinion gears, a fewer number of teeth can be used, but the minimum number of teeth is usually kept above 11. A rule of thumb is, thus, that $z_1 = 15 - 17$ for passenger vehicles and $z_1 = 12 - 17$ for commercial vehicles.

In selecting pressure angle α, its influence on gear strength and NVH characteristics must be taken into consideration. When the pressure angle is large, both bending strength and contact strength of gear teeth increase, but the minimum number of teeth without undercutting also reduces, causing a drop in contact ratio and an increase in noise level. Hence, a pressure angle smaller than standard value is often used to improve NVH performance. The recommended values for helix angle and pressure angle are listed in Table 10.12 [10.1].

TABLE 10.12 Recommended values for the helix angle and pressure angle.

Gear parameters	Passenger EVs	Commercial EVs
Helix angle	25–45° for cars 25–30° for high-end vehicles	18–30° for trucks
Pressure angle	14.5°, 15°, 16°, 16.5°, 17°, 17.5°, 20° for cars	20° for trucks

Reproduced from, Yuan, Y., Electric Vehicle Transmissions: Structure and Control (Beijing: China Machine Press, 2020), ISBN: 978-7-111-66809-1, with the permission of China Machine Press.

10.4.3.4. Selection of Helix Angle and Estimation of Normal Module

The selection of the helix angle β has a large influence on the strength and NVH characteristics of the gearset. With an increase in the helix angle, the contact ratio rises and gear meshing becomes smooth and quiet. Consequently, although the axial force increases, the bending strength of gear teeth decreases, especially when the helix angle exceeds 30°. The preliminary value of the helix angle β can be selected according to Table 10.12.

Next, the normal module m_n can be estimated by

$$m_n = \frac{2A\cos\beta}{z_1 + z_2} \tag{10.204}$$

For standard gears, the normal module is then rounded to the standard value according to the applicable standards or regulations. As mentioned earlier, modified gears of nonstandard modules are often used to improve the strength and NVH performance of the gears. Subsequently, the helix angle β can be updated according to the rounded normal module m_n by

$$\beta = \frac{180}{\pi} \arccos \frac{m_n(z_1 + z_2)}{2A} \tag{10.205}$$

And then the helix angle β should be rounded again to the closest half or integer degree value. The pitch diameters of the pinion and gear as well as the center distance are updated by using the following equations:

$$d_1 = \frac{z_1 m_n}{\cos\beta} \tag{10.206}$$

$$d_2 = \frac{z_2 m_n}{\cos\beta} \tag{10.207}$$

$$A = \frac{d_1 + d_2}{2} \tag{10.208}$$

Finally, the update values of tooth width can be obtained by substituting the updated pitch diameters of the pinion and gear into Equation (10.203).

10.4.3.5. Calculate the Diameters of Base and Addendum Circles

Per the aforementioned pressure angle (de facto normal pressure angle), the transverse pressure angle can be acquired by

$$\alpha_t = \arctan \frac{\tan\alpha}{\cos\beta} \tag{10.209}$$

The base circle diameters of the pinion and gear then become

$$d_{b1} = d_1 \cos\alpha_t \tag{10.210}$$

$$d_{b2} = d_2 \cos\alpha_t \tag{10.211}$$

The addendum is

$$h_a = h_a^* m_n \tag{10.212}$$

where h_a^* is the addendum factor. Let $h_a^* = 1$, then the tip diameters of the pinion and gear are given by

$$d_{a1} = d_1 + 2h_a \tag{10.213}$$

$$d_{a2} = d_2 + 2h_a \tag{10.214}$$

10.4.3.6. Calculate the Base Pitch

The base pitch can be calculated by

$$p_b = \frac{2\pi r_{b1}}{z_1} \tag{10.215}$$

10.4.3.7. Initial Calculation of the Transverse Contact Ratio and Overlap Ratio

The contact ratio is crucial in the design of gearsets for its close relationship with gear noise.

The total contact ratio ε of an involute cylindrical helical gearset is composed of two parts, namely, the transverse contact ratio ε_α and overlap ratio ε_β. The physical meaning of the transverse contact ratio ε_α is the ratio of the length of action L_{act} to the base pitch p_b, namely,

$$\varepsilon_\alpha = \frac{L_{act}}{p_b} \tag{10.216}$$

whereas that the length of action is the length from the start of the tooth contact to the end of contact. For two involute cylindrical helical gears meshing with each other, the transverse contact ratio is given by

$$\varepsilon_\alpha = \frac{0.5\left(\sqrt{d_{a1}^2 - d_{b1}^2} \pm \sqrt{d_{a1}^2 - d_{b1}^2}\right) + \alpha' \sin\alpha_t'}{\pi m_t \cos\alpha_t} \tag{10.217}$$

where

- α' is the engagement angle (also known as the pitch circle pressure angle)
- α_t' is the engagement angle
- α_t is the transverse pressure angle
- m_t is the transverse module
- "+" denotes external gear meshing
- "−" denotes internal gear meshing
- subscripts 1 and 2 denote the pinion and gear, respectively

By ignoring the effect of gear modification temporarily, we have $\alpha' = \alpha$. Thus, the transverse module m_t is given by

$$m_t = \text{transverse}\frac{m_n}{\cos\beta} \tag{10.218}$$

and the transverse engagement angle can be solved by

$$\text{inv}\alpha_t' = \text{inv}\alpha_t + \frac{2(x_{t2} \pm x_{t1})}{z_2 \pm z_1}\tan\alpha_t \tag{10.219}$$

where

$$\alpha_t = \arctan\frac{\tan\alpha_n}{\cos\beta} \tag{10.220}$$

and x_{t1} and x_{t2} are the modification coefficients of the pinion and gear, respectively. Again, by neglecting the effect of the modification coefficient on the engagement angle, we have $\alpha' = \alpha$, which can be substituted into Equation (10.217) to calculate the initial value of the transverse contact ratio.

The physical meaning of the overlap ratio ε_β is the ratio of the effective tooth width and the base pitch:

$$\varepsilon_\beta = \frac{b_f \tan\beta_b}{p_b} \tag{10.221}$$

where

b_f is the effective tooth width
β_b is the base helix angle

Let the effective tooth width b_f be the tooth width of the gear b_2, then the base helix angle β_b is actually the helix angle β, and Equation (10.221) becomes

$$\varepsilon_\beta = \frac{b_2 \tan\beta}{p_b} = \frac{b_2 \sin\beta}{\pi m_n} \tag{10.222}$$

The sum of the transverse contact ratio ε_α and the overlap ratio ε_β is called the total contact ratio ε. Its physical meaning is that the total number of teeth partially meshing with each other alternately changes between ε_1 and ε_2, where ε_1 and ε_2 are the two integers closest to ε. In the meshing action, the total length of the contact lines changes in passing through the field of contact, depending on the values of the transverse contact ratio ε_α and the overlap ratio ε_β. When either of the two ratios is an integer, the sum of the total length of the contact lines remains constant throughout the field of contact. It is generally reckoned that this condition makes the meshing smooth and quiet, especially when the two contact ratios are both integers.

10.4.3.8. Calculation and Determination of the Modification Coefficient

Modified gears are generally used for four effects: avoiding undercutting of the tooth profile while improving the strength and service life of the gears; when the gear ratio is large, increasing the tooth thickness of the pinion and reducing the tooth thickness of the gear so as to make the service life of the pinion and gear equivalent; reducing gear meshing noise; and making up the center distance.

Tooth height modification refers to the distance that the centerline of the rack cutter moves away from or close to the center of the gear from the position where the standard gear is cut along the wheel blank, which is generally expressed as $x \cdot m$, where x is the modification coefficient and m is the module. A positive modification stands for $x > 0$, zero modification for $x = 0$, and negative modification for $x < 0$. The effect of gear modification on its tooth profile is shown in Figure 10.47 [10.1]. For a positively modified gear, its tooth thickness increases, but its tooth tip becomes sharp, while for a negatively modified gear, its tooth thickness decreases, but its tooth tip becomes wide.

FIGURE 10.47 Effect of gear modification on its tooth profile.

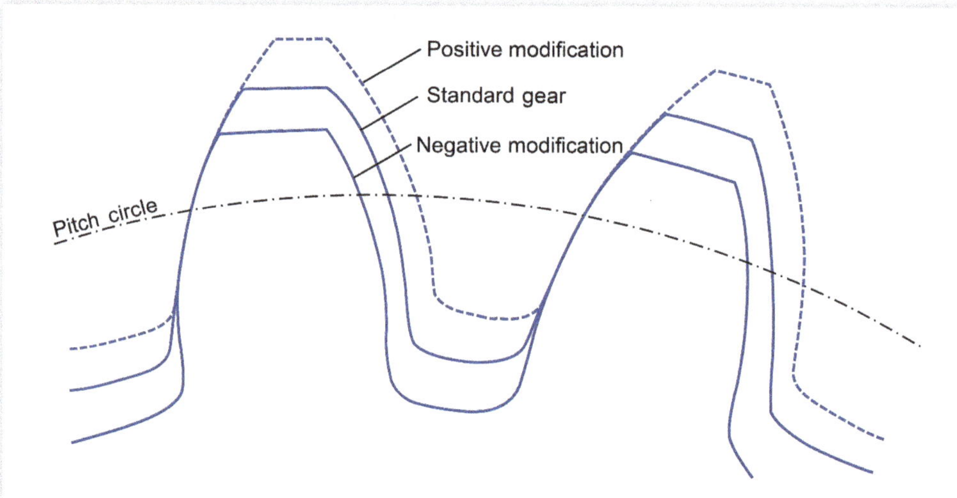

For a helical gearset, the total modification coefficient is the sum of the modification coefficients of the pinion and gear in mesh, namely,

$$x_{tc} = x_{t1} + x_{t2} \tag{10.223}$$

Gear modification can be divided into tooth height modification and tooth lead modification. The tooth height modification can increase the tooth root strength of the pinion but cannot reduce the noise. Its characteristic is that the sum of the total modification coefficient is zero, that is,

$$x_{t1} + x_{t2} = 0 \tag{10.224}$$

On the other hand, tooth lead modification not only increases the strength of the pinion tooth root but also reduces noise. Its characteristic is that the sum of the total displacement coefficient is not zero, that is,

$$x_{t1} + x_{t2} \neq 0 \tag{10.225}$$

When selecting the gear modification coefficient, the following principles should be considered:

1. A large total modification coefficient x_{tc} is preferred for low gears.
2. For the first gear, $x_{tc} > 1$ is recommended to improve the tooth root strength.
3. A smaller x_{tc} should be selected for a higher gear.
4. The total modification coefficient $-0.2 \le x_{tc} \le 0.2$ can be selected for the highest gear to ensure the maximum contact strength, as well as antiadhesive and wear-resistant characteristics.
5. With the drop of the total modification coefficient x_{tc}, the tooth root bending strength weakens, but it is easy to absorb noise and vibration.

In order to select the gear modification coefficient, the equivalent numbers of teeth z_{n1} and z_{n2} of the pinion and gear should be calculated first:

$$z_{n1} = \frac{z_1}{\cos^2 \beta} \tag{10.226}$$

$$z_{n2} = \frac{z_2}{\cos^2 \beta} \tag{10.227}$$

The modification coefficients x_{t1} and x_{t2} of the pinion and gear are then selected according to the chart of the tooth profile factor versus equivalent numbers of teeth at various modification coefficients, in which the areas of the narrow tooth top and the undercut should be avoided. Applicable standards should be complied in selecting the modification coefficients.

Finally, per equivalent numbers of teeth z_{n1} and z_{n2} as well as the modification coefficients x_{t1} and x_{t2}, the tooth profile factor (also called form factor) Y_F can be obtained as well from the aforementioned chart, which is needed for the subsequent calculation of the basic value of tooth root stress.

10.4.3.9. Update of Center Distance, Contact Ratio, and Other Parameters

In order to make up the center distance and adjust the contact ratio, once the modification coefficient is determined, the design parameters such as the center distance and the contact ratio must be updated accordingly.

First, the transverse engagement angle α_t' should be updated using Equation (10.219) with the new modification coefficient values.

Second, because the tooth height modification does not affect the center distance, only the center distance for tooth lead modified gearset needs to be updated with the following equation:

$$A' = A\frac{\cos\alpha_t}{\cos\alpha_t'} = \frac{1}{2}m_t\left(z_1 + z_2\right)\frac{\cos\alpha_t}{\cos\alpha_t'} = \frac{1}{2}m_n\left(z_1 + z_2\right)\frac{\cos\alpha_t}{\cos\beta\cos\alpha_t'} \tag{10.228}$$

where A' is the new value of center distance. The variation coefficient of center distance, therefore, becomes

$$y_t = \frac{A' - A}{m_n} \tag{10.229}$$

The updated values of the addendums for the pinion and gear, respectively, are

$$h_{a1} = m_n\left(h_{an}^* + x_{n1}\right) - \Delta y_t m_t \tag{10.230}$$

$$h_{a2} = m_n\left(h_{an}^* + x_{n2}\right) - \Delta y_t m_t \tag{10.231}$$

where

h_{an}^* is the addendum coefficient, usually $h_{an}^* = h_a^* = 1$
Δy_t is the addendum variation coefficient

$$\Delta y_t = x_{tc} - y_t \tag{10.232}$$

and m_t is the transverse module:

$$m_t = \frac{m_n}{\cos\beta} \tag{10.233}$$

The dedendum of the pinion and gear, respectively, is

$$h_{f1} = m_n\left(h_{an}^* + c_n^* - x_{n1}\right) \tag{10.234}$$

$$h_{f2} = m_n\left(h_{an}^* + c_n^* - x_{n2}\right) \tag{10.235}$$

where c_n^* is the radial clearance factor, $c_n^* = c^* = 0.25$.

The whole depths of the pinion and gear are identical:

$$h = h_{a1} + h_{f1} = h_{a2} + h_{f2} \tag{10.236}$$

The tip diameters of the pinion and gear are given by

$$d_{a1} = d_1 + 2h_{a1} \tag{10.237}$$

$$d_{a2} = d_1 + 2h_{a2} \tag{10.238}$$

The root diameters of the pinion and gear are

$$d_{f1} = d_1 - 2h_{f1} \tag{10.239}$$

$$d_{f2} = d_2 - 2h_{f2} \tag{10.240}$$

Third, the transverse contact ratio ε_α and overlap ratio ε_β after modification can be updated per Equations (10.217) and (10.218).

10.4.3.10. Check the Bending Strength for the Teeth

The bending strength of gear teeth should be checked against the applicable standards.

The checkup of tooth bending strength mainly involves the calculation of tooth root stress. The tooth root stress is obtained by making corrections on the basis of nominal bending stress (i.e., the maximum tensile stress at the root of the tooth profile on the load acting side) with the corresponding coefficients

$$\sigma_F = \sigma_{F0} K_A K_V K_{F\beta} K_{F\alpha} \tag{10.241}$$

where

σ_{F0} is the basic value of tooth root stress
K_A is the coefficient of utilization
K_V is the dynamic load coefficient
$K_{F\beta}$ is the tooth load distribution coefficient calculated by bending strength
$K_{F\alpha}$ is the load distribution coefficient between teeth calculated by bending strength

The bending strength of gear teeth should meet one of the following two conditions: First, the calculated tooth root stress σ_F should not be greater than the allowable tooth root stress σ_{FP}, namely,

$$\sigma_F \le \sigma_{FP} \tag{10.242}$$

Second, the calculated safety factor S_F of bending strength should not be less than the minimum safety factor S_{Fmin} of bending strength, namely,

$$S_F \ge S_{Fmin} \tag{10.243}$$

The allowable tooth root stress σ_{FP} is determined by

$$\sigma_{FP} = \frac{\sigma_{FG}}{S_{Fmin}} \tag{10.244}$$

where σ_{FG} is the bending stress limit of the calculated gear. σ_{FG} can be calculated by

$$\sigma_{FG} = \sigma_{Flim} Y_{ST} Y_{NT} Y_{\delta relT} Y_{RrelT} Y_x \tag{10.245}$$

where

σ_{Flim} is the tooth root bending fatigue limit of the test gear
Y_{ST} is the stress correction coefficient of the test gear
Y_{NT} is the life coefficient calculated by bending strength
$Y_{\delta relT}$ is the relative tooth root fillet sensitivity coefficient
Y_{RrelT} is the relative tooth root surface condition coefficient
Y_x is the size factor calculated for bending strength

Refer to gear design manuals for the calculation method of these coefficients. The minimum safety factor S_{Fmin} of bending strength is generally 2.

If the bending strength of gear teeth does not meet any of these two conditions, then the number of teeth on the pinion and gear should be selected once again according to Section 10.4.3.3, and then these steps should be repeated to iterate until the bending strength of gear teeth meets the two conditions.

10.4.3.11. Contact Strength Checkup of Tooth Face

The contact strength of gear teeth should be checked against the applicable standards.

The checkup of tooth face contact strength mainly involves the allowable contact stress of the pinion and gear for the larger value of the contact stress at pitch points and boundary points in the field of contact of meshing teeth, and they are checked separately. The contact stresses of the pinion and gear can be calculated with the following formulas:

$$\sigma_{H1} = Z_B \sigma_{H0} \sqrt{K_A K_V K_{H\beta} K_{H\alpha}} \tag{10.246}$$

$$\sigma_{H2} = Z_D \sigma_{H0} \sqrt{K_A K_V K_{H\beta} K_{H\alpha}} \tag{10.247}$$

where

K_A is the coefficient of utilization
K_V is the dynamic load factor
$K_{H\beta}$ is the tooth load distribution coefficient for contact strength calculation
$K_{H\alpha}$ is the load distribution coefficient between the teeth for contact strength calculation
Z_B and Z_D are the meshing coefficients of the pinion and gear, respectively
σ_{H0} is the basic value of contact stress at pitch points

Refer to gear design manuals for the calculation method of these coefficients and factors.

The tooth surface contact strength should meet one of the following two conditions. First, the larger value σ_H of the calculated contact stress at pitch points and boundary points in the field of contact of meshing teeth of the pinion and gear should not be greater than its corresponding allowable contact stress σ_{HP}, namely,

$$\sigma_H \leq \sigma_{HP} \tag{10.248}$$

Second, the calculated safety factor S_H of contact strength should not be less than the corresponding minimum safety factor S_{Hmin}, namely,

$$S_H \geq S_{Hmin} \tag{10.249}$$

The allowable contact stress σ_{HP} is determined by

$$\sigma_{HP} = \frac{\sigma_{HG}}{S_{Hmin}} \quad (10.250)$$

where σ_{HG} denotes the contact stress limit of the calculated gear. σ_{HG} can be calculated by

$$\sigma_{HG} = \sigma_{Hlim} Z_{NT} Z_L Z_V Z_R Z_W Z_x \quad (10.251)$$

where

σ_{Hlim} is the contact fatigue limit of the test gear
Z_{NT} is the life coefficient for contact strength calculation
Z_L is the lubricant coefficient
Z_V is the velocity factor
Z_R is the roughness coefficient
Z_W is the work hardening coefficient
Z_x is the size factor for contact strength calculation

Refer to gear design manuals for the calculation method of these coefficients and factors.

The minimum safety factor S_{Hmin} of contact strength is generally 1.50–1.60.

If the tooth face contact strength does not meet any of the two conditions, the teeth numbers of pinion and gear should be selected once again according to Section 10.4.3.3, and the steps should be repeated to iterate until the tooth face contact strength meets the two conditions.

10.4.4. Design of Planetary Gearsets

The most common planetary gearsets used in EV reducers or transmissions can be divided into simple planetary gearsets, double planetary gearsets, and Ravigneaux planetary gearsets, as shown in Figure 10.48 [10.1].

A simple planetary gearset can be defined as a system consisting of a sun gear, a ring gear, and a set of planet gears supported by a planet carrier. The central axis positions of these planet gears on the carrier are relatively fixed to each other, and they mesh with the sun gear and ring gear at the same time.

A double planetary gearset can be defined as a system consisting of a sun gear, a ring gear, and two sets of planet gears supported by a single planet carrier. These planet gears are divided into an inner planet gearset and an outer planet gearset. All mounted on the same carrier, each of inner planet gears meshes with one of outer planet gears. Meanwhile, the inner planetary gearset meshes with the sun gear, while the outer planetary gearset meshes with the ring gear.

A Ravigneaux planetary gearset is composed of two sun gears, a ring gear, and two sets of planet gears supported by a single planet carrier. The two sun gears are divided into a small sun gear and a large sun gear, while the two sets of planet gears are divided into a short planet gearset and a long planet gearset. Each of short planet gears meshes with one of long planet gears, all of which are mounted on the same planet carrier. There are two types of structures in terms of their interconnection. First, the short planet gearset meshes with both small sun gear and ring gear, while the long planet gearset meshes with a large sun gear but not with a ring gear. Second, the short planet gearset meshes with a small sun gear but not with a ring gear, while the long planet gearset meshes with both large sun gear and ring gear. The Ravigneaux planetary gearset is of the first type of structure if not particularly indicated. The second type of structure is omitted here as it consists of similar characteristics.

FIGURE 10.48 Schematic of common planetary gearsets. (a) Simple planetary gearset, (b) double planetary gearset, and (c) Ravigneaux planetary gearset.

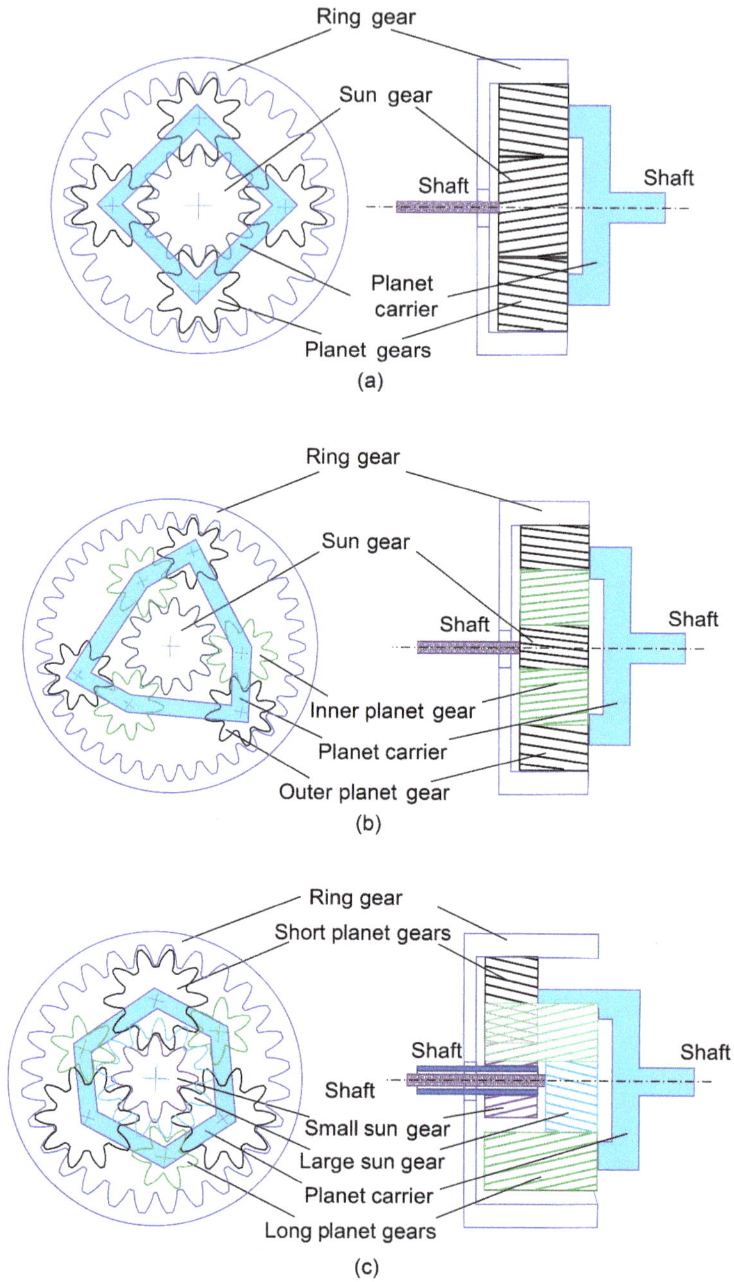

10.4.4.1. Basic Equations of Planetary Gearsets

Each planetary gearset has two DOFs, meaning that when the sun gear, ring gear, or planet carrier is fixed, the system can provide two different ratios according to the combination of the other two members as input and output. Therefore, both simple and double planetary gearsets can theoretically provide six different transmission ratios, as listed in Table 10.13 [10.1].

TABLE 10.13 Transmission ratios of simple and double planetary gearsets.

State or action	Sun gear	Fixed	Fixed	Input	Output	Input	Output
	Planet carrier	Input	Output	Fixed	Fixed	Output	Input
	Ring gear	Output	Input	Output	Input	Fixed	Fixed
Gear ratio	Simple planetary	$K/(K+1)$	$(K+1)/K$	$-K$	$-1/K$	$K+1$	$1/(K+1)$
	Double planetary	$K/(K-1)$	$(K-1)/K$	K	$1/K$	$-K+1$	$-1/(K-1)$

Reproduced from, Yuan, Y., Electric Vehicle Transmissions: Structure and Control (Beijing: China Machine Press, 2020), ISBN: 978-7-111-66809-1, with the permission of China Machine Press.

In Table 10.13, the characteristic parameter $K = \frac{Z_r}{Z_s}$, where Z_r is the number of teeth of the ring gear and Z_s is the number of teeth of the sun gear. It is clear from Table 10.13 that when the planet carrier is fixed, the planetary gearset reverses the direction of rotation from input to output. When the planet carrier acts as the input member, the planetary gearset plays the role of increasing the rotating speed from input to output. On the contrary, when the planet carrier acts as the output member, the planetary gearset serves to decrease the speed.

A lever diagram can be used to visually and conveniently analyze the speed and torque relationship among members in a planetary gearset.

The speed relationships among members of the simple or double planetary gearsets are given by the following formula:

$$\frac{N_s - N_c}{N_r - N_c} = \pm K \tag{10.252}$$

where

N_s is the speed of the sun gear
N_c is the speed of the planet carrier
N_r is the speed of the ring gear

The sign before K should be positive for a simple planetary gearset and negative for a double planetary gearset. Therefore, for a single planetary gear train, we have

$$N_s + KN_r - (K+1)N_c = 0 \tag{10.253}$$

and for a double planetary gearset,

$$N_s - KN_r + (K-1)N_c = 0 \tag{10.254}$$

Apparently, $N_s = N_r = N_c$ is the solution to Equations (10.253) and (10.254). Its physical meaning is that the planetary gearset rotates like a rigid body without any relative movement between its members.

The planet gear speed relative to the planet carrier is given by

$$N_p = (N_r - N_c)\frac{Z_r}{Z_p} = (N_c - N_s)\frac{Z_s}{Z_p} \tag{10.255}$$

where Z_p is the number of teeth of a planet gear.

The torque relationship among the members of a simple planetary gearset is given by the following formula:

$$T_s = \frac{T_r}{K} = \frac{-T_c}{K+1} \tag{10.256}$$

where

T_s is the total torque acting on the sun gear
T_c is the total torque acting on the planet carrier
T_r is the total torque acting on the ring gear

The negative sign means that the torque on the planet carrier is in the opposite direction to the torques of the sun gear and ring gear.

The torque relationship among the members of a double planetary gearset is

$$T_s = \frac{-T_r}{K} = \frac{T_c}{K-1} \tag{10.257}$$

where the negative sign means that the torques on the ring gear are in the opposite direction to the torque on both planet carrier and sun gear.

Per Equations (10.256) and (10.257), we can derive that

$$T_s + T_r = -T_c \tag{10.258}$$

Two or more planetary gearsets can be connected in several ways to provide multiple transmission ratios.

On the other hand, a Ravigneaux planetary gearset also has two DOFs, which means that when the large and small sun gears, ring gear, or planet carrier is fixed, the system can provide two different gear ratios according to the combination of the other two members as input and output. Thus, a Ravigneaux planetary gearset can theoretically provide 12 different transmission ratios, as given in Table 10.14 [10.1].

In Table 10.14, the characteristic parameter $K_1 = \frac{Z_r}{Z_{s1}}$, where Z_r is the number of teeth of the ring gear and Z_{s1} is the number of teeth of the small sun gear; the characteristic parameter $K_2 = \frac{Z_r}{Z_{s2}}$, where Z_{s2} is the number of teeth of the large sun gear.

TABLE 10.14 Transmission ratios of Ravigneaux planetary gearsets.

State or action	Small sun gear	Fixed	Fixed	Input	Output	Input	Output
	Large sun gear	Optional*	Optional*	Optional*	Optional*	Optional*	Optional*
	Planet carrier	Input	Output	Fixed	Fixed	Output	Input
	Ring gear	Output	Input	Output	Input	Fixed	Fixed
Gear ratio		$K_1/(K_1+1)$	$(K_1+1)/K_1$	$-K_1$	$-1/K_1$	K_1+1	$1/(K_1+1)$
State or action	Small sun gear	Optional*	Optional*	Optional*	Optional*	Optional*	Optional*
	Large sun gear	Fixed	Fixed	Input	Output	Input	Output
	Planet carrier	Input	Output	Fixed	Fixed	Output	Input
	Ring gear	Output	Input	Output	Input	Fixed	Fixed
Gear ratio		$K_2/(K_2-1)$	$K_2/(K_2-1)$	$(K_2-1)/K_2$	K_2	$1/K_2$	$-K_2+1$

* "Optional" means that it can operate in any one of input, output, or follow-up modes.

The speed relationship among the members in a Ravigneaux planetary gearset is given by the following formulas:

$$\begin{cases} N_{s1} + K_1 N_r - (K_1+1) N_c = 0 \\ N_{s2} - K_2 N_r + (K_2-1) N_c = 0 \end{cases} \tag{10.259}$$

Obviously, $N_{s1} = N_{s2} = N_r = N_c$ is a solution to Equation (10.259). Its physical meaning is that the Ravigneaux planetary gearset rotates like a rigid body without any relative movement among its members.

The torque relationship among the members in a Ravigneaux planetary gearset is given by the following formulas:

$$\begin{cases} (K_1+K_2+1) T_{s1} + K_2 T_r - (K_2+1) T_c = 0 \\ (K_1+K_2+1) T_{s2} - (K_1+1) T_r + K_1 T_c = 0 \end{cases} \tag{10.260}$$

where both T_{s2} and T_c are acting in the same direction and are opposite to T_{s1}.

10.4.4.2. Number of Planet Gears and Their Spacing

The number of planet gears and their spacing should be determined prior to the design of the planet carrier. The number of planet gears n_p depends on the stress caused by the loads acting on it, as described in the last section. Their spacing is related to the number of planet gears and the number of teeth of the sun gear and ring gear that are meshed with.

In EV reducers or transmissions, a simple planetary gearset generally uses three or more planet gears, and four to six planetary gear systems are the most common. This number is mainly limited by the space available for accommodating planet gears between the ring gear and sun gear, as well as the structure of planet carrier. Two planet gears are only considered when the space does not allow to arrange three large planet gears, typically due to an extremely large gear ratio.

For standard or close-to-standard gears, the number of teeth of each members must satisfy the following relationship:

$$Z_r - Z_s = 2Z_p \tag{10.261}$$

It is obvious that $Z_r - Z_s$ must be an even number, meaning that the number of teeth of the sun gear and ring gear must be either odd or even. Otherwise, there will be a mismatch between them. Since the pitch diameter of a gear is proportional to its number of teeth, the gear meshing at the pitch diameter is completely a tangential contact.

For nonstandard gears, the relationship of Equation (10.261) is not necessarily true. Because of its large number of teeth, the ring gear is easier to be made nonstandard than the sun gear, which is sometimes considered in the design. In this case, $Z_r - Z_s$ can be either an even number or an odd number, and the gear meshing at the pitch diameter is not necessarily a tangential contact.

The planet gears can only be fitted into the planetary gearset to mesh simultaneously with the sun gear and ring gear at a limited number of positions. For a simple planetary gearset, the number of these positions n_{fit} is

$$n_{fit} = Z_r + Z_s \tag{10.262}$$

For double planetary gearsets, the number of these positions n_{fit} is

$$n_{fit} = Z_r - Z_s \tag{10.263}$$

For Ravigneaux planetary gearsets, we have

$$n_{fit} = Z_{s1} + Z_{s2} \tag{10.264}$$

It can be noted from Equation (10.262) that there are only $Z_r + Z_s$ positions within the 360° range of the gears for a simple planetary gearset that allow planet gears to mesh with the sun gear and ring gear simultaneously when assembled. The minimum angle between these positions is, therefore, $360^\circ/(Z_r + Z_s)$, which is called the least-mesh angle. Therefore, for a simple planetary gearset, when planet gears are evenly arranged around the center axis of the planet carrier, $(Z_r + Z_s)/n_p$ is required to be an integer. In other words, the azimuth angle between the planet gears relative to the axis of the planet carrier must be an integral multiple of the least-mesh angle; otherwise, they cannot be installed. For a double planetary gearset, when the planet gears are evenly arranged around the center axis of the planet carrier, $(Z_r - Z_s)/n_p$ is required to be an integer. On the other hand, a Ravigneaux planetary gearset must satisfy all the relations from Equations (10.262) to (10.264).

When the aforementioned conditions for uniform arrangement of planet gears cannot be met, a symmetrical arrangement across the central axis can be taken as an alternative. Generally, when the number of planet gears is an odd number, the planet gears are often evenly distributed relative to the center of the planet carrier; when the number of planet gears is even, the planet gears are often arranged symmetrically across the center of the planet carrier in pairs. For example, given a simple planetary gearset with the numbers of sun gears, planet gears, and ring gears, respectively, being 37, 22, 81 and the number of planet gears being 4, then the least-mesh angle is $360^\circ/(81 + 37) = 3.051^\circ$. Because $(81 + 37)/4 = 29.5$ is not an integer, it is, thus, impossible to distribute the 4 planet gears evenly. Instead, it can be considered to arrange the azimuth angles between adjacent planet gears as 3.051° × 29 = 88.479° and 3.051° × 30 = 91.530°.

In designing a planetary gearset, the numbers of teeth on the gears must be checked against the aforementioned relational expressions. For instance, for a Ravigneaux planetary gearset, if the numbers of teeth Z_{s1} for the small sun gear, Z_{s2} for the large sun gear, Z_{p1} for the short planet gear, Z_{p1} for the long planet gear, and Z_r for the ring gear are 19, 25, 26, 17, and 81, respectively, and the number of planet gears N_p is 4, then the simple planetary gearset composed of the small sun gear, short planetary gear, and ring gear must be checked first. Because

$$\frac{Z_r + Z_{s1}}{n_p} = \frac{81+19}{4} = 25$$

the result is an integer, which, therefore, meets the requirement of a uniform distribution of four planet gears.

Next, the double planetary gearset consisting of a large sun gear, long planetary gear, short planetary gear, and ring gear must be checked. Since

$$\frac{Z_r - Z_{s2}}{n_p} = \frac{81-25}{4} = 14$$

the result is again an integer, which also meets the requirement of a uniform distribution of four planet gears.

Finally, the gear system consisting of the small sun gear, big sun gear, short planet gear, and the long planet gear must be checked. Since

$$\frac{Z_{s1} + Z_{s2}}{n_p} = \frac{19+25}{4} = 11$$

the result is an integer once again, which also satisfies the requirement of a uniform distribution of both four inner planet gears and four outer planet gears. The aforementioned three checkups ensure that the spacing of the Ravigneaux planetary gearset meets the assembly requirements.

In order to reduce the noise and vibration of planetary gearsets, the distance between two adjacent planet gears is usually designed so that one planet gear starts to mesh when its adjacent planet gear and the sun gear mesh halfway. This requires the distance between these two adjacent planet gears on the sun gear to satisfy the following relation:

$$\hat{L}_s = \frac{\pi(2k_s + 1)}{Z_s} \qquad (10.265)$$

where

k_s is an integer

Z_s is the number of teeth on a sun gear

Similarly, the spacing on the ring gear is given by

$$\hat{L}_a = \frac{\pi(2k_r + 1)}{Z_r} \tag{10.266}$$

where
k_r is an integer
Z_r is the number of teeth on a ring gear

Since the ratio of the aforementioned distance between the sun gear and the ring gear must be equal to Z_s/Z_r, the following formula should be held true in the design of a sun gear and ring gear:

$$\frac{2k_s + 1}{2k_r + 1} = \frac{Z_s^2}{Z_r^2} \tag{10.267}$$

In addition, there are other factors affecting gear noise that need to be considered in planet gear design, such as the contact ratio, addendum coefficient, and planetary pitch.

Moreover, in order to make the load evenly shared by all planet gears, it is necessary to control the position tolerances of the planet gear shaft hole on the planet carrier, that is, its coaxiality and parallelism.

10.4.4.3. Transverse Contact Ratio of Planetary Gearsets

Design parameters such as gear blank diameter, center distance, runout, and tolerance of the tooth tip chamfer affect the transverse contact ratio ε_α of a planetary gearset. In design, the following formula can be used to calculate the transverse contact ratio ε_α to examine its change and ensure its extreme value within the acceptable range. The transverse contact ratio ε_α of a planetary gearset is given by

$$\varepsilon_\alpha = \frac{L_{act}}{p_b} \tag{10.268}$$

For external gears,

$$L_{act} = \frac{1}{2}\left[\sqrt{d_o^2 - d_b^2} + \sqrt{D_o^2 - D_b^2} - \sqrt{4C^2 - (d_b + D_b)^2}\right] \tag{10.269}$$

For internal gears,

$$L_{act} = \frac{1}{2}\left[\sqrt{d_o^2 - d_b^2} - \sqrt{D_i^2 - D_b^2} + \sqrt{4C^2 - (D_b - d_b)^2}\right] \tag{10.270}$$

and

$$p_b = p_t \cos\alpha_t \tag{10.271}$$

where

L_{act} is the length of the line of action
p_t is the transverse pitch
p_b is the base pitch
α_t is the transverse pressure angle at the pitch circle diameter
d_o is the addendum circle diameter of the pinion
d_b is the base circle diameter of the pinion
D_o is the addendum circle diameter of the external gear
D_b is the base circle diameter of the gear
D_i is the inner diameter of the internal gear
C is the center distance

Note that when the undercut of the gear profile exceeds the position of the deepest meshing point, Equations (10.269) and (10.270) are no longer valid.

10.4.4.4. Gear Design for Strength and Durability

When a gear pair is entering engagement, the friction force acting on the pinion points to the direction of the pinion side, which tends to reduce the torque produced by a normal force with respect to the pinion center. Therefore, the normal force is a function of the coefficient of friction between the teeth:

$$F_n = \frac{T}{R_b\left(1 - f\tan\phi\right)} \tag{10.272}$$

where

T is the torque transmitted by the gear pair
R_b is the base circle radius of the pinion
f is the coefficient of friction
ϕ is the transverse pressure angle at the pitch diameter

As a gear pair is quitting engagement, the friction force acting on the pinion points away from the pinion side. Both friction force and normal force tend to increase the torque relative to the pinion center. Thus, the normal force is given by

$$F_N = \frac{T}{R_b\left(1 + \mu\tan\phi\right)} \tag{10.273}$$

A couple of planetary gears are sometimes directly connected one after another to obtain three and more transmission ratios. The architecture of such a connection has a significant impact on the loading on the gearset. Hence, it is necessary to comprehensively consider the architecture optimization in addition to the gear design itself. Most of the principles for the strength and durability design of layshaft gears are equally applicable for planetary gearset designs. Refer to the relevant content in the last section for details.

Planetary gearsets mainly have several failure modes such as bending fatigue, contact fatigue, tooth surface bonding, gear tooth wear, and planetary gear rupture under high loads. In some scenarios, the failure of the planetary gearset is not due to the tooth fracture but a result of the fracture between the shaft hole and the tooth root. It is clear that a proper thickness of the material between the shaft hole and the tooth root is key to preventing from the latter type of failure.

Gear wear is a common failure mode, which is usually caused by insufficient surface hardness, dust intrusion, or lubrication design defects. A widely accepted principle of transmission gear design is that no discernible wear is allowed within its design life. Thus, all transmission gears must operate in an enclosure filled with lubricating oil, and their tooth faces are generally hardened by fire to improve wear resistance. The lubrication system design of transmission gears will be covered in Section 10.4.8.

The fatigue failure of gears is usually related to the defects of their supporting design, often accompanied by noise problems. The gear design with good durability must allow the load to be properly distributed on the tooth face. For this reason, the design requirements for the positioning and support of an intermeshing gearset are that the maximum pressure of a contact pattern should fall within the boundaries of gear teeth under all load conditions, and the contact pattern under full load should be distributed on at least two-thirds of the face, with no heavy contact allowed on the edge of the tooth. Hence, the design of the gear supporting structure must prevent excessive displacement and angular deformation under load, otherwise it will cause noise problems and subsequent fatigue failures.

A couple of fatigue failure modes of transmission gears are discussed as follows:

1. Bending fatigue

   Gear bending fatigue is the primary problem in transmission gear design. The selection of parameters such as pitch, pressure angle, and helix angle required for durability and noise often conflicts with each other. For example, design changes made to reduce noise often result in a decrease in durability, while design changes that help the smooth transition of load between gear teeth can adversely affect both durability and noise.

   Although the bending stress of a gearset can be calculated by empirical equations, a more accurate and comprehensive stress distribution can be provided by the finite element analysis (FEA) method commonly used today. When the load and the geometric dimensions of a gear are known, important design data such as the maximum bending stress can be obtained. The maximum bending stress in transmission gears typically ranges from 620 to 900 MPa. Derived from bench and vehicle tests, these data can be used as design guidelines, but there is still no way to know accurate gear life.

   Due to the diversity of road conditions that an EV can experience, the load and speed of transmission gears may change accordingly. This load spectrum usually needs to be acquired with the first prototype mounted on the mule vehicle tested in a proving ground. The time traces of the torque, speed on the output shaft, and corresponding gear of a multispeed transmission (if so equipped) can be decomposed into the torque and speed per each of the gears. In this way, the corresponding rotation cycles can be calculated for different torque intervals. Using this method, a torque histogram can be obtained according to each of the gears involved, which reflects the cycles of rotation corresponding to each of the gears at different torque levels. On the other hand, the number of stress cycles to failure at various torque levels for gear samples can be measured in the test, and the results can be sorted into the so-called S-N curve [10.74], which represents the life of the gear (cycles) under different loads (torque). Since a gear tooth subject to bending is similar to a cantilever beam in terms of the forces acting on it, the torque is proportional to the bending moment on the gear teeth. Thus, Miner's rule can be used to calculate the cumulative damage [10.74]. Note that this is uncertainty in the fatigue life predicted by Miner's rule when it is based on a small number of samples. Hence, a safety factor of 2 for general-purpose gears and 10 for critical gears are usually applied when it comes to their bending fatigue life design.

It is worth noting that the cumulative number of cycles of a sun gear tooth is proportional to its number of revolutions relative to the planet carrier and the number of planet gears. In contrast, the accumulative number of cycles of a planet gear tooth is proportional to its rotational speed relative to the planet carrier, and its mating with sun gear and ring gear simultaneously not only doubles the cycles but also produces a reversal of the stress on it, which is different from the situation of the sun gear and ring gear.

2. Surface contact fatigue
Surface contact fatigue may occur when a gear operates under high contact stress for a long time, which usually starts as pitting near the pitch line.

Spalling, or Hertzian fatigue, will occur under very high compressive loads. Compared with pitting, spalling failure involves a larger area of the gear face. It is generally considered that the spalling starts from the subsurface.

Generally speaking, thicker gears are more prone to compression fatigue failure because when a tooth becomes larger, the bending strength increases faster than the compression strength. Hence, compression fatigue failure hardly occurs on the fine-module gears, which are often used in EV transmissions.

When the elastic modulus of the material is about the same, the compressive stress at the pitch line of an external gearset can be calculated using the Huffaker formula derived based on the Hertz equation describing the contact stress between cylinders [10.75]:

$$\sigma_c = 0.834\sqrt{\frac{F_t E\left(\frac{1}{d_{pp}} \pm \frac{1}{d_{gp}}\right)}{W_{ga}\sin(2\psi_n)}\cos\phi_p} \tag{10.274}$$

where

σ_c is the maximum compression stress
ϕ_p is the helix angle at the operating pitch diameter
F_t is the tangential tooth load at the operating pitch diameter
E is Young's modulus of elasticity
W_{ga} is the active face width in the axial direction
ψ_n is the normal operating pressure angle
d_{pp} is the operating pitch diameter of the pinion
d_{gp} is the operating pitch diameter of the gear
the subtraction sign is for internal gears

Previous tests have shown that the surface fatigue life of the gearset under proper lubrication and cooling conditions is practically infinite when the maximum compressive stress calculated based on Equation (10.274) does not exceed 1380 MPa [10.75].

3. Scoring fatigue
   Usually caused by the joint action of low-speed gear tooth sliding and minor compressive stress, gear scoring failures, in fact, rarely occur in EV transmissions. Fine-module gears are commonly of a great safety margin for scoring failures. For large and heavy-duty transmissions, it is necessary to check the scoring factor to ensure it is 1.5×10^6 or less. It is recommended that the scoring factors at the tips of pinions and gears be calculated. If necessary, the scoring problem can be alleviated by modifying tooth profiles for smooth engagement, changing the outside diameters of gears, or adding antiscoring additives to the lubricating oil [10.75].

   For external meshing gears, the scoring factor is given by

$$\varsigma_k = \sigma_k \frac{\pi N_p (i+1)}{360i} (\rho_k - r \sin\psi_t)^2 \tag{10.275}$$

$$\sigma_k = 476{,}620 \sqrt{\frac{T_p C \sin\psi_n}{W_f L_c Z_k \rho_k (C \sin\psi_t - \rho_k)}} \tag{10.276}$$

$$\rho_k = \sqrt{R_k^2 - (r_k \cos\psi_t)^2} \tag{10.277}$$

where

$i = Z_g/Z_p$
ς is a scoring factor at the tip of either pinion tooth or gear tooth
σ is contact stress at the tip of either pinion tooth or gear tooth
N_p is rotation speed of the pinion
Z is the number of teeth
r is the pitch radius
R is the outer radius
ψ is the pressure angle
C is the center distance
T_p is the pinion torque
ρ is the involute radius of curvature at the pinion tip
W_f is the active face width

In addition, subscriptions $k = p$ or g, p denotes pinion, g denotes gear, t denotes transverse, n denotes normal, o denotes outside, pt denotes pitch, and L_c is the length of the contact path, which can be obtained by

$$L_c = \rho_p + \rho_g - C \cos\psi_t \tag{10.278}$$

10.4.4.5. Selection of Gear Materials

Transmission gears are essentially made of two types of steel.

One type is through-hardened grade of alloy steel, which is carbonitrided to produce a hard surface layer of a depth of 0.13–0.31 mm. The gears are then tempered at 175–220°C for stress relief, with a surface hardness of HRC58–60 and a center hardness of HRC45–60 [10.75]. This type of steel generally has the advantages of low cost and high static strength.

The other type is carburized grade alloy steel. The maximum depth of the carburized layer is limited to 25% of the tooth thickness, with a typical depth of the carburized layer of 0.3 to 0.7 mm for planetary gears and 0.8 to 1.8 mm for layshaft gears. The surface hardness of the finished gear is at least HRC55, and the core hardness is HRC25–40 [10.75]. By controlling the core hardness below HRC40, the fatigue strength of this type of steel can reach a very high level. The most popular steel of this type for EV transmissions includes SAE86 series (e.g., SAE8615 and SAE8620) in the US, the Cr-Mn series (e.g., 16MnCr5, 20MnCr5) in Germany, and 20CrMnTi in China. The mechanical properties of some well-known transmission gear steel are listed in Table 10.15 [10.75].

TABLE 10.15 Mechanical properties of well-known transmission gear steel.

Steel	SAE8620	20MnCr5	40Cr	20CrMo	20CrMnTi
Country	US	Germany	China	China	China
Density (kg/m^3)	7.872×10^3	7.85×10^3	7.87×10^3	7.84×10^3	7.86×10^3
Young's modulus, (GPa)	205	210	211	210	212
Poisson's ratio	0.27	0.28	0.277	0.278	0.289
Thermal expansion coefficient (1/°C)	1.22×10^{-5}	1.26×10^{-5}	1.20×10^{-5}	1.27×10^{-5}	1.27×10^{-5}
Specific heat (J/(kg·K))	477	470	460	460	460
Thermal conductivity, (W/(m·K))	46.6	45.9	44	44	44
Tensile strength (MPa)	≥980	≥1482	≥980	≥885	≥1080
Yield strength (MPa)	≥785	≥1232	≥785	≥685	≥835
Elongation (%)	≥9	≥13	≥9	≥12	≥10
Percent reduction of area (%)	≥40	≥57	≥45	≥50	≥45
Impact energy (J)	≥47	≥47	≥47	≥78	≥55
Impact toughness (J/cm^2)	≥59	≥73	≥59	≥98	≥69
Hardness (HB)	≤214	≤357	≤207	≤197	≤217
Heat treatment specification	Hardening: at 850°C with oil cooling; Tempering at 200°C with air cooling	Hardening: first time at 900°C, second time at 870°C with oil cooling; Tempering at 200°C with oil cooling	Hardening: at 850°C with oil cooling; Tempering at 520°C with water cooling, oil cooling	Hardening: at 880°C, oil cooling; Tempering at 500°C with water cooling, oil cooling	Hardening: first time at 880°C, second time at 870°C with oil cooling; Tempering at 200°C with water cooling, oil cooling

Internal gears generally operate at significantly lower stresses than mating external gears. Some gear manufacturers take advantage of this fact by fabricating internal gears out of pearlitic malleable medium-hardness steel. The gears are machined at the original hardness of 163–207 HB without further heat treatment. This kind of gear has been successfully applied to occasions with low load cycles, such as internal gearsets for reverse only, in spite of their low hardness, less resistance to wear, and vulnerability to pitting failures.

10.4.4.6. Manufacturability of Transmission Gear Design

One of the most important considerations in transmission gear design is to ensure the manufacturability of the gears. In addition, it is also desirable to control the dimensional tolerances within a reasonable limit and to reduce the manufacturing cost as much as possible. Several common gear machining processes are summarized now.

1. Hobbing
   Hobbing is a gear machining process based on the so-called generating method, which is widely used due to its high productivity. The main problem of this process is the concentricity and precision of the addendum and dedendum circles. If they are out of tolerance, the deviation in the center distance would become excessive, causing root interference or insufficient involute contact ratio. Full-topping hobbing technology can, generally, make the addendum circle, tooth profile, and tooth root meet the design requirements.

   There are many factors affecting the concentricity and precision of the addendum and dedendum circles, including the dimensional tolerance of the hob, outer diameter runout, spindle runout, and changes in the lead and pitch, which are often intertwined with each other. For example, the runout of a hob directly affects the addendum and dedendum circles and indirectly increases the lead error of each groove on the hob. Consequently, the deviation of tooth thickness at the index circle increases. In this situation, it is necessary to make a trade-off between the accuracy of the tooth thickness and the outer diameter because they affect the gear backlash and involute contact ratio, respectively.
2. Shaper cutting and shaving
   Shaper cutting and shaving are also processes based on the generating method. Shaper cutting is used for gears that cannot be hobbed due to insufficient runout clearance of the hob. Generally speaking, a wide enough slot (at least 4 mm) on the back of a gear is required for the overtravel of a shaper cutter and the chips between the shaper cutter and the end of recess. Compared with hobbing, shaper cutting produces a more precise tooth profile, but a larger tooth-to-tooth error.

   The shaving process is a kind of gear finishing process, which can further improve the tooth profile and tooth-to-tooth precision on the basis of rough machining and reduce the roughness of the tooth face. This process is mostly used for gears of relatively high precision that have not been quenched and hardened. The main drawback of the shaving process is the angle of cross axis relative to the gear, which is usually between 8 and 12° for external gears. Hence, there should be enough clearance between the end of the gear tooth and the shaft shoulder to ensure the stroke of the shaver along both working axis and the diagonal direction at a certain angle.

   For internal gears with a closed end that are not suitable for broaching, gear shaper cutting is often used for rough machining first and then a disk-shaped shaver is used for finishing. Compared with external gears, internal gears require a shaver of a smaller outer diameter. In addition, because it is very close to the closed end, a smaller cross-axis angle of generally 5–6° should be used.
3. Broaching
   Broaching is a gear machining process based on the shaping method. For internal gears, the broaching process requires that they do not have a closed end. Thus, the flange and the gear must be processed separately, similar to the process, for the important area that needs heat treatment is too close to the gear teeth. If the structure requires an internal gear to transmit all the torque to the sun gear through the flange, electron beam welding or laser welding should be considered to assemble the flange and the gear to minimize the heat-affected zone and ensure the strength of assembly. Compared with the shaper cutting process, the advantage of broaching is that the consistency of manufactured gears is better; the disadvantage of broaching, however, is that the cost of internal gear processing would increase sharply in the event of a broach accident. Hence, it is more suitable for mass production scenarios.

4. Grinding
   Grinding is a finishing process in which the tooth profile is further polished with a grinding wheel. Based on either the generating method or shaping method, only the former is suitable for mass production. Nevertheless, compared with other gear processes, grinding is generally a process of low efficiency and high cost. Hence, it should be avoided whenever possible. Due to the higher speed of EV transmission gears than traditional transmission gears, grinding is becoming more and more common in order to reduce gear noise. In many cases, gear grinding is used to correct the deformation after heat treatment. Grinding after quenching and hardening can further improve the tooth profile precision and surface finish, allowing the tooth profile precision to reach a grade of 2–7 per ISO 1328.
5. Chamfering and tooth relief
   Chamfering of a gear generally refers to the process of removing the edges between the ends of the gear and the top of the tooth. On the other hand, tooth relief actually refers to the removal of a small amount of tooth from the ridge lines where the tooth profile intersects with the tooth tip and both ends. The main effects of chamfering and tooth relief are three-fold: removing the burrs left by the cutting operation at the sharp edges of the gear, as the remaining burrs have a significant impact on the safe and reliable operation of gears and other parts and may even cause transmission failures; eliminating the protrusions at the edges caused by bumps during manufacturing processes to prevent them from affecting the meshing face; and reducing the meshing impact and undercutting, as well as avoiding stress concentration during heat treatment.
   Chamfering and tooth relief must be kept small, otherwise the effective tooth width and contact ratio would be considerably reduced. Although the fine module gearsets commonly used in EV transmission have the advantage of large contact ratio, their tooth relief has a more significant impact on the contact ratio than gearsets of ordinary modules. Therefore, the tooth relief must not be excessive. Generally, it is recommended that the angle between the chamfer and the top land be kept within $30 - 45^{\circ}$ and the chamfer size along the tooth height within 0.1 – 0.15 mm.
   The chamfering of both ends of a gear is generally machined at the same time when the tooth blank is machined with lathe cutting. The tooth tip relief is usually processed directly with a tip relief hob or a tip relief shaper cutter, while end relief is normally carried out by grinding after quenching and hardening treatments.

10.4.4.7. Design Principles for Gear Noise Reduction

In general, the excitation produced by the motion transmission error is one of the major contributors to the gear meshing noise. The motion transmission error is defined as the difference between the actual angular position of the output gear and the angular position it would occupy if the gear pair is fully conjugated:

$$\delta_{TE} = \theta_2 - \frac{Z_1}{Z_2}\theta_1 \tag{10.279}$$

where

δ_{TE} is the motion transmission error
θ_1 is the angle of the driving gear
θ_2 is the angle of the driven gear
Z_1 is the number of teeth of the driving gear
Z_2 is the number of teeth of the driven gear

The angle of an actual driven gear as a function of the angle of a driving gear is compared with that of a fully conjugated driven gear. The motion transmission error is mainly affected by factors such as the cumulative error of the tooth pitch, radial runout error, variation in the base tangent length, and combined tangential and radial errors. The frequency component of the motion transmission error signal can be obtained by fast Fourier transform (FFT), usually including the meshing frequency and its harmonic frequency at which the gear noise is produced.

The gearsets in an EV drivetrain generally operate at higher speeds than their counterparts in the traditional drivetrain of an IC engine-powered vehicle, which often makes the gear noise a prominent issue in the design of EV drivetrains. However, due to various influencing factors of gear noise, it is currently not possible to establish an exact design methodology to ensure the outcome free from excessive or abnormal gear noise.

It is generally reckoned that the gear noise in a vehicle drivetrain comes primarily from the gear body vibrations excited by the mutual collision or friction of the meshing gear pair during torque transmission. Hence, gear noise is highly correlated with tooth contact frequency.

As an important physical quantity related to gear noise, the principle of tooth contact frequency is that the vibrations excited by tooth surface contact is resonated and amplified along the torque transmission path. Thus, the tooth contact frequency usually dominates the frequency content of gear meshing noise. For a layshaft gearset, the tooth contact frequency f refers to the time rate of its meshing, namely,

$$f = \frac{NZ}{60} \tag{10.280}$$

where

N is the rotating speed of a gear

Z is the number of teeth on the same gear

For a planetary gearset, the tooth contact frequency f is defined as the product of the planet carrier speed relative to the reference gear and the number of teeth of any gear in the planetary gearset, namely,

$$f = \frac{\left|N_c - N_{ref}\right| Z_{ref}}{60} \tag{10.281}$$

where

N_{ref} is the rotating speed of a reference gear (either sun gear or ring gear)

Z_{ref} is the number of teeth of the same reference gear

Apparently, the layshaft gearset can be regarded as a special case of Equation (10.281) in which the planet carrier speed is zero.

Despite the complexity in gear noise, the risk of noise problems, however, can be significantly reduced as long as some basic guidelines are followed in the design. Gear design parameters such as module, pressure angle, helix angle, addendum coefficient, tolerance, modification, and gear shafts and their support have significant influence on gear noise. The relevant design principles for gear noise reduction are discussed now.

1. Module
The smaller the module, the lower the noise. Therefore, the smallest possible module is generally selected under the condition of meeting the requirements of bending strength. The module of EV transmission gears can be as fine as 1.5.

The relationship between the module and gear noise is mainly due to its influence on the contact ratio. Since both transverse contact ratio ε_α and overlap ratio ε_β are linear functions of the module, the smaller the module, the more teeth are in contact at the same time, thereby reducing the load on each tooth. For a given transverse contact ratio ε_α and the overlap ratio ε_β, smaller module gears are quieter than larger module gears. This is a result of a reduced slip ratio and involute sensitivity because the meshing action is closer to the pitch circle if the module is smaller. When gear parameters such as contact ratio, center distance, and base circle diameter remain unchanged, the effects of the gear module on the specific sliding, the distance from the start of action to the base circle, the addenda, and the relative gear tooth stiffness of a gear pair can be evaluated by referencing a gear design manual. The reason for the noise generated by sliding is the intermittent elastic deformation and release of the meshing surfaces at the contact frequency [10.75]. It is generally deemed that the deformation rate dips with the decrease in the gear module. Note that a gear with a module of 2.25 is actually more rigid than a gear with a module of 3. The reason is that the tooth height decreases faster than the tooth thickness when the contact ratio is constant. Nevertheless, instead of suppressing gear noise, high stiffness may actually have a negative effect. Therefore, the reason for the gear pair of a smaller module being quieter is not its higher stiffness but its larger transverse contact ratio and overlap ratio due to reduced specific sliding and relative sliding.

2. Pressure angle
A small pressure angle is beneficial for gear noise reduction for three reasons: The transverse contact ratio increases rapidly with the reduction of the pressure angle; the decrease in the pressure angle helps reduce the corresponding meshing speed, thereby cutting down the instantaneous dynamic load at the beginning and end of meshing; and a drop in the pressure angle can bring down the stiffness of gear teeth and the load variation due to dimensional errors, thereby lowering noise.

Note that a small pressure angle is sometimes used to avoid a large helix angle, in order to prevent the gear teeth tip from becoming too narrow, which has nothing to do with noise reduction.

Reducing the pressure angle is also restricted by some factors. A very small pressure angle will cause the starting point of meshing action to be too close to the base circle, which will increase the sliding ratio and, consequently, the bending and compressive stresses. In order to avoid undercutting and make the starting point of action as close to the base circle as possible, the prevailing design rule is to make the radial distance of 0.125–0.25 mm between the lowest contact point and the base circle with the tolerance chain of all dimensions considered, in which the lower bound should be taken for gears of small modules.

3. Helix angle
A large helix angle is generally desirable to reduce noise. In fact, the effect of enlarging the helix angle is similar to that of reducing the pressure angle. In recent years, it has become a trend to reduce noise by increasing the contact ratio, and the increase in the helix angle is conducive to increase in the contact ratio.

However, an excessively large helix angle would also cause the axial load to exceed the axial load-carrying capacity of the bearing. If ball bearings cannot meet the requirements, cylindrical roller bearings would have to be selected, resulting in increased friction loss though.

4. Addendum coefficient
The addendum coefficient is another design variable that has a significant influence on gear noise. As a rule of thumb, the addendum coefficient is usually proportioned to make the angle of approach small and the angle of recess large for a quiet meshing action [10.75].
For helical gears, the contact line along the length of the gear teeth is oblique, so there will be sliding simultaneously on both sides of the pitch diameters on the same gear tooth [10.75]. Thus, the effect of surface finish of the transmission gears on the noise is trivial, as they are generally of high grade of finish and well-lubricated.
For spur gears, because of the direction of the frictional force, the meshing action is quieter in the phase of recess [10.75]. According to Equation (10.273), the normal load on the gear decreases with the friction, making the noise lower than that in the phase of approach when the normal load increases with friction per Equation (10.272).
5. Shaft and support design
The design principle for the shafts and bearings of a transmission is to maximize the stiffness within allowable limits of the transmission size and weight. When loaded, the gears deform in the radial direction, causing the contact ratio to decrease. Besides, as a result of shaft deflection, the gears on it become skewed, causing the load distribution to lean toward one end and, thus, loss of the contact ratio. Hence, the deflection and skewness of the shaft should be analyzed in the design to prevent the reduction of the total contact ratio from generating gear noise.
Drum tooth profile modification on one gear of a gear pair is a widely accepted solution to eliminating end loading caused by deflection. Because the drum modification tends to reduce the contact ratio of helical gearsets, it is generally desirable to have a drum modification as small as possible, which is, in turn, determined by the amount of deflection allowed by the stiffness. It should be noted that the deflection generally inclines to increase the backlash in a layshaft gearset. Hence, the specifications of a backlash only need to avoid too tight meshing (i.e., both ends of the gear teeth being in contact) caused by dimensional tolerances such as center distance, tooth thickness, and runout. Compared to traditional ICE vehicles, an important feature of an EV is the sudden reverse of the torque transmission direction in the drivetrain during switching from the driving mode to the regenerative braking mode. In the most severe case, the change from a forward peak torque to a reverse peak torque can be completed within a few milliseconds. Thus, an excessive backlash will result in gear knocking sound under no-load and torque reversal conditions, instead of making a gear meshing noise. Hence, the nominal center backlash is generally controlled within the range of 0.05–0.2 mm.
Design parameters such as shaft and support stiffness, coaxiality, and radial runout have an impact on gear noise. Bending deformation caused by insufficient stiffness, as well as poor coaxiality and excessive radial runout, will compromise the correct meshing of the gears and generate noise.
In arranging the shaft and its supports, the distance between the supports should be minimized so as to control the deflection of the shaft under loads.

In addition, special attention should be paid to the fatigue strength of splines on the shaft, as the torque direction is frequently switched under the driving conditions of an EV, so as to avoid abnormal noise caused by the wear of splines.

The deformation of transmission housing will also indirectly affect the gear noise through shafts and supports, so its stiffness should also be controlled above a reasonable threshold in the design. In order to maximize the stiffness, the housing should be reinforced using a sufficient number of ribs. The typical values of allowable deflection and the bending angle of the shaft in its axis plane under the maximum input torque are 0.02–0.06 mm and 0.005 rad, respectively. Refer to Section 10.4.9 for details.

6. Gear modification
   For EV transmissions, the driving gear is generally modified to a crown-shaped involute, that is to say, a small portion of the involute at the tip and root of each tooth is removed. When such a modified gear mates with an unmodified gear, it would result in a delayed entry and an early exit [10.75]. Involute modification can be regarded as a tapered backlash. At the point of tooth engagement with an unmodified gear, interference may occur due to deformation or pitch error. Gear modification can generate a clearance at the theoretical start point of engagement to avoid interference. With the theoretical contact point moving down the line of action, the backlash tapers to zero clearance, and the load acting on the tooth varies at a rate depending on both modification taper and the elastic rate of the tooth. It is generally desirable that the taper comes to an end before the next abrupt change in tooth load. The appropriate modification amount of an involute needs to be determined by noise measurements of a large number of samples of gearsets with and without modification [10.75].
7. Dimension tolerances
   Dimension tolerances are an important aspect of transmission gear design. When the number of gears is large enough, the correlation of gear noise to the tolerances assigned to their dimensions can be established by statistical methods, and these tolerances can be continuously updated based on statistical data. The tolerances listed in Table 10.16 can be used as the initial tolerances of transmission gears [10.1, 10.75].

TABLE 10.16 Initial tolerances of transmission gears.

External gears	Lead variation/face width	±0.05%
	Lead variation between any two charted teeth/face width	±0.1%
Internal gears	Lead variation/face width	±0.1%
	Lead variation between any two teeth/face width	≤0.025 mm
Involute	Minus variation, plus variation	−0.005 mm, +0.0075 mm
	Variation of involute form between any two teeth	≤0.025 mm
Spacing	Max. tooth-to-tooth index error	0.015 mm
	Max. index error	0.025 mm
Pitch diameter	Runout	−0.025 mm

10.4.4.8. Trend in Transmission Gear Design

Transmission gears are developing toward "fine and tall" teeth, that is, gears with a low module, a small pressure angle, a large addendum coefficient, and a slim tooth profile are increasingly employed. EV transmissions are no exception. The outstanding feature of this type of gears is that it can increase the transverse contact ratio of the gear teeth for meshing noise reduction. Generally speaking, "fine and tall" gears have an addendum coefficient of no less than 1.3 and a transverse contact ratio of greater than 1.8 [10.1].

Previous studies have shown that when the transverse contact ratio is equal to 2, the gear noise is significantly reduced. This is because in theory, there are exactly two pairs of teeth meshing at the moment, and there is no motion transmission error caused by the change in the number of meshing teeth, with the deformation remaining unchanged. As a result, both vibrations and meshing noise are minimized. In fact, due to manufacturing tolerances, it is impossible to exactly keep the two pairs of teeth in meshing when the transverse contact ratio is equal to 2. Thus, the transverse contact ratio is usually designed to be larger than 2, that is, there are always two or three pairs of teeth in meshing for a certain moment. Such a design can significantly reduce the motion transmission error, deformation, and meshing noise compared to the design with a transverse contact ratio less than 2 [10.1].

The following guidelines are recommended for the design of "fine and tall" gears [10.1]:

1. Mainly used for overspeed and high-speed gears of cars, light vehicles, and main gearboxes of heavy-duty vehicles.
2. Rarely used for low-speed gears with high loads.
3. As the large addendum coefficient makes the tooth top narrow, the width of tooth top is generally larger than $0.3m_n$.
4. Check the width of tooth root slot to avoid interference during meshing.
5. Check the radius of the starting point of meshing action to ensure it is greater than the radius of the starting point of the transition curve at the tooth root during hobbing.
6. The diameter of the starting point of tooth tip relief should be properly controlled. A reduction of 0.2 mm in this diameter can bring down the transverse contact ratio by 0.1.
7. Tooth pitch error should be properly controlled to prevent the loss in tooth root strength.

10.4.5. Design of Clutches

As part of the shift mechanism, multiplate clutches are widely used in automotive transmissions. Most multiplate clutches are wet type, due to their compactness and excellent heat dissipation characteristics. The basic functions of these wet clutches in a transmission are two-fold: locking one of the planetary gear elements to output a certain transmission ratio and changing the power transmission path.

The advantages of a wet clutch mainly are as follows: capable of entering engagement during rotation and locking up both sides, no excessive local stress in the transmission housing caused by lockup of the wet clutch, large torque capacity, ease of torque capacity adjustment by adding or reducing the number of friction plates, and no need of adjusting the clearance due to wear or uneven contact once a proper operating clearance is set.

A wet clutch is primarily composed of friction plates and steel plates. In a typical configuration, the inner clutch hub is splined to friction plates, while the outer clutch hub (or drum) is splined to steel plates for transmitting torque. In contrast, the inner clutch hub is sometimes splined to steel plates, while the outer clutch hub is splined to friction plates. The cylinder piston applies a pressure on the clutch pack through the pressure plate by pumping the pressurized oil into the chamber, so as to compress the assembly to engage the clutch. The reaction plate (or snap ring) at the other end provides the support against the actuation force and the return spring functions to resume the clutch to the disengaged state.

The design of a wet clutch is based on the torque capacity, which is usually dependent on the maximum torque that the wet clutch is capable of being locked or transmitted in different gears. The diameter of a wet clutch is mostly dependent on its torque capacity, despite its relationship with the structural arrangement of the planetary gears and the packaging space of the transmission in the vehicle.

Due to the difference in static and dynamic friction coefficients between the friction plates and the mating steel plates in the clutch pack, static capacity and dynamic capacity are generally different. For the same wet clutch, its static capacity is usually greater than its dynamic capacity. When a wet clutch is used to lock planetary gear elements, its static capacity determines the maximum lockup torque that it can provide; when a wet clutch is employed to transmit torque, its static capacity determines the maximum torque that it can possibly transmit without slippage while its dynamic capacity determines the maximum torque that it is capable of transmitting in the case of slippage.

10.4.5.1. Static Capacity Design of Wet Clutches

The static capacity T_c of a wet clutch is given by

$$T_c = F_{min} f n_f D_m \tag{10.282}$$

where

f is the static friction coefficient
n_f is the number of friction surfaces
D_m is the average effective diameter
F_{min} is the minimum actuation force required by the cylinder piston

$$F_{min} = Ap - F_s - F_v \tag{10.283}$$

where

A is the area of the cylinder piston
p is the net oil pressure on the piston
F_s is the force of return spring
F_v is the viscous resistance

In order to prevent the wet clutch from slipping, the torque capacity is generally designed to be larger than the maximum torque transmitted with a certain safety factor:

$$\varphi_s = \frac{T_c - T_{cmax}}{T_{cmax}} \tag{10.284}$$

where

φ_s is the static safety factor of the wet clutch
T_{cmax} is the maximum torque acting on the clutch

10.4.5.2. Dynamic Capacity Design of Wet Clutches

The dynamic capacity T_{cd} of a wet clutch is given by

$$T_{cd} = F_{min} f_d n_f D_m \tag{10.285}$$

where f_d is the dynamic friction coefficient. The peak dynamic capacity T_{peak} of the wet clutch depends on the maximum torque transmitted under the condition of slip, generally the maximum input torque in the inertia phase of the shift process, namely,

$$T_{peak} = \max(T_{in}) \tag{10.286}$$

where T_{in} is the input torque to the clutch in the inertia phase. Thus, the dynamic safety factor φ_{sd} of the wet clutch is

$$\varphi_{sd} = \frac{T_{cd} - T_{cmax}}{T_{cmax}} \tag{10.287}$$

The average power absorbed by the wet clutch in the torque and inertia phases is

$$P_{avg} = \int T_c \omega_c dt \tag{10.288}$$

where

T_c is the clutch torque
ω_c is the relative angular velocity between driving and driven members

The corresponding power density is

$$D_{pavg} = \frac{P_{avg}}{A_f} \tag{10.289}$$

where A_f is the total area of friction surfaces:

$$A_f = \frac{1}{4}\left(D_o^2 - D_i^2\right)\left(1 - \delta_g\right)n_f \tag{10.290}$$

where

D_o is the outer diameter of the friction surface
D_i is the inner diameter of the friction surface
δ_g is the proportion of oil grooves area on friction surfaces

The peak power density is

$$D_{ppeak} = \frac{P_{peak}}{A_f} \tag{10.291}$$

10.4.5.3. Calculation of Actuation Force

A wet clutch is usually actuated by a cylinder piston with pressurized oil from an oil pump. In order to balance out the centrifugal force produced by the rotating oil in the cylinder for the easy return of the piston to its initial position, a centrifugal balancing chamber is sometimes designed, as shown in Figure 10.49 [10.1]. The actuation force acting on the clutch pack by piston is

$$F_a = F_{pc} - F_{pb} - F_{sp} - F_v - m_c \frac{dv}{d} \tag{10.292}$$

FIGURE 10.49 Schematic of a wet clutch for actuation force calculation.

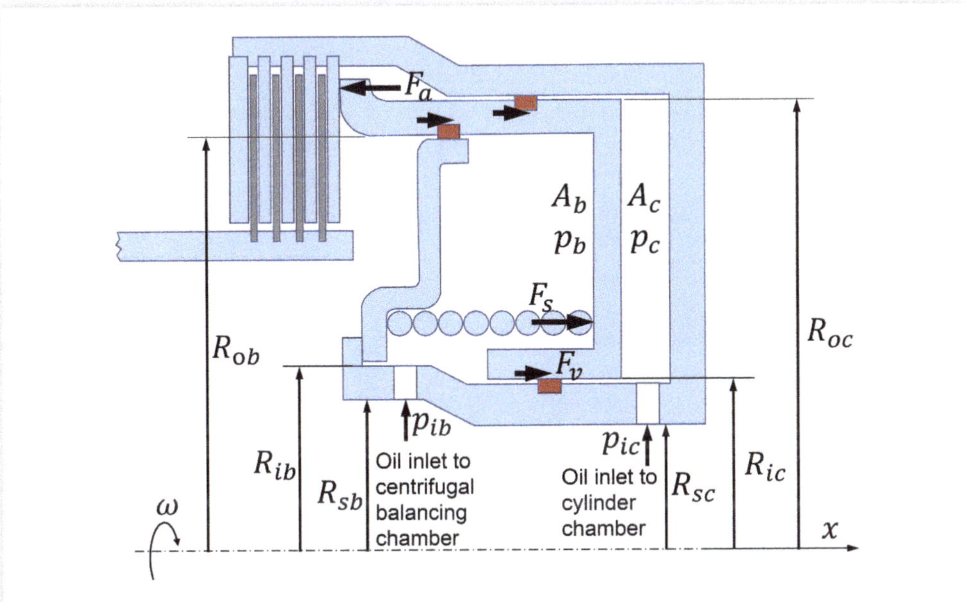

Reproduced from, Yuan, Y., Electric Vehicle Transmissions: Structure and Control (Beijing: China Machine Press, 2020), ISBN: 978-7-111-66809-1, with the permission of China Machine Press.

with

$$F_{pc} = \int_{R_{ic}}^{R_{oc}} 2\pi p_c r dr \tag{10.293}$$

$$F_{pb} = \int_{R_{ib}}^{R_{ob}} 2\pi p_b r dr \tag{10.294}$$

$$F_{sp} = F_0 + K\left(x_0 - x\right) \tag{10.295}$$

$$F_v = Bv \tag{10.296}$$

where

F_{pc} is the resultant force of oil pressure in the cylinder chamber acting on the piston
F_{pb} is the resultant force of oil pressure in the centrifugal balancing chamber acting on the piston
F_{sp} is the force of return spring on the piston
F_v is the resistance acting on the piston from seal rings
R_{ic} is the inner radius of the piston on the cylinder chamber side
R_{oc} is the outer radius of the piston on the cylinder chamber side
R_{ib} is the outer radius of the piston on the centrifugal balancing chamber side
R_{ob} is the force of return spring at its initial position
K is the rate of return spring
x_0 is the initial axial position of the piston
B is the damping coefficient
v is the translational velocity of the piston
m_c is the mass of piston assembly

In addition, p_c is the oil pressure in the cylinder chamber due to the influence of a centrifugal force, which is a function of radial distance r from the axis of rotation, that is, $p_c = p_c(r)$. p_b is the oil pressure in the centrifugal balancing chamber due to the influence of the centrifugal force, which is a function of the radial distance r as well, that is, $p_b = p_b(r)$. Furthermore, the inner radius of the piston on the cylinder chamber side is assumed to be identical to that on the centrifugal balancing chamber side.

Considering the effect of the centrifugal force on the oil pressure in the cylinder chamber as it rotates at an angular velocity ω and ignoring the pressure loss at the oil inlet, the distribution of oil pressure along the radial direction within the cylinder chamber is given by

$$p_c(r) = p_{ic} + \int_{R_{sc}}^{r} \frac{\rho\omega^2}{2}\left(r^2 - R_{sc}^2\right)dr \tag{10.297}$$

Similarly, the distribution of oil pressure along the radial direction within the centrifugal balancing chamber is

$$p_b(r) = p_{ib} + \int_{R_{sb}}^{r} \frac{\rho\omega^2}{2}\left(r^2 - R_{sb}^2\right)dr \tag{10.298}$$

where

p_{ic} is the oil pressure at the inlet of the cylinder chamber
R_{sc} is the radius at the inlet of the cylinder chamber
p_{ib} is the oil pressure at the inlet of the centrifugal balancing chamber
R_{sb} is the radius at the inlet of the centrifugal balancing chamber
ρ is the density of oil

Substitution of Equations (10.297) and (10.298) into Equations (10.293) and (10.294) with an integration in radial reaction yields

$$F_{pc} = p_{ic} A_c + F_{cc} \tag{10.299}$$

$$F_{pb} = p_{ib} A_b + F_{cb} \tag{10.300}$$

$$F_{cc} = \frac{\pi}{4} \rho \omega^2 \left[R_{oc}^4 - 2R_{sc}^2 \left(R_{oc}^2 - R_{ic}^2 \right) - R_{ic}^4 \right] \tag{10.301}$$

$$F_{cb} = \frac{\pi}{4} \rho \omega^2 \left[R_{ob}^4 - 2R_{s2}^2 \left(R_{ob}^2 - R_{ib}^2 \right) - 2R_{ib}^4 + R_{ic}^4 \right] \tag{10.302}$$

where

- F_{cc} is the force acting on the piston from the oil pressure in the cylinder chamber caused by the centrifugal effect
- F_{cb} is the force acting on the piston from the oil pressure in the centrifugal balancing chamber caused by the centrifugal effect
- A_c is the projected contact area between the piston and the oil in the cylinder chamber along the axial direction
- A_b is the projected contact area between the piston and the oil in the centrifugal balancing chamber along the axial direction

10.4.5.4. Clutch Friction Materials

The friction materials of wet clutches can mainly be divided into three types: powder metallurgy materials, asbestos-based materials, and cermet materials. To select a suitable friction material, the most important properties to consider include wear resistance, friction coefficient, and stability and its change with speed. In addition, other properties such as thermal conductivity, elasticity, oil absorption, torque capacity per unit area, and cost must also be considered as secondary factors. The properties of these friction materials are compared in Table 10.17 [10.1].

TABLE 10.17 Comparison of properties of friction materials.

Properties	Powder metallurgy	Asbestos-based	Cermet
Coefficient of friction	0.25–0.5	0.2–0.35	0.4
Stability of coefficient of friction	Good	Fair	Good
Wear resistance	Fair	Poor	Good
Specific heat	Good	Fair	Fair
Thermal conductivity	Good	Fair	Fair
Durability	Good	Fair	Good
Oil absorption	Good	Fair	Fair
Elasticity	Fair	Good	Fair
Cost	Fair	Good	Fair

Reproduced from, Yuan, Y., Electric Vehicle Transmissions: Structure and Control (Beijing: China Machine Press, 2020), ISBN: 978-7-111-66809-1, with the permission of China Machine Press.

10.4.5.5. Reduction of Drag Losses

One of the most important design considerations is the reduction of drag torque in disengaged clutch packs, as it is a major contributor to the parasitic losses in a transmission. The following design principles should generally be considered:

1. Use transmission oil of low viscosity.
2. Use wavy steel plates.
3. Reduce the oil flow rate in the clutch, especially at high speeds.
4. Design proper oil groove patterns on the friction plates, for example, radial and multiparallel oil grooves; increase the depth of oil grooves, the proportion of the oil grooved area, and the number of oil grooves.
5. Increase the clearance between steel and friction plates.

Finally, it is worth noting that the change in the friction material type usually has little effect on the drag torque.

10.4.6. Design of Synchronizers

For a layshaft transmission, its output assembly has a considerable moment of inertia for its direct connection to the vehicle. The output speed changes only slightly during gear shifting, while the input speed alters substantially for the change in the transmission ratio prior to and after the shift. As a result, there exists a speed difference between the input and output assemblies that can be completely eliminated even if the speed of the prime mover, for example, an electric motor, is adjusted in real time. Hence, a synchronizer is generally needed to synchronize the rotation speeds of the input and output assemblies during a gear shift.

From the perspective of architecture of a layshaft gearing system, the gear pair not related to the operating gear usually does not transmit torque, and one of the gears is sleeved on its shaft via a bearing and rotates with the mating gear. Thus, before the gear pair starts to transmit torque, there is a speed difference between the idler gear on the shaft and the shaft itself. The first function of a synchronizer is to eliminate the speed difference between the two through the rotating friction pair installed on the gear and the shaft. Once a synchronization is reached, the teeth on shift sleeve mesh with the teeth on gear to complete the shift operation and start to transmit torque.

10.4.6.1. Types of Synchronizers and Their Characteristics

Synchronizers come in two types, namely, inertial type and Porsche type, with the inertial type being the most widely used in EV transmissions.

The inertial type of synchronizers can be divided into lock pin type and lock ring type. A lock pin synchronizer mainly consists of a sliding sleeve, a hub, a friction cone ring (also called synchronization ring), a friction cone disc, lock pins, detent balls, detent springs, positioning pins, and so on. The advantage of lock pin-type synchronizers is high torque capacity due to the relatively large average radius of their friction cones, with very few parts in the assembly. However, the shortcoming of their excessive axial size has limited their usage in most of the vehicles, except in trucks with a maximum total mass of 6t or more. On the other hand, a lock ring synchronizer is composed of a sliding sleeve, a synchronizer hub, a couple of blocker rings (also called synchronizing rings), a synchronizer cone on the gear, detent balls, detent springs, struts, and so forth. In order to increase the torque capacity, more than one blocker rings are sometimes used, which is referred to as a multicone synchronizer. In spite of their high reliability and durability, lock ring synchronizers generally have an intrinsic drawback of limited torque capacity. Besides, an improper design can lead to clashing, and eventually, the wear and failure of the leading edge of sleeve and gear spline. Hence, they are normally used in the transmissions for passenger vehicles and trucks of a GVW below 6t.

10.4.6.2. Calculation of Shift Force

For a synchronizer to achieve synchronization within a given time interval, the required friction torque T_f should be determined first by

$$T_f = \frac{J_i(\omega_b - \omega_a)}{t_s} \tag{10.303}$$

where J_i is the total moment of inertia of the input assembly, which is the sum of moments of inertia of all components that rely on the synchronizer to change the speed during the shift; ω_a and ω_b are the angular velocities of input assembly prior to and after gear shift, respectively; and t_s is the synchronization time. The recommended values of synchronization time for different types of vehicles are listed in Table 10.18 [10.1].

TABLE 10.18 Recommended values of synchronization time.

Type of vehicles	High gears	Low gears
Passenger vehicles	0.15–0.3	0.5–0.8
Trucks	0.3–0.8	1.0–1.5

Reproduced from, Yuan, Y., Electric Vehicle Transmissions: Structure and Control (Beijing: China Machine Press, 2020), ISBN: 978-7-111-66809-1, with the permission of China Machine Press.

It can be noticed from Equation (10.303) that lowering the difference in angular velocity prior to and after speed synchronization is an effective method to reduce the required friction torque. As the angular velocity of the input assembly of an EV transmission can be controlled by adjusting the speed of the traction motor, the difference in angular velocity between input and output assemblies before synchronization is usually controlled within an appropriate range, such as ≤50 rpm, to greatly reduce the friction torque.

Next, the axial force (also called shift force) acting on the sleeve required for producing the friction torque obtained by Equation (10.303) is given by

$$F_s = \frac{2T_f \sin\alpha}{jfd} \tag{10.304}$$

where

α is the cone angle of the friction cone
j is the number of friction cones
f is the friction coefficient of the friction cone
d is the mean diameter of the friction cone, as shown in Figure 10.50 [10.1]

FIGURE 10.50 Schematic for axial force calculation of a lock ring synchronizer.

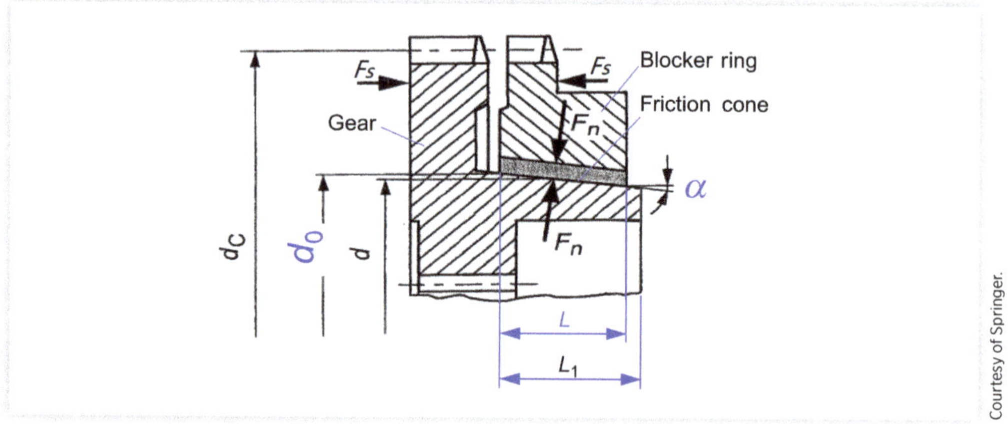

It is clear from Equation (10.304) that the shifting force can be reduced by enlarging the mean diameter of the friction cone, increasing the number of friction cones, or lowering the cone angle of the friction.

Since the gear shift actuating mechanism is usually driven by a power source such as an electric motor, the torque required for shifting from the power source can be calculated based on the aforementioned axial force along with the ratio and mechanical efficiency of the actuating mechanism.

10.4.6.3. Selection of Major Design Parameters

The major design parameters of a lock ring synchronizer include the cone angle α, average diameter d, working axial lengths L and L_1 of the friction cone, shift force F_s, lock angle β, and synchronizer-related clearance δ_1, δ_2, δ_3, etc.

1. Index distance a and proximity distance b

   The index distance a refers to the distance between the centerline of sleeve teeth and the blocker ring teeth when the side of strut is in contact with the side of notch of the blocker ring, as shown in Figure 10.51 [10.1]. In the design, we usually take $a = 0.25t$, where t is the pitch of sleeve teeth.

   Proximity distance b refers to the axial distance between sleeve teeth and blocker ring teeth when the side of strut is in contact with the side of the blocker ring notch before the sleeve moves axially, as shown Figure 10.51. In the design, we normally take $b = 0.2 - 0.3$ mm.

2. Corn angle α

   The smaller the corn angle α, the greater the friction torque. However, if α is too small, the friction surface can become self-locked. In order to avoid self-locking, it is required that $\alpha \geq \arctan f$. For single cone synchronizers, generally $\alpha = 6 - 7^{\circ}$; for multicone synchronizers, $\alpha = 8 - 12^{\circ}$.

3. Mean diameter d of friction cone

   The larger the mean diameter d, the greater the friction torque. However, if d is too large, the radial size of the synchronizer and transmission will be affected. Therefore, d should be made as large as possible if the structure permits. Generally, the mean diameter d of the friction cone can be determined according to the following formula:

FIGURE 10.51 Index and proximity distances of a lock ring synchronizer.

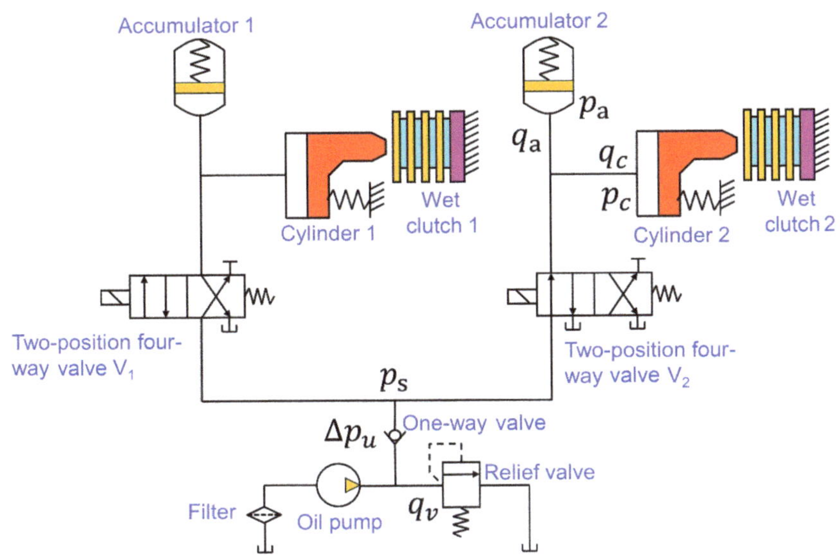

$$d = d_o - L\tan\alpha \tag{10.305}$$

$$d_o = (0.8 - 0.85)d_c \tag{10.306}$$

where

d_o is the diameter of the large end of the blocker ring

d_c is the diameter of the index circle of blocker ring teeth, as shown in Figure 10.50

4. Shift force F_s

   The larger the shift force F_s, the greater the friction torque. However, an excessively large F_s would cause the volume, mass, power consumption, and cost of the shifting power source to increase. Thus, it is usually preferred that $F_s = 80 - 120$ N.

5. Axial working lengths L and L_1 of the friction cone

   The axial working length L of the conical surface of the blocker ring should be less than the axial working length L_1 of the conical surface of gear clutching teeth, that is, $L < L_1$. If not, the conical surface of the blocker ring would be worn out. Eventually, a step can be formed, making a poor contact between the conical surfaces and causing out-of-synchronization engagement. The larger the L, the greater the total area of the friction cone, which is beneficial to reduce the contact pressure on the friction surface. However, an excessively large L can cause the axial length of transmission to increase. Hence, L should be made as small as possible under the premise of meeting the condition of contact pressure on the friction surface. The empirical formula $L = (0.125 - 0.2)d$ can be utilized to determine L.

6. Clearances
   Clearances related to the function of a synchronizer include the clearance δ_1 between the end surface of the strut and the end surface of the blocker ring notch, the clearance δ_2 between the end surface of the sleeve and the end surface of the blocker ring, and the clearance δ_3 between the end surface of the blocker ring and the end surface of gear clutching teeth. When $\delta_1 > \delta_2$, the sleeve teeth can get in touch with the locking surface of the blocker ring before the friction cone surfaces of the blocker ring come into contact, causing the synchronizer to lose its locking function, resulting in out-of-synchronization engagement and shift shock. Therefore, it is required that $\delta_1 < \delta_2$ in the design. In addition, because the friction cone of the blocker ring can be worn out, δ_3 would gradually decrease. When δ_3 is reduced to zero, a clearance would be created between the friction cones, which would cause the friction torque to disappear and a failure in synchronization. Thus, sufficient δ_3 should be left in the design, which is also called reserved backup stroke. Generally, it is recommended that $\delta_1 = 0.5 - 1$ mm, $\delta_2 - \delta_1 = 0.2 - 0.3$ mm, and $\delta_3 = 1.4 - 1.8$ mm.
7. Tread grooves and axial oil grooves
   The tread groove on the inner cone surface of the blocker ring functions to destroy and scrape off the transmission oil film on the friction cone, while the axial oil groove on the inner cone surface of the blocker ring functions to drain the oil as soon as possible. Hence, all these grooves serve the purpose of improving the friction. The rules of thumb in designing these grooves are as follows: First, the size of tread pitch must ensure that the tread clearance is sufficient to accommodate the amount of oil extruded, but an excessively large tread pitch would reduce the contact area of cone surfaces and increase the wear. Hence, the tread pitch is usually 0.6–0.75 mm. Second, the tread angle is generally 60°. Third, the tread depth is recommended to be 0.25–0.4 mm. Fourth, the narrower the tread top width, the faster the oil film can be scraped off. However, if it is too narrow, the contact surface pressure would be too large, resulting in excessive wear. Thus, the tread top width is generally 0.025–0.1 mm. Fifth, the width of axial oil grooves is generally 3 mm, and the groove depth is slightly larger than the bottom diameter of tread. There are usually 6–9 axial oil grooves, depending on the average diameter of the friction cone. The fillet radius at the bottom of oil grooves should be as large as possible to reduce stress concentration.
8. Struts and notches of the blocker ring
   The rotational distance c of struts in the notches of the blocker ring affects the index distance a. It is, therefore, most advantageous for both locking surfaces to be aligned with each other in the locking position. The contact position of both locking surfaces would be poor if c is too small or too large. The rotation distance c can be estimated by

$$c = \frac{tR_1}{4R_2} \tag{10.307}$$

where

R_1 is the outer radius of the blocker ring notch
R_2 is the radius of the indexing circle of sleeve teeth
The width H of the blocker ring notch is calculated using the following formula:

$$H = h + 2c \tag{10.308}$$

where h is the width of struts.

9. Locking angle β
   Note the locking angle defined here is the complementary angle of the teeth chamfer angle (or teeth pointing angle). The locking angle β of lock ring synchronizers serves two functions: the shifting force is transmitted to the synchronization cone via the locking angle slope, and the index torque T_b produced by the tangential component of the shifting force via the locking angle slope causes the blocker ring to rotate an angle, thereby disengaging the tooth end slope of the sleeve, allowing the sleeve to continue moving forward to engage with the gear clutching teeth for the completion of the gear shift. For this reason, before reaching synchronization, it is required that the friction torque T_f for synchronization should always be greater than the index torque T_b, that is,

$$T_f > T_b \tag{10.309}$$

If this condition is not met, the teeth would be engaged before the synchronization is completed, resulting in out-of-synchronization engagement and shift shock.

With the friction between the locking slopes taken into account, the total axial component force ΣF_a acting on the locking surface is

$$\Sigma F_a = F_a + F_{ba} = F_s + F_s f_b \tan\beta \tag{10.310}$$

where

f_b is the friction coefficient between the locking slopes
F_s is the shift force

Hence, the total tangential component force ΣF_t acting on the locking surface is given by

$$\Sigma F_t = F_t - F_{bt} = F\tan\beta - Ff_b \tag{10.311}$$

The index torque T_b can be obtained by

$$T_b = \frac{d_2}{2}\Sigma F_t = \frac{1}{2}Fd_2\left(\tan\beta - f_b\right) \tag{10.312}$$

where d_2 is the diameter of the index circle of the blocker ring.

The friction torque T_f is expressed as

$$T_f = \frac{\left(F + Ff_b\tan\beta\right)fd}{2\sin\alpha} \tag{10.313}$$

Thus, the condition of Equation (10.309) becomes

$$\frac{\left(F + Ff_b\tan\beta\right)fd}{2\sin\alpha} > \frac{1}{2}Fd_2\left(\tan\beta - f_b\right) \tag{10.314}$$

Rearrangement of Equation (10.314) yields

$$\beta < \arctan\left(\frac{fd + d_2 f_b \sin\alpha}{d_2 \sin\alpha - ff_b d}\right) \tag{10.315}$$

The larger the locking angle β, the greater the mean diameter d of the friction cone and the smaller the mean diameter d_c of locking surfaces as well as the cone angle α of the friction cone. Consequently, the locking angle β can, in turn, affect the size of the synchronizer and the satisfaction of locking conditions. Therefore, it is desirable to make β as small as possible. However, if β becomes too small, the index torque would be too low to make a successful engagement. Generally, we use $\beta = 26 - 38^{\circ}$.

10.4.6.4. Breakthrough Load of Detent Balls

When the sliding sleeve is axially displaced from the neutral position under the action of the shift force, the force on the detent ball in the groove increases almost linearly with the axial displacement as long as the detent ball remains in the groove. With a further increase in displacement, the groove thrust on the ball increases to the maximum, and the detent ball begins to move out of the groove. At this moment, the groove thrust on the ball is greatly reduced and the coupling teeth come into contact, that is, the tooth chamfer of sliding sleeve starts to contact with the tooth chamfer of the blocker ring. The breakthrough load (BTL) of the detent ball refers to the groove thrust on the ball, which needs to be overcome in gear shifting, so that the strut comes into contact with the blocker ring to produce index torque. If the BTL drops off too fast with the axial displacement of the sleeve, the blocker ring would be unloaded too fast, which would affect the oil scraping process and cause gear clash; if the BTL drops off too slowly with the axial displacement of the sleeve, the blocker ring may seize, resulting in shift shock. Thus, the rate of detent spring should be correctly selected to provide an appropriate radial force F_a, so that the total BTL meets the aforementioned requirements. The total BTL required during gear shifting can be calculated as follows:

The resultant axial force F_a of a detent ball acting on the sliding sleeve is

$$F_a = F_n \sin\theta + F_f \cos\theta = F_n(\sin\theta + f\cos\theta) \tag{10.316}$$

The resultant radial force of the sliding sleeve acting on the detent ball is

$$F_r = F_n \cos\theta - F_f \sin\theta = F_n(\cos\theta - f\sin\theta) \tag{10.317}$$

Per Equations (10.316) and (10.317), the total BTL of three detent balls around the circumferential is given by

$$F_{a\_total} = 3F_a = 3F_r \frac{f + \tan\theta}{1 - f\tan\theta} \tag{10.318}$$

10.4.7. Design of Shift Actuators

A gear shift actuator refers to a mechanism composed of shift-related components that execute the commands from TCU and complete the shift operation.

The requirements for shift actuators generally include five aspects: fast and accurate performance, that is, capable of achieving fast dynamic position control with high accuracy, good controllability, and clear failure modes; reliable and durable, with a lifespan of no less than the lifespan of vehicle, maintenance-free, and robust enough under various environmental conditions such as temperature, vibration, and dirt when the vehicle is in operation; compact structure, simple installation process, small volume, and high level of integration; low power consumption, avoiding the use of components with high parasitic losses such as oil pumps, wet clutches, and brakes, or minimizing their duty cycles, adopting a speed reduction mechanism with high mechanical efficiency; and low cost and simple design, with no need for parts or components of excessive precision.

There are several shift actuators for EV transmissions, which can generally be divided into three types, namely, electronically controlled hydraulic actuators, electronically controlled pneumatic actuators, and electronically controlled mechanical actuators, which are referred to as hydraulic, pneumatic, and mechanical, respectively, for short.

Hydraulic actuators refer to the actuation mechanisms in which a TCU controls the hydraulic valve using hydraulic oil via a group of solenoid valves, producing a hydraulic force to drive the shift control components such as clutches, brakes, or synchronizers to realize the gear shift.

Pneumatic actuators refer to the actuation mechanisms in which a TCU controls the cylinder using compressed air via a group of solenoid valves, producing a pneumatic force to drive the clutches, brakes, synchronizers, or other shift execution components to realize the shift.

Mechanical actuators refer to the actuation mechanisms in which TCU controls the shift motor to drive the clutches, brakes, synchronizers, or other shift execution components to achieve shifting. When a rotational shift motor is used, the shift execution component is usually driven via a reduction mechanism; when a linear shift motor is used instead, the shift execution component is usually driven directly.

The main characteristics of these actuators are compared in Table 10.19 [10.1].

TABLE 10.19 Comparison of three major types of shift actuators.

Criteria	Hydraulic	Pneumatic	Mechanical
Control accuracy	++++	++++	+++++
Response speed	+++++	+++++	++++
Mass	+++++	+++++	++++
Structure	++	++	+++++
Actuating force	+++++	+++++	++++
Power consumption	++	++++	++++
Shock	+++++	+++++	++++
Sealing requirement	++	++	+++++
Noise	+++++	++	+++++
Temperature influence	++	+++++	+++++
Cost	++	+++++	+++++

Note: +, ++, +++, ++++, and +++++ stand for poor, fair, good, very good, and excellent, respectively.

10.4.7.1. Design of Electrohydraulic Actuators

The power of a hydraulic actuator comes from the actuating piston within a cylinder, with the piston driven by the pressurized hydraulic oil from an oil pump that is usually powered by an electric motor. Taking two-speed hydraulic transmission as a simple example, two of the two-position four-way valves can be utilized to control the piston to shift between 1st and 2nd gears. Its control logic for gear shift is listed in Table 10.20 [10.1].

TABLE 10.20 Shift control logic of a two-speed electrohydraulic transmission.

Gears	State of valves	
	V1	V2
1st	On	Off
2nd	Off	On

Reproduced from, Yuan, Y., Electric Vehicle Transmissions: Structure and Control (Beijing: China Machine Press, 2020), ISBN: 978-7-111-66809-1, with the permission of China Machine Press.

The cylinder and piston of a hydraulic actuator can be designed according to the following steps:

Taking the hydraulic actuator shown in Figure 10.49 as an example, the relationship between the oil pressure and flow rate in cylinder 2 during the shift from the 1st gear to the 2nd gear is given by the following continuity equation:

$$A_c\dot{x}_c + \lambda_l p_c + \frac{V_{c0} + A_c x_c}{\beta}\frac{dp_c}{dt} = q_c \tag{10.319}$$

where

q_c is the flow rate of oil entering the cylinder
A_c is the projected area of the piston–oil contact surface in the cylinder along the axial direction
x_c is the piston displacement
λ_l is the leakage coefficient of cylinder
p_c is the oil pressure in the cylinder
V_{c0} is the initial volume of the high-pressure chamber in the cylinder
β is the bulk elastic modulus of oil

As mentioned earlier in the section about clutch design, when the cylinder rotates with the whole assembly, the oil pressure p_c rises with the increase in the radius due to the centrifugal effect. In this situation, p_c in Equation (10.319) can be approximately replaced by the average oil pressure in the cylinder. Assuming the piston is cylindrical, we have

$$A_c = \pi\left(R_{oc}^2 - R_{ic}^2\right) \tag{10.320}$$

For a positive displacement oil pump commonly used in hydraulic systems, there is a certain relationship between its output flow and the speed of oil pump. With the system leakage ignored, the flow rates in the system follows the relationship of

$$q_p = q_a + q_v + q_r + q_c + q_b \tag{10.321}$$

where q_p, q_a, q_v, q_r, q_c, and q_b are the flow rates at the outlet of oil pump, accumulator, relief valve, oil passage (induced by oil compression), cylinder chamber, and centrifugal balancing chamber, respectively.

The flow rate of the oil pump q_p can be estimated on the basis of the theoretical capacity and the speed of the oil pump.

For a bladder or piston accumulators commonly used in automotive hydraulic systems, by assuming that the filling and draining process is approximately adiabatic, the relationship between the pressure and the flow rate in the bladder or piston is

$$p_a = p_0\left(\frac{V_0}{V_0 - \int q_a dt}\right)^{1.4} \quad (10.322)$$

Equation (10.322) can be rearranged to yield

$$q_a = 1.4V_0 p_0^{-1.4} p_a^{0.4} \frac{dp_a}{dt} \quad (10.323)$$

where

p_0 is the initial pressure in the bladder or piston
V_0 is the initial volume of the bladder or piston

For a pilot relief valve, the relationship between the pressure and flow rate is

$$q_v = \begin{cases} 0 & p_s + \Delta p_u < p_k \\ \pi d_m y C_q \sqrt{\dfrac{2(p_s - p_o)}{\rho}} \sin\alpha & p_s + \Delta p_u \geq p_k \end{cases} \quad (10.324)$$

where

d_m is the mean diameter of the valve port of the relief valve
y is the valve port opening
C_q is the flow coefficient of the valve port
p_o is the outlet pressure of the relief valve, usually $p_o = 0$
α is the half cone angle of the valve core
Δp_u is the pressure drop of the one-way valve
p_k is the opening pressure of the relief valve

For the flow rate induced by oil compression in the oil passage, we have

$$q_r = \frac{V_p}{\beta} \cdot \frac{dp_s}{dt} \quad (10.325)$$

where

V_p is the volume of the oil passage
β is the elastic modulus of oil

For the cylinder and centrifugal balancing chamber, due to the damping of the oil inlet passage to the oil flow, the following relationships hold for the pressure and flow rate:

$$p_s - p_c = R_c q_c + L_c \frac{dq_c}{dt} \tag{10.326}$$

$$p_s - p_b = R_b q_b + L_b \frac{dq_b}{dt} \tag{10.327}$$

where

p_s is the main oil pressure of the system

R_c and R_b are the hydraulic resistance at the inlet oil passage of the cylinder and balancing chamber, respectively

L_c and L_b are the liquid inductance at the inlet oil passage of the cylinder and balancing chamber, respectively

$$R_c = \frac{128\mu l_c}{\pi d_c^4} \tag{10.328}$$

$$L_c = \frac{\rho l_c}{A_c} = \frac{4\rho l_c}{\pi d_c^2} \tag{10.329}$$

$$R_b = \frac{128\mu l_b}{\pi d_b^4} \tag{10.330}$$

$$L_b = \frac{\rho l_b}{A_b} = \frac{4\rho l_b}{\pi d_b^2} \tag{10.331}$$

where

μ is the dynamic viscosity of oil

ρ is the oil density

l_c and l_b are the length of the inlet oil passage of the cylinder and balancing chamber, respectively

d_c and d_b are the diameter of the inlet oil passage of the cylinder and balancing chamber, respectively

A_c and A_b are the cross-sectional areas of the inlet oil passage of the cylinder and balancing chamber, respectively

Per Equation (10.292), the dynamic equation of piston motion is

$$F_{pc} - F_{pb} - F_s - F_v - F_z = m_c \ddot{x}_c \tag{10.332}$$

where

m_c is the mass of the piston assembly

F_{pc} is the force on the piston from the oil pressure in the cylinder

F_{pb} is the force on the piston from the oil pressure in the balancing chamber

F_s is the spring force on the piston

F_v is the resistance acting on the piston from the sealing

F_z is the reaction force on the piston from the clutch

$$F_{pc} = \int_{R_{ic}}^{R_{oc}} 2\pi p_c r dr \tag{10.333}$$

$$F_{pb} = \int_{R_{ic}}^{R_{ob}} 2\pi p_b r dr \tag{10.334}$$

$$F_s = F_0 + K(x_0 - x) \tag{10.335}$$

$$F_v = B\dot{x}_c \tag{10.336}$$

where p_c is the oil pressure in the cylinder; due to the influence of the centrifugal force, p_c is a function of the radial distance r, that is, $p_c = p_c(r)$; p_b is the oil pressure in the balancing chamber; due to the influence of the centrifugal force, p_b is also a the function of the radial distance r, that is, the $p_b = p_b(r)$; R_{ic} is the inner radius of the piston on the cylinder side that is assumed to be the same as the inner radius on the balancing chamber side; R_{oc} is the outer radius of the piston on the cylinder side; R_{ob} is the outer radius of the piston on the balancing chamber side; F_0 is the spring force of the return spring at its initial position; K is the elastic coefficient of the return spring; x_0 is the initial axial position of the piston; and B is the damping coefficient of the piston.

Substitution of Equations (10.333) through (10.336) into Equation (10.332) yields

$$m_c\ddot{x}_c + B\dot{x}_c - Kx + F_0 + Kx_0 + F_z + \int_{R_{ic}}^{R_{ob}} 2\pi p_b r dr - \int_{R_{ic}}^{R_{oc}} 2\pi p_c r dr = 0 \tag{10.337}$$

By combining the continuity equation, that is, Equation (10.319) with Equations (10.320) through (10.331), and carrying out Laplace transform together with Equation (10.337), the dynamic block diagram of the piston–cylinder assembly can be established and solved for the structural dimensions such as the inner radius R_{ic} and the outer radius R_{oc} of the piston and then the structural dimensions such as the outer radius R_{ob} of the balancing chamber.

In the design of the piston–cylinder assembly, particular attention should be paid to the two aspects: Adequate strength and stiffness are essential to prevent excessive deformation and fatigue failure, and proper structural arrangement to prevent the piston from jamming in the cylinder. The reaction plate is generally supported at its outer diameter, with a stress distribution along the radial direction similar to that of a cantilever beam. As a result, it often deforms greatly when loaded. In order to make the contact between the friction plate and the mating steel plate more uniform, the piston can be designed into a convex shape to compensate for the deformation of the reaction plate, piston, and cylinder.

There are commonly three types of seals for the piston–cylinder assembly, namely, O-rings, square rings, and lip rings. Although the seal resistance is insignificant compared with the actuation force, lip rings generally produce smaller resistance than O-rings and square rings.

Traditional seal materials include nitrile rubber, polypropylene, and iron. The service life of polypropylene seals at high temperatures are longer than that of nitrile rubber seals, but they wear more significantly. Iron as a sealing material has good durability at high temperatures and stable friction characteristics, but the leakage is relatively large.

There are usually two types of solenoid valves used in hydraulic actuators, namely, high-speed switch valves and electrohydraulic proportional valves.

High-speed switch valves are also known as high-speed response valves, which are essentially two-position two-way electrohydraulic reversing valves. Functioning only as a switch, they control the flow and pressure drop through the themselves by switching the orifice area between fully closed and fully open, so as to realize the control of output oil pressure. When the flow rate required for clutch or brake engagement or disengagement is small, a high-speed switch valve can be used directly. In places where the required flow rate is relatively large, a high-speed switch valve can be utilized as a pilot valve in combination with a spool valve. High-speed switch valves are generally controlled by the PWM method on the electromagnetic coil voltage. As the duty cycle increases, the output oil pressure becomes higher, thereby controlling the smooth engagement and disengagement of a clutch or a brake.

When a voltage is applied to both ends of the electromagnetic coil of a high-speed switch valve, the coil current gradually increases due to self-inductance, causing the electromagnetic force acting on the moving iron to increase gradually. When the electromagnetic force is greater than the resistance on the moving iron, the moving iron starts to move, resulting in a counter-EMF affecting the coil current again.

When the voltage applied across the coil becomes zero, the coil current gradually decreases, causing the electromagnetic force on the moving iron to drop gradually. When the electromagnetic force is less than the resistance of moving iron, the moving iron stops moving, which produces the counter-electromotive force, causing the coil current to decline gradually to zero.

There is a certain lag in the action of spool relative to the electronic control signal. Such a phenomenon is called the switching response of a solenoid valve, which has a significant impact on the accuracy and sensitivity of system control. The switching response of a solenoid valve generally involves pull-in and release processes, which can be further divided into five stages, namely, pull-in triggering, pull-in motion, position retention, release triggering, and release motion.

The switching response of a solenoid valve is usually characterized by its on and off response times, which can be acquired by tests or simulations under various pump source pressures.

On the other hand, an electrohydraulic proportional valve is actually a bilateral throttle valve controlled by a proportional electromagnet. By controlling the spool to a certain position via the input current, the orifice area is regulated from fully closed to fully open continuously and proportionally to achieve continuous and precise control of the output oil pressure. The current is adjusted by changing the duty cycle of the input voltage signal of the electromagnetic coil using the PWM method. The greater the duty cycle, the higher the current through the solenoid and the oil pressure output, thereby controlling the engagement and disengagement of a clutch or a brake.

For a hydraulic proportional pressure pilot valve, when the input current I is zero, the valve has a minimum control oil pressure p_{smin}, generally $p_{smin} \approx 1$ MPa. When the front cavity oil pressure p_s is higher than a certain value, the valve linearity is better. In addition, the current value I corresponding to the same oil pressure value p_s deviates.

The characteristics of an electrohydraulic proportional valve can be divided into static and dynamic categories. Its static characteristics include hysteresis, linear range and nonlinearity, resolution, and repeat accuracy, and the dynamic indicators include step response and frequency response. Their definitions and selection criteria are listed in Table 10.21 [10.1].

TABLE 10.21 Definitions and selection criteria of characteristics of proportional valves.

Characteristics		Definitions	Selection criteria
Static	Hysteresis error	When the input current of the proportional valve undergoes a reciprocating cycle between the positive and negative rated values, the percentage of the maximum value of the input current difference corresponding to the same output oil pressure or flow rate to the rated current	The smaller the better
	Linear range	The approximate straight part of the oil pressure–current characteristic curve, usually selected as the operating range	The wider the better
	Nonlinearity	The percentage of the maximum current deviation between the characteristic curve and the straight line in the linear range relative to the rated current	The smaller the better
	Resolution	The minimum change in input current required for small changes in output oil pressure or flow rate as a percentage of rated input current	Proper*
	Repeat accuracy	With repeated input current under a certain oil pressure or flow rate, the percentage of the maximum difference between the multiple input currents and the input rated current	The smaller the better
Dynamic	Step response	When the input current is a step signal, the time required for the output oil pressure or flow rate to reach a steady state (greater than 98% of the set value)	The shorter the better
	Frequency response	When a sine disturbance of a certain frequency is added, the complex ratio between the output and the input in a steady state (i.e., the frequency, where the gain is −3 dB and the lag phase angle is −45°)	The higher the better

* As the resolution becomes smaller, the sensitivity of the valve increases and the static performance improves, but if the resolution is too small, the operating stability of the valve worsens.

Reproduced from, Yuan, Y., Electric Vehicle Transmissions: Structure and Control (Beijing: China Machine Press, 2020), ISBN: 978-7-111-66809-1, with the permission of China Machine Press.

Compared with high-speed switch valves, electrohydraulic proportional valves have the advantages of smaller output oil pressure pulsation and lower pressure drop in the valve, but they are more expensive, with slower response and slightly higher requirements for oil filtration.

10.4.7.2. Design of Electropneumatic Actuators

The power of a pneumatic actuator comes from the compressed air in a cylinder, which drives the piston to move the shift components, with the compressed air typically supplied by an air source. Taking a pneumatic two-speed AMT as a simplest example, two of the two-position three-way valves can be used to control the piston in a cylinder to achieve the shift between the 1st and 2nd gears, with the shift control logic listed in Table 10.22 [10.1].

TABLE 10.22 Control logic of a two-speed transmission with two-position three-way valves.

Gears	State of valves	
	S1	S2
1st	Off	On
2nd	On	Off

Reproduced from, Yuan, Y., Electric Vehicle Transmissions: Structure and Control (Beijing: China Machine Press, 2020), ISBN: 978-7-111-66809-1, with the permission of China Machine Press.

Alternatively, four of two-position two-way valves can be used to control the piston in a cylinder to achieve the same shift function, with the shift control logic listed in Table 10.23 [10.1].

TABLE 10.23 Control logic of a two-speed transmission with two-position two-way valves.

Gears	State of valves			
	S1	S2	S3	S4
1st	Off	On	On	Off
2nd	On	Off	Off	On

Reproduced from, Yuan, Y., Electric Vehicle Transmissions: Structure and Control (Beijing: China Machine Press, 2020), ISBN: 978-7-111-66809-1, with the permission of China Machine Press.

These two control principles are also applicable to the control of a clutch or a brake in a similar way.

When a three-position control such as 1st gear, 2nd gear, and neutral gear is required, three of two-position three-way valves can be used to control the piston in a cylinder, with the shift control logic listed in Table 10.24 [10.1].

TABLE 10.24 Control logic of a two-speed transmission with two-position three-way valves.

Gears	State of valves		
	S1	S2	S3
1st	Off	On	Off
2nd	On	Off	Off
Neutral	On	Off	On

Reproduced from, Yuan, Y., Electric Vehicle Transmissions: Structure and Control (Beijing: China Machine Press, 2020), ISBN: 978-7-111-66809-1, with the permission of China Machine Press.

In order to determine the parameters of the piston and cylinder, the example shown in Figure 10.46 is taken here to elaborate the process. First, the relationship between the shifting force and the diameters of the cylinder and piston when the gear is upshifted from 1st to 2nd is

$$F_s = \frac{\pi}{4}\left[p_{in}\left(D^2 - d^2\right) - p_{out}D^2\right](1+\eta) \tag{10.338}$$

where

F_s is the shift force
p_{in} and p_{out} are the pressures at the inlet and outlet of the cylinder, respectively
D is the cylinder diameter
d is the diameter of the piston rod
η is the loss coefficient due to the friction between the cylinder and the piston

Thus, with the shifting force F_s obtained from the aforementioned formula, for example, Equation (10.106), the cylinder diameter D and the piston rod diameter d can be determined accordingly. Second, the air volume V that flows into the cylinder during gear up and the 1st to 2nd upshift is

$$V = \frac{\pi}{4}\left(D^2 - d^2\right)L = Q \cdot t_s \tag{10.339}$$

where

L is the piston travel
Q is the airflow rate
t_s is the shift time

Since the piston travel is generally determined according to the displacement of the synchronizer or clutch during the shift process, the required air volume V can be obtained by Equation (10.141). Finally, the airflow rate is given by

$$Q = \lambda C_a \left(p_{in} - p_{out} \right)^{\varphi} \tag{10.340}$$

where λ is the operating state of the solenoid valve; C_a is the orifice coefficient, which is related to the shape of inlet and outlet ports and the flow state of the gas; and φ is the orifice index, which is related to the shape of inlet and outlet ports. Using Equation (10.340), the airflow rate Q can be obtained and then the shift time t_s is estimated according to Equation (10.339).

The response characteristics of high-speed pneumatic solenoid valves are similar to those of high-speed hydraulic solenoid valves elaborated earlier.

10.4.7.3. Design of Electromechanical Actuators

The power of the electromechanical actuator comes from an electric motor, which is typically a BDC motor or a BLDC motor. The advantages of these motors are five-fold: large torque and speed ranges, good starting and speed regulation performance, and fast response; smooth operation with low impact; simple and compact structure, small size, and convenient mounting and packaging; strong overload capacity, impact resistance, and high reliability; and low energy consumption. However, they also have a few shortcomings: inadequate accuracy in position control; vulnerability of the brush of a BDC motor to wear, affecting the service life; and relatively complicated and costly of BLDC controllers.

In general, an electromechanical shift actuator is mainly composed of an aforementioned actuation motor, a speed reduction mechanism, a shaft, and a terminal actuator such as a shift fork. The speed reduction mechanism usually has two functions: speed reduction and torque amplification, that is, reducing the speed and increasing the torque of the shift motor, and converting the rotary motion output by the shift motor into linear motion and producing adequate shift force on the terminal actuator. Common speed reduction mechanisms include worm gears, ball screws, and screw nuts. Terminal actuators can be shift forks and synchronizers, clutches, release bearings, brakes, etc., which function to change the transmission relationship of the gearset through changes in their positions or states, so as to achieve gear shift. The function of the shift shaft is to provide a certain support to the speed reduction mechanism and terminal actuator, especially to restrain the linear movement of the terminal actuator to ensure its correct direction.

Based on the shift force obtained earlier, the rated power required for the shift motor can be estimated by

$$P_m = \frac{F_s V_f}{\eta_r} \tag{10.341}$$

where

V_f is the linear motion speed of the terminal actuator

η_r is the mechanical efficiency of the shift actuation mechanism

Based on the shift time and shift stroke, the linear motion speed of the terminal actuator can be estimated:

$$V_f = \frac{l_f}{t_s} \tag{10.342}$$

where l_f is the linear motion stroke of the terminal actuator required for the gear shift.

The rated speed required for the shift motor n_m is

$$n_m = \frac{30 i_r V_f}{\pi R_{eq}} \tag{10.343}$$

where

i_r is the reduction ratio of the shift actuation mechanism
R_{eq} is the conversion factor between rotational and linear motion speeds at the output end of the shift actuation mechanism

When a worm gear is used as shift actuation mechanism, i_r is the reduction ratio of the worm screw to the worm gear, then

$$R_{eq} = \frac{m z_1}{2} = \frac{p_z}{2\pi} \tag{10.344}$$

where

m is the module of the worm gear
z_1 is the number of worm heads
p_z is the worm lead

When ball screws or screw nuts are used as the shift actuation mechanism, $i_r = 1$, then

$$R_{eq} = \frac{d_2}{2} \tan \psi = \frac{S}{2\pi} \tag{10.345}$$

where

d_2 is the pitch diameter of tread
ψ is the lead angle of tread
S is the tread lead

More specific design methods of these mechanisms are omitted here. Refer to mechanical design manuals for details.

10.4.8. Design of Hydraulic and Lubrication Systems

The hydraulic and lubrication systems of a drivetrain are generally composed of components such as oil pumps, accumulators, control valves, and orifices, which are connected by hydraulic and lubrication circuits.

10.4.8.1. Oil Pumps

The oil pump in a drivetrain usually functions to supply hydraulic and lubricating oil of a certain pressure and flow rate to the hydraulic components and interfaces where relative motion takes place. In addition to being highly efficient, quiet, and durable, it is also required that the oil pump be designed with only reasonable manufacturing precisions and low mass production costs.

Other than the selection of the oil pump type and its sizing, the design also involves the balance of the axial and lateral forces, the critical speed of cavitation, the volumetric and the mechanical efficiencies, and the oil filter at the inlet.

Oil pumps can be sized according to the desired volumetric efficiency, leak paths, and minimum system pressure. The minimum system pressure mainly depends on the oil pressure required by the hydraulic components, for example, a shift actuator discussed in the last section, or simply the sum of pressure drops in the lubrication circuit if an active lubrication system is employed. The internal leakage in a hydraulic system occurs mostly at the spool valves, clutches, pump seals, and solenoids.

The leakage flow rate at a spool valve, Q (in m^3/s), can be estimated by

$$Q = 0.52 \frac{D}{L} \frac{ph^3}{\mu} \tag{10.346}$$

where

- D is the diameter of the spool valve in m
- L is the length of seal land in m
- p is the differential pressure across the seal land in Pa
- h is the difference between radii of the bore and spool in m
- μ is the dynamic viscosity of oil

Note that the leakage flow rate often reaches the maximum when the oil temperature comes to the highest value. The leakage at the clutches, Q (in m^3/s), can be estimated by making the leakage area equivalent to an orifice:

$$Q = 0.024 d^2 \sqrt{\Delta p} \tag{10.347}$$

where

- d is the orifice diameter equivalent to the leakage area in m
- Δp is the differential pressure across the orifice in Pa

Leakage at the solenoids and at the seal between the pressure port and the suction port of the oil pump can be estimated similarly using Equation (10.346).

Oil pumps commonly used in the drivetrains of EVs are external gear pumps, gerotor pumps, crescent internal gear pumps, or internal gear pumps without a crescent. They are essentially positive displacement gear pumps.

With only a few parts, the gear pump is very compact and simple and is suitable for installation on the input or output shaft of the transmission. The gear pump for EV applications, however, often powered by an BLDC motor, can actually be packaged with great flexibility inside or outside the transmission housing. The crescent internal gear pump generally consists of an inner gear, outer gear, crescent, casing, and cover. Not only being more compact, the crescent internal gear pump is usually more wear-resistant than an external gear pump due to the lower relative speed between the inner and outer gears, as the former has one or two teeth in the inner gear than the latter. Mostly made of cast iron or low carbon steel for wear resistance and easy cutting, the inner and outer gears are usually thicker toothed than transmission gears and are chosen based on the design capacity, rather than strength.

The internal clearance of a gear pump is an important design consideration. If the clearance is too tight, the gears do not rotate smoothly and power consumption would increase; if the clearance is too loose, the leakage becomes severe and the volumetric efficiency would drop. It is, therefore, desired that the outer diameter of the inner gear and the inner diameter of the outer gear be machined to a relatively high precision. At the same time, the inner and outer diameters of the two gears must have sufficient concentricity relative to their respective pitch lines for their proper alignment in the pump casing. In addition, in order to ensure the end face sealing and running clearance of the pump, there are certain tolerance requirements for the width, parallelism, and flatness of the two gears. To ensure the end sealing of the rotor, the flatness of the end cover and the tolerance to the depth of the rotor depression of the pump casing must be strictly controlled. Besides, the tolerance to the shape of the crescent should also be carefully designed so that it fits the tooth top surfaces of the two gears to form a seal between the suction port and pressure port.

The theoretical displacement of the pump, q_c, is given by the displacement motion of the gear mesh and can be expressed as

$$q_c = \frac{\pi}{4}\left(D_1^2 - D_2^2\right)W \tag{10.348}$$

where

D_1 is the outer diameter of the inner gear

D_2 is the pitch diameter of the inner gear minus twice the height of the addendum of the outer gear

W is the face width of the gear

The size of the inner gear can be mainly determined according to the required displacement, but for sizing the outer gear, the requirements for proper formation of the pressure and suction ports should be taken into account. From the perspective of volumetric efficiency, a long crescent is often preferred. However, compromises may be necessary from the standpoint of reducing mechanical losses or improving the flow inside the pump. In addition, for maximized pump displacement, coarse pitch gears are generally employed. For satisfactory suction effect, the gears should fully mesh each other at the end of engagement, and the tooth crest in contact with the crescent should provide an adequate seal.

As far as the typical characteristic curves of a crescent internal gear pump are concerned, it is worth noting that the actual displacement line is parallel to the theoretical displacement line and shifted to the right by a certain amount. Such an offset depends on the internal leakage of the pump, while the two parallel lines imply that the internal leakage is not affected by the speed. The main factors affecting internal leakage include clearances inside the pump, increases in discharge pressure, and decreases in oil viscosity due to elevated temperatures.

The power consumption of the pump is dependent on the mechanical friction losses and the volumetric displacement. To minimize the power consumption, the forthcoming design guidelines should be followed:

1. The operating pressure range of the oil pump can be reduced by enlarging the clutch actuation cylinder.
2. The internal leakage of the transmission should be minimized to reduce the demand for hydraulic oil, primarily by eliminating leakage points in the overall design of the transmission and strictly controlling the tolerances between mating parts.

3. The manufacturing precision and assembly process of the pump parts should be tightly controlled to keep the internal leakage to a minimum, and selective assembly is, thus, generally recommended to keep the end face clearance within limits.
4. The transmission should have sufficient cooling to limit the leakage loss caused by the decrease in oil viscosity.

A gerotor pump consists of components such as the inner rotor, outer rotor, housing, and cover. Its fundamental differences from a crescent internal gear pump are the tooth shape of the rotor and the absence of a crescent. These differences make their pumping action and the manufacturing process of the two rotors and casing very different.

Similar to gear pumps, to ensure proper alignment of the two rotors in the gerotor pump casing and minimize internal leakage, the dimensional tolerances of the rotors must be tightly controlled. The tooth shape of the inner rotor is produced by a gear cutter with the same average diameter as the outer rotor, and the eccentricity between the center of the cutter and the center of the rotor blank is equal to the average eccentricity of the centers of the two rotors installed in the pump casing. The inner rotor has one more tooth than the outer rotor, and the tooth height is equal to 4 times the eccentric distance between the centers of the two rotors. The two rotors mesh with each other with complete disengagement occurring across the diameter of the rotors from the point of complete engagement. With the inner rotor being the driving part and the outer rotor being the driven part, the outer rotor is usually directly guided and positioned by the pump casing and the inner rotor is indirectly guided and positioned. As the inner rotor turns, each of its teeth maintains continuous line contact with a tooth of the outer rotor, with the point of contact moving from full meshing of the tooth sides to maximum separation of the tooth tips. In the section between engagement and disengagement, a continuously expanding volume is formed between two adjacent teeth of the inner rotor and two adjacent teeth of the outer rotor. Similarly, a continuously decreasing volume is formed in the disengagement and engagement sections. In this way, the continuous variation in the displacement with the rotation of the rotor results in a suction effect. The section between engagement and disengagement is the suction port, and the section between disengagement and engagement is the pressure port.

Similar to gear pumps, the gerotor pump has external components such as a casing and a cover, in which oil passages or ports are required to carry oil to the suction port and discharge it from the pressure port. These ports should ideally extend from the point of full engagement to the point of full disengagement between the rotor teeth, depending on displacement. However, they must be shortened by sufficient length in perspective to meet the requirements of tooth pitch and control of face leakage. In order to provide maximum inlet area for oil filling, the suction port is usually eccentric. The increment of displacement between the circumferential teeth of a gerotor pump is relatively large, and the pressure change due to the rapid compression of the oil, as it is delivered to the pressure port, is likely to cause noise. A satisfactory solution to this problem is usually to shorten the length of the initial end of the discharge port and replace it with a section of machined compensating oil groove to control the compressibility of the oil as it enters the discharge area of the pump.

The theoretical displacement of a gerotor pump, q_g, can be calculated by

$$q_g = \frac{\pi}{4}\left(D_{r1}^2 - D_{r2}^2\right)W_r \tag{10.349}$$

where

D_{r1} is the inner diameter of the outer rotor plus twice the eccentricity of the outer rotor
D_{r2} is the inner diameter of the outer rotor minus twice the eccentricity of the outer rotor
W_r is the width of the rotor

10.4.8.2. Control Valves

In a powershift transmission, pulse width-modulated (PWM) solenoid valves are often used to directly control the engagement and disengagement of the clutch. These solenoid valves can be divided into normally closed type and normally open type. The selection of the solenoid valve type can be made according to the gear position being placed on after the transmission is powered off.

Additionally, variable-force solenoid valves (VFSs) are increasingly employed in transmissions. This type of solenoid valve can directly control the line pressure or clutch pressure, making it less susceptible to oil temperature and oil pressure fluctuations, so as to obtain higher-precision clutch actuation flow control and improve the timeliness and smoothness of shifting. Its residual pressure can be controlled between 0 and 13.8 kPa.

Moreover, a common type of control valve, spool valves are often used in conjunction with solenoid valves. Since the spool valves have less surface area that is influenced by the pressure of the flow, they can be actuated using less powerful solenoid valves to save energy. They come in a variety of sizes and shapes but generally have several lands that control hydraulic fluid for different functions. Common spool valves and their typical functions are as follows: pressure regulation valves for blocking oil flow and building up pressure, restriction valves for limiting the pressure inside hydraulic components, on–off valves for switching the direction of oil flow, and fourth, solenoid switch valves for interlock of clutch engagement.

Furthermore, pressure-regulating spool valves build up pressure by introducing flow resistance in the oil circuit, which is necessary for the clutch to engage and prevent the clutch from slipping. When the spool is at its initial position, the flow from the discharge port of the oil pump is blocked by the valve to build up the pressure. With the speed of the oil pump increasing, the valve reaches a force balance and the spool begins to move. At this moment, the oil flow is directed to the most important portion of the system, that is, the hydraulic and lubrication components. When the oil flow rate continues to rise with the increasing oil pump speed, the excessive oil flow is then directed to the inlet of the oil pump.

In order to reduce the power consumption of an oil pump, the hydraulic system nowadays is often designed with variable pressure control, in which the line pressure of the hydraulic system is regulated to an appropriate value in the predetermined range according to the output torque of the traction motor. Although a high line pressure should be avoided as it would incur excessive leakage inside the hydraulic system and unnecessary power consumption, a low line pressure is not desirable either as it may cause clutches to slip.

The spring load tolerance of a pressure-regulating spool valve is a key operating characteristic. High-quality springs should be selected so as to control the load tolerance under its operating length within ±3%. It is always preferred to keep the elastic coefficient of the spring as small as possible. The clearance between the spring and the inner diameter of the valve borehole where it is installed is normally 0.5 mm (one side).

The pressure-regulating spool valve must be designed to withstand a dramatical surge in line pressure due to the speed increase in oil pump by taking the great influence of the oil flow momentum into account. To reduce the surge of line pressure, an orifice can be added to the oil passage downstream the pressure-regulating spool valve or a special pressure port structure can be designed to introduce damping into the system.

In designing the installation position of the valve core and the lands, the sealing length between them should be greater than 1.5 mm, and the distance between the land and the wall of the valve body should be at least 0.75 mm. When the spool reaches the maximum stroke, the pressure difference from the pressure port to the suction port must be less than the minimum line pressure. These design principles can help greatly reduce the amount of leakage.

Spool valves should also be designed to minimize the wear from side loads. The side load is mainly caused by the uneven exposure of the spool around the valve ports. When only the upper half of the spool is exposed to the valve port, the spool is subjected to a downward hydraulic force greater than the upward hydraulic force. The resultant downward force causes the spool to deviate from the centered equilibrium position and deflect downward, which will increase the wear on the valve borehole when the spool moves. In order to reduce wear, some annular grooves can be applied on the spool surface to balance the pressure around the spool and lower the side load, thus reducing the wear.

Another issue to consider in the design of spool valves is the minimization of its hysteresis in response to hydraulic pressure. The draft angle of the valve body wall can affect the hydraulic balance and increase the hysteresis of the valve. The hysteresis due to the draft angle can be minimized by using spool grooves mentioned previously, while both the hysteresis and the wear of the borehole can be significantly reduced by anodizing the spool and borehole.

10.4.8.3. Lubrication System

Consideration should be given to the lubrication system right at the initial stage of transmission design so that it can adapt to the requirements of the different application scenarios, load, life, and efficiency. The arrangement of the lubricating oil inlet and outlet, oil jet, pump, filter, filler, drain, vent plug, and so forth should all be optimized to ensure the best lubrication effect.

The lubrication system of transmission can generally be divided into two types, namely, the pressurized oil circulating method and the oil splashing method. Both lubrication systems can be fully housed in the transmission case or connected via oil lines to an external cooler. The common feature of these two types of lubrication systems is that the lubricating oil can be recycled and reused—in the former the lubricating oil is pumped back to the oil circuit through the oil filter from the oil sump, while in the latter, the lubricating oil flow is gathered to the oil sump through the action of gravity. The appropriate lubrication method can be selected by taking various factors into account, such as gear load, environmental conditions, and expected life. Generally speaking, the pressurized oil circulating lubrication method is mostly used in planetary gear transmissions because of their compact layout, high power density, small gears and bearings, and poor heat dissipation and oil film formation conditions; the oil splashing method is often adopted in layshaft transmissions for their large gears and bearings, along with good heat dissipation and oil film formation conditions.

Based on the theory of tribology, there are mainly three lubrication regimes between two relatively moving surfaces, namely, dry friction, boundary lubrication, and fluid film lubrication, as discussed now.

At first, in the regime of dry friction, the lack of a fluid film between the two surfaces causes the contact actually to only occur at points high enough on the surface of the softer material, resulting in a pressure to balance the total load. Thus, under the action of a normal load, the real contact area is only a small part of its apparent total area, and it increases proportionally with the increase load [10.76-10.78]. Sliding one rough surface over another requires a force that must overcome the frictional forces associated with the yield strength of the asperity points. Friction can be greatly reduced when a very thin coating of the solid lubricating material is applied or absorbed onto the interface. The wear volume is related to the debris generated from asperity contact points during sliding.

Next, in the regime of fluid film lubrication, the relative moving surfaces are completely separated by a liquid or gas lubricant film. At this time, a load-supporting pressure is generally established between the two surfaces due to the effects of hydrodynamic lubrication (HL), squeeze film lubrication, or external pressure film lubrication, as elaborated here. HL occurs when the self-acting pumping effect of the moving surface brings the liquid into the converging wedge-shaped area, thereby forming a fluid film to separate the surfaces. In this fluid film, both pressure and frictional power losses are dependent on lubricant viscosity, geometry, and shear rate due to the relative motion. In transmissions, journal bearings (also known as bushings), axial thrust bearings, and so on are common applications of this type of lubrication. HL also includes a special case, that is, the elastohydrodynamic lubrication (EHL), which is a result of the fact that the pressure in the fluid film is high enough to cause a significant elastic deformation of the lubricated surface. It is different from HL in that the hardness of the material, the change in viscosity with pressure, and the mutual adaptability of the contact surface geometry become more important for EHL. The contact surfaces with good geometrical adaptability are close to each other, with a relatively large area for bearing the load, while the contact surfaces with poor geometric adaptability have a relatively much smaller area for bearing the load, with only one-thousandth of the load-bearing surface area remaining sometimes. EHL can be further categorized into hard EHL and soft EHL. Hard EHL involves situations where the geometry of the contact surfaces is not well-adapted and the material has a high modulus of elasticity. The form of contact mainly includes rolling or rolling combined with sliding. Typical examples of hard EHL include ball bearings, roller bearings, and meshing gears. Due to the poor adaptation of the surface geometry, the loads are concentrated on very small areas where elastic deformation occurs. This type of lubrication is characterized by a very thin fluid film, which leads to localized stresses, which negatively affect the fatigue life of the material. The fluid film pressure is generally as high as 0.5–3 GPa, a thousand times higher than that in most fluid dynamic bearings. The thickness of the fluid film is usually 0.1–0.5 mm, which mainly depends on the viscosity of the lubricating fluid, the shape of the fluid film, and the speed of the fluid entering the contact area. On the other hand, soft EHL involves materials with a lower modulus of elasticity, such as rubber seals and Babbitt journal bearings under heavy loads and low speeds with the use of softer bearing materials. Due to the low contact pressure, its influence on the viscosity of lubricating fluid can be ignored and the difficulty of analysis is lower than that of hard EHL. However, squeeze film lubrication is produced on the surfaces with relative motion such as reciprocation or rotation, or the surfaces subjected to impact load. Since it takes time to extrude the lubricating fluid film out of the contact area, it can withstand much larger loads than the steady, single-directional load. The comparison of its typical values is shown in Table 10.25 [10.1]. Finally, external pressure film lubrication is often produced by injecting lubricating fluid into the contact surface with pressure by an external fluid pump. It is usually used when the hydrodynamic pumping effect is limited, for example, when the lubricating fluid viscosity is low or when the relative speed is slow right after startup.

TABLE 10.25 Typical load bearing capacity under squeeze film lubrication.

Type of load	Load value (MPa)
Steady radial load	1.4–2.8
Steady axial load	2.1–4.8
Impact or dynamic load	17–34

Reproduced from, Yuan, Y., Electric Vehicle Transmissions: Structure and Control (Beijing: China Machine Press, 2020), ISBN: 978-7-111-66809-1, with the permission of China Machine Press.

At last, boundary lubrication is formed when the speed and viscosity of the tribological system in the HL regime eventually become unable to generate enough fluid film pressure to carry the entire load with the increase in load. Consequently, the asperities on the contact surfaces break through the fluid film and start rubbing against each other, increasing the contact area and leading to plastic deformation. The temperature of the asperities in contact increases, and finally, the surface is torn and peeled off on a larger scale. At this time, the friction coefficient of the lubricated area may increase from 0.001 in the case of full fluid film to 0.03–0.1 in the case of mixed fluid film, that is, boundary lubrication. As the load continues to increase, the bearing force of the fluid film will eventually be lost completely, and the friction coefficient may reach 0.2–0.4, which is the typical range of dry sliding friction [10.75].

The knowledge concerning the design of lubrication system is detailed as follows:

1. Commonly used equations for lubrication system design
   When gears, shafts, clutches, and other moving parts in the transmission rotate at high speeds, the lubricating oil in them will be affected by a centrifugal force and a pressure field will be established in a relatively closed space or oil chamber, with the oil pressure rising with increasing distance from the centerline of rotation. By assuming that there are droplet A and droplet B in a rotating field of continuous lubricating oil, with the radii from the centerline of rotation are R_A and R_B, respectively, then the difference of oil pressure between the two droplets is given by

$$\Delta p = p_B - p_A = \frac{1}{2}\rho\omega^2\left(R_A^2 - R_B^2\right) \tag{10.350}$$

   where
   ρ is the density of lubricating oil
   ω is the rotational speed of lubricating oil

   In addition, it is common in the lubrication system that two orifices are arranged in series. In this case, an equivalent diameter, D_{eq}, can be calculated by the following equation:

$$D_{eq} = \frac{D_1 D_2}{\left(D_1^4 + D_2^4\right)^{1/4}} \tag{10.351}$$

   where D_1 and D_2 are the diameters of orifices.

2. Forced oil lubrication system
   The working principle of a forced oil lubrication system for transmissions of EVs is that the lubricating oil collected in the oil pan is sucked into the oil circuit from the oil inlet due to the negative pressure created by the oil pump. With impurities removed by the filter, the oil then enters the pressure-regulating valve, which builds up pressure by applying resistance to the oil flow and, at the same time, controls the output oil pressure to a certain level without fluctuating. The accumulator can improve the time response of the system and further reduce the fluctuation of lubricating oil flow in the system. The control valve precisely controls the oil flow so that it can meet the lubrication requirements of each component without causing the flow to exceed the

demand and incur unnecessary power consumption of the oil pump. In the upstream of the lubricated components, there is usually an orifice through which the flow rate of lubricating oil is finally adjusted, so that different components in the system, including splines, gears, bearings, wet clutches, and wet brakes, can obtain appropriate amount of lubricating oil. The main function of the orifice is to adjust the distribution of the amount of lubricating oil among the various components, so that the flow distribution under all operating conditions can meet the requirements of the transmission.

When an oil pump is employed in the system, the oil level should be designed so that the inlet of the oil pump is not exposed to the air under any road slope conditions, a phenomenon called loss of prime; otherwise, air could be sucked into the oil pump, causing failures such as drop of line pressure, slip of clutches, and air bubbles in the oil circuits.

3. Oil splash lubrication system

   The oil splash lubrication system should meet two requirements: the components that are lubricated by the oil splash due to gear rotation are able to catch enough oil under any designed operating conditions to prevent failure from insufficient lubrication, and under the premise of meeting the aforementioned requirement, it is desired to minimize the quantity of lubricating oil and to lower the oil level, so as to lower the oil churning loss.

   Generally speaking, the amount of oil obtained by each component that needs to be lubricated under different operating conditions can be calculated by establishing a one-dimensional simulation model of the lubrication system, and the churning loss of each component can be solved by establishing a three-dimensional simulation model of the lubrication system.

   For oil splash lubrication, it is generally required that the gear of the final drive has a certain immersion depth in the oil. In principle, the immersion depth should not be kept within 1–3 times the tooth of the final drive gear. If the immersion depth is too small, cooling and lubrication would be insufficient; if the immersion depth is too large, the oil churning loss would be large and the oil temperature would rise too fast. An optimal oil level can be obtained by means of three-dimensional flow field simulation or testing of the transmission. At low temperatures, the transmission is prone to low liquid-level problems because of the shrink of lubricating oil volume and the slow return of oil flow due to the rise of viscosity. At high temperatures, on the other hand, the oil churning loss may increase due to elevated liquid levels, while bubbles may develop and cause transmission failure. Operating conditions such as extreme temperatures, sharp turns, emergency braking, and steep slopes should be considered when simulating or testing the liquid level.

   As far as the depth of the transmission oil pan is concerned, the ground clearance and the influence of the constant velocity universal joint angle for front-wheel-drive models should be taken into account.

4. Selection of transmission oil

   The physical, chemical, and electrical properties of a transmission oil have significant impacts on the performance and durability of transmissions and electric drive systems.

First, the choice of transmission oil should consider the viscosity and its change with temperature. The operating temperature of a transmission is generally within the range of −40 to 135°C, and the viscosity of transmission oil usually declines with the rise in the temperature. When the viscosity drops, the leakage at the oil pumps, seals, and other hydraulic components would increase, so that the oil pump has to speed up to meet the requirements, resulting in increased power consumption of the oil pump. At the same time, the line pressure and clutch actuating pressure decline, which may cause engaged clutches to slip, or excessive clutch slip during shifting, consequently resulting in prolonged shifting time and shortened clutch life. Low-viscosity oil also reduces the thickness of the oil film between bearings and other relative moving surfaces, deteriorating the lubrication conditions, especially under high-temperature and heavy-load conditions. On the other hand, when the viscosity rises with the drop of temperature, the viscous resistance-related loss of the system increases and the dynamic response of the hydraulic system slows down. The latter makes the oil pump prone to malfunction due to possible air suction into the hydraulic system from the inlet. Therefore, the use of transmission oil with low viscosity at low temperatures and high viscosity at high temperatures is recommended to increase clutch capacity and shorten shift time.

Second, the friction characteristics and stability of transmission oil also largely affect the shift time, smoothness, friction work, clutch life, and vibration and noise during shifting. The friction coefficient of the clutch is not only related to the material and microstructure of the friction surface but also related to the base oil and additives of the transmission oil. In addition, the friction characteristics of the clutch are also related to factors such as slip speed, temperatures of friction surface and transmission oil, oil type, and length of usage. For smooth and quiet engagements, it is generally desired that the transmission oil does not cause a rise in the friction coefficient when the clutch slip speed declines.

Third, the antioxidant properties of transmission oil are critical as well. The high temperature generated on the friction surface can cause hydrocarbons in the transmission oil to oxidize, producing organic acids and forming viscous deposits. With the deterioration of the friction characteristics of the transmission oil, its viscosity rises and wear resistance worsens, and thus affecting sealing performance and causing metal corrosion. The antioxidation performance of transmission oil can be improved by selecting appropriate base oil and antioxidation additives.

Fourth, the antiwear properties of transmission oil play a major role in protecting friction surfaces. Surfaces with relative motion, such as clutches, gear pairs, bearings, hydraulic valves, and seals inside the transmission, are subject to wear. Antiwear additives in transmission oils reduce wear and prevent associated component failures.

Fifth, the antifoam properties of the transmission oil are of crucial importance too. In the transmission and electric drive system, the lubrication oil is agitated by moving parts or injected into the liquid surface in the form of oil flow during the circulation process, which often causes foaming. Excessive foaming would overflow from the vent cap and even cause the exposure of the oil pump inlet to the air, leading to oil pump noise, loss of prime, or other failures of hydraulic components. Structural design, chemical composition of oil, oil filter, and other factors have influence on foaming. Particularly, foaming can be reduced with foam inhibitors such as silicon and silicon-containing compounds such as polysiloxanes or acrylic copolymers, which normally have lower surface tension and higher solubility than transmission oils and can be broken down in the oil into tiny droplets.

Sixth, the cooling performance of a transmission oil on friction surfaces depends largely on its thermal conductivity, while the internal volume at extreme temperatures and the compactness of the electric drive system with its transmission are influenced by the volumetric thermal expansion coefficient to a certain extent.

Finally, the physical properties of a transmission oil, such as density, specific heat, bulk modulus, boiling range distribution, gas solubility, shear strength, surface tension, resistivity, and other single-valued properties including freezing point, flash point, ignition point, aniline point, cloud point, heat of vaporization, and corrosion resistance, have different influences on the functions and performances of the drivetrain system.

With continuous improvement in the design of electric drive systems, the electric motor and transmission components are often integrated into the same housing. Meanwhile, a unified lubricating oil is utilized for cooling and lubrication, which puts forward high requirements for the physical, chemical, and electrical characteristics of the lubricating oil, mainly involving the following aspects [10.76, 10.77]:

1. Low electrical conductivity: Since the lubricating oil is in direct contact with the stator winding of the motor, the electrical conductivity of the oil must be strictly controlled to prevent short circuits.
2. High cleanliness: In order to prevent deposits from affecting the normal operation of the system, metal cleaning agents and dispersants are usually added to lubricating oils for system cleanliness, friction characteristic adjustment, wear reduction, acidic product neutralization of oxidation reactions, and rust prevention.
3. Lower liquid dielectric constant, higher dielectric strength, and better copper corrosion resistance than conventional transmission oils are also desired to improve the safety and lifespan of the electric drive system.

The physical properties of several transmission oils are listed in Table 10.26 [10.1].

TABLE 10.26 Physical properties of several transmission oils.

Characteristics	Chevron ATF VI	SAE 80W-90	SAE 75W-90
Color	Red	Amber	Amber
Density (kg/dm^3)	0.849	0.891	0884
Dynamic viscosity at −40°C (MPa·s)	11,700	—	—
Kinematic viscosity at 40°C (mm^2/s)	30.6	132.84	106
Kinematic viscosity 100°C (mm^2/s)	6.0	13.56	17
Viscosity index	147	110	175
Freezing point (°C)	-54	-27	-39
Flash point (°C)	217	240	210

Reproduced from, Yuan, Y., Electric Vehicle Transmissions: Structure and Control (Beijing: China Machine Press, 2020), ISBN: 978-7-111-66809-1, with the permission of China Machine Press.

10.4.9. Design of Housing

The gearsets and other components of a transmission are enclosed in a housing. The transmission housing mainly has the following functions:

1. The lubricating oil is retained within the housing to ensure that the gears, bearings, and other components are fully lubricated; the enclosure can prevent external impurities from entering the interior and deteriorating the cleanliness, so as to ensure the reliability of the transmission.
2. After splashed by the rotating parts, for example, gears, the lubricating oil is directed to the components in need of lubrication by oil guide structures and finally collected by the housing and used repeatedly.

3. Gears, shafts, and other load-carrying components subjected to the load from the prime mover are supported by the housing through bearings, with the housing providing the reaction forces and moments to keep these components in place.
4. Mounted on the vehicle frame via the housing, the transmission assembly is retained in place by the reaction forces and moments that the frame exerts on the housing.

The general requirements for transmission housing design are elaborated as follows:

1. Make the housing as compact and light as possible while ensuring the structural strength and stiffness for bearing the forces and moments from the prime mover.
2. Ensure that the relative positions of components such as shafts and gears change within acceptable ranges.
3. The oil guide structure can direct the lubricating oil to the components in need of lubrication under all operating conditions.
4. Good heat dissipation performance.
5. Decent performance in noise reduction and isolation.
6. Easy to disassemble and repair.
7. Under all operating and allowable ambient conditions, the vent cap on the housing can prevent the internal pressure from exceeding the acceptable level with no lubricant leakage at the vent cap and on any sealing surface of the housing.

10.4.9.1. Overall Structure of Transmission Housing

The structural topology of the transmission housing is usually a split type, that is, it is composed of more than one piece, with the load generally input and output by the ends. The input end of transmission housing can be flanged to the front cover of the traction motor or highly integrated to the traction motor by unifying their shaft and housing, so as to improve compactness and reduce manufacturing costs. The housing shall envelop the internal components of transmission with a minimum volume, with adequate clearance in between. Generally, the clearance should be greater than 15 mm between the gear tooth tip and the inner bottom wall of the transmission; greater than 5 mm between the inner wall and rotating parts, for example, gears; within 3–5 mm between the inner wall and linear moving parts, for example, shift forks; and within 1–2 mm between the oil baffle and other stationary parts [10.1].

The wall thickness of the housing is generally composed of the thickness of the reinforced structure on top of a basic wall thickness. In those places where the stress is concentrated such as bearings and bolts, the wall can be thickened to reduce the stress. In addition, there are usually reinforcing ribs on the housing to enhance the strength and stiffness and reduce its deformation and noise reflection.

10.4.9.2. Loads on the Housing

The loads acting on the housing need to be determined for its structural design. These loads mainly come from four sources: the gravity on the transmission, including mechanical and electrical components and transmission oil; the forces acting on the interior of the housing by bearings, clutches, brakes, and hydraulic system during operation; the forces acting on the exterior of the housing through the transmission mounts; and the assembly load at the bolt connection [10.1]. It is worth noting that the latter two loads would surge when subjected to externally induced impact loads, which can generally be equivalent to multiples of the gravity on the transmission, usually 5 to 10 times the gravity. These impact loads can be obtained from the load spectrum formulated according to the statistical data collected from the vehicle operating on various roads.

A safety factor of at least 2.5 is generally required for static torsional strength of the transmission assembly. Thus, when verifying the strength of the transmission housing, the torque applied to it should be at least 2.5 times the maximum nominal torque in the first gear.

For helical gear transmissions, the load acting on the teeth surface of the gear can be decomposed into the circumferential, radial, and axial components. The corresponding magnitudes, denoted by F_c, F_r, and F_a, are given, respectively, by

$$F_c = \frac{2T}{d} \tag{10.352}$$

$$F_r = \frac{F_c \tan\alpha_n}{\beta} \tag{10.353}$$

$$F_a = \frac{F_c}{\tan\beta} \tag{10.354}$$

where

T is the torque transmitted by the transmission
d is the pitch diameter of the gear
α_n is the normal pressure angle
β is the helix angle at the pitch circle

Since the circumferential force magnitude F_c can be decomposed into a force passing through the gear axis and a moment, the radial resultant force passing through the gear axis is

$$F_{rc} = \sqrt{F_r^2 + F_c^2} \tag{10.355}$$

The radial resultant force is transmitted to the bearing seat through the outer ring of the bearing, and a certain pressure distribution is formed on the surface of the bearing seat on both sides of the radial resultant force line. Studies have shown that the radial pressure distribution can be uniformly simplified into a cosine pressure distribution in the range of 120° [10.1], namely,

$$p(\theta) = \frac{5F_{rc}}{6BR}\cos\frac{3\theta}{2} \tag{10.356}$$

where

B is the width of the bearing seat
R is the radius of the bearing seat
θ is the central angle with respect to the radial resultant force line, with $-60^\circ \le \theta \le 60^\circ$

In addition, the gear also produces an axial force F_a that acts on the housing through the bearing seat, which can be simplified as a uniform pressure distribution on the thrust surface of the bearing seat.

Based on the aforementioned loads, a topology optimization of the housing can be carried out with the help of engineering design software to obtain the optimal topology of the housing. By means of FEA software, the stress distribution and deformation of the housing under various typical operating conditions can be calculated. In general, the maximum stress in the housing should not be higher than the allowable stress of its material, and the deformation at the bearing seat should be less than 0.2–0.25 mm [10.1]. If the stress and deformation do not meet the requirements, the local structure should be strengthened by adding or adjusting ribs and wall thickness.

In addition, the cyclic loads acting on the housing when the transmission is in operation cause cumulative damage to the housing. Hence, the fatigue life of the housing must be analyzed using relevant methods such as nominal stress method (S-N method), local stress–strain method, or multiaxis fatigue life method. An appropriate method can be selected according to the load spectrum for analysis to meet the design life requirements.

10.4.9.3. Wall Thickness and Reinforcement Ribs

The design guidelines for the wall thickness and reinforcement ribs of a transmission housing are as follows:

1. The basic wall thickness of a housing is dependent on its material, generally 3.5–4 mm for die-casting aluminum alloy and 5–6 mm for cast iron [10.1].
2. The reinforcement ribs should always be arranged along the direction of principal stress, so as to reduce the tensile stress on the housing by increasing the supporting cross-sectional area, and their thickness is generally equal to the basic wall thickness, with a symmetrical structure and a uniform thickness.
3. The wall thickness of a bearing seat generally falls in the range of 5.5–7.7 mm, and the reinforcement ribs near it should be arranged radially with the bearing hole as the center, with a height of 3–4 times the wall thickness and a width of 1–2 times the wall thickness.
4. Wide ribs and large radii of the curvature are preferred for the housing wall of longitudinal transmissions, with a width of 1–2 times wall thickness, a radius of curvature 1.2 times the wall thickness, and an angle of 45° to the longitudinal axis.
5. The acoustic characteristics of a transmission can be improved by using densely arranged ribs, with a spacing of 5–15 times the wall thickness.
6. The wall thickness at the bolt bosses generally ranges from 6 to 7 mm.
7. The maximum wall thickness of the housing is usually not more than 5 times the basic wall thickness, so as to avoid casting defects such as air holes and shrinkage cavities.

10.4.9.4. Draft Angle and Fillet

In order to smoothly pull out the housing from the die-casting mold, a certain draft angle must be designed inside and outside the housing, and its size depends on factors such as shape, wall thickness, and demolding depth. A draft angle of 1–3° is commonly used for a housing made of die-casting aluminum alloys, which should be consistent both inside and outside.

The connection between any two walls with a certain angle on the housing should be designed as rounded corners. These rounded corners are not only conducive to the flow and forming of metal during casting but also advantageous in preventing stress concentration and cracks. The size of the fillet can be designed according to the following formula:

$$r = 0.5(s_1 + s_2) \tag{10.357}$$

$$R = r + s_1 \tag{10.358}$$

where

r is the inner radius of the fillet
R is the outer radius of the fillet
s_1 is the thickness of the thicker wall
s_2 is the thickness of the thinner wall

10.4.9.5. Reduction of Housing Vibration and Noise

When the excitation frequency caused by the motion transmission error of gear meshing comes close to the modal frequencies of transmission housing, the system starts to resonate, leading to the amplification of gear tooth excitation and possible transmission whine. Therefore, the modal design and optimization of the housing play a critical role in improving the vibration and noise performance of the transmission. Generally, the modal frequencies of the housing can be increased by adding ribs, adjusting the layout, and optimizing the wall thickness to avoid overlapping with the gear meshing frequency range. The natural frequencies of the housing can be solved by means of FEA software.

In addition, measures should be taken to reduce the noise radiation area of the transmission housing, for example, using small facets with different normal directions, instead of a large flat surface. Applying noise-absorbing materials on the inner wall of the housing can also effectively reduce noise radiation, with a reduction of noise levels up to 4–5 dB.

Recently, the transmission, electric motor, and motor controller are often highly integrated. The transmission and electric motor can share the same housing, or the housings of electric motor and its controller can be made into one piece. Such a housing must meet the respective design requirements of relevant subsystems in structural strength, stiffness, temperature, vibration, noise, etc., which has brought about changes to the structural topology and detailed geometry of the electric powertrain.

10.4.9.6. Oil Passages

For a transmission with an oil pump, oil passages are usually designed inside the housing that transports lubricating oil from the pump to different components for hydraulic actuation, cooling, or lubrication. These oil passages are an integral part of the transmission cooling and lubrication system.

The design of oil passage should meet two requirements: minimize the resistance and pressure drop of the oil flow in it as much as possible because excessive resistance not only increases the power consumption of the oil pump but also affects the response of the hydraulic system, and avoid oil flow leakage in the passages as much as possible. In order to reduce flow resistance, the design of oil passages should avoid sudden changes in their cross sections, cut down the number of turns, and control the roughness of passage walls. For the case where housings are connected by bolts, special attention should be paid to the sealing where the oil channel passes through the joint surface of different housings to prevent leakage.

An external cooler is required for some transmissions to cool down the transmission oil so that it does not exceed the proper temperature range. The external transmission oil cooler is generally installed in front of the radiator and connected to the transmission through oil pipes. In this case, the oil pipes and joints should be designed and arranged properly to avoid leakage.

For the case where the transmission and the electric motor are integrated into the same housing, the seal between the transmission and the motor no longer exists, but the influence of the bearing position change needs to be considered, especially the adaptability of the transmission and the motor to the uniform oil temperature. It is necessary to consider thermal management in a systematic perspective, so as to make full use of the heat generated for the enhancement of system efficiency and reduction of parasitic losses.

10.4.9.7. Material and Manufacturing Process

At present, most of the transmission housings of electric passenger vehicles are made of aluminum alloys. However, magnesium alloys can be employed for further weight reduction of up to 25–30%. Compared to aluminum alloys, magnesium alloys are not only lighter and stronger but also of better characteristics in vibration absorption, casting performance, recycling, and cutting energy consumption. However, magnesium alloys are not as good as aluminum alloys in corrosion resistance, plasticity at room temperature, and flammability. The material properties of aluminum alloy and magnesium alloy are compared in Table 10.27 [10.1].

TABLE 10.27 Comparison of material properties of aluminum and magnesium alloys.

Material	Density (g/cm^3)	Elastic modulus (GPa)	Specific modulus of elasticity (GPa)	Strength (MPa)	Specific strength (MPa)	Thermal conductivity (W/(m·K)	Damping coefficient	Cutting energy consumption index
Aluminum alloy	2.7	70	25.9	200–350	73–128	247	2–5	1.8
Magnesium alloy	1.74	45	25.86	180–300	118–172	157	30–60	1.0

Reproduced from, Yuan, Y., Electric Vehicle Transmissions: Structure and Control (Beijing: China Machine Press, 2020), ISBN: 978-7-111-66809-1, with the permission of China Machine Press.

The manufacturing process is batch size-dependent. Machining and fine carving are generally used for the batch size of up to a dozen, while the sand molded casting process is mostly adopted for a batch size of 100. When the batch size reaches a hundred thousand level, the precision pressure casting process using metal mold is a common practice.

10.4.9.8. Vent Plugs

Lubricated with oil, the transmission usually has oil seals at the places where the input shaft and drive shaft, for example, half shafts, stick out of the housing. These seals cannot take much internal pressure while sealing off the lubricating oil. However, if the housing is completely airtight, the air pressure in the housing changes with temperature, and the rise in its internal pressure can lead to an increase in the contact pressure and local temperature at sealing edges. As a result, wear and leakage may occur at these seals. On the other hand, when a certain vacuum develops in the housing, the contact pressure at the sealing edge declines, and the air, water, and dust outside the housing may be sucked into the housing from the oil seals. Hence, at least one vent plug should be fitted to the housing for air breathing, which not only prevents the lubricating oil, oil foam, oil vapor, or oil mist from escaping but also blocks the water and dust from entering the housing. The vent plug is generally positioned in the area where the transmission oil is least likely to reach when it splashes, often on the top of the transmission.

There are two types of vent plugs, namely, normally closed ones and normally open ones. A normally closed vent plug consists of components such as a plug cap, spring, cover, valve, and plug body. The plug cap is mainly used to prevent water or dust from coming into contact with the internal parts from the top. The spring applies a certain force on the cover, which allows it to open only under a certain internal pressure, for example, 5–10 kPa. The valve provides a smooth contact with the cover to seal off the internal oil as well as the external water or dust. The tread and hex nut on the plug body allow the vent plug to be mounted to the transmission housing.

References

[10.1]. Yuan, Y., *Electric Vehicle Transmissions: Structure and Control* (Beijing: China Machine Press, 2020), ISBN:978-7-111-66809-1.

[10.2]. Cheng, F. and Lin, J., *Modern Electric Machinery Design* (Beijing: Chine Machine Press, 1993), 304-325, ISBN:7-111-03546-1.

[10.3]. Chen, S., *Electric Machine Design*, 2nd ed. (Beijing: Chine Machine Press, 2000), ISBN:7-111-02061-8.

[10.4]. Alger, P., "The Calculation of the Armature Reactance of Synchronous Machines," *Transactions of the American Institute of Electrical Engineers* 47, no. 2 (1928): 493-512, doi:10.1109/T-AIEE.1928.5055008.

[10.5]. Han, L., "Electric Machinery, Chapter 4, Common Problems in AC Motor Theory," School of Electrical Engineering, Chongqing University, accessed August 2023, https://www.doc88.com/p-88861422526233.html.

[10.6]. Zhu, C., Bao, Y., Wang, Y., Ma, J. et al., "Research Progress on Application Status and Property Control of Non-oriented Silicon Steel for Traction Motor of New Energy Vehicles," *Material Reports* 35, no. 23 (2021): 23089-23096, doi:10.11896/cldb. 20070220.

[10.7]. Liu, R., Zhu, J., and Cao, J., "Temperature Field Analysis of Silicon Steel and Amorphous on Permanent Magnet Synchronous Motors Used in Electrical Vehicles," *Journal of Beijing Jiaotong University* 43, no. 5 (2019): 119-125, doi:10.11860/j.issn.1673-0291.20190029.

[10.8]. Makinio, A., Kubota, T., Makabe, M., Chang, C. et al., "FeSiBP Metallic Glasses with High Glass-Forming Ability and Excellent Magnetic Properties," *Materials Science and Engineering B* 148 (2008): 166-170, doi:10.1016/j.mseb.2007.09.010.

[10.9]. Han, X., Tong, W., and Tang, R., "Application of Amorphous Alloy in Motor," *Advanced Technology of Electrical Engineering and Energy* 33, no. 12 (2014): 47-52, doi:10.3969/j.issn.1003-3076.2014.12.010.

[10.10]. Zhu, J., Cao, J., Liu, R., and Ding, Y., "Comparative Analysis of Silicon Steel and Amorphous on the Performance of Permanent Magnet Synchronous Motors on Electric Vehicles," *Transactions of China Electrotechnical Society* 33, no. sup. 2 (2018): 352-358, doi:10.19595/j.cnki.1000-6753.tces.L80576.

[10.11]. Dimier, T., Cossale, M., and Wellerdieck, T., "Comparison of Stator Winding Technologies for High-Speed Motors in Electric Propulsion Systems," in *2020 International Conference on Electrical Machines (ICEM)*, Gothenburg, 2020, 2406-2412, doi:10.1109/ICEM49940.2020.9270943.

[10.12]. Levi, E., *Polyphase Motors: A Direct Approach to Their Design* (New York: John Wiley & Sons, Inc., 1984), ISBN-10:047189866X, ISBN-13:978-0471898665.

[10.13]. Legarra, M., "Thermal and Hydraulic Design of Water-based Cooling Systems for Electric Machines," Ph.D. thesis, Applied Engineering, University of Navarra, Pamplona, Spain, 2017.

[10.14]. Johansson, M., "Evaluation of Sensor Solutions & Motor Speed Control Methods for BLDCM/PMSM in Aerospace Applications," Master thesis, Computer Science, Electrical and Space Engineering, Luleå University of Technology, Luleå, 2017.

[10.15]. Lu, M., "Comparative Study on Power Module Architectures for Modularity and Scalability," *Journal of Electronic Packaging* 142, no. 4 (2020): 040801, doi:10.1115/1.4047472.

[10.16]. Marcinkowski, J., Kempitiya, A., Prabhala V., Kim, M. et al., "Dual-Sided Cooling for Automotive Inverters - Practical Implementation with Power Module," Paper Presented at in *PCIM Europe*, Nuremberg, Germany, May 19–21, 2015.

[10.17]. Wen, Y., Yuan, Y., Ning, H., Zhang, Y. et al., "Review on Short-Circuit Protection Technology of SiC MOSFET," *Transactions of China Electrotechnical Society* 37, no. 10 (2022): 2538-2548, doi:10.19595/j.cnki.1000-6753.tces.211119.

[10.18]. Sheng, K., Dong, Z., and Wu, X., "Review and Prospect of Key Packaging Technologies for Silicon Carbide Power Devices," *Proceedings of the CSEE* 39, no. 19 (2019): 5576-5584, doi:10.13334/j.0258-8013.pcsee.191428.

[10.19]. Yang, Y., Wen, Y., and Li, G., "Review on High-Power IGBT Module and Drive Circuit," *High Voltage Engineering* 44, no. 10 (2018): 3207-3220, doi:10.13336/j.1003-6520.hve.20180925011.

[10.20]. Kang, J., Xin, Z., Chen, J., Wang, H. et al., "Review and Prospect of Short-Circuit Failure and Degradation Mechanism of SiC MOSFET," *Proceedings of the CSEE* 41, no. 3 (2021): 1069-1084, doi:10.13334/j.0258-8013.pcsee.200871.

[10.21]. Wei, J., Liu, S., Yang, L., Fang, J. et al., "Comprehensive Analysis of Electrical Parameters Degradations for SIC Power MOSFETS under Repetitive Short-Circuit Stress," *IEEE Transactions on Electron Devices* 65, no. 12 (2018): 5440-5447, doi:10.1109/TED.2018.2873672.

[10.22]. Wang, Z., Shi, X., Xue, Y., Tolbert, L. et al. "Design and Performance Evaluation of Overcurrent Protection Schemes for Silicon Carbide (SiC) Power MOSFETs," Paper Presented at in *IEEE Energy Conversion Congress and Exposition*, Denver, September 15–19, 2013, doi:10.1109/ECCE.2013.6647436.

[10.23]. Sun, K., Wang, J., Burgos, R., Boroyevich, D. et al., "Design, Analysis, and Discussion of Short Circuit and Overload Gate-Driver Dual-Protection Scheme for 1.2kV, 400A SiC MOSFET Modules," *IEEE Transactions on Power Electronics* 35, no. 3 (2019): 3045-3068, doi:10.1109/TPEL.2019.2930048.

[10.24]. Xue, J., Xin, Z., Chen, J., and Wang, H., "An Improved di/dt-RCD Detection for Short-Circuit Protection of SiC MOSFET," *IEEE Transactions on Power Electronics* 36, no. 1 (2020): 12-17, doi:10.1109/TPEL.2020.3000246.

[10.25]. Zeng, Z., *Packaging Test and System Integration of SiC Power Devices* (Beijing: Science Press, 2020), ISBN:9787030657008.

[10.26]. Wang, J., Shen, Z., Dimarno, C., Burgos, R. et al., "Gate Driver Design for 1.7kV SiC MOSFET Module with Rogowski Current Sensor for Short-Circuit Protection," Paper Presented at in *IEEE Applied Power Electronics Conference and Exposition (APEC)*, Long Beach, May 20–24, 2016, doi:10.1109/APEC.2016.7467921.

[10.27]. Zhang, J., Zhang, T., Feng, Y., Song, M. et al., "Dynamic Characterization Assessment on Series Short Circuit of SiC MOSFET," *Proceedings of the CSEE* 36, no. 12 (2021): 2446-2458, doi:10.19595/j.cnki.1000-6753.tces.201432.

[10.28]. Yao, K., Yano, H., Tadano, H., Iwamuro, N. et al., "Investigations of SiC MOSFET Short-Circuit Failure Mechanisms Using Electrical, Thermal, and Mechanical Stress Analyses," *IEEE Transactions on Electron Devices* 67, no. 17 (2020): 4328-4334, doi:10.1109/TED.2020.3013192.

[10.29]. Yang, Y. and Wen, Y., *High Power IGBT Drive and Protection Technology* (Beijing: Science Press, 2018), ISBN:9787030577030.

[10.30]. Lu, Q., Gao, Z., Teng, S., Lei, T. et al., "Current Limiting Solid State Circuit Breaker Based on MOSEFT and Its Over Voltage Suppression," *Transactions of China Electrotechnical Society* 32, no. 24 (2017): 42-52, doi:10.19595/j.cnki.1000-6753.tces.L70738.

[10.31]. Kazimierezuk, M., *High-frequency Magnetic Components*, 2nd ed. (Hoboken: John Wiley & Sons, 2014), ISBN:978-1-118-71779-0.

[10.32]. Cai, W., Yang, M., Liu, Y., Li, D. et al., "SiC Power Module Packaging Technologies and Prospects," *Automotive Engineering* 44, no. 4 (2022): 638-647, doi:10.19562/j.chinasae.qcgc.2022.04.018.

[10.33]. Zeng, Z., Li, X., Lin, C., Ran, L. et al., "Electric-Thermal-Stress Oriented Multi-objective Optimal Design of Power Module Package," *Proceedings of the Chinese Society of Electrical Engineering* 39, no. 17 (2019): 5161-5171, 5297, doi:10.13334/j.0258-8013.pcsee.180965.

[10.34]. McPherson, B., "Parametric Power Electronic Module Design Techniques for Rapid Analysis, Prototyping and Transition to Manufacturing," Panel Presentation at IEEE 3D-PEIM, 2016.

[10.35]. Jeffrey, C., Pala, V., Brunt, E., Hull, B. et al., "Ultra-Low (1.25mW) On-Resistance 900V SiC 62mm Half-Bridge Power Modules Using New 10mΩ SiC MOSFETs," Paper presented at in *IEEE PCIM, 2016*, Nuremberg, Germany, May 10–12, 2016.

[10.36]. Seal, S., Glover, M., and Mantooth, H., "3-D Wire Bondless Switching Cell Using Flip-Chip-Bonded Silicon Carbide Power Devices," *IEEE Transactions on Power Electronics* 33, no. 10 (2018): 8553-8564, doi:10.1109/TPEL.2017.2782226.

[10.37]. Chen, Z., Yao, Y., Boroyevich, D., Ngo, K. et al., "An Ultra-Fast SiC Phase-Leg Module in Modified Hybrid Packaging Structure," Paper Presented at in *IEEE Energy Conversion Congress and Exposition*, Pittsburgh, September 14–18, 2014, doi:10.1109/ECCE.2014.6953789.

[10.38]. Hoene, E., Ostmann, A., and Marczok, C., "Packaging Very Fast Switching Semiconductors," Paper Presented at in *International Conference on Integrated Power Electronics Systems*, Nuremberg, Germany, February 25–27, 2014.

[10.39]. Huang, Z., Li, Y., Chen, C., and Kang, Y., "An Ultra-Fast and Low Inductive SiC Half-Bridge Module Based on Novel Hybrid Packaging Structure," *Power Electronics* 51, no. 12 (2017): 20-22.

[10.40]. Chen, C., Huang, Z., Chen, L., and Tan, Y., "A Flexible PCB Based 3D Integrated SiC Half-bridge Power Module with Three-Sided Cooling Using Ultra-Low Inductive Hybrid Packaging Structure," *IEEE Transactions on Power Electronics* 34, no. 6 (2019): 5579-5593, doi:10.1109/TPEL.2018.2866404.

[10.41]. Kasko, I., Berberich, S., Gross, M., Beckedahl, P. et al., "High Efficient Approach to Utilize SiC MOSFET Potential in Power Module," Paper Presented in *at International Symposium on Power Semiconductor Devices & ICs*, Sapporo, Japan, 2017, doi:10.23919/ISPSD.2017.7988909.

[10.42]. Buetow, S., Herzer, R., Koenigsmann, G., Rossberg, M. et al., "High Power, High Frequency SiC-MOSFET System with Outstanding Performance, Power Density and Reliability," Paper Presented at in *IEEE ISPSD*, Sapporo, Japan, May 28–June 1, 2017, doi:10.23919/ISPSD.2017.7988973.

[10.43]. Koenigsmann, G., Herzer, R., Buetow, S., and Rossberg, M., "High Power High Frequency Gate Driver for SiC MOSFET Modules," Paper Presented at in *IEEE PCIM Europe*, Nuremberg, Germany, 2016.

[10.44]. Temple, V., Waldron, J., Azotea, J., Francois, M. et al., "High Frequency SiC Majority Carrier Modules," Paper Presented at in *International Exhibition and Conference for Power Electronics, Intelligent Motion, Renewable Energy and Energy Management*, Nuremberg, Germany, May 19–20, 2015.

[10.45]. Chang, H., Bu, J., Kong, G., and Labayen, R., "300A 650V 70m Thin IGBTs with Double-sided Cooling," Paper Presented at in *International Symposium on Power Semiconductor Devices and ICs*, San Diego, May 23–26, 2011, doi:10.1109/ISPSD.2011.5890855.

[10.46]. Weidner, K., Kaspar, M., and Seliger, N., "Planar Interconnect Technology for Power Module System Integration," Paper Presented at in *Proceedings of International Conference on Integrated Power Electronics Systems*, Nuremberg, Germany, March 6–8, 2012.

[10.47]. Gillot, C., Schaeffer, C., Massit, C., and Meysenc, L., "Double-Sided Cooling for High Power IGBT Modules Using Flip Chip Technology," *IEEE Transactions on Components and Packaging Technologies* 24, no. 4 (2001): 698-704, doi:10.1109/6144.974963.

[10.48]. Nakatsu, K., Nishihara, A., Sasaki, K., and Saito, R., "A Novel Direct Water and Double-Sided Cooled Power Module and a Compact Inverter for Electrified Vehicles," Paper Presented at in *European Conference on Power Electronics and Applications*, Lille, France, September 2–6, 2013, doi:10.1109/EPE.2013.6631862.

[10.49]. DiMarino, C., Johnson, M., Mouawad, B., Li, J. et al., "Design of a Novel, High-Density, High-Speed 10kV SiC MOSFET Module," Paper Presented at in *IEEE Energy Conversion Congress and Exposition (ECCE)*, Cincinnati, October 1–5, 2017, doi:10.1109/ECCE.2017.8096699.

[10.50]. Zhu, N., Mantooth, H., Xu, D., Chen, M. et al., "A Solution to Press-Pack Packaging of SiC MOSFETS," *IEEE Transactions on Industrial Electronics* 64, no. 10 (2017): 8224-8234, doi:10.1109/TIE.2017.2686365.

[10.51]. Chang, Y., Luo, H., Iannuzzo, F., Li, H. et al., "Compact Sandwiched Press-Pack SiC Power Module with Low Stray Inductance and Balanced Thermal Stress," *IEEE Transactions on Power Electronics* 35, no. 3 (2019): 2237-2241, doi:10.1109/TPEL.2019.2934709.

[10.52]. Infineon, "Silicon Carbide CoolSiC™ MOSFETs," accessed October 2017, https://www.infineon.com/cms/en/product/power/mosfet/silicon-carbide.

[10.53]. Vagnon, E., Crebier, J., Avenas, Y., and Jeannin, P., "Study and Realization of a Low Force 3D Press-Pack Power Module," Paper Presented at in *IEEE Power Electronics Specialists Conference*, Rhodes, Greece, June 15–19, 2008, doi:10.1109/PESC.2008.4592068.

[10.54]. Vagnon, E., Jeannin, P., Crebier, J., and Avenas, Y., "Bus-Bar-Like Power Module Based on Three-Dimensional Power-Chip-on-Chip Hybrid Integration," *IEEE Transactions on Industry Applications* 46, no. 5 (2010): 2046-2055, doi:10.1109/TIA.2010.2057401.

[10.55]. Herbsommer, J., Noquil, J., Lopez, O. and Jauregui, D., "Innovative 3D Integration of Power MOSFETs for Synchronous Buck Converters," Paper Presented at in *IEEE Applied Power Electronics Conference and Exposition*, Fort Worth, March 6–11, 2011, doi:0.1109/APEC.2011.5744756.

[10.56]. Fraunhofer, "Full SiC Double Sided Busbar Power Module," accessed October 2017, https://www.iisb.fraunhofer.de/content/dam/iisb2014/en/Documents/Research-Areas/Packaging_and_Reliability/Full-SiC-Double-Sided-Busbar-Power-Module_FraunhoferIISB_WWW.pdf.

[10.57]. Kicin, S., Traub, F., Hartmann, S., Bianda, E. et al., "A New Concept of a High-Current Power Module Allowing Paralleling of Many SiC Devices Assembled Exploiting Conventional Packaging Technologies," Paper Presented at in *IEEE.ISPSD*, Prague, Czech Republic, June 12–16, 2016, doi:10.1109/ISPSD.2016.7520879.

[10.58]. Wolfspeed, "Application Consideration for Silicon Carbide MOSFETs," accessed October 2021, http://www.cree.com.

[10.59]. Kelley, R., Ritenour, A., Sheridan, D., Casady, J., "Improved Two-Stage DC-Coupled Gate Driver for Enhancement-Mode SiC JFET," Paper Presented at in *Twenty-Fifth Annual IEEE Applied Power Electronics Conference and Exposition (APEC)*, Palm Springs, February 21–25, 2010, doi:10.1109/APEC.2010.5433483.

[10.60]. Qin, H., Zhu, Z., Wang, D., Xie, H. et al., "A Bridge-Arm Crosstalk Suppression Method Suitable for SiC-Based Converters," *Journal of Nanjing University of Aeronautics and Astronautics* 49, no. 6 (2017): 872-882, doi:10.16356/j.1005-2615.2017.06.018.

[10.61]. Wang, X., "Research on Current Sharing Technology of Paralleled SiC MOSFETs," Master's thesis, College of Electrical Engineering, Zhejiang University, Hangzhou, 2022.

[10.62]. Ke, J., Zhao, Z., Sun, P., Huang, H. et al., "Chips Classification for Suppressing Transient Current Imbalance of Parallel-Connected Silicon Carbide MOSFETs," *IEEE Transactions on Power Electronics* 35, no. 4 (2020): 3963-3972, doi:10.1109/TPEL.2019.2934739.

[10.63]. Watt, G., Mocevic, S., Burgos, R., and Romero, A., "Current Sharing Behavior and Characterization of a 1200 V, 6.5 mΩ SiC Half-Bridge Power Module with Flexible PCB Gate Loop Connection," Paper Presented at in *IEEE Energy Conversion Congress and Exposition (ECCE)*, Baltimore, September 29, 2019, doi:10.1109/ECCE.2019.8912641.

[10.64]. Passmore, B., and Lostetter, A., "A Review of SiC Power Module Packaging Technologies: Attaches, Interconnections, and Advanced Heat Transfer," Paper Presented at in *IEEE International Workshop on Integrated Power Packaging (IWIPP)*, Delft, the Netherlands, April 5–7, 2017, doi:10.1109/IWIPP.2017.7936763.

[10.65]. Wang, D., Cui, Y., and Zhang, X., "Thermal Management Technology for the Power Devices of an EV Motor Controller," *Shanghai Automobile (Version A)* 04 (2022): 4-10, doi:10.3969/j.issn.1007-4554.2022.04.01.

[10.66]. Yvan, A., Laurent, D., and Zoubir, K., "Temperature Measurement of Power Semiconductor Devices by Thermo-Sensitive Electrical Parameters—A Review," *IEEE Transaction on Power Electronics* 27, no. 6 (2012): 3081-3092, doi:10.1109/TPEL.2011.2178433.

[10.67]. Salah, T., Khachroumi, S., and Morel, H., "Characterization, Modeling and Design Parameters Identification of Silicon Carbide Junction Field Effect Transistor for Temperature Sensor Applications," *Sensors* 10, no. 1 (2010): 388-399, doi:10.3390/s100100388.

[10.68]. Nowak, M., Rabkowski, J., and Barlik, R., "Measurement of Temperature Sensitive Parameter Characteristics of Semiconductor Silicon and Silicon Carbide Power Devices," Paper Presented at in *13th International Power Electronics and Motion Control Conference*, Poznan, Poland, September 1–3, 2008, doi:10.1109/EPEPEMC.2008.4635248.

[10.69]. Zhang, Z., Wang, F., Costinett, D., Tolbert, L. et al., "Online Junction Temperature Monitoring Using Turn-off Delay Time for Silicon Carbide Power Devices," Paper Presented at in *IEEE Energy Conversion Congress and Exposition*, Milwaukee, WI, September 18–22, 2016, doi:10.1109/ECCE.2016.7855308.

[10.70]. Gonzalez, J., Alatise, O., Hu, J., Ran, L. et al., "An Investigation of Temperature Sensitive Electrical Parameters for SiC Power MOSFETs," *IEEE Transactions on Power Electronics* 32, no. 10 (2017): 7954-7966, doi:10.1109/TPEL.2016.2631447.

[10.71]. Li, H., Wu, J., Wei, Y., and Yan, H., "Smooth Control of Low-Frequency Junction Temperature Fluctuation for SiC MOSFET Based on Combined Regulation Methods," Paper Presented at in *MEMAT 2022, 2nd International Conference on Mechanical Engineering, Intelligent Manufacturing and Automation Technology*, Guilin, China, January 7–9, 2022.

[10.72]. Yuan, Y., Dual-rotor motor for electric automobile, associated stepless speed change system with planet gear and control method, PCT/CN2011/076434. US Patent 20130190961, June 27, 2011.

[10.73]. Chen, X., Yuan, Y., and Wu, G., "The Study of Control Strategies and Simulation for EVT Power Train Applying to EV," Paper Presented at in *CEVS-17*, Beijing, July 6–8, 2013.

[10.74]. Lee, Y., Pan, J., Hathaway, R., and Barkey, M., *Fatigue Testing and Analysis (Theory and Practice)* (Amsterdam: Elsevier, 2005), doi:10.1016/B978-0-7506-7719-6.X5000-3.

[10.75]. DeVincent, E., *Design Practices: Passenger Car Automatic Transmissions*, 4th ed. (Warrendale: SAE International, 2012), ISBN:978-0-7680-7531-1.

[10.76]. Khonsari, M. and Booser, E., *Applied Tribology, Bearing Design and Lubrication* (Hoboken: John Wiley & Sons, 2001), doi:10.1002/9781118700280.

[10.77]. Tang, T., Devlin, M., Mathur, N., Henly, T. et al., "Lubricants for (Hybrid) Electric Transmissions," *SAE Int. J. Fuels Lubr.* 6, no. 2 (2013): 289-294, doi:https://doi.org/10.4271/2013-01-0298.

[10.78]. Gahagan, M., "Lubricant Technology for Hybrid Electric Automatic Transmissions," SAE Technical Paper 2017-01-2358 (2017), doi:https://doi.org/10.4271/2017-01-2358.

11

Suspension System Design

Suspension is a lattice of components that connect the vehicle body to the wheels. It not only transfers torques and forces between the vehicle body and the wheels but also allows relative motion between the two. For optimal handling and riding comfort, all the relative motions between the vehicle body and wheels should be restrained, except the translational motion in the vertical direction. Composed of springs, shock absorbers, and linkages, the functions of suspension are mainly two-fold: damping the impact loads from the road surface to the body and attenuating the resultant vibrations to ensure a smooth ride, and providing desired road-holding and handling characteristics on uneven roads or in changing loading conditions to ensure stability, particularly at high speeds.

In addition to providing a smooth ride, proper damping to vibrations, and good handling, the suspension is also required to stabilize the vehicle body during acceleration or braking with a minimal pitch movement, to keep an appropriate roll angle in cornering, and to be compact and lightweight with adequate strength and lifespan.

To design suspension for a smooth ride, the following guidelines are generally followed: making the natural frequency of the vibration system composed of sprung mass and springs as low as possible; working out a proper match between the natural frequencies of front and rear suspensions, with the natural frequency of rear suspension slightly higher than that of front suspension for sedans; avoiding the suspension hitting the vehicle body directly; minimizing the variation of body height due to the change in sprung mass with the suspension of nonlinear characteristics; and isolating the impact load from the road to the vehicle body using rubber liners at the hinges of the guide rod system.

The vehicle body is subject to vertical vibrations on uneven roads due to the action of suspension springs. To ensure ride comfort and stable handling, the following design guidelines are recommended: finding a proper damping ratio to suppress the vibrations and resonance of the vehicle body and wheels so that the amplitude continuously drops to zero, and selecting an appropriate suspension type and parameters not only to minimize the positioning angle of the kingpin so as to prevent the front wheels from swinging and coordinate the movements of wheels and linkages as the wheels bounce up and down but also to acquire a slight understeer characteristics in cornering.

Traditionally, most of vehicles use passive suspension in which the characteristics of components such as stiffness and damping ratio are fixed. However, in recent years, active suspension is increasingly employed by high-end vehicles in which the stiffness and damping of suspension can be adjusted dynamically and adaptively according to the changes in parameters such as the state of motion of a vehicle, road condition, and loading, so that the suspension system can always provide the optimal damping effect and reduce the body pitch and roll in braking and cornering. An active suspension system that includes a power source is referred to as a full active suspension, while an active suspension system that does not include a power source is a semiactive suspension.

11.1. Types of Vehicle Suspension System

In general, the vehicle suspension system can be divided into nonindependent suspension and independent suspension. As illustrated in Figure 11.1, nonindependent suspension (Figure 11.1a) features a single axle that connects both left and right wheels, and the vehicle body connects onto the top of axle via suspension; on the contrary, independent suspension (Figure 11.1b) features a separate suspension that connects each wheel to the vehicle body.

Compared with independent suspension, nonindependent suspension is structurally simpler, with a fewer number of parts and, therefore, lower costs. In addition, it is also easy to fabricate and repair, with good reliability. However, it comes with several drawbacks: the left and right wheels influence each other, causing the vehicle body and axle to roll sideways as one of the wheels hit a pothole or bump; steering can be interfered as one of the wheels bounces, or as both left and right wheels bounce in the opposite directions when the vehicle is driven straight forward, or as a vehicle is cornering under the influence of a centrifugal force; and extra space above the suspension spring is required for its travel. Hence, nonindependent suspension is mostly utilized in trucks, buses, or the rear of passenger vehicles.

On the other hand, independent suspension offers the following advantages: light unsprung mass; only a small space required for accommodating suspension; suspension springs are only subject to a vertical force, allowing a reduction of spring stiffness and, thus, a lowered vibration frequency of the vehicle body and improved smoothness; when used in the front of a vehicle, the electric powertrain can be positioned lower, helping to drop the vehicle centroid and enhance the handling stability; and reduced vehicle body roll and vibrations, with improved road-holding characteristics on undulating roads. Nevertheless, independent suspension has a more complex and costly structure and is relatively hard to repair. It is often used in passenger vehicles, light-duty trucks, and off-road vehicles.

FIGURE 11.1 Schematic of nonindependent suspension and independent suspension. (a) Nonindependent suspension and (b) independent suspension.

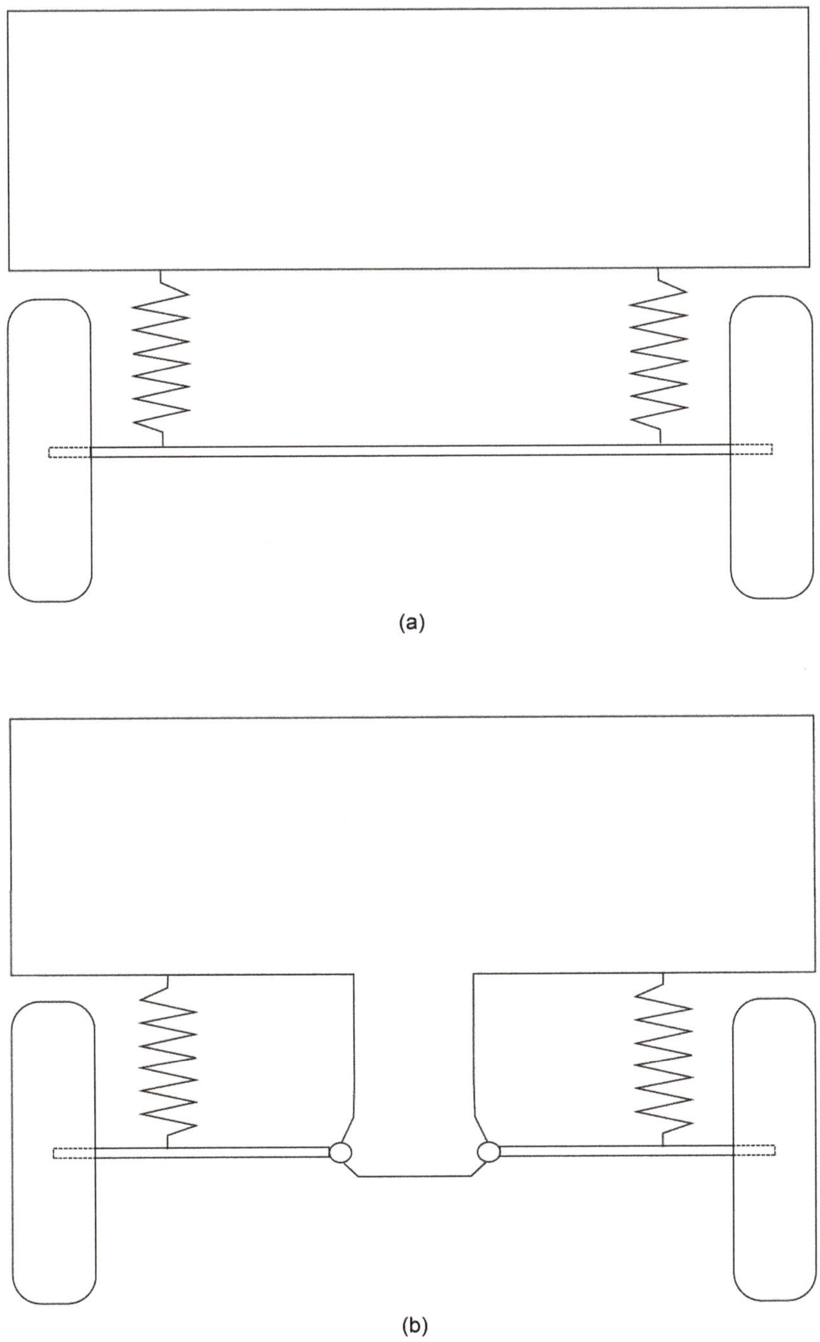

11.1.1. Types of Nonindependent Suspension

The nonindependent suspension frequently utilized in EVs are torsion beam and leaf spring suspensions.

The torsion beam suspension, also known as twist-beam suspension, features an H-shaped structure that connects to the two trailing arms together with a cross member to provide the roll stiffness. The structure attaches to the vehicle body on each side via rubber bushings in the front and shock absorbers in the rear. The cross member remains rigid under a lateral force to resist the body roll; however, it twists as the two trailing arms move relative to each other vertically on uneven roads. Torsion beam suspension is simple, reliable, light-weight, and compact, allowing maximization of vertical space for passengers or cargo. In addition, it features a high load-carrying capacity and low cost. Nevertheless, it behaves poorly in dealing with wheel bounces, with intrinsic oversteer characteristics. As a result, the vehicle body is subject to severe roll motions on rough roads, leading to rollover in the worst scenario. In addition, the bouncing up of one side wheel would cause the other side wheel to partially come off the ground, resulting in wear or damage on tire tread and rim at their outer edges. If a torsional beam rear suspension is used for an RWD vehicle, not only roll stability, ride comfort, and tire longevity but also trafficability can be affected due to possible loss of traction. Although an addition of Watt's linkage helps improve lateral stiffness, it still does not offer adequate control over the wheel toe and inclination.

As for leaf spring suspension, several leaves of progressively shorter length can be stacked on top of each other for increased stiffness. Leaf springs are typically fixed to the axle with a U-bolt at the center of the slender arc-shaped spring of a rectangular cross section. The front of the leaf spring assembly is connected to the chassis via a bushing in the eye of the master leaf spring, while the rear is connected to the chassis through an eye in a shackle with a short swing arm. The shackle takes up the elongation of the leaf spring assembly under compression and prevents the parts at the attaching points from damage by allowing the eye to move around inside with a certain flexibility under heavy or impact loading. Leaf spring suspension is simple, reliable, inexpensive, and easy to repair. It serves not only as a springy and guiding mechanism but also as a damping element, thanks to the friction on the contact surfaces between the leaves. However, leaf spring suspension is generally very stiff and heavy, resulting in a poor ride comfort. In the absence of a subframe, the entire leaf spring suspension is actually unsprung mass. Its relatively large longitudinal dimension is not favorable to the reduction of front and rear overhangs. Besides, the leaf spring is prone to wear at the bushings and eyes where it connects to the vehicle frame. For performance improvement and weight reduction, alternative leaf spring suspensions of variable-and single-cross section leaf springs should be considered.

11.1.2. Types of Independent Suspension

The types of independent suspension often used in EVs mainly include MacPherson, double-wishbone, multilink, and transverse leaf springs. These types of independent suspension can be evaluated by the following characteristics.

1. Roll center height

   The roll center is defined as the point at which the lateral force is transferred from the axle to the sprung mass, which is actually the instantaneous center of rotation of the vehicle body with respect to the ground when it rolls in the transverse vertical plane that passes through the centers of left and right wheels. The roll center height is the vertical distance from the roll center to the ground. When the roll center is high, its distance to the centroid of the vehicle body shortens, resulting a smaller roll moment and roll angle. However, when the roll center is excessively high, the wheel track changes more drastically as the vehicle body rolls, causing tread wear to increase.

2. Suspension roll stiffness
   Suspension roll stiffness affects the roll angle of the vehicle body about its roll axis under the action of a lateral force when a vehicle is making a steady-state circular motion. Thus, it impacts the handling stability and smoothness of a vehicle.
3. Suspension lateral compliance
   Suspension lateral compliance has an influence on the handling stability of a vehicle because an excessive lateral compliance of the suspension for steering wheels can result in steering wheel shimmy.
4. Changes in wheel alignment parameters
   As a wheel bounces up and down, the wheel alignment parameters such as kingpin inclination, castor, camber, and toe change accordingly. However, a large change in castor may cause steering wheel shimmy, while a large change in camber can affect the driving stability of a vehicle in linear motion, the wheel track, and the rate of tire wear.
5. Space occupation
   When a suspension occupies a large space, the arrangement of other subsystems or components, for example, battery packs, electric powertrain, and cargo space, would be affected.

The characteristics of MacPherson, double-wishbone, and multilink suspensions are compared in Table 11.1.

TABLE 11.1 Comparison of several independent suspensions.

	Guiding mechanism		
Characteristics	**MacPherson**	**Double-wishbone**	**Multilink**
Roll center height	++++	+++++	+++++
Change in wheel track	+++++	+++++	+++++
Suspension roll stiffness	++++ (No stabilizer bar is required)	+++++	+++++
Suspension lateral compliance	+++++	+++++	+++++
Changes in wheel alignment parameters	++++	+++++	+++++
Space occupation	+++++	+++	++
Structural simplicity	+++++	+++	++
Cost	+++++	+++	++

Note: ++, +++, ++++, and +++++ stand for fair, good, very good, and excellent, respectively.

11.2. Fundamental Parameters of the Suspension System

In the design of a suspension system for an EV, several basic parameters must be determined on the system level.

11.2.1. Static Deflection of Suspension

Static deflection refers to the ratio of load carried on the suspension per unit stiffness when a fully loaded vehicle is in standstill, namely,

$$\delta_s = \frac{F_{zf}}{K} \tag{11.1}$$

where

δ_s is the static deflection of suspension
F_{zf} is the full load carried by suspension
K is the stiffness of suspension

If the nonlinearity in suspension elasticity is negligible, the static deflections of front and rear suspensions are, respectively,

$$\delta_{s1} = \frac{m_1 g}{K_1} \tag{11.2}$$

$$\delta_{s2} = \frac{m_2 g}{K_2} \tag{11.3}$$

where

m and K are the sprung mass and stiffness of suspension
subscriptions 1 and 2 denote front and rear suspensions, respectively

In selecting the static deflection of suspensions, it is generally recommended that the static deflection of front suspension be slightly larger than that of rear suspension for electric passenger vehicles, that is, $\delta_{s1} = (1.1 - 1.25)\delta_{s2}$, in order to prevent the vehicle body from excessive angular vibrations in the longitudinal direction. For electric trucks, it is recommended to have $\delta_{s1} = (1.25 - 1.67)\delta_{s2}$ with an emphasis on the ride comfort of the driver because of the difference between front and rear axle loads. Nevertheless, for electric microcars, a higher natural frequency is sometimes selected for the front suspension than the rear suspension in order to improve the ride comfort in rear seats. The statistical data of typical static deflection for the front and rear suspensions for various types of vehicles are listed in Table 11.2.

TABLE 11.2 Statistical data of static deflection for suspensions per type of vehicle.

	Typical static deflection (cm)	
Type of vehicle	Front suspension	Rear suspension
Regular cars	12–24	10–18
Luxury cars	20–30	15–26
Trucks	6–11	5–9
Buses	7–15	
Off-road vehicles	6–13	

11.2.2. Natural Frequency of Suspension

The natural frequencies of front and rear suspension systems including their corresponding sprung masses are major parameters that affect the ride comfort. As the sprung mass at the front and rear suspensions is nearly identical, the vibrations at the front and the rear are very well decoupled. As a result, the natural frequencies of front and rear suspension systems are approximately

$$\omega_1 = \frac{\sqrt{K_1 / m_1}}{2\pi} \tag{11.4}$$

$$\omega_2 = \frac{\sqrt{K_2 / m_2}}{2\pi} \tag{11.5}$$

where K_1 and K_2 are the stiffness of front and rear suspensions, respectively.

Substitution of Equations (11.2) and (11.3) into Equations (11.4) and (11.5) yields

$$\omega_1 \approx \frac{0.5}{\sqrt{\delta_{s1}}} \tag{11.6}$$

$$\omega_2 \approx \frac{0.5}{\sqrt{\delta_{s2}}} \tag{11.7}$$

It is obvious that the static deflection of suspension has a direct impact on the natural frequency of vehicle body vibrations. Thus, it is crucial to select a proper static deflection of suspension to ensure a satisfactory ride comfort.

The ride quality requirement varies with the type of vehicles, so as the natural frequencies for the front and rear suspensions. The statistical data of typical natural frequencies for the front and rear suspensions for various types of vehicles are listed in Table 11.3.

TABLE 11.3 Statistical data of natural frequencies for suspensions per type of vehicle.

	Typical natural frequency (Hz)	
Type of vehicle	**Front suspension**	**Rear suspension**
Regular cars	1.00–1.44	1.18–1.58
Luxury cars	0.80–1.15	0.98–1.30
Trucks	1.50–2.10	1.67–2.23
Buses	1.29–1.89	
Off-road vehicles	1.39–2.04	

11.2.3. Dynamic Deflection of Suspension

The dynamic deflection of suspension refers to the vertical displacement of the wheel center with respect to the vehicle body when the suspension is compressed from the static balancing position to the maximum deflection allowed by the structure under full loads, which usually corresponds to the buffer being compressed by 1/3–1/2 of its unloaded height. An adequate dynamic deflection of suspension is desirable in the design to prevent the buffer from damage when driving on poor roads. The statistical data of typical dynamic deflection for the front and rear suspensions for various types of vehicles are listed in Table 11.4.

TABLE 11.4 Statistical data of dynamic deflection for suspensions per type of vehicle.

	Typical dynamic deflection (cm)	
Type of vehicle	**Front suspension**	**Rear suspension**
Regular cars	8–11	10–14
Luxury cars		
Trucks	6–9	6–8
Buses	5–8	
Off-road vehicles	7–13	

11.2.4. Elastic Characteristics of Suspensions

The elastic characteristic of a suspension refers to the relationship between the vertical external force F_z on the suspension and the resultant displacement of the wheel center δ with respect to the vehicle body.

The elastic characteristics of a suspension can be either linear or nonlinear. The characteristics of a typical nonlinear suspension are illustrated in Figure 11.2 [11.1], where S_r is the recovery stroke of buffer and S_c is the compression stroke of buffer. Thus, the buffer is fully recovered at point A and is detached from the bracket at point B. Since the slope of F_z versus δ curve represents suspension stiffness, it can be noticed that suspension stiffness varies linearly between B and D around C, corresponding to a relatively small stiffness of the primary spring and smooth ride. Hence, it is ideal to design the suspension stiffness so that point C corresponds to the rated full load of the suspension. When the vertical load is much lighter than the full load (from points A to B) or much greater than the full load (from points D to E), the stiffness is nonlinear. Under compression, the buffer starts to come in contact with the elastic bracket at point D, while the curve DE reflects the elastic characteristics in this stage. The steep increase in stiffness allows the nonlinear suspension to consume more dynamic energy in the limited dynamic deflection range than the linear suspension, which reduces the likelihood of buffer damage.

The leaf spring suspension without secondary leaves is an example of linear suspension, which is rarely used in EVs. On the contrary, most of the electric passenger vehicles have nonlinear suspensions to reduce the roll angle in cornering and the pitch angle in accelerating or braking, as well as the likelihood of impact on the vehicle frame. On the other hand, most of the electric buses and trucks also employ nonlinear suspensions to the vibration frequency and change in body height due to substantial change in sprung mass under unloaded and fully loaded conditions.

FIGURE 11.2 Characteristics of a typical nonlinear suspension.

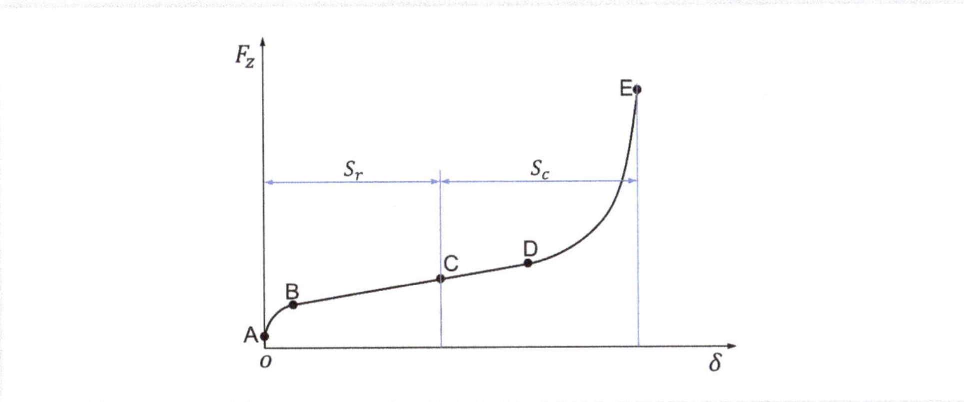

Courtesy of Butterworth-Heinemann.

11.2.5. Roll Stiffness of Front and Rear Suspensions

Based on the definition given in Section 5.6.1, roll stiffness is the elastic recovery moment that the suspension acts on the vehicle body per unit roll angle produced on sprung mass. Due to its direct impact on the roll angle of sprung mass, proper roll stiffness is essential because excessive roll stiffness can result in large sideslip angles of tires, oversteer tendency at rear wheels, and insufficient rollover feeling, while inadequate roll stiffness can cause an uncomfortable ride and feeling of unsafety. At a lateral acceleration of 0.4*g*, it is generally required that the roll angle fall in the range of 2.5–4° for a sedan and 6–7° for a truck. In addition, roll stiffness also affects steering characteristics. A vehicle can be made slightly understeer by letting the sideslip angle at front wheels be 1–3° larger than that at rear wheels and, subsequently, the roll stiffness of front suspension larger than that of rear suspension. For example, the front-to-rear roll stiffness ratio usually ranges from 1.4 to 2.6 for a sedan.

11.2.6. Stiffness of Primary and Secondary Leaf Springs

A nonlinear suspension consisting of primary and secondary leaf springs can be used as the rear suspension of trucks and pickups. The typical elastic characteristics of such suspension is shown in Figure 11.3 [11.2]. The secondary leaf spring does not come into operation until the vertical load rises to a certain threshold. With the secondary leaf spring coming into contact with the bracket, it starts to share the load together with the primary leaf spring.

FIGURE 11.3 Characteristics of a typical nonlinear rear leaf spring suspensions.

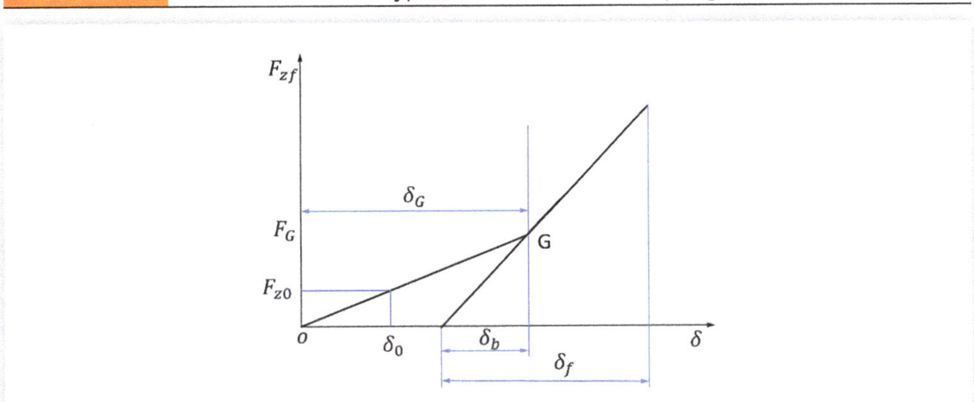

Courtesy of Wangyu Wang.

Depending on the type of vehicles, the following two methodologies can be taken in the design of stiffness of primary and secondary leaf springs.

1. For vehicles that frequently operate nearly half of their rated loading, it is desired to keep the vibration frequencies of front and rear suspensions unchanged before and after the secondary leaf spring kicks in. Thus, it is recommended that the secondary leaf spring start to take the load at the average loads in unload and fully conditions, namely,

$$F_G = \frac{F_0 + F_{zf}}{2} \tag{11.8}$$

   where F_0 is the vertical load on the suspension when the vehicle is unloaded.

   Additionally, to make the frequency corresponding to the averaged load between F_0 and F_G equal to that corresponding to the averaged load between F_G and F_{zf}, the stiffness of secondary leaf spring must satisfy the following relationship:

$$K_s = \frac{2\lambda - 2}{\lambda + 3} K_p \tag{11.9}$$

   where

   $\lambda = F_0/F_{zf}$

   K_p and K_s are the stiffness of primary and secondary leaf springs, respectively

2. For vehicles often operating in either unloaded or fully loaded conditions, it is preferred to keep vibration frequencies unchanged in these two loading conditions. This can be achieved by making the suspension deflection upon the kick-in of the secondary leaf spring equal to that when the vehicle is unloaded. Thus, we have

$$F_G = \sqrt{F_0 F_{zf}} \tag{11.10}$$

3. The stiffness of the secondary leaf spring must satisfy the following relationship:

$$K_s = \left(\sqrt{\lambda} - 1\right) K_p \tag{11.11}$$

   It is worth noting that this design methodology can cause a considerable change in the vibration frequency before and after the secondary leaf spring comes into contact with the bracket.

11.3. Design of Elastic Elements

Elastic elements commonly used in suspensions can be of four types, namely, coil springs, torsion bar springs, leaf springs, and air springs.

11.3.1. Coil Springs

Coil springs are widely used in light-duty vehicles, particularly in passenger vehicles, due to their simple structure, high specific energy capacity, and good manufacturability. In addition, coil springs are very compatible as they allow installation of shock absorbers, stroke limiters, or guide columns inside them. The stiffness of a coil spring can be made variable by using spring structures of variable pitch, wires of variable diameter, or their combination.

Coil springs absorb the energy mainly by shear deformation of spring wires. Thus, the shear stress on the surface of spring wire can be expressed by

$$\tau = k\frac{8PD_m}{\pi d^3} \tag{11.12}$$

where P is the load along the axis of the coil spring, D_m is the mean diameter of the coil spring, d is the diameter of the spring wire, and k is a correction factor that takes the relatively high shear stresses on the inner side of the coil spring into account, which can be calculated using the Wahl formula:

$$k = \frac{4C-1}{4C-4} + \frac{0.165}{C} \tag{11.13}$$

where C is the spring index, with $C = D_m/d$.

In the design, shear stress must not exceed the allowable shear stress $[\tau]$, which is typically related to the allowable tensile stress as $[\tau] = 0.63[\sigma]$.

Under the action of an axial load P, the deflection of a coil spring is

$$\delta = \frac{8NPD_m^3}{Gd^4} \tag{11.14}$$

where

N is the effective number of coils

G is the shear modulus of elasticity, usually $G = 8.3 \times 10^4$ MPa

The stiffness of coil spring is

$$K = \frac{P}{\delta} = \frac{Gd^4}{8ND_m^3} \tag{11.15}$$

The natural frequency of coil spring suspension, ω_0, is given by

$$\omega_0 = \sqrt{\frac{K}{m}} = \sqrt{\frac{Kg}{P}} \tag{11.16}$$

where m is the sprung mass.

The damping force of the shock absorber can be determined based on the critical damping coefficient, C_{cr}

$$C_{cr} = 2\sqrt{mK} = 2\sqrt{\frac{PK}{g}} \tag{11.17}$$

The load corresponding to the extreme compression position of a coil spring in operation, P_{max}, can be calculated by

$$P_{max} = P_r + K\left(H_r - H_{max}\right) \tag{11.18}$$

where

P_r and H_r are the rated load of the suspension and the corresponding height of the coil spring, respectively

H_{max} is the height of the coil spring at its extreme compression position

The corresponding shear stress τ_{max} can be obtained by using Equation (11.12), and its value must be checked against the allowable shear stress to make sure that $\tau_{max} \le [\tau]$.

Moreover, the cycle life of a coil spring, n_c, can be estimated by

$$n_c = \left(\frac{1.808}{\zeta}\right)^{\frac{1}{0.13}} \tag{11.19}$$

and

$$\zeta = \frac{0.74\left(\tau_c - \tau_e\right)}{1.48[\sigma] - \left(\tau_c + \tau_e\right)} \tag{11.20}$$

where τ_c and τ_e are the shear stresses corresponding to the extreme extrusion and compression positions of the coil spring, respectively.

The free height of the coil spring is given by

$$H_0 = H_r + \frac{P_r}{K} \tag{11.21}$$

and the minimum operating height of coil spring is

$$H_{min} = H_s + \gamma N d \tag{11.22}$$

where

H_s is the fully compressed height of the coil spring

γ is a factor that is associated with the spring index C

Finally, to prevent the coil spring from buckling under large loading, its relative deflection δ/H_0 must be examined to ensure that it is smaller than the critical value $(\delta/H_0)_{cr}$, which is given by

$$(\delta / H_0)_{cr} = 0.811\left(1+\sqrt{1-6.89\left(\frac{C_0}{\lambda}\right)^2}\right) \tag{11.23}$$

where λ is the ratio of free height to mean diameter, that is, $\lambda = H_0/D_m$, and C_0 is a coefficient ranging from 0.5 to 2.0, depending on the way the coil spring is supported.

11.3.2. Torsion Bar Springs

Torsion bar springs are used extensively in the suspensions of EVs for their lightweight structure and high mass specific energy. In addition, the fact that torsion bar springs attach to the vehicle frame helps reduce the unsprung mass and improve ride smoothness. When used in a front suspension, torsion bar springs can be arranged longitudinally to make room for the half shafts of the front axle to swing. Nevertheless, the design of suspension with torsion bar springs can be complicated as they must work with a guide mechanism. The strict requirements for their materials and processes often result in high manufacturing costs.

The cross section of a torsion bar spring can take various shapes, such as circle, ring, rectangle, or a combination of multiple blades of a rectangular cross section. The mechanical characteristics of torsion bar springs of various cross sections are listed in Table 11.5 [11.3], where T is the torque acting on the torsion bar; L is the action length; D is the outer diameter; d is the inner diameter of the ring cross section; w is the width of the rectangular cross section; t is the thickness of the rectangular cross section; G is the shear modulus of elasticity, with $G = 7.60 \times 10^4$ MPa for $D \geq 16$ mm and $G = 7.93 \times 10^4$ MPa for $D < 16$ mm; K_τ and K_C are the correction factors for stress and stiffness of the rectangular cross section, respectively, which can be obtained from Figure 11.4 per aspect ratio w/t [11.3].

TABLE 11.5 Mechanical characteristics of torsion bar springs.

	Shape of cross section		
Characteristics	**Circle**	**Ring**	**Rectangle**
Torsion angle θ (rad)	$\frac{32TL}{\pi D^4 G}$	$\frac{32TL}{\pi G(D^4-d^4)}$	$\frac{TL}{K_C t^3 w G}$
	$\frac{2\tau L}{DG}$	$\frac{2\tau L}{DG}$	$\frac{K_\tau \tau L}{K_C t G}$
Torsional stiffness C_τ (N·m/rad)	$\frac{\pi D^4 G}{32L}$	$\frac{\pi G(D^4-d^4)}{32L}$	$\frac{K_C t^3 w G}{L}$
Maximum shear stress τ (MPa)	$\frac{16T}{\pi D^3}$	$\frac{16T}{\pi G(D^3-d^3)}$	$\frac{T}{K_\tau t^2 w}$
	$\frac{\theta DG}{2L}$	$\frac{\theta DG}{2L}$	$\frac{K_C \theta t G}{K_\tau L}$
Specific stress τ/θ (MPa/rad)	$\frac{DG}{2L}$	$\frac{DG}{2L}$	$\frac{K_C t G}{K_\tau L}$

FIGURE 11.4 Correction factors versus aspect ratio for torsion bars of a rectangular cross section.

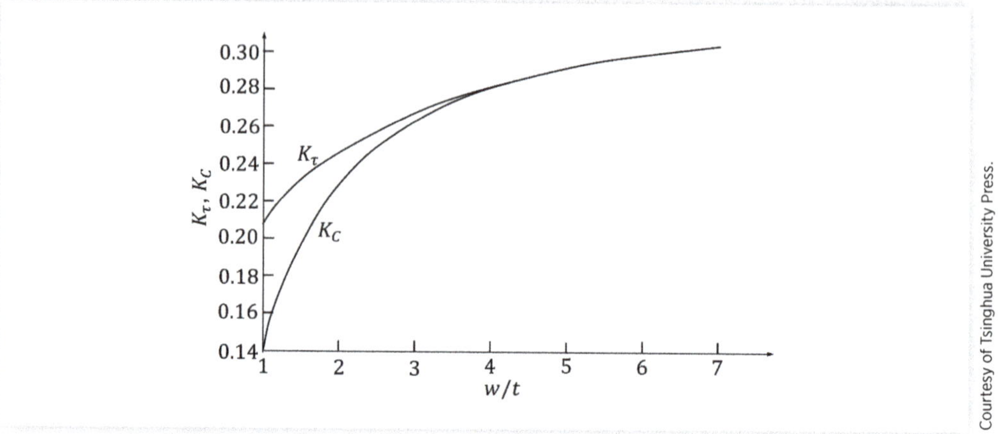

The schematic of a torsion bar arm with torsional deflection under pure axial torque is illustrated in Figure 11.5 [11.3]. The vertical stiffness of a torsion bar suspension at the end of torsion bar, C_v, is given by

$$C_v = \frac{dP}{dy} = \frac{1+(\alpha+\beta)\tan\alpha}{\cos^2\alpha}\frac{C_\tau}{R^2} = C_{st}\frac{C_\tau}{R^2} \tag{11.24}$$

where C_{st} is called the stiffness factor, with $C_{st} = \frac{1+(\alpha+\beta)\tan\alpha}{\cos^2\alpha}$. The stiffness factor C_{st} is a function of α and sinα for various angle β. The vertical stiffness of a torsion bar suspension at the end of torsion bar, C_v, reaches the minimum when the stiffness factor, C_{st}, falls on the curve of $C_{st} = C_{st}\left(C_v = C_{v,min}\right) = \frac{1}{2\sin^2\alpha + 1}$.

FIGURE 11.5 Schematic of a torsion bar arm with torsional deflection.

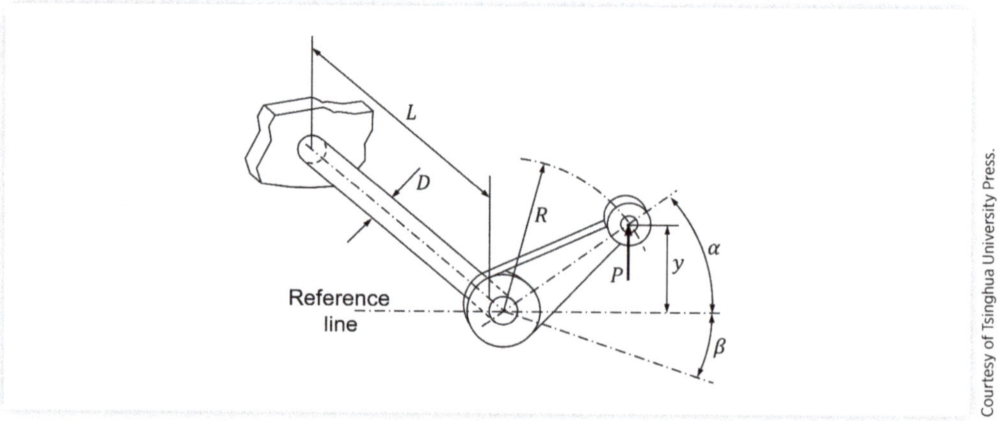

The torsional stiffness of such torsion bar arm is given by

$$C_\tau = \frac{T}{\theta} = \frac{T}{\alpha+\beta} = \frac{PR\cos\alpha}{\alpha+\beta} \tag{11.25}$$

Rearranging Equation (11.25), we have

$$P = \frac{\alpha+\beta}{\cos\alpha}\frac{C_\tau}{R} = C_L\frac{C_\tau}{R} \tag{11.26}$$

where C_L is called the load factor, $C_L = \dfrac{\alpha+\beta}{\cos\alpha}$.

An equivalent static deflection can be defined by

$$\delta_{se} = \frac{P}{C_s} = \frac{\cos\alpha}{\dfrac{1}{\alpha+\beta}+\tan\alpha}R = C_{se}R \tag{11.27}$$

where

C_s is the stiffness of suspension under the load P

C_{se} is called the equivalent static deflection factor, with $C_{se} = \dfrac{\cos\alpha}{\dfrac{1}{\alpha+\beta}+\tan\alpha}$

The load factor C_L is the function of sinα or y/R for various angle β. The torsional stiffness of such torsion bar arm, C_τ, reaches the minimum when the load factor, C_L, falls on the curve $C_L = C_L\left(C_\tau = C_{\tau,min}\right) = \dfrac{-3\sin\alpha}{2\sin^2\alpha+1}$, while the equivalent static deflection, δ_{se}, reaches maximum when the load factor, C_L, falls on the curve of $C_L = C_L(\delta_{se} = \delta_{se,\,max})$.

The equivalent static deflection factor, C_{se}, is the function of sinα for various angle β. The stiffness of suspension, C_s, reaches the minimum when the equivalent static deflection factor, C_{se}, falls on the curve $C_{se} = C_{se}(C_s = C_{s,\,min}) = 3\sin\alpha$, while the equivalent static deflection, δ_{se}, reaches maximum when the equivalent static deflection factor, C_{se}, falls on the curve of $C_{se} = C_{se}(\delta_{se} = \delta_{se,\,max})$.

Note that the values of angle α and angle β in Equations (11.24) through (11.27) are all in radian.

When the load P is known, the equivalent static deflection δ_{se} and then β can be solved using Equation (11.27). Substituting β into Equation (11.26) yields the torsional stiffness of the torsion bar, C_τ.

Thus, the maximum torque acting on the torsion bar at the extreme position of suspension can be obtained by

$$T_{max} = C_\tau\left(\beta + \frac{1}{\sin\left(R\sin\alpha + h_c\right)}\right) \tag{11.28}$$

where h_c is the maximum deflection of the end of torsion bar arm under maximum compression load with respect to its position under design load.

With torque acting on the torsion bar being $T = PR \cos \alpha$, the dimensions of the cross section and the action length of the torsion bar can be worked out according to the shape of its cross section and the equations in Table 11.5.

The end of a torsion bar can be formed into a triangular spline, quartet, or hexagon for connecting to other parts, with a triangular spline of 90° being the most frequently used for its reliable force transmission and easy shot peening process. The diameter of the end portion of a torsion bar is generally 1.15–1.6 times the diameter of the torsion bar itself, with the length of the spline no shorter than 0.4–1.3 times diameter of the end portion.

To minimize stress concentration, a smooth transition is needed between the end portion and the middle portion of the torsion bar. The transition can be made in two methods: using a large radius arc with $R_t \geq 4d$ and using a corn of 30° angle with a radius $r = 1.5d$ at the corner where the tapered surface connects to the torsion bar.

The action length of a torsion bar, L, can be calculated by

$$L = L_{tot} + 2L_e - 2L_{end} - 2L_t \tag{11.29}$$

where

L_{tot} is the total length of the torsion bar
L_e is the equivalent action length of the transition portion
L_{end} is the length of the end portion
L_t is the length of the transition portion

For a large-radius arc transition,

$$L_e = \left(\frac{1}{\chi} + \frac{1}{\chi^2} + \frac{1}{\chi^3} \right) \frac{L_t}{3} \tag{11.30}$$

and for a corn transition,

$$L_e = \left(8 + 10\chi + 15\chi^2 + \frac{15\chi^3}{\sqrt{\chi - 1} \tan \sqrt{\chi - 1}} \right) \frac{L_t}{48\chi^3} \tag{11.31}$$

where $\chi = D/d$.

Torsion bars are generally made of spring steel with an allowable shear stress of over 800 MPa, which can be further strengthened to 1000–1250 MPa by quenching, shot peening, and pretwisting. In the same direction as its operation, pretwisting should be conducted 4–5 times in sequence with a residual angular deflection of no more than 2° and the direction of pretwisting. A clear label on the torsion bars is generally required to prevent it from mistaken interchange between the left and right ones.

11.3.3. Leaf Springs

Leaf springs are one of the oldest forms of springs commonly used for traditional vehicles, which are sometimes used for EVs as an alternative for their simplicity, low cost, and the capability of taking up loads from all directions. In leaf spring suspensions, one or more narrow, arc-shaped, thin plates are so attached to the axle and chassis that the leaf springs can deform vertically for storing the kinetic energy from vibrations of wheels excited by uneven road surfaces. Transverse leaf springs are used from time to time to take up lateral loads for enhanced handling and stability.

A common leaf spring suspension is shown in Figure 11.6 [11.4]. The leaf spring can be either symmetric with $l_1 = l_2$ or asymmetrical with $l_1 \neq l_2$.

FIGURE 11.6 Leaf spring suspension and its basic dimensions.

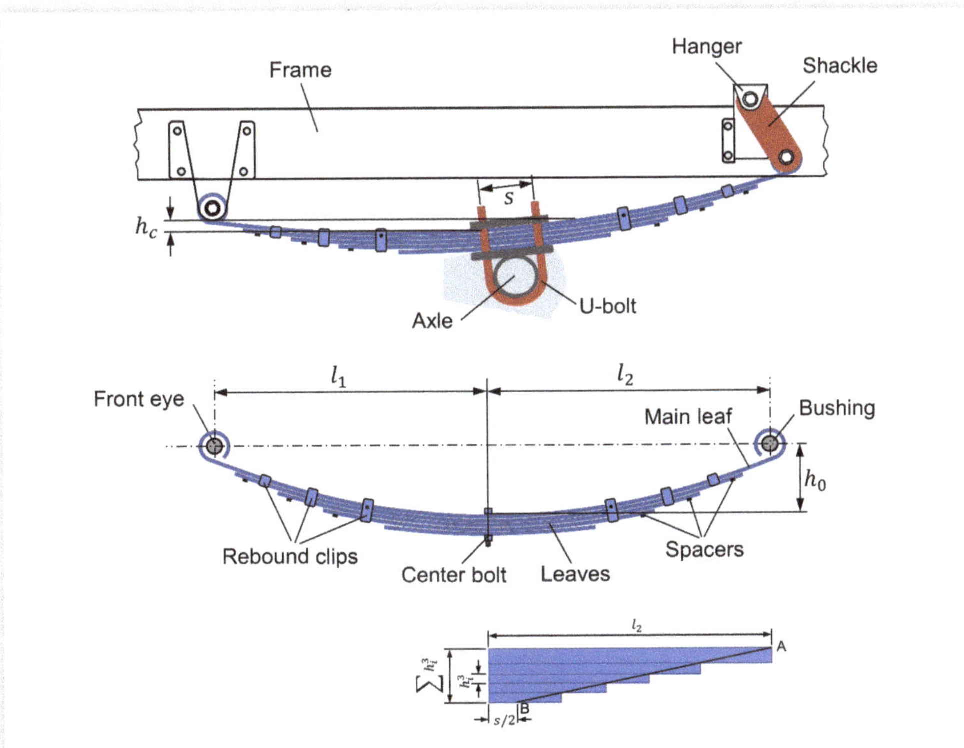

Courtesy of TestingAutos.com.

In general, the following steps can be taken in the design of leaf spring suspension.

1. Calculation of load on the leaf spring suspension
The load on a leaf spring suspension is actually the sprung mass, which is the difference between the axle load and unsprung mass. Thus, the load on the front leaf spring suspension, F_{sf}, is given by

$$F_{sf} = F_{zf} - m_{uf} g \tag{11.32}$$

where

F_{zf} is the vertical load on the front axle
m_{uf} is the unsprung mass of the front suspension

Similarly, the load on the rear leaf spring suspension, F_{sr}, is given by

$$F_{sr} = F_{zr} - m_{ur}g \tag{11.33}$$

where

F_{zr} is the vertical load on the rear axle

m_{ur} is the unsprung mass of the rear suspension

2. Determine the length of the main leaf
   The length of the main leaf refers to the distance between the centers of the front and rear eyes when the leaf spring is straightened. For trucks, the length of the main leaf, l, is typically 0.26–0.35L for the front suspension and 0.35–0.45L for the rear suspension, where L is the wheelbase. A longer length l is always preferred if it is allowed by packaging space.
3. Determine spring camber under full load and in free state
   The spring camber under full load, h_c, is measured to the top of the main leaf immediately below the center of the end eyes. This dimension affects the height of the vehicle body or cargo box. In general, $h_c = 10 - 25$ mm.

   The spring camber in its free state can be calculated by

$$h_0 = \delta_s + h_c + \Delta \tag{11.34}$$

where δ_s is the static deflection and Δ is the change in the spring camber caused by the clamping of the U-bolt of the leaf spring assembly,

$$\Delta = \frac{s(h_c + \delta_s)(3l - s)}{2l^2} \tag{11.35}$$

The radius of the curvature of the leaf spring assembly in its free state is given by

$$R_0 = \frac{l^2}{8h_0} \tag{11.36}$$

4. Estimate the average thickness of leaf springs
   The average thickness of leaf springs is given by [11.2]

$$h_{avg} = \frac{(l - ks)^2 \lambda[\sigma_b]}{6E\delta_s} \tag{11.37}$$

where k is the ineffective length factor considering the section of the leaf spring clamped by the U-bolt, and $k = 0$ for flexible clamping and $k = 0.5$ for rigid clamping; s is the center-to-center distance between threaded parts of the U-bolt; λ is the deflection augmentation factor, and $\lambda = 2.885/(n_m/n_\Sigma + 2)$, with n_m and n_Σ being the number of overlapping main leaves and the estimated total number of leaves, respectively; $[\sigma_b]$ and E are the allowable bending strength and elastic modulus of the leaf spring material; and δ_s is the static deflection of the leaf spring suspension.

5. Select the width of leaf springs
   The width of leaf springs b is usually selected within the range of (6 – 10) h_{avg}. Excessively wide leaf springs for the front suspension may affect the maximum steerable angle of the vehicle.
6. Select the thickness of leaf springs
   Leaf springs of equal thickness should be used wherever possible; however, leaf springs of variable thickness are sometimes adopted to strengthen the main leaf or eyes. It is preferred not to use more than three different thicknesses in a leaf spring suspension and the thickest leaf is no more than 50% thicker than the thinnest leaf for equalizing their lifespans.
7. Select the cross section of leaf springs
   Although a rectangular cross section is standard, there are quite a few different cross sections available, which can reduce the weight by up to 14% and increase the fatigue life by up to 30% [11.3].
8. Determine the number of leaves
   The number of leaves falls commonly within the range of 6–14 but can go as many as 20 for heavy-duty trucks. The use of a reduced number of leaves is obviously advantageous for manufacturing and assembly, but this can result in unequal fatigue life among leaves.
9. Estimate the length of each of the leaves
   Based on the lengths of the main leaf l_1 and l_2, the center-to-center distance of the U-bolt s, and the thickness of each of the leaves $h_1, h_2, h_3, \ldots, h_n$, the length of each of the leaves can be estimated using the graphic method. With the right portion of the main leaf on the top and $s/2$ on the bottom, all the right portions of leaves can be drawn as an elongated rectangle with one on top of another, as shown in Figure 11.6. Note that the height of the rectangle for the ith leaf is the cubic of its thickness, that is, h_i^3, and as a result, the total height of the trapezoid is $\sum h_i^3$. Consequently, points A and B can be drawn and connected with a straight line and then the length of each leaf can be obtained by finding the intersection between the line and the top side of each leaf [11.2].

The design variables mentioned previously are subject to checkups and examinations using FEA and design optimization methods. Design iterations are usually needed to reach an optimal design that can meet strength, stiffness, and other requirements.

In recent years, leaf spring suspension with a fewer number of leaves is increasingly used in engineering practice, which features only one to three leaves of identical width and length, with thickness parabolically or linearly varied from the center to ends. Compared with the spring of all equal thickness leaves, this type of leaf spring suspension can be made equal strength along the length direction, with a 20–40% weight reduction [11.2]. Plastic pieces can be placed between leaves to reduce the friction and noise.

For a parabolic leaf spring, the thickness h_x tapers to the end according to the following formula:

$$h_x = h_c\sqrt{\frac{x}{l_2}} \tag{11.38}$$

where

h_c is the thickness at the center of the leaf
x is the distance from the end of leaf
l_2 is the distance from the center to the right end when the leaf is straightened

The stiffness of a single piece of leaf spring is given by

$$K = \frac{6\zeta J_2 E}{(\eta + 1) l_2^3} \tag{11.39}$$

where

ζ is a correction factor, with $\zeta = 0.92$
the coefficient $\eta = 1 - (h_e/h_c)^3$
$J_2 = bh_c^3 / 12$

For a linear leaf spring, thickness h_x tapers to the end according to the following formula:

$$h_x = h_e + (h_c - h_e)\frac{x}{l_2} \tag{11.40}$$

The corresponding stiffness of a single piece of leaf spring can be calculated using Equation (11.39), but the coefficient η must be substituted by

$$\eta = -\frac{3}{2(1-\epsilon)^3}\left[2\ln\epsilon + (1-\varepsilon)(3-\varepsilon)\right] - 1 \tag{11.41}$$

where $\varepsilon = h_e/h_c$.

For a parabolic or linear leaf spring suspension, its total stiffness is the sum of the stiffness of all leaves, while the load is shared by all leaves as well. A relatively large width is recommended in order to enhance the transverse stiffness, which generally falls within the range of 75–100 mm. The thickness is usually greater than 8 mm at the ends and 12–20 mm at the center, to prevent quenching crack and ensure adequate shear strength [11.2].

11.3.4. Air Springs

Air spring suspensions are generally composed of elastic elements, shock absorbers, and guide mechanisms, along with a body height adjustment mechanism in most scenarios. The elastic elements are convoluted bellows or rolling lobes (diaphragms) made of textile-reinforced rubber that are inflated by compressed air.

The cross section of a bellow-type air spring is illustrated in Figure 11.7a [11.5]. It usually has two or three bellows, of which the inner layer is made of air tight rubber and the outer layer made of oil-resistant rubber. A chord waist, typically made of steel, is used between the bellows to prevent their expansion in the radial direction and rubbing against each other.

The cross section of a rolling diaphragm-type air spring is shown in Figure 11.7b [11.5]. Different from bellow-type air springs, the rubber diaphragm is allowed to expand radially during its operation.

FIGURE 11.7 Two types of air springs. (a) The bellow type and (b) the rolling diaphragm type.

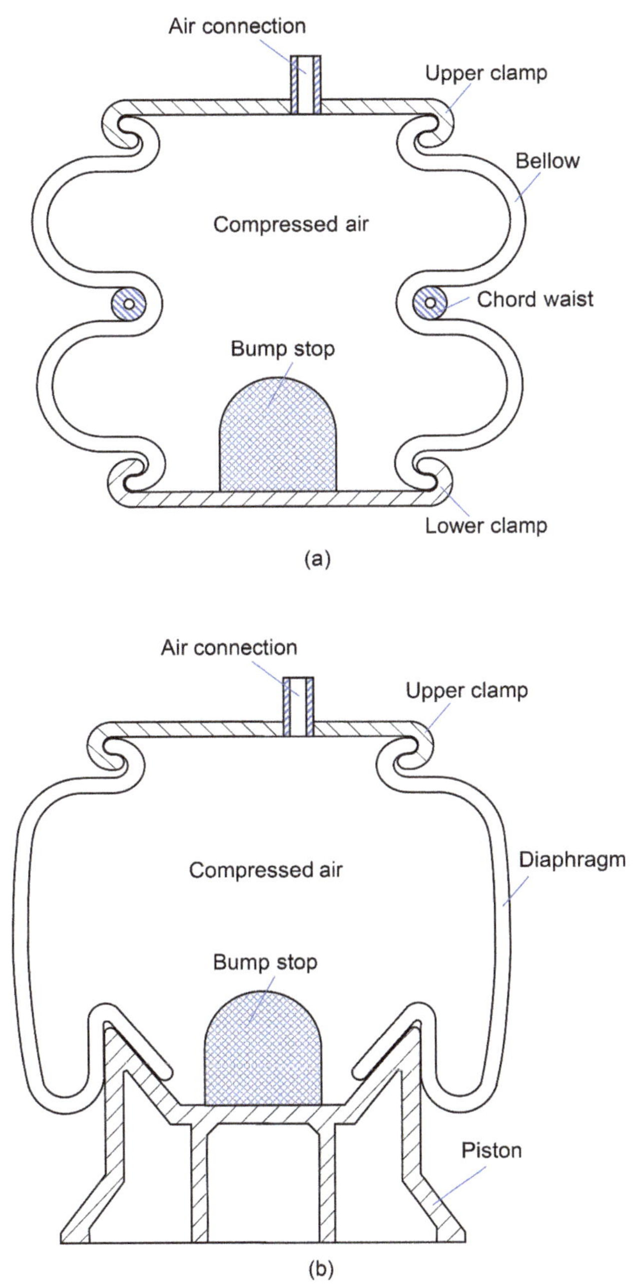

In the design of air springs, with the load F and the allowable air pressure for the bellow or diaphragm known, the effective pressure carrying area, A_e, is given by

$$A_e = \frac{F}{p - p_a} \tag{11.42}$$

where

p is the air pressure inside the bellow or diaphragm
p_a is the standard atmospheric pressure, that is, 101.325 kPa

When a bellow or diaphragm is subjected to a load, its internal volume changes according to the equation of state for a real gas:

$$p = p_0 \left(\frac{V_0}{V} \right)^m \tag{11.43}$$

where p_0 and V_0 are the air pressure and volume within the bellow or diaphragm in the static equilibrium position, respectively, and m is the gas constant. In general, $m = 1$ for an isothermal process when the load varies slowly, $m = 1.4$ for an adiabatic process when the load fluctuates, and $m = 1.33$ can generally be assumed in engineering designs.

Noting that $\mathrm{d}V = -A_e \mathrm{d}z_{as}$, the stiffness of an air spring can be derived from Equations (11.42) and (11.43):

$$K = \frac{\mathrm{d}F}{\mathrm{d}z_{as}} = \left[p_0 \left(\frac{V_0}{V} \right)^m - p_a \right] \frac{\mathrm{d}A_e}{z_{as}} + m p_0 A_e^2 \frac{V_0^m}{V^{m+1}} \tag{11.44}$$

where z_{as} is the stroke of air spring. It is evident that the stiffness of an air spring can be decreased by enlarging V_0, which is often restricted by the packaging space. One or multiple auxiliary air chambers can sometimes be used for the purpose, which usually has a volume no more than three times of the volume of the air spring itself.

The typical stiffness characteristics of an air spring in comparison to a steel spring including jounce bumper are shown in Figure 11.8. In most of the operating ranges around the static equilibrium position of the suspension, the air spring can offer stiffness characteristics of better linearity than a steel spring, while dramatically higher stiffness with prominent nonlinearity is also available when approaching the maximum jounce or rebound points of the suspension travel. In addition, stiffness characteristics can be adjusted according to the wheel load by using active suspension control.

In general, active air suspensions can be classified into semiactive and fully active suspensions. Also referred to as adaptive suspension, the semiactive suspension alters the damping coefficient of its shock absorber with no need for a power supply, while the fully active suspension is capable of adjusting its stiffness characteristics and the damping coefficient using a power supply.

FIGURE 11.8 Stiffness characteristics of an air spring.

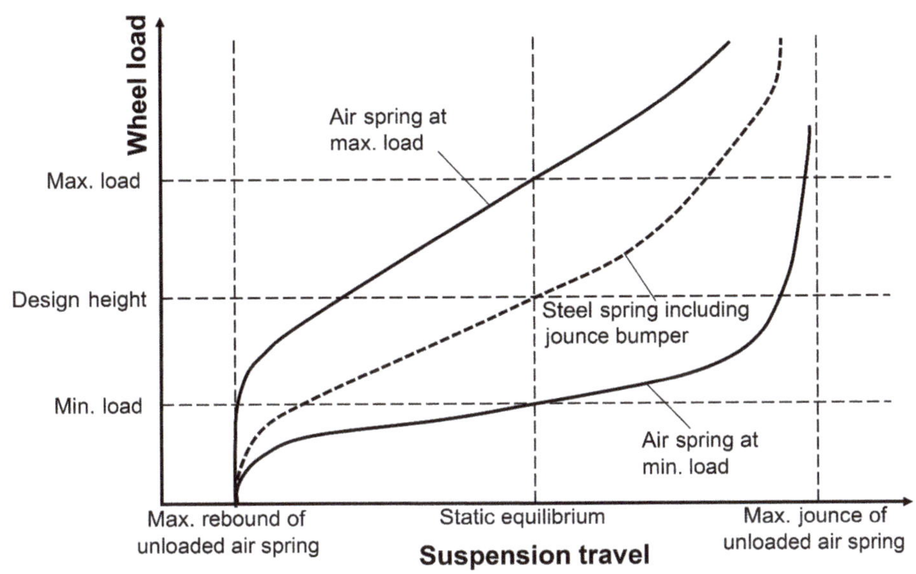

As air springs are not capable of carrying lateral forces or torques, it is thus necessary to properly design guide mechanisms, such as leaf springs, trailing arms, or wishbones, for the air suspension.

A body height adjustment mechanism working together with air springs, typically electronically controlled, is illustrated in Figure 11.9 [11.6]. With such an electronically controlled air suspension (ECAS), when the vehicle is traveling on a road, the acceleration sensor constantly detects the excitation from the road surface that the wheel is experiencing and the signal is processed in the controller together with other signals such as the vehicle speed and the operating mode, for example, driving off, acceleration, coasting, or braking. Consequently, the controller sends the signals to the solenoid and shock absorber for altering the stiffness of the air spring, the damping coefficient of the shock absorber, and the height of the vehicle body. The body height can be adjusted by inflating or deflating the air spring. In general, the air suspension for each of the wheels can be controlled independently so that the vehicle body can be kept level or at desired posture for ride comfort enhancement.

Furthermore, in a more advanced design, road unevenness in front of the vehicle can be detected continually, allowing a predictive regulation of suspension firmness and damping characteristics for minimizing the movement of the vehicle body.

FIGURE 11.9 Body height adjustment mechanism.

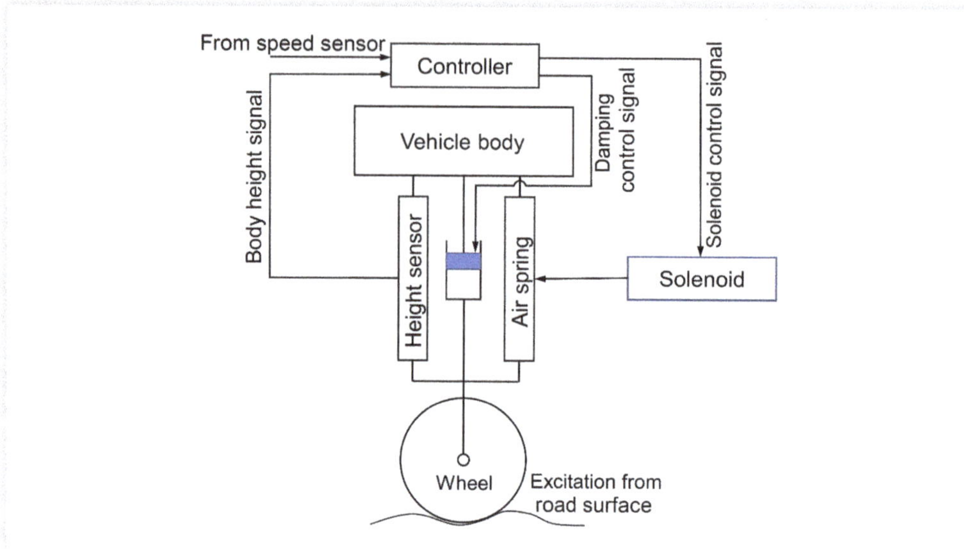

Courtesy of IEEE.

An air suspension is usually configurated in either a two-bellow per axle or a four-bellow per axle form, as shown in Figure 11.10a [11.7] and 11.10b, respectively. The latter offers better damping effects than the former as the vibration from the road can be almost completely absorbed in the latter configuration in comparison to roughly 70% in the former configuration. Besides, the latter is also advantageous in load-carrying capacity, body height adjustment capability, and reliability over the former, although at an increased cost and structural complexity. The four-bellow per axle configuration has been gaining popularity due to aforementioned advantages and a remarkable decline in its cost in recently years.

FIGURE 11.10 Typical configurations of air suspension. (a) Two-bellow per axle and (b) four-bellow per axle.

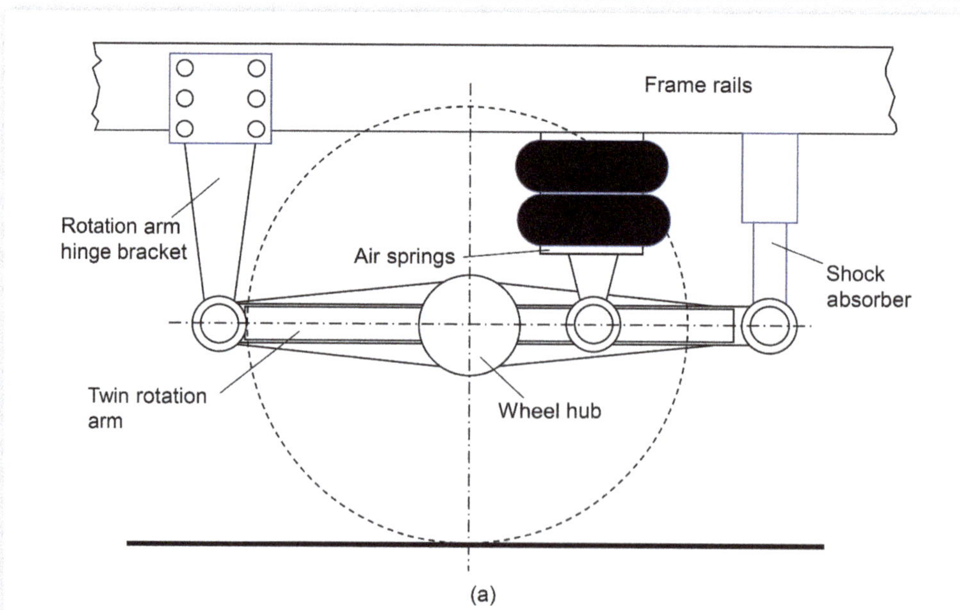

Courtesy of MDPI.

FIGURE 11.10 **(Continued)** Typical configurations of air suspension. (a) Two-bellow per axle and (b) four-bellow per axle.

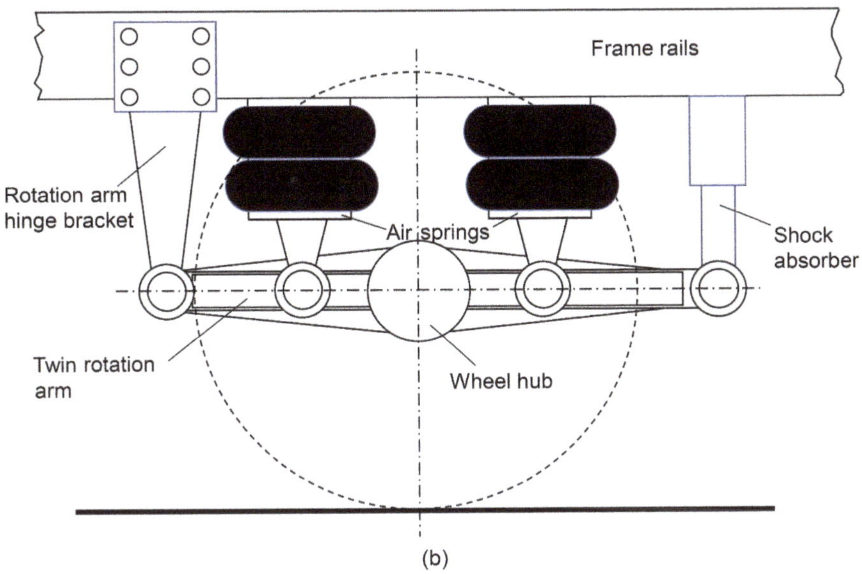

11.4. Design of Shock Absorbers

Shock absorbers are a part of suspension that act primarily to dampen out the vertical vibrations of the vehicle body quickly by absorbing the kinetic energy and converting it into heat or even electricity. It also functions as a secondary device to soften the impact on the suspension only next to the elastic elements such as springs and tires.

11.4.1. Hydraulic Shock Absorbers

Today, hydraulic shock absorbers are either twin-tube or mono-tube type, with the twin-tube type being most commonly used in EVs.

As shown in Figure 11.11 [11.8], a twin-tube shock absorber mainly consists of a working chamber W, a compensation chamber C, a pressure tube T, a reserve cylinder Cl, a dust shield D, piston valves 1 and 2, and base valves 3 and 4. The working chamber W is separated into chamber W_1 and chamber W_2 by the piston, both of which are fully filled with fluid, while the compensation chamber C is usually only half-filled for the fluid to expand under high temperatures. The piston displaces vertically in working chamber W with the bouncing wheel, forcing fluid to flow through the orifices, thereby converting the kinetic energy into heat, which is then dissipated into the system. When the wheel bounces downward, the suspension is stretched and the piston moves upward. Thus, the pressure in the chamber W_1 rises, forcing the fluid to flow through piston valve 1 into the chamber W_2. As a result, a damping force is generated, with some fluid entering chamber C through the clearance between the piston rod R and the rod guide G, as well as through the returning orifice O, and some fluid entering the chamber W_2 through base valve 4 in the meanwhile; when the wheel bounces downward, the suspension is compressed and the piston moves downward. The fluid flows into the chamber W_1 through piston valve 2 from the chamber W_2, and some fluid enters the chamber C through base valve 3 from the chamber W_2 at the same time.

FIGURE 11.11 Main components of a twin-tube shock absorber.

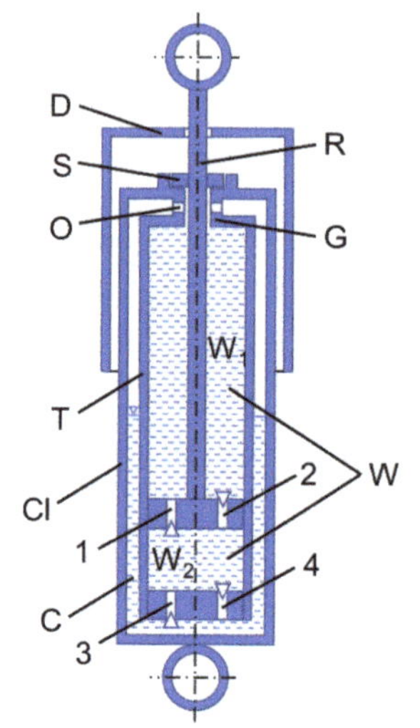

The damping force of a shock absorber, F, can generally be expressed as

$$F = Cv^k \tag{11.45}$$

where

C is the damping coefficient of the shock absorber
v is the velocity of the piston with respect to the cylinder
k is a constant

Usually, $k = 1$ before the opening of the unloader valve.

In addition, the behavior of a shock absorber can be expressed as the force–velocity and force–stroke characteristics, as shown in Figure 11.12 in the left side and the right side, respectively [11.1]. It is clear from the force–velocity characteristics that the damping force increases with the increase in moving velocity with a certain nonlinearity. The total work in the full cycle including both compression and rebound strokes is actually the area enclosed by the curves corresponding to specified moving velocity in the force–stroke characteristics.

FIGURE 11.12 Force-velocity and force-stroke characteristics of a shock absorber.

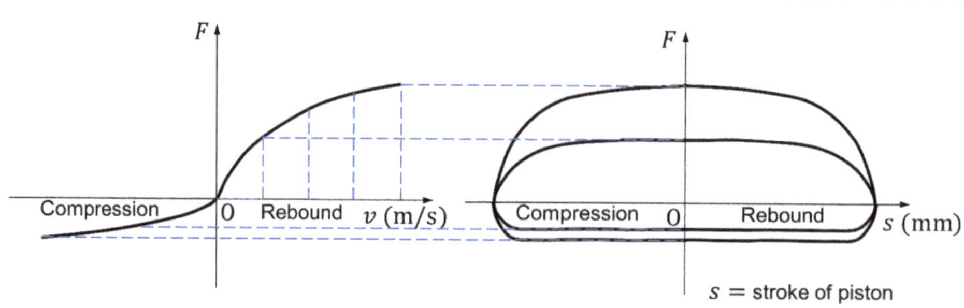

As discussed in Section 6.2.2, the ride comfort is closely related to the damping ratio, which determines the decay rate of vibrations of sprung mass. The recommended damping ratios for achieving good ride smoothness are listed in Table 11.6 [11.3].

TABLE 11.6 Recommended damping ratio for suspensions per type of vehicle.

Type of vehicle		Damping ratio ξ	
		Front suspension	Rear suspension
Cars	Steel spring	0.4	0.2
	Air spring	0.8	0.6
Trucks	Steel spring	0.4	0.3
	Air spring	0.8	0.6

Courtesy of Tsinghua University Press.

By examining the definition of the damping ratio ξ and the natural frequency for sprung mass ω_0 in Equation (6.58), the damping coefficient C can, therefore, be obtained by

$$C = 2\xi\sqrt{m_s K} = 2\xi m_s \omega_0 \tag{11.46}$$

where

m_s is the sprung mass

K is the stiffness of the suspension spring

The shock absorber should be mounted vertically wherever possible. If not, by assuming that the angle between the shock absorber and vertical line is θ, as illustrated in Figure 11.13 [11.2], the damping coefficient can be calculated by

$$C = \frac{2\xi b^2 m_s \omega_0}{b_1^2 \cos^2\theta} \tag{11.47}$$

where b and b_1 are dimensions, as shown in Figure 11.13.

FIGURE 11.13 Mounting a shock absorber at an angle.

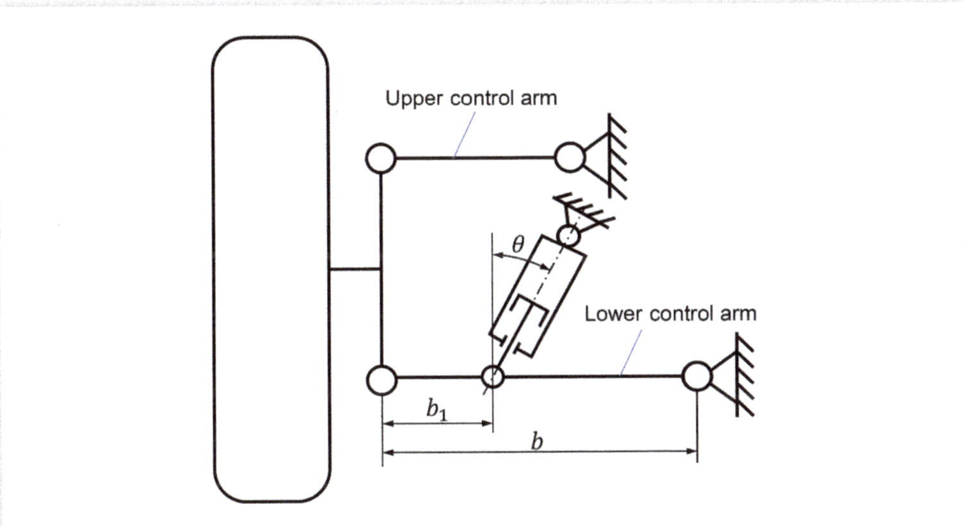

Courtesy of Wangyu Wang.

In order to reduce the impact force on the vehicle body, the unloader valve is opened when the piston velocity reaches a certain value during its oscillation, which is called unloading velocity v_u:

$$v_u = \frac{b_1 \omega_0 A \cos\theta}{b} \tag{11.48}$$

where A is the amplitude of vehicle body oscillation, typically $A = \pm 0.04$ m. The unloading velocity v_u generally ranges from 0.15 to 0.3 m/s. Thus, the maximum unloading force is given by

$$F_u = C_r v_u \tag{11.49}$$

where C_r is the damping coefficient in the rebound stroke. Finally, the diameter of the working chamber D can be estimated by

$$D = 2\sqrt{\frac{F_u}{\pi\left(1-\lambda^2\right)[p]}} \tag{11.50}$$

where λ is the ratio of rod diameter to cylinder diameter, in general $\lambda = 0.4 - 0.5$ for twin-tube shock absorbers, and $[p]$ is the allowable operating pressure of the working chamber, usually $[p] = 3 - 4$ MPa. Standard values of the diameter of the working chamber, D, should be used whenever applicable, while the diameter of the reserve cylinder is generally $1.35 - 1.50D$, with a wall thickness of 2 mm.

11.4.2. Magnetorheological Shock Absorbers

Magnetorheological (MR) shock absorbers utilize MR fluids instead of hydraulic oils in a construction similar to the above-mentioned hydraulic shock absorbers. MR fluids are a class of smart fluid whose properties, for example, viscosity, can be altered by applying an electric field or a magnetic field [11.9]. With magnetically polarizable particles of typically 3–10 μm in diameter dispersed in a carrier medium such as silicone or mineral oil, MR fluids are essentially Newtonian fluids in absence of a magnetic field; however, they behave as non-Newtonian fluids, which can reach a large yield shear stress between 50 and 100 kPa when a magnetic field of 150–250 kA/m is applied [11.10], typically using an electromagnet. Thus, the damping characteristics of the shock absorber can be varied continuously by adjusting the power of the electromagnet, as the result of increased fluid viscosity with the intensity of the magnetic field. The fundamental strategy is to control the yield point shear stress of the MR fluid by altering the electric current.

Based on the hysteresis loops of an MR shock absorber at various input voltages, for a given velocity or displacement, the damping force grows rapidly in the beginning and gradually becomes saturated with the increase of input voltage. Since the MR fluids react to an external stimulus in a few milliseconds, the MR shock absorber is capable of altering its stiffness up to 100 Hz over a broad range of temperature, for example, from −40 to 150°C [11.11].

The behavior of an MR fluid can be described by the following Bingham's model [11.12]:

$$\tau = \mu\dot{\gamma} + \tau_H(H)\mathrm{sgn}(\dot{\gamma}) \tag{11.51}$$

where

τ is the shear stress
μ is the MR fluid viscosity when no magnetic field applied
$\dot{\gamma}$ is the shear–strain rate
τ_H is the yield stress as a function of magnetic field strength H

However, as a quasistatic plastic model, Bingham's model cannot accurately depict the nonlinear and hysteresis behavior of MR fluids under dynamic loading. Other improved models, such as biviscous models and Bouc–Wen models, are available to improve the accuracy in predicting the dynamic behavior of MR shock absorbers [11.12]. Particularly, in a biviscous model, the yield stress of an MR fluid can be described in terms of its preyield and postyield conditions [11.12]:

$$\tau = \begin{cases} \mu_{pre}\dot{\gamma} & |\tau| < \tau_0 \\ \mu_{pos}\dot{\gamma} + \tau_H(H) & |\tau| \geq \tau_0 \end{cases} \tag{11.52}$$

where

τ_0 is the static yield stress
μ is the MR fluid viscosity when no magnetic field is applied

The interested reader is referred to the cited references for a detailed description of other models.

As a mature technology, MR shock absorbers have been widely applied in semiactive and active suspensions today due to their quick response, high reliability, and low power consumption, although at an extra cost.

11.4.3. Electrorheological Shock Absorbers

Similar to the behavior of MR fluids in a magnetic field, electrorheological (ER) fluids are another class of smart fluids whose viscosity can be changed sharply in an electric field when its intensity exceeds a threshold.

The shear stress in an ER fluid is given by

$$\tau = \tau_0 + \tau_E = \mu_0 \frac{dv}{dy} + k_r E^2 \tag{11.53}$$

where τ_0 is the shear stress of the ER fluid without excitation of an electric field, τ_E is the shear stress of the ER fluid induced by the ER effect, μ_0 is the viscosity of the ER fluid without excitation of an electric field, $\frac{dv}{dy}$ is the shear rate of ER fluid without excitation of an electric field, E is the intensity of electric field, and k_r is a factor related to the polarizability, volumetric concentration of dielectric particles, the activation of additives.

The ER effect of an ER fluid can also be expressed by so-called apparent viscosity that is defined by

$$\mu = \mu_0 + \mu_E \tag{11.54}$$

where μ_E is the change in the viscosity of the ER fluid induced by the ER effect, and

$$\mu_E = \frac{\tau_E}{dv / dy} \tag{11.55}$$

Although an ER fluid require a similar level of power for polarization of particles as an MR fluid, the former is deemed better than the latter as it does not cause the fine particles to attach and exert extra force on the walls of the container.

11.4.4. Electromagnetic Shock Absorbers

In most shock absorbers, the kinetic energy of vehicle body vibration is converted into heat and dissipated without being used; however, in electromagnetic shock absorbers, kinetic energy can be converted into electrical energy and stored in energy storage devices, for example, a battery in the form of chemical energy, for later use, which is, therefore, sometimes called regenerative shock absorbers. These active shock absorbers can be self-powered and function as sensors and actuators simultaneously.

At present, the electromagnetic shock absorbers are mainly based on the piezoelectric material, the electromagnetic generator, or their combination. However, as shock absorbers for vehicle suspension applications need to operate in a relatively low frequency range of 1–30 Hz and a large displacement up to 100 mm, electromagnetic generators are reckoned to be more advantageous than piezoelectric material transducers [11.13].

In general, electromagnetic shock absorbers can be categorized into direct and indirect drive types. In a direct drive-type design, magnets and stator windings are attached to the vehicle body and wheel assembly directly. The relative motion between the magnets and stator windings caused by the road unevenness induces an AC electric current in the stator winding circuit. On the other hand, in an indirect drive-type design, the bidirectional linear motion of vehicle body vibration with respect to the wheels is often converted into one directional rotational motion first, mainly for speed augmentation, through a variety of mechanisms, such ball screw, rack and pinion, hydraulic, pneumatic, or their variants [11.13] before being converted into electrical energy using a rotational DC generator. They are more complex than the direct drive-type design in structure due to the additional mechanism required. It is also worth noting that the direct drive type has an AC output, and indirect drive type has a DC output; thus, the former needs rectification before a DC-DC converter for charging the battery, while the latter does not need rectification. Moreover, hybrid electromagnetic shock absorbers are also proposed by combining direct and indirect drive types of design [11.13]. Nevertheless, their structural complexity, reliability, and costs make it difficult to be used in vehicle suspensions.

For an electromagnetic shock absorber making use of a linear electric generator for actuation and energy regeneration, according to Faraday's law of induction, the electromotive potential in the stator windings is given by

$$U = Blv \tag{11.56}$$

where
- B is the magnetic field intensity of the permanent magnet stack
- l is the total length of the stator winding
- v is the magnet speed with respect to the stator winding

Thus, regenerative power can be expressed by

$$P = \frac{U^2}{R} \tag{11.57}$$

where R is the total electrical resistance of the stator windings. In order to maximize the energy regeneration, the magnet arrangement should be optimized to improve the magnetic field intensity. A common approach is to separate magnets by spacers, and it is found that the energy regeneration can be maximized by using magnets and spacers of equal thickness.

Another magnet arrangement method is called Halbach array, in which the polarity orientation of each magnet is turned 90° from the adjacent ones, instead of being opposite to each other. It is found that the Halbach array has the highest magnetic flux density compared to the conventional longitudinal and transverse layouts with or without spacers [11.13].

The damping effect of an electromagnetic shock absorber is entirely or partially contributed by the energy regeneration device, typically the electric generator. The equivalent damping coefficient of the electromagnetic generator is given by [11.13]

$$C_{eq} = \frac{B^2 l^2}{R} \tag{11.58}$$

For a linear electric generator-based direct drive shock absorber, multiple strong magnet arrays and a stator winding of sufficient length may be necessary to reach satisfactory damping performance, which are, however, often subject to the limit in packaging space. Additional hydraulic or pneumatic shock absorbers and MR fluids can be employed to solve the problem.

A compromise can be made between energy harvesting, road holding, and ride comfort. This multiobjective optimization problem can be solved to find the best trade-off by using control algorithms such as maximum induced power control (MIPC), mode switching between electric drive and regeneration, skyhook control, and fuzzy control [11.13].

11.5. Guide Mechanism Design of Independent Suspension

For independent suspension, the guide mechanism carries all loads in a bouncing system in the form of forces and moments, except the vertical force, and largely determines the wheel trajectories and the change in the alignment angle. The general requirements for the design of a guide mechanism are as follows: forming a proper roll center, roll axis, and pitch center, with resistance to pitching both forward in braking and backward in accelerating; minimizing the force at all hinges and elastic deformation of rubber elements for better guide accuracy; reliably transmitting all forces and moments but vertical forces with adequate fatigue strength and lifespan; ensuring the alignment parameters and their changes with bouncing wheels are within limits, with a change in the wheel track indiscernible for the front and no more than ±4 mm for the rear, and no longitudinal acceleration is induced by bouncing wheels; and ensuring a minimal roll angle during cornering, preferably less than 6–7° under the action of a lateral acceleration of 0.4g, with the front wheel leaning to the same direction as the vehicle body for the enhancement of the understeer effect.

11.5.1. Design Considerations on Alignment Parameters

11.5.1.1. Wheel Track

The wheel track changes almost inevitably when wheels bounce up and down. Its effect is similar to adding a sideslip angle to the wheels, and the resultant lateral force can cause the vehicle to deviate from the desired travel direction, leading to an increase in rolling resistance and an impact on steering maneuver. Hence, it is generally desirable to minimize the change in the wheel track in wheel bounce for a satisfactory driving performance. This can be accomplished by making the roll center of suspension as close to the ground level as possible. The reason is that the bouncing velocity of the wheel at the point that touches the ground does not have a horizontal component, and thus, the change in the wheel track is theoretically zero. Nevertheless, for given lateral acceleration and roll stiffness, the roll moment rises with the decrease in the roll center, causing the roll angle of the vehicle body to increase. Hence, a compromise may be necessary in the design of the arm length of a guide mechanism. The change in the wheel track of double-wishbone suspension is ordinarily smaller than that of MacPherson suspension.

11.5.1.2. Camber

For a sedan, the ideal camber angle for avoidance of uneven tire wear ranges between 5 and 10′ (i.e., 1/12–1/6°). In practice, the suspension is generally designed in a way that the camber angle is around the ideal value when the vehicle is unloaded. As a result, the camber angle would become slightly negative when the vehicle is loaded, which can compensate the positive camber angle of outer wheels in cornering and help the outer wheels to better carry lateral loads. Thus, it is required that the camber angle increases as the wheels fall down and decreases as the wheels bounce up. A double-wishbone suspension can better meet these requirements than a MacPherson suspension.

In cornering, the average camber angle of the inner and outer wheels is $\gamma = (\gamma_i + \gamma_o)/2$. A roll camber coefficient can be defined as the rate of change in the camber angle with respect to the change in the body roll angle, namely, $d\gamma/d\Phi$, as shown in Figure 11.14 [11.1]. The roll camber coefficient is about 0.85 for a MacPherson suspension and 0.80 for a double-wishbone suspension.

FIGURE 11.14 Change in the camber angle with respect to the change in the body roll angle.

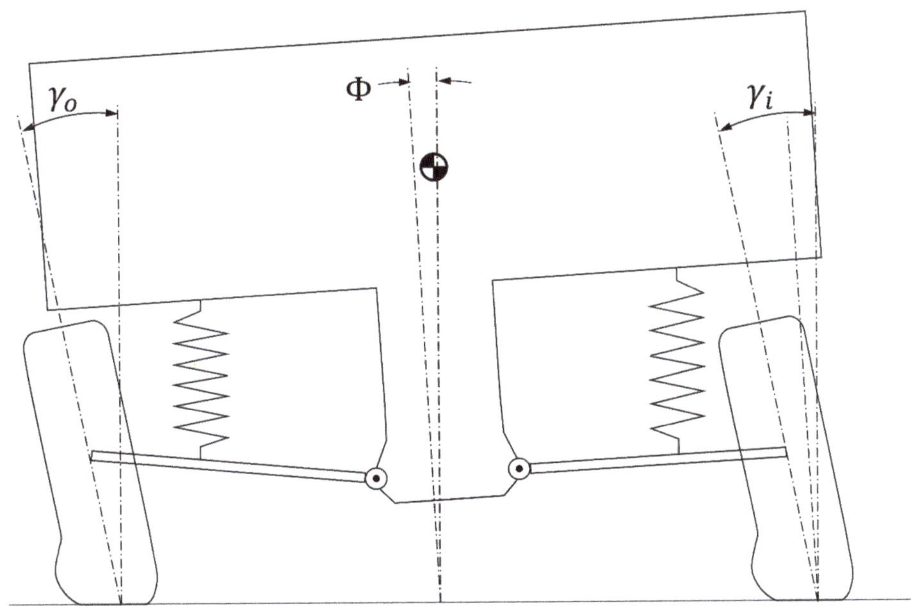

Courtesy of Butterworth-Heinemann.

11.5.1.3. Toe

Although typically ranging from 1.0 to 4.5 mm, the toe-in value of modern cars tends to shrink. When the front wheel camber is negative, a negative toe, or toe-out, is often used to ensure a straightforward motion of wheels with elastic deformation under the action of tractive force.

Additionally, in the steering system matching with an independent suspension system, the tie rod is generally segmented into the inner rod and the outer rod to prevent wheel bouncing from affecting the normal operation of the other wheel and avoid motion interference. The connection point of inner and outer rods is positioned according to the structure type of suspension. A good design of the connection point should not only produce adequate understeer but also prevent the bouncing wheel from generating a significant bouncing of the other wheel that compromises driving safety. When a tie rod is mounted in the rear of the front axle, the position of the connection point may vary, as shown in Figure 11.15 [11.3]. With point A being the ideal position of the connection point, points B and C are the positions for a tie rod being relatively too short and too long, while points D and E are for being relatively too high or too low.

FIGURE 11.15 Positioning of the connection point for the tie rod.

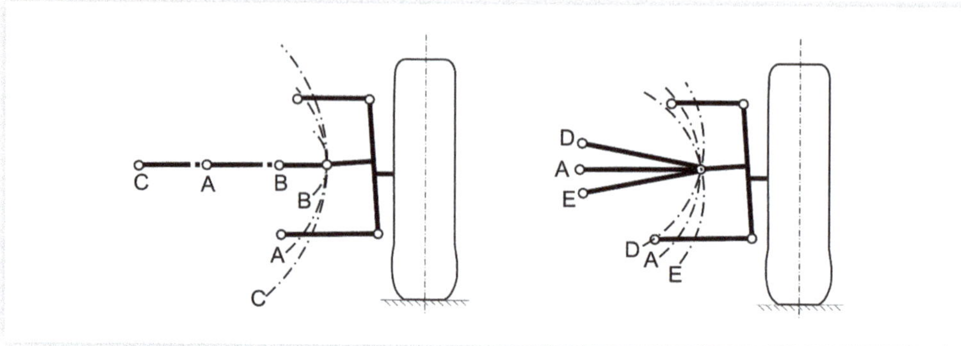

Courtesy of Tsinghua University Press.

It is generally desired to design the suspension so that the toe does not vary with the bouncing wheel, in order to prevent its adverse consequences such as an increase in rolling resistance, tire tread wear, and inability of maintaining a straightforward motion. The influence of the connection point position of the tie rod on the toe angle is also illustrated. It is a common practice to design the connection point so that the front axle tends to change to a weak negative toe in the presence of wheel bouncing.

11.5.2. Design of Roll Center

For a McPherson strut, in order to find the roll center, point A should be located first, which is, in fact, the intersection of the extended line of the lower arm and the line perpendicular to the moving direction of the piston passing point G where the suspension attaches to the vehicle body. Then, by the connecting point A to the tire–road contact point T with a line, the roll center C_r is its intersection with the centerline of the vehicle, as illustrated in Figure 11.16 [11.1].

FIGURE 11.16 Roll center of a McPherson strut suspension.

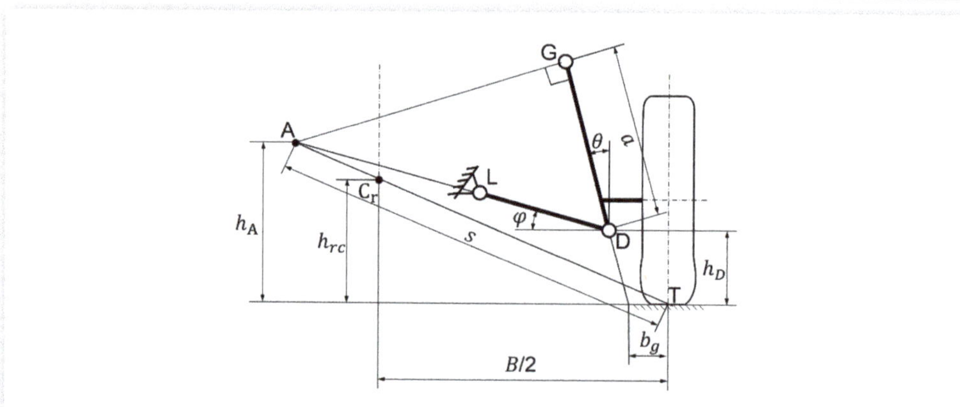

Courtesy of Butterworth-Heinemann.

The height of the roll center, h_{rc}, can be computed by

$$h_{rc} = \frac{B}{2} \frac{h_A}{s\cos\varphi + h_D \tan\theta + b_g} \tag{11.59}$$

where

$$s = \frac{a}{\sin(\theta + \varphi)}$$

$$h_A = s \sin \varphi + h_D$$

For a double-wishbone suspension, point A can be located at the intersection of the extended lines that connect the inner and outer pivot points of the upper and lower arms. Then, the roll center C_r can be found by using the same method as used for the McPherson strut. If the upper and lower arms are parallel, the roll center C_r is simply at the point where the centerline of the vehicle intersects with the line passing the point T, that is, parallel to these arms, as shown in Figure 11.17 [11.1].

FIGURE 11.17 Roll center of a double-wishbone suspension.

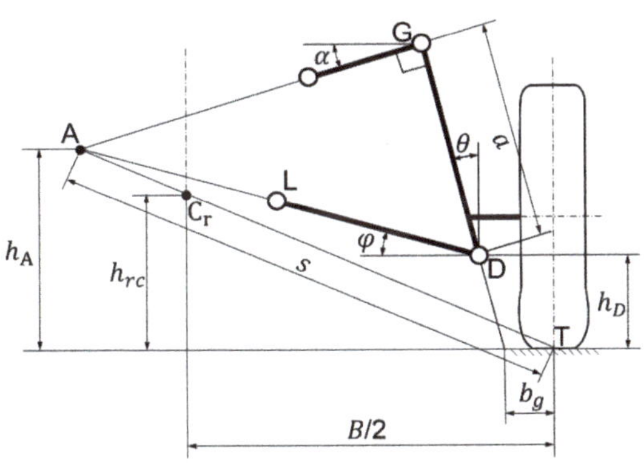

Similarly, the height of the roll center, h_{rc}, can be calculated using Equation (11.59) with

$$s = \frac{a\sin\left(\frac{\pi}{2} + \theta - \alpha\right)}{\sin(\alpha + \varphi)}$$

$$h_A = s \sin \varphi + h_D$$

In suspension design, the roll moment and the force proportion supported by spring and linkages can be adjusted by setting the roll center to proper height. In fact, as the height of the roll center is raised, both roll moment and roll angle drop, and the force acting on the spring and stabilizer bar decreases. Hence, it is generally desired to set a relatively high roll center as long as it does not cause too much tire tread wear due to excessive change in the wheel track as a result of wheel bouncing up and down. The recommended height of the roll center is 0–120 mm for an independent front suspension and 80–150 mm for an independent rear suspension.

The imaginary line connecting the front and rear roll centers is called the roll axis. For a vehicle with independent suspensions, it is often desired to have the roll axis parallel to the ground to minimize the change in front and rear axle loads in cornering for a neutral steering characteristic.

11.5.3. Design of the Pitch Center

For a McPherson strut suspension, the pitch center is located at the intersection of the line passing joint D in parallel to the axis of the lower control arm and the line passing point G in perpendicular to the direction of piston motion, as shown in Figure 11.18 [11.14]. Depending on the orientation of the axis of the lower control arm, the pitch center O_p can possibly be located either in the front or in the rear of front wheels, and the latter is generally desired to minimize the forward pitch in braking. The reason is that such a design makes the kingpin castor angle increase during the compression stroke of suspension. As a special case, when the angle between the axis of the lower control arm and the horizontal line is equal to the kingpin castor angle, the latter would be a constant, with the kingpin axis only undergoing a translational motion with no forward or backward pitch at all during wheel bouncing.

FIGURE 11.18 Pitch center of a McPherson strut suspension.

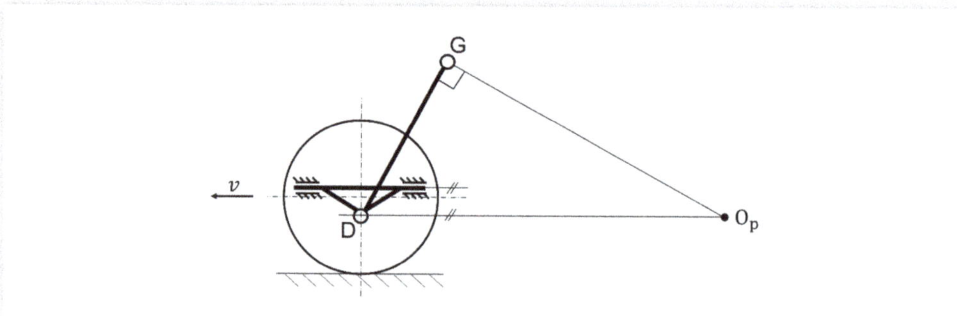

For a double-wishbone suspension, the pitch center is located at the intersection of the lines passing joints D and G in parallel to the rotating axes of lower and upper arms P and Q, respectively, as shown in Figure 11.19 [11.14].

FIGURE 11.19 Pitch center of a double-wishbone suspension.

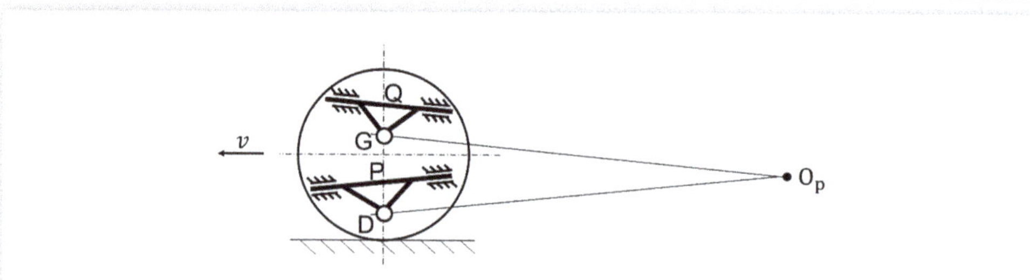

In order to minimize the forward pitch of the vehicle body during braking, the front and rear pitch centers must be positioned between the front and rear axles. On the other hand, the backward pitch during acceleration can be reduced by two means: using a single axle to drive the vehicle or setting the pitch center higher than the wheel center for an independent suspension.

11.5.4. Guide Mechanism Design for McPherson Suspension

A schematic of the guide mechanism for a McPherson suspension is shown in Figure 11.20 [11.1]. The transverse force acting on the piston rod from the bearing sleeve, F_1, can be obtained by solving the force and moment equilibrium equations of the system.

$$F_1 = \frac{bc\left(F_g - 0.5m_{sf}g\right)}{(a+d)(c-d)} \tag{11.60}$$

where
F_g is the static loads on the front wheel
m_{sf} is the sprung mass on the front suspension
g is the gravitational acceleration
a, b, c, and d are dimensions depicted in Figure 11.21 [11.2]

FIGURE 11.20 Guide mechanism for a McPherson suspension.

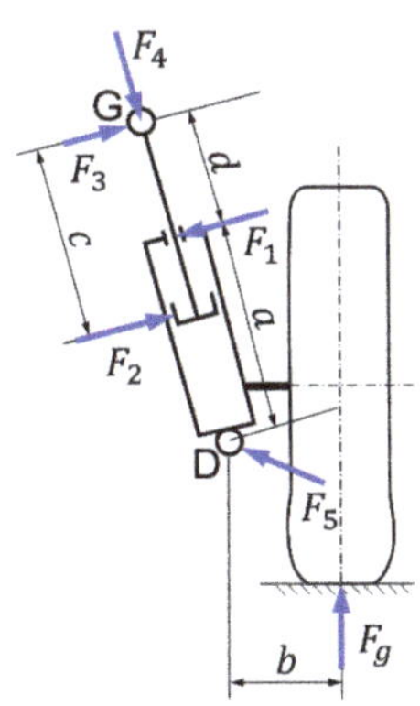

Courtesy of Butterworth-Heinemann.

Thus, the friction force between the piston rod and bearing sleeve, fF_1, increases with the transverse force F_1, causing a detriment to the riding conform. Although the increase in dimension $a + d$ or a decrease in dimension b helps reduce the transverse force F_1, a difficulty in suspension packaging may occur. On the contrary, offsetting the spring axis from the axis of the shock absorber can be a more effective approach to reducing the transverse force F_1, as shown in Figure 11.21. With this configuration, the transverse force, F_1, becomes

$$F_1 = \frac{bc\left(F_g - 0.5m_{sf}g\right)}{(a+d)(c-d)} - \frac{F_s\epsilon}{c-d} \tag{11.61}$$

where

F_s is the spring force along its axis

ϵ is the offset between the axes of the coil spring and shock absorber

FIGURE 11.21 Guide mechanism for a McPherson suspension with an offset spring axis.

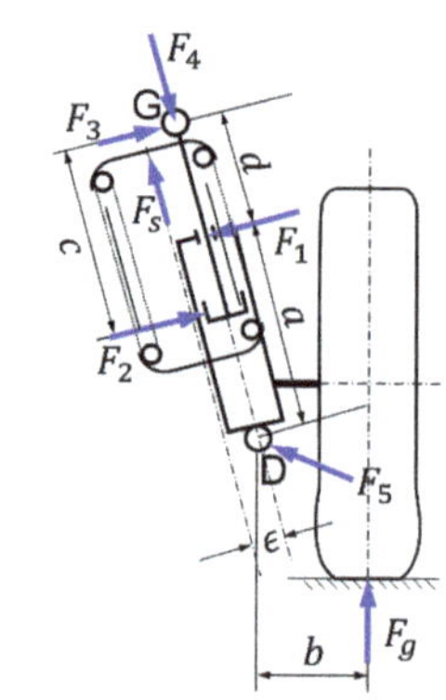

Courtesy of Wangyu Wang.

In order to reduce the transverse force F_1 even further, the bottom of coil spring is sometimes positioned as close as possible to the wheel, leaving an angle between the axes of the coil spring and shock absorber. The kingpin pivot, that is, point D, can be moved to inside the wheel to not only minimize the dimension b but also reduce the kingpin offset at the ground to enhance braking stability, with a negative kingpin offset preferred in some designs.

In the design of a lower control arm for a McPherson strut suspension, it is recommended its length should be increased as much as possible for the enhancing the lifespan of tires and handling stability if there is no interference with other components and there is space available for its packaging. Thus, all design parameters such as the wheel track t_r, camber angle γ, kingpin axis inclination angle ψ, and castor angle χ decrease with the increase in the length of the lower control arm L_l, resulting in an improved handling stability.

11.5.5. Guide Mechanism Design for Double-Wishbone Suspension

The orientation of upper and lower control arms in a double-wishbone suspension largely affects the handling, ride, and packaging of various of chassis components.

In the longitudinal plane of symmetry, the angles of upper and lower arms with respect to the horizontal line, β_u and β_l, respectively, have significant effects on the change in the castor angle χ in braking. For the configurations with $\beta_u > 0$, the castor angle χ decreases in the compression stroke, which is not desired because this worsens the forward pitch in braking. For the configuration of $\beta_u = \beta_l = 0$, the castor angle χ does not change, but no moment is produced on the control arms to work against forward pitch motion in braking, which is, therefore, rarely adopted either. As a result, the configurations corresponding to $\beta_u < 0$ are widely used as a moment can be created to minimize the forward pitch in braking.

In the vertical transverse plane, the configuration of upper and lower control arms critically affects the position of roll centers. As shown in Figure 11.22 [11.1], the roll center C_r can be located at either side of the wheel, depending on the directions of arm axes. The angular positions of upper and lower arms can be determined according to the principles mentioned in the design of roll centers earlier. The configuration shown in Figure 11.22c is generally not acceptable as the roll center C_r is below the ground level.

FIGURE 11.22 Angular positions of control arms in the vertical transverse plane for a double-wishbone suspension. (a) Roll center above the ground, (b) roll center on the ground, and (c) roll center below the ground.

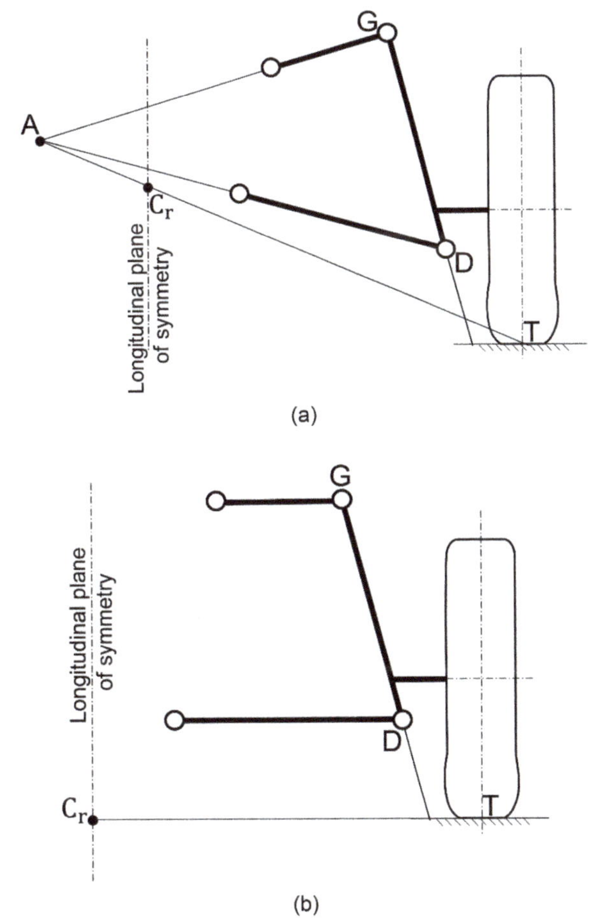

FIGURE 11.22 (Continued) Angular positions of control arms in the vertical transverse plane for a double-wishbone suspension. (a) Roll center above the ground, (b) roll center on the ground, and (c) roll center below the ground.

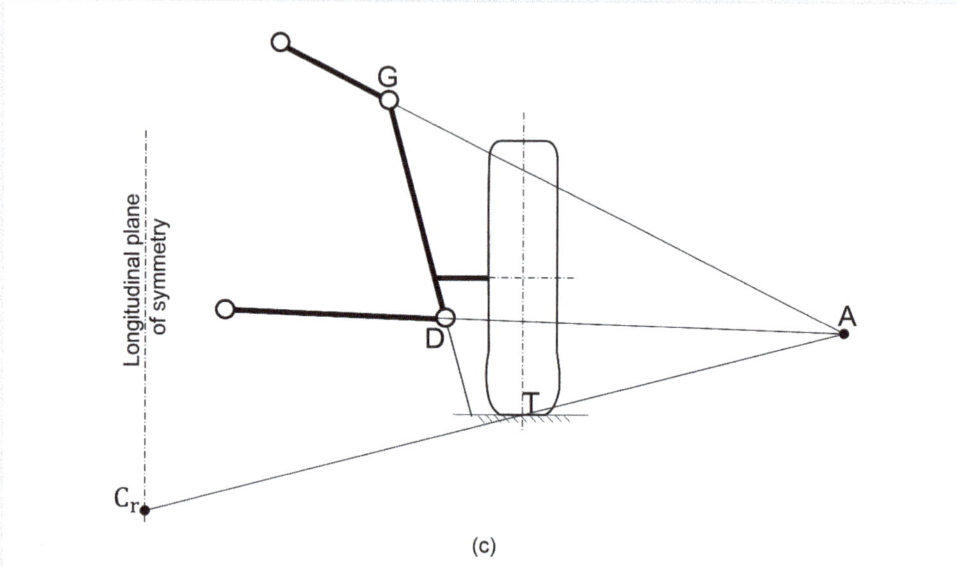

In the horizontal plane, the upper and lower arms generally have three angular positions, as shown in Figure 11.23 [11.2]. The lower arm axis is usually positioned with its front end farther away from the vehicle centerline (i.e., $\alpha_l > 0$) for packaging powertrain and driveline components, allowing the wheel to bounce backward when hitting a raised road obstacle and, thus, reduce the impact on the vehicle body. In contrast, the upper arm axis can be arranged as $\alpha_u > 0$, $\alpha_u = 0$, or $\alpha_u < 0$, as shown in Figure 11.23a, 11.23b, and 11.23c, respectively. When $\alpha_u > 0$, the castor angle does not increase much with wheel bouncing and even decreases when $\alpha_l < \alpha_u$. When $\alpha_u = 0$ or $\alpha_u < 0$, the castor angle increases with wheel bouncing, which is generally beneficial to the suppression of the forward pitch in braking. However, an overlarge castor angle may cause wheel shimming or change in steering effort due to excessive reaction torque at the sprung mass mount. For a car, it is generally desired that the change in the castor angle ranges between 10 and 40″ for every 1 cm compression of suspension with wheel bouncing starting from an initial caster angle within −1 to 2°.

The lengths of upper and lower control arms largely affect the alignment parameters of wheels. It is a general rule of thumb to make the upper arm shorter than the lower arm for the desired characteristics of suspension motion as well as the convenience of powertrain packaging. The effects of the vertical displacement of the tire–ground contact point, z, on the motion characteristics of a double-wishbone suspension for various ratios of upper arm length to lower arm length, L_u/L_l, indicate that the change in wheel alignment parameters is minimal when the ratio L_u/L_l is around 1.0, while the change in the wheel track is the smallest when the ratio L_u/L_l is about 0.6. Hence, it is recommended that the ratio L_u/L_l be selected in the range of 0.66–0.7 for improving the balance between the handling stability and tire life.

FIGURE 11.23 Angular positions of control arms in the horizontal plane for a double-wishbone suspension. (a) $\alpha_u > 0$, (b) $\alpha_u = 0$, and (c) $\alpha_u < 0$.

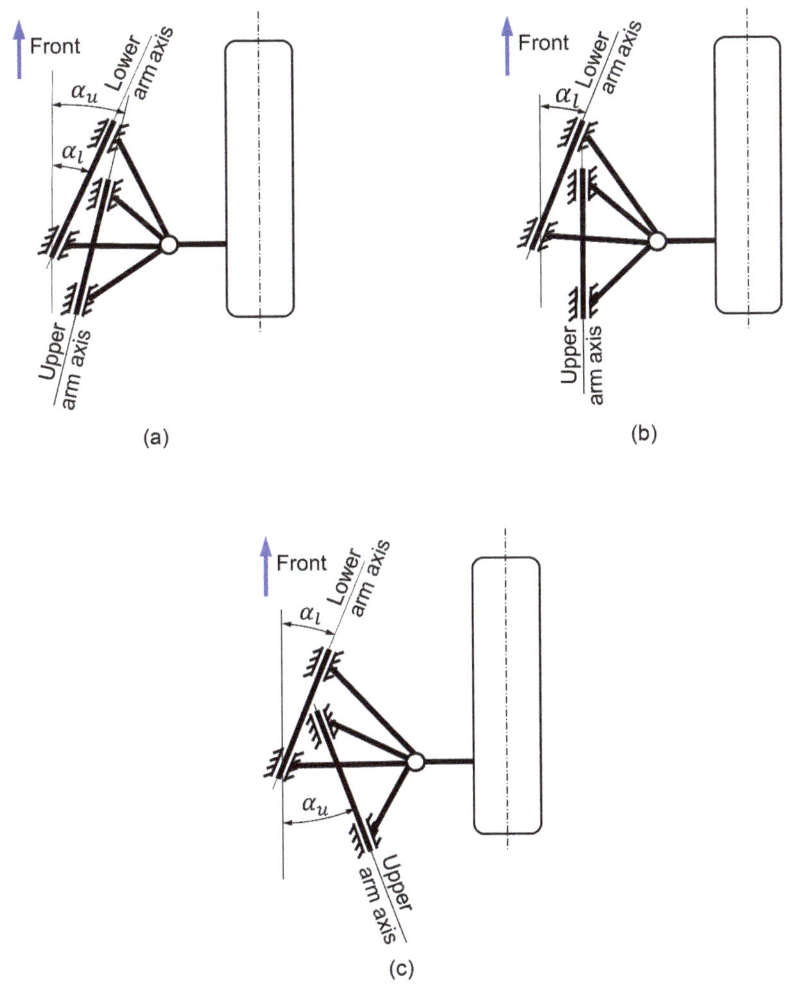

Courtesy of Wangyu Wang.

11.6. Design of Stabilizer Bars

As a component in many suspensions, stabilizer bars, or antiroll bars, are widely used in modern vehicles to increase the roll stiffness to improve their driving stability. The schematic of a stabilizer bar is given in Figure 11.24 [11.1]. The effect of a stabilizer bar is mainly two-fold: reducing the body roll during cornering and enhancing understeer characteristics. However, an uncomfortable ride can occur on rough roads because of increased vertical stiffness at the wheels when a stabilizer bar is used.

FIGURE 11.24 Schematic of a stabilizer bar.

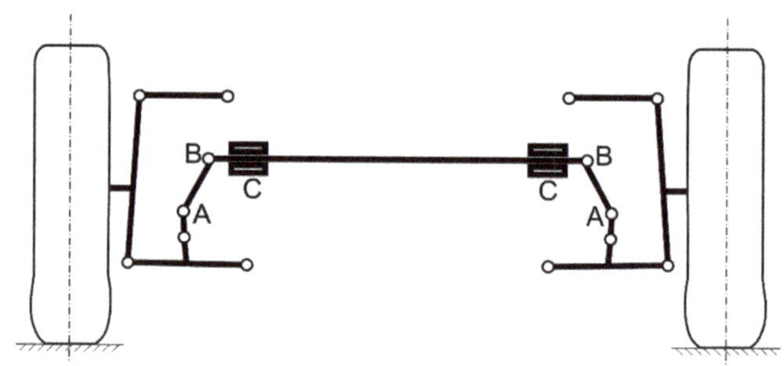

Courtesy of Butterworth-Heinemann.

Although some vehicles only use a stabilizer bar in the front suspension, most vehicles employ two stabilizer bars, with one in the front and the other in the rear. It is generally not recommended to use a stabilizer bar only in the rear suspension as it leads to oversteer.

For a nonlinear suspension with an integral axle, the roll stiffness of its stabilizer bar is identical to the equivalent roll stiffness at the wheels.

For an independent suspension, however, the roll stiffness of its stabilizer bar, $K_{\Phi b}$, is related to the equivalent roll stiffness at the wheels, $K_{\Phi w}$, by

$$K_{\Phi b} = K_{\Phi w}\left(\frac{L_b}{B}\right)^2\left(\frac{\delta_w}{\delta_b}\right)^2 \tag{11.62}$$

where

L_b is the distance between the ends of the stabilizer bar
B is the wheel track
δ_w is the vertical deflection of the left and right wheels
δ_b is the vertical deflection at the end of the stabilizer bar (point A), as shown in Figure 11.25 [11.3]

FIGURE 11.25 Dimensions of a stabilizer bar.

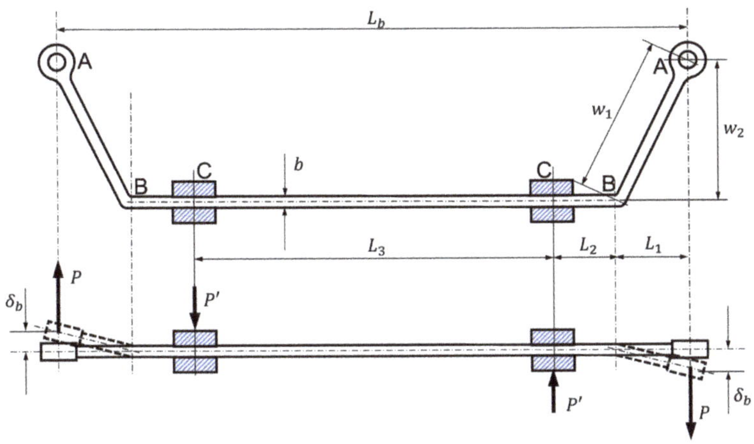

It is assumed that the left and right wheels are subjected to vertical forces of equal magnitude in opposite directions. When rubber bushings are used at the end points A, the roll stiffness of a stabilizer bar would drop approximately by 15–30%. The vertical deflection at the ends of the stabilizer bar, δ_b, is given by

$$\delta_b = \left[w_1^3 - L_1^3 + \frac{L_b}{2}(L_1 + L_2)^2 + 2w_2^2(2L_2 + L_3) \right] \frac{P}{3EI} \tag{11.63}$$

where

P is the vertical force acting on the ends of the stabilizer bar
E is the elastic modulus of material, typically $E = 2.06 \times 10^5$ MPa
I is the moment of inertia of the stabilizer bar section, $I = \pi d^4/64$
d is the diameter of the stabilizer bar
L_1, L_2, L_3, w_1, and w_2 are dimensions, as illustrated in Figure 11.25

Thus, the angular stiffness of the stabilizer bar becomes

$$K_{\Phi b} = \frac{PL_b^2}{2\delta_b} = \frac{3EIL_b^2}{2\left[w_1^3 - L_1^3 + \frac{L_b}{2}(L_1 + L_2)^2 + 2w_2^2(2L_2 + L_3) \right]} \tag{11.64}$$

When the angular stiffness of the stabilizer bar $K_{\Phi b}$ is known, the moment of inertia I can be solved using Equation (11.64), and subsequently, the diameter d can be obtained.

In general, the maximum stress in a stabilizer bar is located at the inner side of the cross section B (Figure 11.25), depending on the fillet radius R_f there. It is recommended that the maximum torsional stress τ should not exceed its allowable value $[\tau]$, with $[\tau] = 700$ MPa, that is,

$$\tau = \frac{16K_c w_2 P}{\pi d^3} \leq [\tau] \tag{11.65}$$

where

K_c is the curvature coefficient, $K_c = \frac{4C-1}{4C-4} + \frac{0.615}{C}$

C is the spring index, $C = (2R_f + d)/d$

The minimum fillet radius R_f can be solved accordingly, which is generally recommended that $R_f \geq 1.25d$. In addition, when the points of support, C, are close to the middle of the stabilizer bar, the principle stress may also be high on the cross section at C, due to the joint action of bending and torsional moments, which needs to be checked as well.

The bending stress on the cross section at B under the action of bending moment Pw_1 is given by

$$\sigma = \frac{Pw_1}{32\pi d^3} \tag{11.66}$$

which should not exceed the allowable value of 1250 MPa.

Stabilizer bars can be made of the same materials as that of torsion bars described in Section 11.3.2, with no need for pretwist processing. A higher grade of materials may be necessary if the stabilizer bar also functions as a guide mechanism.

11.7. Design of Other Suspension Components

The structural design of other suspension components including control arms, trust rods, and joints is discussed in this section.

11.7.1. Control Arms

Utilized for connecting the wheel to the vehicle frame or the body, control arms are often subject to complex loading conditions such as tractive force, braking force, lateral force, and moments. Hence, a monoshell or clamshell cross section is commonly used to enhance their stiffness. In order to transfer the longitudinal force from the wheel, an adequate distance between pivot points, s, is required to minimize the force on the attachment to the vehicle body, as shown in Figure 11.26 [11.2]. Today, control arms are often made of aluminum die casting alloy, instead of stamping or die forging steel, for lightweight purposes.

FIGURE 11.26 Distance between pivot points for an independent suspension.

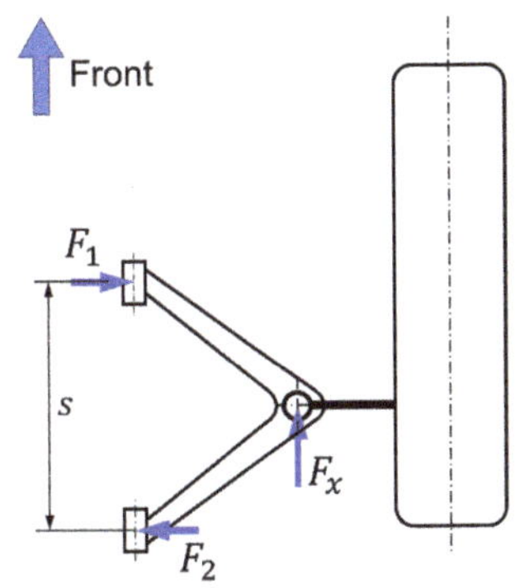

Courtesy of Wangyu Wang.

11.7.2. Thrust Rods

Similar to control arms, thrust rods are used to connect the axle to the vehicle body or frame. For only transmitting simple tension or compression force, thrust rods can be made of simple steel tubes with joints at the ends, which need to have adequate bending strength. To further enhance their bending strength, steel plates can be formed into boxy or "I"-shaped cross section.

In addition, a threaded connection is sometimes used for thrust rods that are required to have their length adjustable or connected together at an adjustable angle to compensate the errors in fabricating and installation processes for easy assembly.

11.7.3. Joints

At the ends of control arms or thrust rods, either hinge joints or ball joints are used to make connections to other vehicle components. In general, the requirements for the joints are as follows: low friction, maintenance free, a certain elasticity, and good noise insulation. Rubber or plastic bushings are often utilized inside the joints to satisfy these requirements.

The selection of the joint type is dependent on the motion required for the joints. For example, for the joint at the inner side of a steering wheel with double-wishbone suspension, a hinge joint is suitable since it allows the wheel to spin about the kingpin axis in cornering and to bounce in the vertical direction at the same time.

References

[11.1]. Reimpell, J., Stoll, H., and Betzler, J., *The Automotive Chassis: Engineering Principles*, 2nd ed. (Oxford, UK: Butterworth-Heinemann, 2001), ISBN:0750650540.

[11.2]. Wang, W., *Automobile Design*, 3rd ed. (Beijing: China Machin Press, 2000), ISBN:7111076133.

[11.3]. Liu, W., *Automobile Design* (Beijing: Tsinghua Press, 2001), ISBN:9787302045298.

[11.4]. TestingAutos.com, "When Do Leaf Springs Need to Be Replaced?," accessed August 2023, https://www.testingautos.com/car_care/leaf-springs.html.

[11.5]. Heisler, H., *Advanced Vehicle Technology*, 2nd ed. (Burlington: Elsevier, 2002), ISBN:0750651319.

[11.6]. Zhao, R., Xie, W., Wong, P. et al., "Robust Ride Height Control for Active Air Suspension Systems with Multiple Unmodeled Dynamics and Parametric Uncertainties," *IEEE Access* 7 (2019): 59185-59199, doi:10.1109/ACCESS.2019.2913451.

[11.7]. De Melo, F., Pereira, A., and Morais, A., "The Simulation of an Automotive Air Spring Suspension Using a Pseudo-Dynamic Procedure," *Applied Sciences* 8, no. 7: 1049, doi:10.3390/app8071049.

[11.8]. Gillespie, T., *Fundamentals of Vehicle Dynamics* (Warrendale: Society of Automotive Engineers, Inc., 1992), 21-118, ISBN-10:1560911999, ISBN-13:978-1560911999.

[11.9]. Michele D., and Giovanna, B., Introduction to Magnetorheological Fluids, in Olabi, A.G. (ed.), *Encyclopedia of Smart Materials 5* (Amsterdam: Elsevier, 2022), 187-202, doi:10.1016/B978-0-12-803581-8.11744-8.

[11.10]. Cortés-Ramirez, J., Villarreal-González, L., and Martinez, M., "Characterization, Modeling and Simulation of Magnetorheological Damper Behavior under Triangular Excitation," in Arai, E. and Arai, T. (eds), *Mechatronics for Safety, Security and Dependability in a New Era* (Amsterdam: Elsevier, 2007), 353-358, doi:10.1016/B978-008044963-0/50072-2.

[11.11]. Patil, R., and Gawade, S., "Design and Static Magnetic Analysis of Electromagnetic Regenerative Shock Absorber," 2012, accessed October 19, 2022, https://www.academia.edu/31005481.

[11.12]. Rossi, A., Orsini, F., Scorza, A., Botta, F. et al., "A Review on Parametric Dynamic Models of Magnetorheological Dampers and Their Characterization Methods," *Actuators* 7, no. 2 (2018): 16, doi:10.3390/act7020016.

[11.13]. Zhang, R., Wang, X., and John, S., "A Comprehensive Review of the Techniques on Regenerative Shock Absorber Systems," *Energies* 11, no. 5 (2018): 1167, doi:10.3390/en11051167.

[11.14]. Bastow, D., Howard, G., and Whitehead, J., *Car Suspension and Handling*, 4th ed. (Warrendale: Society of Automotive Engineers, Inc., 2004), ISBN:0-7680-0872-6.

12

Steering System Design

The steering system of an EV controls the direction of its motion. It is critical that the vehicle follows the desired course, with a coordinated angular relationship between steerable wheels.

Because of the friction between the road and the tires of steerable wheels, effort is needed to turn the steering wheel. In order to minimize such an effort, a power steering system has almost become standard equipment, which generally uses the power of an electric motor to assist in swiveling the steering wheels.

The general requirements for the steering system are as follows [12.1-12.3]:

1. To reduce the tire wear and enhance the handling stability, all wheels of a vehicle in cornering must turn about the instant center of rotation without any sideslips.
2. The steerable wheels must be able to align themselves to the center position in a stable manner after making a turn when the driver releases the driving wheel.
3. No vibrations or shimmy for steerable wheels under any driving conditions, particularly without any shimmy oscillation caused by interference between the steering and suspension guiding mechanisms.
4. Easy to maneuver, with minimized impact on the steering wheel from the road obstacles that the steerable wheels hit.
5. Fast and precise response to the steering wheel to ensure a high maneuverability of the vehicle, with an adjustable mechanism for eliminating the influence of component wear of the steering system.

6. Equipped with mechanisms that can prevent the driver from injury when the steering system is crumpled in a crash.
7. Check the correctness of motion to ensure that the steerable wheels turn in the same direction as the steering wheel.

The components of a basic steering system are illustrated in Figure 12.1 [12.4].

FIGURE 12.1 The components of a basic steering system.

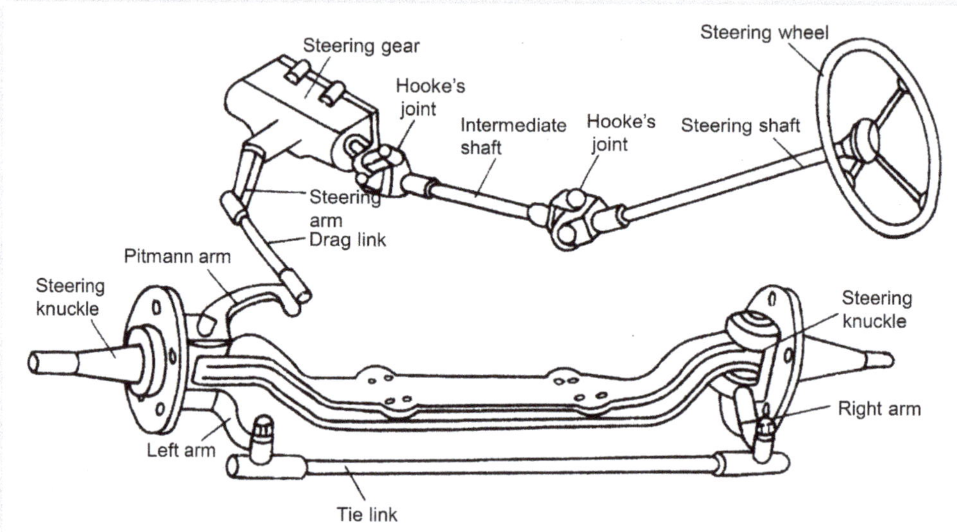

Courtesy of Jianwen Zhang.

12.1. Trapezoidal Steering Mechanism

A trapezoidal linkage used to connect the steerable wheels typically conforms to a variation of Ackermann steering geometry, which allows all the wheels of a vehicle to turn about the instant center of rotation in a rolling motion without any sideslip. The trapezoidal linkage mechanism can be differentiated into the integral type and split type, with the former suitable for nonindependent suspensions and the latter suitable for independent suspensions.

12.1.1. Design of Integral-Type Trapezoidal Linkage

An integral-type trapezoidal linkage consists of a track rod, trapezoid arm, and front axle, as shown in Figure 12.2 [12.5]. With a simple structure and low manufacturing cost, this type of mechanism allows easy toe alignment. However, when one of the steerable wheels starts to bounce up and down, the other one will be affected. The track rod can be arranged either in the front or in the rear of the front axle, preferably at a height no lower than the axle to prevent it from damage caused by road hazards.

FIGURE 12.2 A sketch of an integral-type trapezoidal linkage.

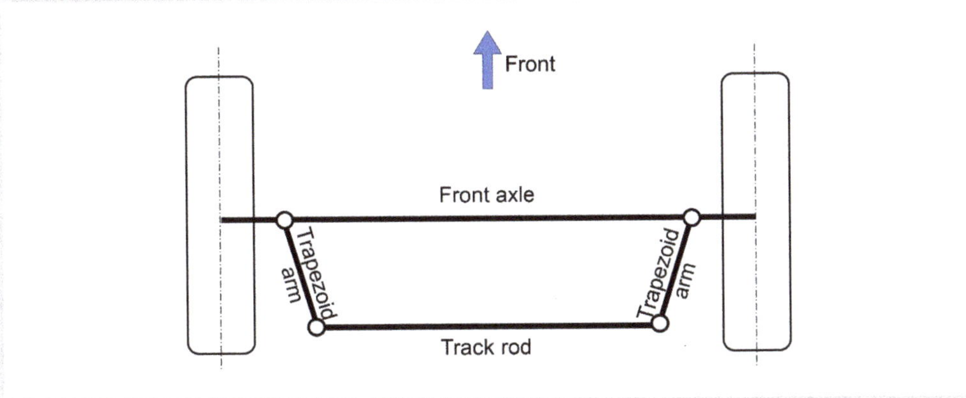

By ignoring the sideslip angle of tires, the basic geometry of a two-axle vehicle in cornering can be simplified by a graph, as shown in Figure 12.3 [12.6]. With the instant center of rotation of the vehicle at O, the trapezoidal mechanism must warrant the following relationship between the steering angles of outer and inner wheels, δ_o and δ_i, respectively [12.1-12.3]:

$$\cot\delta_o - \cot\delta_i = \frac{B_k}{L} \tag{12.1}$$

where

B_k is the distance between the points where the kingpin centerlines intersect with ground

L is the wheelbase

FIGURE 12.3 Basic geometry of a two-axle vehicle in a turn.

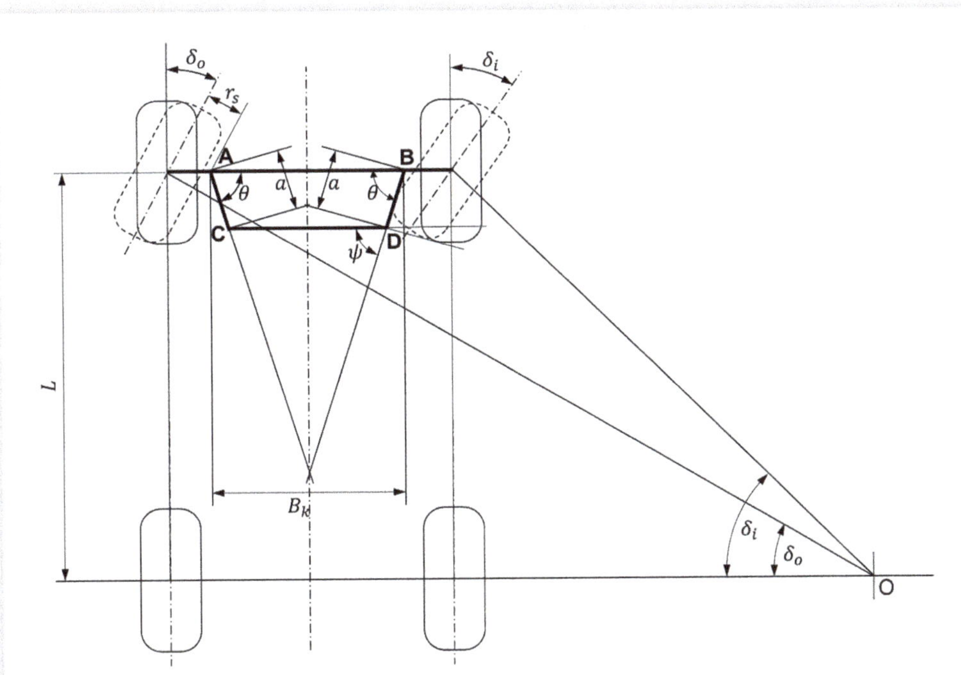

In fact, the trapezoidal mechanism can only approximate the relationship of Equation (12.2). Using the steering angle of the outer wheel, δ_o, as the independent control variable, the dependent control variable δ_i can be solved by cosine law:

$$\delta_{ia} = \theta - \arccos\frac{\frac{B_k}{a}\left[2\cos\theta - \cos(\theta+\delta_o) - \cos 2\theta\right]}{\sqrt{\left(\frac{B_k}{a}\right)^2 - 2\frac{B_k}{a}\cos(\theta+\delta_o)+1}} - \frac{\sin(\theta+\delta_o)}{\sqrt{\left(\frac{B_k}{a}\right)^2 - 2\frac{B_k}{a}\cos(\theta+\delta_o)+1}} \tag{12.2}$$

where

a is the length of the trapezoidal arm

θ is the bottom angle of the trapezoid

In the design, it is preferred that the actual steering angle of the inner wheel, δ_{ia}, be as close as possible to its theoretical value, δ_i, particularly at small steering angles that are used most frequently, in order to reduce tire wear at high speeds. This criterion can be loosened for large steering angles under low-speed operation due to their infrequent usage.

Hence, an objective function with weighting factors can be constructed to solve the optimization problem for proper design variables a and θ:

$$f(x) = \sum_{\delta_o=0}^{\delta_{omax}} w(\delta_o)\frac{\delta_{ia}(\delta_o) - \delta_i(\delta_o)}{\delta_i(\delta_o)} \tag{12.3}$$

where x is the design variable, with $x = [a\ \ \theta]$; $w(\delta_o)$ is the weighting factor; and δ_{omax} is the maximum angle of the outer steerable wheel, with $\delta_{omax} = \arcsin\frac{L}{R_{min} - r_s}$, in which R_{min} is the minimum turning radius of the vehicle and r_s is the kingpin offset at the ground. For a vehicle with a steering angle most often within 10° and rarely exceeding 20°, the weighting factor can be assigned as follows:

$$w(\delta_o) = \begin{cases} 1.5 & 0^\circ < \delta_o \le 10^\circ \\ 1.0 & 10^\circ < \delta_o \le 20^\circ \\ 0.5 & 20^\circ < \delta_o \le \delta_{omax} \end{cases} \tag{12.4}$$

The following constrains are usually applied to prevent excessive steering force on the track rod because the trapezoidal arm is too short or the bottom angle of trapezoid is too small, or to prevent the difficulty in packaging the trapezoidal mechanism:

$$\begin{cases} a_{min} \le a \le a_{max} \\ \theta \ge \theta_{min} \end{cases} \tag{12.5}$$

where $a_{min} = 0.11B_k$, $a_{max} = 0.15B_k$, and $\theta_{min} = 70^\circ$ can be used as typical values.

Furthermore, the transmission angle of a four-bar linkage, ψ, must be greater than a lower bound, ψ_{min}, typically $\psi_{min} = 40^{\circ}$. Thus, the constrain for the minimum transmission angle can be added:

$$\frac{2a}{B_k} - \frac{\cos(\theta + \delta_{omax}) + \cos\psi_{min} - 2\cos\theta}{(\cos\psi_{min} - \cos\theta)\cos\theta} \leq 0 \tag{12.6}$$

where ψ_{min} is the minimum transmission angle that depends on the variables a and θ.

The nonlinear constrained optimization problem defined by Equations (12.3), (12.5), and (12.6) can be solved by a complex method. As a result, the typical solution domains suitable for vehicles requiring a relatively small θ_{min} and a large ψ_{min}, a relatively large θ_{min} and a small ψ_{min}, or in between the above two can be solved.

12.1.2. Design of Split-Type Trapezoidal Linkage

In a split-type trapezoidal linkage, the track rod is split into a Pitman arm, drag link, and center link, which are connected by ball joints, as illustrated in Figure 12.4 [12.3]. When the vehicle is traveling in a straight line, both Pitman arms and steering arms on the left and right sides must be symmetric with respect to the center link and connected to the drag links at the right angle. In addition, by arranging the drag link in line with the control arm axis of the suspension guiding mechanism, the wheel shimmy induced by bouncing wheels can be minimized. This type of mechanism can isolate the vertical vibration and prevent it from affecting the wheel on the other side. Nevertheless, the complexity in such a linkage with multiple rods and joints makes it difficult to adjust the toe and expensive to manufacture.

FIGURE 12.4 A sketch of a split-type trapezoidal linkage.

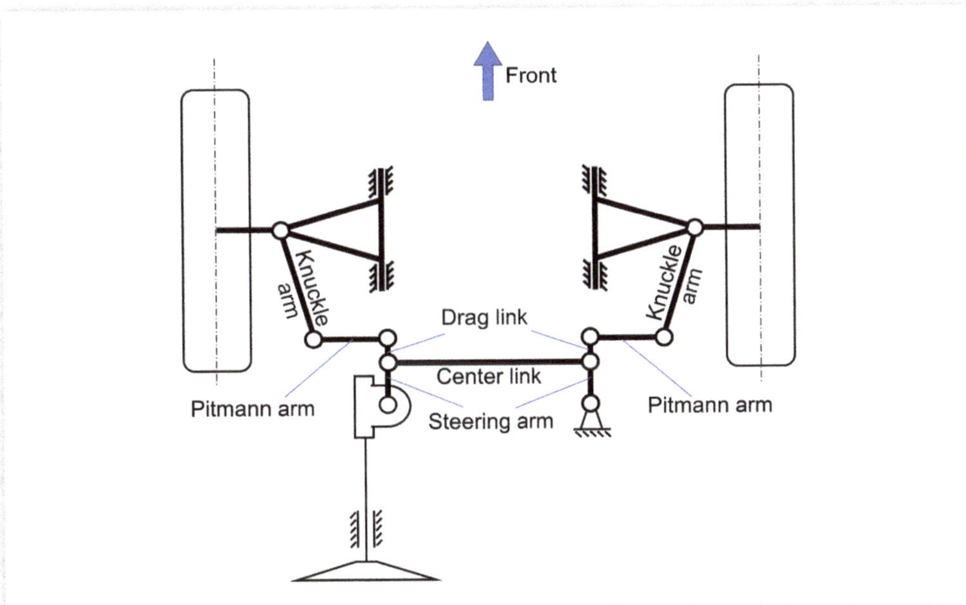

Courtesy of Wangyu Wang.

The split point of the track rod should be positioned according to the configuration of independent suspension. The split point can be determined by using Kennedy's theorem. Refer to mechanical engineering principles for details. A graphic method is often used, with the steps elaborated in the following for an example of MacPherson strut front suspension, as shown in Figure 12.5 [12.7].

FIGURE 12.5 The graphic method for determining the split point of the track rod.

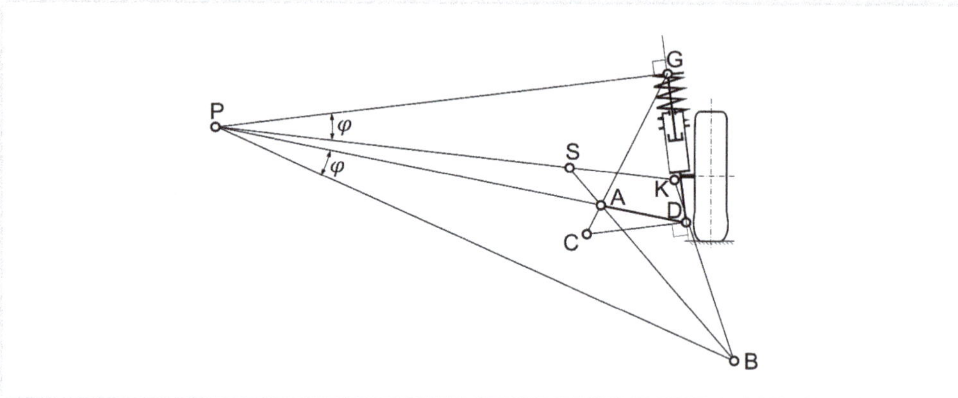

Courtesy of Butterworth-Heinemann.

1. On the plane perpendicular to the control arm axis, draw a line perpendicular to the centerline of the shock absorber cylinder from the strut top-mounting point G and an extension line along the control arm DA (A is the end point of the control arm), and then find the point of their intersection P.
2. Draw a line perpendicular to the centerline of the shock absorber cylinder from point D and a line GA with its extension, and then find the point of their intersection C.
3. Draw a line $\overline{KP}$ from the knuckle joint, that is, point K, and find the angle $\overline{KPG}$, which is equal to φ.
4. Draw a line from the point P with an angle φ below the line $\overline{DP}$ and find its intersection B with the extension of line KD, that is, angle $\overline{DPG} = \varphi$.
5. Draw a line $\overline{BA}$ and its extension, and then find its point of intersection S with line KP.
6. Point S is the theoretical position of the split point projected on the plane perpendicular to the control arm axis.

Note the split point obtained by the graphic method does not take the angle of steerable wheels into account. In fact, the position of point S may vary with the angle of steerable wheels, which can be examined with the same approach. If the variation of point S is very minimal for all extreme steerable wheel angles, then the bouncing steerable wheels will not affect the steering, regardless of operating in a straight or curved road.

12.2. Design of Steering Gears

The steering gear is a device used for converting the rotary motion of the steering wheel into the straight-line motion of a trapezoidal mechanism, as discussed previously. There are several different types of steering gears in terms of their structure, including rack and pinion, recirculating ball, and worm and sector.

12.2.1. Motion Transmission in a Steering System

The motion transmission in a steering system is generally characterized by the steering ratio, the forward and backward efficiencies, and the steering effort.

12.2.1.1. Steering Ratio

The steering ratio represents the sensitivity of steerable wheels in response to the turning of the steering wheel. Denoted by i_s, this ratio refers to the angular velocity ratio of the steering wheel with respect to the Pitman arm on the same side, namely,

$$i_s = \frac{\omega_s}{\omega_p} = \frac{\mathrm{d}\varphi_s}{\mathrm{d}\varphi_p} = i_{sg} i_{sp} \tag{12.7}$$

where

ω and $\mathrm{d}\varphi$ are the angular velocity and incremental angle, respectively
subscripts s and p stand for the steering wheel and Pitman arm, respectively
i_{sg} is the angular velocity ratio of the steering wheel to the steering arm
i_{sp} is the angular velocity ratio of the steering arm to the Pitman arm

The smaller the steering ratio i_s, the more responsive the steering system. Note that Equation (12.7) is not applicable to systems with rack and pinion steering gears.

12.2.1.2. Forward and Backward Efficiencies

The forward efficiency of a steering system, denoted by η_+, is defined as the percentage of the ratio of the net power output from the Pitman arm with respect to the power input from steering shaft, that is,

$$\eta_+ = \frac{100\left(P_s - P_f\right)}{P_s}\% \tag{12.8}$$

where

P_s is the power input from the sheering shaft
P_f is the power loss in the steering system

On the other hand, the backward efficiency, η_-, is defined as

$$\eta_- = \frac{100\left(P_p - P_{fb}\right)}{P_p}\% \tag{12.9}$$

where P_p is the power output from the Pitman arm and P_{fb} is the power loss in the steering system when the power is transmitted in the backward direction, typically $P_{fb} \neq P_f$.

In general, a high forward efficiency is desired in order to make it easy for the driver to turn the steering wheel. In contrast, the steering system with an excessively high backward efficiency would aggravate the feedback of road unevenness to the hands and cause discomfort and fatigue of the driver. Nevertheless, an excessively low backward efficiency would result in difficulties in aligning the steering wheel to the on-center position by itself after turning. Hence, a proper level of backward efficiency is crucial to make a balance between these situations.

Among different types of steering gears, both rack and pinion gear and recirculating ball gear have high forward and backward efficiencies, while the forward and backward efficiencies of worm and sector gear are both very low.

The use of needle bearings, instead of journal bearings, in the steering gears helps improve the forward efficiency considerably.

12.2.1.3. Steering Effort

Steering effort refers to the effort in turning the steering wheel to overcome the resistance between the tires and the road in cornering, which can be expressed by the ratio of the combined forces acting on both steerable wheels from the centers of tire–road contact patches to the force needed to turn the steering wheel from the driver:

$$i_f = \frac{D_s T_k}{r_s T_s} \tag{12.10}$$

where

i_f is the force ratio in the steering system
D_s is the diameter of the steering wheel
r_s is the kingpin offset
T_k is the torque on the steering knuckle
T_s is the torque effort on the steering wheel from the driver

The larger the ratio i_f, the lower the steering effort. Note that the steering wheel diameter must be selected according to applicable standards or regulations. The kingpin offset r_s is typically within 0.4–0.6 tread width of the tire for passenger cars and 4–6 cm for trucks.

By applying the law of conservation of energy, the following equation can be derived from Equation (12.7):

$$i_s = \frac{2T_k}{T_s \eta_+} \tag{12.11}$$

The torque on the steering knuckle, T_k, required for turning the steerable wheels is mainly dependent on factors such as the axle load on steering wheels, resistance from the road surface, inflation pressure of tires, and the friction resistance in the steering system, which can be estimated by the empirical equation:

$$T_k = \frac{f}{3}\sqrt{\frac{F_{z1}^3}{p_t}} \tag{12.12}$$

where

f is the coefficient of sliding friction between the tire and the road, generally $f = 0.7$
F_{z1} is the axle load on the steering wheels
p_t is the inflation pressure of tires

Substitution of Equation (12.11) into Equation (12.10) yields

$$i_f = \frac{D_s \eta_+}{2r_s} i_s \tag{12.13}$$

Equation (12.13) implies that the steering effort increases with the drop in the steering ratio. In other words, the steering wheel becomes increasingly difficult to turn as the steering sensitivity increases.

Thus, the tangential force from the hand of the driver acting on the rim of the steering wheel can be obtained by

$$F_s = \frac{2L_{sa}T_k}{L_{pa}D_s i_s \eta_+} \tag{12.14}$$

where

L_{sa} is the length of the steering arm
L_{pa} is the length of the Pitman arm

Note the force given by Equation (12.14) is the maximum force required to make a turn when the vehicle is stationary. Due to the physical limit of a human driver, the force F_s should normally not exceed 700 N. If not, the steering ratio i_s must be modified to satisfy this condition.

To resolve the contradiction between the steering sensitivity and steering effort, steering gears of a variable ratio can be utilized. For a vehicle without power steering, a large steering ratio is usually preferred for operation at low speeds to reduce the steering effort and to avoid overloading the steering shaft, while a small steering ratio is beneficial to acquire a responsive and robust steering feeling at high speeds. On the contrary, for a vehicle equipped with a power steering system, a small steering ratio is desired to enhance the maneuverability by reducing the maximum steering angle. Note that the steering ratio corresponding to the on-center position of the steering wheel is generally kept above 15–16 in order to prevent the steering from overly sensitive in controlling the direction of vehicle motion.

12.2.2. Rack and Pinion Steering Gears

The fundamental principle of rack and pinion steering gears is illustrated in Figure 12.6. As the driver turns the steering wheel, the pinion that is typically integrated to the steering shaft turns accordingly, which moves the rack, that is, a linear gear that meshes with the pinion, converting the circular motion into the transverse motion of the tie rod. Rack and pinion gears are widely used for their simplicity, compactness, low cost, and a high transmission efficiency of up to 90%. There is no need for a steering arm and drag link, as shown in Figure 12.1, making it possible to increase the maximum angle of steerable wheels. The housing is often made of lightweight materials such as aluminum or magnesium alloys. A spring on the back of the rack near the pinion can produce an adjustable pressing force on the pinion to eliminate the backlash between their teeth produced by teeth wear. However, due to its high efficiency in reverse motion transmission, rack and pinion gears feature a high degree of feedback, causing the road impacts to be transmitted back to the steering wheel and, thus, affecting the comfort and safety in the operation of drivers.

FIGURE 12.6 Fundamental principle of rack and pinion steering gears.

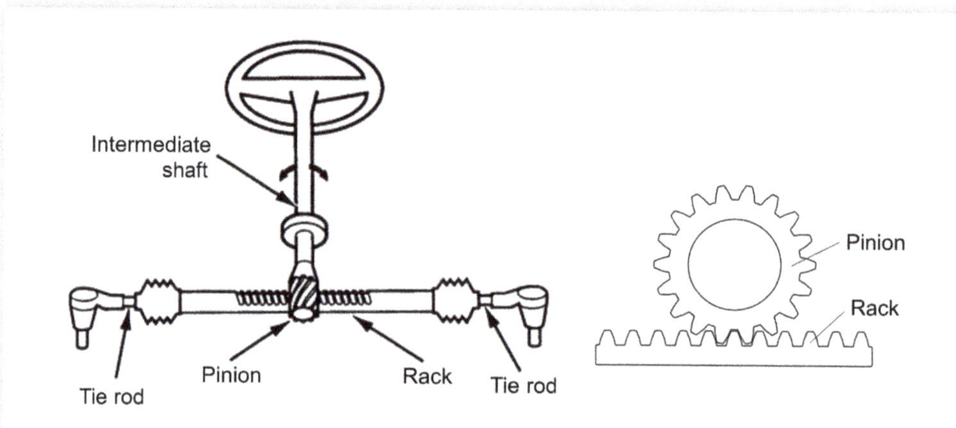

Besides, helical rack and pinion gears are often used instead of spur rack and pinion gears to reduce the noise and vibrations for their higher contact ratio and advantage in packaging as a result of the angle between rack and pinion axes. Usually with 5–7 teeth, most of the pinions have a module of 2–3 mm, a pressure angle of 20°, and a helix angle between 9 and 15°. The number of teeth for the rack is determined by the stroke corresponding to the maximum steering angle of the steerable wheels. The steering rack and pinion are generally made of aluminum alloys or steel, while the housing is often made of aluminum alloys.

In terms of steering torque input from the pinion relative to the force output at the rack, there are four arrangements mostly used in practice, as illustrated in Figure 12.7 [12.7]. When the force output from both ends of the rack, regardless of the input from the center (Figure 12.7a) or side (Figure 12.7b), the length of tie rods will be limited, causing difficulties in avoidance of motion interference with a suspension guiding mechanism. The side input, single-side output configuration (Figure 12.7c) is often used in minitrucks for its simplicity and low cost. The side input, center output arrangement (Figure 12.7d) would be advantageous as the increased length of its tie rods helps reduce their swing angle with bouncing wheels.

FIGURE 12.7 Four arrangements of rack and pinion steering gears. (a) Central input, double-sided outputs, (b) side input, double-sided outputs, (c) side input, single-sided output, and (d) side input, central outputs.

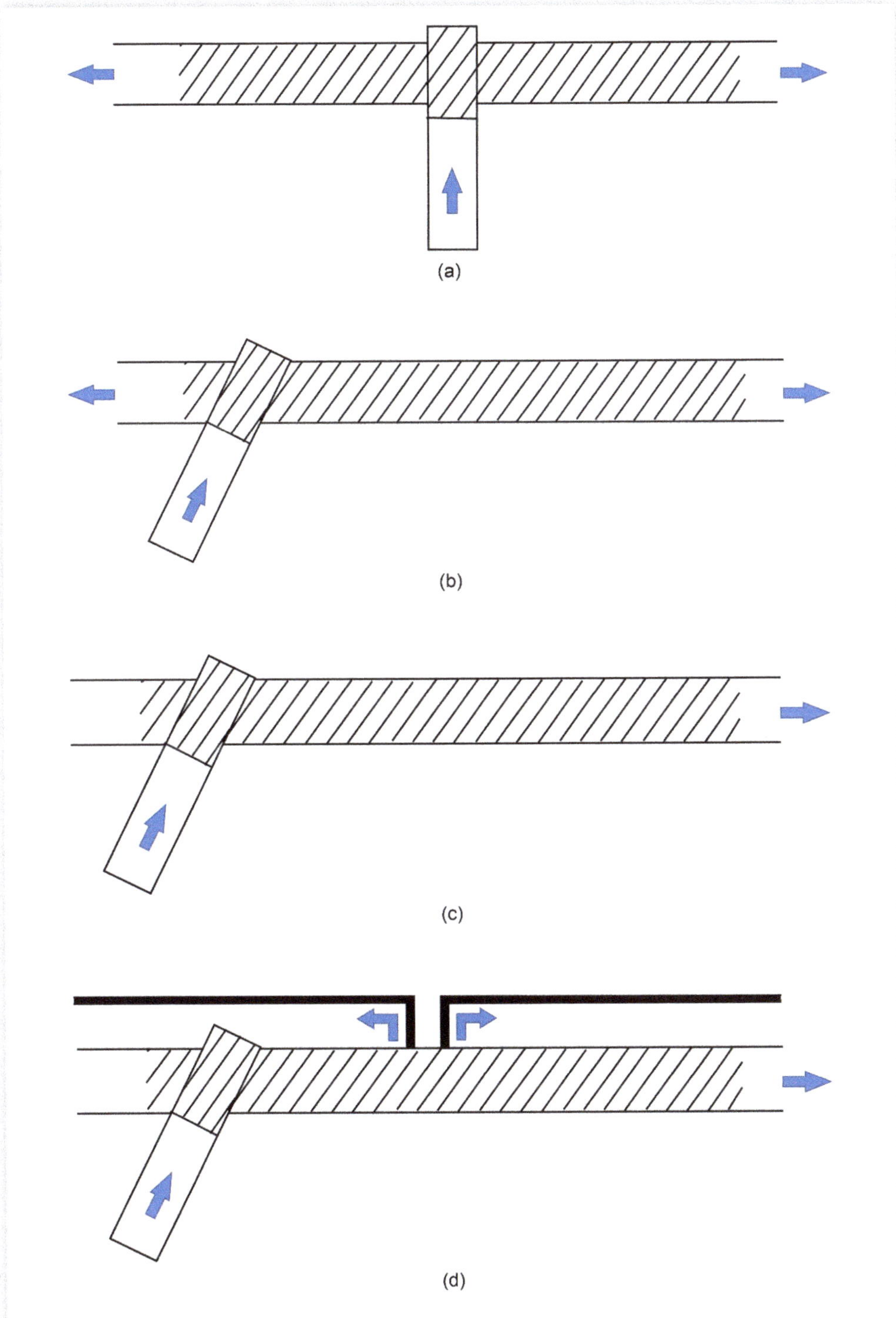

Courtesy of Butterworth-Heinemann.

The rack can be made into round, V-shape, or Y-shape cross sections. Although round racks are easy to fabricate, the V-shape or Y-shape racks can not only save the material and reduce the weight but also prevent the racks from rotating about their axes.

Rack and pinion steering gears can be positioned either in the front or in the rear of the front axle, on the same side with or on the opposite side to the trapezoidal linkage, as shown in Figure 12.8.

FIGURE 12.8 Positions of rack and pinion steering gears with respect to the front axle. (a) In the rear of front axle and on the same side with trapezoidal linkage, (b) in the front of front axle and on the opposite side to trapezoidal linkage, (c) in the front of front axle and on the same side with trapezoidal linkage, and (d) in the rear of front axle and on the opposite side to trapezoidal linkage.

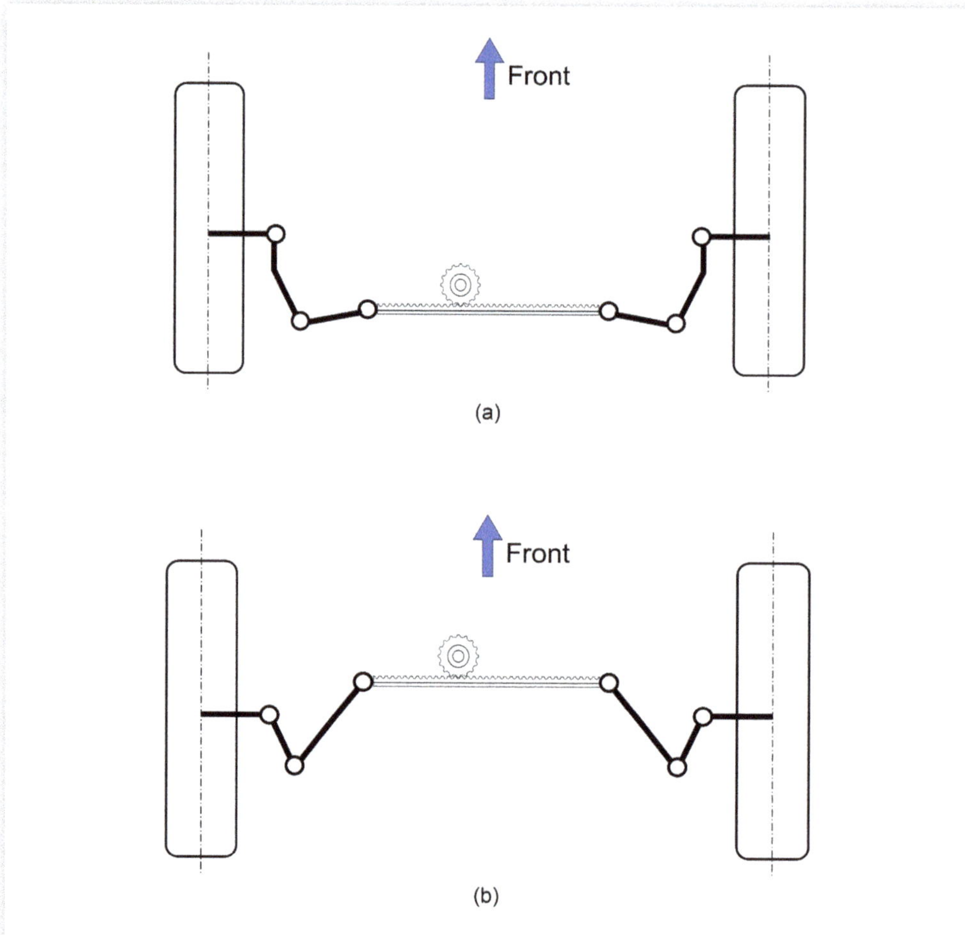

FIGURE 12.8 **(Continued)** Positions of rack and pinion steering gears with respect to the front axle. (a) In the rear of front axle and on the same side with trapezoidal linkage, (b) in the front of front axle and on the opposite side to trapezoidal linkage, (c) in the front of front axle and on the same side with trapezoidal linkage, and (d) in the rear of front axle and on the opposite side to trapezoidal linkage.

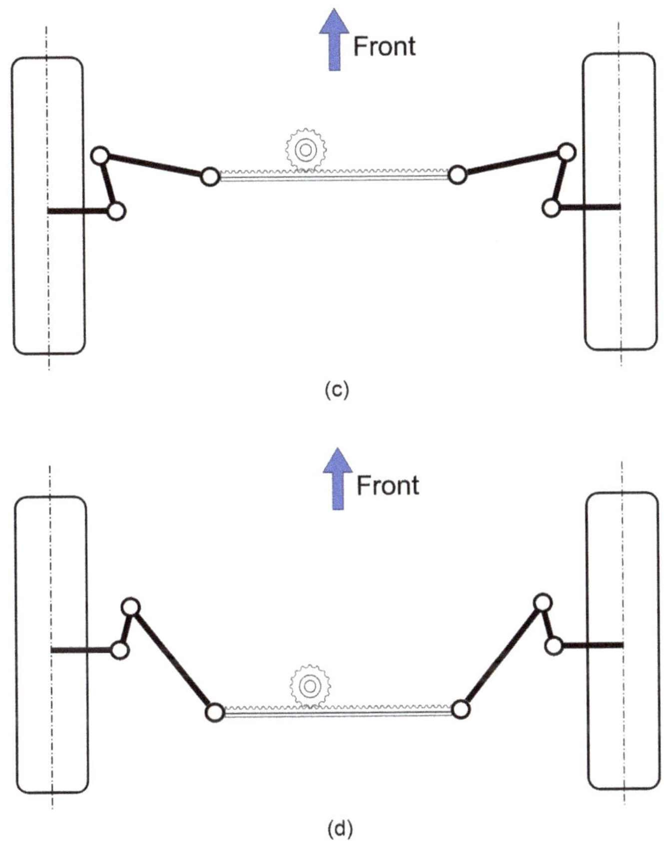

In order to reduce the steering effort without worsening the sensitivity in response to the steering action of the driver, the ratio of the rack and the pinion can be made variable. For the rack and pinion to mesh each other properly, the pitch of the rack must be identical to the pitch of the pinion at all the contact points, that is,

$$p_r = p_p \tag{12.15}$$

where the pitch of the rack $p_r = \pi m_r \cos \alpha_r$ and the pitch of the pinion $p_p = \pi m_p \cos \alpha_p$. Thus, the condition of Equation (12.15) becomes

$$m_r \cos\alpha_r = m_p \cos\alpha_p \tag{12.16}$$

where

- m_r is the module of the rack
- α_r is the pressure angle of the rack
- m_p is the module of the pinion
- α_p is the pressure angle of the pinion

It is obvious from Equation (12.16) that a standard pinion of a module m_p and pressure angle α_p can mesh with a rack with a variable module m_r and pressure angle α_r as long as the condition in Equation (12.16) is satisfied. By making both pressure angle and module the largest in the middle of the rack and decrease toward both ends, the meshing radius on the pinion decreases accordingly. The pressure angle of a rack typically ranges from 12 to 35°. Hence, the distance that the rack moves decreases as the meshing comes from the middle to the end of the rack for a given angular displacement of the steering wheel or pinion.

The backlash between motion transmission pairs, for example, the rack and pinion, in the steering gears has a significant influence on the stability in straight-line operations of a vehicle because such a backlash can cause the steerable wheels to deviate from the original direction under the action of a random lateral force from an uneven road. Therefore, it is generally desired that the backlash within the steering gears be minimized when the steering wheel operates around its on-center position. In practice, however, due to more frequent usage, the gear pairs corresponding to the on-center position of the steering wheel wear more severely, resulting in a larger backlash. By designing a proper adjustment mechanism, the backlash for the steering wheel operating around its on-center position can be kept minimal, while the backlash at its off-center position can be reduced.

As a safety-critical component, the design of the steering rack and pinion must be carefully checked to ensure an adequate margin in its bending and contact strength. A safety factor of 2.5–3.5 is generally used in the design of steering components.

12.3. Design of Power Steering Systems

Although traditional power steering systems come in the form of hydraulic power steering and then electrohydraulic power steering as a more recent upgrade, most of the EVs today use electric power steering (EPS) for its high efficiency, structural simplicity, compactness, and relatively low manufacturing cost.

In an EPS system based on rack and pinion steering gears, the steering is assisted by an electric motor mounted to the steering column, or directly to the steering gear, or even to the steering rack. Sensors are used to detect the direction and angular position of the steering wheel, and then the direction and magnitude of the assisting force is determined by the EPS controlling program. Consequently, the electric motor of EPS applies the assisting force, which generally varies with the steering angle.

The requirements for an EPS system usually include the following: First, a certain kinematic proportion must be maintained between the angle of steerable wheels and the steering angle; second, a proper road feeling is needed, meaning that the force acting on the hands of the driver increases or decreases as the road force on steerable wheels increases or decreases; third, power steering should set in when the tangential hand force acting on the steering wheel reaches 25–190 N; fourth, the steering wheel should be able to align itself back to the on-center position after turning; fifth, fast response to the turning action of the driver; and finally, a mechanical backup steering can take over if EPS fails to operate.

12.3.1. Types of EPS

Some most common types of EPS used in EVs nowadays are as follows:

1. Column EPS
   In this configuration, the EPS motor is mounted at the top of the steering shaft, with the former driving the latter directly. Located inside the dashboard, the motor is less

likely to be affected by water, dust, and extreme environmental conditions. Hence, the column EPS generally features a simple structure and low cost. Nevertheless, it is infamously not sensitive enough in response to the steering operation of the driver, primarily because of the increased inertia from the strengthened steering shaft and universal joints for mounting the motor, as well as the additional friction brought by the motor for its connection to the steering shaft. As a result, the column EPS is typically used only in mini EVs.

2. Single-pinion EPS
   The motor applies the assisting force to the pinion gear directly, thus giving a better steering feel than the column EPS. However, as the motor is mounted in front of the pedals, it could be pushed toward the legs of the driver and cause injuries in a front-end collision. Dual-pinion EPS can be a good alternative to avoid this drawback.
3. Dual-pinion EPS
   The electric motor drives the second pinion, which is added to the steering rack, and the motor is located away from the steering column. Thus, such a configuration can substantially lower the risk of leg injuries caused by front-end collision and enhance the steering feel due to the separation of the motor from the steering column. However, it is not suitable for applications in heavy-duty vehicles due to its limited power.
4. Parallel axis EPS
   Mounted in parallel to the steering rack, the electric motor drives the rack via a cogged belt and a reduction gearbox. The gearbox and the cogged belt allow the assisting torque from the electric motor to be delivered to the steering rack more powerfully and precisely. However, it is generally more costly than other aforementioned types due to the increase in system complexity.

12.3.2. Design of Boost Characteristics

Based on the vehicle speed and the status of the steering wheel, the EPS control unit selects the appropriate one from the following three operating modes:

1. Boost mode: This is the most fundamental mode in which the EPS operates by default. The electric motor may need to provide an assisting torque during the turning of the steering wheel to reduce the steering effort based on the boost characteristics.
2. Aligning mode: Upon completion of the turn, the motor should assist the steering system to return to the on-center, straight forward position.
3. Damping mode: When the vehicle operates at high speeds, a proper damping torque is needed from the motor to stabilize the steering system from being affected by disturbance from the road.

The boost characteristics determine the amount of assisting torque to be exerted on the steering mechanism. In general, the requirements for boost characteristics are as follows:

1. With the increase in the vehicle speed, the required assisting torque decreases and the speed interval corresponding to zero assisting torque extends accordingly.
2. The magnitude of assisting torque varies with the input torque from the steering wheel, with little assisting torque corresponding to low input torque.
3. A left–right symmetry in the boost characteristic curves must be kept above 85%, given the tolerances in component fabrication and assembly.

4. The aligning process following the turn and the consequent steering operating must be kept stable.
5. An upper limit of assisting torque is needed to protect the motor from overloading and the consequent system failure.
6. The transition between characteristic curves in adjacent speed intervals must be smooth to avoid shocks in the system.

Three basic types of boost characteristics are discussed as follows:

1. Linear characteristics
   The assisting torque is a linear function of the input torque from the steering wheel, and the slope varies with the vehicle speed. Such a characteristic can be expressed as [12.1-12.3]

$$T_a = \begin{cases} 0 & 0 \le T_s < T_{s0} \\ k(v) \cdot (T_s - T_{s0}) & T_{s0} \le T_s < T_{smax} \\ T_{amax} & T_s \ge T_{smax} \end{cases} \tag{12.17}$$

   where
   - T_a is the assisting torque
   - T_s is the steering wheel torque
   - T_{s0} is the minimum steering wheel torque above which an assisting torque is needed
   - v is the vehicle speed
   - $k(v)$ is the slope of assisting torque increasing with vehicle speed
   - T_{smax} is the maximum input torque from the steering wheel in the presence of an assisting torque
   - T_{amax} is the upper limit of assisting torque

2. Polyline characteristics
   The assisting torque is a polylinear function of the input torque from the steering wheel. Not only varying with the vehicle speed, the slope also changes abruptly with the input torque across a certain level. Such a characteristics can be expressed as

$$T_a = \begin{cases} 0 & 0 \le T_s < T_{s0} \\ k_0(v) \cdot (T_s - T_{s0}) & T_{s0} \le T_s < T_{s1} \\ k_0(v) \cdot (T_{s1} - T_{s0}) + k_1(v) \cdot (T_s - T_{s1}) & T_{s0} \le T_s < T_{smax} \\ T_{amax} & T_s \ge T_{smax} \end{cases} \tag{12.18}$$

   where
   - $k_0(v)$ and $k_1(v)$ are the different slopes of assisting torque increasing with vehicle speed
   - T_{s1} is the demarcation input torque corresponding to different slopes of assisting torque

3. Parabolic characteristics
   The assisting torque is a parabolic function of the input torque from the steering wheel. The parabolic characteristics can be expressed as

$$T_a = \begin{cases} 0 & 0 \le T_s < T_{s0} \\ k_1(v)\cdot(T_s - T_{s0})^2 + k_0(v)\cdot(T_s - T_{s0}) & T_{s0} \le T_s < T_{smax} \\ T_{amax} & T_s \ge T_{smax} \end{cases} \tag{12.19}$$

Apparently, due to the nature of such a parabolic function, the slope increases with the increase in input torque. Hence, this is the characteristic closest to the ideal one among the three types.

Note that when the input torque is smaller than a certain level, that is, $T_s < T_{s0}$, all the boost characteristics discussed earlier come with a dead zone in which no boost torque is produced by the electric motor.

After a turn, the steering wheel will need to align itself to the on-center position. This can be detected by the angular position sensor or the torque sensor in the system. Obviously, the time rate of the steering wheel angle changing from positive to negative signifies the entry into the aligning process. As discussed in Chapter 5, the presence of a sideslip angle and camber angle produces a self-aligning torque, which enables the steering system to align itself when the driver releases the steering wheel. With the increase in vehicle speed, the lateral adhesion coefficient at the tire–road interface drops and the self-aligning torque rises. However, there are two scenarios that need the EPS to intervene: the steering wheel can align by itself or in a timely manner when the self-aligning torque is inadequate, and an overshoot occurs in the angular position of the steering wheel due to an excessively large self-aligning torque. For the first scenario, the EPS system needs to switch to the aligning mode with a good steering feel; for the second scenario, the system should switch to the damping mode to prevent the overshoot, with the damping force usually growing with increasing vehicle speed.

A brushless DC motor is commonly used in the EPS systems for its high efficiency, good speed control characteristics, high reliability, and maintenance-free operation. Recently, dual three-phase motors have been increasingly applied in the EPS system to further improve its reliability and functional safety, which can make the system operatable even in case of the failure in one of the three-phase windings.

The steering angle and torque sensors can either be contact or contactless type, in the form of stand-alone ones or integrated assembly. Mainly potentiometer-based, the contact type of sensors is subject to the influence of temperature and wear, and prone to noise problems or failures if frequently operated on poor roads. However, they are still often used due to their simplicity and low cost. In contrast, the contactless types of sensors can be inductive, photoelectric, capacitive, or Hall effect-based, which are typically of high accuracy, low hysteresis, reliability, and free from wear and noise. Although a little pricy, they are utilized in almost all of mainstream EPS systems today.

12.4. Design of Other Steering System Components and Features

Other important components in a steering system include ball joints, tie rods, Pitman arms, and steering dampers.

12.4.1. Ball Joints

The failure mode of ball joints is mainly the wear on the ball surface. Thus, the contact stresses should be checked by

$$\sigma = \frac{F}{A} \tag{12.20}$$

where F is the force acting on the ball and A is the projected area of the portion of the ball surface that bears the load on the plane passing the ball center and perpendicular to the force. The allowable contact stress is typically 25–30 MPa.

The initial value of the ball diameter can be selected as given in Table 12.1 [12.3].

TABLE 12.1 Suggested ball diameter based on the load on steerable wheels.

Load on steerable wheels (kN)	0–6	6–9	9–12.5	12.5–16	16–24	24–34	34–49	49–70	70–100
Ball diameter (mm)	20	22	25	27	30	35	40	45	50

Courtesy of Wangyu Wang.

Ball joints are often made of alloy structural steel or liquid carbonitriding steel.

12.4.2. Tie Rods

Lightweight and adequately stiff tie rods are generally desired. In order to fit the packaging requirements, it is sometimes necessary to make them into a curved shape, causing a reduction in their stiffness. The calculation formula for the stability of compression rods in the mechanics of materials should be used to check the design. A safety factor of stability no less than 1.5–2.5 is normally used.

Traditionally made of seamless low-carbon steel pipes, tie rods made of aluminum alloys are used increasingly for weight reduction purposes in EVs.

12.4.3. Pitman Arms

The Pitman arm is subject to the combined action of bending and torsional moments from the force acting on the ball joint. The unified strength theory should be used to check its stresses at the critical section at the root of the arm:

$$\sigma = F\sqrt{\frac{d^2}{I_b^2} + \frac{4e^2}{I_t^2}} \tag{12.21}$$

where

I_b is the section modulus in bending

I_t is the section modulus in torsion, that is, Polar modulus

d and e are dimensions as shown in Figure 12.9 [12.3]

FIGURE 12.9 Dimensions for Pitman arm design.

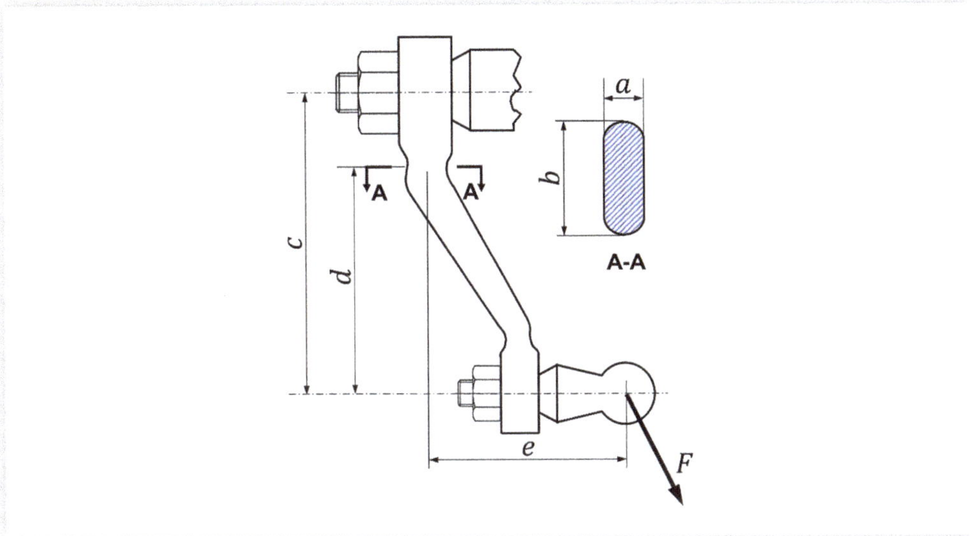

Courtesy of Wangyu Wang.

It is required that $\sigma \leq \sigma_y/n_s$, where σ_y is the yield strength of the material and n_s is the safety factor, typically $n_s = 1.7 - 2.4$.

Since the Pitman arm splines to its shaft, the extrusion stress and tangential stress of the spline must be checked for durability.

12.4.4. Steering Dampers

Steering dampers, or steering stabilizers, are used in most of the EVs to attenuate vibrations from shimmy steerable wheels or the impact from uneven roads.

The steering damper makes use of the viscosity of liquid in a cylindrical device to dampen out the vibrations and impacts. With a structure similar to the suspension damper illustrated in Figure 11.25, the steering damper is often mounted in an almost horizontal position, with one end bolted to the center link of the steering system and the other end bolted to the vehicle frame.

The damping characteristics of the steering damper in its stretch stroke is generally symmetric to that in compression stroke [12.1-12.3], in order to reduce the swing motion of steerable wheels to the left or right.

12.4.5. Other Safety-Related Components

Safety-related components in a steering system include the steering column and shaft, as well as airbags.

The design of the steering column and shaft must comply with the applicable standards or regulations. Countermeasures must be taken to prevent the driver from injuries in front collisions when the assembly of the steering column and shaft is pushed against the driver. It is generally required that the rearward motion of the assembly should not exceed 127 mm (5 in.) in a 48 km/h front crash. In addition, the force on the steering wheel must not exceed 11,121 N (2,500 lbf) if the mannequin torso hits the steering wheel at a speed of 6.7 m/s (15 mph).

A collapsible steering column is used in most of the vehicles to absorb kinetic energy, usually in a combined form of plastic deformation, elastic deformation, and friction of the components. A basic design of collapsible steering columns makes use of an inner sleeve and an outer sleeve to the shaft, with a number of steel bearings pressed in between each sleeve. The bearings are bonded to the sleeves with a certain resin. If a specific level of pressure is reached in a crash, the resin shatters, allowing the sleeves to compress telescopically. Alternatively, the steel bearings can be made into a mesh design that folds in a concertina fashion upon receipt of an impact.

Moreover, the introduction of universal joints to the assembly of the steering column and the shaft helps deflect the impact force toward the driver during a front-end collision. The universal joint connects shafts at an angle, thus mitigating the driver's contact with the steering assembly.

A front crumple zone must be considered to absorb the kinetic energy of a front-end crash. A rule of thumb is to mount the steering gears behind the front axle on the front of bulkhead, at the rear of the front crumple zone.

Front airbags are generally fitted as standard in most of the EVs, with rear and side airbags remaining as an option, except for luxury models, which can protect the driver's face from hitting the steering wheel in a front-end collision.

12.5. Steer-by-Wire

Steering-by-wire aims at the removal of all the mechanical components between the steering wheel and steerable wheels, such as steering column, shaft, and gears. It comes with the following advantages: the interior and front compartment design can be simplified, leaving extra space for flexible utilization; the weight of the steering system can be reduced, making the vehicle lighter; the steering ratio can be made fully variable in a wide range because its adjustment is no longer limited by the gear reduction mechanism; no physical feedback from the road to the driver makes it possible to eliminate the rubber bushings in the system, which enhances the steering response and precision; the risk of driver injuries due to the backward motion of the steering mechanism in front-end collisions can be minimized; the mechanical constraints on the installation of a steering mechanism are loosened, making it easy to modularize and serialize the steering mechanisms; the steering system characteristics, for example, response sensitivity and steering feel, can be adjusted to fit the needs of various driving conditions; and it allows an easy conversion into a further simplified steering system for autonomous driving applications.

A steer-by-wire system primarily consists of a steering wheel module, a road wheel module, and an ECU, as illustrated in Figure 12.10 [12.8].

The steering wheel module contains the steering wheel assembly, a steering angle sensor, and a feedback motor. The angular motion of the steering wheel is detected by the steering angle sensor, and the signal is processed by the ECU to identify the intention of the driver. Then, the ECU sends a control signal to the feedback motor that produces an aligning torque to the steering wheel that emulates the road feel.

The road wheel module is composed of a road wheel steering assembly, a pinion angle sensor, and a steering actuator. The pinion angle sensor detects the angular position of the steerable wheels and feedback the signal to the ECU. The steering actuator receives the command from ECU and adjusts the angle of steerable wheels to realize the steering intention of the driver.

FIGURE 12.10 Sketch of a steer-by-wire system.

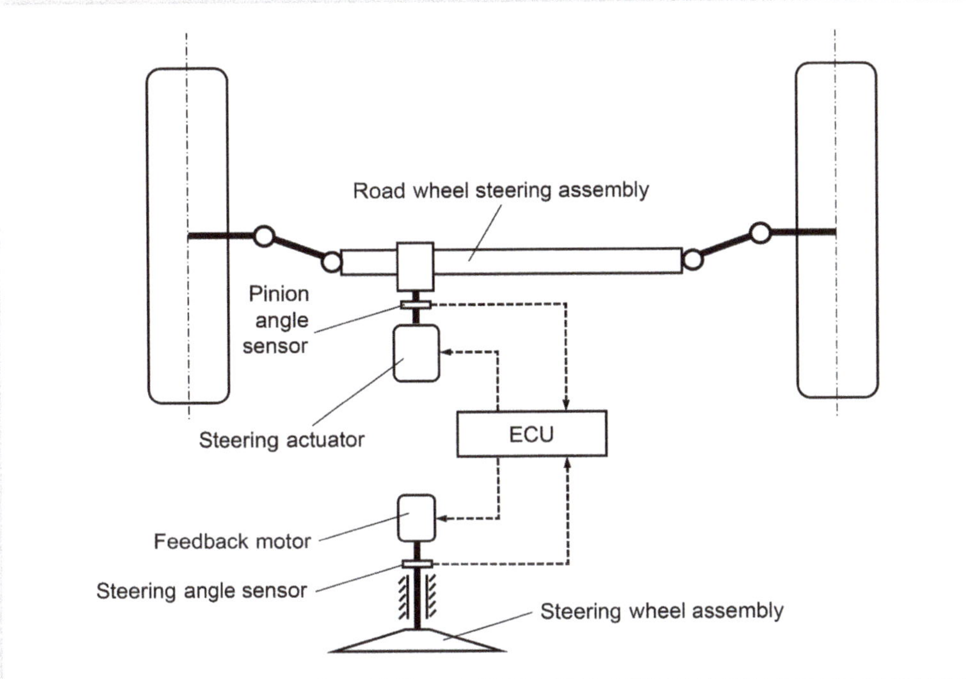

Courtesy of Applied Sciences.

The ECU processes the signals from the sensors and identifies the condition of vehicle motion. It also sends commands to the feedback motor and steering actuator to coordinate their operation. If any instability in the vehicle operation or any misoperation of the driver is identified, the ECU will execute stability control or replace with a correct maneuver. The ECU normally features a fail-safe design that allows the steer-by-wire system to remain operatable in a safe manner. Some standards or regulations require a mechanical backup system to function in case of electronic hardware or software failures. A high level of redundancy is generally deemed necessary to make it fault-proof to avoid accidents because of the safety-critical nature of the steer-by-wire system. For example, three identical ECUs can be used in parallel to compare their results in order to improve the system reliability, at a price of increased system complexity and cost though.

Due to safety, reliability, and cost concerns, the steer-by-wire system is barely used in mass produced EVs yet. In addition, the lack of a proper road feel is another challenge to tackle since the feedback motor is not up to the job of mimicking all the real-world road disturbances.

12.6. All-Wheel Steering

All-wheel steering mainly aims at reducing the turning radius and enhancing the stability and handling in cornering. It typically features an added spindle drive that turns the rear wheels to offer added maneuverability. One of the key problems here is how to convey the motion of steering wheels to rear wheels. Traditionally, a mechanical connection between the steering wheel and the front and rear wheels is used. However, an electric connection has become a more attractive alternative, thanks to its simple structure, lightweight, low cost, and packaging flexibility. Besides, the mechanical, hydraulic, and electrohydraulic spindle drives have largely been replaced by electric ones for the convenience and precision in control design and execution.

12.6.1. Basic Principles of All-Wheel Steering

In an all-wheel steering system, the rear wheels must be turned to an angle that varies with the steering angle and the operating condition of the vehicle.

12.6.1.1. Opposite-Direction Steer

This is sometimes called counterphase steer. When a steering angle is large or vehicle speed is low, the rear wheels are turned in the opposite direction to the front wheels. In the meanwhile, the rear wheel angle increases with the steering angle in a certain range, with the maximum rear wheel angle usually around 5° from the straight-ahead position. As a result, the maneuverability and agility of the vehicle can be enhanced, and the turning radius can be reduced. The inner wheel difference, that is, the difference in the turning radius of the front and rear wheels, decreases accordingly, as illustrated in Figure 12.11. However, the rear wheels tend to bear against the side of the curb, which could become a serious problem at very large steering angles.

FIGURE 12.11 Effect of opposite-direction steering on a four-wheel steering vehicle comparing to a two-wheel steering vehicle.

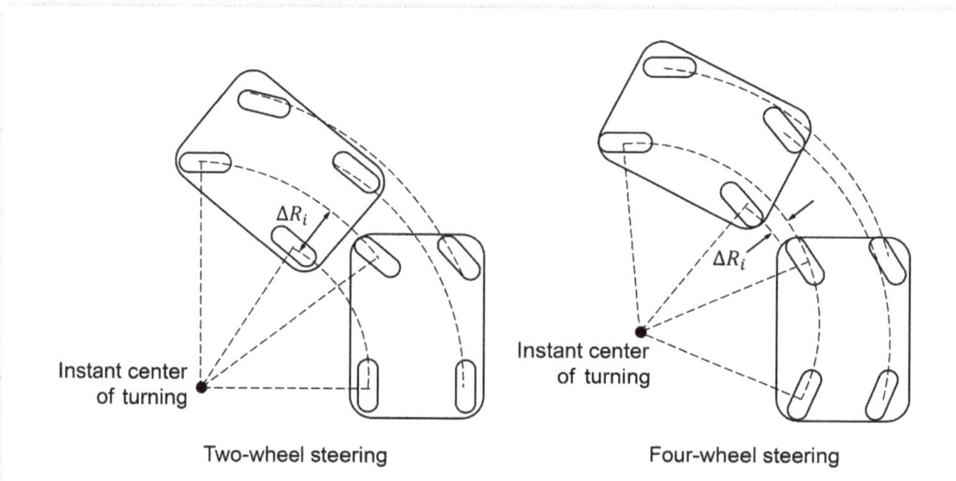

12.6.1.2. Same-Direction Steer

This is sometimes called in-phase steer. When the steering angle is small or vehicle speed is high, the rear wheels are turned in the same direction as the front wheels, with the maximum rear wheel angle around 1° from the straight-ahead position. Thus, the steering response and stability can be improved, and the yaw rate and tendency to roll over can be reduced dramatically due to the introduction of understeer characteristics to the vehicle. The turning of a vehicle is actually a combination of two motions, namely, the revolution of the centroid of mass around the instant center of turning that changes the forward direction and the rotational motion around the centroid of mass. In the case of two-wheel steering, the turning of front wheels results in sideslip angles at the front wheels and the centripetal force, causing the vehicle body to spin. When the vehicle body deviates from the initial direction, sideslip angles occur at rear wheels also, producing the centripetal force, as shown in Figure 12.12a. Nevertheless, with increasing vehicle speed, the centrifugal force increases and so does the sideslip angles at front wheels. Consequently, the rotational motion of the vehicle becomes stronger, leading to instability such as excessive spin or sideway skid. It is desired that the motion of the vehicle body follows the same direction of vehicle motion without excessive rotation. By using four-wheel steering, sideslip angles at rear wheels are produced, which balance the centripetal force at the front wheels. As a result, the rotational motion of the vehicle body is suppressed and made in line with the direction of vehicle motion with improved stability, as shown in Figure 12.12b.

FIGURE 12.12 Effect of same-direction steering on a four-wheel steering vehicle comparing to a two-wheel steering vehicle. (a) Two-wheel steering and (b) four-wheel steering.

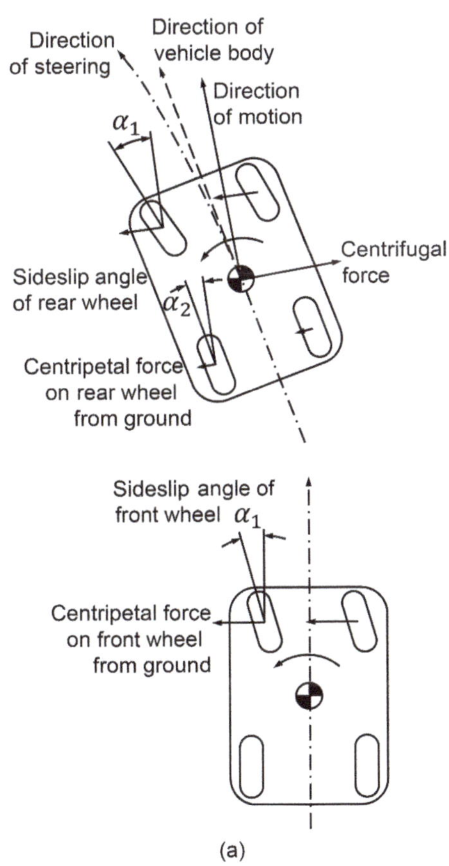

FIGURE 12.12 **(Continued)** Effect of same-direction steering on a four-wheel steering vehicle comparing to a two-wheel steering vehicle. (a) Two-wheel steering and (b) four-wheel steering.

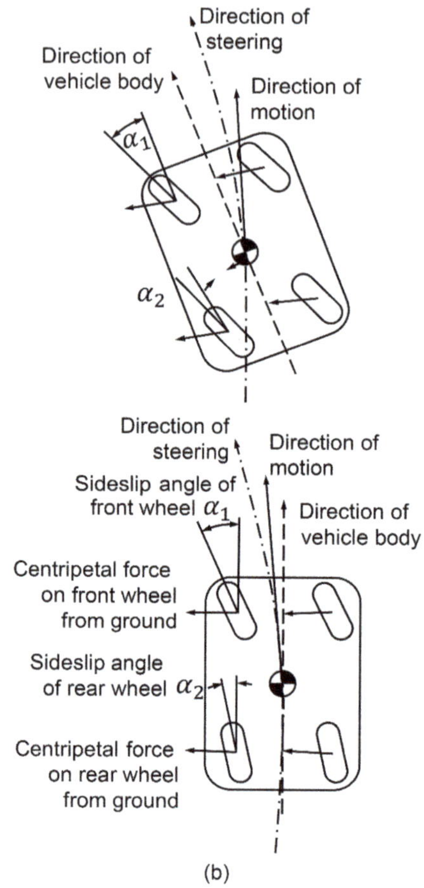

12.6.2. Mechanism of All-Wheel Steering

As illustrated in Figure 12.13, a typical all-wheel steering system consists of the following components.

1. A steering angle sensor: It functions to detect the steering angle that is needed to identify the steering intention of the driver.
2. Front and rear wheel angle sensors: Mostly Hall effect-based, they are used to detect the angle of road wheels with respect to the straight-ahead position.
3. A vehicle speed sensor: Typically, photoelectric-based, the sensor detects the vehicle speed in the form of the pulsed signal.
4. A yaw rate sensor: Often mounted on the vehicle body near the centroid of the vehicle, the yaw rate sensor is typically based on piezo jet principles.

5. An ECU: It processes the signals from these sensors and outputs the commands to the front and rear e-motors to actuate the front and rear steering assemblies. It is mainly composed of hardware such as an input signal conditioning circuit, a microprocessor, an output signal processing circuit, a power supply circuit, and software such as control program and software platform. It must also have effective means of anti-interference and fault self-diagnosis to warrant the reliability of the control system.
6. E-motors: The front e-motor outputs a proper torque to assist the steering, while the rear e-motor, typically a step motor, outputs a proper angle based on the commands from ECU. A step motor is capable of producing a precise angular displacement according to the digital pulsed signal with a fast response in start–stop, forward or reverse, and speed change. By working with an angle feedback mechanism, a simple and robust closed-loop control system can be formed with self-locking and torque-holding functions and strong anti-interference ability.

FIGURE 12.13 Sketch of a typical all-wheel steering system.

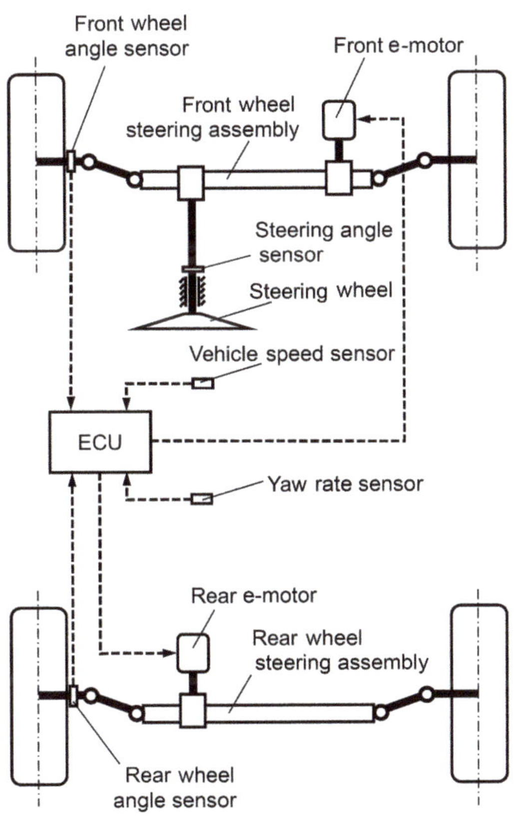

References

[12.1]. Heinz, H., *Advanced Vehicle Technology*, 2nd ed. (Burlington: Elsevier, 2002), doi:10.1016/B978-0-7506-5131-8.X5000-3.

[12.2]. Liu, W., *Automobile Design* (Beijing: Tsinghua Press, 2001), ISBN:9787302045298.

[12.3]. Wang, W., *Automobile Design*, 3rd ed. (Beijing: China Machin Press, 2000), ISBN:7111076133.

[12.4]. Wang, X., *Automotive Chassis Design* (Beijing: Tsinghua University Press, 2010), ISBN:978-7-302-21471-7.

[12.5]. Gillespie, T., *Fundamentals of Vehicle Dynamics* (Warrendale: Society of Automotive Engineers, Inc., 1992), 21-118, ISBN-10:1560911999, ISBN-13:978-1560911999.

[12.6]. Bastow, D., Howard, G., and Whitehead, J., *Car Suspension and Handling*, 4th ed. (Warrendale: Society of Automotive Engineers, Inc., 2004), ISBN:0-7680-0872-7.

[12.7]. Reimpell, J., Stoll, H., and Betzler, J., *The Automotive Chassis: Engineering Principles*, 2nd ed. (Oxford, UK: Butterworth-Heinemann, 2001), ISBN:0750650540.

[12.8]. Iqbal, J., Zuhaib, K., Han, C., Khan, A. et al., "Adaptive Global Fast Sliding Mode Control for Steer-by-Wire System Road Vehicles," *Applied Sciences* 7, no. 7: 738, doi:10.3390/app7070738.

13

Braking System Design

The braking system in an EV generally offers four functions: reducing the speed of the vehicle until it stops with a proper deceleration, maintaining a proper stable speed when driving downhill, recovering as much as possible kinetic energy under the premise of safety, and parking the vehicle reliably in situ, even on the ramp of a certain slope.

Thus, there are two types of brakes in an EV, namely, service brakes and parking brakes. In fact, the former works with the electric powertrain to provide the first three functions, while the latter is used for the fourth function. Nevertheless, some vehicles are also equipped with emergency brakes and auxiliary brakes to ensure safety in case of any service brake failure or overload during a continuous usage in a long downhill road.

The requirements for the braking system are generally as follows:

1. Adequate braking capacity, which is typically evaluated by the deceleration and braking distance for a given initial speed and the maximum slope for a vehicle to park safely and subjected to applicable standards and regulations.
2. High reliability, which is typically warranted by a redundant brake pipeline capable of providing at least 30% of the regular braking capacity in case of any failure in the on-duty pipeline.
3. Large brake factor, with minimized influence from heat and wet.
4. No loss in handling or directional stability when the brake pedal is depressed at any speed.

5. Quick response to the depression of the brake pedal.
6. Water or mud is prevented from entering the friction surfaces as much as possible, with a gap adjustment mechanism to minimize its influence on brake performance and warning signal if any failure occurs.
7. Good thermal stability, with a long-lifespan brake pad.
8. Low noise and easy to maneuver.

In general, to quantify the mechanical advantage of a brake system that can be taken to minimize the braking effort, a nondimensional brake factor, C_b, can be defined by

$$C_b = \frac{T_b}{F_{in} R_c} \tag{13.1}$$

where

T_b is the braking torque output from the brake
F_{in} is the force input from the brake pedal
R_c is the effective radius of the friction force on the disc or drum

The brake factor often declines as the friction interface becomes hot or wet due to its influence on the friction coefficient. A good design should aim at reducing the sensitivity of the brake factor to these conditions.

13.1. Determination of Braking Torques

A reasonable proportion of front and rear braking torques is necessary to ensure a good braking effectiveness. Per Equations (4.25) and (4.26), if f is the adhesion coefficient for the front and rear wheels to lock simultaneously, the ratio of the front to rear braking torque is given by [13.1]

$$\frac{T_{b1}}{T_{b2}} = \frac{L_2 + fh_g}{L_1 - fh_g} \tag{13.2}$$

where

T_{b1} and T_{b2} are the braking torque at the front and rear wheels, respectively
L_1 and L_2 are the distance from the vehicle center of mass to the front and rear axles, respectively
h_g is the height of the vehicle center of mass

Once the ratio T_{b1}/T_{b2} is obtained from Equation (13.1), the maximum value of T_{b1} can be solved for the vehicle in emergency braking until wheel locks and slides on a well-paved road, and then the maximum value of T_{b2} can be calculated accordingly.

During an emergency braking, since the rear wheels are often locked and slide, the braking force at the rear axle is

$$F_{b2} = \frac{m_a g f L_1}{L + f h_g} \tag{13.3}$$

where
m_a is the total mass of the vehicle
L is the wheelbase

Accordingly, the braking torque at the rear axle is

$$T_{b2} = \frac{m_a g f L_1 R_d}{L + f h_g} \tag{13.4}$$

where R_d is the dynamic radius of rear wheels.

13.2. Design of Service Brakes

There are two types of service brakes often used in EVs, namely, disc brakes and drum brakes.

13.2.1. Disc Brakes

A disc brake utilizes the caliper to squeeze pairs of pads against a disc (sometimes called a rotor) to create friction. The brake pads are the friction parts that are mounted inside the caliper. Although mounted on the stub axle, the caliper does not rotate with the wheel.

The major advantages of disc brakes are as follows:

1. High thermal stability: Less prone to brake fade or deviation in the direction because the rise in temperature has little effect on the axial expansion of the disc, and the radial expansion of the disc has no influence on the braking performance.
2. Good wet braking performance: Not much performance decay when the brake gets wet, and it recovers quickly from immersion after a couple of applications as water can be wiped out easily by the pads and centrifugal effects.
3. Braking torque is independent on the direction of traveling of the vehicle.
4. Facilitating a dual-circuit braking system for higher reliability and safety.
5. Uniform pressure and wear on the pads make them less prone to premature failures.
6. It is easy to make the gaps between pads and disc self-adjustable, and the small gaps of only 0.05–0.15 mm shortens the response time of the brake.
7. Compact, lightweight, good heat dissipation, and easy replacement of brake pads.

However, disc brakes come with the following drawbacks:

1. A power-assisting device is needed to reduce the braking effort.
2. They are noisier than drum brakes and hard to prevent the ingress of dust and rust due to their open structure.
3. Relatively small-sized pads cause them to wear fast, demanding highly durable pad materials.
4. They are not as effective as the drum brakes to be used as a parking brake.
5. With the brake pedal released, the friction pads may remain in contact with the disc, generating a slight drag to the wheel and, thus, reducing the driving range a bit.

13.2.1.1. Types of Disc Brakes

Disc brakes can be divided into two types, namely, floating calipers and fixed calipers, depending on whether the calipers are fixed or not in the axial direction.

A floating caliper is allowed to slide along the axis of rotation with respect to the disc, with the pad on the other side of the hydraulic piston being fixed on the caliper. The contact area between the brake pads and the disc is quite small, with the angle φ generally ranging only 30–50°, as shown in Figure 13.1a. Under the action of brake fluid, the hydraulic piston pushes the pairs of pads against the disc when the brake pedal is depressed by the driver and the friction torque between the pads and disc produces a resistance to the rotating wheel. The advantages of a floating caliper are as follows: The only hydraulic piston is on the inner side of the disc, allowing the brake to be mounted closely to the wheel hub with a small axial dimension; both good cooling condition for the hydraulic piston and no need of pipelines across the disc help reducing the risk of vaporization of brake fluid; and low lost due to the simple structure.

FIGURE 13.1 Disc brakes. (a) Floating caliper and (b) fixed caliper.

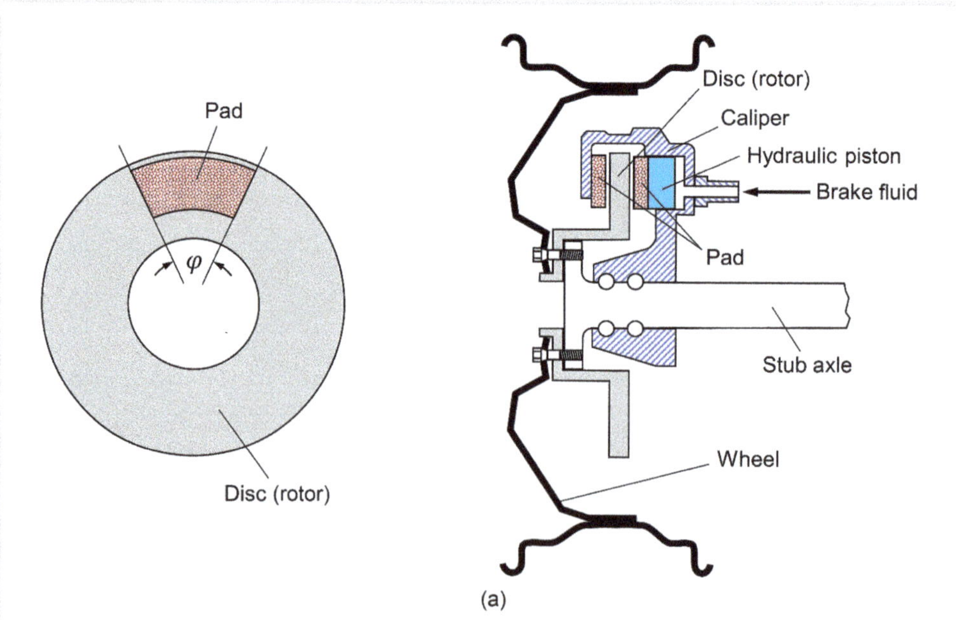

Courtesy of Wangyu Wang.

FIGURE 13.1 (Continued) Disc brakes. (a) Floating caliper and (b) fixed caliper.

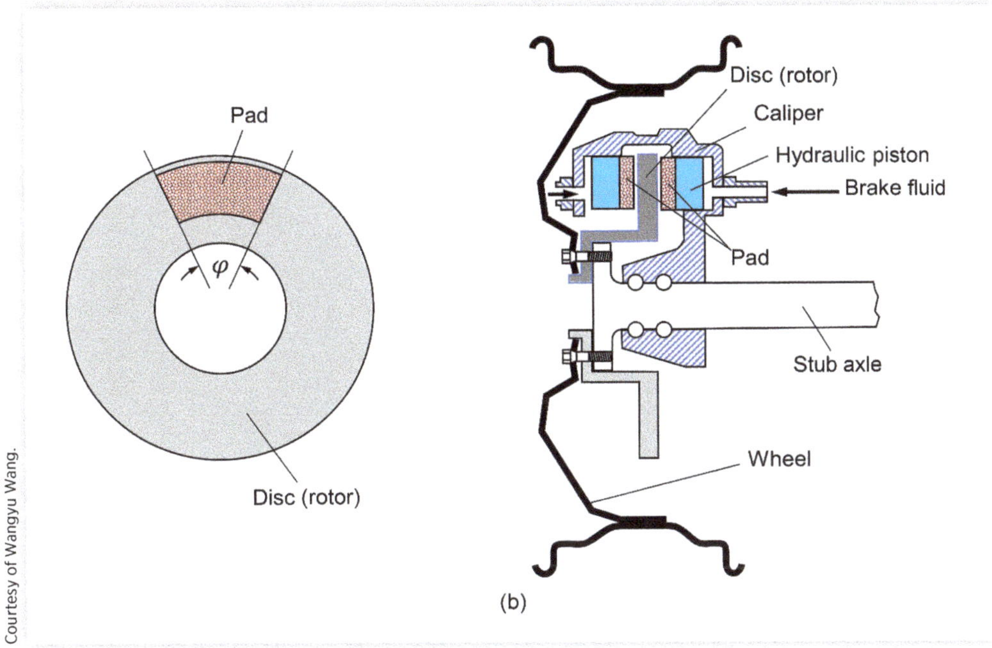

The fixed caliper type of design features one or more pairs of hydraulic pistons, as shown in Figure 13.1b. When the brake is applied, hydraulic pistons push the pads to clamp against the disc from each side to create the resistance torque. One advantage of fixed calipers is that they can be made highly stiff because there are no sliding parts other than the pistons and pads. However, they have several disadvantages: Pipelines or fluid passages across the disc are needed, thus increasing both the axial and radial dimensions of the brake and the difficulties in packaging; the brake is prone to the failure caused by fluid vaporization in pipelines due to the heat generation in braking processes; the brake performance is less tolerant of disc imperfection; and the cost is higher than the cost of floating calipers due to the increased number of pistons and system complexity.

13.2.1.2. Sizing of Disc Brakes

The major design parameters for a disc brake involves the disc diameter, disc thickness, inner diameter, outer diameter, and working area of brake pads.

The disc diameter is limited by the size of the rim, thus typically ranging from 70 to 79% of the rim diameter. A value near the upper limit is recommended for a vehicle with a GVW over 2000 kg, in order to increase the effective radius of braking action and reduce the clamping force, pressure, and temperature on the pads.

The disc can be made into either a solid piece or casted with some ventilation holes in it. Solid discs generally have a thickness of 10–20 mm, and discs with holes are mostly 20–30 mm thick. Disc thickness should be properly selected to minimize the mass of the disc and the rise of temperature in the braking operation.

It is recommended that the ratio of outer to inner diameters be kept below 1.5. When the ratio becomes too large, uneven brake pad wear would occur due to the large difference in circumferential velocities between the inner and outer edges, thus resulting in a decreased contact area and significant change in braking torque. The area of the brake pad for each side of the disc can be estimated based on the vehicle mass per unit pad area, with the recommended value within the range of 1.6–3.5 kg/cm^2.

By assuming the friction surface of braking pads is completely and uniformly in contact with the disc, the braking torque is given by

$$T_{b1} = 2 f_b F_c R_c \tag{13.5}$$

where
- f_b is the friction coefficient
- F_c is the clamping force
- R_c is the action radius of the clamping force, sometimes referred to as effective radius:

$$R_c = \frac{2\left(R_2^3 - R_1^3\right)}{3\left(R_2^2 - R_1^2\right)} \tag{13.6}$$

13.2.1.3. Detailed Design of Disc Brakes

1. Discs
   Commonly made of pearlitic gray cast irons, the discs are usually machined to the tolerances, as listed in Table 13.1 [13.1].

TABLE 13.1 Recommended tolerances for machining discs.

Tolerances	Flatness	Roughness	Parallelism*	Runout
Recommended values (mm)	12	0.7–1.3	50	30

* Between two friction faces.

Courtesy of Wangyu Wang.

2. Brake pads
   The design of brake pads must take specific energy dissipation rates and unit area friction forces into consideration.

   The specific energy dissipation rate refers to the energy that is dissipated by unit area of pads in unit time, which reflects the energy loads on braking pads in the braking process. The specific energy dissipation rates for a single brake of the front and rear wheels in a two-axle vehicle, e_1 and e_2, can be calculated, respectively, by

$$e_1 = \frac{\delta m_a \left(V_0^2 - V^2\right)\beta}{4 A_1 t} \tag{13.7}$$

$$e_2 = \frac{\delta m_a \left(V_0^2 - V^2\right)(1-\beta)}{4 A_2 \Delta t} \tag{13.8}$$

where

δ is the rotating mass conversion factor
m_a is the total mass of the vehicle
V_0 is the initial speed before braking
V is the final speed after braking
β is proportioning ratio of the front and rear brake forces
A_1 is the friction face area of front pads
A_2 is the friction face area of rear pads
Δt is the duration of braking

It is evident that $\Delta t = (V_0 - V)/a_b$, where a_b is the deceleration in braking.

It is generally recommended that the specific energy dissipation rate should not exceed 6.0 W/mm^2 at a deceleration of 0.6g and an initial speed of 100 km/h for a sedan, 80 km/h for a truck of 3.5t total mass or lighter, or 65 km/h for a truck of 3.5t total mass or heavier. A specific energy dissipation rate higher than the recommended value may result in excessive wear to the pads.

The unit area friction force, or specific friction force, refers to the friction force per unit area of the friction face. The specific friction force of the brake for a single wheel, f_b, is given by

$$f_b = \frac{T_b}{R_c A} \tag{13.9}$$

where

T_b is the braking torque of a single brake
R_c is the effective radius of the brake pad
A is the friction area of a single brake pad

To prevent excessive wear of the brake pads, it is usually required that for a typical large deceleration, the specific friction force should not exceed an upper limit, which is obviously dependent on the characteristics of the friction material.

3. Mounting position of calipers
The caliper can be mounted either in the front or in the rear of the axle. As shown in Figure 13.2 [13.2], the combined load on wheel bearings in braking, F_L, can be reduced by mounting the calipers in the rear of the axle. The ground force acting on the wheel, F_w, is a combination of the braking force F_b and normal force F_z. Thus, the load on wheel bearings is the combination of the friction force F_p acting on the disc from pads and the force F_w. It is clear that the load F_L is smaller when the calipers are mounted in the rear of the axle than when mounted in the front of the axle. Nevertheless, mounting the calipers in the front of the axle has an advantage of avoiding the tires throwing mud into the calipers.

FIGURE 13.2 Influence of the caliper position on wheel bearing load. (a) Mounted in the front of axle and (b) mounted in the rear of axle.

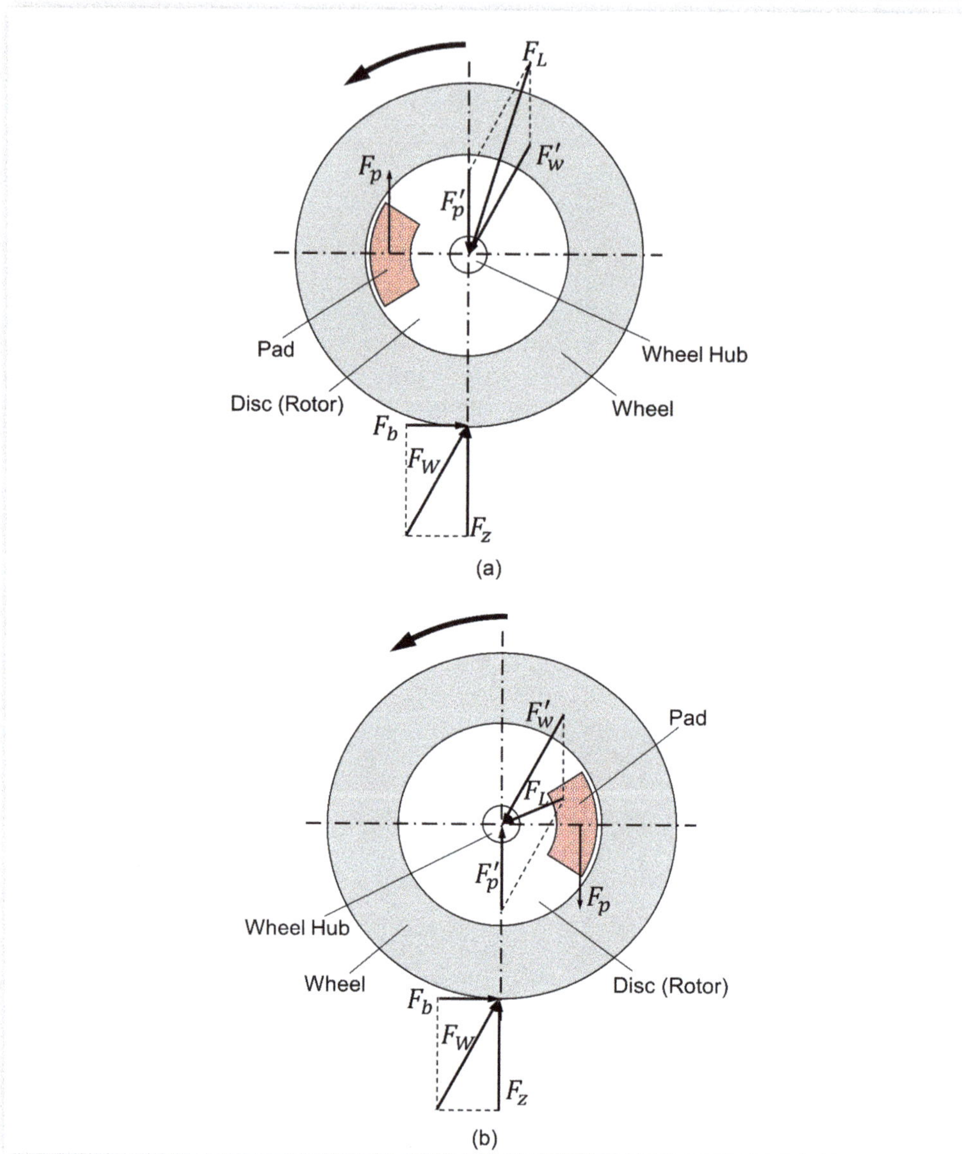

4. Clearance adjustment mechanism
   A clearance between the brake pads and the disc is necessary to ensure the free rotation of the wheel when the brake pedal is not depressed. It is generally desired to keep the clearance as small as possible, as a large clearance deteriorates the braking performance. Hence, an automatic clearance adjustment mechanism is often used to keep the clearance at the initial level.

In the routine design of such self-adjuster mechanisms for a new disc brake, there exists a clearance between the pads and the disc when the brake is not applied. When the brake with worn pads is applied, the piston is pushed further from its initial position by the hydraulic force to a point where the pads are clamping the disc again. At this moment, the piston first causes an ultimate elastic deformation of the seal and then displaces relative to the seal toward the disc to compensate the reduced thickness of the pads. After the brake pedal is released, this elastic deformation of the seal vanishes, pulling the piston back slightly off the disc to resume a normal clearance between the pads and the disc.

13.2.2. Drum Brakes

A drum brake uses a set of shoes that press outward against a rotating brake drum to create friction. The brake shoes usually press on the inner surface of the cylindrical-shaped drum.

The most distinctive advantage of a drum brake is its large brake factor due to the self-servo effect. In other words, a small actuation force can produce a large brake force, which makes it particularly suitable for applications in large and heavy-duty vehicles. In addition, its drum enclosure can largely keep the dust and dirt out of the friction surface and reduce their effect on brake factor. Drum brakes are also well-known for their low costs in manufacturing.

However, since the brake shoes are enclosed in the drum, drum brakes tend to heat up quickly in a continued braking operation due to poor heat dissipation. The high temperature may cause complex deformations of the drums and brake shoes, thus deteriorating the braking effectiveness and producing vibrations. Besides, drum brakes are prone to be slippery when they get wet, dramatically reducing their braking effectiveness. The brake wear debris tends to remain in the drum and accumulate, also reducing their braking effectiveness.

Recently, there has been a renewed interest in using drum brakes for the rear wheels of RWD or four-wheel drive EVs. This is mainly driven by three reasons: A drum brake imposes less drag to the wheel than a disc brake in which the pads can be naturally pulled against the rotor to create a slight resistance, a drum brake makes it easy to integrate a parking brake or emergency brake, without the need for separate mechanisms, further reducing weight and increasing driving range, and a drum brake is inexpensive and almost maintenance-free since the electric motor does a majority of the braking through regeneration.

13.2.2.1. Types of Drum Brakes

Based on the number and position of the fixed pivot, the shape and number of expenders, or whether there is interaction between shoes, the drum brakes are divided into leading/trailing, twin leading shoe, twin trailing shoe, dual-twin leading shoe, uni-servo, and duo-servo. Each of these types may offer different brake factors. For a given coefficient of friction, the duo-servo generally has the greatest brake factor and the twin trailing has the smallest. The brake factors of twin leading shoes and leading/trailing shoes lie in between those of duo-servo and twin trailing shoes, with the former generally larger than the latter. Note that the brake factor is affected not only by the type of brake shoes but also by the contact condition between the shoes and drum. For a given configuration, the brake factor is usually larger when the shoe and drum come into contact at the ends of the shoe than at the middle of the shoe. Hence, shoe adjustment plays a critical role to ensure proper contact between the shoes and drum for maximizing the brake factor.

1. Leading/trailing shoes
   This is the basic type of drum brake design, featuring a leading shoe and a trailing shoe, as shown in Figure 13.3 [13.3]. Each shoe is pivoted about its anchor point, with both anchor points being located at the same end of each shoe. The expander can be either a cam or wedge, which can be either floating or fixed. The floating expander can equalize the force acting on both shoes, but the fixed expander cannot.

FIGURE 13.3 A leading/trailing shoe type of drum brake.

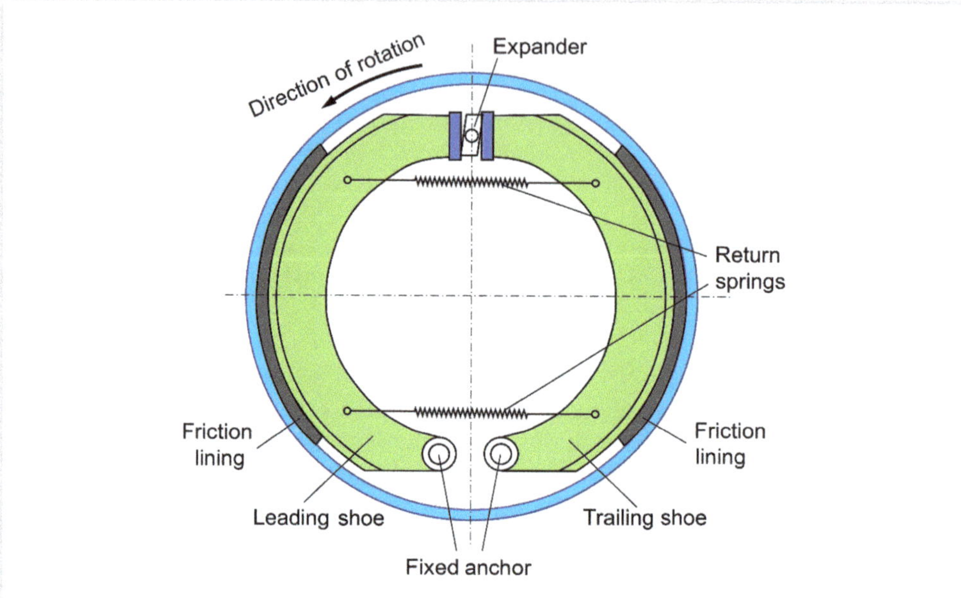

Due to the friction force acting on the leading shoe from the drum, the leading shoe is pulled into the friction surface of the drum and, thus, exhibiting a self-servo effect. As a result, the braking force is greatly increased. The trailing shoe, however, tends to be pushed away from the friction surface of the drum, causing the braking force to drop.

An advantage of leading/trailing shoe drum brakes is that they work equally effective whether the vehicle is traveling forward or in reverse. They are simple, inexpensive, and easy to adjust the clearance between the shoe and drum. With an average level of brake factor, they also have drawbacks of unequal pressure, wear, and lifespan of the shoes, and both shoes sharing the same wheel cylinder and hydraulic circuit.

Leading/trailing shoe drum brakes are commonly used in rear wheel applications.

2. Twin leading shoes
   This type of drum brake consists of two leading shoes that pivot about fixed anchor points, each located at the opposite end of the other. Besides, each shoe has its own wheel cylinder located at the opposite end to its anchor point. Because of the self-servo effect of both leading shoes, this type of drum brake is very effective at braking in the forward direction. In addition, both shoes have equal pressure, wear, and lifespan, and it is easy to adjust the clearance between the shoes and the drum. However, the

brake factor is poor when braking in the reverse direction, and powerful springs are necessary to pull the shoes back when the braking force is not needed.

A twin leading shoe type of drum brake can be used in front wheels of EVs for two reasons: It is not suitable for rear wheel applications because the mutually symmetrical wheel cylinders make it difficult to attach an additional actuator for the parking brake, and the front wheels generally require larger braking force than rear wheels, while the very large brake factor in retarding forward motion makes it a good choice.

3. Twin trailing shoes
Two trailing shoes are used in this configuration, each of which has its own wheel cylinder and a fixed anchor point located at opposite ends. Due to the low brake factor in both forward and reverse directions, this type of drum brake is rarely employed.

4. Dual-twin leading shoes
This configuration features pistons that displace in both directions, allowing both shoes to always act as leading ones, regardless of the direction of travel.

With a very high and constant brake factor in both forward and reverse directions, this type of shoe fits an actuation mechanism with dual-hydraulic circuits. In case of any failure in one of the circuits, the brake becomes a leading/trailing type but is still functionable. The equal pressure on both shoes results in their undistinguishable wear and lifespan. Nonetheless, this type of shoes comes with disadvantages of the complexity from dual-bidirectional cylinders and the difficulties in adjusting the clearance between shoes and the drum.

This type of shoe is widely used for its high reliability and braking effectiveness.

5. Uni-servo
Only one anchor point is used for both shoes in this configuration, with a push rod connecting the shoes in the bottom that is actuated by a single wheel cylinder. With the increase in hydraulic pressure in the wheel cylinder, the wheel cylinder pushes against the primary shoe, moving the primary and secondary shoes in an arc. The end of the secondary shoe is anchored against a fixed anchor, resulting in a levered force applied to the brake shoes.

When the vehicle is traveling forward, both shoes are leading, giving the highest brake factor in all types of drum brakes. However, in the reverse motion, both shoes become trailing ones, yielding a poor brake factor. The unequal pressure on shoes causes differences in wear and lifespan between the shoes. The single wheel cylinder does not accommodate dual-hydraulic circuits. The linkage at the bottom of shoes makes it hard to adjust the clearance between the shoes and the drum.

Consequently, this type of shoe is only used on front wheels of some trucks.

6. Duo-servo
A single cylinder with two pistons is used to activate the brake shoes in this configuration. There is an adjustable push rod floating in the bottom that links the shoes, allowing them to move in the direction of the drum. The anchor point at the top stops the shoes when they rotate toward it, while allowing the shoes to move away from it. This configuration creates the equal braking force in both forward and reverse directions by making use of the servo effect. A small force on the primary shoe can produce a large force on the secondary shoe through the bottom push rod. However, the unequal pressure on the primary and secondary shoes results in differences in their wear and lifespan. The single cylinder also makes it hard to work with an actuation system with dual-hydraulic circuits. It is also not easy to adjust the clearance between the shoes and the drum.

This type of brake shoes is often used for commercial vehicles that require especially large braking forces.

13.2.2.2. Sizing of Drum Brakes

Design parameters of a drum brake mainly include the inner diameter of the drum, the distance from the brake center to the line of action of expanding force, the anchor position of brake shoes, and the lead angle, wrap angle, and width of friction lining, as illustrated in Figure 13.4 [13.3].

FIGURE 13.4 Design parameters of a drum brake.

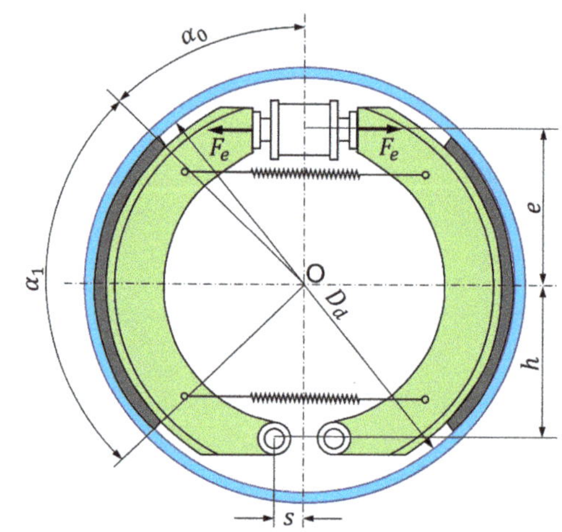

For a given input force from the brake pedal, a large inner diameter of the drum is generally desired for maximizing the brake force and cooling capacity. However, the increase in the inner diameter of the drum is limited by the inner diameter of the rim. A clearance of at least 20 mm is generally recommended between the drum and the rim to prevent the heat radiation from damaging tires. The inner diameter ratio of the drum to the rim, D_d/D_r, typically falls in the range of 0.64–0.74 for passenger vehicles and 0.70–0.83 for trucks. In addition, an adequate thickness of the drum wall is also preferred to ensure its stiffness and heat capacity for reducing the temperature rise in braking. Relevant standards or regulations may apply in the design of the inner diameter of brake drums.

A large distance from the brake center to the line of action of expanding force, e, is generally preferred to improve the brake factor as long as there is enough room for packaging the wheel cylinder in the drum. A value of $e = 0.4D_d$ can be used as a good guess in the preliminary design.

Provided no interference between the ends of two brake shoes, the anchor points of brake shoes are usually positioned with dimension h maximized and dimension s minimized. A value of $h = 0.4D_d$ is usually used in the preliminary design.

As far as the friction lining is concerned, its wrap angle, α, is generally selected in the range of 90–100° to minimize the wear and temperature rise and maximize the brake factor. In reality, since the friction lining applies lower pressure on the drum near each of the ends than in the middle, the excessively large wrap angle does not help much in reducing the pressure. Instead, a large wrap angle may cause unsmooth or self-locking problems in braking. Hence, a wrap angle of over 120° is rarely used in design.

The lead angle α_0 is often determined by centering the friction lining to the brake shoe, that is, $\alpha_0 = (180^{\circ} - \alpha_1)/2$. As an alternative, it is sometimes centered to the area of maximum pressure though, in order to enhance the uniformity of the wear and brake factor.

The width of friction lining, b_f, can be determined according to the statistical data of the total area of friction lining for a single drum brake, A_f, as listed in Table 13.2 [13.1].

TABLE 13.2 Statistical data of the total area of friction lining for a single drum brake.

Type of vehicle	Total mass of vehicle (t)	Total area of friction lining in a single brake A_f (dm^2)
Passenger vehicles	0.9–1.5	1.0–2.0
	1.5–2.5	2.0–3.0
Trucks and buses	1.0–1.5	1.2–2.0
	1.5–2.5	1.5–2.5
	2.5–3.5	2.5–4.0
	3.5–7.0	3.0–6.5
	7.0–12.0	5.5–10.0
	12.0–17.0	6.0–15.0

Courtesy of Wangyu Wang.

The width of friction lining, b_f, is given by

$$b_f = \frac{2A_f}{\alpha_1 D_d} \tag{13.10}$$

The width of friction lining b_f affects the lifespan of friction lining. A narrow friction lining shortens its lifespan due to accelerated wear, while a broad friction lining causes its mass and cost to increase and uneven contact between the lining and drum.

13.2.2.3. Detailed Design of Drum Brakes

Once the basic sizes of a drum brake are determined based on the method explained in the previous section, a bunch of calculations are needed to come up with other detailed design parameters and to verify them.

1. Determination of other design parameters
   Based on the discussion in the previous section, a brake shoe is of either one DOF if its ends fixed to an anchor pin or two DOF, for example, dual-twin leading shoes.

   In the design of drum brakes, the pressure of the brake shoe acting on the inner surface of the brake drum can be solved using FEA at the required accuracies.

   A theoretical pressure distribution on the friction lining can be obtained in a simple model on the basis of two assumptions: both the drum and the brake shoe are rigid, with the friction lining being the only deformable parts in the system, and the pressure and deformation satisfy Hooke's law.

At first, a brake shoe system with one DOF is discussed. For the friction lining and brake shoe system with one DOF, as shown in Figure 13.5 [13.1], it can be proved that the pressure distribution is a sine function of the angle α:

$$p = p_{max} \sin\alpha \tag{13.11}$$

where α is the angle with respect to the line OO′ connecting the center of the drum to the anchor point and p_{max} is the maximum pressure that occurs on the radial line perpendicular to the line OO′.

FIGURE 13.5 Friction lining and brake shoe system with one DOF.

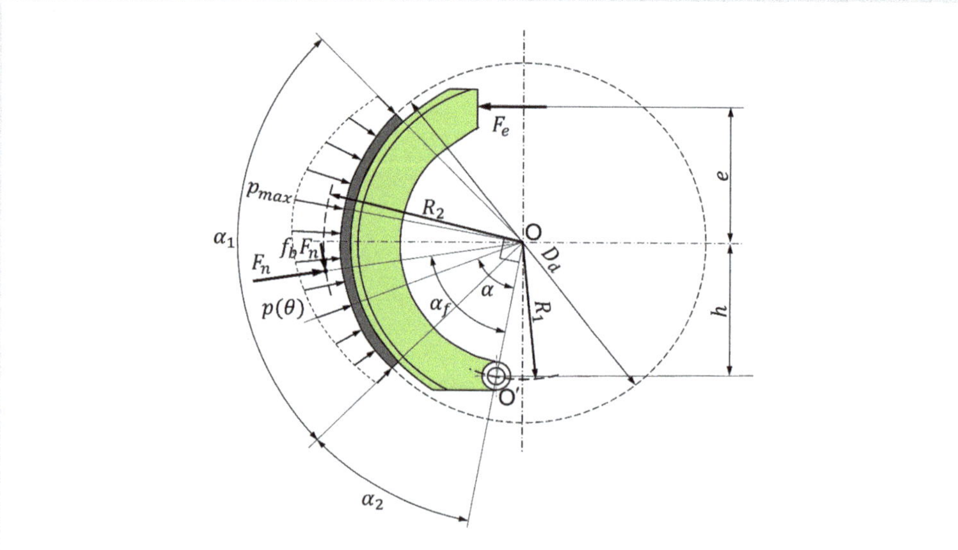

Courtesy of Wangyu Wang.

The unevenness of the pressure distribution can be quantified by the ratio Δ_p:

$$\Delta_p = \frac{p_{max}}{\bar{p}} \tag{13.12}$$

where $\bar{p}$ is the average pressure corresponding to the same braking torque.

For a leading shoe, the combined normal force reacting from the drum on the shoe is

$$F_n = \frac{(e+h)F_e}{R_1\left(\sin\alpha_f + f_b\cos\alpha_f\right) - f_b R_2} \tag{13.13}$$

where

F_e is the expanding force

R_1 is the radius of the anchor point

R_2 is the radius of line of action of the combined friction force

α_f is the angle from the line of action of the combined normal force on the friction lining to the line OO′

We have

$$R_2 = \frac{2D_d(\cos\alpha_2 - \cos\alpha_1)}{\sqrt{(\cos 2\alpha_2 - \cos 2\alpha_1)^2 + (2\alpha_2 - 2\alpha_1 + \sin 2\alpha_2 - \sin 2\alpha_1)^2}} \quad (13.14)$$

Thus, the braking torque produced by the assembly of single shoe and friction lining is given by

$$T_b = \frac{f_b(e+h)R_2F_e}{R_1(\sin\alpha_f + f_b\cos\alpha_f) - f_bR_2} \quad (13.15)$$

Similarly, for a trailing shoe, the braking torque produced by the assembly of single shoe and friction lining is given by

$$T_b = \frac{f_b(e+h)R_2F_e}{R_1(\sin\alpha_f - f_b\cos\alpha_f) + f_bR_2} \quad (13.16)$$

In addition, the leading shoe may lock itself if not designed properly, which can quickly lead to mechanical failure of other parts of the brake system. Therefore, as an important design step, the following condition must be checked:

$$\frac{R_1\sin\alpha_f}{R_2 - R_1\cos\alpha_f} > f_b \quad (13.17)$$

If the inequality in Equation (13.17) holds, then the drum brake will not lock up. The maximum pressure on the friction lining of a leading shoe is given by

$$p_{max} = \frac{4(e+h)R_2F_e}{(\cos\alpha_2 - \cos\alpha_1)\left[R_1(\sin\alpha_f + f_b\cos\alpha_f) - f_bR_2\right]bD_d^2} \quad (13.18)$$

where

α_1 is the angle from the beginning to the end of the friction lining
α_2 is the angle from the line OO′ to the lower end of the friction lining
b is the width of the friction lining

Hence, the unevenness ratio of pressure distribution becomes

$$\Delta_p = \frac{\alpha_1 - \alpha_2}{\cos\alpha_2 - \cos\alpha_1} \quad (13.19)$$

Lastly, for the friction lining and brake shoe system with two DOFs, the pressure distribution is a sine function of the angle α':

$$p = p_{max} \sin\alpha' \tag{13.20}$$

where α' is the angle with respect to the line OO′ connecting the center of the drum to the instantaneous velocity center O′. Note that the maximum pressure p_{max} varies with the location of the instant center of rotation, so does the pressure distribution. Consequently, the maximum pressure p_{max} does not stay at a fixed angular position.

2. Design of major mechanical components
First, the design of drums is addressed. In addition to adequate strength, stiffness, and heat capacity, a drum also needs to have a high friction coefficient for mating with the friction lining material.

Drums can be either casted or assembled by parts made of different materials. Mostly made of gray cast iron, the casted drums are easy to process and wear-resistant, and have a large heat capacity. Ribs are often added to their outer circumferences to enhance stiffness and heat dissipation. The empirical value of drum wall thickness generally falls into a range of 7–12 mm for a passenger vehicle and 13–18 mm for a truck, and a precise wall thickness can be determined using FEA methods. The assembled drums are made of steel, aluminum, or gray iron parts that are processed by stamping or casting, or even their combination, to further reduce the weight, improve the wear resistance, and elevate the friction coefficient.

Second, as far as the brake shoes for the passenger vehicle or light-duty truck applications, they are often made of rolling T-shaped steel or welding steel sheets. One or two radial grooves are often made on the web section of a brake shoe to lower its stiffness, which helps reduce the squeak during braking and the uneven wear of friction lining. In contrast, the brake shoes for heavy-duty trucks are mostly made of cast iron or cast steel with an I-shaped, E-shaped, or U-shaped cross section. The webs and flanges usually have a thickness of 3–5 mm for a passenger vehicle and 5–8 mm for a truck. A roller or a gasket is sometimes used at the end near the expander to improve the efficiency, reduce the wear, and increase the lifespan of the brake shoe.

Third, the friction lining material must meet the following requirements: First of all, a stable friction coefficient that does not vary significantly with temperature, pressure, and relative velocity. Next, good wear resistance that minimizes the wear of both friction lining and mating surface, with the latter typically not exceeding 10% of the former. Furthermore, small compression and expansion ratios that avoid the drop in braking sensitivity due to changes in the displacement of the master cylinder and the stroke of brake pedal, and prevent the friction lining from continual drag on the drum and get stuck as the clearance vanishes under expansion. In addition, adequate extrusion strength, impact strength, and shear strength. In addition, limited thermal conductivity, with the back plate temperature not exceeding 190°C after putting friction lining on a heating plate of 300°C for 30 min, in order to prevent the seals and dust cover from premature failures and the brake fluid from evaporation. Lastly, no harm to human health and the environment, and low noise. At present, semimetallic, nonasbestos organic polymer, ceramic-based composite, and carbon fiber-reinforced carbon (CFRC) composite are used as brake friction materials. Semimetallic materials use steel wools and porous iron powders as the main constituent. They have been widely used for their good thermal stability and excellent wear resistance. However, they are prone to noise and rust problems, and their great hardness can result in excessive wear of the mating parts. In nonasbestos organic polymer friction materials, fibers such as carbon fibers, aramid fibers, glass fibers, ceramic fibers, or hybrid fibers are utilized to reinforce the base material, which is typically an organic material like a resin. In addition to its relatively large specific gravity, the main drawback of this type of materials is its mediocre thermal resistance due to the presence of resins. In ceramic-based composite materials, modified resins and rubbers are used as binders, and a variety of synthetic organic and inorganic materials are employed as friction performance modifiers. With the advantages of no noise, no rust, long life, good high temperature resistance, moderate specific gravity, reasonable price, and no harm to the environment, it has been widely used for their all-around characteristics. CFRC is a composite material consisting of carbon fiber reinforcement in a matrix of graphite. It has advantages of small specific gravity, high strength, low thermal expansion, and excellent high temperature resistance. Nevertheless, it has not been widely applied in EVs yet due to their high costs.

The friction lining can be either riveted or bonded to the brake shoe. The riveted friction lining features low noise, while the bonded friction provides a thicker lining for wear before its replacement in spite of the more complicated bonding and replacement processes.

Fourth, on the clearance adjustment mechanism, a clearance between the friction lining and the drum is a must to ensure the free rotation of the wheel when the brake is not applied, similar to disc brakes. It is also desired to keep the clearance as small as possible, as a large clearance makes the brake spongy and sluggish. Hence, an automatic clearance adjustment mechanism is often used to maintain a proper clearance even when the friction lining is worn out. In a typical design of such self-adjuster mechanisms, the adjuster mechanism is attached to the secondary shoe. It is a mechanical device that is made up of three main components, namely, an adjuster assembly, an adjuster lever, and a cable guide. The adjuster assembly comes with an adjusting screw that is a threaded device like a bolt and nut. The head of the bolt is a notched wheel with a cylindrical pin. A washer and slotted cap fit over the pin and engage the web section of the secondary shoe. The nut is also a slotted cap that threads onto the bolt and engages the primary shoe. In addition, the adjuster lever moves on a pivot to engage and turns the notched wheel. Moreover, the adjusting cable is attached to the adjuster lever and the anchor pin to move the adjuster lever at its pivot point. There are springs and retainers to hold the mechanism in place. The adjustment takes place when the vehicle is in reverse and the brakes are applied or when the parking brake is engaged. The adjusting cable attached to the anchor pin pulls the lever mechanism based on the movement of the secondary shoe. When the brakes are applied, the adjuster lever cannot move the adjuster screw because the secondary shoe is engaged by the brake pressure. The adjuster lever mechanism must store the adjustment and turn the notched wheel when the brake is released.

13.3. Design of Park Brakes

The park brakes are usually applied to the rear wheels.

For a vehicle parked on a slope facing uphill, the brake force acting on the rear wheel from the ground is given by [13.1-13.2]

$$F_{b2} = fF_2 = m_a g f\left(\frac{L_1}{L}\cos\theta + \frac{h_g}{L}\sin\theta\right) \tag{13.21}$$

where

f is the tire–ground adhesion coefficient
F_2 is the normal force acting on the rear wheel
L is the wheelbase
h_g is the center-of-mass height of the vehicle
L_1 is the distance from the center of mass to the front axle

For a vehicle parked on a slope facing downhill, the brake force acting on the rear wheel from the ground is given by

$$F_{b2} = fF_2 = m_a g f\left(\frac{L_1}{L}\cos\theta - \frac{h_g}{L}\sin\theta\right) \tag{13.22}$$

There is an upper limit for the slope that the vehicle can park and remain balance on. The maximum slope angle for uphill parking is given by

$$\theta_{max} = \arctan\frac{fL_1}{L - fh_g} \tag{13.23}$$

Similarly, the maximum slope angle for downhill parking is given by

$$\theta_{max} = \arctan\frac{fL_1}{L + fh_g} \tag{13.24}$$

The maximum slope angles must be no less than the specified values of applicable standards and regulations.

13.4. Brake Actuation Mechanism

The brake actuation mechanism receives the braking command from the driver or the vehicle controller and executes the command to produce the braking force. The braking command can be a force transmitted through a mechanical linkage, a fluid pressure through a hydraulic or pneumatic circuit, or simply an electronic signal through a wire.

13.4.1. Types of Brake Actuation Mechanism

Nowadays, the mechanical linkage type of brake actuation mechanisms has been primarily phased out due to the difficulties in proportioning the braking forces between wheels, except for low speed, neighborhood EVs.

On the contrary, hydraulic actuation mechanisms are widely used in EVs that are not equipped with a source of compressed air. The working pressure of as high as 10–20 MPa makes it possible to mount the compact wheel cylinder inside the brake for activating an expander or a piston. In addition, they also come with other advantages of fast response, fairly high efficiency, and lightweight. However, a hydraulic actuation mechanism may cause brake effectiveness to decay or fail when exposed to excessive heat due to the evaporation of brake fluid.

Furthermore, a pneumatic actuation mechanism can be a wise alternative if a source of compressed air is available on a vehicle such as a heavy-duty truck or a bus. In general, a pneumatic actuation mechanism is highly reliable, easy to maneuver, and convenient to maintain or repair. However, the need of air compressors, air tanks, and valves increases the complexity, mass, and cost of the system. Its 0.3–0.9 sec of lag time results in a long braking distance, often calling for additional relay valves along the pipelines if the brakes are distant from the brake chamber or air tank. Its low operating pressure of 0.5–0.7 MPa usually leads to a brake chamber of large size, which can only be mounted outside the brake. In this case, a linkage, cam, or wedge to actuate the brake is needed, causing unsprung mass to increase. A loud noise is made when the brake chamber discharges the compressed air.

13.4.2. Power Brakes

In order to reduce the braking effort of a driver, power brakes are equipped by most EVs today, in which a power-assisting mechanism is used to amplify the actuation force exerted on the brake pedal. In a typical power-assisting mechanism, an electrically powered vacuum pump is used to maintain a constant vacuum that the vacuum booster needs in a braking system, as illustrated in Figure 13.6. When the brake pedal is depressed, a poppet valve (not shown) opens and air rushes into a pressure chamber on the driver's side of the booster. The pressure exerted by this air against the vacuum pushes a piston, thus assisting the pressure exerted by the driver on the pedal. The piston, in turn, exerts pressure on the master cylinder, from which brake fluid is forced to act on the brakes.

FIGURE 13.6 A power-assisting mechanism with an electrically powered vacuum pump.

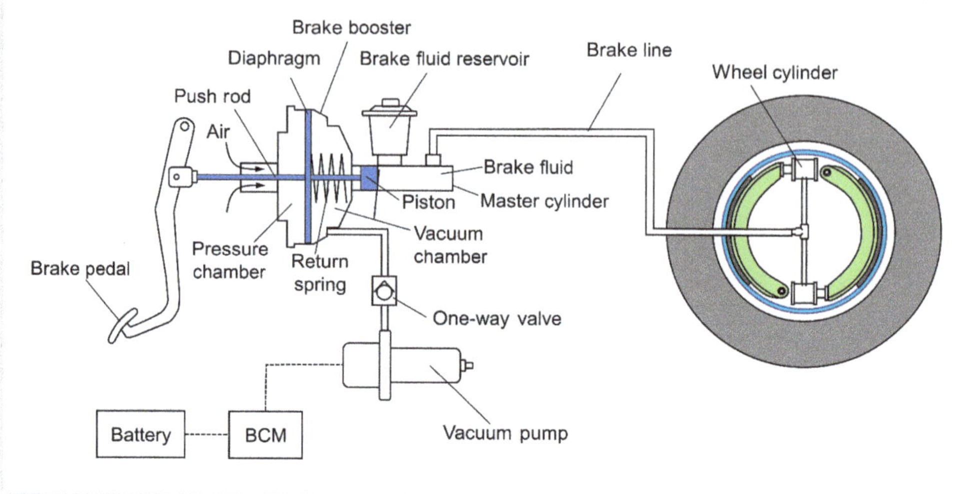

13.4.3. Layouts of Brake Line Circuit

For an EV, two or more independent brake line circuits are often used to improve the reliability of the brake system. In case of any failure in one of the circuits, the rest of the system can still provide the braking functionality, often at a reduced performance though. As illustrated in Figure 13.7 [13.1], for a two-axle vehicle with two independent brake line circuits, there are usually five types of brake line layout, as elaborated as follows.

1. II type: The front wheels share the same brake line circuit, and the rear wheels share the other, as shown in Figure 13.7a. This type of layout is widely used in trucks for its simplicity and low cost. However, in case of failure in the brake circuit for rear wheels, any braking operation may cause the front wheel to lock up, and the vehicle would become uncontrollable due to the loss of both steering and braking.
2. X type: One front wheel shares the same brake line circuit with the rear wheel on the other side, as shown in Figure 13.7b. This type of layout is widely used in passenger vehicles for its simplicity and fail-safe characteristics as 50% of braking capacity remains in case of any failure in one of the braking circuits. However, if any fault breaks the lateral symmetry in braking forces, the front wheels would start to spin about the kingpin on the side of the larger braking force, and the vehicle would become unstable. Hence, it would be advantageous to use this type of layout on a vehicle with a negative kingpin offset, in which the nonsymmetric left–right braking force reverses the spin direction of front wheels and, thus, enhances the vehicle stability.
3. HI type: Half of the front wheel cylinders share the same brake line circuit with all rear wheel cylinders, and the other front wheel cylinders share the other brake line circuit, as shown in Figure 13.7c. This type of layout, as well as the LL and HH types, features an increased level of complexity. It allows a high portion of braking capacity to remain functionable if any failure occurs only in the circuit for the front wheels. Its weakness is that the rear wheels are prone to lock up under emergency braking.

FIGURE 13.7 Various brake line layouts. (a) II type, (b) X type, (c) HI type, (d) LL type, and (e) HH type.

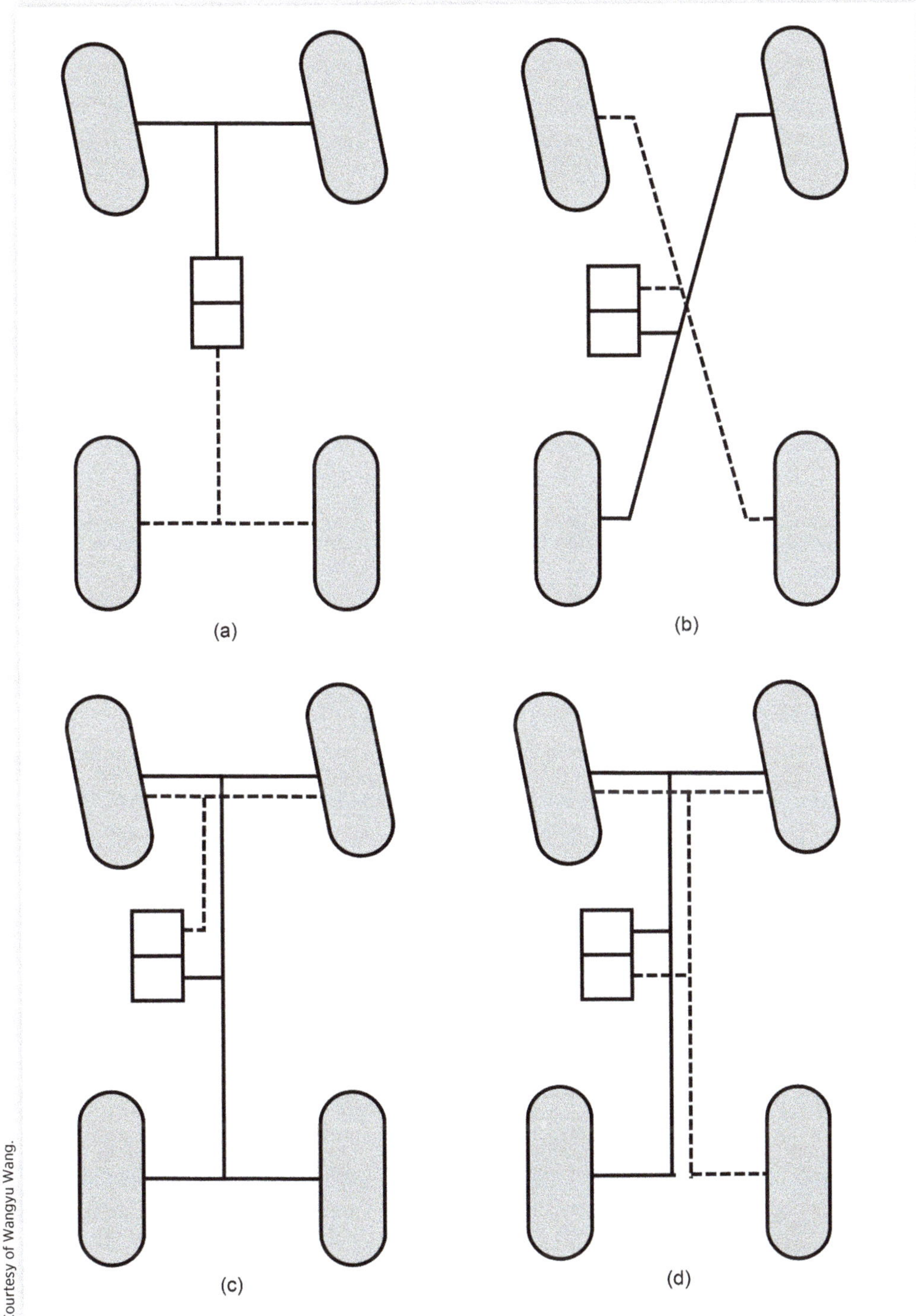

Courtesy of Wangyu Wang.

FIGURE 13.7 (Continued) Various brake line layouts. (a) II type, (b) X type, (c) HI type, (d) LL type, and (e) HH type.

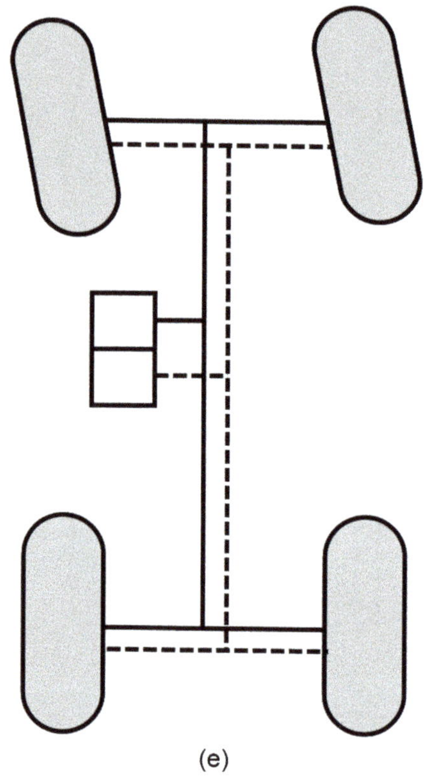

Courtesy of Wangyu Wang.

4. LL type: Half of the front wheel cylinders share the same brake line circuit with the rear wheel cylinders on one side, as shown in Figure 13.7d. If any failure occurs in one of the circuits, the front to rear proportion of braking forces remains the same, with roughly 50% of the braking capacity left, thus improving the fail-safe feature of the vehicle. It also has the same drawback as the HI type that the rear wheels are prone to lock up under emergency braking.
5. HH type: Half of the front wheel cylinders share the same brake line circuit with half of the rear wheel cylinders, as shown in Figure 13.7e. Similar to the LL type, if any failure occurs in one of the circuits, the front to rear proportions of braking forces remain unchanged, with nearly 50% of the braking capacity still available, thus improving the fail-safe feature of the vehicle.

13.4.4. Design of Hydraulic Actuation System Components

The design of hydraulic actuation system components mainly involves the diameter of the wheel cylinder, diameter of the master cylinder, and brake pedal force and stroke.

13.4.4.1. Diameter of the Wheel Cylinder

The diameter of the wheel cylinder, D_{wc}, is related to the actuation force, F_a, and the brake line pressure, p_b, by

$$D_{wc} = \sqrt{\frac{4F_a}{\pi p_b}} \tag{13.25}$$

where F_a is the clamping force F_c in the case of disc brakes or the expanding force F_e for drum brakes. Although an elevated brake line pressure helps reduce the size of the system, it may impose high requirements for tightness at the connectors. A brake line pressure of 10–12 MPa is typically used for drum brakes, although a higher value may be used for disc brakes. If there is applicable standards or regulations, then the diameter of the wheel cylinder must be selected from a series of specified values.

13.4.4.2. Diameter of the Master Cylinder

Since the working volume of the jth wheel cylinder, V_j, is given by

$$V_j = \frac{\pi}{4}\sum_{j=1}^{m} D_j^2 S_j \tag{13.26}$$

where

D_j is the diameter of the jth wheel piston
S_j is the maximum stroke of the jth wheel piston
m is the total number of wheel pistons in the jth wheel cylinder

A value of $S_j = 2.0 - 2.5$ mm can be assumed in the preliminary design.

Thus, the total working volume of wheel cylinders, V_w, is

$$V_w = \sum_{i=1}^{n} V_i \tag{13.27}$$

where n is the total number of wheel cylinders.

Then, the working volume of the master cylinder, V_m, is given by

$$V_m = V_w + V_h \tag{13.28}$$

where V_h is the total volume of hoses counting in the expansion under the maximum brake line pressure. In general, $V_h = 0.1V_w$ can be assumed for passenger vehicles and $V_h = 0.3V_w$ for trucks in a preliminary design.

Finally, the diameter of the mass cylinder, D_m, can be obtained by

$$D_m = \sqrt{\frac{4V_m}{\pi S_m}} \tag{13.29}$$

where S_m is the stroke of the mass cylinder. Usually, $S_m = (0.8 - 1.2)D_m$. The diameter of the mass cylinder must be selected from a series of values specified by any applicable standards or regulations.

13.4.4.3. Brake Pedal Force and Stroke

The brake pedal force is computed by

$$F_p = \frac{\pi D_m^2 p_b}{4 i_p \eta_{pm}} \tag{13.30}$$

where

i_p is the force ratio of the brake pedal to the master cylinder
η_{pm} is the mechanical efficiency from the brake pedal to the master cylinder

Generally, $\eta_{pm} = 0.82 - 0.86$.

The recommended pedal force should fall into the range of 200–500 N, with the maximum value of 500 N for a passenger vehicle and 700 N for a truck.

On the other hand, the brake pedal stroke can be estimated by

$$S_p = \left(S_m + S_{m0} + \delta_m + S_w\right) i_p \tag{13.31}$$

where S_{m0} is the dead stroke of the master piston from the extreme position where no braking force is produced to the point where the piston completely blocks the bypass hole on the master cylinder, δ_m is the gap between the push rod and the piston (typically, $\delta_m = 1.5 - 2.0$ mm), and S_w is the increased stroke due to the allowable wear of brake pads or friction linings. In general, $S_w = (0.4 - 0.6)S_p$.

With the stroke due to allowable wear of brake pads or friction linings taken into account, the brake pedal stroke must not exceed 100–150 mm for passenger vehicles and 180 mm for trucks. When used as park brakes, the hand force and stroke should be kept below 400 N and 160 mm for passenger vehicles, and 600 N and 220 mm for trucks, respectively.

With the brake pedal released, a residual pressure of 0.05–0.14 MPa in brake lines is generally desired to prevent the air ingress. The return spring for the master piston should be calculated accordingly.

13.4.5. Braking Force Adjustment Mechanism

Demands for braking force adjustment come for two reasons: To ensure the front wheels lock up earlier than rear wheels, the brake system of a fixed front-to-rear force proportion generally has a poor adhesion utilization; both ideal and actual proportioning lines, that is, the I-curve and line β, vary with the loading conditions of a vehicle. Thus, various mechanisms are used to adjust the proportion so as to make it as close to the I-curve as possible. In fact, most of the mechanisms are designed to adjust the brake pressure for rear brakes.

13.4.5.1. Brake Pressure Regulator

Brake pressure regulators are an open-loop control devices that help improve the adhesion utilization and handling stability during braking.

For a typical brake pressure regulator installed in the rear brake line, its purpose is to limit the pressure acting on the rear brakes so that the risk of rear wheel locking and skidding is reduced. With a spring-loaded plunger enclosed in a valve body, full pressure p_1 acts on all the brakes initially when the brake is applied because the pressure is not high enough to overcome the spring. Hence, at this moment, $p_1 = p_2$. As pressure p_1 rises, the valve closes and shuts off the fluid flow to rear brakes once a predetermined pressure p_s is reached. Subsequently, any further increase in pressure p_1 affects the front brakes only. Therefore, although the rear brake pressure p_2 is proportioned to the front brake pressure p_1 when $p_1 < p_s$, the rear brake pressure p_2 would remain at $p_2 = p_s$ without further rise, thus enhancing the adhesion utilization and preventing rear wheels from locking.

Brake pressure regulators are suitable for vehicles of a short wheelbase and high CG that have a large axle load shift. For those vehicles of large load variations, such as cargo trucks, proportional valves can be used instead of brake pressure regulators as the fixed pressure limit of the latter for rear brakes cannot fit both I-curves for the same vehicle operating in full-load and that in no-load conditions.

13.4.5.2. ABS

Different from the aforementioned open-loop control devices such as a brake pressure regulator or a proportional valve, the ABS is a closed-loop control system usually consisting of four-wheel speed sensors, a brake control module, a hydraulic or pneumatic control modulator, and a warning light. In some configurations, a Doppler radar speed sensor and a piezoelectric accelerometer may also be utilized.

The wheel speed sensor is a device for sensing the wheel speed. It is mainly composed of a ferromagnetic toothed reluctor ring (also called the tone wheel) and a sensor that produces a voltage signal for the control module to process. Either a magnet or a Hall effect-based, the sensor detects the speed by counting the number of pulses of the voltage signal in a certain period of time due to the variation in the magnetic field induced by the rotation of the tone wheel.

The hydraulic or pneumatic control modulator consists of an oil pump, an accumulator, and several solenoids for regulating the brake pressure to each of the wheels, as shown in Figure 13.8. Located between the master cylinder and the wheel cylinders, the hydraulic control modulator can increase, decrease, or maintain the brake pressure by controlling the brake fluid flow between the master cylinder (or accumulator) and the wheel cylinders.

FIGURE 13.8 A hydraulic or pneumatic control modulator.

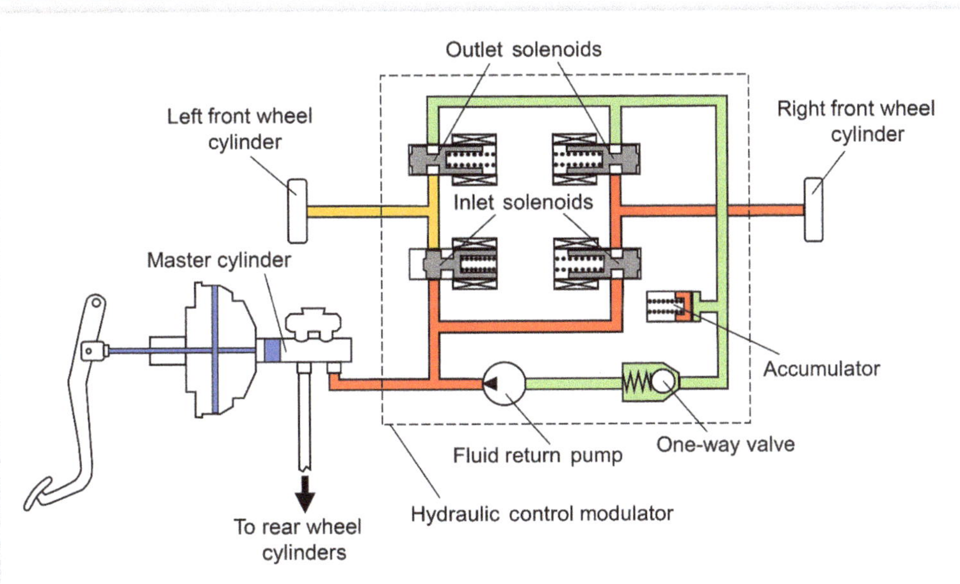

In a normal braking operation, the ABS is not activated, with no current going to any of the solenoids. Inlet solenoids are open, allowing the brake fluid from the master cylinder to enter the wheel cylinders. In the meanwhile, outlet solenoids are closed to isolate the brake fluid from the return lines, resulting in a pressure rise in wheel cylinders and a consequent increase in the braking force. When the brake pedal is released, the brake fluid flows back to the master cylinder from the wheel cylinders.

In emergency braking, if any wheel is about to lock, the control module would activate the ABS to operate in one of the following three modes, in order to prevent the wheel from locking up: Hold mode: When an excessive brake force and slip ratio occur at the left front wheel, the inlet solenoid for that wheel is energized to close off the brake line from the master cylinder to the left front wheel cylinder. This prevents the pressure from further rising, should the driver depress the brake pedal harder. Decrease mode: When the left front wheel operates in the hold mode and the slip ratio continues to soar, the outlet solenoid for that wheel is energized to open up the brake line, causing the pressure to drop. Increase mode: With the pressure at the left front wheel dropping, the slip ratio declines to the acceptable range and the brake force needs to increase again, which can be accomplished by de-energizing both the inlet and outlet solenoids to the left front wheel to pass the pressure from the master cylinder right through to the brake.

The brake line is often referred to as control channel if the brake pressure in it can be independently adjusted. A control channel can be used alone by a single wheel or be shared by two wheels. There are also select-high and select-low operating principles. For a control channel shared by two wheels, the select-high operating principle aims at preventing the wheel of a larger adhesion force from locking or sliding, while the select-low operating principle aims at preventing the wheel of a lower adhesion force from locking or sliding.

In general, an ABS can be configurated with one to four control channels. However, the ABS with one or two control channels are rarely used in practice due to their poor performance in directional stability, handling, and braking distance.

The three-channel ABS comes with either three- or four-wheel speed sensors. The brake pressure is usually independently controlled for each of the front wheels, with shared control for rear wheels. In a three-channel, four-sensor ABS for an RWD vehicle, the rear wheels share the same brake pressure control using a single hydraulic pressure modulator. The two sensors for rear wheels allow the select-low operating principle to be implemented, ensuring identical brake force exerted at both left and right rear wheels under various operating conditions. Moreover, a three-channel, three-sensor ABS for an RWD vehicle is a simplified configuration. Only one speed sensor is mounted in the rear axle to detect the average speed of rear wheels, enabling an approximate select-low operating principle. On the other hand, in a three-channel, four-sensor ABS in a diagonal arrangement, although there is a hydraulic pressure modulator for each of the four wheels, the two hydraulic pressure modulators for rear wheels share the same select-low control principle. Thus, it is virtually a three-channel ABS. In the presence of a large difference between the adhesion forces at both left and right rear wheels, the directional stability of the vehicle can be improved by adopting the select-low operating principle, at a cost of reduced adhesion utilization for the wheel with a high adhesion coefficient though. This is acceptable for the vehicles whose rear wheels provide a relatively small portion of the total brake force due to the load shift toward the front in an emergency braking. The independent pressure control for the front wheels offers two advantages: maximizing the adhesion utilization for both front wheels and the total braking force, thereby shortening the braking distance; allowing the front wheels to retain large lateral adhesion forces and, thus, good steering capability. Consequently, even if the braking forces at two front wheels become unbalanced, its influence on the directional stability of the vehicle can be mitigated by the steering maneuver of the driver. For these reasons, the three-channel ABS is widely used in electric passenger vehicles.

The four-channel, four-sensor ABS can be made in either a diagonal or front-rear arrangement. With independent brake pressure control for all wheels, this type of ABS can maximize the adhesion utilization of each wheel, resulting in good directional stability, handling, and shortest braking distance. Nonetheless, if there is a large difference in adhesion forces between left and right wheels, a yaw moment is produced by the braking forces from the ground. As a result, the vehicle would deviate toward the side with a larger braking force, compromising the directional stability. This problem cannot be easily corrected by steering maneuver of the driver. The four-channel ABS is normally used for vehicles equipped with a TCS as both systems can share the wheel speed sensors.

13.5. Brake-by-Wire

Brake-by-wire is a technology that allows control of brakes through electrical means instead of mechanically connecting the brake pedal to brake actuators. It can be used either to supplement ordinary service brake or as a stand-alone brake system if allowed by applicable regulations. It can also be used as electric park brake in a similar way. The ultimate goal of this technology is to replace components of a traditional brake system such as master cylinders, vacuum servos, pumps, pipes, hoses, and fluids with electronic sensors and actuators. The state-of-the art electrohydraulic actuators and pedal–master cylinder assembly will be transformed to electromechanical actuators and novel human–machine interfaces such as brake feel emulators. Eventually, the development of fully autonomous vehicles will further simplify the system by abolishing the brake feel emulators.

13.5.1. Electrohydraulic Brakes

An electrohydraulic brake (EHB) is a type of brake-by-wire system that uses hydraulic components for brake actuation without connecting them to the brake pedal. In comparison to a traditional hydraulic brake system, the EHB eliminates the vacuum brake booster and replaces the hydraulic brake pedal with an electronic brake pedal to work with the pedal feel emulator. Thus, the brake force is decoupled with the pedal force. In addition, it replaces the brake pressure regulators and ABS modules with an integrated brake module, which consists of a pump driven by an electric motor, an accumulator, and a control unit, to produce and store brake pressure for independent adjustment of braking torque on each of the wheels. The diagram of a typical EHB is provided in Figure 13.9.

FIGURE 13.9 Diagram of a typical EHB.

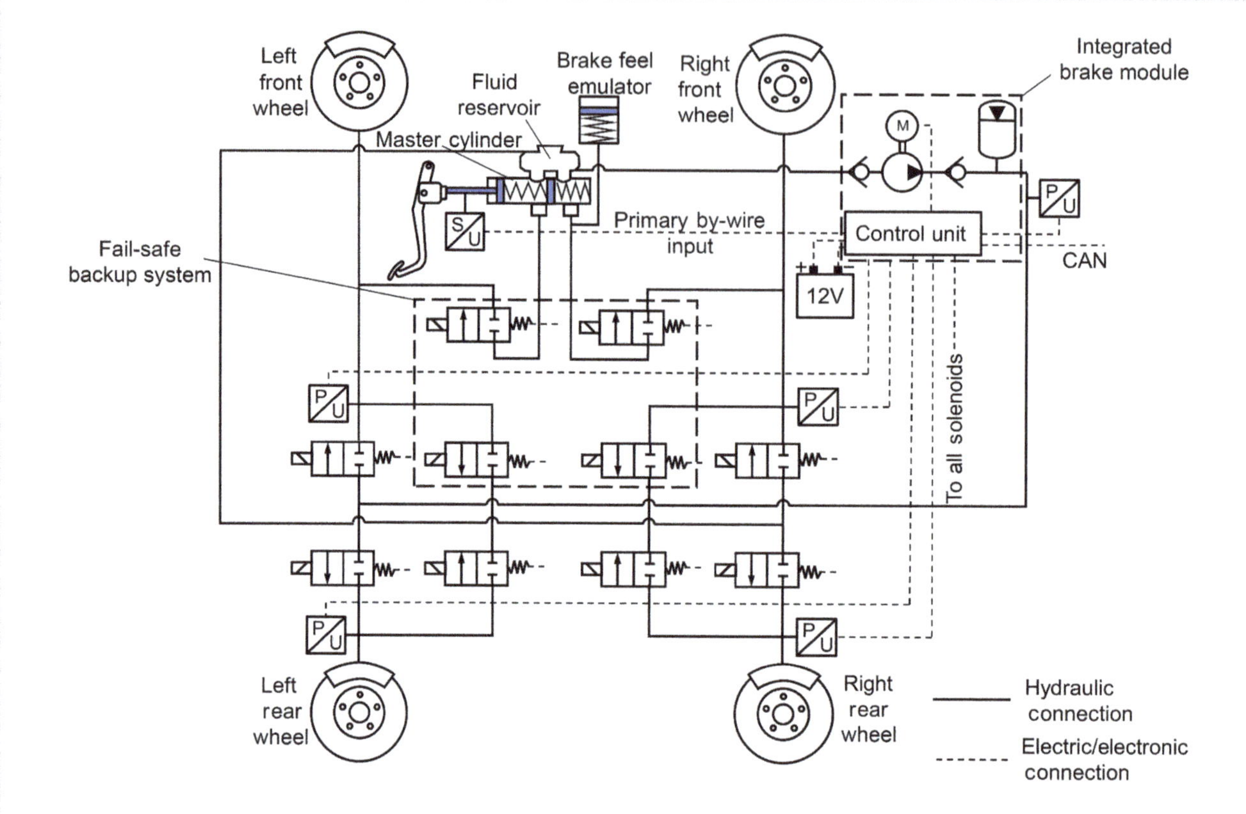

Compared to the traditional hydraulic brake system, the advantages of EHBs are primarily as follows:

1. A large DOF in braking force proportioning, resulting in enhanced adhesion utilization and braking effectiveness.
2. Easier and more reliable control, allowing compensation of changes in system characteristics after long-term or emergency use to make the relationship between brake pressure and pedal stroke always consistent.
3. Largely reduced braking noise.
4. A more compact package with no need of a vacuum booster.
5. A smooth braking process and better pedal feeling without kicking back.
6. A reduced number of parts and components as well as occupation of space with a modularized design and improved design flexibility.
7. Allowing automatic application of the service brake and parking brake to fit the needs in various operating conditions such as stop-and-go and red light hold.

13.5.2. Electromechanical Brakes

An electromechanical brake (EMB) is a type of brake-by-wire system that completely eliminates hydraulic components and brake fluid. Instead, it uses the electric motor mounted on each of the wheels for brake actuation. Figure 13.10 is a typical EMB configuration. It consists of a pedal module and a brake module on each of the wheels. The pedal module has the brake pedal, brake fell emulator, sensors, and a central ECU integrated together. The central ECU processes the signal from brake pedal sensors to determine the intention of the driver, and the brake feel emulator provides proper feedback to the driver. The braker module integrates the caliper, disc, electric motor, reducer, wheel sensor, and wheel ECU. For the rear wheels, the electric motor and reducer unit are integrated with a park brake mechanism. The central ECU determines the brake force that should be applied at each wheel based on the brake pedal sensors and wheel speed sensors. Then, the actuation command signal is sent to the wheel ECU via the EMB bus. Each wheel ECU executes the command by actuating the motor, reducer, and caliper assembly, independently from other wheels. The power needed for driving the electric motor comes from two batteries, which each serves as the backup for the other. Each battery supplies power to the two diagonal wheels, in order to retain half of the brake capacity in case of failure in one of them.

FIGURE 13.10 Diagram of a typical EMB.

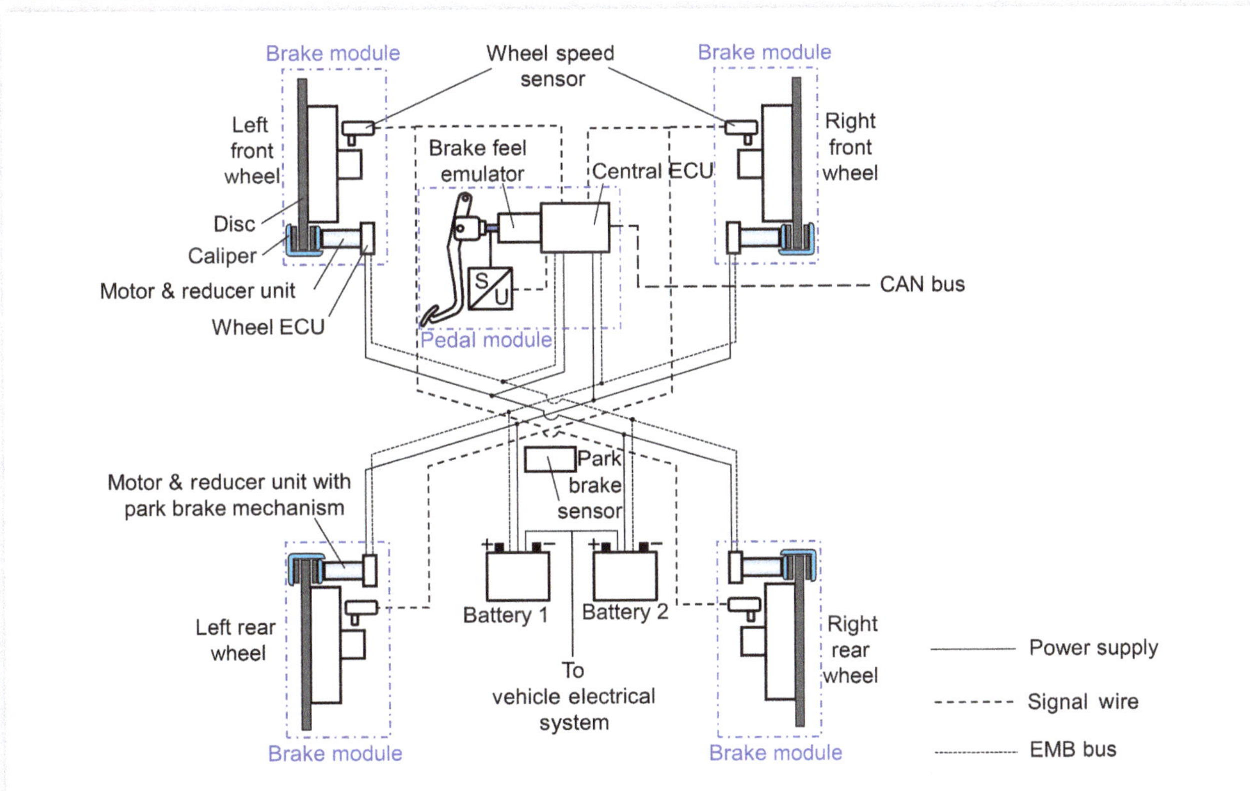

Either a brushless permanent magnet motor (BLDC) or an SRM can be used for the actuators. It is claimed that the electromechanical components in an EMB have much higher output-to-weight ratio than hydraulic or pneumatic components in a traditional brake system [13.1]. Due to the large brake force demand in an EMB, a high electrical power, for example, the maximum power of 1 kW and a stand-by power of 30 W, is required for the actuator [13.1].

The EMB offers the following advantages:

1. Significant shortened braking distance due to a response time typically less than 100 ms.
2. No need of a boost mechanism for reducing pedal effort.
3. No hydraulic components and brake fluid, which reduces system weight and wipes out the probability of environment pollution due to fluid leakage.
4. No vibration and noise feedback to the driver in the ABS mode.
5. Easy integration of functions, such as electric parking brake, antilocking, and brake force proportioning, to the system by direct addition of codes to the control unit.

However, the development of EMB technology is facing the following challenges:

1. To get rid of the hydraulic backup braking system, an extremely high system reliability is required, including a stable power supply, high fault tolerance of bus communication, and anti-interference ability of electronic circuits.
2. Insufficient braking force due to limited volume of the wheel hub for accommodating a powerful brake motor, making it necessary to adopt a system with higher voltage (such as 48 V) for raising the motor power.
3. As an unsprung mass close to the friction surface of a brake, the wheel motor must be able to withstand the intense vibration and high temperature of up to several hundred degrees Celsius. For example, due to space constraints, the braking motor can only be of permanent magnet type. However, the best magnetic material, NdFeB, is only allowed to operate under 80°C, and its magnetism disappears at 310°C. Furthermore, some semiconductors in the EMB that need to work near the brake pads cannot withstand the high temperatures there, and it is also difficult to configure a cooling system due to volume restrictions.

For these reasons, few EMBs have been successfully developed and put into mass production to date. However, the EMB will no doubtfully be the future of the brake system for EVs.

References

[13.1]. Wang, W., *Automobile Design*, 3rd ed. (Beijing: China Machin Press, 2000), ISBN:7111076133.

[13.2]. Liu, W., *Automobile Design* (Beijing: Tsinghua Press, 2001), ISBN:9787302045298.

[13.3]. Limpert, R., *Brake Design and Safety* (Warrendale: Society of Automotive Engineers, Inc.), 1992, ISBN-10:1560912618, ISBN-13:978-1560912614.

14

Chassis and Body Design

As the underpart of the EV, a chassis is the structurally independent framework on which the energy system, electric powertrain and driveline, suspension, and vehicle body are all suitably attached to or suspended from, including the wheels and sometimes the driver's seat. Thus, it is essentially the complete vehicle minus the vehicle body.

The vehicle body is the outer covering of the vehicle, typically designating the portion of the vehicle mounted on the chassis, including fenders, bumpers, windshields, doors, and windows.

The chassis and body design are of crucial importance to the vehicle, with the former significantly impacting its safety, trafficability, stability, and ride comfort, while the latter substantially affecting its safety, fuel economy, noise, vibration, and exterior appearance. Since many subsystems of the chassis, such as the battery system, powertrain, suspension, steering, and braking, have been discussed in previous chapters, the emphasis on chassis design will be herein placed on the major framework for different types of vehicles. As for the body, the discussion in this chapter will be focused primarily on the main structure of body design.

14.1. Requirements for the Chassis and Body

The design of the chassis and body for EVs should generally meet the following requirements:

1. Adequate strength and stiffness to bear the loads including the vehicle components and subsystems, passengers, and cargos and to keep the deformations within acceptable limits.
2. Low aerodynamic drag to minimize energy consumption.
3. High aesthetic attraction to please people of all ages or certain ages.
4. Lightweight but greatly resistant to impact, with reasonable crumple in an accident for minimizing forces on the driver, passengers, and pedestrians.
5. No vibration or noise, especially those excited by rotating components or road unevenness.
6. High resistance to chemical and electrochemical corrosions due to exposure to wet, salty, solar, or extreme cold environments or vibrations.
7. Good all-round visibility with minimized blind spots.

14.2. Design of an EV Chassis

14.2.1. Analysis of Load Cases

Since the chassis is the main load-bearing structure in the EV, the first stage in its design is the analysis of load cases. In general, the following five load cases should be taken into consideration:

1. Bending about y-axis: caused by the load in the x-z plane due to the weight of occupants, cargos, and vehicle components and subsystems distributed along the longitudinal direction.
2. Torsion about x-axis: resulted by the upward and downward loads at the centerline of each axle due to the road unevenness such as potholes or a bump.
3. Combined bending and torsion: generated by the collective effect of aforementioned two load cases.
4. Fore and aft loading: produced by the inertia forces in acceleration and deceleration of the vehicle.
5. Lateral loading: caused by centrifugal forces for the vehicle in cornering, which are balanced by the forces at the tire–road interface.

Note that the coordinate system and axis convention shown in Figure 8.12 are followed herewith. Since all these load cases boil down to the reaction forces at the tire–road interface, it is therefore necessary to determine the extreme and average values of the reaction forces on the tires from the road.

In generally, the reaction force in a certain direction, F, can be expressed as

$$F = F_s + F_d \tag{14.1}$$

where
F_s is the static force
F_d is the dynamic force

The static force is that for the vehicle in standstill or in uniform linear motion, while the dynamic force is produced as the vehicle diverts from standstill and uniform linear motion, thus highly dependent on the road unevenness. The static strength of the chassis and body is designed to take the forces corresponding to the worst road surface scenario, while endurance is designed with forces corresponding to the road surface.

14.2.1.1. Maximum Vertical Reaction Forces on Tires

The maximum vertical reaction forces on tires from the road for the design of static strength and endurance are given, respectively, by [14.1]

$$F_{vs} = k_s F_{z0} \tag{14.2}$$

$$F_{ve} = k_e F_{z0} \tag{14.3}$$

where
k_s and k_e are dynamic load factors for static strength and endurance, respectively
F_{z0} is the full static vertical load on the wheel

Both dynamic load factors are the functions of vertical static stiffness of the tire, K_t, which is related to its inflation pressure. Hence, the maximum vertical forces can be reduced by lowering tire pressure.

14.2.1.2. Maximum Lateral Reaction Forces on Tires

When a vehicle moves linearly on the road, the lateral force acting on the tires from the ground are stochastic in direction, which can be can be expressed by [14.1]

$$F_{ls} = \pm \mu_s F_{z0} \tag{14.4}$$

$$F_{le} = \pm \mu_w F_{z0} \tag{14.5}$$

where μ_s and μ_w are lateral force factors for static strength and endurance, respectively. Both lateral force factors are dependent on F_{z0}.

As the lateral force is associated with the vertical force, both of them must be taken into account in determining the maximum bending moment on the axle for the calculation of static strength. In the worst-case scenario, the lateral force is in such a direction that it produces a moment on the axle adding up to the moment generated by the vertical force. Since the lateral force does not reach the maximum at the same time as the vertical force, there are four load cases that need to be considered, as listed in Table 14.1 and shown in Figure 14.1 [14.1]. Note that in load cases 1 and 2, the bending moments produced by lateral and vertical forces are in the same direction for the wheels on both sides, while in the load cases 3 and 4, this is only true for the left wheel. As a matter of fact, the bending stresses in the axle resulted by the lateral force are lower for a vehicle in cornering than in linear motion.

TABLE 14.1 Four load cases related to the maximum bending moments on the axle.

	Left wheel		Right wheel	
Load case	Vertical	Lateral	Vertical	Lateral
1	F_{vs}	F_{le}	F_{vs}	$-F_{le}$
2	F_{ve}	F_{ls}	F_{ve}	$-F_{ls}$
3	F_{vs}	F_{le}	F_{vs}	F_{le}
4	F_{ve}	F_{ls}	F_{ve}	F_{ls}

Courtesy of Xiaofeng Wang.

FIGURE 14.1 Four load cases related to the maximum bending moments on the axle. (a) Load case 1, (b) load case 2, (c) load case 3, and (d) load case 4.

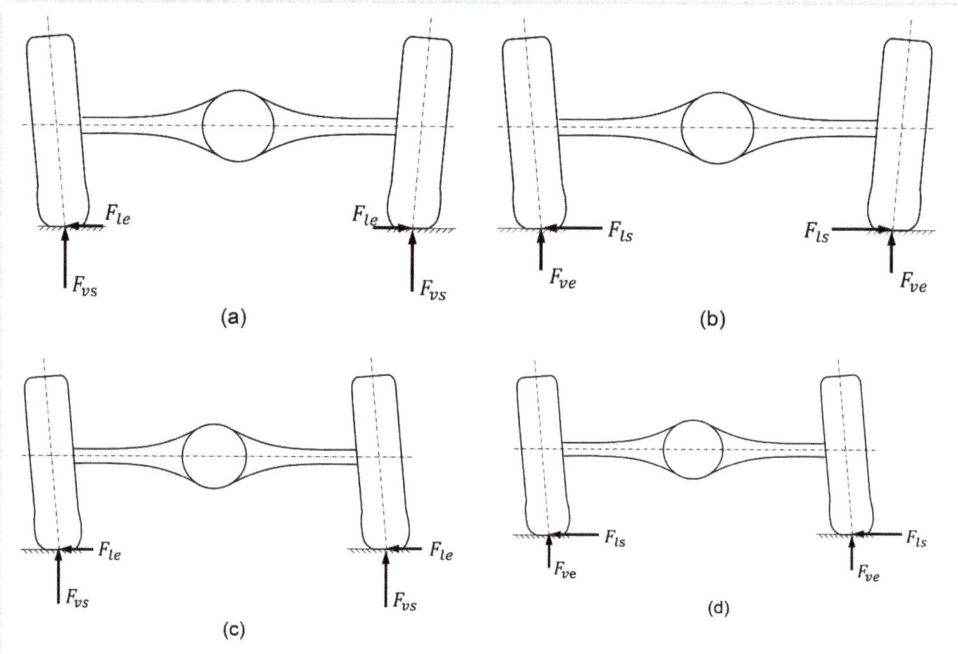

Courtesy of Xiaofeng Wang.

14.2.1.3. Maximum Braking Force

The vertical forces on the front and rear axles of a vehicle in braking are given, respectively, by [14.1]

$$F_{zf} = \frac{m}{L}\left(gL_r + ah_g\right) \tag{14.6}$$

$$F_{zr} = \frac{m}{L}\left(gL_f - ah_g\right) \tag{14.7}$$

where a is the vehicle deceleration.

Hence, the maximum braking force acting on the front axle from the ground is

$$F_{xf} = \mu F_{zf} = \frac{\mu m}{L}\left(gL_r + ah_g\right) \tag{14.8}$$

where μ is the tire–road adhesion coefficient.

The worst load case for the front axle in braking is the combination of vertical force and braking force at the maximum vehicle deceleration allowed by the tire–road adhesion. Or, the maximum vertical load and longitudinal braking load on the front axle can be approximated by

$$F_{zfmax} = 1.25F_{zf0} \tag{14.9}$$

$$F_{xfmax} = k_e F_{zf0} \tag{14.10}$$

where F_{zf0} is the full static load on the front axle. Similarly, the maximum vertical load and longitudinal braking load on the rear axle can be approximated by

$$F_{zrmax} = 0.8F_{zr0} \tag{14.11}$$

$$F_{xrmax} = F_{zr0} \tag{14.12}$$

where F_{zr0} is the full static load on the rear axle.

14.2.1.4. Maximum Tractive Force

For a vehicle in acceleration, the vertical load decreases on the front axle and increases on the rear axle. Thus, the maximum vertical load and longitudinal tractive load on the front axle can be approximated by [14.1]

$$F_{zfmax} = 0.8F_{zf0} \tag{14.13}$$

$$F_{xfmax} = F_{zf0} \tag{14.14}$$

The maximum vertical load and longitudinal tractive load on the rear axle can be approximated by

$$F_{zrmax} = 1.1F_{zr0} \tag{14.15}$$

$$F_{xrmax} = k_e F_{zr0} \tag{14.16}$$

14.2.2. Design of Vehicle Frames

The general requirements for the design of vehicle frames are as follows:

1. Adequate strength to take the worst static loads and typical loading sequence within the specified service life.
2. Adequate bending stiffness to keep the deflection of the frame within the maximum allowable limits, which is normally 10 mm for trucks.
3. Proper torsional stiffness that is usually small in the middle and large at the ends of the frame, with a preferred stiffness.
4. Lightweight enough to keep the mass of the frame as low as possible, generally within 10% of the curb weight.

14.2.2.1. Types of Vehicle Frames

The most common frames used in EVs are ladder and perimeter types.

A ladder frame consists of two longitudinal spars and several cross members, as well as mounts and brackets for mounting the body and other subsystems, as shown in Figure 14.2. Spars can be parallel to each other or arranged at a certain angle relative to each other. When subject to torsional loading, spar frames allow bending and twisting at the same time. The cross members can take different shapes, such as X-shaped, K-shaped, and tubular. Spars usually have variable lengths, with increased section height in the heavily loaded areas. Usually with a channel cross section, spars can have a closed cross section at least for part of their length. Tubular spars with a round cross section are sometimes used on sports cars for a better ratio of mass and stiffness. Frame parts can be connected by bolts, rivets, or welding. Common in small-scale production, bolted frames are also sometimes used in modern heavy trucks for easy maintenance and repair. Riveted frames are usually used in most trucks, with welded frames only found in light and super-heavy dump trucks.

FIGURE 14.2 Schematic of a ladder frame.

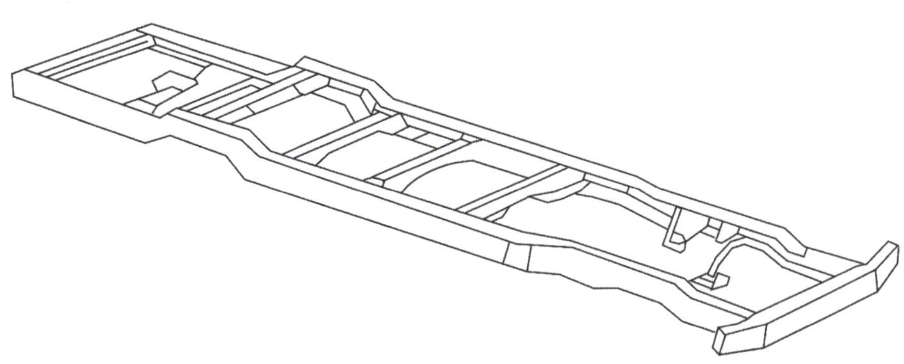

The distinctive feature of a perimeter frame is that the distance between spars is increased in the middle and narrowed at both ends, as illustrated in Figure 14.3 [14.2]. In fact, in the middle, spars can be made directly behind the door threshold when the body is installed, which is very conducive to improving the safety for the occupants in a side collision. At the weak points, that is, the places of transition from the narrow distance between the spars to the broad one, box-shaped reinforcements, often known as torque boxes, are often adopted. When subject to loading, a certain elastic deformation can occur in the torque box, allowing an angle to be formed between the spar in the middle and that at either ends. Hence, perimeter frames have the advantage of subsiding the impacts from the uneven road surface and the subsequent noise. The narrow distance at both ends helps make up adequate space for the turning of front wheels and avoid an excessively large rear wheel track. In addition, with this type of frame, the floor of the vehicle body can be significantly lowered and even placed completely between the spars, thus reducing the overall height of the vehicle. Owing to these advantages, perimeter frames are often utilized in electric passenger vehicles, despite their complex structure and relatively high costs. Note that perimeter frames are different from monocoque or unibody construction often found in traditional vehicles where the external shell, that is, vehicle body, provides a significant portion of its strength.

FIGURE 14.3 Schematic of a perimeter frame.

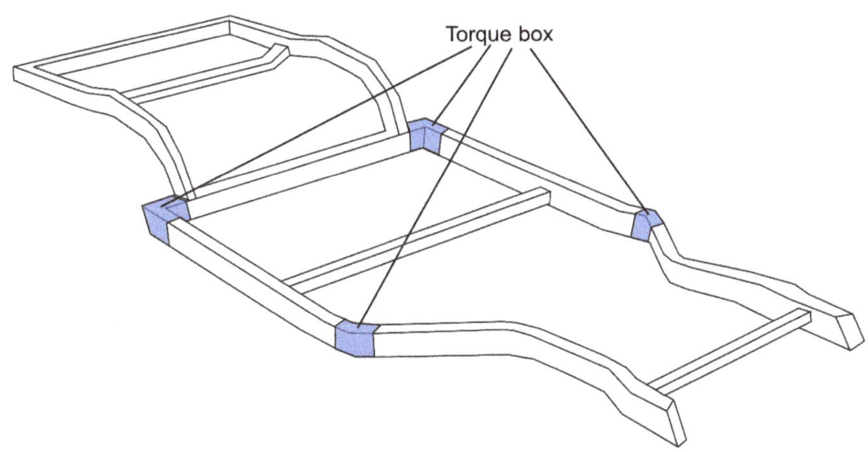

14.2.2.2. Design of Spars and Cross Members

Spars are the longitudinal beams in a vehicle frame that carry the most of the load. They often have a length almost equal to that of the vehicle, usually ranging from 1.4 to 1.7 wheelbases. The design of spars should be kept as simple as possible, with little variation in the cross section along the longitudinal direction, to simplify the manufacturing process. The height of the spar can sometimes be made large in the middle portion and small at the ends with smooth transition in between to bring the stress of each cross section to the same level. Spars can take different cross sections, such as channel, tubular, channel flitch plate, top-hat, double-hat, circular, and I-shaped. The spars of the channel cross section are widely used due to their high bending stiffness, easy craftsmanship, and simple assembly, despite their poor torsional stiffness. In contrast, the tubular cross section can be employed, instead when large bending and torsional stiffness are essential.

On the other hand, cross members, also known as traverses, are mainly used to connect the left and right spars to form a framework, giving the vehicle frame adequate torsional stiffness. Moreover, the cross members may also serve to support the subsystems such as battery packs, electric propulsion systems, and steering systems. In general, four or five cross members are used for a truck, normally with a variety of shapes for different purposes. K- and X-shaped cross members are sometimes used instead of straight ones to improve stiffness. Their cross sections can take one of the following shapes: channel, top-hat, I-shaped, double-hat, circular, and Z-shaped. The channel and tubular shapes are most frequently used, with the latter being more rigid torsional than the former.

The way of connecting the cross members to the spars can affect the local stresses around the bolt or rivet holes. There are usually three forms of connection, as follows:

1. Cross members are connected to the upper and lower flanges of a spar. This form of connection is favorable to increasing the torsional stiffness of the frame, but large stresses in the flanges of the spar can be resulted when the frame is subjected to torsion. Therefore, this form of connection is generally used for the cross members at the front and rear ends.
2. Cross members are connected to the web of a spar. This kind of connection is often adopted for cross members in the middle of spars since it usually does not cause large stresses in spars, despite poor stiffness.
3. Cross members are connected to the web and one of the flanges of a spar. This method of connection offers the two aforementioned advantages, but the forces acting on the spars can be transmitted to the cross members directly. Local reinforcement may become necessary to improve the stiffness of spars so as to minimize such a force transmission.

14.2.2.3. Sizing of Vehicle Frames

The width of a vehicle frame refers to the outside distance between left and right spar webs. The frame width is generally determined according to the overall width of the vehicle, as discussed in Section 8.4. The frame in the front should be wide enough to accommodate the electric drive system, especially when it is transversely mounted, as well as the front wheels at their maximum turning angle, or toe-out on turns. The frame width in the rear may depend on the wheel track, width of tires, width of leaf springs, and so on.

A large frame width is normally desired for the enhancement of lateral stability and reduction in the overhang length of mounting brackets. In passenger vehicles, the frame width is often small in the front to make room for wheels and independent suspension. For trucks, however, an equal width is preferred most of the time to simplify the manufacturing process and reduce the costs, although designs of wide front narrow rear or narrow front wide rear are used from time to time as well.

The width of vehicle frames should comply with the standards or regulations if applicable.

14.2.2.4. Frame Design for Strength and Stiffness

The vehicle frame is subjected to complex loading conditions that are closely related to the state of the vehicle, the road, and their interaction, as summarized in Table 14.2.

TABLE 14.2 Loading conditions on the vehicle frame and its deformation.

Vehicle state or road condition	Loading on vehicle frame and its deformation	Impact factors
Vehicle in standstill	Static vertical load equal to total sprung mass	Total sprung mass
Fairly high speed on level road	Vertical load and bending deformation in transverse symmetry	Static vertical load and vertical acceleration
Uneven road	Frame leans together with the body, producing torsional deformation	Road unevenness, stiffness of the frame and suspension
Accelerating or decelerating	Longitudinal force, causing redistribution of vertical loads on front and rear frames	Acceleration or deceleration
Cornering	Lateral force on frame	Total sprung mass, turning radius, and vehicle speed
Wheel hitting road humps	Horizontal and vertical forces on the frame	Total sprung mass, vehicle speed, and hump height
Force from vehicle subsystem	Additional local torque on the frame	Distance from the action line of force to the center of the spar cross section

Under the action of the spatial system of forces, the behavior of the vehicle frame can be solved using the FEA method. At least four loading conditions must be examined in the design of a vehicle frame, namely, vertical bending, lateral bending, longitudinal torsion, and horizontal lozenging, as illustrated in Figure 14.4 [14.3].

FIGURE 14.4 Loading conditions need to be examined in the design of vehicle frames. (a) Vertical bending, (b) lateral bending, (c) longitudinal torsion, and (d) horizontal lozenging.

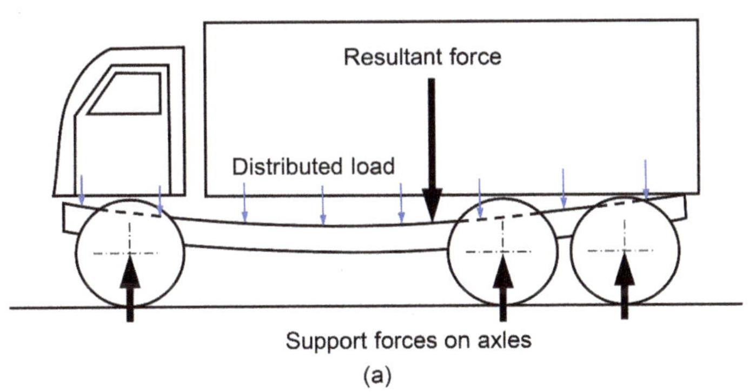

FIGURE 14.4 (Continued) Loading conditions need to be examined in the design of vehicle frames. (a) Vertical bending, (b) lateral bending, (c) longitudinal torsion, and (d) horizontal lozenging.

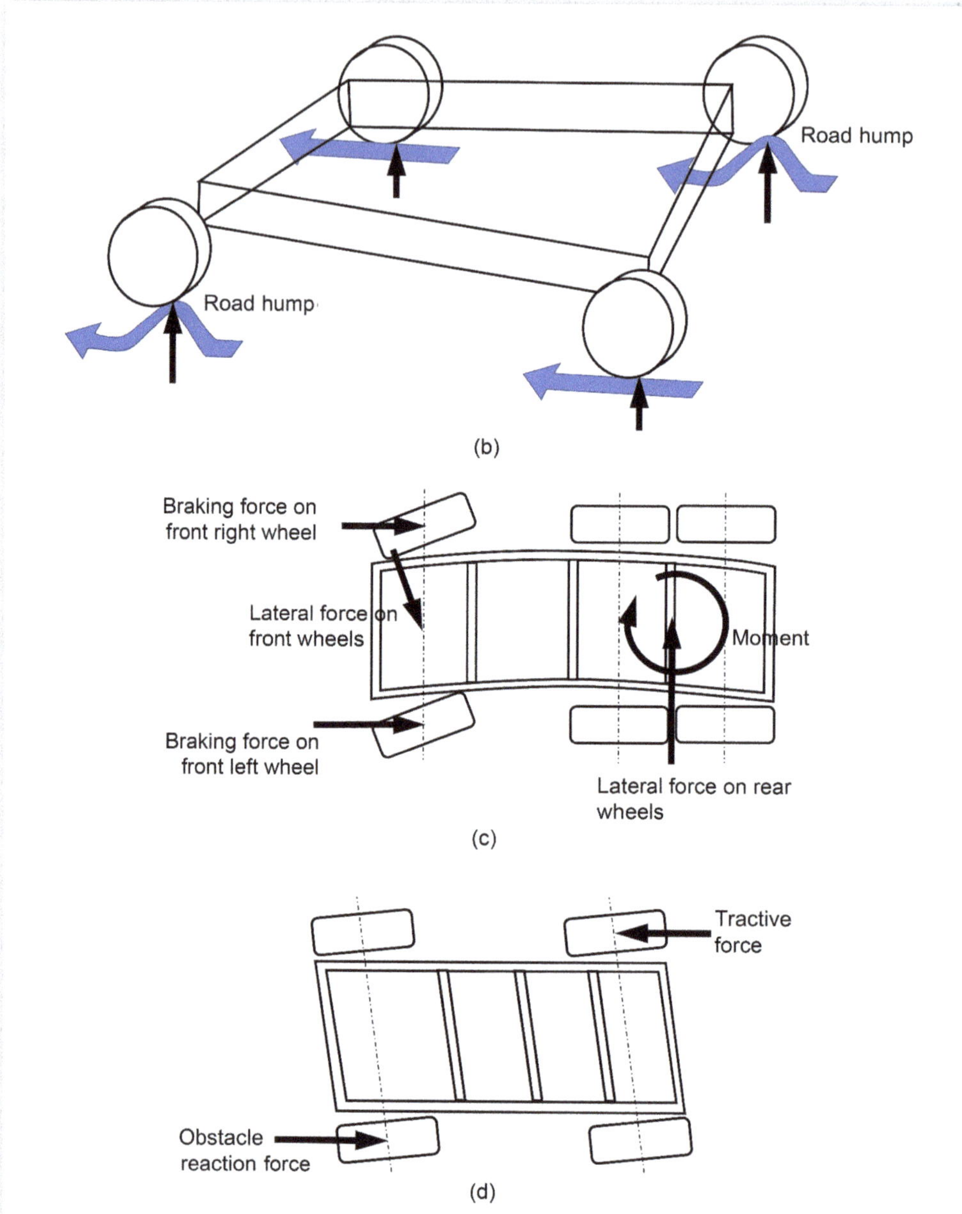

In design, torsional stiffness is often made different for meeting the specific requirements of the loads carried by different portions of the frame, although high torsional stiffness is generally desired. For the frame in the front of the vehicle, high torsional stiffness helps reduce the deflections of the mounting points of steering system components, in particular the motion interference between the steering linkage and suspension. For independent front suspensions, excessive deflections in the vehicle frame can result in an unpredictable consequence on the motion of left and right wheels, deteriorating the handling even more significantly. For a cargo truck, high tortional stiffness and strength in the rear portion of the frame are also beneficial to minimize the deflection and likelihood of structure failure under the larger share of cargo load as compared to the front. Moreover, as the middle of the frame usually carries most of the load from the battery pack, it is also desired to be adequately stiff and sturdy. In addition, it is also necessary to keep the lower order natural frequencies of the vehicle frame out of the range of the road excitation frequency to prevent resonance. Finally, a section of the frame with a certain deflection, preferably in the middle of the frame, is sometimes favored to avoid local structural damage by buffering the impacts and stress concentration.

The torsional stiffness of a vehicle frame is related to the relative torsional angle between the front and rear axles, φ, by

$$\varphi = \frac{\varphi_r}{\frac{K_{\varphi f}}{K_{\varphi s}} + 1} \tag{14.17}$$

where

φ_r is the road unevenness quantified by the angle between the two front or rear wheels
$K_{\varphi f}$ is the torsional stiffness of the vehicle frame
$K_{\varphi s}$ is the angular stiffness of the suspension including wheels, as illustrated in Figure 14.5 [14.4]

For a vehicle of a high-torsional stiffness frame and independent suspension, $\frac{K_{\varphi f}}{K_{\varphi s}} \gg 1$, and φ can be considerably larger than φ_r.

FIGURE 14.5 Torsional stiffness of the vehicle frame versus relative torsional angle between front and rear axles.

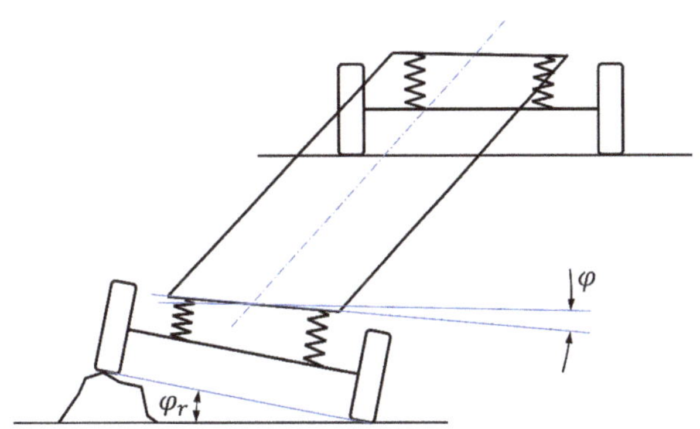

For a passenger vehicle, its framework is generally designed with large torsional stiffness. Thus, the deflection under bending moment corresponding to the nominal loading and the torsion angle under a certain torsional moment of the framework need to be examined [14.4]. In contrast, for a truck, a frame of relatively low stiffness is often employed. Consequently, the stresses in frame components caused by dynamic loading in two circumstances need to be examined. First, the fully loaded vehicle is traveling at high speeds on the road surface of little unevenness. Second, the fully loaded vehicle is traveling at low speeds with the wheel under lighter axle loading rolling over a bump of 0.3 m high on a poor road [14.4]. Special attentions must be paid to the stresses at critical sections where the cross members connect to spars, where the dimension of spars changes, or where the load is applied. Note that the vertical loads on a frame may come from assemblies, components, the cab, the cargo container, or the cargo mass attached to it. The reaction forces acting on the frame from the front and rear suspensions should also be taken into account.

For a simple estimation, the dynamic loading on a vehicle frame can be taken into account by multiplying a dynamic load factor k_d to the maximum bending moment, T_{bmax}, acting on the spars, with $k_d = 2.5 - 4.0$ in general. To prevent fatigue failures, a factor of safety, N_f, can be applied to that, that is,

$$T_{dbmax} = k_d N_f T_{bmax} \tag{14.18}$$

The bending stress is given by

$$\sigma_b = \frac{T_{dbmax}}{\psi_b} \tag{14.19}$$

where ψ_b is the shape factor for bending of beams. The maximum bending moment obtained from Equation (14.19) should be kept within the fatigue limit of the material.

Moreover, the maximum deflection of the spars is often kept within a certain limit to prevent excessive deflection from affecting the normal operation of the vehicle. By assuming the vehicle frame to be a simply supported beam with a span equal to the wheelbase L, then its deflection, δ, under a concentrated load, F, in the middle of the beam is given by

$$\delta = \frac{FL^3}{48EJ} \tag{14.20}$$

where

E is the elastic modulus of the beam material in MPa
J is the moment of inertia of the beam section in cm^4
L is in m
F is in N

A simple rule of thumb is to design the rails so that

$$\delta \le \frac{10^5 F}{576E} \tag{14.21}$$

where the deflection is in mm.

Today, the FEA method is widely used in the calculation of stress and deflection in the structure of the vehicle frame. With the static loads and constrains properly applied to the structure according to their actual positions of acting points, the results can be obtained for any nodes of interest under various loading conditions of bending, torsional, or their combination.

To take the effect of dynamic loading on the fatigue strength into account, a more precise approach is to compute the accumulated damage to the frame structure based on load spectrum data collected from real-world vehicle tests. These data, generally stochastic in nature, can be imported into the FEA software package and imposed to the structure as excitations. As a result, the dynamic response of the frame structure to the excitations can be obtained in the form of vibration modes, modal frequencies, dynamic stresses, deformations, etc. For example, the first three orders of the natural frequencies and mode shapes of a vehicle frame are shown in Figure 14.6 [14.4]. The dynamic stresses and their mean squares can be obtained accordingly for evaluating the fatigue life of the frame structure.

FIGURE 14.6 The first three orders of the natural frequencies and mode shapes of a vehicle frame. (a) 1st-order bending, (b) 1st-order torsion, (c) 2nd-order bending, (d) 2nd-order torsion, (e) 3rd-order bending, and (f) 3rd-order torsion.

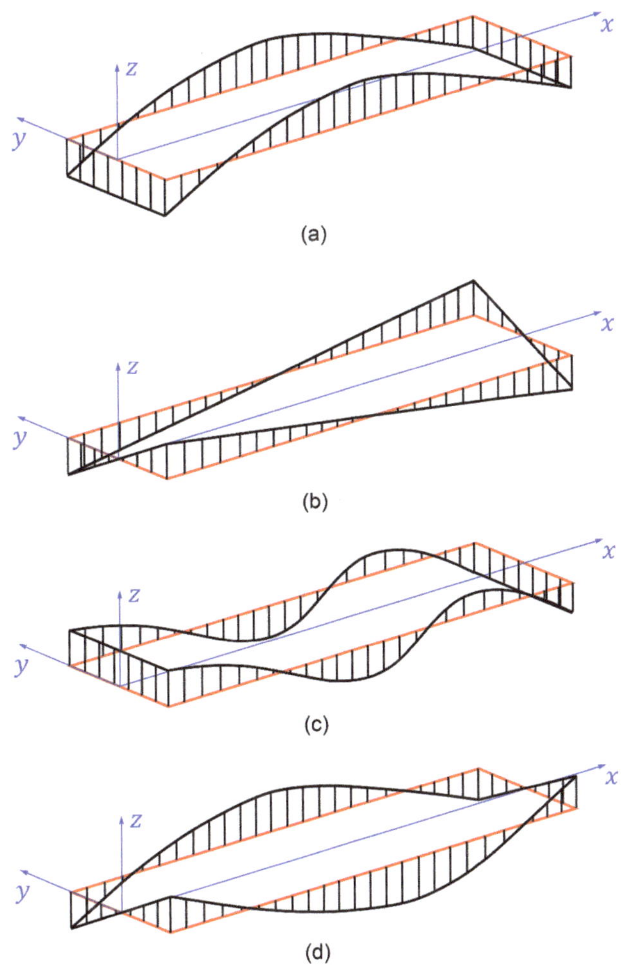

FIGURE 14.6 (Continued) The first three orders of the natural frequencies and mode shapes of a vehicle frame. (a) 1st-order bending, (b) 1st-order torsion, (c) 2nd-order bending, (d) 2nd-order torsion, (e) 3rd-order bending, and (f) 3rd-order torsion.

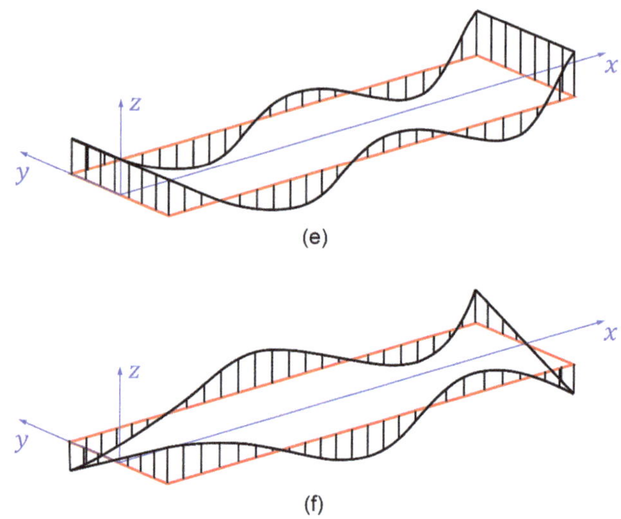

Courtesy of Tsinghua University Press & Learn Mechanical Engineering.

14.2.2.5. Skateboard Chassis

A skateboard chassis is a new type of platform specifically used for EVs, which must be complemented by a non-load-bearing vehicle body to form the entire vehicle. The vehicle body is typically attached on top of the chassis through several mounting points. As a load-carrying base structure, the skateboard chassis houses batteries, electric drive systems, drivetrains, steering systems, braking systems, and suspension into a compact, skateboard-like configuration.

Today, a state-of-the-art skateboard chassis is often designed with the following features:

1. CTC technology: Instead of battery modules, the battery cells are integrated directly into the chassis, together with BMS and thermal management system. Although not an enabling technology to skateboard chassis, CTC is reckoned to be particularly promising and suitable for the application.
2. X-by-wire technology: The mechanical linkages in traditional driving control systems, such as steering or braking, are replaced by electronic ones. This enables a complete decoupling of a vehicle body from its skateboard chassis with no need for any forces or torques being transmitted through their interfaces.
3. Domain-centralized electric and electronic architecture (E/E): This allows a dramatic cut in the number of ECUs from the traditional decentralized E/E architecture, where a one-to-one mapping between vehicular functions and ECUs is a norm. In addition, the communication loads on the vehicle network can be mitigated in comparison to the decentralized E/E architecture when one function is implemented by several coordinated ECUs. Moreover, the hardware costs of ECUs and wiring harnesses, as well as the software complexity and number of software variants, can be reduced also.

4. Decentralized electric drive system: Although a centralized electric drive system can well-satisfy the functional and performance requirements of an EV and is still utilized frequently in skateboard chassis, a decentralized electric drive system, mostly consisting of two or four in-wheel or wheel-side electric motors, is increasingly adopted, in order to simplify the driveline, free up space, and control each wheel independently for a faster response.
5. Modular and compact construction: There is a trend toward making the chassis as flat as possible by incorporating the in-wheel motor, steering, braking, and suspension subassemblies into a compact assembly, sometimes known as a corner module, which can be easily removed and replaced in case that a service becomes necessary.

The advantages of a skateboard chassis are twofold: the complexity and costs involved in the design and manufacturing of EVs can be lowered dramatically, and one such platform can fit a variety of vehicle bodies of different sizes or stylings, even for different types of vehicles or different vehicle manufacturers.

Nevertheless, the skateboard chassis also comes with some drawbacks:

1. The design of the vehicle body is subjected to additional constrains, as its strength and stiffness must properly match that of the pre-existing chassis to ensure collision safety.
2. The flat skateboard chassis imposes a severe limitation on the strokes of the spring and shock absorber, which may leave inadequate design space for tuning the suspension to achieve the optimal performance. As a result, the skateboard chassis may not be suitable for vehicle models requiring an ultimate suspension performance.
3. The wheelbase is somewhat fixed for a particular skateboard chassis, which poses a challenge to fit different vehicle models to the same chassis in view of the fact that the vehicle is, as a matter of fact, largely characterized by its wheelbase today.

14.2.3. Vehicle Frame Testing

The validation of vehicle frame design generally involves rig tests on stress, stiffness, reliability, and durability, as well as proving ground and road tests on endurance of the vehicle.

14.2.3.1. Stress Tests

Stress distribution can be obtained indirectly by measuring the deformation using strain gauge foils on various parts of the frame of the test vehicle under static and dynamic conditions on a road simulation test bed or on a proving ground.

14.2.3.2. Stiffness Tests

Stiffness tests usually involve bending and torsional stiffness tests of the vehicle frame.

In a bending stiffness test, the frame is rigidly supported at the front and rear axles and is loaded according to the actual condition of the vehicle in operation. The deformations are measured using displacement sensors.

In a torsional stiffness test, the frame is usually loaded by applying clockwise and counterclockwise torques at the mounting points of front shock brackets. The frame must be properly fastened to the test bed to avoid the effect of the test bed on the stiffness of the frame. The deformations of the frame are also measured by displacement sensors.

14.2.3.3. Reliability and Durability Rig Tests

These bending and torsional fatigue tests of the frames are carried out on a fatigue test bench, which applies a dynamic loading on the frame either using repeated constant amplitude or variable amplitude based on road load spectrum data.

14.2.3.4. Modal Tests

In modal tests, the vehicle frame is suspended freely in the air using a rubber rope positioned right at the node of its low-order mode shapes. The frame is excited at a fixed point of its lower part, preferably away from the nodes of mode shapes. An electrodynamic shaker is often used to excite the frame in the z direction, and the response is acquired by a three-axis accelerometer.

14.2.3.5. Proving Ground and Road Endurance Tests

These tests are generally conducted using a fully loaded vehicle on a grooving ground or roadways for the validation of the bending and torsional fatigue strength. Cumulative damage to the frame and the fatigue life of a vehicle can also be predicted in virtual proving ground using computer simulation software packages to save the time and costs for building and testing a real vehicle.

14.3. Design of an EV Body

The body of an EV consists of a body-in-white (BIW) and accessories. Traditionally, the BIW refers to the body shell before painting and installation of powertrain subassembly, any chassis subassemblies, or trims such as glass, seats, handles, and locks. It is generally a welded assembly of sheet metals; however, lightweight materials such as aluminum, carbon fiber, or other composite materials are increasingly utilized in EVs, with processes of riveting, laser brazing, clinching, and bonding other than welding for joining components together. On the other hand, body accessories are components or devices attached to the vehicle body, such as windshield glass, windshield wiper, defroster, lighting device, horn, air conditioning, and ventilation.

14.3.1. Body Styling of EVs

As an important means of transport, EVs must not only provide the functions and performance in moving people or cargos to the destination but also provide artistic quality and aesthetic attraction to everyone or at least the segment of targeted customers. Body styling is therefore an essential part of body design, in which some fundamental aesthetic principles must be obeyed, as elaborated in the following subsections.

14.3.1.1. Sense of Integrity

The entire body must be harmonious and coherent, with coordinated proportions among all major parts or design elements and also with well-organized line patterns. The proportions involving the overall length, width, and height of the vehicle body, the dimensions of the front face and tail, as well as doors and windows, are all critical to body styling. By repeating one or several proportions or using proportions based on one or a number of sequences can often yield a concordant and congruous result. In particular, based on classic aesthetic theory, the golden ratio is often used in the body styling of EVs. For example, the statistical data of the wheelbase versus overall length of the production EVs from different regions of the world, as shown in Figure 14.7, indicates that the golden ratio is approximately followed, despite trivial deviations resulting from some extremely proportioned vehicles. In addition, proper organization of line patterns can give an extraordinary sense of integrity. Repeated design features, such as side windows and their frames of a bus, offer a feeling of harmony. Sometimes, scaled curves can be employed as the profiles of body parts of a variety of sizes to create an appealing look. Moreover, the use of radial, parallel, and perpendicular lines; curves of equal curvature; or similar shapes can also produce a pleasant appearance.

FIGURE 14.7 Wheelbase versus overall length for major production models of passenger EVs.

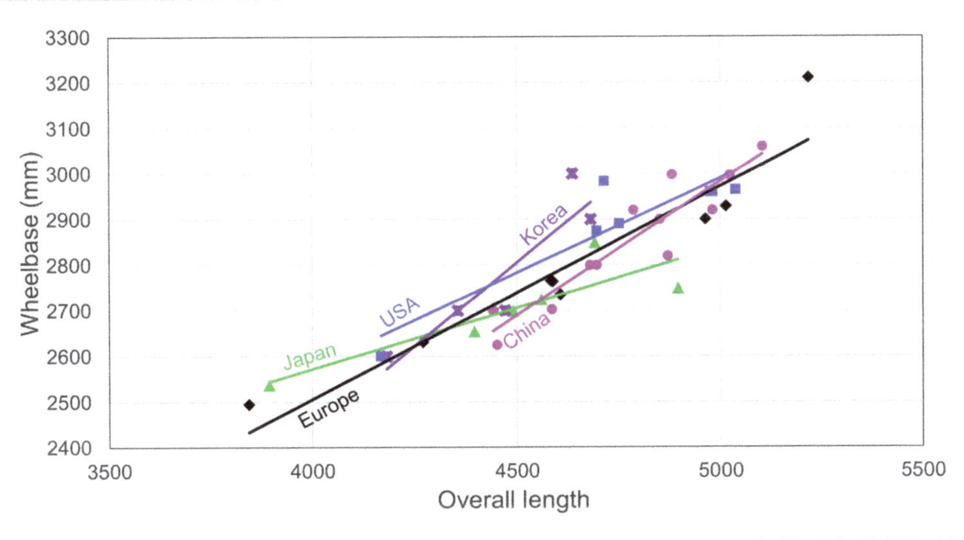

Courtesy of Tsinghua University Press & Learn Mechanical Engineering.

14.3.1.2. Dynamic Sensation

Since the EV is a mobile device, it is generally desired to create the dynamic sensation out of its body style. The first method is to shape the vehicle body by mimicking insects, for example, beetles, or, animals such as horse, cheetah, and ram can be made into decorative markers or logos. Besides, lively and streamlined profiles and smooth surfaces can enhance the dynamic sensation of a vehicle body. The mathematical meaning of a smooth surface is that there is only a unique curvature for every point on the surface of a vehicle body section in a given direction. A smooth surface can be obtained readily using the computer-aided engineering (CAE) methodology. Another way of improving the dynamic sensation of a vehicle body is emphasizing the horizontal dividing lines and subsiding the vertical ones. Horizontal decorative strips, embossed lines, or two-tone colored bodies separated by a horizontal line are utilized from time to time in practice to give the vehicles a sense of motion. To weaken the vertical dividing lines, it is often desired to reduce the number of vertical pillars and their widths, to increase the distance between them, or to use a pair of eye-catching horizontal plated strips along the top edge and sill of side windows. Brown glasses are sometimes utilized for side windows to cover up the vertical pillars and visually integrated them into a harmonious whole.

14.3.2. Body Construction

The normal body construction of an electric car includes major components such as front panel, rear quarter panel, roof, floor, A pillar, B pillar, and C pillar from the front to the rear.

14.3.2.1. Unibody Construction

In most passenger vehicles today, the vehicle body is unified with and structurally integrated to the chassis mainly for weight reduction, which is referred to as the unibody or monocoque construction. The distinct feature of the unibody construction is that the loads are supported by the external cover, not by the frame.

The unibody construction offers the advantage of substantially reduced vehicle mass and height. Hence, the energy economy can be improved, and the floor can be lowered for easy access to the cabin. Nevertheless, it comes with two primary drawbacks: the noise and vibration from suspension and driveline can transmit into the cabin directly, causing cavity resonance and, thus, deteriorating the ride comfort, and the body is prone to large deformation when four wheels are not uniformly loaded due to the relatively low stiffness of the structure.

14.3.2.2. Body-on-Frame Construction

In contrast to the unibody construction, nearly all buses, trucks, and most pickups use a structural frame separate from its body, which is known as the body-on-frame construction. With this the construction, the majority of loads on the wheels and suspension are carried by frames, while only a small amount of loading, often caused by the deformation of the frame, is borne by the body. Multiple elastic rubber pads can be used between the vehicle body and the frame to absorb such deformation.

The body-on-frame construction offers a few advantages: The manufacturing process can be simplified by separating the assembly of the chassis and body; the impact loading from roads can be largely absorbed by frames, thus damping out vibrations and noise transmitted into the body; vehicle frames can protect the body in a collision; and the independent chassis enables platformization, which makes it easy to develop multiple vehicle models using the same chassis.

However, the shortcomings of the body-on-frame construction are also worth noting: The needs for adequate strength and stiffness of vehicle frames often result an excessive vehicle mass, and the frame sandwiched between the chassis and floor inevitably results in an elevated vehicle height.

14.3.3. Aerodynamic Design of Vehicle Bodies

Due to the relatively low energy density of the on-board energy source, the aerodynamic design of an EV is crucial to its energy economy and driving range.

14.3.3.1. Ideal Aerodynamic Body Shapes

The aerodynamic drag to a vehicle is mainly dependent on its frontal area and drag coefficient C_D. Since the frontal area is closely related to the height and width of the vehicle body, which are not subject to much reduction, the drag coefficient C_D, therefore, becomes the major parameter in slashing the aerodynamic resistance.

Several ideal aerodynamic body shapes for a ground vehicle are summarized in Table 14.3 in terms of the minimal achievable drag coefficient C_D, according to the drag measurements performed by W. Klemperer in 1922 [14.5].

TABLE 14.3 Minimal drag coefficients of ideal aerodynamic body shapes.

Diagram of body shape	Description of body shape	Minimal C_D
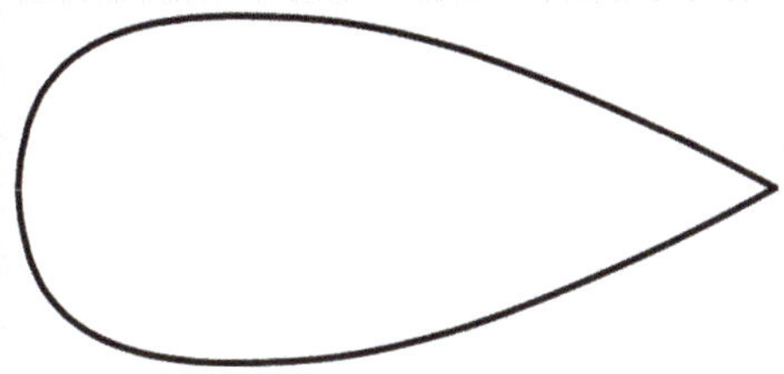	Streamlined body without wheels	0.05
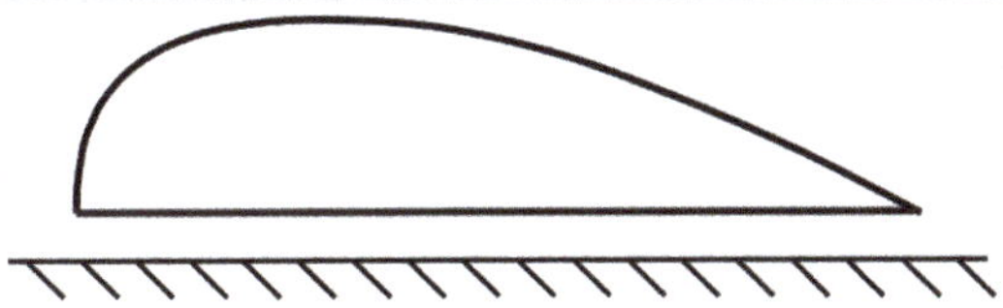	Streamlined half-body without wheels	0.13
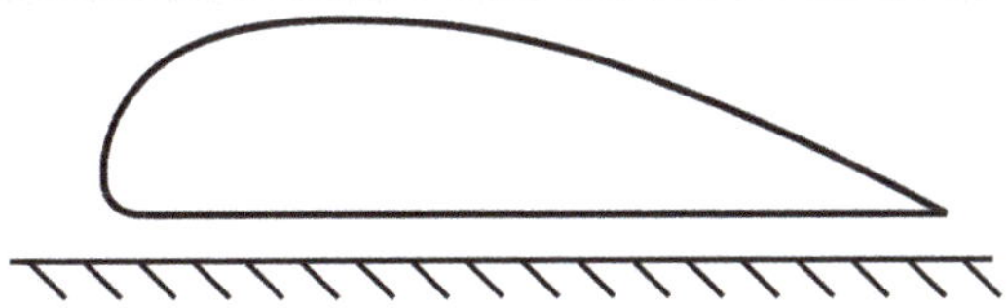	Front-rounded streamlined half-body without wheels	0.09
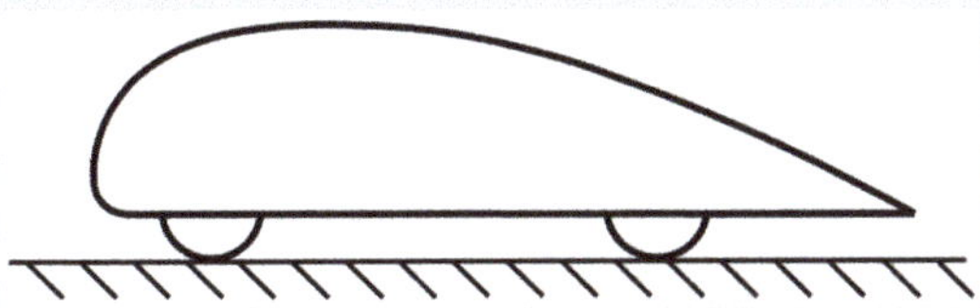	Front-rounded streamlined half-body with wheels	0.15

The aforementioned ideal aerodynamic shapes give us a good reference to the minimum drag coefficient, which can possibly be achieved for a ground vehicle. Nevertheless, such ideal shapes are generally not only impractical but also unattractive. For aerodynamic and safety reasons, straight lines, polylines, and sharp turns are rarely used in an EV. Rather, smooth curves and transitions at large angles are employed instead to make the body shape closely resemble the ideal ones. Additionally, the vehicle body can be kept as streamlined as possible by embedding exterior parts, such as door handles, mirrors, and light bulbs in the doors or the vehicle body, to further reduce the drag coefficient.

14.3.3.2. Types of Aerodynamic Drag

The aerodynamic drag on the vehicle body consists of two different components, namely, pressure drag and skin friction drag. The pressure drag is caused by pressure forces acting normal to the body surface, while the skin friction drag is a result of the viscous friction force acting tangential to the body surface.

The pressure drag comes mainly from the pressure difference, exterior accessories, interior air flow, and lift induction. The pressure difference drag is a result of the higher pressure in the front where air is compressed by the vehicle body in motion and the lower pressure in the rear where air is entrained back as vortices. Another form of pressure difference drag is produced by air flow separation from the body surface. As the most significant portion of the aerodynamic drag, the pressure difference-related drag usually takes up over 58% of the total drag. Besides, the exterior accessories-related drag is caused by the accessories that protrude from the body surface, which takes up roughly over 10–14% of the total drag.

Moreover, the interior air flow-related drag refers to the drag produced by the air circulation inside the cabin and front compartment, which also accounts for above 10–12% of the total drag. Finally, associated with the pressure redistribution in the production of a lift force due to trailing vortices, the lift-induced drag accounts for approximately 6–8% of the total drag.

On the other hand, the skin friction drag is generated by the viscosity of the air and the velocity gradient of the airflow along the direction normal to the body surface, which is usually about 8–10% of the total drag. An excessive velocity gradient can cause the transition of airflow from laminar to turbulent, further increasing the drag.

14.3.3.3. Design Principles for Aerodynamic Drag Reduction

The general rules for minimizing the aerodynamic drag coefficient are mainly as follows:

1. Front end

   With the vehicle moving forward, the air flow is divided into upward, downward, leftward, and rightward streams by the front end of the body. A round and smooth front end would have the advantage of separating the flow without inducing any vortices and, thus, keeping the drag coefficient low. In addition, mounting a spoiler under the front end can also reduce the drag coefficient, mainly by bringing down the lift-induced drag of the underbody.

   In order to reduce the aerodynamic drag, the front grille, like the one in traditional ICE-powered vehicles, is minimized or completely eliminated in an EV. In fact, the cooling requirement of an EV can be satisfied by a totally different radiator layout and a front grille is no longer a necessity. However, since the front grille has been the key part in the design language of vehicle body styling, it is sometimes preferred to retain it with part of the mesh closed or even a faux one. Today, the design of the front fascia for EVs is still evolving, mainly from mostly operational to strictly ornamental.

2. Front windshield

   The angle of the windshield has considerable effects on the drag coefficient. In a certain range, the smaller the windshield angle, the lower the drag coefficient. However, if the angle becomes too small, either the cabin size or the body proportion can be adversely affected. A great deal of test data has indicated that a windshield angle of 25–30° allows an optimal balance among all design targets.

3. Roof

   Compared to a flat roof profile, an increase in the roof profile depth helps reduce the drag coefficient to a certain extent. The mechanism behind this phenomenon is deemed to be the increased base pressure possibly due to enhanced mixing of longitudinal vortices shed from the upper side edges of the rear body and increased effective boat-tailing angle in the near-wake [14.6]. However, the roof profile depth must be carefully designed to balance the aerodynamic drag with the vehicle dimensions, as the frontal area becomes larger at the same time.

4. Rear windshield

   If the rear windshield curves downward sharply from the roof, it becomes difficult to control the airflow. Thus, large wake vortices may develop, leading to a rise in the drag coefficient. A smooth transition from the rear windshield to the tail is therefore recommended, such as the fastback body style, with the angle of the rear windshield made as small as possible, in order to reduce the drag coefficient.

14.3.3.4. Design Principles for Aerodynamic Stability

As the vehicle moves forward, it is subject to three aerodynamic forces, namely, resistance force F_D, side force F_S, and lift force F_L, as well as three aerodynamic moments, namely, roll moment M_R, pitch moment M_P, and yaw moment M_Y. Hence, the joint action point of all of these forces and moments is called the aerodynamic center, which is dependent on the exterior shape of the vehicle.

Under the action of an upward lift force, the road–tire adhesion declines, causing the driving force to decline and worsening the acceleration. At high speeds, the upward lift force on the front axle can reach as high as 20–25% of the total mass. In cornering, gusty wind, or other interference from the road, the handling and stability of the vehicle are also compromised. These effects are particularly sensitive for lightweight vehicles with a CG leaning to the rear. Hence, it is required in engineering practice that the lift force of a road vehicle be negative, that is, a down force, which is exactly opposite to that of an airplane.

To reduce the lift force on a vehicle, the following measures are often taken:

1. Use a negative angle of attack. The angle of attack is defined as the angle between the oncoming flow and the reference line of the body, as shown in Figure 14.8 [14.4]. For a vehicle body, the oncoming flow is assumed to be parallel to the road surface, and the chord line is taken as the reference line. The chord line is simply the straight line connecting the intersections of the camber line with the leading edge and trailing edge, while the camber line is an imaginary line that lies halfway between the upper surface and lower surface of the body. In general, a negative angle of attack can be obtained by lowering the hood and making the tail bulky and upturned. A negative angle of attack helps reduce the amount of airflow entering the bottom of the vehicle and prevent blockage to the underfloor airflow. As a result, the air flow is directed to the low pressure region behind the vehicle at a certain velocity, thus reducing the lift force.

FIGURE 14.8 Camber line, chord line, and angle of attack for a vehicle.

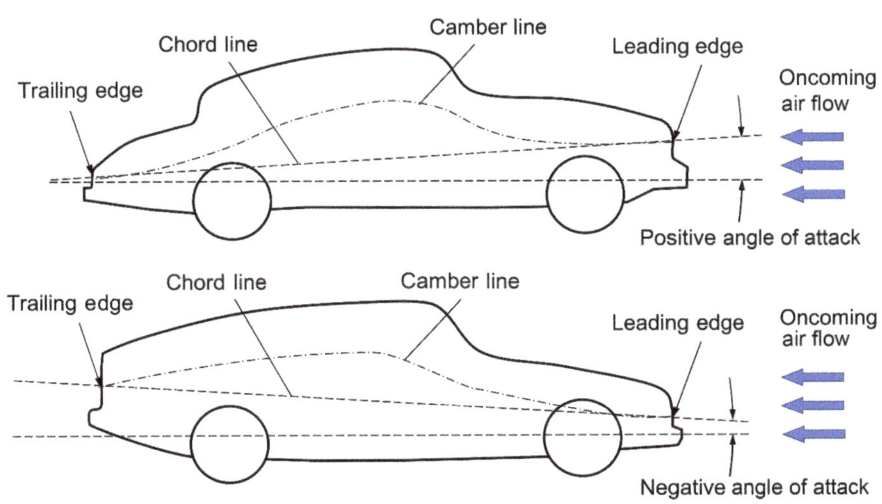

2. Mount a spoiler to the bottom in the front end of the vehicle. A spoiler is an automotive aerodynamics device used for nullifying unfavorable air movement across a body of vehicle in motion. When fitted at the bottom of the nose or front bumper of a vehicle, it is often called an air dam or front splitter. They can cut the amount of underfloor airflow and maintain a certain flow velocity toward the rear and side, thus slashing the lift coefficient. The lift coefficient is defined as the ratio of lift force to the product of surface area and dynamic pressure, that is, $C_L = F_L / \left(\frac{1}{2}\rho V^2 A\right)$. Listed in Table 14.4 are examples of the spoiler effect on the lift coefficient of a car [14.4].
3. Raise the rear portion of the vehicle floor at an angle toward to the rear and raise the floor slightly toward both sides to divert the underfloor airflow. Consequently, the underfloor airflow velocity is increased and the lift force is reduced.

TABLE 14.4 An example of the effects of spoilers on lift coefficients of a car.

Model	Schematics	Lift coefficients		
		Front axle	Rear axle	Total
Baseline		0.301	0.126	0.427
Baseline with rear spoiler		0.313	0.007	0.320
Baseline with both front and rear spoilers		0.280	0.016	0.296

Courtesy of Tsinghua University Press.

The proper position of the aerodynamic center with respect to the CG is of significance to the aerodynamic stability of the vehicle. When the aerodynamic center is located in the front of CG, a yaw moment is produced in such a direction that the action of lateral wind is strengthened; in contrast, when the aerodynamic center is located in the rear of CG, a yaw moment is created in such a direction that tends to weaken the action of lateral wind. Hence, a rule of thumb in body styling is to make the lateral projection area of the front body as small as possible so as to move the aerodynamic center toward the rear.

14.3.3.5. Trend of Drag Coefficients of EVs

The statistical data of the drag coefficients of production electric passenger vehicles are presented in Figure 14.9. Clearly, the average has been ever decreasing in the past decade, as indicated by the trendline. Today, with the lowest drag coefficient approaching 0.2, a further drop in the drag coefficient is very likely in a continued pursuit of energy economy and driving range.

The drag coefficients of electric trucks have lowered substantially as well. Currently, the lowest drag coefficient might soon reach 0.35 for heavy-duty electric trucks and 0.3 for electric pickups.

FIGURE 14.9 Statistical data of the drag coefficients of production EVs.

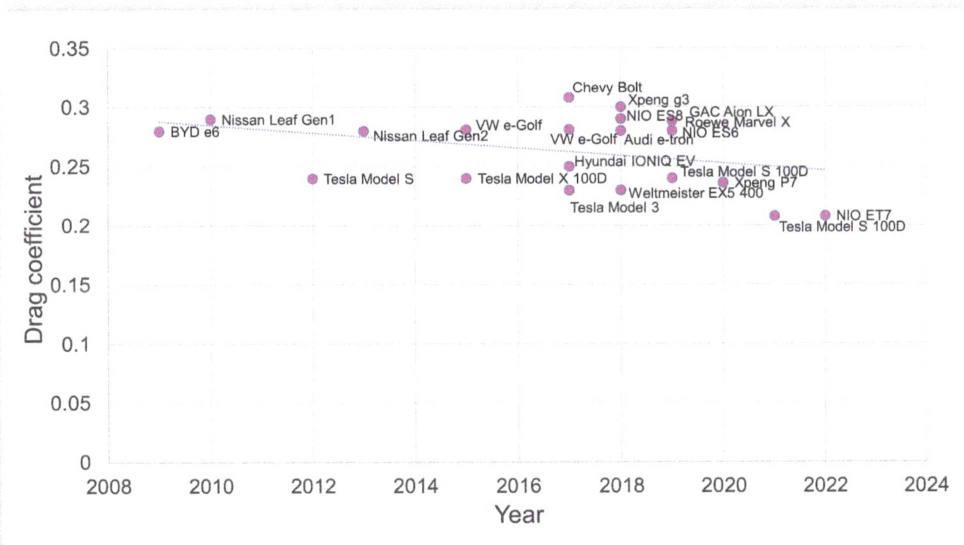

14.3.4. Structural Design of Vehicle Bodies

The requirements for the structure of the vehicle body are as follows:

1. Adequate strength and stiffness for the design life: If the stiffness is inadequate, large deformations can occur at body openings such as window frames, door frames, front compartment, and trunk, causing cracks in glasses or a door stuck in its frame. In addition, a structure of inadequate stiffness is prone to resonance and noise associated with its low natural frequency, thus weakening the connection strength of structural joints. Body parts of low strength is susceptible to failures such as cracks or fracture.

2. Lightweight: Since the vehicle body takes up a large percentage of the vehicle mass, its lightweight design helps reduce energy consumption significantly.
3. Good manufacturability: The body of an EV is often made of different types of materials for weight reduction. It, therefore, becomes increasingly important to reduce the complexity and costs involved in the manufacturing, repair, and replacement process of the body parts and their interconnections. In particular, the vulnerable body parts prone to damage in collisions should be made separate to facilitate the repair.

14.3.4.1. Design Principles for Crash Safety

To ensure crash safety, it is generally desired to design the vehicle body so that it deforms in a predetermined way under the impact of a crash, while the deformation energy can balance out the kinetic energy of the impact. The structure of the vehicle body must be able to minimize the force and acceleration acting on the occupants in order to avoid their injuries. In a crash, the body structure must preserve at least the minimum survival space to keep injury and fatality levels as low as possible and make it possible for the occupant to escape or to be rescued.

In a vehicle crash, the occupants are mainly injured by the acceleration or deceleration of the vehicle and the subsequent forces. Such an acceleration or deceleration can last for a certain duration, causing body and even brain injuries to the occupants. Metrics such as severity index (SI) and head injury criterion (HIC) have been established and widely used to evaluate the severity of a crash.

The SI is a measure of the severity of a crash or series of crashes, which is defined by

$$SI = \int_0^T a^{2.5} \mathrm{d}t \tag{14.22}$$

where

a is the resultant head acceleration in a crash
T is the complete duration of deceleration

The HIC is a measure of the likelihood of head injury arising from an impact, which is defined as

$$HIC = \max_{t_1, t_1} \left[(t_2 - t_1) \left(\frac{1}{t_2 - t_1} \int_{t_1}^{t_2} a(t) \mathrm{d}t \right)^{2.5} \right] \tag{14.23}$$

where t_1 and t_2 are the initial and final times chosen to maximize HIC, and the time duration $t_2 - t_1$ is usually 15 ms, with an upper limit of 36 ms.

It is generally recommended to keep both SI and HIC within 1000 for the structural design of a vehicle body to reduce the probability of a severe head injury.

The vehicle body design must also comply with applicable motor vehicle safety standards, which are established primarily for protecting the occupants from impacts during and after crashes. These standards impose requirements on the vehicle or its parts and components to pass specific validation tests. Under specified loading conditions, a design is only acceptable when the structural response is equal to or better than the specified criteria.

Motor vehicle safety standards generally give detailed requirements for the test conditions and results, mainly involving the vehicular structure, door assemblies, bumper assemblies, and seat assemblies. For example, in the FMVSS of the US, the vehicular structure-related tests include three impact tests and a roof crush resistance test, as elaborated briefly in the following:

1. Front barrier crash
   In this test, the test vehicle impacts a fixed collision barrier that is perpendicular to the line of travel or at any angle up to 30° in either direction from what is perpendicular to the line of travel when the vehicle is traveling longitudinally forward at any speed up to and including 48 km/h.
2. Rear moving barrier impact
   The test vehicle is impacted from the rear by a barrier that is moving at any speed up to and including 80 km/h (50 mph).
3. Side moving deformable barrier impact
   The test vehicle is impacted from the side by a barrier that is moving at any speed up to and including 54 km/h.
4. Postimpact test static rollover
   The test vehicle is rotated on its longitudinal axis to each successive increment of 90° after each impact test specified.

   The following requirements are examined after the aforementioned four impact tests. First, electrolyte spillage from propulsion batteries: Not more than 5.0 L of the electrolyte from propulsion batteries shall spill outside the passenger compartment, and no visible trace of the electrolyte shall spill into the passenger compartment. Spillage is measured from the time the vehicle ceases motion after a barrier impact test until 30 min thereafter, and throughout any static rollover after a barrier impact test. Second, electric energy storage/conversion device retention: During and after each test, electric energy storage/conversion devices shall remain attached to the vehicle by at least one component anchorage, bracket, or any structure that transfers loads from the device to the vehicle structure, and electric energy storage/conversion devices located outside the occupant compartment shall not enter the occupant compartment. Third, electrical safety: After a test, each high voltage source in a vehicle must meet one of the following requirements, that is, the electrical isolation requirements of the subparagraph, the voltage-level requirements of the subparagraph, or the physical barrier protection requirements of the subparagraph. The subparagraphs are omitted here. The interested reader is referred to the cited references for detailed subparagraphs [14.7].
5. Roof crush resistance
   In order to limit the roof protruding into the passenger compartment in a rollover accident, the roof of a vehicle must meet certain static load requirements; that is, the loading device has a rigid lower surface with a rectangular (76 cm × 183 cm) plane, with the loading direction of 25° in the front view and 5° in the side view. The loading force is 1.5 times the vehicle weight or 22,240 N, whichever is smaller. It is required that the movement of the test device not exceed 127 mm. Both the left and right front corners of the vehicle must meet this requirement, but a vehicle only needs to be tested in any one of the two locations [14.8].

To meet the crash safety and occupant protection requirements, two design principles must be followed in the design of vehicle bodies: The specified portion of the vehicle body should be able to deform in the collision for crash energy absorption, thus reducing the acceleration and impact force as well as keeping the SI and HIC within specific values, and the deformation of the vehicle body must be properly controlled to ensure the integrity of the cabin, in order to provide a survival space for occupants. In other words, the front compartment and the rear trunk are usually designed with a relatively low stiffness for crash energy absorption, while the cabin should have a high stiffness for occupant protection.

Both metals and composites can absorb kinetic energy in a collision, but the mechanisms are different. The metals absorb the energy through plastic deformation, while composites through fragmenting. The behavior of metals in a crash can be predicted accurately using FEA methods, whereas it is much more difficult to predict the behavior of composites.

14.3.4.2. Lightweight Design Principles

Often used to assess the lightweight level of a BIW, the lightweight index, L, is defined as

$$L = \frac{1000 m_{BIW}}{K_T A} \tag{14.24}$$

where m_{BIW} is the mass of BIW without doors, hood, and trunk cover, K_T is the static torsional stiffness of BIW with body accessories such as windshields and subframes mounted in place, and A is the projected area of BIW depending on the wheelbase and track.

The lower the lightweight index, the better the lightweight design. A good lightweight design of an electric passenger vehicle today has a lightweight index, normally falling in the range of 2.0–2.5.

The material properties of several frequently used lightweight materials are compared with steel, as in Table 14.5 [14.9].

TABLE 14.5 Comparison of serval frequently used lightweight materials with steel.

Material	Density (kg/m^3)	Fracture strength (MPa)	Young's module (GPa)
Mild steel	7850	465	207
Stainless steel (FSM 1)	7855	980	185
Aluminum alloy (DTD 5050B)	2810	500	71
Magnesium alloy (AX 31)	1780	185	45
Carbon fiber-reinforced plastic (58%)	1500	1050	189
Glass-reinforced plastic	2000	1240	48.2

14.3.4.3. Noise and Vibration Considerations in Vehicle Body Design

When subjected to the excitation from road unevenness, the body structural panels may vibrate and radiate noises, mainly in the range of 20–200 Hz. These noises are amplified by resonance through the cavity of the passenger compartment. However, due to the large number of structural modes involved and the significance of phase relationships between the movements of panels in the frequency range of interest, it is often resorted to combine the method of FEA and testing in corrective structural modifications [14.3].

The FEA methods have been well-established for solving the modes of body panel vibrations and the acoustic field in the cavity of the passenger compartment. The resonance in the enclosure of the passenger compartment can be characterized by the acoustic vibration modes related to the pressure distribution in the form of mode shapes and natural frequency. For forced vibrations, the total pressure at any internal position is dependent on the way of excitation of every internal mode, while the acoustic modes in the cavity are coupled to the vibrations of the surrounding structure through boundary conditions. Hence, the internal sound pressure, or noise, obviously depends on the acoustic characteristics of the cavity and the dynamic properties of the structure [14.3].

The frequency and mode shape of the resonance in the passenger compartment can be predicted by using FEA methods. If the pressure variation in the transverse direction of the vehicle body is neglected, a two-dimensional analysis can be carried out to give a simplified result in the longitudinal vertical cross section of the passenger compartment. In the initial stage of design, the body panels are often assumed rigid, and the frequency and mode shape can be predicted as long as the geometry of the passenger compartment is provided to the FEA model as input data. Consequently, the first longitudinal acoustic mode in the passenger compartment of a passenger vehicle, the nodal line (i.e., the line of zero acoustic pressure), and the relative phase of acoustic pressure can be solved [14.3].

To diagnose a noise problem in the passenger compartment of a vehicle, the following steps are usually taken to identify the body panels that generate the noise [14.3]:

1. The dominant frequency of low-frequency noise is measured in a road test.
2. The vibrations of surrounding body panels are measured, while excitations of the dominant frequency are applied to the vehicle in a laboratory.
3. The vibrations of these body panels are used as the inputs to the FEA model in acoustic simulations for calculating the amplitude and phase of the noise produced by each panel in the passenger compartment.
4. The contribution of each panel to the noise is then displayed in a polar coordinate system.
5. Noise is actually the resultant of vectors of sound pressure produced by each of the panels.

The polar plot of sound vectors provides useful information on the phase of acoustic waves generated by each of the panels. For example, sound pressure can be attenuated by dampening the movement of the panel that produces a sound wave in phase with the total sound wave [14.3].

14.3.4.4. Novel Body Materials and Processes of EVs

For vehicle bodies made of aluminum alloys, advanced connection technologies such as resistance spot welding (RSW), flow drill screw (FDS), cold metal transfer (CMT), self-piercing riveting (SPR), monobolt, structure adhesive, and laser welding have been employed to enhance the strength of connections between body parts [14.10]. In some EV models, over 96% of the vehicle body in weight is made of aluminum alloys. The aluminum alloy BIW offer a number of advantages: reducing the energy consumption in the manufacturing process, slashing emission of greenhouse gases, lowering the weight and improving the freedom of configuration, bringing down the energy consumption and enhancing the driving range, cutting down the braking distance and kinetic energy in collision, shortening the acceleration time, and increasing the effective payload of a vehicle.

Novel body construction technologies are also emerging, among which the Audi Space Frame (ASF) and BMW LifeDrive technologies are elaborated as follows:

1. ASF
   ASF is a novel body construction technology that Audi AG developed for weight reduction. Although initially only aluminum alloys are used, other lightweight materials, such as CFRP and carbon fiber, are also employed lately for further improvement of structural strength and weight reduction.

   In the early stage of ASF application, both frames and body panels are made of aluminum alloys, with very high strength and stiffness, thus providing excellent protection in collisions. The aluminum–alloy frames are formed using die-casting or extrusion and connected by clinching, stud welding, laser brazing, and metal inert gas (MIG) welding. The body outer cover is made of stamped aluminum alloy plates, which is 20–25% thicker than their steel counterparts. The outer cover and the frame are connected using punch rivets, with a 30% higher strength over spot welding connections. With a tortional stiffness of up to 32,000 N·m/deg, the body frames made of aluminum alloys offer excellent energy absorption performance. Hence, the crumple zone at the front wrinkling in the event of a collision can absorb a lot of impact energy, thus protecting the seating area. Furthermore, compared to steel, aluminum alloys allow higher DOF in selecting complex shapes of the die-casted parts.

   In recent years, a high proportion of aluminum and CFRP is utilized for the body shell, with CFRP mainly for the central tunnel, rear wall, and the area around B pillar.
2. BMW LifeDrive
   LifeDrive is a body-on-frame construction developed by BMW AG, which is composed of two separate functional units: The upper module, called "Life," is a high-strength and lightweight passenger cabin made from CFRP, which is 50% lighter than a steel cabin or 30% lighter than an aluminum cabin. The lower module, called "Drive," is a skateboard-like chassis mostly made of aluminum alloys in which the electric powertrain, battery, suspension, steering, and brake systems are packed, with crash protection structures as well.

14.3.5. Vehicle Body Testing

A large number of tests must be carried out on the vehicle body for design and process validation, as described briefly in the following paragraphs.

14.3.5.1. Measurements of Overall Stiffness

The static measurement of the overall stiffness of the body structure is generally conducted on a test rigs. In addition to the torsional and bending stiffness, the shear stiffness across the frames of body openings is also measured by applying a torque on the front axle at an angle to the forward direction, and the result is presented by the distance variations between the diagonal corner points of these frames.

In the tests, the vehicle body should be supported, and the forces and torques acting on it should be applied according to the actual loading condition. The result of measurement is usually presented by the vertical deflection at the measuring points a certain distance apart from each other in the c from the front to the rear.

14.3.5.2. Body Fatigue Strength Tests

The body fatigue strength test is carried out on a multiaxis fatigue test bench that applies dynamic random loading on the center of each wheel based on the large amount of data collected from road tests. The relationship between the fatigue failure at some critical parts of the vehicle body and the dynamic loading can be obtained.

14.3.5.3. BIW Modal Tests

In the modal test, the BIW is usually suspended under free conditions by elastic cords or set on air bags. Upon hammer excitation, the response of the BIW is detected with accelerometers using a single-input multi-output (SISO) or multi-input multi-output (MIMO) method. The MIMO method is deemed more reliable in capturing all the vibrational modes than the SISO method.

14.3.5.4. Other Body-Related Tests

Other BIW tests include air tightness, waterproof, thermal insulation, and corrosion tests.

The major parts of a vehicle body such as bumpers, glasses, doors, locks, and seats are often tested separately.

Other vehicle body-related tests, for example, aerodynamic performance tests, collision tests, and interior noise tests, are generally performed on the whole vehicle, instead of the BIW, due to the significant contributions from other portions of the vehicle on these performances.

References

[14.1]. Wang, X., *Automotive Chassis Design* (Beijing: Tsinghua University Press, 2010), ISBN:978-7-302-21471-7.

[14.2]. qcwxjs.com, "Introduction to Vehicle Models and Types of Frames," accessed August 2023, https://www.qcwxjs.com/cheshen/14324.html.

[14.3]. Sidelko, W., "An Objective Approach to Highway Truck Frame Design," SAE Technical Paper 660162 (1966), doi:https://doi.org/10.4271/660162.

[14.4]. Liu, W., *Automobile Design* (Beijing: Tsinghua Press, 2001), ISBN:9787302045298.

[14.5]. Klemperer, W., "Air Resistance Studies on Automobile Models," *Z. Flugtech. Mot.* 13 (1922): 287-289.

[14.6]. Holt, J., Garry, K., and Velikov, S., "A Wind Tunnel Investigation into the Effects of Roof Curvature on the Aerodynamic Drag Experienced by a Light Goods Vehicle," *International Journal of Vehicle Design* 67, no. 1 (2015): 45-62, doi:10.1504/IJVD.2015.066478.

[14.7]. U.S. Federal Motor Vehicle Safety Standards, "Electric-Powered Vehicles: Electrolyte Spillage and Electrical Shock Protection," FMVSS Standard No. 305, Title 49, Part 571, Section CFR 571.305, 2019.

[14.8]. U.S. Federal Motor Vehicle Safety Standards, "Roof Crush Resistance," FMVSS Standard No. 216, Title 49, Part 571, Section CFR 571.216, 2009.

[14.9]. Larminie, J. and Lowry, J., *Electric Vehicle Technology Explained*, 2nd ed. (Hoboken: John Wiley & Sons, 2012), doi:10.1002/0470090707.

[14.10]. Brünger, E., Engler, O., and Hirsch, J., "Al-Mg-Si Sheet Alloys for Autobody Applications, Virtual Fabrication of Aluminum Products," Chapter I-6 in Hirsch, J. (ed.), *Virtual Fabrication of Aluminium Products* (New York: Wiley Press, 2006), ISBN:3-527-31363-X.

15

Control System Design

An EV is generally controlled by a distributed hierarchical system. On the top of the system, a VCU is the master controller that identifies the intention of the driver, makes a decision based on all the information that is either stored in the system or received from sensors and lower-level control units, and then sends commands to lower-level control units for them to execute, as illustrated in Figure 15.1. The vehicle-level control ultimately aims at realizing the driver intent with optimized safety, dynamic performance, energy economy, and comfort. Traditionally, this is an open-loop control system that requires feedback from the sense organs of a driver to close the control loop for making adjustments on forward or reverse propelling, steering, braking, and parking. With rapid growth of intelligence in the VCU control program, some misoperations of a driver can be intervened or corrected in real time by active control of steering or braking systems.

FIGURE 15.1 A distributed hierarchical system for vehicle control.

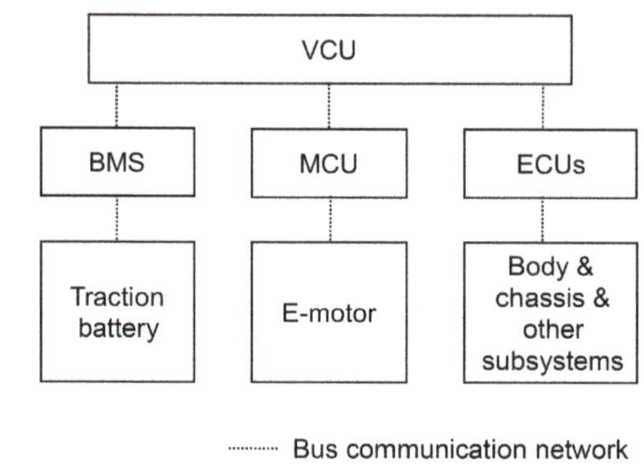

For a fully autonomous EV, however, the VCU needs a high degree of intelligence and a great deal of information to implement closed-loop control. The decisions on route planning, energy management, and motion control are made on the basis of the road, vehicle, and environmental conditions that are identified according to the signals from sensors, data from V2X communication network, high-precision maps, position information, and so on. Once the decision is made, the VCU issues the commands to the corresponding lower-level control units for execution. The resulted change in the state of vehicle motion is continuously monitored and fed back to the VCU for any further corrections in the commands to achieve optimal velocity and trajectory on the road.

The electric MCU includes the high-voltage module, for example, the inverter, and the low-voltage module, for example, the microcontroller and control circuit. The MCU is used to control the motor operation, usually in a closed-loop manner. It receives instructions from the VCU and controls the operating mode of the motor, for example, driving or regenerative braking, as well as the motor torque or speed based on its own control algorithm.

The BMS is a distributed control system consisting of sensors, wires, fuses, and electrical circuit boards for monitoring and controlling the battery down to the cell level. The BMS is primarily used for SOC estimation, charge and discharge process control, parameter monitoring, online calibration, and fault handling. It communicates with the VCU and other controllers through CAN bus and wires to transmit information and control commands.

The ECUs for body, chassis, and other subsystems mainly involve subsystems such as e-motor cooling, gear shift panel, EPS motor, A/C, instrument cluster, infotainment, monitor host, calibration tools, lights, doors, and windows. The instrument cluster usually displays the data related to the motor and the battery, such as voltage, current, SOC, remaining mileage, current mileage, accumulated mileage, temperature, warnings, and indicator lights. The infotainment system, such as the radio, phone, audio, video, maps, positioning and navigation, and network services, allows the driver and passengers to receive additional information and entertainment normally by interacting with one or more LCD screens. The calibration and monitoring system is used for online monitoring of the operating parameters as well as online calibration and optimization of the control parameters of the VCU.

The common E/E architecture of an EV is shown in Figure 15.2. Today, over 100–300 microcontrollers and 200 million lines of the control code are used for a highly intelligent EV.

FIGURE 15.2 A typical E/E architecture of today's EV.

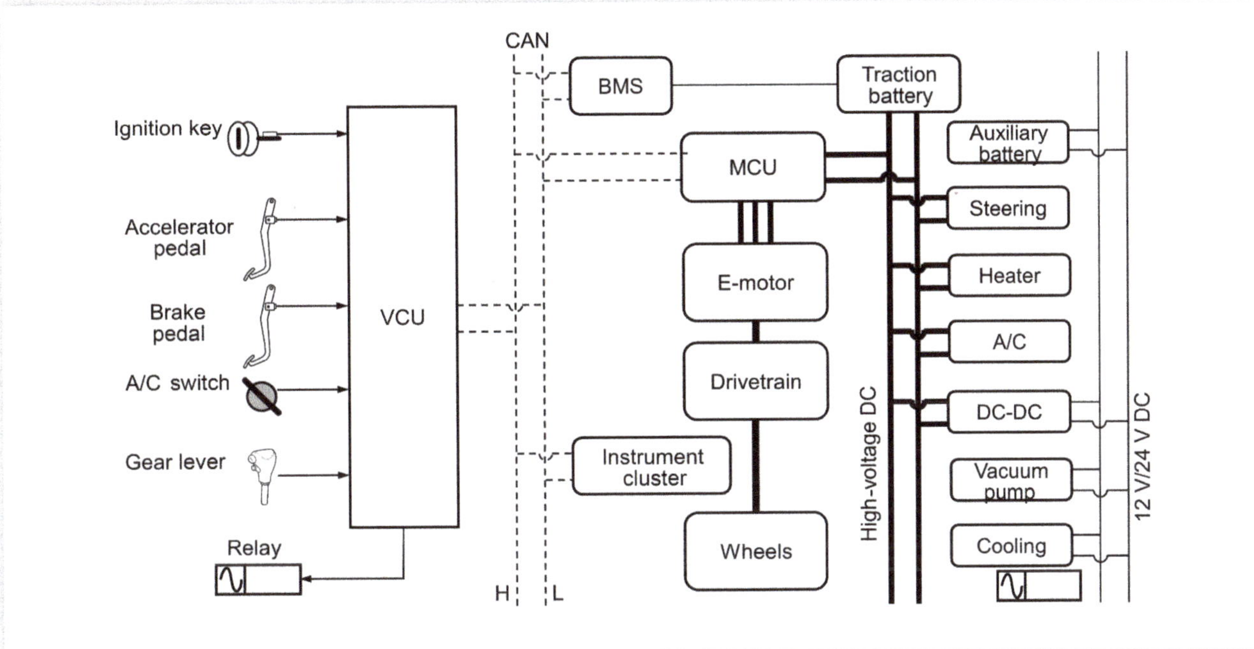

The basic functions of a VCU are as follows:

1. Vehicle-level energy management and optimization
   In an EV, the battery supplies power not only to e-motors but also to the electric auxiliaries. Hence, in order to maximize the driving range, the VCU handles energy management of the vehicle by optimizing the utilization of energy. When the battery SOC drops below a certain threshold, the VCU will issue commands to some electric auxiliaries to limit their output power.
2. Vehicle driving control
   The e-motor must output the driving or braking torque according to the intention of the driver. Thus, the VCU should identify the intention of the driver properly, mainly based on the opening of the accelerator or brake pedal and then send commands to the MCU through the CAN bus. The MCU generally controls the motor to output a certain driving torque that is proportional to the pedal opening. Besides, the VCU receives feedback information from the vehicle subsystems and provides the driver with selected and necessary information for decision-making. It also sends control commands to vehicle subsystems in order to coordinate their outputs for driving the vehicle. The electric drive control-related functionality of the VCU primarily involves the wake-up, shutdown, fault diagnosis, failure handling, and safety assurance of the electric drive system. The VCU should also support calibration online using dedicated programs on a personal computer.
3. Regenerative braking control
   In braking, the VCU coordinates the electric and mechanical braking torques to maximize the energy recovery under the premise of safety. At the same time, the

VCU sends a command to the MCU, which controls the e-motor to act as a generator, converting the kinetic energy of the vehicle back into electricity. Meanwhile, this electric energy is stored in the battery if allowed by the SOC condition. In this process, the VCU judges whether the regenerative braking can be performed in real time based on pedal signals and the battery SOC.

4. Vehicle network management
   The control units in an EV are networked using CAN, LIN, or FlexRay bus for communication, among which CAN bus is the most popular. As a node in the CAN bus, the VCU is not only the center of information collection and processing but also a communication control hub that does network status monitoring, network node management, network fault diagnosis and handling, and so on.
5. Monitor and display of the vehicle operational status
   Through the CAN bus, the signals from all the sensors in vehicle subsystems are transmitted to the VCU, where the signals are processed and analyzed in real time to monitor the operational status of the vehicle. Then, the selected information of each subsystem is sent to the on-board display for notifying the driver. The contents for display include motor speed, vehicle speed, battery SOC, and fault information.
6. Fault diagnosis and handling
   Vehicle subsystems are continuously monitored for fault diagnosis. If any fault is detected, the corresponding fault code is sent to the VCU, and some selected faults are displayed to the driver or serviceman using fault indicators. The VCU carries out corresponding safety protection actions based on the type and severity of the fault in a timely manner. For less-serious faults, the VCU imposes a limp home mode, which allows the vehicle to be driven at a low speed to a nearby maintenance station for checkup and repair. For serious faults, the VCU sends out commands to shut down the high-voltage power supplies to ensure safety.
7. Charging management
   The VCU communicates with the external charger using an appropriate protocol and manages the connectivity of the charging circuitry. It controls the charging process such as the starting and ending time and charging current. It also monitors the charging process and reports the charging status by sending corresponding information to the display in the vehicle or even mobile devices. When the charging is complete, it shuts off the charging module of the vehicle.

15.1. Hardware Architecture of a VCU

The VCU hardware usually includes modules of microcontroller, analog input and output, digital signal conditioning, relay driver, high-speed CAN bus interface, integrated instrument interface, and power supply, for managing, coordinating, and monitoring each link from energy storage to electric drive systems to improve the energy utilization efficiency as well as to ensure the safety and reliability of the vehicle, as shown in Figure 15.3.

FIGURE 15.3 The hardware architecture of an EV.

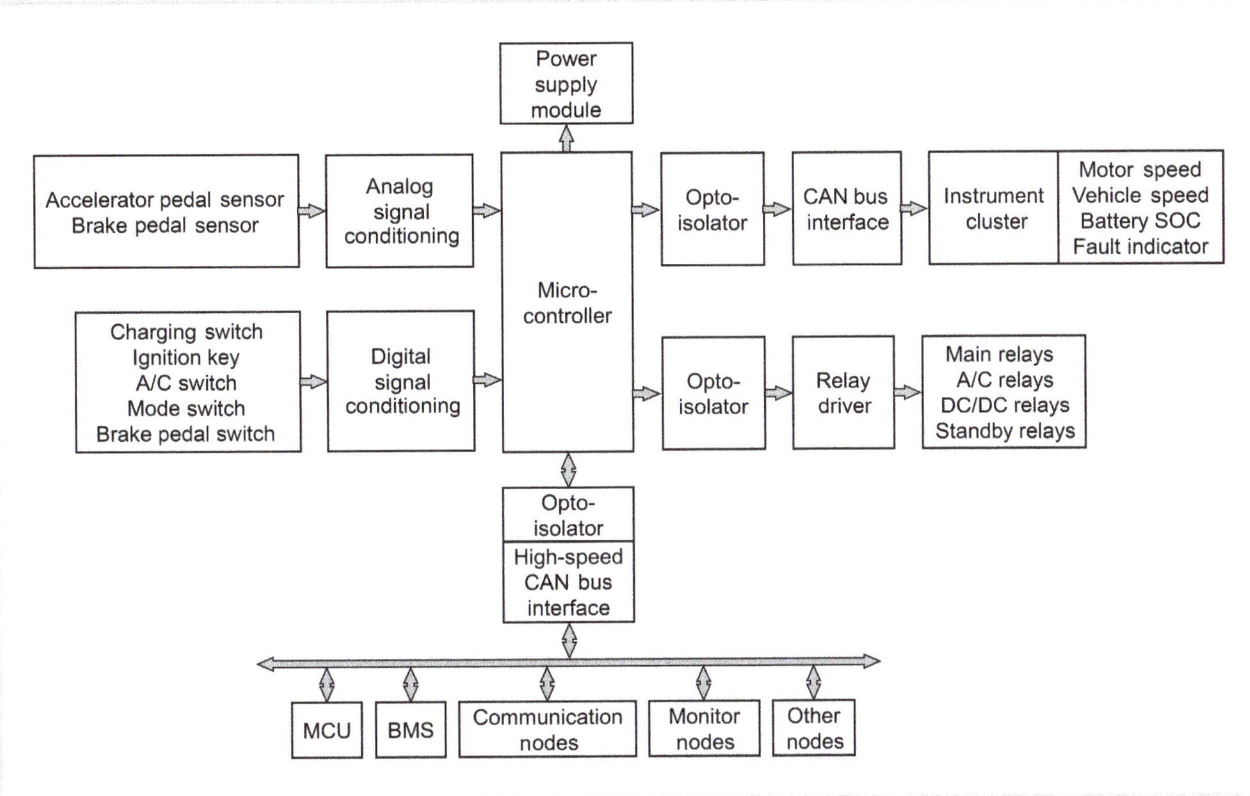

The function of each hardware module in the VCU is elaborated as follows:

1. Power supply module
   The module not only provides isolated power supply for the microprocessor and each input and output module but also monitors the battery voltage.
2. Analog input and output module
   This module can collect analog signals and then convert them into analog voltage signals, usually in a range of 0–5 V, before output.
3. Digital signal conditioning module
   This module is used for voltage-level conversions and conditioning of the digital signals, with one end connecting to a plurality of digital sensors and the other end connecting to the microcontroller.
4. Relay driver module
   This module functions to drive multiple relays, with one end connecting to the microcontroller through an optoelectronic isolator and the other end connecting to multiple relays.
5. High-speed CAN bus interface module
   This module provides a high-speed CAN bus interface, with one end connecting to the microcontroller through an optoelectronic isolator and the other end connecting to the high-speed CAN bus.

6. Pulse signal input and output module
   With this module, pulse signals are collected and conditioned to a certain frequency and amplitude. Meanwhile, it can output a PWM signal of a certain frequency and amplitude.
7. Data and fault storage module
   The vehicle characteristic parameters, calibration data, and fault codes can be stored in the memory of the module.

15.2. Power-up and Shutdown Procedures

15.2.1. Low-Voltage Power-up Procedure

Once the vehicle is turned on by a key or a button, the ignition relay is powered on by the low-voltage battery, usually 12 or 24 V, and the voltage (or POWER-ON signal) will wake up the VCU first. Subsequently, the VCU will wake up the BMS and other systems. An example of power-up steps are as follows:

1. The VCU wakes up the BMS through the POWER-ON signal, and the BMS undergoes low-voltage power-on and self-check and feeds back the self-check status and fault-level information through the CAN bus.
2. The BMS can recognize the control pilot function (i.e., CP signal) sent by the charging gun and wake up the BMS. Then, the BMS is powered on normally and completes the self-check.
3. The BMS can recognize the +12 V wake-up signal of the on-board charger. Once this signal is detected, the BMS can power on normally and complete the self-check.

15.2.2. Low-Voltage Shutdown Procedure

There are two circumstances of low-voltage shutdown:

First, the low-voltage shutdown of BMS controlled by a VCU follows the subsequent steps:

1. The VCU sends a shutdown command to the BMS and waits for the BMS to switch off both main positive and main negative relays. Then, the BMS feeds back the HIGH-VOLTAGE OFF signal to the VCU.
2. Upon receipt of the HIGH-VOLTAGE OFF signal, the VCU will command the BMS to power off, and then the BMS enters a power-off delay procedure.
3. If the VCU does not receive the HIGH-VOLTAGE OFF feedback signal from the BMS within a certain time, then the BMS enters the power-off delay procedure after the diminishing of the POWER-ON signal and controls the main positive and main negative relays to switch off by itself.

Second, the low-voltage shutdown of BMS controlled by the charging wake-up signal follows the following steps:

1. Upon completion of charging, the charger should turn off the 12 V charging ignition signal.
2. The BMS enters the power-off delay control procedure and disconnects the charging relay and the main negative relay.
3. If the charger cannot turn off the low-voltage output voltage, the BMS should switch off the charging relay and the main negative relay by itself.

Note that the BMS itself must have a constant power supply input to complete the power-off delay control procedure after the VCU switches off the power supply relay of the BMS or turns off the low-voltage output after charging.

15.2.3. High-Voltage Power-up Procedure during Discharge

When the battery system operates in the discharge mode, the BMS can be powered up by following the subsequent steps:

1. After powered on, the BMS completes the system initialization within 700 ms. The system initialization time is defined as the time from power-up to sending out the first valid CAN message.
2. The BMS sends out a BMS READY message via CAN after its power-up with nothing abnormal detected in the self-check, otherwise the BMS status would be NOT READY. With the self-check completed in system initialization, the BMS then receives the power-on command sent by the VCU for precharging. The maximum allowable time for precharging is 2 sec, and the minimum time interval between two precharges is 500 ms.
3. Upon receipt of the battery disconnection command from VCU, the BMS switches off the positive relay first and then the negative relay, with an interval of 200 ms in between.
4. The BMS decides whether the relay is switched off at a delay or immediately based on the severity of the fault. Consult the BMS faults and classification from the battery manufacturer for details.
5. After the ignition signal is off, the BMS waits for the switch-off command from the VCU. If no command is received, the BMS turns itself off after 3 sec.

15.2.4. High-Voltage Shutdown Procedure during Discharge

When the battery system operates in the discharge mode, the BMS can be powered off by following the following steps:

1. If the BMS receives the power-off command from the VCU, it switches off the main positive relay first and then the main negative relay thereafter.
2. If the BMS does not receive the power-off command from the VCU and receives a 4th class fault signal from the VCU instead, then the BMS switches off the main positive relay after 5 sec of delay and then the main negative relay thereafter.

3. If the BMS receives a 3rd class fault signal instead of a 4th class fault signal from the VCU, then the BMS switches off the main positive relay after 30 sec of delay and then the main negative relay thereafter.
4. If the BMS does not receive either 3rd or 4th class fault signal from the VCU, then its sends a report message to the VCU.

15.2.5. High-Voltage Power-up Procedure during Charge

When the battery system operates in the charge mode, the BMS can be powered up by following the steps:

1. Once the BMS is activated and powered up, it will undergo self-check for fault diagnosis. If any fault is detected, then high-voltage power-up is not allowed, and a fault signal is sent to the VCU. If no fault is detected, a diagnosis on charging fault is carried out.
2. If any charging fault is detected, then high-voltage power-up will not be allowed, and a fault signal will be sent to the VCU. If no charging fault is detected, then check slow charging connectivity and fast charging connectivity at the same time.
3. If there is no slow charging connectivity, go back to charging fault diagnosis. If slow charging connectivity is confirmed, then check whether the battery is charged at low temperatures. If not low temperature, then enter the standard slow charging mode. If it is low-temperature charging, then switch on the main negative relay. After a delay of 300 ms, switch on the main positive relay and then enter the heat-up procedure until the completion of heat-up. Enter the standard slow charging mode thereafter.
4. If there is no fast charging connectivity, go back to charging fault diagnosis. If fast charging connectivity is confirmed, then check the readiness of the international charging protocol. If not ready, go back to charging fault diagnosis. If ready, then check whether the battery is charged at low temperatures. If not low temperature, then enter the standard fast charging mode. If it is low-temperature charging, then switch on the main negative relay. After a delay of 300 ms, switch on the main positive relay and then enter the heat-up procedure until the completion of heat-up. Enter the standard fast charging mode thereafter.

15.2.6. High-Voltage Shutdown Procedure during Charge

When the battery system operates in the charge mode, the BMS can be powered off by following the steps:

1. When the BMS operates in the charging state, detect if there is any disconnection in charging hardware. If the charging hardware is disconnected, check if there is any requested delay for turning the charging off. If the charging hardware is not disconnected, check if there is any fault in the charging protocol. If there is a fault in the charging protocol, check if there is any requested delay for turning the charging off.

2. If there is no fault in the charging protocol, check if the battery is full or if there is any serious fault in the BMS charging module. If the battery is full or there is a serious fault in the BMS charging module, then check if there is any requested delay for turning it off. If the battery is not full or there is no serious fault in the BMS charging module, then continue to charge the battery with limited current.
3. If there is any requested delay for turning the charging off, then check if the battery is in the state of slow charging or fast charging. If there is no requested delay for turning the charging off, then go back to check if there is any requested delay for turning the charging off again.
4. If the battery is in the state of slow charging, then switch off the relay for slow charging and then switch off the main negative relay to complete the high-voltage shutdown.
5. If the battery is in the state of fast charging, then switch off the relay for fast charging and then switch off the main negative relay to complete the high-voltage shutdown.

15.3. Battery Management Strategy

A battery energy control model is shown in Figure 15.4. In this model, the battery SOC, accelerator pedal opening, and current motor speed are used as the input parameters for battery energy management and usage control. After data processing and analysis, the driving range in the current battery state is predicted and output to the driver. The speed and acceleration of the vehicle are controlled by adjusting the output torque of the e-motor to ensure the safe and rational use of battery energy. Consequently, different control targets, such as minimization of travel time or energy consumption for a given course, can be achieved by adjusting the driving mode as well as allocating and utilizing the battery energy in a different way. Hence, various information such as the state of the vehicle and battery, intention of the driver, and path information and planning must be integrated in the model for making control decisions.

FIGURE 15.4 A battery energy control model.

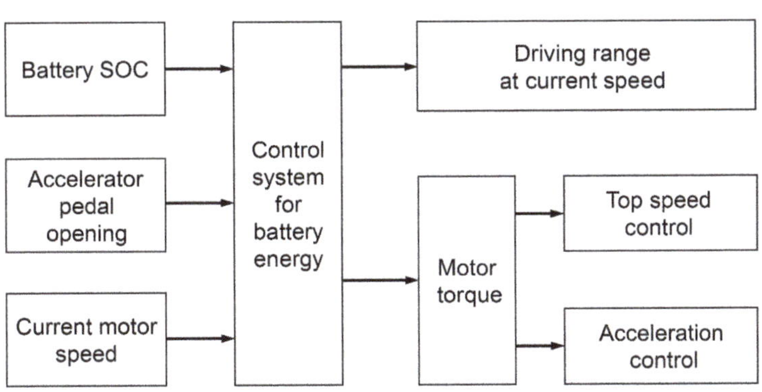

15.3.1. Power Management

When the battery charge is too low or the e-motor overheats, the output power of the e-motor must be limited to protect the battery and e-motor from damage due to overloading. The state of the battery can be recognized by using parameters such as SOC, voltages of signal cells, and the total voltage of the battery pack.

In a battery system, if its SOC drops below a safe value, its voltage declines accordingly. To meet a given power demand, the current will have to increase, which can cause permanent damage to the battery system due to its overdischarge or cause the trip of protection device in the high-voltage system, which may subsequently lead to road accidents. Therefore, in these circumstances, the power draw from the battery must be limited in the form of restricting the output torque of the e-motor.

Since the power of the battery system mainly supplies to the e-motor and high-voltage auxiliary systems, the e-motor power must not exceed the difference between the maximum discharge power of the battery and the total power that the auxiliary systems consume. Hence, the limit for the power supplied to the e-motor, P_{lmt}, is given by

$$P_{lmt} = \min\left[P_e / \eta_e, \left(P_{b,max} - P_{aux}\right)\right] \tag{15.1}$$

where

P_e is the power demand from e-motor output based on the accelerator pedal opening
η_e is the efficiency of the e-motor system
$P_{b,max}$ is the maximum discharge power of the battery system
P_{aux} is the total battery power consumption by auxiliary systems, including the cooling and systems

The maximum discharge power of the battery system, $P_{b,max}$, is primarily a function of SOC of the battery system and its temperature T_b:

$$P_{b,\max} = f\left(SOC, T_b\right) \tag{15.2}$$

When the SOC drops below a certain limit, the limp home mode will be activated, which allows the vehicle to go home slowly.

Similarly, the battery system can also be protected by setting a maximum discharge current, $I_{b,max}$:

$$I_{b,max} = f\left(SOC, T_b\right) \tag{15.3}$$

The maximum discharge current can be determined on the basis of discharge characteristics of the battery, which are usually available from the battery manufacturer. An exemplary strategy for battery protection is twofold: automatically switch the VCU to the limp home mode in which the discharge current is limited if SOC drops below 0.3 and cut off the e-motor output if SOC drops below 0.1 and the total voltage drops below $2.85N_cV_c$ (N_c is the total number of cells in the battery pack and V_c is the voltage of a single cell), or the voltage of any single cell drops below 2.5 V, or the battery temperature exceeds 60°C for 20 sec consecutively.

On the other hand, when the e-motor is overloaded, its heat generation increases, resulting in a rise of temperature. Consequently, the thermal protection program may be triggered to turn the e-motor system off if temperature exceeds a certain level in its windings and the torque command can simply be cleared out, otherwise the motor can be burned up if the climb in temperature is not halted timely. Thus, both the overload multiplier and overload time must be restricted, which are normally addressed in the power limiting strategy. The overload multiple of the output torque has a linear relationship with the accelerator pedal opening. When the accelerator pedal opening exceeds the set threshold, the motor is overloaded to meet rapid acceleration needs of the driver. The overload multiple can be obtained from the curves of torque demand for acceleration. The overload time is determined according to the motor temperature characteristics based on the feedback from the motor temperature sensors. The VCU keeps the output torque of the motor below a certain level based on the set temperature and duration thresholds of the e-motor.

15.3.2. BTM Strategy

The BTM takes different strategies for the vehicle being charged or being driven on a road.

For the vehicle equipped with a PTC heater being charged, the general control strategies are as follows:

1. Ignore the A/C switch signal.
2. Turn on the A/C if the battery temperature exceeds a high temperature threshold T_h, and turn it off if the battery temperature drops below a normal temperature threshold T_n.
3. Turn on the PTC if the battery temperature drops below a low temperature threshold T_l, and turn it off if the battery temperature rises above an acceptable low temperature threshold T_{la}. Turn on and off the water pump accordingly if the battery pack is water-cooled.

For the vehicle equipped with a PTC heater being driven on a road, the general control strategies are as follows:

1. If the A/C is switched off, turn the A/C on if the battery temperature is 35°C or higher, and then turn the A/C off if the battery temperature is 33°C or lower. Turn on and off the water pump accordingly if the battery pack is water-cooled.
2. If the A/C is switched on, leave the A/C on if the battery temperature is 35°C or higher, and then turn the A/C off if the battery temperature is 33°C or lower. Turn on and off the water pump accordingly if the battery pack is water-cooled.
3. Turn on the PTC heater if the battery temperature is 0°C or lower, and turn the PTC heater off if the battery temperature is 10°C or higher. Turn on the water pump if the battery pack is water-cooled.

15.4. Control of Electric Motors

15.4.1. Control of Three-Phase IMs

A three-phase IM can be equivalent to a T-shaped circuit, as shown in Figure 15.5, where R_s and $L_{s\sigma}$ are the resistance and leakage reactance of stator windings, respectively;

L_m is the equivalent excitation reactance of three-phase windings; R_s and $L_{s\sigma}$ are the resistance and leakage reactance of rotor windings, respectively; and s is the slip, which is defined as

$$s = \frac{\omega_s - \omega_r}{\omega_s} \tag{15.4}$$

Traditionally, IMs are controlled using the variable voltage and variable frequency (VVVF) method.

FIGURE 15.5 The T-shaped equivalent circuit of an IM.

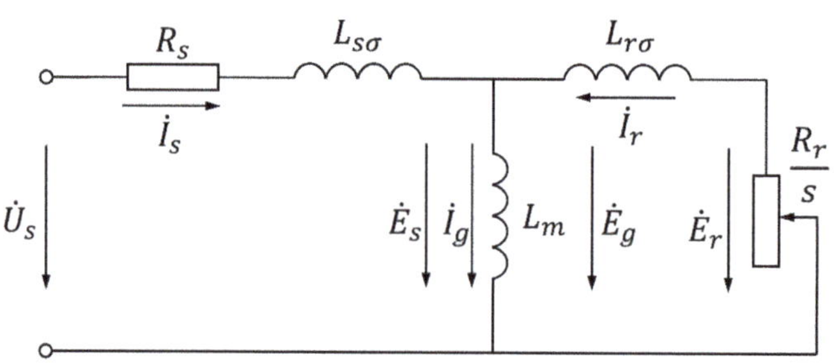

15.4.1.1. VVVF Speed Control of IMs

The electromagnetic torque of IMs, T_e, can be expressed by

$$T_e = 3p_0\Psi_g^2 \frac{s\omega_s R_r}{R_r^2 + (s\omega_s L_{r\sigma})^2} \tag{15.5}$$

or

$$T_e = 3p_0\Psi_g^2 \frac{\omega_f R_r}{R_r^2 + (\omega_f L_{r\sigma})^2} \tag{15.6}$$

where

p_0 is the number of pole pairs
Ψ_g is the air gap flux vector
ω_s is the angular frequency of the power supply to the stator, that is, the angular speed of the air gap circular rotating magnetic field
ω_f is the slip frequency

The electromagnetic torque T_e as a function of slip s depicted by Equation (15.5) is generally called the mechanical characteristics of IMs. When the amplitude of air gap magnetic flux Ψ_g is kept constant, under a certain supply frequency in stator windings, the electromagnetic torque is dependent only on the slip frequency, as indicated by Equation (15.6). Thus, the electromagnetic torque can be controlled by the slip frequency, which is the principle of VVVF speed control.

In an open-loop control system of AC speed regulation, the air gap magnetic flux Ψ_g is kept constant by making E_g/f_s a constant, where E_g is the induced EMF and f_s is the frequency of stator current. Hence, Equation (15.6) can be rewritten as

$$T_e = \frac{3p_0}{4\pi^2}\left(\frac{E_g}{f_s}\right)^2 \frac{\omega_f R_r}{R_r^2 + \left(\omega_f L_{r\sigma}\right)^2} \tag{15.7}$$

In practice, the external voltage imposed on stator windings, U_s, can be adjusted to retain a constant E_g/f_s.

On the other hand, in a closed-loop control system of AC speed regulation, the air gap magnetic flux Ψ_g is kept constant by controlling the excitation current I_g, which is actually accomplished by adjusting the current in state windings.

The fundamental weakness of the VVVF control methodology is that Equation (15.7) is only valid under steady-state conditions, which does not hold under dynamic conditions. In fact, a poor instantaneous performance would be observed if the VVVF control method is used under dynamic conditions. This leads to more-effective schemes such as FOC and DTC, which are elaborated now.

15.4.1.2. FOC of IMs

The basic idea of the FOC scheme, also known as the space vector control scheme, is to transform the magnetomotive force generated by the three-phase current in stator windings into that generated by the two-phase current in a stationary coordinate system first and then transform the two-phase current in the stationary coordinate system into a two-phase current in a rotating coordinate system again in order to decouple the excitation and torque generation current components. In the first step, by keeping the power and magnetomotive force unchanged, the three-phase current in the ABC coordinate system is transformed into the two-phase current in the stationary $\alpha\beta$ coordinate system by the following Clarke transformation:

$$\begin{bmatrix} i_\alpha \\ i_\beta \end{bmatrix} = \sqrt{\frac{2}{3}} \begin{bmatrix} 1 & -\frac{1}{2} & -\frac{1}{2} \\ 0 & \frac{\sqrt{3}}{2} & -\frac{\sqrt{3}}{2} \end{bmatrix} \begin{bmatrix} i_A \\ i_B \\ i_C \end{bmatrix} \tag{15.8}$$

The inverse Clarke transformation needed for the backward calculation of the three-phase current is

$$\begin{bmatrix} i_A \\ i_B \\ i_C \end{bmatrix} = \sqrt{\frac{2}{3}} \begin{bmatrix} 1 & 0 \\ -\frac{1}{2} & \frac{\sqrt{3}}{2} \\ -\frac{1}{2} & -\frac{\sqrt{3}}{2} \end{bmatrix} \begin{bmatrix} i_\alpha \\ i_\beta \end{bmatrix} \tag{15.9}$$

In the second step, the two-phase current in the stationary $\alpha\beta$ coordinate system is transformed into the two-phase current in a rotating dq reference frame using the following Park transformation:

$$\begin{bmatrix} i_d \\ i_q \end{bmatrix} = \begin{bmatrix} \cos\theta & \sin\theta \\ -\sin\theta & \cos\theta \end{bmatrix} \begin{bmatrix} i_\alpha \\ i_\beta \end{bmatrix} \tag{15.10}$$

where i_d and i_q are excitation current and torque generation current, respectively. Consequently, the time-varying stator vector current is decomposed into DCs i_d and i_q, which are time-invariant as they are oriented to the rotor flux. Thus, the magnetic flux and torque can be controlled separately by adjusting i_d and i_q, respectively.

Accordingly, the inverse Park transformation is

$$\begin{bmatrix} i_\alpha \\ i_\beta \end{bmatrix} = \begin{bmatrix} \cos\theta & -\sin\theta \\ \sin\theta & \cos\theta \end{bmatrix} \begin{bmatrix} i_d \\ i_q \end{bmatrix} \tag{15.11}$$

In the dq reference frame, the electromagnetic torque is given by

$$T_e = p_0 \frac{L_m^2}{L_r} i_d i_q \tag{15.12}$$

where

L_m is the equivalent excitation inductance of the stator and the rotor

L_r is the equivalent self-inductance of the rotor

Although the d axis can be either oriented to the rotor flux, stator flux, or air gap, the rotor flux-oriented control scheme is mostly utilized in practice. It is obvious from Equation (15.12) that the constant torque control can be implemented by adjusting the torque generation current i_q only and keeping the excitation current i_d at the rated value for operating below the rated speed, while the constant power control can be achieved by adjusting i_d and i_q so that both $i_d\omega_r$ and $T_e\omega_r$ are constants.

The block diagram of a rotor flux-oriented control system of IMs for EV traction applications is illustrated in Figure 15.6. Both torque and flux are closed-loop-controlled, and the signal to the three-phase inverter is produced using space vector pulsed width modulation (SVPWM).

The implementation of the flux-oriented control algorithm requires updated values of characteristic variables of IMs, for example, R_s, R_r, $L_{s\sigma}$, $L_{r\sigma}$, and L_m, since resistances are dependent on the temperature and reluctances are dependent on the magnetic field saturation. The dynamic characteristics of an IM can be deteriorated if these variables provided to the controller deviate from their true values.

FIGURE 15.6 The block diagram of a rotor flux-oriented control system for IMs.

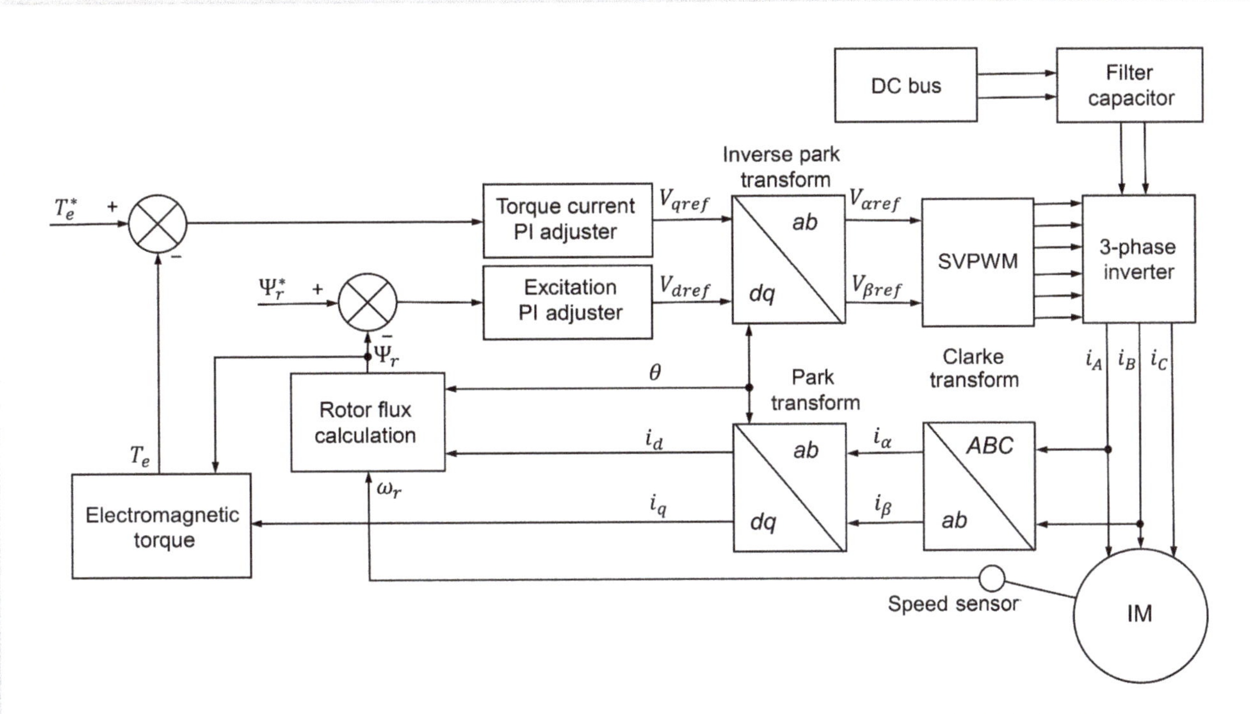

15.4.1.3. DTC of IMs

In the DTC scheme, the stator flux and torque are directly employed as control variables with no need of flux orientation, vector transformation, or current control, so that the dynamic response of the system can be improved further due to the simple and fast nature of the control algorithm.

The electromagnetic torque can be expressed by

$$T_e = p_0 \frac{L_m}{L_s' L_r} \Psi_r \Psi_q \tag{15.13}$$

where

L_s' is the instantaneous inductance of the stator
Ψ_r is the rotor flux vector
Ψ_q is the component on the q-axis of the stator flux vector, with

$$\Psi_q = \Psi_s \sin \delta_{sr} \tag{15.14}$$

In Equation (15.14), the load angle $\delta_{sr} = \delta_s - \delta_r$, where δ_s and δ_r are the space electric angle of the stator and rotor flux vectors with respect to the A-axis, as illustrated in Figure 15.7.

FIGURE 15.7 Coordinate systems for DTC of an IM.

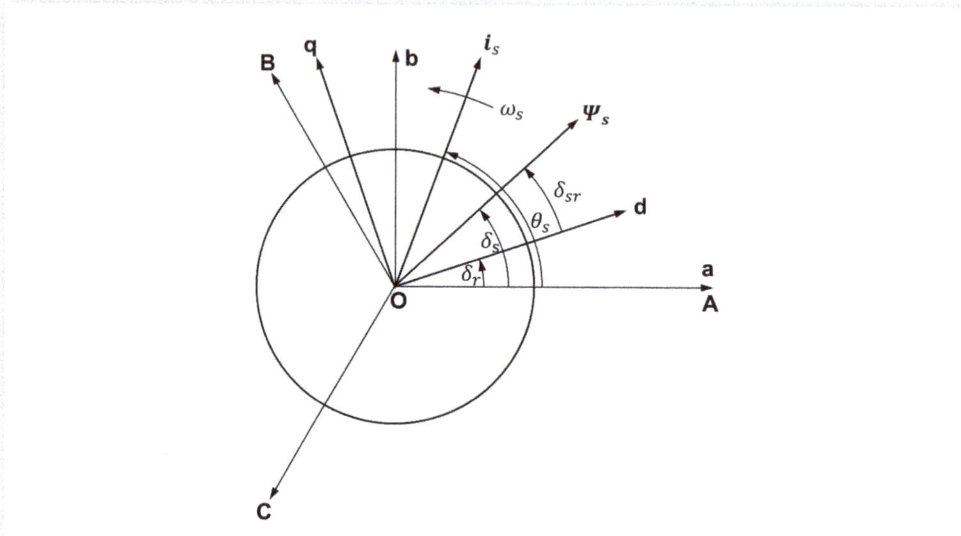

Based on Equation (15.13), it is evident that the electromagnetic torque is only dependent on Ψ_q if the amplitude of Ψ_r is kept constant. Thus, the electromagnetic torque can be controlled by adjusting Ψ_s and δ_{sr} per Equation (15.23).

The block diagram of a DTC system of IMs for EV traction applications is illustrated in Figure 15.8. Both torque and stator flux hysteresis comparators are usually a two-value bang-bang controller, which is an effective nonlinear controller. Not relying on the mathematical model or affected by parameter variations of IMs, the bang-bang controller is, therefore, highly robust. Moreover, it does not need complex transformations or accurate estimation of the stator flux vector; the control algorithm is improved in its simplicity and response time. Nevertheless, the output errors of the estimators may result in torque fluctuations, particularly at low speeds.

FIGURE 15.8 Block diagram of a DTC system for IMs.

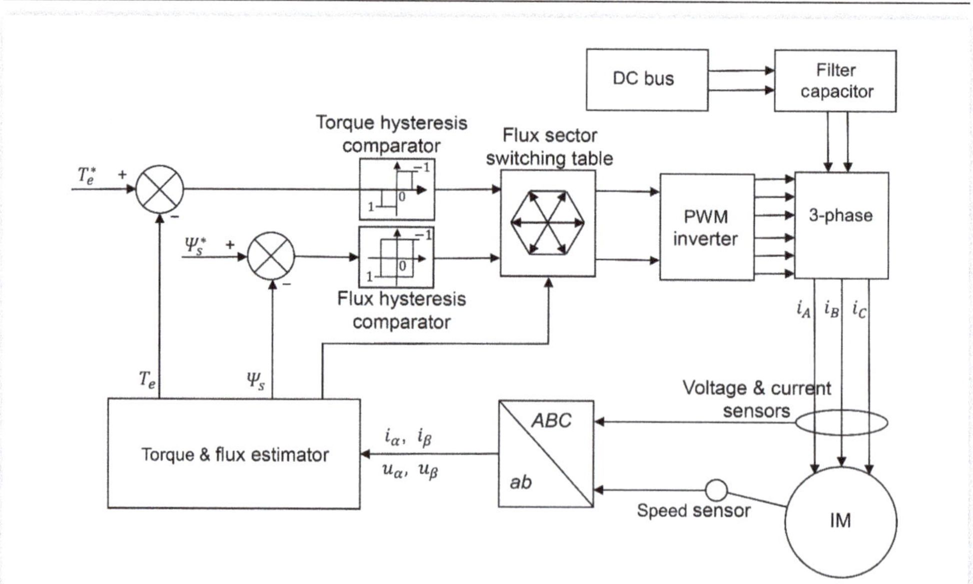

15.4.2. Control of Three-Phase PMSMs

The PMSMs are typically controlled with rotor flux-oriented space vector schemes. For the IPM type of PMSMs that are most frequently used for EVs, the stator flux can be decomposed into components in d-axis and q-axis, respectively:

$$\Psi_d = L_d i_d + \Psi_f \tag{15.15}$$

and

$$\Psi_q = L_q i_q \tag{15.16}$$

where

L_d is the direct axis synchronous inductor

L_q is the quadrature synchronous inductance, as illustrated in Figure 15.9

FIGURE 15.9 Equivalent physical model of an IPM type PMSM.

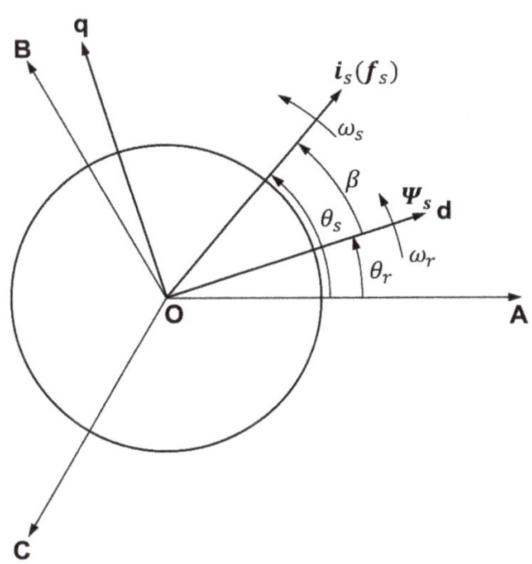

The voltage components in the d-axis and q-axis are given, respectively, by

$$u_d = R_s i_d + L_d \frac{di_d}{dt} - \omega_r L_q i_q \tag{15.17}$$

and

$$u_q = R_s i_q + L_q \frac{di_q}{dt} + \omega_r \left(L_d i_d + \Psi_f\right) \tag{15.18}$$

where

R_s is the stator phase resistance

ω_r is the electric angular speed of the rotor

Ψ_f is the excitation flux linkage vector of the rotor

The electromagnetic torque can be expressed by

$$T_e = p_0\left[\boldsymbol{\Psi}_f \boldsymbol{i}_s \sin\beta + \frac{1}{2}\left(L_d - L_q\right)i_s^2 \sin 2\beta\right] \tag{15.19}$$

where β is the phase angle that the stator current vector $\boldsymbol{i}_s$ leads the rotor excitation flux vector $\boldsymbol{\Psi}_f$. Hence, in Equation (15.19), the first term on the right side is the excitation torque produced by rotor current under the action of the excitation flux of permanent magnets, while the second term is the reluctance torque created by the difference between direct axis reluctance and quadrature axis reluctance. Because we have $L_d < L_q$ for the IPM type of PMSMs, the reluctance torque can be taken advantage of by adjusting the angle β. For example, the torque can be increased by letting $\frac{\pi}{2} < \beta < \pi$ for the PMSM operating in the constant torque region, while the torque can be augmented and the speed range broadened by adjusting the angle β for operating in the constant power region.

In addition, in the dq coordinate system,

$$i_d = i_s \cos\beta \tag{15.20}$$

and

$$i_q = i_s \sin\beta \tag{15.21}$$

Equation (15.19) can be rewritten as

$$T_e = p_0\left[\Psi_f i_q + \left(L_d - L_q\right)i_d i_q\right] \tag{15.22}$$

15.5. Driving Intent and Accelerator Pedal Strategy

For nonautonomous EVs, the torque demand for acceleration is one of the most important driving intentions that need to be determined.

The torque demand for acceleration is estimated according to the opening of the accelerator pedal and the external characteristics of electric traction motors. The relationship between the accelerator pedal opening and the torque demand for acceleration has a significant impact on the dynamic performance and ride comfort, as shown in Figure 15.10 [15.1]. Curve 1 is a hard pedal strategy, which is characterized by good pedal feel at medium and high loads but poor pedal feel at low loads; Curve 3 is a soft pedal strategy, which offers a good pedal feel at low loads, with a weak acceleration though; and Curve 2 is a linear pedal strategy with a pedal feel between Curves 1 and 3, which is the most frequently used pedal strategy. The torque demand for acceleration is usually converted and plotted on the external characteristic curve of the e-motor. Based on the aforementioned linear pedal strategy, the torque demand for acceleration, T_e, using the simplest conversion method is given by

$$T_e = \beta T_{emax}(n_e) \tag{15.23}$$

where

β is the opening of the accelerator pedal
n_e is the speed of the e-motor
$T_{emax}(n_e)$ is the maximum output torque of the e-motor at the speed n_e

FIGURE 15.10 Typical pedal strategies.

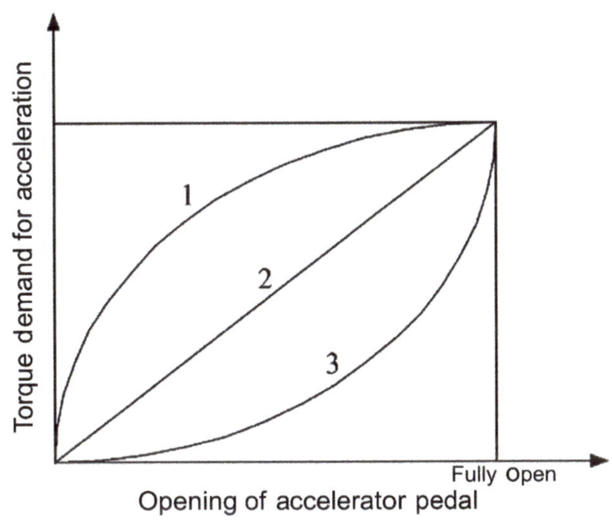

Courtesy of SAIC.

It is obvious that the rated characteristics are not taken into account in this conversion method. Since the e-motor outputs a constant torque when its speed is smaller than the base speed and a constant power when its speed is higher than the base speed, we have

$$T_{emax}(n_{eb}) = \begin{cases} T_p & \text{if } n_e \le n_{eb} \\ \dfrac{T_p n_{eb}}{n_e} & \text{if } n_e > n_{eb} \end{cases} \tag{15.24}$$

where

n_{eb} is the base speed of the e-motor
T_p is the peak torque of the e-motor

Note that the base speed is the maximum speed at which an e-motor can operate under constant torque characteristics or the minimum speed to operate at rated power. Hence, for the peak characteristics of an e-motor, the constant torque and constant power regions are not always demarcated at the base speed. The conversion method presented in Equation (15.24) may not allow a full utilization of the motor power. In order to fully utilize the motor power, a more reasonable conversion method can be employed, instead of the linear pedal strategy:

$$T_e = \begin{cases} \beta T_p & \text{if } n_e \le n_{ep} + \dfrac{\beta T_{emax}(n_e) - T_p}{T_r - T_p}(n_e - n_{ep}) \\ \beta T_p \dfrac{n_{ep}}{n_e} + \dfrac{n_e - n_{ep}}{n_e} \dfrac{\beta T_p^2 (1-\beta)}{T_p - T_r} & \text{if } n_e > n_{ep} + \dfrac{\beta T_{emax}(n_e) - T_p}{T_r - T_p}(n_e - n_{ep}) \end{cases} \tag{15.25}$$

where

n_{ep} is the speed demarcating the constant torque and constant power regions of the peak characteristics of the motor
T_r is the rated torque

In order to make the pedal feel of EVs similar to traditional vehicles, EVs are often designed with a creep function as it moves off from a standstill. It is actually the default speed for the vehicle operating in either forward or reverse when the foot is taken off the brake pedal while the accelerator pedal is not depressed. For this to happen, the e-motor must output a certain torque to the wheels so that the vehicle can move slowly on flat roads or prevent it from rolling back on a slope. It is generally recommended that the creep be kept below a speed of 6 km/h.

On the other hand, when the accelerator pedal is released and the vehicle is coasting at a high speed, the e-motor produces braking torque to simulate the driving feel of the engine braking mode of a traditional vehicle. In such a regenerative braking mode, the e-motor converts the kinetic energy of the vehicle into electricity and stores it in the traction battery. The torque demand for acceleration by adopting a linear pedal strategy with both creep function at low speeds and regenerative braking function at high speeds can simply be proportioned by the accelerator pedal opening. Apparently, the e-motor outputs a certain positive torque (driving torque) in the low-speed region and outputs a certain negative torque (braking torque) in the high-speed region. These curves can be simply mirrored to the second quadrant for driving the vehicle in the reverse direction.

15.6. Launch Control Strategy

For EVs, in the moment when the brake pedal is released but the accelerator pedal is not depressed, a launch control strategy must be taken to ensure the safety and comfort in the process. The basic requirements for launch control generally include the following: making the process as smooth as possible, providing a proper creep when the vehicle is on a level road, and preventing rolling backward when going up or rolling forward when going down if launching on a steep slope. As illustrated in Figure 15.11 [15.1], to satisfy these requirements, the e-motor must produce a torque for the vehicle to creep that diminishes with the increasing speed of the vehicle and vanishes when the vehicle reaches the maximum creep speed. At high vehicle speeds, the motor generally produces a regenerative braking torque to recover the kinetic energy of the vehicle in coasting. While the vehicle starts to roll backward on a slope, the torque needed to stop such a motion must increase with the increase in rolling back speed.

FIGURE 15.11 E-motor torque control strategy in the launch of an EV.

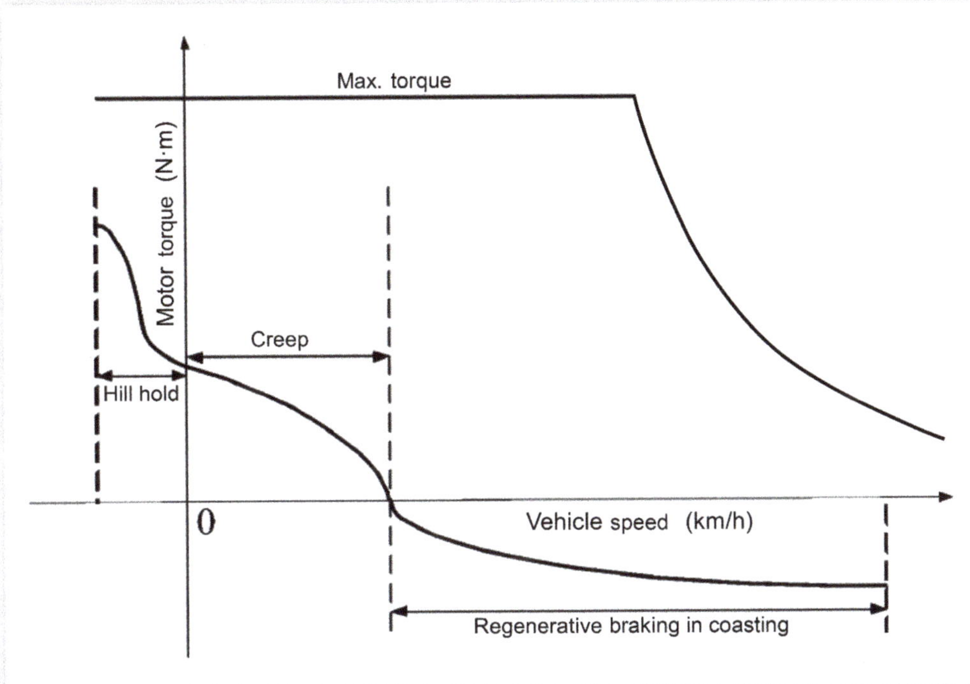

Courtesy of SAIC.

15.6.1. Torque Compensation Control Strategy

In order to meet the requirements of smooth launch with creep and avoid the torque pulsation of the e-motor, an inertia link is added to the system before the e-motor outputs any torque [15.2]:

$$T_e^*(s) = \frac{T_e(s)}{Ts+1} \tag{15.26}$$

where

T is the time constant
T_e is the output torque of the e-motor
s is the complex variable of Laplace transform

Based on the vehicle driving equation, that is, Equation (1.32), we have

$$\frac{dv}{dt} = \frac{\dfrac{T_e i_g i_0 \eta_T}{R_d} - mg\left(f\cos\alpha + \sin\alpha\right) - \dfrac{1}{2}\rho C_D A v^2}{\delta m} \tag{15.27}$$

As the aerodynamic drag is minimal during launch, the term $\frac{1}{2}\rho C_D A v^2$ can be ignored. Thus, by combining Equation (15.26), Equation (15.27) becomes

$$v(t)=\int\left[\frac{T_e^*(t)i_g i_0 \eta_T}{R_d \delta m}-\frac{mg(f\cos\alpha+\sin\alpha)}{\delta m}\right]dt=C_1\int T_e^*(t)dt-C_2 \tag{15.28}$$

where

$$C_1=\frac{i_g i_0 \eta_T}{R_d \delta m}$$

$$C_2=\int\left[\frac{mg(f\cos\alpha+\sin\alpha)}{\delta m}\right]dt=\frac{mg(f\cos\alpha+\sin\alpha)t}{\delta m}$$

Taking the Laplace transform of Equation (15.28) yields

$$v(s)=\frac{C_1 T_e(s)}{s^2 T+s}-\frac{C_2}{s^2} \tag{15.29}$$

During launch, the deviation of the actual vehicle speed $v_a(i)$ from the target vehicle speed v_t at a certain time t_k for $k = 1, 2, 3, \ldots, n$ is

$$\Delta v(k)=v_t-v_a(k) \tag{15.30}$$

Per Equations (15.29) and (15.30), we have

$$T_e(k)=T_e(k-1)+\Delta T_e(k) \tag{15.31}$$

where

$T_e(k)$ is the torque command issued by the VCU at the time t_k

$\Delta T_e(k)$ is the compensation torque at the same time

In addition, under the condition that the drive wheels do not slip, the following linear relationship between the vehicle speed and the motor speed holds:

$$v=\frac{3\pi n_e R_d}{25i} \tag{15.32}$$

where

n_e is the motor speed

i is the total transmission ratio ($i = i_g i_0$)

Hence, the deviation of the actual motor speed from the target motor speed can be obtained from Equations (15.30) and (15.32):

$$\Delta n(k)=n_t-n_a(k) \tag{15.33}$$

Based on Equations (15.32) and (15.33), the deviation of the actual vehicle speed and the target vehicle speed during creep can be obtained from the deviation of corresponding motor speeds. The VCU calculates the compensation torque demand at the moment according to the motor speed deviation and then sends the output torque command calculated by Equation (15.31) to the e-motor. Driven by the updated motor torque, the vehicle acceleration will change, resulting in changes in both the vehicle speed and the motor speed. The feedback of the updated motor speed signal to the VCU would allow calculating torque demand for the next command cycle, so as to realize the closed-loop control of the launch process. Due to the nonlinear characteristics of the system, it is suggested to calculate the aforementioned compensation torque using the incremental form of the proportional integral (PI) algorithm. To meet the requirements of launching the vehicle on a slope with a creep, the torque increment $\Delta T_e(k)$ at the kth moment is given by

$$\Delta T_e(k) = k_p\left[n_{err}(k) - n_{err}(k-1)\right] + k_p \frac{t_s}{T_i} n_{err}(k) \tag{15.34}$$

where

n_{err} is the deviation value of motor speed
t_s is the sampling time of the control system
T_i is the integral time constant
k_p is the proportional coefficient

In Equation (15.34), the value of the proportional coefficient k_p affects the response speed and amplitude of the compensation torque, and the value of integral time constant T_i is used to adjust the range of compensation torque variation at any time.

Finally, the output torque command of VCU at the kth moment can be derived from Equations (15.31) and (15.34) [15.2]:

$$T_e(k) = T_e(k-1)k_p\left[n_{err}(k) - n_{err}(k-1)\right] + k_p \frac{t_s}{T_i} n_{err}(k) \tag{15.35}$$

When the gradient is greater than 10%, it is necessary to strengthen the role of the proportional part of the incremental PI algorithm in the early stage of launch from standstill, in order to improve the response speed and amplitude of torque compensation and avoid long-term and excessive speed deviations as well as the overflow of integral control variables. Therefore, by setting the motor speed deviation threshold $\varepsilon > 0$, the following judgments can be made according to the absolute value of the motor speed deviation $|n_{err}(k)|$.

When $|n_{err}(k)| > 0$, a prolonged duration of the launch process would imply that this is in the initial stage of a long uphill slope. In order to speed up the system response and prevent the vehicle from rolling back, the proportional control part is changed to $k_p = k_c|e(k)|$, where $e(k)$ is the deviation value of the input variable at the kth sampling time and k_c is the incremental strengthening factor.

When $|n_{err}(k)| \leq 0$, the vehicle is in the initial stage of launch on a gentle slope. The proportional coefficient k_p remains unchanged, and the torque compensation for creep is controlled by the aforementioned basic incremental PI algorithm to ensure the precision of speed control.

Furthermore, to ensure a smooth ride in creep away from standstill, the jerk of the vehicle must be controlled within an acceptable range, typically no more than 10 m/s^3.

15.6.2. Self-Adaptive Control Strategy

The launch process can be divided into two stages [15.3, 15.4]: first, preparation of control parameters, and second, application of control parameters, until the end of the process indicated by a change in accelerator pedal opening.

In the first stage, since the gradient of the slope is unknown, it is difficult to make the e-motor output the correct torque for launch. The control strategy here is to switch the e-motor into the closed-loop speed control mode first and operate in the stall condition. As a result, the stall torque of the e-motor would hold the vehicle on the slope without rolling backward or forward. At the moment, the torque of the e-motor can automatically adjust its direction and magnitude to overcome the resistance for launch until the motor reaches the speed given by the speed command. In this process, the resistance for launch is used to identify the conditions of the slope on which the vehicle is traveling, including uphill, downhill or flat road, and the gradient of the slope. In this stage, the motor speed command n_0 is set to zero and the stall torque T_{e0} must be able to not only overcome the slope resistance corresponding to the maximum gradeability α_{max} but also meet the jerk requirements. Note that the deviation of the actual motor speed from the target speed, if any, is related to the stall characteristics of the motor. Thus, the target function for system control in this stage is

$$n_0 = 0 \tag{15.36}$$

with the following constraint conditions:

$$\begin{cases} T_{e0} > \dfrac{mg\left(f\cos\alpha_{max} + \sin\alpha_{max}\right)R_d}{i_g i_0 \eta_T} \\ j = \ddot{v} = \dfrac{i_g i_0 \eta_t}{\delta m R_d}\dfrac{\mathrm{d}T_e}{\mathrm{d}t} \le j_{max} \end{cases} \tag{15.37}$$

In the second stage, the e-motor operates in the closed-loop torque control mode. According to the gradient estimated in the first stage, the resistance and the e-motor torque required to overcome such a resistance can be calculated corresponding to zero vehicle speed under the current slope. Then, the e-motor torque is sent to the motor controller as a torque command to keep the vehicle stationary.

At this moment, if any accelerator pedal signal is detected, then it is an indication that the driver is taking over the control and the second stage comes to the end. After that, the torque command for the e-motor, including the direction and magnitude of the torque, can be determined according to the accelerator pedal signal and the identified slope data.

Based on different situations such as uphill, downhill or flat road, the algorithms for the estimation of gradient and torque command are different, as discussed in the following section.

15.6.2.1. Launching on Uphill Roads

In the first stage, the e-motor is commanded to operate in stalling under the closed-loop speed control mode and the target motor speed is set to zero.

Let F_{tb} be the driving force at the wheels coming from the e-motor through the drive-train when the rolling back speed of the vehicle is zero, then the vehicle acceleration is

$$a = \begin{cases} \dfrac{F_t - F_s + F_f}{\delta m}, & F_t < F_{tb} \\ 0, & F_{tb} \le F_t \le F_s + F_f \\ \dfrac{F_t - F_s - F_f}{\delta m}, & F_t > F_s + F_f \end{cases} \tag{15.38}$$

where

F_t is the driving force at wheels
F_s is the gradient resistance
F_f is the rolling resistance
δ is the rotational mass conversion fact
m is the vehicle mass

Since the vehicle is on an uphill slope and the target of the closed-loop speed control is zero, the motor is bound to increase the forward torque rapidly from zero to produce the wheel driving force for overcoming the resistances during launch. However, before the driving force increases from zero to F_{tb}, the vehicle starts to roll back. Let t_1 be the moment when the rolling back speed drops to zero, then the distance of rolling back is given by

$$s = \left| \int_0^{t_1} \int_0^{t} \frac{F_t - F_s + F_f}{\delta m} dt dt \right| \tag{15.39}$$

Only after $F_t > F_{tb}$, the vehicle stops rolling back and the driving force at wheels is in balance with the resistances, with the motor output torque being

$$T_{e1} = \frac{mg(f \cos\alpha + \sin\alpha) R_d}{i_g i_0 \eta_T} \tag{15.40}$$

In this process, the e-motor outputs a positive torque. Hence, once a positive torque is detected by the VCU, it can be identified that the vehicle is moving off uphill. The gradient of the slope can be estimated per Equation (15.40):

$$\alpha = \arcsin \frac{i_g i_0 \eta_T T_{e1}}{mgR_d \sqrt{1 + f^2}} - \arctan f \tag{15.41}$$

Subsequently, in the second stage, the e-motor is switched into the closed-loop torque control mode and the torque command is given by

$$T_{e2} = \frac{mg\left[\sin(1+e)\alpha + f\cos(1+e)\alpha + \frac{1}{2}\rho C_D A v^2\right]R_d}{i_g i_0 \eta_T} \tag{15.42}$$

where

e is the error in slope identification

α is the slope angle ($\alpha \geq 0$)

15.6.2.2. Launching on Downhill Roads

In the first stage, as the e-motor operates in the closed-loop speed control mode, it can be reckoned that the vehicle is traveling on a downhill slope if a negative output torque is detected. Due to the stall torque, the driving force at the wheel would increase rapidly from zero and its direction is opposite to the direction of the vehicle movement for overcoming the resistances during launch. At this moment, the vehicle acceleration is

$$a = \frac{F_t + F_s - F_f}{\delta m} \tag{15.43}$$

Before the driving force becomes large enough to overcome the resistances, the vehicle may start rolling forward. Let t_1 be the moment when the rolling forward speed drops to zero, then the distance of rolling forward is given by

$$s = \left| \int_0^{t_1}\int_0^{t} \frac{F_t + F_s - F_f}{\delta m} dtdt \right| \tag{15.44}$$

Since the vehicle is on a downhill slope and the target of the closed-loop speed control is zero, the motor is bound to increase the reverse torque rapidly from zero to produce the wheel driving force for overcoming the resistances during launch. The motor torque for balancing the vehicle is given by

$$T_{e1} = \frac{mg(f\cos\alpha - \sin\alpha)R_d}{i_g i_0 \eta_T} \tag{15.45}$$

Note that T_{e1} is a negative value, virtually a braking torque. For slope recognition, the gradient of the slope is given by

$$\alpha = \arctan f - \arcsin\frac{i_g i_0 \eta_T T_{e1}}{mgR_d\sqrt{1+f^2}} \tag{15.46}$$

Similarly, $\alpha \geq 0$.

In the subsequent second stage, the e-motor is switched to a closed-loop torque control mode and the torque command can be estimated by

$$T_{e2} = \frac{mg\left[\sin(1+e)\alpha - f\cos(1+e)\alpha + \frac{1}{2}\rho C_D A v^2\right]R_d}{i_g i_0 \eta_T} \tag{15.47}$$

15.6.2.3. Launching on Level Roads

When the motor operates in stall under the closed-loop speed control mode, if neither reverse nor forward speed is detected and the motor torque is very minimal, then it can be recognized that the vehicle is on a flat road or a very small slope that is not enough to cause the vehicle to move. In the first stage, there is no significant stall torque or driving force at wheels required for overcoming the resistance. Therefore, the vehicle acceleration is

$$a = \begin{cases} 0 & F_t \le F_s + F_f \\ \dfrac{F_t - F_s - F_f}{\delta m}, & F_t > F_s + F_f \end{cases} \tag{15.48}$$

In the first stage, the slope can be identified using Equation (15.41). In the second stage, the motor is switched to the closed-loop speed torque mode, and the torque command can be computed using Equation (15.42).

In summary, the logical block diagram of the adaptive control strategy involves the following seven steps [15.3, 15.4]:

1. After releasing the brake pedal, detect whether there is any accelerator pedal signal. If yes, calculate the torque command directly according to the accelerator pedal opening, and the system hands over the torque control in the launch process to the driver.
2. If there is no accelerator pedal signal, enter the first stage of the launch process and make the motor work in stalling under the closed-loop speed control mode, with the speed command set as $n_0 = 0$.
3. Detect the motor torque. If it is positive, the vehicle is identified as launching on an uphill road. When the motor output torque is stabilized, use Equation (15.41) to estimate the gradient and then use Equation (15.42) to calculate the torque command for the motor based on the estimated gradient.
4. If the motor torque is negative, the vehicle is identified as launching on a downhill road. When the motor output torque is stabilized, use Equation (15.46) to estimate the gradient and then use Equation (15.47) to calculate the torque command for the motor based on the estimated gradient.
5. If the motor torque is zero, the vehicle is identified as launching on a level road, the gradient and the torque command are calculated using Equations (15.41) and (15.42), respectively.
6. Set the motor to operate in the closed-loop torque control mode and use the torque command obtained from steps 4 to 6.
7. Detect the accelerator pedal signal again. If there is still no accelerator pedal signal, return to continue the detection repeatedly. If there is an accelerator pedal signal, then hand over the launch control to the driver by calculating the torque command based on the accelerator pedal opening signal and the identified gradient information.

15.7. Electric Drive System Control Strategy

15.7.1. E-Motor Control Strategy

The e-motor control strategy is dependent on the layout of the electric drive system in an EV. Among the variety of layouts discussed in Section 8.9.2, four most frequently used ones are discussed here in detail: single motor driving a single axle, dual-motors driving a single axle, dual-motors driving dual-axles, and four motors driving four wheels. In addition, the e-motor control strategy can also be quite different if a multispeed transmission is used, instead of a single-speed transmission. The control strategy for an e-motor mating with a single-speed transmission is discussed in this section, while the control strategy related to multispeed transmissions and CVTs will be discussed in Section 15.7.2.

15.7.1.1. Single Motor Driving a Single Axle

When the vehicle is driven by a single e-motor through a single speed transmission in either FF or RR configurations, the control strategy is quite straightforward. The motor speed n_e can be detected by the signals from speed sensors, for example, a resolver. The torque demand for motor, T_e, can be calculated according to the accelerator pedal opening β by

$$T_e(\beta) = \begin{cases} \beta T_{epk}(n_e) & \text{if } 0 \le n_e < n_{eb} \\ \dfrac{\beta P_{epk}(n_e)}{2\pi n_e} & \text{if } n_{eb} \le n_e < n_{emax} \end{cases} \tag{15.49}$$

where

$T_{epk}(n_e)$ is the peak motor torque at the speed n_e
$P_{epk}(n_e)$ is the peak motor power at the speed n_e
n_{eb} is the base speed of the motor
n_{emax} is the maximum motor speed

In order to prevent the excessive motor torque pulsation caused by the jitter of the accelerator pedal, the voltage signal from the accelerator pedal sensor needs to be filtered. This signal is periodically sampled through real-time interruption to the MCU, and then a certain number of the latest sampling values are used for processing. With the maximum and minimum values removed first from these values, an average is then taken among the rest of the values, which will be used as the accelerator pedal opening, usually in percentage, under the current sampling period.

In practice, as the requirements for driving torque and its response are different in driving forward and backward, different driving torque strategies are often designed according to their respective characteristics.

1. Forward driving torque strategy
   In a normal forward driving scenario, it is generally desired to give the driver a good driving torque control under small load conditions, so that the vehicle can gain driving stability. In addition, it is also preferred that the vehicle can have decent acceleration response near 50–60% of the accelerator pedal opening. Hence, a strategy of prorated

torque response from the motor can be adopted, with the relationship between the response torque from the motor T_{er} and the demand for motor torque T_e under the premise of fully guaranteeing the vehicle dynamic performances as follows:

$$T_{er} = \begin{cases} T_e & \text{if } 0 \le n_e < n_{elmt} \\ T_e\left(n_{emax} - n_e\right)/\left(n_{emax} - n_{elmt}\right) & \text{if } n_{elmt} \le n_e < n_{emax} \\ 0 & \text{if } n_e \ge n_{emax} \end{cases} \tag{15.50}$$

where n_{elmt} is a motor speed limit, which can be determined in the vehicle calibration.

2. Backward driving torque strategy
In reality, the vehicle speed is usually quite low when it is driven in reverse. To ensure safety, the torque response from the motor can be prorated according to the motor speed so that the torque declines quickly with the increasing speed in reverse. Thus, the relationship between the torque response from the motor T_{er} and the demand for motor torque T_e can take the following strategy:

$$T_{er} = \begin{cases} T_e\left(n_{emax} - n_e\right)/n_{emax} & \text{if } 0 \le n_e < n_{emax} \\ 0 & \text{if } n_e \ge n_{emax} \end{cases} \tag{15.51}$$

where

n_e is the current motor speed in the reverse direction

n_{emax} is the maximum motor speed in the reverse direction

15.7.1.2. Dual-Motors Driving a Single Axle

This architecture is used in either an FF or an RR format. Dual-motors can either mate with the same gear reduction mechanism in the single-speed transmission and output from the shared differential, or each mate with a separate gear reduction mechanism without differential. The discussion here will be focused on the control strategy in the former architecture since the control strategy in the latter architecture is similar to that of four motors driving four wheels and will be discussed later.

Furthermore, the operation of dual-motors can be decoupled if at least one decoupling mechanism, such as a clutch, is used for one of the motors. However, when both motors are coupled to the drivetrain for driving the vehicle, their operating speeds becomes dependent on each other.

When the dual-motors drive a single axle, the driving torque can come from one of the motors or both motors. However, the pedal strategy is made by first assuming that both motors drive the vehicle to find the maximum torque.

Thus, the maximum output torque, as a function of the torque and speed of both motors, can be computed by

$$\begin{aligned} T_{omax}\left(n_o\right) &= T_{omax}\left(T_{e1}, n_{e1}, T_{e2}, n_{e2}\right) \\ &= i_{g1} T_{e1max}\left(n_{e1}\right) \eta_{g1}\left(T_{e1}, n_{e1}\right) + i_{g2} T_{e2max}\left(n_{e2}\right) \eta_{g2}\left(T_{e2}, n_{e2}\right) \end{aligned} \tag{15.52}$$

where

η_{g1} and η_{g2} are the efficiencies of the gear reduction mechanisms mating with the motor 1 and motor 2, respectively

i_{g1} and i_{g2} are the ratios of these gear reduction mechanism, respectively

In Equation (15.52),

$$T_{e1max}\left(n_{e1}\right)=\begin{cases}T_{e1pk}\left(n_{e1}\right) & \text{if } 0\le n_{e1}<n_{eb1}\\ \dfrac{P_{e1pk}\left(n_{e1}\right)}{2\pi n_{e1}} & \text{if } n_{eb1}\le n_{e1}<n_{e1max}\end{cases} \tag{15.53}$$

and

$$T_{e2max}\left(n_{e2}\right)=\begin{cases}T_{e2pk}\left(n_{e2}\right) & \text{if } 0\le n_{e2}<n_{eb2}\\ \dfrac{P_{e2pk}\left(n_{e2}\right)}{2\pi n_{e2}} & \text{if } n_{eb2}\le n_{e2}<n_{e2max}\end{cases} \tag{15.54}$$

Additionally, due to the aforementioned coupling between the speeds of dual-motors, Equations (15.52) through (15.54) are subjected to the constraint of

$$n_o=\frac{n_{e1}}{i_{g1}}=\frac{n_{e2}}{i_{g2}} \tag{15.55}$$

as the output speed for motor 1 and its gear reduction is identical to that of motor 2 and its gear reduction, with i_{g1} and i_{g2} being their respective gear reduction ratios.

If the characteristics of both motors and gear reduction ratios are known, the systems of equations—Equations (15.52) through (15.55)—can be solved readily. For a given output speed n_o, we have

$$n_{e1}=n_o i_{g1} \tag{15.56}$$

and

$$n_{e2}=n_o i_{g2} \tag{15.57}$$

With T_{e1max} and T_{e2max} obtained from Equations (15.53) and (15.54), respectively, the maximum output torque $T_{omax}(n_o)$ can be solved from Equation (15.52).

Thus, the pedal strategy can be made according to the output torque of the electric drive system. In its simplest format, the output torque demand $T_o(n_o)$, as a function of the accelerator pedal opening β and the output speed n_o, can be assigned as a direct proportion to the accelerator pedal opening β:

$$T_o\left(\beta,n_o\right)=\beta T_{omax}\left(n_o\right) \tag{15.58}$$

where n_{omax} is the maximum output speed, $0\% \le \beta \le 100\%$, and $0\le n_o\le n_{omax}$.

Since the output torque comes from both motors, we have

$$\begin{aligned} T_o(\beta, n_o) &= T_o(T_{e1}, n_{e1}, T_{e2}, n_{e2}) \\ &= i_{g1} T_{e1}(n_{e1}) \eta_{g1}(T_{e1}, n_{e1}) + i_{g2} T_{e2}(n_{e2}) \eta_{g2}(T_{e2}, n_{e2}) \end{aligned} \tag{15.59}$$

For a given accelerator pedal opening β and output speed n_o, there could be multiple combinations of motor operating points (T_{e1}, n_{e1}) and (T_{e2}, n_{e2}) that can produce the output torque $T_o(\beta, n_o)$. Evidently, the combination that yields the highest electric drive system efficiency should be selected for the optimal energy economy.

Since the electric drive system efficiency is given by

$$\eta_{sys}(T_{e1}, n_{e1}, T_{e2}, n_{e2}) = \frac{T_{e1} n_{e1} \eta_{g1} + T_{e2} n_{e2} \eta_{g2}}{\dfrac{T_{e1} n_{e1}}{\eta_{e1}} + \dfrac{T_{e2} n_{e2}}{\eta_{e2}}} \tag{15.60}$$

among the multiple solutions of Equation (15.59), an optimal combination of motor operating points (T'_{e1}, n'_{e1}) and (T'_{e2}, n'_{e2}) can, therefore, be found so that

$$\eta_{sys}(T'_{e1}, n'_{e2}, T'_{e2}, n'_{e2}) = \max\left[\eta_{sys}(T_{e1}, n_{e1}, T_{e2}, n_{e2})\right] \tag{15.61}$$

By mapping through the two-dimensional space of variables (β, n_o) in the range of 0% $\leq \beta \leq$ 100% and $0 \leq n_o \leq n_{omax}$, all the optimal motor operating points (T'_{e1}, n'_{e1}) and (T'_{e2}, n'_{e2}) can be obtained and utilized to render the pedal strategy of Equation (15.58).

For a drivetrain system not equipped with a decoupling mechanism, when only one motor is producing the torque for driving the vehicle, the other motor may produce a certain drag torque to the system. The drag torque can be determined in vehicle calibration and compensated accordingly.

When the torque comes from one of the motors (preferably from the one that does not have a decoupling mechanism for the sake of energy saving), the demand for motor torque can be calculated in a similar way to the single motor driving single axle scenario mentioned earlier in Equation (15.49).

15.7.1.3. Dual-Motors Driving Dual-Axles

This is the most frequently used architecture for four-wheel drive EVs today, essentially with one motor driving the front axle and the other driving the rear. Different types of motors are often used, for example, an IM for the front and a PMSM for the rear, or the other way around, mainly aiming at improving the energy efficiency of the vehicle. Besides, motors of different characteristics in torque, power, or high-efficiency regions are also matched, with each fitting the needs of certain driving conditions taking advantage of its characteristics. The drivetrain for mating with one motor may also have a different reduction ratio from the other for the same purpose.

Similar to the last section, in the case of both motors driving the vehicle together, assuming that the front and rear wheels are of the same size with no slip at the tire–road interface, the maximum total torque at the wheels, as a function of the torque and speed of both motors, can be computed by

$$T_{wmax}(n_w) = T_{wmax}(T_{e1}, n_{e1}, T_{e2}, n_{e2}) = i_{g1}i_{o1}T_{e1max}(n_{e1})\eta_{t1}(T_{e1}, n_{e1}) + i_{g2}i_{o2}T_{e2max}(n_{e2})\eta_{t2}(T_{e2}, n_{e2}) \tag{15.62}$$

where

η_{t1} and η_{t2} are the efficiencies of drivetrains mating with motor 1 and motor 2, respectively
i_{o1} and i_{o2} are the respective final drive ratios

In Equation (15.43), $T_{e1max}(n_{e1})$ and $T_{e2max}(n_{e2})$ can be solved using Equations (15.53) and (15.54). Nevertheless, the motor speeds are subject to the constraint of

$$n_w = \frac{n_{e1}}{i_{g1}i_{o1}} = \frac{n_{e2}}{i_{g2}i_{o2}} \tag{15.63}$$

as the front and rear drivetrains, possibly of different final drive ratios, are coupled by the ground.

If the characteristics of both motors and drivetrain ratios are known, for a given wheel speed n_w, we have

$$n_{e1} = i_{g1}i_{o1}n_w \tag{15.64}$$

and

$$n_{e2} = i_{g2}i_{o2}n_w \tag{15.65}$$

With T_{e1max} and T_{e2max} obtained from Equations (15.53) and (15.54), respectively, the maximum wheel torque $T_{wmax}(n_w)$ can be solved from Equation (15.62).

Thus, the pedal strategy can be made according to the total torque at the wheels of the vehicle. In its simplest format, the output torque demand $T_w(n_w)$, as a function of the accelerator pedal opening β and the wheel speed n_w, can be assigned as a direct proportion to the accelerator pedal opening β:

$$T_w(\beta, n_w) = \beta T_{wmax}(n_w) \tag{15.66}$$

where n_{wmax} is the maximum wheel speed, $0\% \le \beta \le 100\%$, and $0 \le n_w \le n_{wmax}$.

Since the total torque at the wheels comes from both motors, we have

$$\begin{aligned} T_w(\beta, n_w) &= T_w(T_{e1}, n_{e1}, T_{e2}, n_{e2}) \\ &= i_{g1}i_{o1}T_{e1}(n_{e1})\eta_{t1}(T_{e1}, n_{e1}) + i_{g2}i_{o2}T_{e2}(n_{e2})\eta_{t2}(T_{e2}, n_{e2}) \end{aligned} \tag{15.67}$$

For a given accelerator pedal opening β and wheel speed n_w, there could be multiple combinations of motor operating points (T_{e1}, n_{e1}) and (T_{e2}, n_{e2}) that can produce the total torque at the wheels $T_w(\beta, n_w)$. Apparently, the combination that yields the highest electric drive system efficiency should be selected for the optimal energy economy.

Since the electric drive system efficiency is given by

$$\eta_{sys}\left(T_{e1},n_{e1},T_{e2},n_{e2}\right)=\frac{T_{e1}n_{e1}\eta_{t1}+T_{e2}n_{e2}\eta_{t2}}{\dfrac{T_{e1}n_{e1}}{\eta_{e1}}+\dfrac{T_{e2}n_{e2}}{\eta_{e2}}} \tag{15.68}$$

among the multiple solutions of Equation (15.66), an optimal combination of motor operating points $\left(T_{e1}^{'},n_{e1}^{'}\right)$ and $\left(T_{e2}^{'},n_{e2}^{'}\right)$ can, therefore, be found so that

$$\eta_{sys}\left(T_{e1}^{'},n_{e2}^{'},T_{e2}^{'},n_{e2}^{'}\right)=\max\left[\eta_{sys}\left(T_{e1},n_{e1},T_{e2},n_{e2}\right)\right] \tag{15.69}$$

By mapping through the two-dimensional space of variables (β, n_w) in the range of 0% $\leq \beta \leq$ 100% and $0 \leq n_w \leq n_{wmax}$, all the optimal motor operating points $\left(T_{e1}^{'},n_{e1}^{'}\right)$ and $\left(T_{e2}^{'},n_{e2}^{'}\right)$ can be obtained and utilized to render the pedal strategy of Equation (15.66).

If none of the drivetrain systems is equipped with any decoupling mechanism, with only one motor producing the torque for driving the vehicle, the other motor may produce a certain drag torque to the vehicle. The drag torque can be determined in vehicle calibration and compensated accordingly.

15.7.1.4. Four Motors Driving Four Wheels

In this architecture, each of the wheels is driven by a motor independently. Due to the complex nature in controlling such a system, some fundamental assumptions are made in the following discussion. First, the motor, its drivetrain, and the wheel on the left is identical to that on the right in all aspects of its characteristics. Second, the full utilization of external characteristics of the motors is supported by the tire-road adhesion. And third, the ground forces acting on all the wheels produced by motor torques are all in the longitudinal direction.

1. Accelerator pedal strategy

   The pedal strategy is made by first assuming that all motors drive the vehicle to find the maximum torque and no slip at any wheels.

   Thus, the maximum output torque, as a function of the torque and speed of each of the motors, can be computed by

$$T_{wmax}\left(n_w\right)=T_{wmax}\left(T_{e1},n_{e1},T_{e2},n_{e2}\right)=\sum_{j=1}^{4}i_{g,j}T_{emax,j}\left(n_{e,j}\right)\eta_{t,j}\left(T_{emax,j},n_{e,j}\right) \tag{15.70}$$

   where j = 1, 2, 3, 4 labeling the motors for driving front left, front right, rear left, and rear right wheels, respectively, and

$$T_{emax,j}\left(n_{e,j}\right)=\begin{cases}T_{epk,j}\left(n_{e,j}\right) & \text{if } 0\leq n_{e,j}<n_{eb,j}\\ \dfrac{P_{epk,j}\left(n_{e,j}\right)}{2\pi n_{e,j}} & \text{if } n_{eb,j}\leq n_{e,j}<n_{emax,j}\end{cases} \tag{15.71}$$

The motor speeds are subject to the constraint of

$$n_w = \frac{n_{e,j}}{i_{g,j}} \quad \text{for } j = 1,2,3,4 \tag{15.72}$$

If the characteristics of all motors and gear reduction ratios are known, for a given wheel speed n_w, we have

$$n_{e,j} = i_{g,j} n_w \quad \text{for } j = 1,2,3,4 \tag{15.73}$$

With $T_{emax,j}$ obtained from Equation (15.52), the maximum wheel torque $T_{wmax}(n_w)$ can be solved from Equation (15.70).

Thus, the pedal strategy can be made according to the total torque at the wheels of the vehicle. In its simplest format, the output torque demand, $T_w(n_w)$, as a function of the accelerator pedal opening β and the wheel speed n_w, can be assigned as a direct proportion to the accelerator pedal opening β:

$$T_w(\beta, n_w) = \beta T_{wmax}(n_w) \tag{15.74}$$

where n_{wmax} is the maximum wheel speed, $0\% \le \beta \le 100\%$, and $0 \le n_w \le n_{wmax}$.

Since the total torque at the wheels comes from all the motors, we have

$$T_w(\beta, n_w) = T_w(T_{e,j}, n_{e,j}) = \sum_{j=1}^{4} i_{g,j} T_{e,j}(n_{e,j}) \eta_{g,j}(T_{e,j}, n_{e,j}) \tag{15.75}$$

For a given accelerator pedal opening β and wheel speed n_w, there could be multiple combinations of motor operating points ($T_{e,j}$, $n_{e,j}$) for j = 1, 2, 3, 4 that can produce the total torque at the wheels $T_w(\beta, n_w)$. Apparently, the combination that yields the highest system efficiency should be selected for the optimal energy economy.

Since the electric drive system efficiency is given by

$$\eta_{sys}(T_{e,j}, n_{e,j}) = \frac{\sum_{j=1}^{4} T_{e,j} n_{e,j} \eta_{g,j}}{\sum_{j=1}^{4} \frac{T_{e,j} n_{e,j}}{\eta_{e,j}}} \tag{15.76}$$

among the multiple solutions of Equation (15.76), an optimal combination of motor operating points $\left(T'_{e,j}, n'_{e,j}\right)$ for j = 1, 2, 3, 4 can, therefore, be found so that

$$\eta_{sys}\left(T'_{e,j}, n'_{e,j}\right) = \max\left[\eta_{sys}\left(T_{e,j}, n_{e,j}\right)\right] \tag{15.77}$$

By mapping through the two-dimensional space of variables (β, n_w) in the range of $0\% \le \beta \le 100\%$ and $0 \le n_w \le n_{wmax}$, all the optimal motor operating points $\left(T'_{e,j}, n'_{e,j}\right)$ for j = 1, 2, 3, 4 can be obtained and utilized to render the pedal strategy of Equation (15.74).

To ensure the lateral stability of the vehicle, additional constraints may be imposed to the motor torque at each of the wheels. For example, if a yaw moment T_{yaw} is required for lateral stability control, then the following constraint must be satisfied [15.5]:

$$\frac{\left(T_{e,2}i_{g,2} - T_{e,1}i_{g,1}\right)t_{rf} + \left(T_{e,4}i_{g,4} - T_{e,3}i_{g,3}\right)t_{rr}}{R} = T_{yaw} \tag{15.78}$$

where

t_{rf} and t_{rr} are the wheel tracks at the front and rear, respectively

R is the wheel radius

Moreover, to prevent the slip at any of the wheels, the following constraints would become necessary [15.5]:

$$T_{e,j} \le \min\left(T_{emax,j}, \frac{f_j F_{z,j} R}{i_{g,j}}\right) \quad \text{for } j = 1,2,3,4 \tag{15.79}$$

where

f_j is the adhesion coefficient at the jth wheel

$F_{z,j}$ is the vertical force at the jth wheel

Finally, if the moment produced by driving forces of rear wheels is greater than that of front wheels, the vehicle may suddenly spin out. To avoid this undesired motion, the following constraint must also be applied [15.6]:

$$\sum_{j=1}^{2}\left(\frac{i_{g,j}F_{z,j}}{f_j F_{z,j} R}\right)^2 \ge \sum_{j=3}^{4}\left(\frac{i_{g,j}F_{z,j}}{f_j F_{z,j} R}\right)^2 \tag{15.80}$$

The optimization problem described by Equation (15.77) with constraints defined by Equations (15.78), (15.79), and (15.80) can be solved using a numerical method.

2. Failsafe control strategy
If any failures occur in one or more motors, the vehicle may deviate from its course, leading to potential accidents. Hence, a proper failsafe control strategy is needed to ensure the safe operation of the vehicle.
The recommended failsafe control strategy reduces the output torque of the failed motors with limiting factors [15.5]:

$$T_{e,j} = \lambda_j T_{emax,j} \tag{15.81}$$

where λ_j is the limiting factor for the jth motor. The value of λ_j falls between 0 and 1, with $\lambda_j = 1$ if the motor is in a perfectly normal condition and $\lambda_j = 0$ if the motor is in total failure and unable to output any torque.

If only one motor completely fails, then set its limiting factor to zero [15.5].

If two motors on the same side fail completely, then set the limiting factors for all the four motors to zero to avoid unbalanced driving forces, which implies that the vehicle cannot limp home; if two motors not on the same side fail completely, then set the limiting factors of failed motors to zero [15.5].

If three out of the four motors fail completely, then set the limiting factors for all the four motors to zero, with no limp home mode provided.

15.7.2. Gear Shift Control Strategy

For a nonautonomous vehicle, the gear shift control strategy is needed for the vehicles equipped with multispeed transmissions only. Nevertheless, for a fully autonomous vehicle, the gear shift can actually be broadened to any shifts among the gears of "Park," "Reverse," "Neutral," and "Drive," although there can be no physical shift lever equipped in the vehicle. In this section, the gear shift will stick to its original meaning of shift between forward gears.

15.7.2.1. Gear Shift Decision-Making

Gear shift decision-making refers to determining the gear that the vehicle should be operated in based on the intention of the driver, operational status of the vehicle, and road conditions, aiming at the optimal energy economy or dynamic performance. The result of the gear shift decision-making is the shift schedule.

15.7.2.2. Forms of Shift Schedules

The shift of a transmission from one gear to another is determined by the shift logic selection routine, while the shift schedule must be determined before being used as the input of the shift logic selection routine.

Generally, the shift schedule can be expressed as

$$\beta_{ij} = f(N_o, P_{sl}, T, P_{tcase}) \tag{15.82}$$

where

β_{ij} is the accelerator pedal opening when the ith gear shifts to the jth gear
N_o is the output shaft speed
P_{sl} is the shift handle position
T is the transmission temperature
P_{tcase} is the transfer case position.

For EV transmissions with the shift lever position in D (i.e., forward gear) and their temperature within a normal range, the shift schedule Equation (15.81) can be simplified as

$$\beta_{ij} = f(N_o) \tag{15.83}$$

which is called the two-parameter shift schedule.

In order to improve the dynamic performance and energy economy, a three-parameter shift schedule based on accelerator pedal opening, vehicle speed, and vehicle acceleration is also proposed [15.5], namely,

$$\beta_{ij} = f(V, \dot{V}) \tag{15.84}$$

In view of the corresponding relationship between the output shaft speed and the vehicle speed, the three-parameter shift schedule of Equation (15.84) can also be expressed as

$$\beta_{ij} = f\left(N_o, \dot{N}_o\right) \tag{15.85}$$

where $\dot{N}_o$ is the acceleration of the output shaft. Although some studies have shown that the three-parameter shift schedule has the advantages of improving dynamic performance and fuel economy, smoothing gear shift schedule surfaces, and fitting better with the shifting experience and habits of a driver, its application in EVs has not been fully studied.

In addition, a four-parameter shift schedule has been suggested based on accelerator pedal opening, vehicle speed, vehicle acceleration, and SOC [15.6].

$$\beta_{ij} = f\left(V, \dot{V}, SOC\right) \tag{15.86}$$

When the battery SOC is low, its maximum discharge power decreases, thus affecting the maximum output power of the motor. It is claimed by some researchers, when the accelerator pedal opening is large, that the four-parameter shift schedule can help improve the dynamic performance and shorten the acceleration time.

15.7.2.3. Shift Schedule for Optimal Dynamic Performance

The two-parameter shift schedule is taken as an example herein to illustrate the formulation steps of the shift schedule for optimal dynamic performance.

The vehicle acceleration in each gear is taken as the basis for selecting the shift point, while the driving force in each gear is sometimes used as the basis for selecting the shift point instead for the sake of simplicity. The former method is adopted here. With the influence of the road gradient ignored, the vehicle acceleration in the ith gear can be derived from the driving equation, that is, Equation (1.32):

$$a_i = \frac{1}{\delta m}\left(\frac{T_e i_{gi} i_0 \eta_T}{r} - mgf\cos\alpha - \frac{1}{2}\rho C_D A V^2\right) \tag{15.87}$$

The vehicle acceleration in the next higher gear, j^{th} gear (thus, $j = i + 1$), can also be obtained by substituting the subscript j for i in Equation (15.87). These acceleration curves for various accelerator pedal openings in the gears of i and j can be plotted and then the shift points for optimal dynamic performance can be obtained by solving the following equation for the intersections of these curves:

$$\frac{1}{\delta_i}\left(\frac{T_{ei} i_{gi} i_0 \eta_T}{r} - mgf\cos\alpha - \frac{1}{2}\rho C_D A V^2\right) = \frac{1}{\delta_j}\left(\frac{T_{ej} i_{gj} i_0 \eta_T}{r} - mgf\cos\alpha - \frac{1}{2}\rho C_D A V^2\right) \tag{15.88}$$

The vehicle speed for upshifting from the ith gear to the jth gear under different accelerator pedal opening β can be obtained by solving Equation (15.88) in combination with Equation (15.1) or Equation (15.3). It is worth noting that the upshift speed V_{ij} must satisfy

$$V_{jb}(\beta) < V_{ij}(\beta) < V_{imax} \tag{15.89}$$

where

V_{imax} is the maximum vehicle speed in the ith gear, usually corresponding to the peak speed of the motor

$V_{jb}(\beta)$ is the vehicle speed in the jth gear corresponding to the base speed of the motor

The engineering significance of Equation (15.89) is that the selection of the shift points must not only avoid the motor overspeed in lower gears but also ensure the continuity of the driving force and power of the vehicle. The shift points that meet the requirements of optimal dynamic performance are located between the two straight lines $V_{jc}(\beta)$ and V_{imax}, which constitute the gear shift region for optimal dynamic performance, that is, dynamic shift region.

It is worth pointing out that the motor drive torques before and after gear shifting are generally not equal, that is,

$$T_{ei} \neq T_{ej} \tag{15.90}$$

Clearly, the motor torque in the higher gear is greater than that in the lower gear in order to maintain the same acceleration before and after shifting.

The upshift schedule curve can be obtained by connecting the shift points (V_{ij}, β) solved by Equation (15.89) at different accelerator pedal openings and applying linear interpolation between the shift points. Then, the vehicle speeds of downshift points, $V_{ji}(\beta)$, can be obtained by taking a certain differential speed, $\Delta V(\beta)$, from the vehicle speed of each upshift point:

$$V_{ji}(\beta) = V_{ij}(\beta) - \Delta V(\beta) \tag{15.91}$$

By connecting the downshift points (V_{ji}, β), the downshift schedule curve can be obtained. In engineering practice, in order to avoid cyclic shifting and to keep the calculation simple, the differential speed ΔV for downshift is often a constant that does not change with the accelerator pedal opening β, with a value generally between 2 and 8 km/h. The downshifting vehicle speed $V_{ji}(\beta)$ must also fall into the shift region for optimal dynamic performance.

15.7.2.4. Shift Schedule for Optimal Energy Economy

Theoretically, the shift points for optimal energy economy should be determined by taking the energy consumption of the vehicle in each gear into account. Since the losses in a battery system at a normal discharge rate are usually small and negligible, the calculations are often simplified by only considering the efficiency of the electric drive system. Again, taking the two-parameter shift schedule as an example, we can express the efficiency of the electric drive system as a function of its torque and speed

$$\eta_{mc} = f(T_e, N_e) \tag{15.92}$$

In view of the relationship between the motor and vehicle speeds, the efficiency of the electric drive system as a function of the accelerator pedal opening and the vehicle speed can be obtained by combining Equation (15.92) with Equation (15.1) or Equation (15.3):

$$\eta_{mc} = f(\beta, V) \tag{15.93}$$

Based on Equation (15.90), the efficiencies of the electric drive system in the ith gear and the next higher gear, j^{th} gear, are, respectively, given by

$$\eta_{mci} = f_1(\beta, V) \tag{15.94}$$

and

$$\eta_{mcj} = f_2(\beta, V) \tag{15.95}$$

The shift points for optimal energy economy must satisfy the relationship $\eta_{mci} = \eta_{mcj}$, or

$$f_1(\beta, V) = f_2(\beta, V) \tag{15.96}$$

The upshift point (V, β) can be obtained by solving Equation (15.96) for vehicle speed V at different accelerator pedal opening β and then the upshift schedule can be obtained by connecting these upshift points. If Equation (15.96) has no solution, then the end points of the efficiency curves of the electric drive system for the ith gear can be taken as the upshift point.

In order to avoid frequent downshifts and upshifts (also known as cyclic shifts) due to drops in vehicle speed or fluctuations in accelerator pedal opening, the equal differential speed method represented by Equation (15.91) is generally used to calculate the downshift speed, with a differential speed ΔV of 2–8 km/h [15.7]. Another way of calculating the upshift and downshift curves is to translate $\Delta V/2$ to the left and right from the center point solved from Equation (15.96), respectively, with $\Delta V/2$ being 1–4 km/h [15.7].

In addition, the other method to prevent cyclic shifting is to delay the shift, usually a constant delay is applied. Hence, the actual shifting timing is given by

$$t_a = t_s + \Delta t \tag{15.97}$$

where

Δt is the shift delay

t_s is the shift timing at which the downshift schedule is satisfied

15.7.2.5. Comprehensive Shift Schedule

Since the shift schedule for optimal dynamic performance can result in an excessive energy consumption and the shift schedule for optimal energy economy can lead to an abrupt change in the driving force, a comprehensive shift schedule is, therefore, proposed to take both dynamic performance and energy economy into account. It can be made by going through the following steps:

1. Obtain shift schedules for optimal dynamic performance and for optimal energy economy, respectively, according to aforementioned methods.
2. If there is an overlap between the economical upshift curve and the dynamic shift region, the highest accelerator pedal opening of the overlapped portion is taken as the dividing point. For the accelerator pedal opening smaller than this point, the shift schedule for optimal energy economy is used; otherwise, the shift schedule for optimal dynamic performance is adopted [15.7].
3. If there is no overlap between the economical upshifting curve and the dynamic shift region, take the accelerator pedal opening of the economical upshifting curve closest to the shift point of the dynamical shifting area as the dividing point. For the accelerator pedal opening smaller than this point, the shift schedule for optimal energy economy is used; otherwise, the shift schedule for optimal dynamic performance is adopted [15.7].

It is asserted that the comprehensive shift schedule allows the vehicle not only to reach an energy economy level very close to the shift schedule for optimal energy economy but also to attain dynamic performance comparable to the shift schedule for optimal dynamic performance [15.7].

15.7.2.6. Shift Schedule Correction under Special Operating Conditions

The output of the shift schedule needs to be corrected in two circumstances, namely, power up at speed and operation at low voltage.

In the practical control procedure, the decision-making procedure of the shift schedule usually involves a relatively complex algorithm. In a simple form of expression, the sequential upshift from the ith gear to the jth gear (i.e., $j = i + 1$) needs to follow the condition of $C_{i \to j}$, while downshift sequential from the jth gear to the ith gear follows the condition $C_{j \to i}$.

The conditions $C_{i \to j}$ and $C_{j \to i}$ are generally logical expressions formulated according to the shift schedule and related state variables. For upshift, these state variables mainly involve the accelerator pedal opening, the output shaft speed, and the current and maximum speeds of the motor in the ith gear to the jth gear. In general, upshifts are often needed in accelerating or releasing the accelerator pedal, and downshifts are required in downhill, coasting, and other operating conditions.

The following principles are usually considered in shift process control:

1. Avoid exceeding the maximum motor speed.
2. When the time change rate of the output shaft speed is smaller than the acceleration value corresponding to the torque change of the output shaft caused by the upshift, the slope detection procedure must be carried out.
3. Under ESP traction control conditions, an alternative shift schedule is sometimes used for limiting the accelerator pedal opening value.

Both road gradient and curvature may affect the shift schedule.

The driving resistance is larger on an uphill road than that on a horizontal road surface due to gradient resistance. If the same shift schedule is adopted, cyclic shifts may occur within a specific speed range, worsening the ride comfort [15.3, 15.4]. Hence, the following principles need to be considered in formulating the shift schedule for uphill driving:

1. Avoid cyclic shifts by increasing the delay in each shift.
2. Make the motor operate in the constant power region by selecting a lower gear, in order to maximize the power to drive the vehicle and overcome the large driving resistance.
3. If the gradient is large and the current vehicle acceleration cannot be guaranteed after the upshift, then the upshift should be abandoned to prevent the tractive force from dropping. For this reason, a slope detection procedure must be invoked to determine whether the required acceleration can be met. The method is to detect the current vehicle acceleration and compare it with the predicted acceleration in the next higher gear based on the difference between the output torques in the two gears. If the predicted acceleration is lower, then the upshift is inhibited. The calculation formula is

$$\alpha_o - \frac{T_{oi} - T_{oj}}{I} < \alpha_{or} \tag{15.98}$$

where

α_o is the angular acceleration of the output shaft
α_{or} is the required angular acceleration of the output shaft
T_{oi} is the output torque in the current gear
T_{oj} is the output torque in the next higher gear
I is the moment of inertia of the vehicle equivalent to the output shaft

The equivalent moment of inertia I can be estimated by the following formula:

$$I = 91.189 \frac{mV^2}{n_o^2} \tag{15.99}$$

where

m is the vehicle mass in kg
V is the vehicle speed in m/s
n_0 is the output shaft speed in rpm

On a downhill road, the driving resistance decreases, leading to a drop in the accelerator pedal opening and an upshift. If the motor enters the constant torque region after upshift, the regenerative braking power of the motor would be reduced, thus affecting the effectiveness of motor braking and energy recovery. Therefore, the following principles are often implemented in the formulation of the shift schedule for downhill driving:

1. Inhibit upshift to prevent the motor from operating in the constant torque region. In a downhill road, the vehicle speed may increase continuously due to the release of potential energy. By inhibiting upshifts, a large wheel braking torque can be maintained, and the effectiveness of motor braking and energy recovery can be improved.
2. Downshift is allowed if the vehicle is currently operating in a high gear.

Slope detection can simply be carried out using the relationship between the longitudinal acceleration of the vehicle and tractive force (or accelerator pedal opening), with its algorithm omitted here.

On the other hand, since the vehicle speed may change when driving on a curve, a regular shift schedule may result in frequent shifts. In order to avoid the impact on the ride comfort, the shift schedule may be subject to revision when driving into a curve. The basic principle is to increase the corresponding vehicle speed at each upshift point and reduce that at each downshift point.

15.7.2.7. Inhibited Shift Operation

For EVs equipped with a park gear, that is, with Gear "P," since the shift from "P" to "R," "N," or "D" is usually carried out when the vehicle is in standstill, such a shift is not subject to any inhibitions.

In addition, since an EV without a park gear is generally parked in "N," its shift to "R" or "D" is not inhibited either when the vehicle is in standstill.

Moreover, when driving in "D," it is not inhibited to shift to "N" and then from "N" back to "D." When backing up the vehicle in "R," shift to "N" is not inhibited, but when the vehicle speed is higher than a certain threshold (typically 16–24 km/h), shifting from "N" to "R" gear will be inhibited to prevent damage to the transmission parts.

Finally, when operating in "D," the current gear should be maintained in the following circumstances to prevent the wrong gear from being selected:

1. When the fault count value exceeds a certain value.
2. When the second test is planned to confirm the fault shown the first time.

15.7.2.8. Other Actions Related to Failsafe

The following actions are recommended to ensure that the drivetrain is failsafe:

1. The current gear should be kept unchanged in any of the following cases: if a fault in the shift mechanism is detected; if the battery voltage is low and the input/output integrity may be lost; if the accelerator pedal is faulty, for example, wrong data with the accelerator pedal opening; and if the speed check fault count reaches a certain threshold.
2. The torque demand corresponding to the accelerator pedal opening is limited if the torque management fault count reaches a certain value.

The failsafe logic can be divided into the following two parts:

1. For the part that is sensitive to voltage with a quick response, some electronic hardware functions may be affected by unreliable input data during low voltage. These hardware functions include park, reverse, neutral, and drive (PRND) gear lever data, ignition switch operating status, and watchdog. Hence, the following failsafe actions may need to be taken: inhibit any change of shift lever position, inhibit setting with PRND data error, and inhibit watchdog tests.
2. Some electronic hardware functions are not controllable or the input data are unreliable during low battery voltage. These hardware functions may include solenoid valves and shift actuation motors. The recommended failsafe actions include the following: end any ongoing shift; keep the output gear of current shift schedule unchanged if the low-voltage duration is short; inhibit the change in the shift lever position if the low-voltage duration is short, otherwise set the shift lever position to "N"; inhibit the in-gear transmission ratio check; inhibit the speed check in the neutral gear; and inhibit the drive circuit continuity test.

The action to recover from a low-voltage-tolerant condition to a normal state is controlled by a low-voltage recovery procedure to avoid possible interference between shift elements.

This action usually involves two sequences: replace reverse gear and all other forward gears with neutral gear and then inhibit the speed check in the neutral gear.

15.7.2.9. Gear Shift Control Strategy of a Clutchless AMT

The shift control principle of a clutchless AMT for EV applications is elaborated here [15.8]. The maneuver of the driver is first detected by the sensor signals from shift lever position, accelerator pedal, and brake pedal signals. In addition, motor speed and output shaft speed are also detected. Based on the information, the TCU makes a gear decision according to preset shift schedules and determines whether to upshift or downshift. If a shift decision is made, then the TCU enters the shift procedure by sending a command of the operating mode to the MCU. In the different stages of the shifting process, the motor is switched to the torque closed-loop mode with the target torque of T_e^*, free-running mode, or speed closed-loop mode with the target speed of n_e^*. At the same time, the TCU sends a current to drive the shift actuator, motor, and the actuator then completes actions such as disengaging, engaging, or putting the current gear on-hold.

The block diagram of the shift control procedure for a clutchless AMT is shown in Figure 15.12 [15.9]. Due to the small moment of inertia of its rotor, the traction motor usually has a fast response in its torque and speed outputs, enabling the gear shift by simply coordinating the motions of synchronizers and the traction motor. The shifting process is usually divided into five phases, namely, off-loading, disengaging, speed synchronizing, engaging, and load restoring. The control strategies for these phases are discussed in detail as follows:

FIGURE 15.12 Block diagram of the shift control procedure for a clutchless AMT.

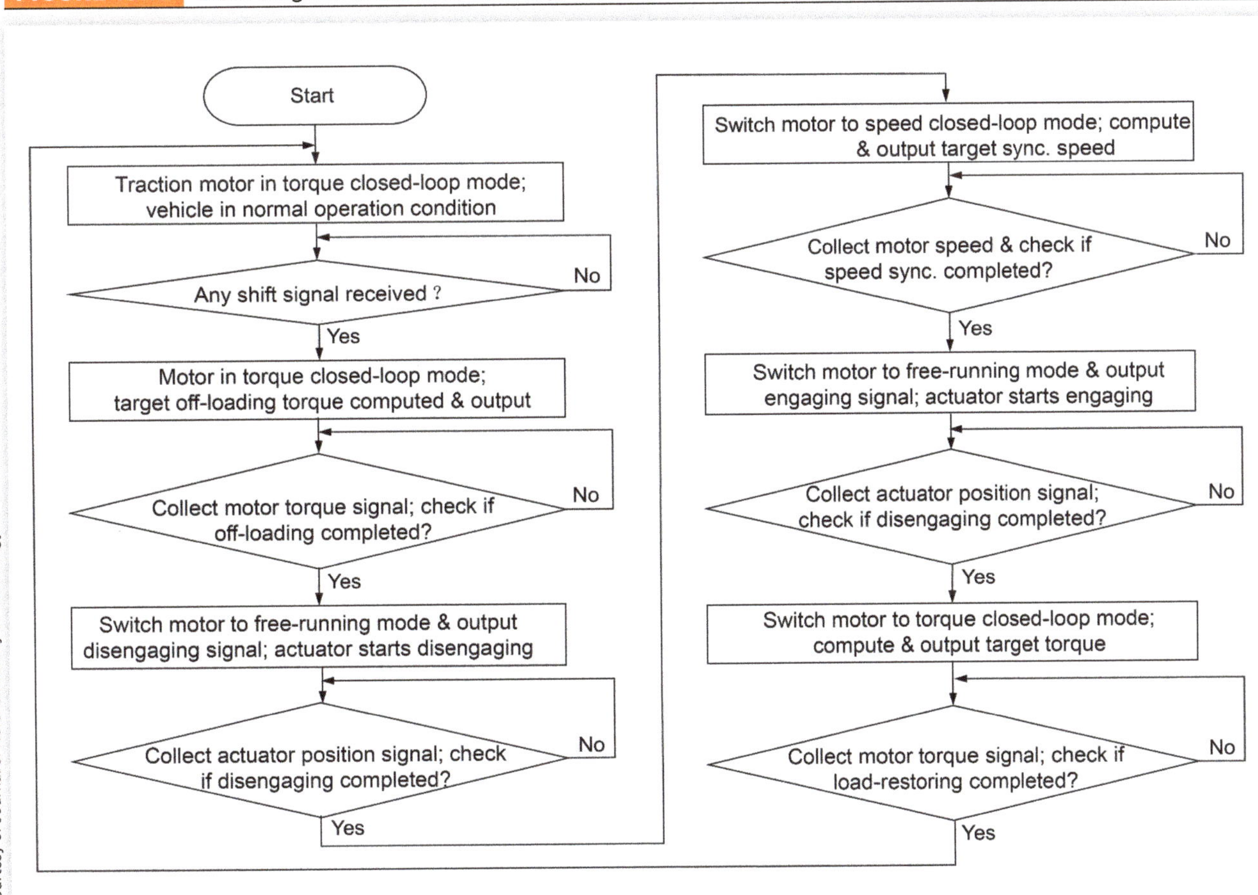

1. Off-loading
   Controlled by MCU, the motor initially operates in the torque closed-loop control mode. Based on the target torque for off-loading from the MCU, the output torque of the motor drops. Since the shift jerk at this stage is proportional to the torque change rate of the motor, the torque change rate must be kept within a certain limit in order to avoid shock jerk, which is given by

$$\frac{\mathrm{d}T_e}{\mathrm{d}t} \leq \frac{I_{ve} j_s}{R_d \eta_T i_g i_o} \tag{15.100}$$

   where

   T_e is the output torque of the traction motor
   I_{ve} is the equivalent moment of inertia of the vehicle mass
   j_s is the shift jerk
   R_d is the dynamic radius of tire
   η_T is the transmission efficiency
   i_g is the ratio of gearbox
   i_o is the final drive ratio

   It is generally desired that $|j_s| \leq 10$ m/s^3.
2. Disengaging
   When the target off-loading torque is reached, the motor is switched from the torque closed-loop control mode to the free-running mode, so that the shift actuator can complete the shifting under the no-load condition. To shorten the shift time, the control strategy in this stage is to drive the actuation motor with its peak power, so as to disengage the gear at the maximum speed.
3. Speed synchronizing
   The control goal of this phase is to quickly synchronize the driving part of the synchronizer to the driven parts through active speed regulation of the motor, so as to minimize the speed difference across the synchronizer, thereby reducing the shift jerk and the wear of the synchronizer. The motor has to be switched from the free-running mode to the speed closed-loop control mode. Since the motor speed is equal to the input shaft speed, the target motor speed is given by

$$n_{i,tar} = \frac{n_{o,cur} i_{g,tar}}{i_{g,cur}} \pm \Delta n_i \tag{15.101}$$

   where

   $n_{i,tar}$ is the target input speed
   $n_{o,cur}$ is the current output speed
   $i_{g,tar}$ is the ratio of target gear
   $i_{g,cur}$ is the ratio of current gear
   Δn_i is the correction amount for input speed

The introduction of speed correction amount is because the synchronization speed changes with the vehicle speed during synchronization. To take both the decrease in motor speed during the period from the completion of traction motor speed regulation to the beginning of synchronization and the influence of motor overshoot into consideration, a speed correction amount of +50 to +100 rpm is generally recommended.

4. Engaging
When the difference between the current and target speeds of the motor is less than a certain threshold, the motor is switched from the speed closed-loop mode to the free-running mode, so that the shift actuator can engage the synchronizer to complete the gear shifting under no-load conditions. This process can be divided into the following three subphases:

Subphase 1 starts from the beginning of the engaging phase and ends just before the synchronization phase is reached. Under the action of the axial force from the shift fork, the engagement sleeve starts to move from the neutral position, but the blocker ring has yet come into contact with the conical surface of the gear. To shorten the shift time, the control strategy is to make the actuation motor move at the maximum speed to reach the synchronization as quick as possible.

In subphase 2, two cases may occur during the engaging process based on the relationship between sleeve speed ω_s and blocker ring speed ω_b. In the first case, since $\omega_s \le \omega_b$, there is usually no tooth knocking during the shifting process because the blocker ring can directly enter the correct locking position. In the second case, since $\omega_s > \omega_b$, tooth knocking may occur as the blocker ring must be rotated forward relative to the sleeve by one more tooth width before entering the correct locking position. At the time, the sleeve must overcome the clearance in the strut before the blocker ring begins to contact the conical surface of the gear with a certain differential rotational speed. Consequently, a friction torque is generated, which drives the blocker ring to rotate rapidly relative to the sleeve by an angle so that the side surface of the blocker ring is in contact with the side surface of the strut. The tooth ends of the sleeve and blocker ring just interfere with each other and cannot be engaged. Under the action of the axial force of shift fork, a friction torque M_f is generated between the blocker ring and the conical surface of the gear, and an index torque M_i on the blocker ring is produced by the sleeve. When the synchronization is not completed, the structure of the synchronizer ensures that $M_f > M_i$, and the synchronizer remains locked. Under the action of the friction torque, the speeds of the blocker ring and sleeve tend to be synchronized, and the friction torque gradually decreases. In this subphase, if the engaging speed is excessively fast, the sleeve may jump over the current subphase and engage directly with the gear, resulting in tooth knocking. This problem can be avoided by setting a limit on the engaging speed, that is, making $t_1 \ge t_2$, where t_1 is the time for the sleeve to move one axial chamfer length of the blocker ring L, and t_2 is the time for the blocker ring to rotate one tooth width. It can be proved that t_1 and t_2 are given by

$$t_1 = \frac{L + \delta_2 - \delta_1}{v_s} \tag{15.102}$$

$$t_2 = \frac{60}{z\Delta n} \tag{15.103}$$

where

L is the axial length of the blocker ring and sliding sleeve
δ_1 and δ_2 are the clearances of the strut and sleeve, respectively
v_s is the axial speed of the sleeve
z is the number of teeth of the blocker ring
Δn is the differential speed between the driving and driven part of the synchronizer

Thus, the condition to avoid tooth knocking in the synchronization process is

$$v_s \le \frac{(L + \delta_2 - \delta_1)z\Delta n}{60} \tag{15.104}$$

Nevertheless, it is deemed that when $\Delta n < 20$ rpm, the engaging speed can go beyond the restriction imposed by Equation (15.104) because of the small moment of inertia generated by the traction motor during the engagement process and its limited impact on the vehicle.

In subphase 3, the friction torque vanishes after synchronization is achieved. Meanwhile, the index torque pushes the blocker ring to turn an angle with respect to the sleeve, allowing the sleeve to engage with the gear. Pushed by the axial force from the shift fork, the sleeve continues to move and engages with the target gear.

5. Load restoring
Finally, in the load restoring phase, the motor is switched from the free-running mode to the torque closed-loop control mode and its output torque is restored to the target value corresponding to the target gear ratio. At the same time, the shift actuator is kept in the current position without moving. The shifting process is complete upon attainment of the target output torque.

By assuming that the shift time is short and the vehicle speeds before and after shifting are approximately equal, the target value of torque recovery can be estimated from the transmission ratios before and after shifting by

$$T_t = T_c \frac{i_c}{i_t} \tag{15.105}$$

where

T_t is the target output torque of the motor
T_c is the current output torque of the motor
i_c is the transmission ratio of the current gear
i_t is the transmission ratio of the target gear

The torque recovery process can be controlled by a linear recovery strategy or an intelligent strategy based on fuzzy logic, as elaborated here.

As for the linear torque recovery strategy, a constant change rate of output torque is usually adopted, that is, the torque increases linearly until it reaches the target value:

$$\Delta T = K\Delta t \tag{15.106}$$

where K is the rate of torque recovery. The larger the K, the faster the torque recovery. Under the condition that the shift jerk is satisfied, a value of K as large as possible is usually taken for a quick recovery of the torque. However, the appropriate value of K may vary under different operating conditions such as whether driving on a flat road or grading on a slope, or accelerating for overtaking. Hence, it is difficult to meet the requirements of all operating conditions with a universal K value.

Alternatively, the intelligent torque recovery strategy based on fuzzy logic is simply to restore the torque quickly if the torque demand is strong and slowly if ride comfort demand is strong. To implement this principle, the language sets for the variables are usually set as accelerator pedal opening $\alpha \in \{\text{small (S), medium small (MS), medium (M), Big (B), Very Big (VB)}\}$; rate of accelerator pedal opening change $\Delta\alpha \in \{\text{negative(N), small(S), medium(M), big(B)}\}$; and torque recovery rate $K \in \{\text{small(S), medium(M), big(B)}\}$.

The torque recovery rate K can be determined by using the fuzzy control rules, as shown in Table 15.1 [15.8].

TABLE 15.1 Torque recovery rate K for the traction motor.

		α				
K		**S**	**MS**	**M**	**B**	**VB**
$\Delta\alpha$	N	S	S	S	S	S
	S	S	S	S	M	B
	M	S	M	M	B	B
	B	S	M	B	B	B

Courtesy of Chongqing University.

15.7.2.10. Gear Shift Control Strategy of a Powershift Transmission

A powershift transmission generally involves at least two clutches or brakes or a combination of them, and the gear shift is made by gradually reducing the load transmitted via the off-going clutch or brake and at the same time, gradually increasing the load transmitted through the on-coming clutch or brake. A brake is hereinafter collectively referred to as a clutch for simplicity.

The shifting process can be divided into four types based on the pedal and road conditions, as listed in Table 15.2 [15.10].

TABLE 15.2 Type of shifts for a powershift transmission.

Type of shifts	Typical operating condition	Probability of occurrence	Shift process sequence
Power-on upshift	Accelerating with accelerator pedal depressed	Most upshifts fall into this category	1. Torque phase 2. Inertia phase
Power-off upshift	1. Accelerating with accelerator pedal released 2. Accelerating during coasting downhill	Few upshifts fall into this category	1. Inertia phase 2. Torque phase
Power-on downshift	Decelerating in driving uphill with accelerator pedal depressed	Few downshifts fall into this category	1. Inertia phase 2. Torque phase
Power-off downshift	1. Depressing brake pedal 2. Coasting uphill without depressing accelerator pedal	Most downshifts fall into this category	1. Torque phase 2. Inertia phase

Courtesy of University of Technology, Sydney.

Thus, the type of shifts should be determined first.

Before entering the shifting process, preparation is needed to prefill the actuator, for example, a cylinder, with oil to eliminate the internal clearance of the mechanism and to make the system ready for a rapid response to the shift command.

In the torque phase, the friction torque transmitted by the off-going clutch is gradually transferred to the on-coming clutch. Initially in the no-slip condition, the off-going clutch starts to slip according to the target torque, while the on-coming clutch begins to engage according to the acceleration requirement of the vehicle. At the same time, the decrease in the driving force on the wheels and the fluctuation in vehicle acceleration are minimized by proper control of pressure in both clutches. The control will move on to the inertia phase once the target torque is reached.

In the inertia phase, the motor speed is adjusted so that the on-coming clutch can reach a speed synchronization. When the rotational speeds of driving and driven sides of the clutch become identical, the clutch pressure is raised to the highest value to lock it. The torque applied to the on-coming clutch is kept as constant as possible to minimize shift jerk.

The detailed control strategies are elaborated as follows according to the type of the shifts:

At first, in the torque phase of power-on upshifts, the timing for torque switching is a key factor that affects the shift quality [15.10]. Shift jerks can be resulted by too early engagement of the on-coming clutch, while a torque hole may occur if the engagement is too late. Hence, a good approach is to maintain a little slip during the disengagement of the off-going clutch before engaging the on-coming clutch, with a recommended slip speed of 5–12 rpm [15.10]. In the inertia phase, the motor speed is reduced to synchronize the speed of the on-coming clutch, while a decline in motor torque helps curtail the shift jerk. It is generally recommended to use closed-loop speed control to make the motor speed decrease according to a predetermined curve, but this method may bring about changes in the clutch torque, which, in turn, causes the output shaft torque to fluctuate. It is difficult to adjust the control parameters to meets the requirements of shift smoothness. Another approach is to use a simple segmented torque control algorithm to synchronize the motor speed with the on-coming clutch, then increase the on-coming clutch pressure to the maximum value and restore the motor torque to the required value. This completes the shifting process. The speed synchronization process is divided into three steps: First, when $|n_e - n_{c2}| \leq \Delta n_1$, the motor torque gradually decreases. Second, when $\Delta n_1 \leq |n_e - n_{c2}| \leq \Delta n_2$, the motor output torque remains at the set value. And third, when $|n_e - n_{c2}| > \Delta n_2$, the motor torque starts to increase again, where n_e is the motor speed and n_{c2} is the speed of on-coming clutch. Both the thresholds Δn_1 and Δn_2 need to be determined by calibration. Since the torque output decreases inevitably with decreasing motor torque, a minimum motor torque should be set to prevent the excessive torque hole [15.10].

Next, in the power-off upshift scenario, since no power is transmitted to the vehicle, the problem of torque interruption does not exist and the control is thus simpler than in the power-on upshift scenario. As shown in Table 15.2, the power-off upshift control starts with the inertia phase for speed synchronization, followed by torque phase for torque switching [15.10]. The upshift is started by decreasing the pressure of the off-going clutch. Meanwhile, the pressure of the on-coming clutch is boosted, allowing the clutch to slip. Speed synchronization begins after a short delay, which takes two steps: First, when $|n_e - n_{c2}| \geq \Delta n_1$, the pressure of the on-coming clutch increases gradually. Second, when $\Delta n_1 < |n_e - n_{c2}| < \Delta n_1$, the torques of both on-coming and off-going clutches are retained at the set levels. When $|n_e - n_{c2}| \leq \Delta n_2$, speed synchronization is complete, indicating the end of the inertia phase and the beginning of the torque phase. The motor torque remains constant throughout the inertia phase [15.10]. In the torque phase, torque switching starts with the compensated slip control of the off-going clutch to avoid shift jerk, similar to that in the power-on upshift control. The torque phase ends when the pressure of the off-going clutch drops to zero and the pressure of the on-coming clutch reaches the maximum value [15.10].

Additionally, the power-on downshift starts from the inertia phase, which is exactly the opposite to the power-on upshift. At the beginning, as the pressure of the off-going clutch drops, the pressure of the on-coming clutch ramps up and the slip occurs. The motor speed is increased to synchronize with the on-coming clutch. If lower than the peak value, the motor torque can be increased to shorten the time of the inertia phase. The algorithm is the same as that of power-on upshift control, but the thresholds Δn_3 and Δn_4 need to be selected and calibrated. When the motor speed is synchronized with the on-coming clutch speed, the inertia phase ends and the torque phase begins [15.10]. In the torque phase, the off-going clutch pressure drops and the on-coming clutch pressure rises. During the torque switching process, the same feedback compensation control as in the power-on upshift control is adopted to control the slip of the off-going clutch toward a given target value to ensure smooth shifting. When the off-going clutch pressure drops to zero, the on-coming clutch pressure rises to the maximum value and the shifting process ends. The motor torque then returns to a level that satisfies the intention of the driver [15.10].

Finally, the power-off downshift is the same as the power-on upshift control in that they both start with the torque phase and subsequently the inertia phase. Since there is no power transmission during shifting, its control is much simpler than that of power-on downshift [15.10]. The torque phase is started by decreasing the pressure of the off-going clutch, allowing it to slip. At the same time, the pressure of the on-coming clutch rises. The slip compensation control strategy is used to restrict the slip speed within a given range. The motor speed is increased to synchronize with the on-coming clutch speed [15.10]. The torque phase ends and the inertia phase begins when the off-going clutch pressure drops to zero. The aforementioned simple segmented torque control algorithm for power-on upshift control is employed to synchronize the motor speed with the on-coming clutch speed. When the on-coming clutch pressure reaches the maximum value, the motor torque is restored to the requested value [15.10]. A variety of control strategies, such as feedforward PID closed-loop slip control [15.11], fuzzy logic [15.12], PSO [15.13], linear quadratic optimization [15.14], and dynamic programming [15.15], have been proposed for the optimization of the shifting process, that is, to find the best compromise solution for the problem generally involving multiple objectives like shift jerk, shift time, and clutch lifespan. The interested reader is referred to the cited references for a detailed description of the control strategies.

15.7.2.11. Control Strategy of the Dual-Rotor EVT Drive System

Based on the architecture and structure of the dual-rotor EVT drive system discussed in Chapter 10, its control system is mainly composed of the VCU, MCU, and pedal signals [15.16]. Based on the accelerator or brake pedal signal and other necessary information such as the battery SOC and vehicle speed, the operating mode, that is, driving or regenerative, is determined first, and then the corresponding torque and speed commands for next command cycle are computed and output to the inner and outer motor subsystems for execution.

Next, the output torques $T_{e1}$and T_{e2} of the inner and outer rotors are determined according to the current motor speed and the number of gear teeth in the planetary gear set, and then the planet carrier speed ω_c can be calculated according to the vehicle speed and the dynamic radius of wheels. Subsequently, based on the efficiency maps of the inner and outer motor subsystems, the optimal speeds of the inner and outer rotors can be computed in real time [15.16].

A double closed-loop control method is proposed for controlling the EVT system with the torque of one rotor and the speed of the other rotor being set as the targets simultaneously [15.16]. The closed-loop torque control and the closed-loop speed control are employed for the inner and outer motor subsystems, respectively. The common field-oriented vector control technology (FOC) is utilized for both inner and outer subsystems, essentially controlling the excitation current i_d and torque current i_q, respectively, in the rotating coordinate system of the rotor magnetic field. Good linear characteristics can be obtained by applying the classic PI linear regulator to the system. For the inverter control, the space vector PWM method (SVPWM) is recommended for its maturity [15.17].

The control strategies for the dual-rotor EVT can be classified into targeting either for optimal energy economy or for optimal dynamic performance, depending on which has the higher priority [15.18].

First, the dual-rotor motor EVT propulsion system allows optimum energy economy control. Assuming that at a certain moment, the angular velocity of the carrier corresponding to the vehicle speed is ω_c and the carrier torque corresponding to the accelerator pedal opening is T_c, then we have

$$T_1 = \frac{T_2}{K} = \frac{T_c}{K+1} \tag{15.107}$$

where K is the characteristic parameter of the planetary gear set [15.18]. Based on Equation (15.107), the output torques T_1 and T_2 of the inner and outer motor subsystems can be obtained, respectively. On the other hand, the rotational speeds of the inner and outer motor subsystems should satisfy the following relation:

$$\omega_1 + K\omega_2 = (K+1)\omega_c \tag{15.108}$$

Theoretically, there are an infinite number of combinations of ω_1 and ω_2, and the operating points that satisfy the requirements of relationships Equations (15.107) and (15.108) falling on the horizontal lines of $T = T_1$ and $T = T_2$, respectively [15.18]. The EVT system efficiency η_{sys} is given by

$$\eta_{sys} = \frac{\left[T_1\omega_1 + T_2\omega_2\right]\eta_T}{\dfrac{T_1\omega_1}{\eta_1(T_1,\omega_1)} + \dfrac{T_2\omega_2}{\eta_2(T_2,\omega_2)}} \tag{15.109}$$

where η_T is the total mechanical efficiency of the planetary gear set, final drive, and differential. By assuming that with the variation in operating points of the inner and outer motor subsystems, the variation in η_T is much smaller than that of η_1 and η_2, then a set of optimal operating points $A(T_1, \omega_{1,opt})$ and $B(T_2, \omega_{2,opt})$ can always be found among the numerous combinations of ω_1 and ω_2, such that

$$\max\left[\eta_{sys}\right]=\frac{\left[T_1\omega_{1,opt}+T_2\omega_{2,opt}\right]\eta_T}{\dfrac{T_1\omega_{1,opt}}{\eta_1\left(T_1,\omega_{1,opt}\right)}+\dfrac{T_2\omega_{2,opt}}{\eta_2\left(T_2,\omega_{2,opt}\right)}} \tag{15.110}$$

where
η_1 is the inner motor subsystem efficiency
η_2 is the outer motor subsystem efficiency

Once the target torque T_1 of the inner motor and target speed $\omega_{2,opt}$ of outer motor subsystems are determined, the excitation current components i_{d1} and i_{d2} and torque current component i_{q1} and i_{q2} of the inner and outer three-phase windings can be determined according to the vector control algorithm [15.18].

Alternatively, the control strategy for optimal dynamic performance for the dual-rotor motor EVT drive system is also feasible. The system can realize rapid response to the abrupt change of system power demand so as to meet the requirements of the driver for the dynamic performance of vehicle. Assuming that at a certain moment, the angular velocity of the planet carrier corresponding to the vehicle speed is ω_c and the inner and outer motors operate at the points $A(T_1, \omega_1)$ and $B(T_2, \omega_2)$, respectively, if the accelerator pedal opening suddenly increases and the corresponding planet carrier torque increases to $T_c^{'}$, then there are the following two scenarios [15.18]:

In the first scenario, based on Equation (15.104), the updated torques $T_1^{'}$ and $T_2^{'}$ of the inner and outer motor subsystems corresponding to $T_c^{'}$ are obtained and the new operating point $A'\left(T_1^{'},\omega_1^{'}\right)$ and $B'\left(T_2^{'},\omega_2^{'}\right)$ can be found in the operating regions of the inner and outer motor subsystems that simultaneously satisfy

$$\begin{cases}\omega_1^{'}=\omega_1\\ \omega_2^{'}=\omega_2\end{cases} \tag{15.111}$$

This implies that the requirements for dynamic performance can be met by increasing the respective torques only without changing the speeds of inner and outer rotors. In this scenario, $T_1^{'}$ is taken as the target for closed-loop torque control of the inner motor subsystem and $\omega_2^{'}=\omega_2$ is taken as the speed closed-loop control target for the outer motor subsystem, so as to achieve the instantaneous response to the demand for dynamic performance. Since this process does not involve any change in motor speeds, the system can respond quickly. In the subsequent command cycle, the target value for the closed-loop speed control is estimated and updated according to the vehicle speed change rate obtained by calibration [15.18].

In the second scenario, based on Equation (15.107), the updated torques $T_1^{'}$ and $T_2^{'}$ of the inner and outer motor subsystems corresponding to $T_c^{'}$ are obtained, respectively, but the new operating points $A'\left(T_1^{'},\omega_1^{'}\right)$ and $B'\left(T_2^{'},\omega_2^{'}\right)$ cannot be found in the operating regions of the inner and outer motor subsystems that simultaneously satisfy Equation (15.111). In other words, at least one of the new operating points does not fall into the operable regions of motor subsystems. For example, if the new operating point $A'\left(T_1^{'},\omega_1^{'}\right)$ of the inner motor subsystem falls outside its operable region, then the operating point must be adjusted to A″$\left(T_1'', \omega_1''\right)$ to make it operable, with $\omega_1'' < \omega_1'$, which means that the inner rotor must be decelerated. According to Equation (15.108), the outer rotor must be accelerated, that is, $\omega_2'' > \omega_2'$. In this scenario, T_1' is taken as target for the closed-loop torque control of the inner motor subsystem and $\omega_2'' > \omega_2$ is taken as the target for the closed-loop speed control of the outer motor subsystem, in realizing fast response to the demand on dynamic performance. However, due to the change of motor speed, the closed-loop speed control of the outer motor requires a certain response time, and the overall response to the demand on dynamic performance is not as fast as that in the first scenario [15.18].

15.7.2.12. Quasi-Stall Operating Principle of Dual-Motor Drive Systems

The dual-rotor motor EVT drive system discussed earlier can be further extended to a general dual-motor drive system composed of two separate motors coupled by a single planetary gear set [15.19]. Research on the operating modes of such a system indicates that under some operating conditions, the optimal system efficiency can often be realized by having one of the motors operate in the stall condition, just like locking up one of the motor connected elements, for example, sun gear or ring gear, using a brake [15.19]. The mechanism behind this phenomenon is that the stalled motor only consumes very little power and the other motor can be assigned to operate at high efficiency points, thereby maximizing the system efficiency.

Based on this idea, a "quasi-stall" control strategy is proposed [15.19], which allows simplification of the general dual-motor system architecture, with the clutch and brake eliminated. The "quasi-stall" control strategy refers to one of the motors operating in the stall or near-stall condition, allowing the system efficiency to approach the level of a more-complicated system equipped with clutches or brakes or a combination of both, while still satisfying the demand for dynamic performance. Since a motor cannot operate in the stall condition for a long time, its operating point can be adjusted in real time according to the temperature rise. Two adjustment strategies have been suggested: when any one of the two motors can meet the power requirements of the current operating condition, the two motors can be made to operate in the stall condition, in turn. Second, when only one of the motors is capable of meeting the power requirements of the current operating condition, the operating point of the stalling motor can be deviated from the stall condition for a period of time and then return to the stall condition when the motor temperature drops below a certain threshold. In summary, such a simplified system architecture offers a high system efficiency using "quasi-stall" control strategy at a low cost and system complexity due to the elimination of clutches and brakes.

15.8. Regenerative Braking Control Strategy

The amount of energy recovered by regenerative braking of an EV is restricted by many factors. The primary factors are elaborated as follows.

1. In the braking process, the kinetic energy of the vehicle can only possibly be recovered by applying motor braking torque to the drive wheels through a drivetrain and obviously cannot be recovered by driven wheels on which no e-motors are mounted to generate the braking torques. However, as discussed in Section 4.5, the front and rear braking force proportion requires a certain braking

force from the mechanical braking system in driven wheels. This inevitably results in the loss of some kinetic energy, which can otherwise be recovered.

2. Whether a regenerative braking can work or not depends on the SOC of the battery system. If the SOC is higher than a certain threshold, then regenerative braking is inhibited, in order to protect the battery system.
3. Limited by the battery charge rate, the current of regenerative braking should not exceed the maximum allowable charging current of the battery system.
4. Limited by the power rating of the motor, the maximum torque generated by motor braking is subject to the upper limit corresponding to the operating speed at that time.
5. In a harsh braking, the regenerative braking torque of the motor may not meet the torque requirements, and additional mechanical braking torque becomes necessary to supplement that.

In addition to proportioning the electric and mechanical braking torques, the regenerative braking control must also take the ride comfort during the introduction and cancelation of electric braking torque into consideration.

Based on the input signal of brake pedal sensor and the operational state of the vehicle, the regenerative braking process is divided into three phases as follows:

1. Soft entry phase: When the brake pedal is depressed, a signal from the brake pedal sensor is sent to the VCU. Once the VCU detects the signal, it will command the motor to increase its braking torque gradually with time, in order to eliminate the discomfort caused by the sudden addition of motor braking torque. The electric braking torque from the e-motor should decrease with the battery SOC and drop to zero when the SOC is higher than a certain threshold. Hence, the electric braking torque is actually a function of time and battery SOC in this phase.
2. Deep regenerative phase: As the braking action of the driver continues, the demand for braking torque increases. Braking effectiveness can be improved by adding a large electric braking torque in this stage, which is a function of the battery SOC. This is, therefore, a critical period of the regenerative braking for energy recovery.
3. Soft withdrawal phase: In this phase of the braking process, both the vehicle and motor speeds decrease considerably. The efficiency of motor power generation becomes very low. With the mechanical braking torque becoming dominant, the electric braking torque withdraws. However, the electric braking torque should be gradually reduced to zero without any abrupt changes as the vehicle speed decreases. In addition, as the vehicle control quits braking and enters driving, the driving torque demand from the accelerator pedal cannot be responded to immediately, in order to ensure a smooth ride of the vehicle. Instead, a buffer zone is generally set for a certain period of time in which the electric braking torque is zero and no accelerator pedal demand is responded to. A diagram of electric braking torque versus vehicle speed in a braking process is given Figure 15.13.

In the deep regenerative phase, the electric braking torque, T_{be}, can be calculated using an algorithm similar to the following equation:

$$T_{be} = f(SOC) = \begin{cases} T_{bemax} & \text{if } SOC < 50\% \\ (T_{bemax} - 50)(95\% - SOC)/45\% + 50 & \text{if } 50\% \le SOC < 95\% \\ 0 & \text{if } SOC \ge 95\% \end{cases} \quad (15.112)$$

where T_{bemax} is the maximum electric braking torque.

FIGURE 15.13 Electric braking torque versus vehicle speed in a braking process.

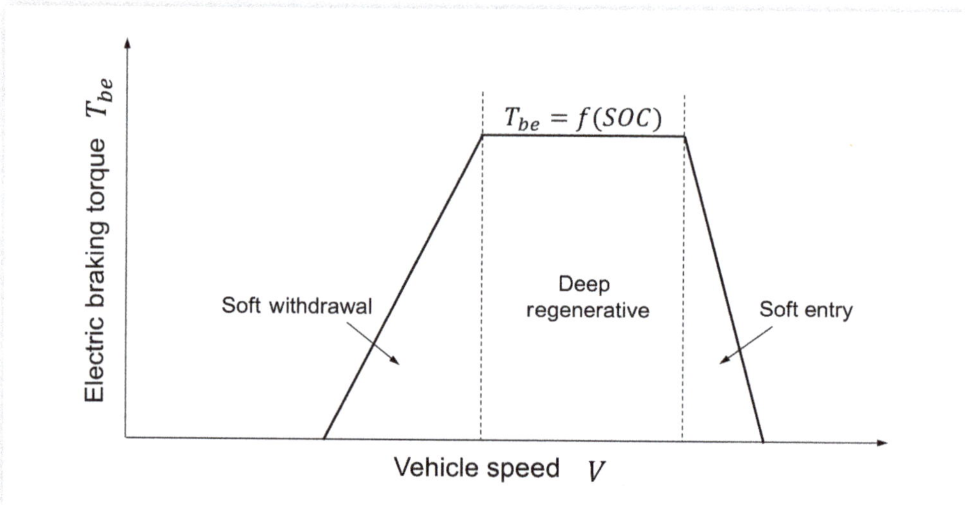

The coordination of regenerative braking with ABS is extremely important to maximize the energy recovery under the premise of safety. The coordination control strategies proposed to date can be generalized into four fundamental types: cancel the regenerative braking torque upon the kick-in of ABS, maintain the regenerative braking torque for a certain duration before tapering off after the ABS kicks in, compute the maximum regenerative braking torque allowed by the tire–road adhesion in real time and then adjust the motor to produce that torque based on the braking torque demand of the driver without interfering with the control of ABS once the ABS is initiated, and use the regenerative braking torque to meet all the required braking torque demand once the ABS is activated, without any mechanical (friction) torque.

To ensure vehicle stability and maximize energy recovery, many coordinative control strategies of regenerative braking with ABS have been proposed, but unfortunately, few of them have been proven with experiments. Hence, a three-level hierarchical controller is presented herein as an exception since it is claimed that the kinetic energy recovered in the NEDC is increased by about 80% compared with parallel braking control. As shown in Figure 15.14 [15.20], the top-level control functions to proportion the front and rear wheel slip ratios based on the maneuver of the driver and the road status; the middle-level control serves to manipulate the actuators to follow the requested slip ratio based on the outputs of top-level control program; and the bottom-level control is responsible to execute the braking torque commands from middle-level control.

FIGURE 15.14 A three-level hierarchical controller for coordinative regenerative braking with ABS.

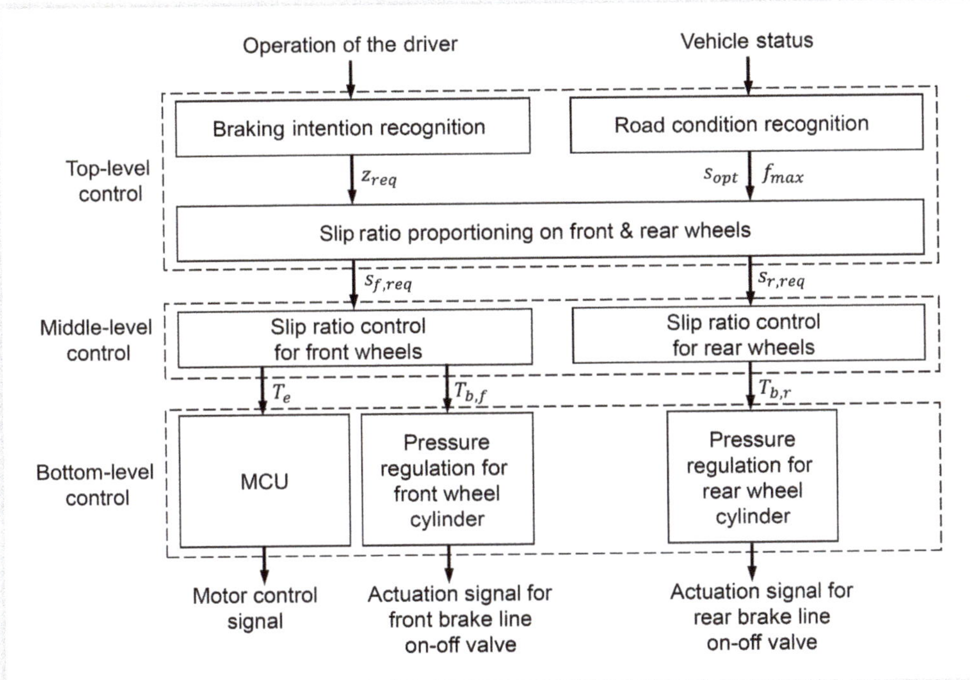

Courtesy of Journal of Tsinghua University (Sci. & Tech.).

FIGURE 15.15 Braking force proportioning strategy for an FWD EV.

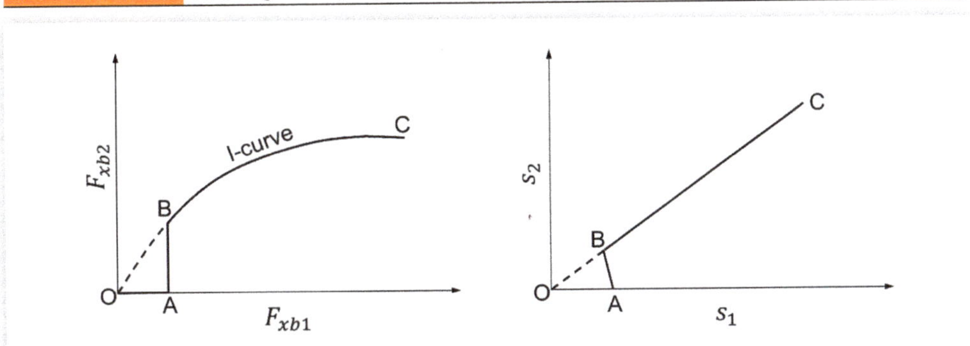

Courtesy of Journal of Tsinghua University (Sci. & Tech.).

In this approach, the regenerative braking torque is assumed to be applied only on the front wheels for an FWD EV. In order to maximize the energy recovery, the front and rear braking forces are proportioned by following lines OA and AB and curve BC, as shown in Figure 15.15 [15.20]. The motor reaches the maximum allowable torque in braking at point A, corresponding to the braking ratio z_A. With a continued increase in braking torque request, the front braking force remains unchanged, and the rear braking force is increased to satisfy the requirement, until curve I is reached. If the braking ratio exceeds that at point B, that is, z_B, then curve I is followed to ensure the vehicle stability and safety. The strategy for proportioning the slip ratio can be expressed as [15.20]:

$$s_1 = \begin{cases} \dfrac{z_{req}}{\left(\rho_1 + z_{req}h_g / L\right)\mu(s_1)} & z_{req} \le z_A \\ \dfrac{z_A}{\left(\rho_1 + z_{req}h_g / L\right)\mu(s_1)} & z_A < z_{req} \le z_B \\ \dfrac{1}{\mu(z_{req})} & z_{req} > z_B \end{cases} \tag{15.113}$$

$$s_2 = \begin{cases} 0 & z_{req} \le z_A \\ \dfrac{z_{req} - z_A}{\left(\rho_2 - z_{req}h_g / L\right)\mu(s_2)} & z_A < z_{req} \le z_B \\ \dfrac{1}{\mu(z_{req})} & z_{req} > z_B \end{cases} \tag{15.114}$$

where

ρ_1 and ρ_2 are the axle load proportion to the front and rear axles, respectively
μ is the adhesion coefficient, as a function of the slip ratio
h_g is the height of the mass centroid of the vehicle
L is the wheelbase
z_{req} is the requested braking ratio
z_A and z_B are the braking ratios at point A and point B, respectively:

$$z_A = \frac{k_\mu T_{emax} i_g i_0}{mgR} \tag{15.115}$$

$$z_B = \frac{\sqrt{s_1^2 + 4z_A h_g / L} - s_2}{2h_g / L} \tag{15.116}$$

In Equations (15.115) and (15.116), T_{emax} is the maximum allowable motor torque depending on the external characteristics and other factors such as the malfunction level of the motor, battery SOC and voltage, and vehicle speed; i_g is the transmission ratio; i_0 is the final drive ratio; m is the vehicle mass; R is the wheel radius; and k_μ is the adhesion coefficient factor defined by [15.20]

$$k_\mu = \begin{cases} 1 & 0 \le \mu_{max} \le 0.4 \\ 5(\mu_{max} - 0.4) & 0.4 < \mu_{max} < 0.6 \\ 1 & \mu_{max} \ge 0.6 \end{cases} \tag{15.117}$$

where

μ_{max} is the maximum adhesion coefficient of the road surface

s_1 and s_2 are the slip ratios of the front and rear wheels under the braking condition, defined as follows:

$$s_1 = \frac{\omega_v - \omega_1}{\omega_1} \tag{15.118}$$

$$s_2 = \frac{\omega_v - \omega_2}{\omega_2} \tag{15.119}$$

In braking, the kinetic equations for the front wheel, rear wheel, and the vehicle are as follows [15.20]:

$$\begin{cases} \dot{\omega}_1 = \dfrac{F_{z1}R\mu(s_1) - T_{r1} - T_{f1} - T_e i_g i_0}{J_1} \\ \dot{\omega}_2 = \dfrac{F_{z2}R\mu(s_2) - T_{r2} - T_{f2}}{J_2} \\ \dot{\omega}_v = \dfrac{-F_{z1}\mu(s_1) - F_{z2}\mu(s_2) - F_a}{mR} \end{cases} \tag{15.120}$$

where

ω_1 and ω_2 are the front and rear wheel speeds, respectively

$\omega_v = v/R$, with v being the vehicle speed

T_{r1} and T_{r2} are the rolling resistance moment of the front and rear wheels, respectively

T_{f1} and T_{f2} are the friction braking torques on the front and rear wheels, respectively

T_e is the regenerative braking torque of the motor

J_1 and J_2 are the rotational inertia of the front and rear wheels, respectively

F_a is the aerodynamic drag to the vehicle

Based on Equations (15.119) and (15.120), the kinetic equation for the front wheels can be expressed as [15.20]

$$\dot{s}_1 = g_1 + b_e u_e + b_{f1} u_{f1} \tag{15.121}$$

where

$$g_1 = \frac{R}{J_1 v}\left[T_{r1} - F_{z1}R\mu(s_1)\right] - \frac{1-s_1}{mv}\left[F_a + F_{z1}\mu(s_1) + F_{z2}\mu(s_2)\right]$$

$$b_e = \frac{R i_g i_0}{J_1}$$

$$b_{f1} = \frac{R}{J_1}$$

$$u_e = \frac{T_e}{v}$$

$$u_{f1} = \frac{T_{f1}}{v}$$

For the dual-input first-order nonlinear system described by Equation (15.121), a sliding mode control (SMC) is proposed with the sliding surface of [15.20]:

$$S_1 = s_1 - s_{1req} \tag{15.122}$$

where s_{1req} is the slip ratio proportioned to the front wheel as the result from the top-level controller. The control law for the dual-input system depicted by Equation (15.121) is designed as [15.20]

$$u_e = \frac{-a\left(\hat{g}_1 - \dot{s}_{1req}\right) - k_e \text{sat}\left(S_1\right)}{b_e} \tag{15.123}$$

$$u_{f1} = \frac{-(1-a)\left(\hat{g}_1 - \dot{s}_{1req}\right) - k_{f1} \text{sat}\left(S_1\right)}{b_{f1}} \tag{15.124}$$

where

$\hat{g}_1$ is the estimate of g_1

sat is a saturation function, with

$$\text{sat}\left(S_1, \phi\right) = \begin{cases} 1 & S_1 \geq \phi \\ S_1 & \left|S_1\right| < \phi \\ -1 & S_1 \leq -\phi \end{cases} \tag{15.125}$$

and

$$\left|g_1 - \hat{g}_1\right| \leq G_1 \tag{15.126}$$

where

G_1 is a known bounded function

ϕ is the saturation value

The parameter a in Equations (15.124) and (15.125) reflects the ratio of regenerative to friction braking torques, with its maximum given by [15.20]

$$a_{max} = \min\left[\frac{T_{emax}\hat{b}_e}{v\left(\hat{g}_1 - \dot{s}_{1req}\right)}, 1\right] \tag{15.127}$$

Although highly robust, SMC comes with an intrinsic chattering problem. In order to eliminate chattering, a fuzzy parameter adapter is developed for dynamically changing the sliding gains. By defining $\bar{a} = a/a_{max}$, the parameters k_e, k_{f1}, and $\bar{a}$ are the outputs of the fuzzy parameter adapter with the requested slip ratio for the front wheel, s_{1req}, and the maximum regenerative braking torque allowed by the system, T_{emax}, serving as inputs. Using the fuzzy logic reasoning rules listed in Table 15.3, the surfaces corresponding to the parameters k_e, k_{f1}, and $\bar{a}$ can be plotted against T_{emax} and s_{1req} in the form of 2-D surfaces [15.20].

TABLE 15.3 Fuzzy logic reasoning rules.

Rule no.	s_{1req}	T_{emax}	$\bar{a}$	k_e	k_n
1	Small	—	Large	0	0
2	Large	Fairly large	Fairly small	Fairly large	Fairly small
3	Large	Fairly small	Fairly large	Fairly small	Fairly large

In the bottom-level control, the friction braking torque is adjusted by means of high-speed on-off valves for wheel cylinders. The mathematically model for such as system is [15.20]

$$Q_1 = C_v A_1 \sqrt{\frac{2(p_a - p_w)}{\rho}} \tag{15.128}$$

$$Q_2 = C_v A_2 \sqrt{\frac{2p_w}{\rho}} \tag{15.129}$$

$$Q_1 - Q_2 = A_p \dot{x}_p \tag{15.130}$$

$$x_p = c_1 p_w^{c_2} / A_p \tag{15.131}$$

where

Q_1 and Q_2 are the flow rate at booster and relief valves, respectively
C_v is the valve flow coefficient
A_1 and A_2 are the cross-sectional areas of booster and relief valves, respectively
ρ is the brake fluid density
p_a is the pressure of the accumulator
p_w is the pressure of the wheel cylinder
x_p is the piston stroke
A_p is the piston area
c_1 and c_2 are the characteristic coefficients of the wheel cylinder

Thus, the change rate of cylinder pressure can be derived from Equations (15.128) through (15.131) [15.20]:

$$\dot{p}_w = \begin{cases} \dfrac{C_v A_1 p_w^{1-b}}{c_1 c_2} \sqrt{\dfrac{2(p_a - p_w)}{\rho}} & \text{for increaing pressure} \\ -\dfrac{C_v A_2 p_w^{\frac{3}{2}-b}}{c_1 c_2} \sqrt{\dfrac{2}{\rho}} & \text{for decreaing pressure} \end{cases} \tag{15.132}$$

In an instruction cycle, the wheel cylinder pressure can be assumed constant, the time for opening the booster and relief valves, respectively, are [15.20]

$$t_1 = \frac{c_1 c_2 \cdot \max\left(0,\ p_w^{req} - p_w\right)}{C_v A_1 p_w^{1-b}} \sqrt{\frac{\rho}{2\left(p_a - p_w\right)}} \tag{15.133}$$

$$t_2 = \frac{c_1 c_2 \cdot \max\left(0,\ p_w - p_w^{req}\right)}{C_v A_2 p_w^{\frac{3}{2}-b}} \sqrt{\frac{\rho}{2}} \tag{15.134}$$

where p_w^{req} is the target pressure value.

15.9. Electronic Differential Control

Electronic differential control (EDC) aims at a safe and robust methodology for controlling all the wheels of an EV with each wheel driven by an independent in-wheel or wheel-side motor that can replace the functionality of a traditional mechanical differential.

The targets of EDC are essentially twofold: minimize the wheel slip and reduce the tire wear, and minimize and stabilize the sideslip angle at the mass centroid of the vehicle and yaw rate in cornering. In such an EDC problem, the steering angle and the accelerator pedal opening are the input quantities from the driver or an autonomous driving control unit, while the wheel velocities, the slip ratio, the sideslip angle at the mass centroid of vehicle, yaw rate, and wheel trajectories are the output quantities subjected to observation and optimization.

To solve this control problem, a model capable of relating the output quantities to the input quantities becomes necessary. Such a model must not only take all the major factors affecting the dynamics of each wheel into account but also allow real-time solution to the model for yielding command instructions in an adequately fast manner in response to the highly random, transient, and frequent disturbances. To simplify the problem, the following discussion will be restricted to the scope of a four-wheel vehicle with only two steerable wheels in the front connected by the same steering mechanism. In other words, the front wheels are not independently steerable and their steering angles are always identical.

15.9.1. Two-DOF Model Based on the Ackermann Steering Principle

By assuming that a rigid vehicle moves on a flat road with equal operating conditions on all the wheels and only translational motions in the horizontal plane, a simple two-DOF model based on the Ackermann steering principle is given by

$$\begin{cases} v_1 = \left(1 + \frac{t_r}{L}\tan\delta\right)v \\ v_2 = \left(1 - \frac{t_r}{L}\tan\delta\right)v \\ v_3 = \left(1 + \frac{t_r}{2L}\tan\delta\right)v \\ v_4 = \left(1 - \frac{t_r}{2L}\tan\delta\right)v \end{cases} \tag{15.135}$$

where

t_r is the wheel track

L is the wheelbase

δ is the steering angle

v is the mass centroid velocity of the vehicle

the subscripts 1, 2, 3, and 4 denote the left front, right front, left rear, and right rear wheels, respectively, as shown in Figure 15.16 [15.21].

FIGURE 15.16 A simple two-DOF model based on the Ackermann steering principle.

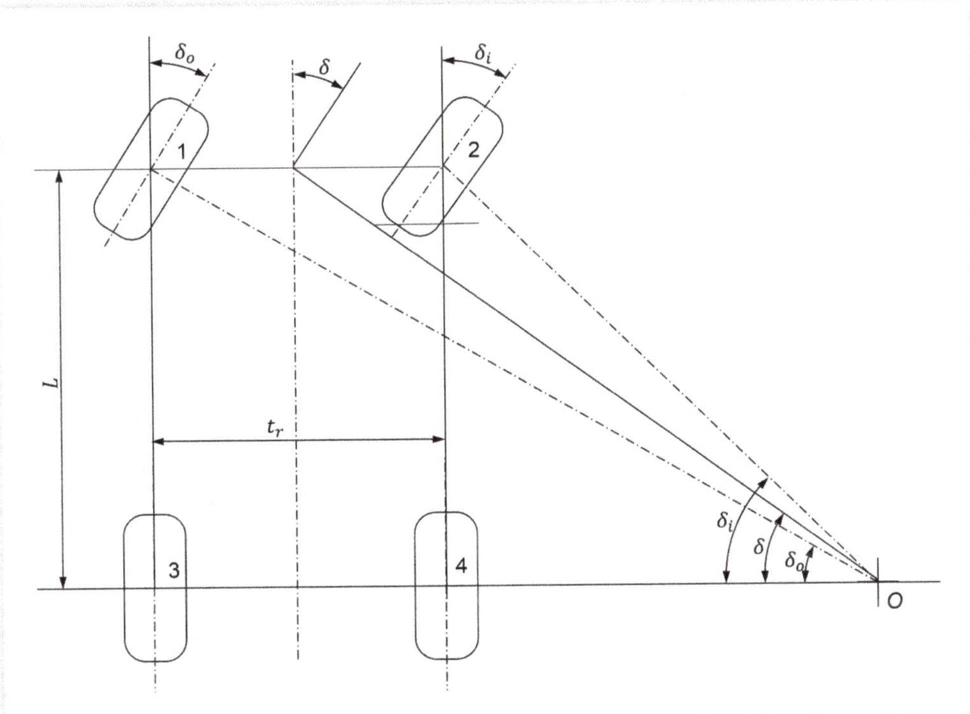

For an all-wheel-drive vehicle with each wheel independently driven by a motor, Equation (15.30) can only yield valid results for a rigid vehicle of all wheels operating in exactly the same condition. However, any nonideal conditions, such as an uneven road, body roll, and wheel slip can introduce disturbances to the steering angles at all the wheels, resulting in the following deviations of wheel velocities from their ideal values [15.22]:

$$\begin{cases} \Delta v_1 = \tan\Delta\delta_1\left[1+\tan\delta\tan(\delta+\Delta\delta_1)\right]v\dfrac{t_r}{L} \\ \Delta v_2 = -\tan\Delta\delta_2\left[1+\tan\delta\tan(\delta+\Delta\delta_2)\right]v\dfrac{t_r}{L} \\ \Delta v_3 = \tan\Delta\delta_3\left[1+\tan\delta\tan(\delta+\Delta\delta_3)\right]v\dfrac{t_r}{2L} \\ \Delta v_4 = -\tan\Delta\delta_4\left[1+\tan\delta\tan(\delta+\Delta\delta_4)\right]v\dfrac{t_r}{2L} \end{cases} \tag{15.136}$$

It is clear from Equation (15.136) that the deviations of wheel velocities are nonlinearly related to the disturbances to the steering angles. Hence, the results from the Ackermann principle-based model may lead to uncoordinated movements among the wheels.

15.9.2. Three-DOF Steering Model

By taking yaw motion in the horizontal plane into consideration, a more sophisticated three-DOF steering model can be established. In the coordinate system shown in Figure 15.17 [15.22], the combined forces acting on the vehicle in cornering are

$$F_x = m\frac{dv}{dt}\sin\delta = F_{1x}\sin\delta_o - F_{1y}\cos\delta_o + F_{2x}\sin\delta_i - F_{2y}\cos\delta_i + F_{3x} + F_{4x} \tag{15.137}$$

$$F_y = -m\frac{dv}{dt}\cos\delta = -F_{1x}\cos\delta_o - F_{1y}\sin\delta_o - F_{2x}\cos\delta_i + F_{2y}\sin\delta_i - F_{3y} - F_{4y} \tag{15.138}$$

and the equation of yaw motion is [15.23]

$$I_z\omega_r = \left(F_{1x}\sin\delta_o - F_{1y}\cos\delta_o\right)L_f + \left(F_{2x}\sin\delta_i - F_{2y}\cos\delta_i\right)L_f + F_{3y}L_f + F_{4yL_f} \tag{15.139}$$

where

m is the vehicle mass
I_z is the moment of inertia of the vehicle about vertical axis
ω_r is the yaw rate

FIGURE 15.17 A three-DOF model and the coordinate system.

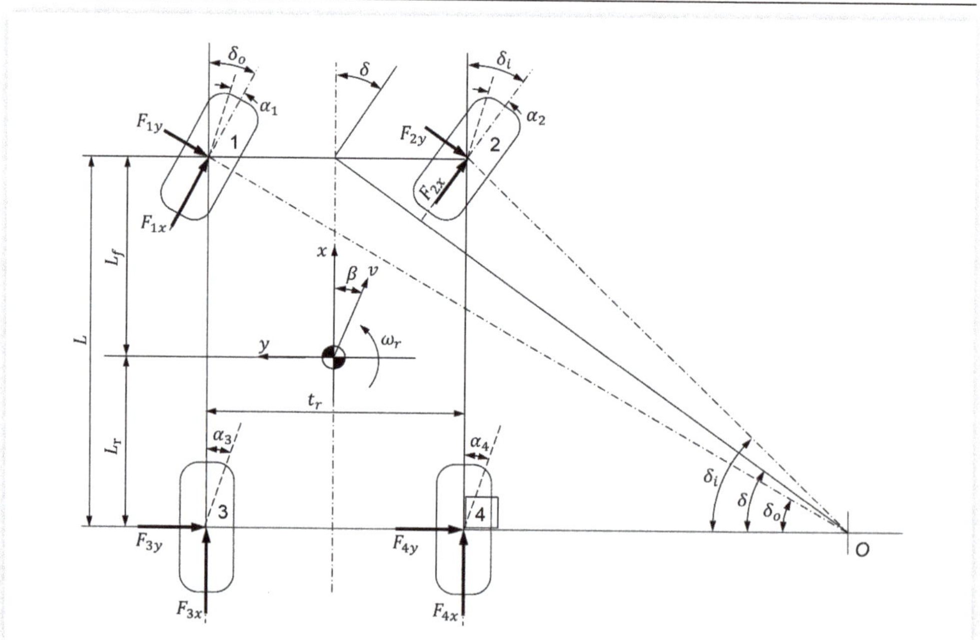

Courtesy of Inderscience Enterprises Ltd.

Due to the constraints imposed by the chassis, the wheel velocities can be expressed by the sideslip angle at the mass centroid of the vehicle β [15.23]:

$$\begin{bmatrix} v_{1x} \\ v_{1y} \\ v_{2x} \\ v_{3y} \end{bmatrix} = \begin{bmatrix} v_{3x} \\ v_{2y} \\ v_{4x} \\ v_{4y} \end{bmatrix} = \begin{bmatrix} \cos\beta & -0.5t_r \\ -\sin\beta & L_f \\ \cos\beta & 0.5t_r \\ \sin\beta & -L_r \end{bmatrix} \begin{bmatrix} v \\ \omega_r \end{bmatrix} \tag{15.140}$$

The sideslip angles of tires due to the elastic deformation in the lateral direction can be calculated by [15.23]

$$\begin{cases} \alpha_1 = \delta_o - \arctan\left(\dfrac{v_{1y}}{v_{1x}}\right) \\ \alpha_2 = \delta_i - \arctan\left(\dfrac{v_{2y}}{v_{2x}}\right) \\ \alpha_3 = \arctan\left(\dfrac{v_{3y}}{v_{3x}}\right) \\ \alpha_4 = \arctan\left(\dfrac{v_{4y}}{v_{4x}}\right) \end{cases} \tag{15.141}$$

The centripetal force acting on the mass centroid of vehicle in cornering is [15.23]

$$F_c = mv\left(\dot{\beta} + \omega_r\right) \tag{15.142}$$

The sideslip angle at the mass centroid of the vehicle is given by [15.23]

$$\beta = \arctan\left(-\frac{v_y}{v_x}\right) \tag{15.143}$$

The vertical loads on the wheels of a vehicle in cornering change accordingly, which can be computed by [15.23]

$$\begin{cases} F_{z1} = \dfrac{L_r}{2L}\left(mg + \dfrac{2h_g F_c}{t_r}\right) \\ F_{z2} = \dfrac{L_r}{2L}\left(mg - \dfrac{2h_g F_c}{t_r}\right) \\ F_{z3} = \dfrac{L_f}{2L}\left(mg + \dfrac{2h_g F_c}{t_r}\right) \\ F_{z4} = \dfrac{L_f}{2L}\left(mg - \dfrac{2h_g F_c}{t_r}\right) \end{cases} \tag{15.144}$$

For the four independently driven wheels, the slip ratio of the jth wheel, s_j, is given by

$$s_j = \begin{cases} \dfrac{\omega_j R - v_{xj}}{\omega_j R} & \text{if } \omega_j R \geq v_{xj} \\ \dfrac{v_{xj} - \omega_j R}{v_{xj}} & \text{if } \omega_j R < v_{xj} \end{cases} \tag{15.145}$$

where
- ω_j is the angular speed of the jth wheel
- v_{xj} is the longitudinal velocity of the jth wheel center
- R is the wheel radius

The longitudinal velocity of the jth wheel center v_{xj} for $j = 1, 2, 3, 4$ can be obtained by [15.24]

$$\begin{cases} v_{x1} = \cos\alpha_1 \sqrt{\left(v_x - \dfrac{t_r\beta}{2}\right)^2 + \left(v_y + L_f\beta\right)^2} \\ v_{x2} = \cos\alpha_2 \sqrt{\left(v_x + \dfrac{t_r\beta}{2}\right)^2 + \left(v_y + L_f\beta\right)^2} \\ v_{x3} = \cos\alpha_3 \sqrt{\left(v_x - \dfrac{t_r\beta}{2}\right)^2 + \left(v_y - L_r\beta\right)^2} \\ v_{x4} = \cos\alpha_4 \sqrt{\left(v_x + \dfrac{t_r\beta}{2}\right)^2 + \left(v_y - L_r\beta\right)^2} \end{cases} \tag{15.146}$$

Thus, the disturbance to the slip ratio of the jth wheel, Δs_j, can be derived from Equation (15.145):

$$\Delta s_j = \begin{cases} \dfrac{v_x \Delta\omega_j}{\omega_j\left(\omega_j + \Delta\omega_j\right)R} & \text{if } \omega_j R \geq v_{xj} \\ \dfrac{-\Delta\omega_j R}{v_x} & \text{if } \omega_j R < v_{xj} \end{cases} \tag{15.147}$$

The disturbance to the yaw rate as a result of disturbances to the steering angles at all the wheels of a vehicle in cornering can be derived from Equation (15.139) [15.23]:

$$\begin{aligned} \Delta\omega_r = \frac{L_f}{I_z}\Big\{ & F_{1x}\left[\sin\delta_o\left(\cos\Delta\delta_o - 1\right) + \cos\delta_o \sin\Delta\delta_o\right] - F_{1y}\left[\cos\delta_o\left(\cos\Delta\delta_o - 1\right) - \sin\delta_o \sin\Delta\delta_o\right] \\ & + F_{2x}\left[\sin\delta_i\left(\cos\Delta\delta_i - 1\right) + \cos\delta_i \sin\Delta\delta_i\right] - F_{2y}\left[\cos\delta_i\left(\cos\Delta\delta_i - 1\right) - \sin\delta_i \sin\Delta\delta_i\right]\Big\} \end{aligned} \tag{15.148}$$

It is inferable from Equation (15.148) that the disturbance to the yaw rate can be reduced by dynamically adjusting the driving forces at the front wheels, so as to keep the vehicle at a desired attitude.

It is worth noting the three-DOF model is far from acceptable for real-world applications in terms of accuracy and response time. A number of more-sophisticated models have been proposed to date, but none of them has the proven effectiveness, safety, and robustness under all-around operating conditions. Continued work is needed to bring this control approach to a practically applicable level.

15.9.3. Motor Speed-Based EDC

In this approach, all the motors operate in closed-loop speed control. The principle of the motor speed-based EDC strategy is illustrated in Figure 15.18. Using the vehicle velocity v and target steering angle δ detected by the sensors as the inputs, the theoretical wheel velocities, v_1, v_2, v_3, and v_4, can be computed by the aforementioned analytical steering model. Then, the difference between the measured and theoretical wheel velocities for each of the wheels are used as control variables for the in-wheel or wheel-side motors after being converted into armature currents, in order to coordinate the motion of all wheels.

FIGURE 15.18 Principle of motor speed-based EDC strategy.

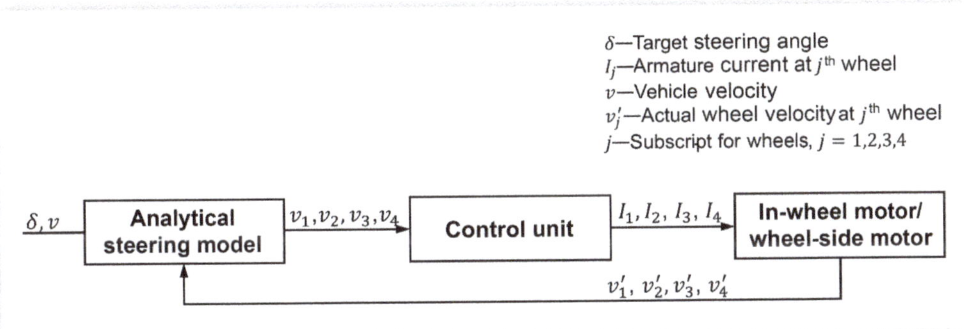

15.9.4. Motor Torque-Based EDC

In this approach, all the motors operate in closed-loop toque control. The principle of the motor torque-based EDC strategy is illustrated in Figure 15.19. The disturbance values of the slip ratio, sideslip angle at the mass centroid of the vehicle, and yaw rate can be selected as control variables. By converting the selected disturbance values into armature currents, the torques from all the motors can be adjusted accordingly. As the input quantity, the target steering angle is converted into the disturbance values of the slip ratio by a road surface recognition model and to the sideslip angle at the mass centroid of the vehicle and yaw rate by a two-DOF model. In addition, the vehicle dynamics model is used to compute the actual sideslip angle at the mass centroid of the vehicle and actual yaw rate according to the measured motor torque.

In addition, a road surface recognition model can be established on the basis of extensive experimental data to acquire the relationship between the adhesion coefficient and the slip ratio for typical road surfaces.

FIGURE 15.19 Principle of the motor torque-based EDC strategy.

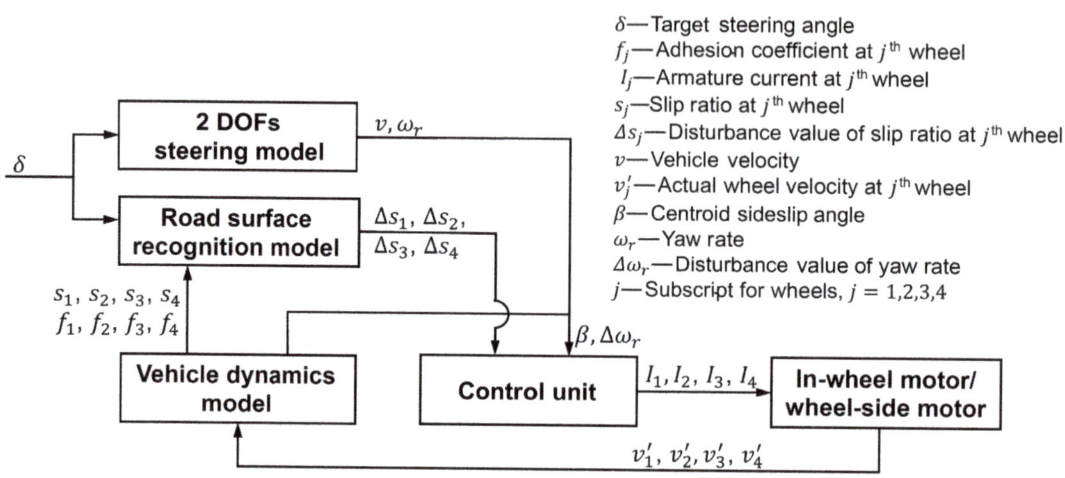

A nonlinear regression model proposed by Burckhardt is often used for curve-fitting these data [15.25]:

$$f(s) = C_1\left(1 - e^{-C_2 s}\right) - C_3 s \tag{15.149}$$

where s is the slip ratio and coefficients C_1, C_2, and C_3 are listed in Table 15.4 [15.25].

TABLE 15.4 Coefficients and results of the road surface recognition model.

Road surface	C_1	C_2	C_3	s_{opt}	f_{max}
Asphalt, dry	1.281	23.993	0.520	0.170	1.171
Cement, dry	1.196	25.166	0.539	0.160	1.092
Asphalt, wet	0.856	33.821	0.345	0.130	0.800
Pebbles, wet	0.400	33.708	0.120	0.140	0.340
Snow	0.195	94.129	0.065	0.065	0.190
Ice	0.050	306.39	0.001	0.030	0.500

The optimal slip ratios and corresponding maximum adhesion coefficients in Table 15.4 are obtained by

$$s_{opt} = \frac{1}{C_2}\ln\frac{C_1 C_2}{C_3} \tag{15.150}$$

$$f_{max} = \frac{C_1 C_3}{C_2}\left(1 + \ln \frac{C_1 C_2}{C_3}\right) \quad (15.151)$$

It is evident that the optimal slip ratios and corresponding maximum adhesion coefficients are dramatically different under different road conditions.

In a fuzzy-logic-based road surface recognition method, the real-time slip ratio and longitudinal adhesion utilization coefficient are utilized as inputs to the controller. By fuzzification, logic inference, and clarification, the similarity coefficients with respect to the six typical curves are obtained as outputs, allowing further calculation of the current adhesion coefficient and optimal slip ratio [15.26].

In the fuzzification process, due to the sharp variation in the adhesion coefficient for a slip ratio lower than 0.2, three fuzzy subsets, namely, small, medium, and large, are selected, with triangular membership functions. The longitudinal adhesion coefficient is fuzzified in three different cases of the slip ratio, namely, small, medium, and large [15.25].

In the logic inference process, 13 fuzzy logic reasoning rules are made, as listed in Table 15.5 [15.25].

TABLE 15.5 Fuzzy logic reasoning rules for logic inference.

	Inputs		Outputs					
Rule no.	Slip ratio	Adhesion coefficient	Ice	Snow	Pebbles, wet	Asphalt, wet	Cement, dry	Asphalt, dry
1	Large	Ice	Similar	Fairly similar	Not similar	Not similar	Not similar	Not similar
2	Large	Snow	Fairly similar	Very similar	Somewhat similar	Not similar	Not similar	Not similar
3	Large	Pebbles, wet	Somewhat similar	Similar	Very similar	Somewhat similar	Not similar	Not similar
4	Large	Asphalt, wet	Not similar	Somewhat similar	Somewhat similar	Very similar	Somewhat similar	Somewhat similar
5	Large	Cement, dry	Not similar	Not similar	Not similar	Similar	Very similar	Similar
6	Large	Asphalt, dry	Not similar	Not similar	Fairly similar	Very similar	Similar	Fairly similar
7	Medium	Ice	Similar	Somewhat similar	Not similar	Not similar	Not similar	Not similar
8	Medium	Snow	Fairly similar	Very similar	Somewhat similar	Not similar	Not similar	Not similar
9	Medium	Pebbles, wet	Somewhat similar	Similar	Very similar	Somewhat similar	Not similar	Not similar
10	Medium	Asphalt, wet	Not similar	Somewhat similar	Somewhat similar	Very similar	Fairly similar	Somewhat similar
11	Medium	Cement, dry	Not similar	Not similar	Not similar	Similar	Very similar	Similar
12	Medium	Asphalt, dry	Not similar	Not similar	Somewhat similar	Very similar	Similar	Fairly similar
13	Small	Cement, dry	Not similar	Not similar	Not similar	Very similar	Very similar	Very similar

Courtesy of Gang Li.

In the clarification process, Gaussian membership functions are employed as it offers better results than the triangular membership functions [15.25]. Based on the outputs of the fuzzy logic inference, the six weight factors, x_j, for $j = 1, 2, 3, 4, 5, 6$, are obtained using the bisector of area method.

Finally, the optimal slip ratio and the maximum adhesion coefficient are given by

$$s_{opt} = \frac{\sum_{j=1}^{6} x_j s_{opt,j}}{\sum_{j=1}^{6} x_j} \tag{15.152}$$

$$f_{max} = \frac{\sum_{j=1}^{6} x_j f_{max,j}}{\sum_{j=1}^{6} x_j} \tag{15.153}$$

where $s_{opt,j}$ and $f_{max,j}$ are the optimal slip ratio and the maximum adhesion coefficient for the jth out of six road surfaces, respectively, listed in Table 15.5 [15.25].

A two-DOF steering model is employed for obtaining the desired dynamic parameters such as sideslip angle β and yaw rate ω_r based on the target steering angle δ. By recalling Equation (5.17), we have

$$\begin{cases} \left(\dot{\beta} + \omega_r\right) mv = \dfrac{\omega_r}{v_x}\left(C_{\alpha 1} L_f - C_{\alpha 2} L_r\right) + \left(C_{\alpha 1} + C_{\alpha 2}\right)\beta - C_{\alpha 1}\delta \\ I_z \dot{\omega}_r = \dfrac{\omega_r}{v_x}\left(C_{\alpha 1} L_f^2 + C_{\alpha 2} L_r^2\right) + \left(C_{\alpha 1} L_f - C_{\alpha 2} L_r\right)\beta - C_{\alpha 1} L_f \delta \end{cases} \tag{15.154}$$

where $C_{\alpha 1}$ and $C_{\alpha 2}$ are the sideslip stiffness of the front and rear tries, respectively.

15.9.5. Disturbance Rejection Control Methodology

Wheels are subjected to highly frequent disturbances from the road as a vehicle moves. These disturbances have significant effects on the operation of the motors and the vehicle, resulting in unstable motor speeds, wheel slips, deviation from expected trajectories, inappropriate yaw rate, and so forth.

Various control schemes have been proposed to improve the disturbance rejection capability of the multimotor system, such as PID [15.26], fuzzy logic [15.27], neural network [15.28], SMC [15.29], PSO [15.30], and PSO-BP neural network [15.31]. The interested reader is referred to the cited references for detailed description of the methods.

15.10. Control Strategies for Auxiliary Systems

The auxiliary systems of an EV may include the water pump, fan, PTC heater, A/C, vacuum pump, and thermal management system for batteries.

The water pump is turned on right after the vehicle is powered up.

The fan is turned on if the e-motor temperature exceeds 60°C or the temperature of MCU exceeds 45°C, and turned off if the e-motor temperature is lower than 45°C or the temperature of MCU is lower than 30°C.

When the vehicle is in the READY state, the PTC heater is turned on by closing the PTC relay if the following conditions are satisfied simultaneously: there is a request for PTC, no faults in the PTC relay, the battery SOC is higher than 5%, and no fault message from the battery pack.

When the vehicle is in the READY state, the VCU sends an enable signal and the target speed command to the compressor of the A/C if the following conditions are satisfied simultaneously: there is a request for A/C on, the evaporator temperature is higher than 2°C, the battery SOC is higher than 5%, and no fault message from the A/C compressor. If the evaporator temperature drops below 2°C, then VCU disables the compressor and enables it again if the evaporator temperature rises back to 4°C or above.

When the vehicle operates on a plain, the vacuum pump is turned on if the pressure in the vacuum chamber of the brake booster is higher than $-0.45p_{amb}$ (p_{amb} is the ambient pressure) and turned off if the pressure is lower than $-0.75p_{amb}$ of the ambient pressure; when the vehicle operates on a plateau, the vacuum pump is turned on if the pressure is higher than $-0.45p_{amb}$ and turned off if the pressure is lower than the higher value of $-0.75p_{amb}$ and p_{vl} (p_{vl} is the lowest achievable pressure of the vacuum pump).

15.11. Trend of VCU Technology and E/E Architecture

There is a trend to integrate the ever-increasing number of ECUs in today's EV to the functional domain controller-based architecture or zonal architecture with a centralized computer. The state-of-the-art distributed E/E architecture will almost certainly be replaced by the vehicle-centralized E/E architecture in the future. Some hybrid designs of the distributed and centralized E/E architectures are being developed before evolving to the ultimate form.

Although the E/E architecture keeps transforming, it is generally agreed that future architecture will have the following distinctive features:

1. Centralized computing: Service-oriented system architecture (SOA) will become the mainstream, providing a high-performance real-time computing platform for software. Hence, a central computer will form a collaborative solution with multiaccess edge computing (MEC) and cloud computing to avoid unlimited growth in the demand for computing power at the vehicle end.
2. Decoupled software and hardware: Hardware will no longer be exclusive to a certain function, and rather it will become a common resource that can be called by any on-board functions. This will allow the stand-alone safety-redundant hardware required by functional safety to be achieved in an economical way.
3. Platform standardization: Perhaps only one platform will be needed to cover all vehicle models from low to high end, which will bring down the costs for developing and maintaining multiple platforms. With the progress in the commonality, standardization, and interoperability, the platform will likely be shared by a number of vehicle manufacturers.

4. Function customization: Intelligence is the core element of future brand differentiation, which is mainly achieved by adding software functions. In the future, the vehicle delivered by a manufacturer will not be a functionally solidified product, but rather a continuously evolving one that features an open application platform.

References

[15.1]. Dou, G., Liu, F., Cheng, H., Ke, X. et al., "Development Practice of Driving Control Strategy for a Battery Electric Sedan," *Shanghai Automobile* 05 (2010): 8-11.

[15.2]. Mo, X., Zhu, Y., and Yang, H., "A Study on Motor Speed Control for Ramp Start Up of Electric Vehicle," *Computer Simulation* 35, no. 4 (2018): 107-111.

[15.3]. Chen, S., Qin, D., Hu, M., and Wei, H., "Ramp Adaptive Launch Control Strategy of Pure Electric Vehicles," *Journal of Chongqing University* 35, no. 9 (2012): 1-7, doi:10.11835/j.issn.1000-582X.2012.09.001.

[15.4]. Chen, S., "Study on Parameter Matching and Integrated Control Strategy for Pure Electrical Vehicle Driveline," Doctoral thesis, College of Mechanical Engineering, Chongqing University, Chongqing, 2013, doi:10.7666/d.D426550.

[15.5]. Wang, B., Luo, Y., Fan, J., and Li, K., "A Study on Driving Force Distribution of Four-Wheel-Independent Drive Electric Vehicle Based on Control Allocation," *Automotive Engineering* 32, no. 2 (2010): 128-132.

[15.6]. Hattori, Y., Koibuchi, K., and Yokoyama, T., "Force and Moment Control with Nonlinear Optimum Distribution for Vehicle Dynamics," in *Proceedings of 6th International Symposium on Advanced Vehicle Control (AVEC 02)*, Hiroshima, Japan, 2002, 595-600.

[15.7]. Chen, S., Qin, D., Hu, M., and Hu, J., "Pure Electric Vehicle AMT Comprehensive Shift Strategy Considering Dynamics and Economic Performances," *Chinese Journal of Mechanical Engineering* 24, no. 19 (2013): 2687-2692, doi:10.3969/j.issn.1004-132X.2013.19.025.

[15.8]. Chen, X., "Study on Shifting Control Strategy for Two-Speed Automatic Mechanical Transmission of Pure Electric Vehicle," Master's thesis, College of Mechanical Engineering, Chongqing University, Chongqing, 2013, doi:10.7666/d.D353731.

[15.9]. Zhang, B., Tang, B., Zhou, D., and Wu, X., "Study of Coordinated Control Strategy of Automatic Shift Based on AMT," *Journal of Hefei University of Technology (Natural Science)* 39, no. 9 (2016): 1178-1183, doi:10.3969/j.issn.1003-5060.2016.09.006.

[15.10]. Zhu, B., "Research of Two Speed DCT Electric Power-Train and Control System," Ph.D. thesis, Engineering and Information Technology Department, University of Technology, Sydney, 2015.

[15.11]. Dai, Z., Xu, X., Liu, Y., and Wang, S., "Closed-Loop Slip Control of Clutches in Hydraulic Automatic Transmission," *Automotive Engineering* 34, no. 8 (2012): 718-722, doi:10.19562/j.chinasae.qcgc.2012.08.011.

[15.12]. Sun, X. and Wu, G., "Shifting Control Strategy Simulation for Dual Clutch Transmission Vehicle," *Journal of Southeast University* 41, no. 4 (2011): 729-733, doi:10.3969/j.issn.1001-0505.2011.04.014.

[15.13]. Gu, Q. and Cheng, X., "Coordinated Control of Upshift Power of Double Clutch Transmission for Electric Vehicle Based on Particle Swarm Optimization," *Transaction of the Chinese Society of Agricultural Engineering* 28, no. 8 (2012): 84-91, doi:10.3969/j.issn.1002-6819.2012.08.013.

[15.14]. Li, J., Chen, J., Wang, Y., and Wang, Y., "Shift Process Optimal Control of Combined Clutch in Electric Vehicles," *Journal of Hunan University (Natural Sciences)* 43, no. 2 (2016): 22-27, doi:10.16339/j.cnki.hdxbzkb.2016.02.004.

[15.15]. Song, X., Zulkefli, M., Sun, Z., and Miao, H., "Automotive Transmission Clutch Fill Control Using a Customized Dynamic Programming Method," *Journal of Dynamic Systems, Measurement and Control* 133, no. 5 (2011): 054503, doi:10.1115/1.4003797.

[15.16]. Chen, X., Yuan, Y., and Wu, G., "The Study of Control Strategies and Simulation for EVT Power Train Applying to EV," Paper Presented at in *CEVS-17*, Beijing, July 6–8, 2013.

[15.17]. Wang, C., Xia, J., and Sun, Y., *Modern Motor Control Technology* (Beijing: China Machine Press, 2008), ISBN:9787111249016.

[15.18]. Yuan, Y., *Electric Vehicle Transmissions: Structure and Control* (Beijing: China Machine Press, 2020), ISBN:978-7-111-66809-1.

[15.19]. Wang, L., Yuan, Y., and Shi, L., "Improvements of Energy Efficiency under Electric Driving Mode for Plug-In Hybrid Electric Vehicles," Paper Presented at in *2018SAECCE-EV011*, Shanghai, November 6, 2018.

[15.20]. Zhou, L., Luo, Y., Li, K., and Lian, X., "Braking Control of Electric Vehicles Coordinating Regenerative Brake and Anti-lock Brake," *Journal of Tsinghua University (Sci. & Tech.)* 49, no. 5 (2009): 105-110, doi:10.16511/j.cnki.qhdxxb.2009.05.017.

[15.21]. Gillespie, T., *Fundamentals of Vehicle Dynamics* (Warrendale: Society of Automotive Engineers, Inc.), 1992, 21-118, ISBN-10:1560911999, ISBN-13:978-1560911999.

[15.22]. Hudha, K. and Jamaluddin, H., "Hardware-in-the-Loop Simulation of Automatic Steering Control for Double Lane Change and Sine Steer Manoeuvres," *International Journal of Vehicle Autonomous Systems* 10, no. 2 (2012): 67-104, doi:10.1504/IJVAS.2012.047695.

[15.23]. Yao, F., Lin, X., Wu, Z., and Li, G., "Summary of Research on Electronic Differential Control Technology of Electric Vehicle," *Acta Automatica Sinica* 47, no. 8 (2019): 1785-1798, doi:10.16383/j.aas.c190293.

[15.24]. Wang, Y., Yan, S., Lv, B., and Yang, J., "Simulation on Multifunction Steering System of Wheel Drive Electric Vehicle," *Journal of Mechanical & Electrical Engineering* 35, no. 10 (2018): 1128-1132, doi:10.19508/j.cnki.1672-4801.2019.02.014.

[15.25]. Li, G., "Study on Stability and Energy Saving Control for X-by-Wire Electric Vehicle with Four-Wheel Independent Drive In-Wheel Motors," Doctoral thesis, College of Automotive Engineering, Jilin University, 2013.

[15.26]. Fahrwerktechnik, R. and Burckhardt, M., *Radschlupf-Regelsysteme* (Würzburg: Vogel Verlag, 1993), ISBN 10:3802304772, ISBN 13:9783802304774.

[15.27]. Tan, H., and Liao, W. "Application of Fuzzy PID in Steering Control of Wheeled Robot," *Journal of Hunan University of Technology* 32, no. 1, (2008): 81-87, doi:10.3969/j.issn.1674-117X.2018.01.015.

[15.28]. Lu, W., Zhang, Y., and Zhang, H., "Research on P-Fuzzy PID Control of Electronic Differential Speed System of Hub Motor Drive Vehicle," *Machinery Manufacturing and Automation* 46, no. 6 (2017): 193-196, doi:10.19344/j.cnki.issn1671-5276.2017.06.053.

[15.29]. Zhai, L., Dong, S., and Luo, K., "Steering Electronic Differential Control of Vehicles Independently Driven by Four-Wheel Hub Motors," *Journal of Beijing Institute of Technology* 30, no. 8 (2010): 901-905, doi:10.15918/j.tbit1001-0645.2010.08.011.

[15.30]. Yang, Y., Zhao, H., Wang, W., and Xu, X., "Research on Electronic Differential Speed Control of Double-Hub Motor Driven Electric Vehicle," *Tractor and Agricultural Transport Vehicle* 45, no. 5 (2008): 34-39+50.

[15.31]. Guan, P. and Huang, Q., "Design and Research of Electronic Differential Speed Controller for Hub Electric Vehicle," *Computer Simulation* 35, no. 9 (2008): 155-160.

Index

A

Acceleration capacity, 19–20
Acceleration resistance, 17
Acceleration time, 228
Ackermann steering principle, 644–646
Adhesion
  and braking forces, 64
  effect of sideslip, 67–68
Adhesion coefficient, 77, 80
  and adhesion condition, 20–21
  different types of tires, 67
  effect of initial speed, 73
  effect of road surface, 66–67
  effect of speed, 67
  factor, 640
  hard-surface roads
    effect of sideslip on adhesion, 67–68
    longitudinal behavior, 64–67
  minimal adhesion coefficient requirements, 24–26
  road surfaces, 21
Adhesion condition, 20, 21
Adhesion-related forces, 21–24
Adhesion utilization, 82
Aerodynamic design of vehicle bodies
  aerodynamic drag reduction, design principles for
    front end, 574
    front windshield, 574
    rear windshield, 574
    roof, 574
  aerodynamic drag, types of, 573–574
  aerodynamic stability, design principles for, 575–577
  drag coefficients of EVs, 577
  ideal aerodynamic body shapes, 572–573
Aggressive brake proportioning strategy, 92
Air-cooled battery systems, 273
Air gap flux density, 306, 307
Air resistance
  air resistance coefficient and frontal area, of various types of EVs, 15
  body components/features to drag coefficient, 15
  designing EVs, 14–15
  drag coefficients of popular EV models, 14
  normal pressure distribution, 12–13
Air springs
  bellow type, 472, 473
  body height adjustment mechanism, 475, 476
  rolling diaphragm type, 472, 473
  stiffness characteristics, 474, 475
  typical configurations of air suspension, 476–477
Aligning torque, 98, 101–102
All-wheel drive (AWD) EVs, 94–95
All-wheel steering, 519–520
  basic principles
    opposite-direction steer, 520
    same-direction steer, 520–522
  mechanism
    ECU, 523
    e-motors, 523
    front and rear wheel angle sensors, 522
    steering angle sensor, 522
    vehicle speed sensor, 522
    yaw rate sensor, 522
Amplitude–frequency characteristics, 117, 177
  dynamic load on wheels, 178
  sprung mass, 176
  sprung mass acceleration, 177
  suspension deflection, 177–178
  unsprung mass, 176
Analog active drive, 369
Analog quantity acquisition module, 355
Antifade performance of brakes, 70
Anti-lock brake system (ABS), 64, 69, 84–86, 139, 549–551, 638, 639
Apparent power, 290
Approach angle γ_1, 187
Armature current, 294
Asymmetrical mass imbalance, 160
Audi Space Frame (ASF), 582
Automated mechanical transmissions (AMTs), 52
Automatic transmissions (ATs), 375
Average curb weight, 226
AWD vehicle overcoming steps, 206–208
Axial flux AC motors
  flux path size, optimization of, 301–302
  outer-to-inner diameter ratio, selection of, 300–301
Axial flux motors, 289, 290, 293–295
Axle load distribution, 227

B

Ball joints, 516
Batteries, state estimation
  SOC, 279–281
  SOH, 281–282
  SOP, 282
  SOS, 282–283
Battery cells
  battery pack, 261
  parallel connection, 262–264
  serial connection, 262–263
  series–parallel connection, 264–265
  type, 260
Battery electric vehicles (BEVs), 32, 35
Battery management strategy, 593
  BTM strategy, 595
  power management, 594–595
Battery modules
  cylindrical cells, 268–269
  pouch cells, 270–271
  prismatic cells, 269–270
Battery packs
  architecture, 261–262
  electrical topology
    battery cells connected in parallel, 263–264
    battery cells connected in series, 262–263
    battery cells connected in series–parallel, 264–265
    guidelines for design of pack topology, 265–266
  structural design, 267–268
    integration of battery modules, 268–271
    trends in battery system integration, 272–273

Battery system
battery cell type, 260
battery pack
architecture, 261–262
BMS, 276–284
BTMS, 273–275
electrical topology, 262–266
structural design, 267–273
battery system parameters, selection of, 232–234
battery type selection, 231–232
fundamental requirements, 259–260
integration
CTB technology, 273
CTC technology, 272–273
CTP technology, 272
large-module approach, 272
layout, 248–249
packing, 234–236
Battery thermal management (BTM), 44, 595
Battery thermal management system (BTMS), 260, 273–275
Bending fatigue, 400–401
Bending stress, 566
BMS
cell equalization, 278–279
functions, 276
state estimation of batteries, 279–283
topology of BMS controllers, 276–278
BMW LifeDrive, 582
Body aerodynamics, 43–44
Body, EVs
aerodynamic design of vehicle bodies
aerodynamic drag reduction, design principles for, 574
aerodynamic drag, types of, 573–574
aerodynamic stability, design principles for, 575–577
drag coefficients of EVs, 577
ideal aerodynamic body shapes, 572–573
body construction
body-on-frame construction, 572
unibody construction, 572
body styling
dynamic sensation, 571
sense of integrity, 570–571
structural design of vehicle bodies, 577–578
crash safety, design principles, 578–580
lightweight design principles, 580
noise and vibration considerations, 580–581
novel body materials and processes of EVs, 581–582
vehicle body testing
BIW modal tests, 583
body fatigue strength tests, 583
body-related tests, 583
measurements of overall stiffness, 582
Body-in-white (BIW), 570, 583
Body-on-frame construction, 572
Brake actuation mechanism
braking force adjustment mechanism, 548–551
hydraulic actuation system components, 546–548
layouts of brake line circuit, 544–546
power brakes, 543–544
types, 543
Brake-by-wire
electrohydraulic brakes, 551–552
electromechanical brakes, 553–554
Brake line circuit
HH type, 545, 546
HI type, 544–546
II type, 544–546
LL type, 545, 546
X type, 544–546
Brake pedal force, 548
Brake pressure regulator, 548–549
Brake proportioning
FWD EVs, 91–92
RWD EVs, 92–93
Brake torque, 85
Braking
directional stability
braking deviation, 71–72
sideways skid and loss of steering, 72–74
wheel force analysis, 62–68
Braking efficiency
and adhesion utilization, 81–82
and constancy, 70–71
analysis of stopping distance, 69–70
stopping distance and deceleration, 68–69
factor, 71
Braking force
and adhesion, 64
adjustment mechanism
ABS, 549–551
brake pressure regulator, 548–549
coefficient, 65, 66, 68
control strategies, 142
front and rear proportioning, 74–75
ABS, 84–86
actual distribution, 78
adhesion utilization and braking efficiency, 81–82
braking process of EVs, with fixed brake proportioning, 78–81
fixed proportioning, 77–78
f-lines, 79
ideal curve, 76–77
normal reaction forces, 75–76
requirements for brake proportioning, 83–84
r-lines, 79
front and rear tractive force proportion, 137
ground, 63
maximum, 558–559
proportioning strategy for FWD EV, 639
stability control, 137
tire circumference, 63–64
Braking-in-a-turn tests, 87
Braking performance
definition, 61
evaluation indicators, 61–62
parameters, 230
regulations of different regions, 62
rig tests, 88
road tests, 86–87
Braking ratio, 82
Braking system design
brake actuation mechanism
braking force adjustment mechanism, 548–551
hydraulic actuation system components, 546–548
layouts of brake line circuit, 544–546
power brakes, 543–544
types, 543
brake-by-wire
electrohydraulic brakes, 551–552
electromechanical brakes, 553–554
braking torques, determination of, 526–527
park brakes, 542
service brakes
disc brakes, 527–533
drum brakes, 533–541
Braking torques, 526–527
Breakthrough load (BTL), 422
Brushed (BDC) motors, 236
Brushless DC (BLDC) motors, 237
Brushless permanent magnet motor, 554

C

Camber angle, 98, 102–104
Camber stiffness, 103
Camber thrust, 103
Carrying capacity, 226–227
Carter coefficient, 304–305, 332
Cartesian coordinate system, 97
Cauer model, 372
Cell equalization, 278–279
Cell supervising circuit (CSC), 276
Cell-to-body (CTB) technology, 273
Cell-to-chassis (CTC) technology, 272–273
Cell-to-pack (CTP) technology, 272
Center distance factor, 380, 381
Centralized powertrain system layout, 250
Charge capacity, 233
Chassis, EVs
analysis of load cases, 556–557
maximum braking force, 558–559
maximum lateral reaction forces on tires, 557–558

maximum tractive force, 559
maximum vertical reaction forces on tires, 557
vehicle frame design
sizing, 562
skateboard chassis, 568–569
spars and cross members, 561–562
strength and stiffness, 562–568
types, 560–561
vehicle frame testing
modal tests, 570
proving ground and road endurance tests, 570
reliability and durability rig tests, 570
stiffness tests, 569
stress tests, 569
Circumferential forces, 2
Clarke transformation, 597
Clutches, 410–411
actuation force, calculation of, 413–415
clutch friction materials, 415
drag losses, reduction of, 416
wet clutches
dynamic capacity design of, 412
static capacity design of, 411
Coherence function, 165
Coil springs, 463–465
Cold braking tests, 87
Column EPS, 512–513
Communication module, 356
Compliance camber change, 129
Compliance steer
aligning torque, 128–129
lateral force, 127–128
steering system stiffness, 135
Composite phase change materials (CPCM), 275
Computer-aided engineering (CAE), 571
Conservative brake proportioning strategy, 91
Constancy of braking efficiency, 70–71
Continuously variable transmissions (CVTs), 375
Control system design
auxiliary systems, strategies for, 652–653
basic functions of VCU
charging management, 588
fault diagnosis and handling, 588
monitor and display of vehicle operational status, 588
regenerative braking control, 587–588
vehicle driving control, 587
vehicle-level energy management and optimization, 587
vehicle network management, 588
battery management strategy, 593
BTM strategy, 595
power management, 594–595
driving intent and accelerator pedal strategy, 602–604
E/E architecture of EV, 586, 587
electric drive system
e-motor control strategy, 612–620
gear shift control strategy, 620–636
electric motors
three-phase IMs, 595–600
three-phase PMSMs, 601–602
electronic differential control
disturbance rejection control methodology, 652
motor speed-based EDC, 649
motor torque-based EDC, 649–652
three-DOF steering model, 646–649
two-DOF model based on Ackermann steering principle, 644–646
hardware architecture of VCU, 588–590
high-voltage power-up procedure
charge, 592
discharge, 591
high-voltage shutdown procedure
charge, 592–593
discharge, 591–592
launch control strategy
e-motor torque, 604, 605
self-adaptive control strategy, 608–611
torque compensation control strategy, 605–607
low-voltage power-up procedure, 590
low-voltage shutdown procedure, 590–591
regenerative braking, 636–644
braking force proportioning strategy for FWD EV, 639
deep regenerative phase, 637
electric braking torque *vs.* vehicle speed, 638
primary factors, 636–637
soft entry phase, 637
soft withdrawal phase, 637
three-level hierarchical controller, 639
VCU technology and E/E architecture, 653–654
Control valves, 436–437
Cooling technology, 371
Coreless transformers (CLTs), 358
Corn transition, 468
Coulomb's law of friction, 189
Coupled inductor method, 368
Curb weight, 226
Cylindrical cells, 268–269, 274

D

Damping ratio, 112, 115
Deceleration sensing proportioning valve (DSPV), 84
Departure angle γ_2, 187
Digital active drive, 369
Digital human models (DHMs), 255
Dimensional deviation-related excitation, 157–159
Dimensional parameters
front and rear overhangs, 225
overall dimension of vehicle, 223
truck, 225
wheelbase, 223–224
wheel tracks, 224
Direct-axis armature reaction reactance, 314
Direct bonding copper (DBC) liner, 362
Directional stability, in braking
braking deviation, 71–72
sideways skid and loss of steering, 72
effect of initial speed and adhesion coefficient, 73
effect of wheel lockup sequence, 73
front- and rear-wheel skids, 73–74
Direct torque control (DTC), 237, 599–600
Disc brakes
advantages, 527
detailed design
brake pads, 530–531
clearance adjustment mechanism, 532
discs, 530
mounting position of calipers, 531–532
disadvantages, 528
sizing, 529–530
types
fixed caliper, 528, 529
floating calipers, 528, 529
Discharging efficiency, 37
Distributed powertrain system layout, 250
Double-layer full-pitch coil sides, 317
Double-layer stacked windings, 319
Double-layer wave windings, 319
Double pulse tester (DPT), 373
Double-sided cooling, 359–360
Double-wishbone suspension
guide mechanism design, 490–493
pitch center, 488
roll center, 487
Drawbar pull of EVs, 191
calculation, 197
soft ground resistance to pneumatic tire, 196–197
soft ground resistance to rigid rolling wheel, 193–195
traction force, 192–193
Driver boards, 357
basic push-pull output circuit with FETs, 358
gate driver circuit of full bridge topology, 358
Driveshaft-related excitation, 155–156
Drivetrain, 43, 374–375
basic characteristic parameters, determination of, 380–381
clutches, design of, 410–411
actuation force, calculation of, 413–415

clutch friction materials, 415
drag losses, reduction of, 416
dynamic capacity design of wet clutches, 412
static capacity design of wet clutches, 411
driveshaft-related excitation, 155–156
EVT, 377, 378
gear shift-related excitation, 156–157
general layout design, 379–380
handling
cornering stability control, tractive and braking forces, 137–139
electronic control systems, 139–142
tractive force effect on cornering characteristics, 135–137
housing, design of, 442–443
draft angle and fillet, 445–446
loads on housing, 443–445
material and manufacturing process, 447
oil passages, 446–447
overall structure of transmission housing, 443
reduction of housing vibration and noise, 446
vent plugs, 447–448
wall thickness and reinforcement ribs, 445
hydraulic and lubrication systems, design of
antifoam properties of transmission oil, 441
antioxidant properties of transmission oil, 441
antiwear properties of transmission oil, 441
boundary lubrication, 439
control valves, 436–437
cooling performance of transmission oil, 441
equations for lubrication system design, 439
forced oil lubrication system, 439–440
friction characteristics and stability of transmission oil, 441
lubrication system of transmission, 437
oil pumps, 432–435
oil splash lubrication system, 440
physical properties of transmission oil, 442
selection of transmission oil, 440
typical load bearing capacity, squeeze film lubrication, 438
layshaft gearsets, design of
base pitch, 384
bending strength for teeth, 389–390
calculation and determination of modification coefficient, 386–387
center distance, initial selection of, 381
contact strength checkup of tooth face, 390–391
diameters of base and addendum circles, 383–384
initial calculation of transverse contact ratio and overlap ratio, 384–386
initial selection of tooth width, number of teeth and pressure angle, 382
pitch diameters, 381–382
selection of helix angle and estimation of normal module, 382–383
update of center distance, contact ratio and parameters, 387–389
mechanical efficiency, 4–5
planetary gearsets, design of, 391
basic equations, 393–395
gear design for strength and durability, 399–402
gear materials, selection of, 402–403
gear noise reduction, design principles for, 405–409
number of planet gears and spacing, 395–398
schematic, 392
transmission gear design, manufacturability of, 403–405
transmission gear design, trend in, 409–410
transverse contact ratio, 398–399
shift actuators, design of
characteristics, 423
electrohydraulic actuators, design of, 424–429
electromechanical actuators, design of, 431–432
electropneumatic actuators, design of, 429–431
synchronizers, design of
breakthrough load of detent balls, 422
major design parameters, selection of, 418–422
shift force, calculation of, 417–418
types of synchronizers and characteristics, 416–417
typical two-speed AMT, 375
typical two-speed DCTs, 375, 376
typical two-speed PSTs, 376, 377
Driving force, 18–19
Driving torque, 136
Drum brakes, 533
design parameters, 536
detailed design
determination of design parameters, 537–540
mechanical components, 540–541
single, 537
sizing, 536–537
types, 533
dual-twin leading shoes, 535
duo-servo, 535
leading/trailing shoes, 534
twin leading shoes, 534–535
twin trailing shoes, 535
uni-servo, 535
Dual-air gap axial flux motor, 295
Dual-mass system, 172, 173
transfer characteristics, 174–177
two dominant modes, 173
Dual-pinion EPS, 513
Dynamic current sharing, 368–369
analog active drive, 369
digital active drive, 369
Dynamic deflection of suspension, 460
Dynamic performance
equilibrium of power, 26–28
indicators, 1–2
multispeed transmissions, 52
resistance forces, 6
acceleration resistance, 17
air resistance, 12–15
grading resistance, 16
rolling resistance, 6–12
rig tests, 29
road tests, 28
tire–ground adhesion
adhesion coefficient, 21
adhesion coefficient and adhesion condition, 20–21
adhesion-related forces, 21–24
minimal adhesion coefficient requirements, 24–26
tractive effort, 2
in different gears, 5–6
electric motors, characteristics of, 3–4
mechanical efficiency of drivetrain, 4–5
wheel, 3
wheel radius, 5
vehicle dynamic equation, 17–18
acceleration capacity, 19–20
equilibrium diagrams of driving force and running resistance, 18–19
gradeability, 20

E

ECE Safety Regulations, 83
Effective axial length of armature l_i, 307
Elastic characteristics of suspensions, 460–461
Elastic elements
air springs, 472–477
coil springs, 463–465
leaf springs, 468–472
torsion bar springs, 465–468
Elastic resistance force, 197
Elastohydrodynamic lubrication (EHL), 438
Electrical reactance
leakage reactance, 315–320
magnetizing reactance, 313–315

Electrical resistance, 311–313
Electric drive system, 287–288
  configurations, 288–289
  drivetrains, design of, 374–375
    basic characteristic parameters, determination of, 380–381
    clutches, design of, 410–416
    EVT, 377, 378
    general layout design, 379–380
    housing, design of, 442–448
    hydraulic and lubrication systems, design of, 432–442
    layshaft gearsets, design of, 381–391
    planetary gearsets, design of, 391–410
    shift actuators, design of, 423–432
    synchronizers, design of, 416–422
    two-speed AMT, 375
    two-speed DCTs, 375, 376
    two-speed PSTs, 376, 377
  electric motor controllers, design of
    circuit schematic, 354
    driver boards, 357–358
    high-voltage main circuits, 359–360
    main control boards, 355–357
    SiC motor controllers, design guidelines for, 360–374
  electric motors
    basic design parameters and equations, 290–295
    electromagnetic design, 311–328
    guidelines for PMSMs, 342–354
    losses, temperature rise and cooling, 329–342
    magnetic circuit, 302–311
    sizing, 295–302
    topology, 289–290
  e-motor control strategy
    dual-motors driving dual-axles, 615–617
    dual-motors driving single axle, 613–615
    four motors driving four wheels, 617–620
    single motor driving single axle, 612–613
  gear shift control strategy
    actions related to failsafe, 626–627
    clutchless AMT, 627–631
    comprehensive shift schedule, 623–624
    dual-rotor EVT drive system, 633–636
    forms of shift schedules, 620–621
    gear shift decision-making, 620
    inhibited shift operation, 626
    optimal dynamic performance, shift schedule for, 621–622
    optimal energy economy, shift schedule for, 622–623
    powershift transmission, 631–633
    quasi-stall operating principle of dual-motor drive systems, 636
    special operating conditions, shift schedule correction under, 624–626
  layout
    basic principles, 221–222
    electric buses, 222
    electric cars, 221
    electric trucks, 222
    number of axles and axle load, 220–221
  mounts, 244–245
  selection of
    electric drive system mounts, 244–245
    electric motor selection, 236–240
    extreme motor speed, 244
    maximum motor speed, 244
    parameters, 240–244
    peak power and torque, 242–244
    rated power and torque, 240–242
    system voltage, 244
Electric MCU, 586
Electric motor
  basic design parameters and equations, 290
    axial flux motors, 293–295
    radial flux motors, 291–293
  characteristics, 3–4, 240
  combined efficiency map, 36
  control
    three-phase IMs, 595–600
    three-phase PMSMs, 601–602
  cooling, 341–342
  electrical losses, 333
  electromagnetic design
    electrical reactance, calculation of, 313–320
    electrical resistance, calculation of, 311–313
    magnetic saturation on winding design, 325–326
    skin effect on rotor winding design, 324–325
    skin effect on stator winding design, 320–324
    types of winding, 326–328
  energy consumption, 37
  iron losses, 329–331
  key efficiency indicators, 43
  load stray losses, 333–334
  magnetic circuit, design of, 302–304
    excitation current and no-load characteristics, 310–311
    magnetic potential drop, in air gap, 304–307
    magnetic potential drop, teeth, 307–309
    magnetic potential drop, yoke, 309–310
    pole flux leakage, 310
  mechanical losses, 333
  no-load stray losses, 331–332
  operating modes, 36
  peak power, 48
  PMSM rotors, design of, 342–343
    air gap length, selection of, 344–345
    design of air gap structure on rotor surface, 344
    iron core materials, selection of, 347
    magnetic path design of IPM type of rotors, 343–344
    permanent magnetic materials, selection of, 346–347
    weight reduction, slots for, 345
  PMSM stators, design of
    aspect ratio, selection of, 351–352
    cooling structure, design of, 350–351
    electrical loading, selection of, 352
    iron core, 348
    pole count, selection of, 352
    slot opening width, selection of, 352
    slot-to-pole ratio, selection of, 352
    slot type, selection of, 352–354
    windings, design of, 348–349
  regenerative braking, characteristics on, 90
  selection
    BDC motors, 236
    BLDC motors, 237
    IMs, 238
    PMSMs, 237–238
    SRMs, 239–240
  selection of electric motor power, 47–48
  sizing
    axial flux AC motors, 300–302
    motor speed, selection of, 295–296
    radial flux AC motors, 298–300
    specific electromagnetic loading, selection of, 296–298
  steady-state temperature rise, calculation of, 335–339
  topology, 289–290
  torque ripples
    current sensor errors, 154
    nonsinusoidal magnetic field, 152–153
    resolver, 155
    switching frequency of inverter, 153–154
  transient temperature rise, calculation of, 339–341
Electric motor controllers
  circuit schematic, 354
  driver boards, design of, 357
    basic push-pull output circuit with FETs, 358
    gate driver circuit of full bridge topology, 358
  high-voltage main circuits, design of
    double-sided cooling, 359–360
    single-sided cooling, 359
    triple-or-above-sided cooling, 360

main control boards, design of
analog quantity acquisition module, 355
communication module, 356
fault handling module, 355–356
power supply module, 355
PWM module, 357
relay module, 356
speed detection module, 356–357
SiC motor controllers, design guidelines for, 360
current sharing, parallel paths, 366–369
EMC, improvement of, 365–366
low-stray inductance packaging, 362–364
short-circuit protection, 361–362
suppression of bridge arm crosstalk, 364
thermal management, 369–374
Electric power steering (EPS), 512
column EPS, 512–513
dual-pinion EPS, 513
parallel axis EPS, 513
single-pinion EPS, 513
Electric variable transmission (EVT), 376, 377
advantages, 378
schematic, 378
Electrochemical impedance spectroscopy (EIS), 279
Electrohydraulic actuators, 424–429
Electrohydraulic brake (EHB), 551–552
Electromagnetic compatibility (EMC) design principles
BMS power supply, 283–284
interface circuit, 284
PCB design, 284
sensitive signal acquisition circuit, 284
software filtering, 284
Electromagnetic design
electrical reactance, calculation of
leakage reactance, 315–320
magnetizing reactance, 313–315
electrical resistance, calculation of, 311–313
magnetic saturation on winding design, 325–326
skin effect
on rotor winding design, 324–325
on stator winding design, 320–324
types of winding, 326–328
Electromagnetic shock absorbers, 482–483
Electromagnetic torque, 294, 596, 598, 602
Electromechanical actuators, 431–432
Electromechanical brake (EMB), 553–554
Electromotive force (EMF), 262
Electronically controlled air suspension (ECAS), 475
Electronic differential control (EDC)
disturbance rejection control methodology, 652
motor speed-based EDC, 649
motor torque-based EDC, 649–652
three-DOF steering model, 646–649
two-DOF model based on Ackermann steering principle, 644–646
Electronic stability control (ESC), 140
Electronic traction system (ETS), 137
Electropneumatic actuators, 429–431
Electrorheological shock absorbers, 481–482
E-motor control strategy
dual-motors driving
dual-axles, 615–617
single axle, 613–615
four motors driving four wheels
accelerator pedal strategy, 617–619
failsafe control strategy, 619–620
single motor driving single axle
backward driving torque strategy, 613
forward driving torque strategy, 612–613
End-winding leakage per phase, 318
Energy consumption, 37
balance maps, 37, 38
of EVs, 229
Energy economy
vs. acceleration for EV, 50
estimation, 36–37
factors
body aerodynamics, 43–44
drivetrain, 43
electric motor system, 43
energy consumption balance maps, 37, 38
reduction in vehicle weight and size, 42
thermal management, 44–45
tire, 44
well-to-wheel energy balance maps, 39–42
FTP-7, 32–35
HWFET, 32–35
indicators, 31–35
NEDC, 32–35
parameters, 229
of popular BEV and FCV models, 35
SC03, 32–35
testing, 45–46
US06, 32–35
WLTC, 32–35
Energy utilization index, 186
Equation of motion, 2
Equilibrium of power, 26–28
Equivalent polar arc coefficient a_p, 305–307
Equivalent static deflection, 467
Excitation sources
drivetrain, 155–157
electric motors, 152–155
road roughness, 161–165
wheels and tires, 157–161
Extended Kalman filter (EKF), 281
Extra-urban driving cycle (EUDC), 32
Extreme motor speed, 244

F

Fast Fourier transform (FFT), 406
Fault handling module, 355–356
Fault under load (FUL), 361
Federal Motor Vehicle Safety Standard (FMVSS), 83
Field effect tubes (FETs), 357, 358
Field-oriented control (FOC), 237, 597–599
Final drive ratio, 54–55
Finite element analysis (FEA), 400
First-order dominant mode, 173
Flexible printed circuit boards (FPCBs), 271
Force analysis, 75
Forced oil lubrication system, 439–440
Foster model, 372
Fourier transform, 116, 163, 164, 169
Four-speed transmission, 55
Friction torque, 417
Front and rear proportioning, of braking force, 74–75
ABS, 84–86
actual distribution, 78
adhesion utilization and braking efficiency, 81–82
braking process of EVs, with fixed brake proportioning, 78–81
fixed proportioning, 77–78
f-lines, 79
ideal curve, 76–77
normal reaction forces, 75–76
requirements for brake proportioning
ECE Safety Regulations, 83
FMVSS, 83
variable brake proportioning, 84
r-lines, 79
Front and rear wheel angle sensors, 522
Front barrier crash, 579
Front suspension, 245–246
Front wheel angle
instantaneous response
equation of motion and solution, 111–115
stability conditions, 115–116
steady-state response
ratio of turning radii, 108–109
sideslip angles, 110–111
static margin, 109–110
yaw velocity gain, 106–108
vehicle response, 104–106
Front-wheel drive (FWD), 21, 22, 24–26
Fuel cell vehicle (FCV), 32, 35, 42
Fuzzy logic reasoning rules, 643, 651
FWD vehicle overcoming steps, 202–204

G

GA optimization algorithm
- calculation of fitness function, 58
- coding, 57
- crossover operation, 57
- initialization of population, 57
- judgement of termination conditions, 58
- mutation operation, 58
- selection of population size, 57
- selection operation, 58

Gear noise reduction, design principles, 405–406
- addendum coefficient, 408
- dimension tolerances, 409
- gear modification, 409
- helix angle, 407–408
- module, 407
- pressure angle, 407
- shaft and support design, 408–409

Gear shift control strategy
- actions related to failsafe, 626–627
- clutchless AMT
  - block diagram, 627
  - disengaging, 628
  - engaging, 629–630
  - load restoring, 630–631
  - off-loading, 628
  - speed synchronizing, 628–629
- dual-rotor EVT drive system, 633–636
- gear shift decision-making, 620
- inhibited shift operation, 626
- powershift transmission, 631–633
- quasi-stall operating principle of dual-motor drive systems, 636
- shift schedule
  - comprehensive, 623–624
  - correction under special operating conditions, 624–626
  - forms of, 620–621
  - for optimal dynamic performance, 621–622
  - for optimal energy economy, 622–623

Gear shift-related excitation, 156–157
Global sideslip angle, 105
Gradeability, 20, 50, 228
Grading resistance, 16
Gross combined vehicle weight (GCVW), 226
Gross vehicle weight rating (GVWR), 83
Ground braking force, 63
Ground-tread interface, 192
Grouped single-layer concentric windings, 319

H

Hall-effect sensors, 356
Handling and stability
- drivetrain system effect
  - cornering stability control, tractive and braking forces, 137–139
  - electronic control systems, 139–142
  - tractive force effect on cornering characteristics, 135–137
- instantaneous response to input of front wheel angle
  - equation of motion and solution, 111–115
  - stability conditions, 115–116
- rollover
  - quasistatic rollover analysis of rigid vehicle, 143–144
  - quasistatic rollover analysis of vehicle with suspension, 145
  - transient rollover analysis, 146–147
- steady-state response to input of front wheel angle
  - ratio of turning radii, 108–109
  - sideslip angles, 110–111
  - static margin, 109–110
  - yaw velocity gain, 106–108
- steering system effect
  - compliance steer, 135
  - functionality, 129–132
  - fundamental handling characteristics, 132
  - movement interference, 135
  - on-center handling characteristics, 133–134
  - steering effort characteristics, 132–133
- suspension effects, 117–118
  - compliance camber change, 129
  - compliance steer related to aligning torque, 128–129
  - compliance steer related to lateral force, 127–128
  - roll steer, 126–127
  - suspension-related cornering behavior, 118–125
- tires, cornering characteristics of
  - aligning torque, 101–102
  - behavior of tire with camber angle, 102–104
  - coordinate system, 97–98
  - cornering behavior, factors affecting, 100–101
  - lateral forces in cornering, 99–100
  - road-related factors, 101
  - tire-related factors, 100–101
- vehicle handling tests
  - low-speed steering effort tests, 147
  - parameters, 147
  - pulse steering tests, 149
  - self-aligning tests, 149
  - steady-state steering characteristics tests, 148
  - transient yaw response tests, 148
- vehicle response to input of front wheel angle, 104–106
- yaw rate frequency response characteristics, 116–117

Hard switching fault (HSF), 361
Head injury criterion (HIC), 578
High-speed handling, 132, 134
High-temperature-resistant packaging, 370
High-voltage main circuits
- double-sided cooling, 359–360
- single-sided cooling, 359
- triple-or-above-sided cooling, 360

High-voltage power-up procedure
- charge, 592
- discharge, 591

High-voltage shutdown procedure
- charge, 592–593
- discharge, 591–592

Hot braking tests, 87
Housing design, 442–443
- draft angle and fillet, 445–446
- loads on housing, 443–445
- material and manufacturing process, 447
- oil passages, 446–447
- overall structure of transmission housing, 443
- reduction of housing vibration and noise, 446
- requirements, 443
- vent plugs, 447–448
- wall thickness and reinforcement ribs, 445

Hydraulic actuation system components, 546
- brake pedal force and stroke, 548
- diameter of
  - master cylinder, 547
  - wheel cylinder, 547

Hydraulic actuators, 423
Hydraulic shock absorbers, 477–480
Hydrodynamic lubrication (HL), 438
Hysteresis-related losses, 329

I

IGBTs, 355, 357, 359, 361, 362, 369, 373
Independent suspension, 120, 454, 455
- alignment parameters, design considerations
  - camber, 484–485
  - toe, 484–486
  - wheel track, 484
- guide mechanism design
  - double-wishbone suspension, 490–493
  - McPherson suspension, 489–490
- pitch center, design of, 488–489
- pivot points, 497
- roll center, design of, 486–488
- roll center height, 456–457
- space occupation, 457
- suspension lateral compliance, 457
- suspension roll stiffness, 457
- wheel alignment parameters, changes in, 457

Induction motors (IMs), 238
DTC, 599–600
FOC, 597–599
rotor flux-oriented control system, 599
T-shaped equivalent circuit, 596
VVVF speed control, 596–597
Instantaneous response, front wheel angle
equation of motion and solution, 111–113
circular eigenfrequency ω_0 of yaw rate fluctuation, 114
damping ratio ξ, 115
peak time t_p, 114
rise time t_r, 113–114
stability conditions, 115–116
Integral-type trapezoidal linkage, 500–502
Interior permanent magnet (IPM), 335, 336, 601
schematic, 335
thermal circuit diagram, 336
type of rotors, 343
Inverse Park transformation, 598
Iron core materials, 347

K

Kalman filter, 281
Kingpin position-related geometric parameters, 130

L

Ladder frame, 560
Laplace transform, 605, 606
Lap winding, 327
Large-module approach, 272
Large-radius arc transition, 468
Largest speed ratio, 50–51
Lateral forces, 99–100
Lateral reaction force, 98
Layout optimization method, 367–368
Leaf spring suspensions, 468–469
average thickness, 470
and basic dimensions, 469
calculation of load, 469–470
cross section of leaf springs, 471
length of each of the leaves, 471
length of main leaf, 470
linear leaf spring, 472
number of leaves, 471
spring camber, full load and in free state, 470
thickness of leaf springs, 471
width of leaf springs, 471
Leakage reactance, 315–320, 323
Lithium ferrophosphate (LFP) batteries, 231
Load factor, 467
Lockup analysis, 85
Logarithmic decrement, 169
Longitudinal friction coefficient, 10, 11
Longitudinal slip ratio, 10
Low electrical conductivity, 442
Low rolling resistance (LRR), 9
Low-speed steering effort tests, 147
Low-stray inductance packaging, 362–364
Low-viscosity oil, 441
Low-voltage power-up procedure, 590
Low-voltage shutdown procedure, 590–591
Lubrication system
boundary lubrication, 439
design
equations, 439
forced oil, 439–440
oil splash, 440
dry friction, 437
fluid film lubrication, 438
oil splashing method, 437
pressurized oil circulating method, 437
transmission oil
antifoam properties, 441
antiwear properties, 441
choice of, 440
cooling performance, 441
friction characteristics and stability, 441
physical properties, 442
typical load bearing capacity, squeeze film lubrication, 438

M

Magnetic circuit, 302–304
excitation current and no-load characteristics, 310–311
magnetic potential drop
air gap, 304–307
teeth, 307–309
yoke, 309–310
pole flux leakage, 310
Magnetizing reactance, 313–315
Magnetorheological shock absorbers, 480–481
Main control boards
analog quantity acquisition module, 355
communication module, 356
fault handling module, 355–356
power supply module, 355
PWM module, 357
relay module, 356
speed detection module, 356–357
Manual transmissions (MTs), 374, 375
Mass imbalance, 160
Mass imbalance-related excitation, 159–160
Mass-related parameters
axle load distribution, 227
carrying capacity, 226–227
curb weight, GVW and GCVW, 226
payload per unit curb weight, 227
Material utilization factor, 294
Maximum braking force, 558–559
Maximum motor speed, 244
Maximum tractive force, 559
McPherson suspension, 489–490
Mean tangential electromagnetic stress, 295
Mechanical actuators, 423
Mechanical efficiency of drivetrain, 4–5
Miles per gallon of gasoline equivalent (MPGe), 32
Minimal adhesion coefficient requirements, 24–26
Minimum ground clearance h_{min}, 187
Minimum turning radius r_{min}, 188
Modified Coulomb counting method, 280
Motor speed-based EDC, 649
Motor torque-based EDC, 649–652
Multiconstrained dynamic (MCD), 282
Multi-input multi-output (MIMO), 582

N

Natural circular eigenfrequency, 111
Natural frequency of suspension, 459
Net tractive force, 7
Neutral steer, 106, 108
Neutral steer point (NSP), 109, 110
New European Driving Cycle (NEDC), 32–35
Nickel manganese cobalt (NMC) batteries, 231
Noise, vibration, and harshness (NVH) quality, 12
Nonindependent suspension, 119, 454–456
Nonlinear rear leaf spring suspensions, 461
Nonsinusoidal magnetic field, 152–153
Normal force, 98

O

Oil pumps, 432–435
Oil splash lubrication system, 440
On-center handling characteristics, 133–134
Opposite-direction steer, 520
Overall vehicle layout design, 247–248
battery system layout, 248–249
brake system layout, 253
coordinate system, 248
layout of luggage space and spare tire, 254
layout of operator pedals, 253–254
layout of suspensions, 252
mounting positions of seat belts and airbags, 257
powertrain system layout, 248–252
sizing of vehicle body, 256–257
steering system layout, 252
vehicle interior layout, 254–256
Overcharge safety device (OSD), 269, 270
Oversteer, 106, 110
Overturning moment, 98

P

Packing efficiencies, 235, 236
Parallel axis EPS, 513
Parallel braking, 89, 90, 93
Parallel-side slots, 354
Parameter screening method, 367
Park brakes, 542

Park transformation, 598
Particle swarm optimization (PSO), 56–57
Passive cell equalization circuit, 278
Payload per unit curb weight, 227
Pear-shaped slots, 354
Performance-related parameters
  dynamic parameters
    acceleration time, 228
    gradeability, 228
    specific power and specific torque, 228
    top speed, 228
  energy economy parameters, 229
  trafficability and handling parameters, 229–231
Perimeter frame, 561
Permanent magnetic materials
  alnico alloys (AlNiCo), 346
  ferrite, 346
  NdFeB, 346
  physical properties, 346
  samarium cobalt (SmCo), 346–347
Permanent magnet synchronous motors (PMSMs), 237–238
  control of three-phase PMSMs, 601–602
  equivalent physical model, 601
  PMSM rotors, design of, 342–343
    air gap length, selection of, 344–345
    design of air gap structure on rotor surface, 344
    iron core materials, selection of, 347
    magnetic path design of IPM type of rotors, 343–344
    permanent magnetic materials, selection of, 346–347
    weight reduction, slots for, 345
  PMSM stators, design of
    aspect ratio, selection of, 351–352
    cooling structure, design of, 350–351
    electrical loading, selection of, 352
    iron core, 348
    pole count, selection of, 352
    slot opening width, selection of, 352
    slot-to-pole ratio, selection of, 352
    slot type, selection of, 352–354
    windings, design of, 348–349
Per-unit direct-axis armature reaction reactance, 314
Per-unit magnetizing reactance, 313
Phase change material (PCM), 44
Phase resistance of stator winding, 312
Pitch center
  double-wishbone suspension, 488
  McPherson strut suspension, 488
Pitman arms, 516–517
Planar vehicle model, 166
Planetary gearsets, 391
  basic equations, 393–395
  gear design for strength and durability, 399–400
    bending fatigue, 400–401
    scoring fatigue, 402
    surface contact fatigue, 401
  gear materials, selection of, 402–403
  gear noise reduction, design principles for, 405–406
    addendum coefficient, 408
    dimension tolerances, 409
    gear modification, 409
    helix angle, 407–408
    module, 407
    pressure angle, 407
    shaft and support design, 408–409
  number of planet gears and spacing, 395–398
  schematic, 392
  transmission gear design, manufacturability of, 403
    broaching, 404
    chamfering and tooth relief, 405
    grinding, 405
    hobbing, 404
    shaper cutting and shaving, 404
  transmission gear design, trend in, 409–410
  transverse contact ratio, 398–399
Plate brake tester, 88
Pneumatic actuators, 423
Pneumatic control modulator, 549
Pneumatic tire, 196–197
Pole flux leakage, 310
Positive lateral acceleration, 110
Positive sideslip angle, 110
Positive temperature coefficient (PTC), 367
Postimpact test static rollover, 579
Pouch cells, 270–271, 274
Power balance equation, 27
Power brakes, 543–544
Power management, 594–595
Power shift transmissions (PSTs), 375–377
Power spectral density (PSD), 161–163
Power steering system, 131
  boost characteristics, design of
    aligning mode, 513
    boost mode, 513
    damping mode, 513
    linear characteristics, 514
    parabolic characteristics, 515
    polyline characteristics, 514
    requirements, 513–514
  column EPS, 512–513
  dual-pinion EPS, 513
  parallel axis EPS, 513
  single-pinion EPS, 513
Power supply module, 355
Powertrain system layout, 248–252
Prismatic cells, 269–270, 274
Propulsion power, 26
Pulse steering tests, 149
Pulse-width modulation (PWM), 153, 357

Q

Quadrature-axis armature reaction reactance, 314
Quarter vehicle model, 167
Quasistatic rollover analysis
  of rigid vehicle, 143–144
  of vehicle with suspension, 145

R

Rack and pinion steering gears, 507–512
Radial flux AC motors
  air gap length, selection of, 299–300
  armature diameter, determination of, 300
  aspect ratio, selection of, 298–299
Radial flux motors, 289–293
Ramp angle β, 187
Ratio of turning radii, 108–109
Ravigneaux planetary gearset, 394, 395
Rear aerodynamic lift coefficients, 23
Rear moving barrier impact, 579
Rear suspension, 246
Rear-wheel-drive (RWD) EVs, 25, 26
  parallel braking-based brake proportioning strategy, 92, 93
  series braking-based brake proportioning strategy, 93
Regenerative braking, 88
  brake proportioning for maximization of regenerative energy
    AWD EVs, 94–95
    FWD EVs, 91–92
    RWD EVs, 92–93
  control strategy, 636–644
  electric motor characteristics on, 90
  mechanical and electrical compound braking, 89–90
Relay module, 356
Resistance forces, 6
  acceleration resistance, 17
  air resistance, 12–15
  grading resistance, 16
  rolling resistance, 6–12
Resistance power, 26
Ride comfort
  excitation sources
    drivetrain, 155–157
    electric motors, 152–155
    road roughness, 161–165
    wheels and tires, 157–161
  human perception of vibrations, 179–180
  parameters, 231
  ride quality, evaluation of
    alternative evaluation method, 181
    basic evaluation method, 180–181
    sound quality evaluation, 181–182
  system parameters on vibration response, 179

vehicle response to excitations
dynamic model of vehicle, 165–168
frequency response to road excitation, 169–171
fundamental behavior of sprung mass, 168–169
statistical characteristics of system response, 171–178
Ride quality, 151, 152
alternative evaluation method, 181
basic evaluation method, 180–181
sound quality evaluation, 181–182
Rig tests, 29
plate brake tester, 88
roller test stands, 88
Road excitation
amplitude–frequency characteristics of sprung mass, 170
frequency response, 169–171
statistical characteristics of system response, 178
Road load, 16
Road-related factors, 101
Road roughness, 161–165
Road slope, 16
Road tests, 28, 86–87
braking-in-a-turn tests, 87
cold braking tests, 87
hot braking tests, 87
Roll center
double-wishbone suspension, 487
height, 487
McPherson strut suspension, 486
Roller test stands, 88
Rolling resistance, 6
effect of load and inflation pressure, 10
moment, 98
net tractive force, 7
pressure distribution, 7, 8
road surface, effect of, 9
rolling friction coefficient, 8
rolling resistance coefficients, 9
rolling resistant force, 7
rolling resistant moment, 7
speed, effect of, 8–9
tire slip, effect of, 10–12
Roll motion
independent suspension, 120
nonindependent suspension, 119
Rollover
quasistatic rollover analysis of rigid vehicle, 143–144
quasistatic rollover analysis of vehicle with suspension, 145
transient rollover analysis, 146–147
Roll steer, 126–127
Roll stiffness, 119–120
Roof crush resistance, 579
Root-mean-square (RMS), 180, 181
RWD vehicle overcoming steps, 204–206

S

Same-direction steer, 520–522
Saturation coefficient K_s, 306
Saw profile diagram, 53, 54
Scoring fatigue, 402
Second-order dominant mode, 173
Selection of dynamic parameters
electric motor power, 47–48
final drive ratio, determination of, 54–55
largest speed ratio, 50–51
number of gears and intermediate speed ratios, 51–54
smallest speed ratio, 49–50
speed ratios, optimization of, 55–56
GA optimization algorithm, 57–58
PSO optimization algorithm, 56–57
Self-adaptive control strategy
downhill roads, launching on, 610–611
level roads, launching on, 611
uphill roads, launching on, 608–610
Self-aligning tests, 149
Serial braking, 89, 90
Series–parallel wired battery cells, 264, 265
Service brakes
disc brakes, 527–533
drum brakes, 533–541
Severity index (SI), 578
Shear deformation, 188–190, 199
Shear stress, 188–190, 199, 209
Shift actuators
characteristics, 423
design of
electrohydraulic actuators, 424–429
electromechanical actuators, 431–432
electropneumatic actuators, 429–431
Shift force, 417–419
Shift jerk, 157
Shock absorbers
electromagnetic, 482–483
electrorheological, 481–482
hydraulic, 477–480
magnetorheological, 480–481
Short-circuit protection, 361–362
SiC MOSFET, 355, 357, 359, 361–367, 369, 371–374
SiC motor controllers, 360
current sharing, 366
dynamic, 368–369
steady-state, 367–368
improvement of EMC, 365–366
low-stray inductance packaging, 362–364
short-circuit protection, 361–362
suppression of bridge arm crosstalk, 364
thermal management, 369
cooling technology, 371
high-temperature-resistant packaging, 370
strategies, 373–374
system modeling and junction temperature extraction, 371–373
Side moving deformable barrier impact, 579
Sideslip angle, 98, 106
difference in, 110–111
front and rear wheels, 106
global, 105, 111
positive, 110
total, 118
Simplified stator winding, 322
Single-air gap axial flux motor, 295
Single-input multi-output (SISO), 583
Single-layer chain windings, 319
Single-layer winding, 317
Single-pinion EPS, 513
Single-sided cooling, 359
Skateboard chassis
advantages, 569
CTC technology, 568
decentralized electric drive system, 569
disadvantages, 569
domain-centralized electric and electronic architecture, 568
modular and compact construction, 569
X-by-wire technology, 568
Sliding mode control (SMC), 642
Slip ratio, 10, 11, 85
vs. braking force coefficient, 66, 68
vs. lateral force coefficient, 68
Slot electrical potential star vectogram, 327
Smallest speed ratio, 49–50
Soft ground resistance
pneumatic tire, 196–197
rigid driven wheel, 193
rigid rolling wheel, 193–195
Sound quality evaluation, 181–182
Space vector pulsed width modulation (SVPWM), 598
Specific electromagnetic loading, 296–298
Speed detection module, 356–357
Split-type trapezoidal linkage, 503–504
Sport utility vehicles (SUVs), 44
Sprung mass
amplitude–frequency characteristics, 170, 177
fundamental behavior, 168–169
Stability control
braking force, 137
inner and outer tractive force proportion, 138–139
tractive force, 137
Stabilizer bars
angular stiffness, 495
dimensions, 495
maximum stress, 496
schematic, 494
Standard deviation, 171
State estimation of batteries
SOC, 279–281
SOH, 281–282
SOP, 282
SOS, 282–283

State of charge (SOC), 28, 276–281
State of health (SOH), 276, 277, 281–282
State of power (SOP), 276, 282
State of safety (SOS), 276, 282–283
Static deflection of suspension, 458
Static margin, 109–110
Statistical characteristics of system response, 171–174
  amplitude-frequency characteristics
    dynamic load on wheels, 178
    sprung mass acceleration, 177
    suspension deflection, 177–178
  road excitation, 178
  transfer characteristics of dual-mass system, 174–177
Steady-state current sharing
  coupled inductor method, 368
  layout optimization method, 367–368
  parameter screening method, 367
Steady-state response, front wheel angle
  ratio of turning radii, 108–109
  sideslip angles, 110–111
  static margin, 109–110
  yaw velocity gain, 106–108
Steady-state steering characteristics tests, 148
Steer angle, 114
Steer-by-wire, 518–519
Steering angle sensor, 522
Steering dampers, 517
Steering effort, 506–507
Steering gears
  motion transmission
    forward and backward efficiencies, 505–506
    steering effort, 506–507
    steering ratio, 505
  rack and pinion, 507–512
Steering ratio, 505
Steering-related handling criteria, 134
Steering system
  all-wheel steering, 519–520
    basic principles, 520–522
    mechanism, 522–523
  basic, 500
  compliance steer, 135
  components and features
    ball joints, 516
    Pitman arms, 516–517
    safety-related components, 517–518
    steering dampers, 517
    tie rods, 516
  functionality, 129–132
  fundamental handling characteristics, 132
  general requirements, 499–500
  layout, 252
  movement interference, 135
  on-center handling characteristics, 133–134
  power steering systems, design of
    boost characteristics, 513–515
    EPS, types of, 512–513
  steer-by-wire, 518–519
  steering effort characteristics, 132–133
  steering gears, design of
    motion transmission, 505–507
    rack and pinion, 507–512
  trapezoidal steering mechanism
    integral-type trapezoidal linkage, design of, 500–502
    split-type trapezoidal linkage, design of, 503–504
Steering wheel force-related criteria, 132
Stiffness tests, 569
Stiffness variation-related excitation, 161
Subsidence index, 190
Surface contact fatigue, 401
Surface permanent magnet (SPM), 342, 343
Suspension-related cornering behavior
  camber change in cornering, 124–125
  effect of load shift on cornering stiffness, 123–124
  load shift, 122–123
  roll angle of body, 120–122
  roll axis and roll centers, 119
  roll stiffness, 119–120
  single wishbone independent suspension, 118
Suspension system design
  elastic elements, design of
    air springs, 472–477
    coil springs, 463–465
    leaf springs, 468–472
    torsion bar springs, 465–468
  fundamental parameters
    dynamic deflection, 460
    elastic characteristics, 460–461
    natural frequency, 459
    roll stiffness of front and rear suspensions, 461
    static deflection, 458
    stiffness of primary and secondary leaf springs, 461–462
  guide mechanism design of independent suspension
    alignment parameters, 484–486
    double-wishbone suspension, 490–493
    McPherson suspension, 489–490
    pitch center, 488–489
    roll center, 486–488
  shock absorbers
    electromagnetic shock absorbers, 482–483
    electrorheological shock absorbers, 481–482
    hydraulic shock absorbers, 477–480
    magnetorheological shock absorbers, 480–481
  stabilizer bars
    angular stiffness, 495
    dimensions, 495
    maximum stress, 496
    schematic, 494
  suspension components
    control arms, 496–497
    joints, 497
    thrust rods, 497
  types of vehicle
    independent suspension, 454–457
    nonindependent suspension, 454–456
Switched reluctance motors (SRMs), 239–240
Synchronizers
  breakthrough load of detent balls, 422
  major design parameters, selection of
    axial working lengths L and L_1 of friction cone, 419
    clearances, 420
    corn angle α, 418
    index distance a and proximity distance b, 418, 419
    locking angle β, 421–422
    mean diameter d of friction cone, 418, 419
    shift force F_s, 419
    struts and notches of blocker ring, 420
    tread grooves and axial oil grooves, 420
  shift force, calculation of, 417–418
  types of synchronizers and characteristics, 416–417
System voltage, 244

T

Tangential reaction force, 24
Thermal conductivity, 338
Thermal management, 44–45, 369
  cooling technology, 371
  high-temperature-resistant packaging, 370
  strategies
    combined methodology, 374
    dead time adjustment, 373–374
    gate voltage adjustment, 374
    switching frequency adjustment, 374
  system modeling and junction temperature extraction, 371–373
Thermal resistance, 336, 337
Three-DOF steering model, 646–649
Three-motor four-wheel-drive powertrain layouts, 251
Three-parameter shift schedule, 621
Three-phase IMs, 595–596
  DTC of IMs, 599–600
  FOC of IMs, 597–599
  VVVF speed control of IMs, 596–597
Three-phase PMSMs, 601–602
Three-speed transmission, 55
Tie rods, 516
Tire circumference, 63–64

Tire coordinate system, 97, 98
Tire deformation, 99
Tire–ground adhesion
adhesion coefficient, 21
adhesion coefficient and adhesion condition, 20–21
adhesion-related forces, 21–24
minimal adhesion coefficient requirements, 24–26
Tire-related factors, 100–101
Tire–road adhesion coefficient, 559
Tone-to-noise ratio (TNR), 182
Torque compensation control strategy, 605–607
Torque fluctuation rate, 152–154
Torque hole and overshoot, 156
Torque interruption, 156
Torque management, 156
Torque ripples
current sensor errors, 154
nonsinusoidal magnetic field, 152–153
resolver, 155
switching frequency of inverter, 153–154
Torsional deflection, 466
Torsional stiffness, 467, 565
Torsion bar springs
action length of torsion bar, 468
correction factors *vs.* aspect ratio, 466
mechanical characteristics, 465
torsional stiffness of such torsion bar arm, 467
torsion bar arm with torsional deflection, 466
Total brake force, 77
Total eddy current loss, 332
Total sideslip angle, 118
Total tone-to-noise ratio (total TNR), 182
Traction coefficient, 11, 186
Traction control system (TCS), 137, 139–140
Traction efficiency, 186
Traction force, 192–193
Tractive effort, 2
in different gears, 5–6
electric motors, characteristics of, 3–4
force, 17
mechanical efficiency of drivetrain, 4–5
wheel, 3
wheel radius, 5
Tractive trafficability, 197–199
Trafficability
AWD vehicle overcoming steps, 206–208
drawbar pull of EVs, 191
calculation, 197
soft ground resistance to pneumatic tire, 196–197
soft ground resistance to rigid rolling wheel, 193–195
traction force, 192–193
FWD vehicle overcoming steps, 202–204
geometric parameters, 229–230
indices and geometric parameters for trafficability evaluation
ground bearing capacity, 185–186
structural and geometric parameters, 186–188
obstacle clearance
front end, 200–201
underbody, 201–202
physical properties of soft ground
density of semifluid mud and snow, 191
shear stress and shear deformation, 188–190
subsidence and vertical loads, 190–191
RWD vehicle overcoming steps, 204–206
testing
brief descriptions, 208
measurements of soil parameters, 209–210
tractive, 197–199
trench obstacle capability, evaluation of, 208
Transient rollover analysis, 146–147
Transient yaw response tests, 148
Translational acceleration, 105
Transmission control unit (TCU), 375
Transmission gear design
manufacturability of, 403
broaching, 404
chamfering and tooth relief, 405
grinding, 405
hobbing, 404
shaper cutting and shaving, 404
trend in, 409–410
Transmission housing, 443
Transverse contact ratio, 398–399
Trapezoidal steering mechanism
integral-type trapezoidal linkage, 500–502
split-type trapezoidal linkage, 503–504
Triple-or-above-sided cooling, 360
True power, 290
Turning path circle, 188
Twisted-end transposition, for whole-coil windings, 321
Two degrees of freedom (DOFs), 104
Two-parameter shift schedule, 620
Two-speed electrohydraulic transmission, 424
Two-speed transmission, 55
Two-wheel planar vehicle model, 104

U

Understeer, 106, 108
Ungrouped single-layer concentric windings, 319
Unibody construction, 572
Unscented Kalman filter (UKF), 281
Unsprung mass, 176
Urban driving cycles (UDCs), 32

V

Vapor compression cycle (VCC), 44
Variable brake proportioning, 84
Variable-force solenoid valves (VFSs), 436
Variable-frequency drive (VFD), 237
Variable voltage and variable frequency (VVVF), 596–597
Vehicle body
structural design, 577–578
crash safety, design principles, 578–580
lightweight design principles, 580
noise and vibration considerations, 580–581
novel body materials and processes of EVs, 581–582
testing
BIW modal tests, 583
body fatigue strength tests, 583
body-related tests, 583
measurements of overall stiffness, 582
Vehicle control unit (VCU), 585
analog input and output module, 589
basic functions
charging management, 588
fault diagnosis and handling, 588
monitor and display of vehicle operational status, 588
regenerative braking control, 587–588
vehicle driving control, 587
vehicle-level energy management and optimization, 587
vehicle network management, 588
data and fault storage module, 590
digital signal conditioning module, 589
distributed hierarchical system, 586
hardware architecture, 588, 589
high-speed CAN bus interface module, 589
power supply module, 589
pulse signal input and output module, 590
relay driver module, 589
VCU technology and E/E architecture
centralized computing, 653
decoupled software and hardware, 653
function customization, 654
platform standardization, 653
Vehicle dynamic equation, 17–18
acceleration capacity, 19–20
equilibrium diagrams of driving force and running resistance, 18–19
gradeability, 20
Vehicle frame design
1st-order bending, 567, 568
1st-order torsion, 567, 568
loading conditions, 563, 564
2nd-order bending, 567, 568

2nd-order torsion, 567, 568
sizing, 562
skateboard chassis, 568–569
spars and cross members, 561–562
strength and stiffness, 562–568
3rd-order bending, 567, 568
3rd-order torsion, 567, 568
torsional stiffness, 565
types, 560–561
Vehicle frame testing
modal tests, 570
proving ground and road endurance tests, 570
reliability and durability rig tests, 570
stiffness tests, 569
stress tests, 569
Vehicle handling tests
low-speed steering effort tests, 147
parameters, 147
pulse steering tests, 149
self-aligning tests, 149
steady-state steering characteristics tests, 148
transient yaw response tests, 148
Vehicle-level design
battery system, selection of
battery system parameters, selection of, 232–234
battery type selection, 231–232
packing, 234–236
body styles of buses
articulated, 217, 218
dog-nose, 217, 218
flat-nose, 217, 218
body styles of cars, 212
convertible, 213–216
coupe, 213–216
MPV, 213–217
sedan, 213–216
sports car, 213–217
station wagon, 213–216
SUV, 213–217
body styles of trucks, 218
characteristics of commercial truck body styles, 220
flat-nose, 219, 220
long-nose, 219, 220
pickup, 219
short-nose, 219, 220
classification
electric buses, 212
electric cars, 211–212
trucks, 212
electric drive system layout
basic principles, 221–222
electric buses, 222
electric cars, 221
electric trucks, 222
number of axles and axle load, 220–221
electric drive systems, selection of
electric drive system mounts, 244–245
electric motor selection, 236–240
parameters, 240–244
motion correctness check, 258
overall design parameters, selection of
dimensional parameters, 223–225
mass-related parameters, 226–227
performance-related parameters, 228–231
overall vehicle layout design, 247–248
battery system layout, 248–249
brake system layout, 253
coordinate system, 248
layout of luggage space and spare tire, 254
layout of operator pedals, 253–254
layout of suspensions, 252
mounting positions of seat belts and airbags, 257
powertrain system layout, 248–252
sizing of vehicle body, 256–257
steering system layout, 252
vehicle interior layout, 254–256
suspensions, selection of
front suspension, 245–246
rear suspension, 246
tires, selection of, 247
Vehicle speed sensor, 522
Vehicle suspension system
independent suspension, 454–457
nonindependent suspension, 454–456
Vertical load, 65
Vibration dose value (VDV), 181

W

Wave winding, 327
Wet clutches
dynamic capacity design of, 412
static capacity design of, 411
Wheel force analysis, braking, 62
adhesion, 64
adhesion coefficient on hard-surface roads, 64–68
ground braking force, 63
tire circumference, 63–64
Wheel radius, 5
Wheels and tires
dimensional deviation-related excitation, 157–159
mass imbalance-related excitation, 159–160
stiffness variation-related excitation, 161
Wide-open throttle (WOT), 1
World Harmonized Light-Duty Vehicles Test Cycle (WLTC), 32–35
Woven wire rods for transposition, 321

Y

Yaw moment and brake force
front axle skid and control, 140–141
rear axle skid and control, 141–142
Yaw moment control, 138
Yaw rate frequency response, 116–117
Yaw rate sensor, 522
Yaw velocity gain
definition, 106
as function of vehicle speed, 108
neutral steer, 106
oversteer, 106
understeer, 106

About the Author

Dr. Yiqing Yuan is a chief expert at the National New Energy Vehicle Technology Innovation Center (NEVC) of China. He has worked in the areas of electric vehicles, chassis, hybrid and electric powertrains, and their control. Holding a PhD from the Mechanical Engineering Department of the Virginia Polytechnic Institute and State University, he was a product engineer and then a lead engineer at the Advanced Vehicle Engineering (AVE) division of Chrysler developing powertrain products. After that, he served as the deputy director and professor at the Electric Vehicle R&D Center of the Shenzhen Institute of Advanced Technology of Chinese Academy of Science (CAS); he was concurrently the deputy general manager of the Electric Vehicle R&D Center of CAS. Before he joined NEVC, he was a distinguished professor at the Clean Energy Engineering Center of Tongji University. From the year 2020 to 2023, he was honored as a distinguished professor at Jiangsu University, and a chair professor at the Fujian Institute of Engineering.

Courtesy of Yiqing Yuan.

www.ingramcontent.com/pod-product-compliance
Lightning Source LLC
LaVergne TN
LVHW071342210926
848486LV00003B/211
* 9 7 8 1 4 6 8 6 0 7 7 3 4 *